AF344391

# THERMAL STRESS AND STRAIN IN MICROELECTRONICS PACKAGING

# THERMAL STRESS AND STRAIN IN MICROELECTRONICS PACKAGING

*Edited by John H. Lau*

VAN NOSTRAND REINHOLD
New York

Library of Congress Catalog Card Number 92-43285
ISBN 0–442–01058–3



I(T)P Van Nostrand Reinhold is a division of International Thomson Publishing. ITP logo is a trademark under license.

Printed in the United States of America

Van Nostrand Reinhold
115 Fifth Avenue
New York, New York 10003

International Thomson Publishing
Berkshire House
168–173 High Holborn
London WC1V 7AA, England

Thomas Nelson Australia
102 Dodds Street
South Melbourne 3205
Victoria, Australia

Nelson Canada
1120 Birchmount Road
Scarborough, Ontario MIK 5G4, Canada

16 15 14 13 12 11 10 9 8 7 6 5 4 3 2 1

**Library of Congress Cataloging-in-Publication Data**
Thermal stress and strain in microelectronics packaging / edited by
  John H. Lau.
     p.  cm.
  Includes bibliographical references and index.
  ISBN 0–442–01058–3
  1. Electronic packaging.  2. Microelectronic packaging.
3. Thermal stresses.  I. Lau, John H.
TK7870.15.T48  1993
621.381′046—dc20                    92-43285
                                        CIP

# Contents

# Preface

Microelectronics packaging and interconnection have experienced exciting growth stimulated by the recognition that systems, not just silicon, provide the solution to evolving applications. In order to have a high density/performance/yield/quality/reliability, low cost, and light weight system, a more precise understanding of the system behavior is required. Mechanical and thermal phenomena are among the least understood and most complex of the many phenomena encountered in microelectronics packaging systems and are found on the critical path of nearly every design and process in the electronics industry.

The last decade has witnessed an explosive growth in the research and development efforts devoted to determining the mechanical and thermal behaviors of microelectronics packaging. With the advance of very large scale integration technologies, thousands to tens of thousands of devices can be fabricated on a silicon chip. At the same time, demands to further reduce packaging signal delay and increase packaging density between communicating circuits have led to the use of very high power dissipation single-chip modules and multi-chip modules. The result of these developments has been a rapid growth in module level heat flux within the personal, workstation, midrange, mainframe, and super computers. Thus, thermal (temperature, stress, and strain) management is vital for microelectronics packaging designs and analyses. How to determine the temperature distribution in the electronics components and systems is outside the scope of this book, which focuses on the determination of stress and strain distributions in the electronics packaging.

An electronics package is a composite structure that undergoes thermal loadings. Due to the thermal expansion mismatch of different parts of the package, thermal stresses and strains can occur and concentrate with very large values at a small area inside the packaging system while it is being manufactured and while it is being used. The determination of thermal stresses and strains in electronics packaging is a very difficult task because of its geometry and materials construction. Fortunately, we are now beginning to obtain useful insight and new understanding of the thermo-mechanical behaviors of microelectronics packaging. Some useful closed-form solutions, efficient simulation schemes, and practical experimental techniques have also been obtained. These results already have been disclosed in diverse journals or, more incidentally, in the proceedings of many conferences, symposia, and workshops whose primary emphasis is materials science or electronics packaging and interconnection. Consequently, there is no single source of information devoted to the state of the art of thermo-mechanical methods and responses for packaging. This book aims to remedy this deficiency and to present, in one volume, a timely summary of progress in all aspects of this fascinating field.

This book is divided into eight basic parts. The first part (Chapter 1) presents the fundamental equations of linear as well as nonlinear thermo-mechanics for microelectronics packaging. The second part of the book discusses thermal stresses in multilayer structures. In Chapter 2, Peter Hall presents a simplified method of calculating the thermal stresses in multilayer structures using spreadsheet software, which is available to any personal computer, such as LOTUS 1-2-3. Chapter 3, by Wan-Lee Yin, examines the thermal stresses in anisotropic layers by the variational method. In Chapter 4, An-Yu Kuo and Kuan-Luen Chen determine the transient thermal stresses in multilayer devices by the hybrid finite element method. Rajen Chanchani and Peter Hall explore the temperature dependence of in-plane expansivity (thermal coefficient of expansion) for several microelectronics packaging materials in Chapter 5, and in Chapter 6, Goran Matijasevic, Chen Yu Wang, and Chin Lee examine the thermally induced stresses in the die (chip) due to die-attachment and the methods to reduce them.

The third part of the book addresses thermomechanical measurement techniques for microelectronics packaging. James Sweet (Chapter 7) reports on the use of Si test chips with piezoresistive stress sensors for measurement of the components of the stress tensor at the chip surface. In Chapter 8, Arkady Voloshin introduces an enhanced Moiré interferometry method for measurements of displacement field resulting from the thermal loading on the microelectronics package. In Chapter 9, C.-P. Yeh, C. Ume, R. E. Fulton, K. W. Wyatt, and J. W. Stafford determine the printed circuit board warpages by a shadow Moiré method and then validate the results by correlating them with finite element results.

The fourth part of this book looks at the thermal stress-induced voids in thin-film metallizations. Quanxin Guo, Leon Keer, and Yip-Wah Chung (Chapter 10) present a stress-induced voiding and cracking model based on the thermally activated grain boundary and surface diffusion and fracture mechanics. In Chapter 11, M. A. Korhonen, P. Børgesen, and Che-Yu Li examine the formation of tensile thermal stresses in the metallizations (pure aluminum metallizations on oxidized silicon substrates) during cooldown from elevated temperatures, and on the relaxation of them by void growth.

The fifth part of this book addresses the thermal stress and strain in plastic packages. Ephraim Suhir (Chapter 12) evaluates the effect of plastic package geometry and materials on its residual bow (warpage) and designs a plastic package with sufficiently low residual warpages. In Chapter 13, Michel Mermet-Guyennet presents a failure criterion for moisture-induced stresses in plastic packages and compares the thermal stress analysis results from finite element method and piezoresistive technique. Suresh Golwalkar (Chapter 14) reviews in detail the mechanisms involved in the temperature–moisture interactions that challenge the structural integrity of the plastic packages.

The sixth part of this book presents the thermal stress and strain in bulk solders and surface mount solder joints. In Chapter 15, Peter Hacke, Arnold Sprecher, and Hans Conrad measure the stress–strain hysteresis loops and the growth of fatigue cracks of the near-eutectic 63wt%Sn/37wt%Pb solder joints in shear. Ryohei Satoh (Chapter 16) estimates the thermal fatigue life of actual solder joints (63wt%Sn/37wt%Pb, 5wt%Sn/95wt%Pb, and 96.5wt%Sn/3.5wt%Ag) by using a crack propagation model with thermal strain range. In Chapter 17, Jenq-Gong Duh, Kuo-Chuan Liu, and Bi-Shiou Chiou examine the microstructural development of Sn/Pb solder joints subjected to aging and thermal cycling. Jim Lynch and Alberto Boetti look at leadless chip carriers mounted upon thick film alumina substrates and printed circuit boards in Chapter 18, and in Chapter 19, Ronald Ross, Jr. and Liang-Chi Wen explore the complex systems-level creep–fatigue interactions involved in leaded chip carrier solder joints.

The seventh part of the book examines the thermal stress and strain in plated-through-hole (PTH) composite structures with and without solders. In Chapter 20, Donald Barker and Abhijit Dasgupta determine the critical thermal stresses of an anisotropy PTH with very complex geometries by the finite element method. John Lau, Ravi Subrahmanyan, Steve Erasmus, Sherman Leung, and Che-Yu Li (Chapter 21) evaluate the effects of the global and local thermal expansion mismatches of a ceramic pin grad array soldered to an orthotropic epoxy PCB on the PTH copper and solder joint reliability, respectively, by fatigue experiments and finite element methods.

The last part of this book presents two very important subjects in microelectronics packaging. Chapter 22, by Michael Pecht and Pradeep Lall, discusses the thermomechanical considerations in the manufacture and design of wirebond interconnects, and presents guidelines for design of reliable wirebond interconnects. Xuning Shan and Michael Pecht (Chapter 23) examine the forms, reaction mechanisms, thermodynamics, and kinetics of corrosion, discuss the governing equations for generic corrosion processes, and present general approaches to model and simulate various forms of corrosion in microelectronics packages.

For whom is this book intended? Undoubtedly it will be of interest to three groups of specialists: (1) those who are active or intend to become active in research and development of microelectronics packaging; (2) those who have encountered practical packaging problems and wish to understand and learn more methods of solving such problems; and (3) those who have to choose a high performance and cost effective packaging technique for their interconnect system.

I hope this book will serve as a valuable source of reference to all those faced with the challenging problems created by the ever more increasing heat fluxes within microelectronics packaging. I also hope that it will aid in stimulating further research and development on useful closed-form solutions, efficient simulation schemes, and practical experimental techniques and more sound use of thermomechanical methods for packaging designs and analyses. The organizations that learn how to find the stresses and strains by thermomechanical methods and to minimize them by changing the geometry and materials of their interconnect systems have the potential to make major advances in microelectronics packaging and to gain great benefits in cost, performance, quality, size, and weight.

John H. Lau, PhD, PE
*Hewlett-Packard Company*

# Acknowledgments

Development and preparation of *Thermal Stress and Strain in Micro-electronics Packaging* was facilitated by the efforts of a number of dedicated people at Van Nostrand Reinhold (VNR) and Keyword Publishing Services. I would like to thank them all, with special mention to Alan Chesterton of Keyword for his effective coordination of the publication process and to Marjorie Spencer of VNR for her unswerving support and for solving many problems that arose during the book's preparation. My special thanks to Stephen Chapman of VNR who made my dream of this book come true by effectively sponsoring the project. It has been a great pleasure and fruitful experience to work with them.

The material in this book has clearly been derived from many sources including individuals, companies, and organizations, and the various contributing authors have attempted to acknowledge, in the appropriate parts of the book, the assistance that they have been given. It would be quite impossible for them to express their thanks to everyone concerned for their cooperation in producing this book, but on their behalf, I would like to extend due gratitude.

Each chapter of the book was reviewed by at least three individuals who are experts in microelectronics packaging and related areas. According to their specialities, each individual reviewed at least three chapters of the book. These reviewers are Dr. Tom Chung, Tandem Computers, Inc., Professor James Clum, SUNY Binghamton, Professor H. Donald Conway, Cornell University, Professor J. Eischen, North Carolina State University, Dr. Darrel Frear, Sandia National Laboratories, Dr. Donald Helling, Hughes Aircraft,

Professor Tai-Ran Hsu, San José State University, Professor Klod Kokini, Purdue University at West Lafayette, Professor Y. C. Lee, University of Colorado at Boulder, Dr. Z. Mei, University of California at Berkeley, Professor Robert Miller, University of Illinois at Urbana, Dr. Larry Moresco, Fujitsu Computer Packaging Technology, Inc., Dr. Tsung-Yu Pan, Ford Scientific Research Laboratories, Dr. Yi-Hsin Pao, Ford Scientific Research Laboratories, Dr. Jason Pei, Digital Equipment Corporation, Dr. Robert Riddle, Lawrence Livermore National Laboratories, Dr. Charles Schmidt, Stanford Research Institute, Dr. G. Scott, AT&T Bell Laboratories, Professor W. H. Stevenson, Purdue University at West Lafayette, Professor Donald Stone, University of Wisconsin at Madison, and Dr. Frank Wu, International Business Machines Corporation. I want to thank them for their many helpful comments and constructive suggestions that added significantly to this book.

I express my deep appreciation to the 43 contributing authors, all experts in their respective fields, for their many helpful suggestions and cooperation in responding to requests for revisions. Their depth of knowledge, dedication, and patience has been demonstrated throughout the process of preparing this book. Their brief technical biographies are presented at the end of this book.

Lastly, I want to thank my employer, Hewlett-Packard, for providing an excellent environment in which completing this book was possible. I also want to thank my managers, Steve Erasmus and Anita Danford, for their trust, respect, and support of my real work at HP. Finally, I want to thank my daughter (Judy) and my wife (Teresa) for their love, consideration, and patience by allowing me to work on many weekends for this private project. Their simple belief that I am making my small contribution to mankind was strong motivation for me, and to them I have dedicated my efforts on this book.

John H. Lau, PhD, PE
*Palo Alto, California*

# THERMAL STRESS AND STRAIN IN MICROELECTRONICS PACKAGING

# 1

# Thermomechanics for Electronics Packaging

*John H. Lau*

## 1.1 INTRODUCTION

With very few exceptions (for example, Invar, an iron–nickel alloy), substances expand when their temperature is raised and contract when cooled. The deformation (expansion or contraction) due to temperature change in the absence of mechanical loads is called thermal strain. The thermal strain is not exactly linear with temperature change (for example, see Chapter 5), but for first-order approximation and small temperature changes, this strain can be described as proportional to the temperature change. This proportionality is expressed by the *coefficient of linear thermal expansion*, which is defined as the change in length that a bar of unit length undergoes when its temperature is changed by one degree.

Thermal stresses occur when any portion of the thermal expansion or contraction in a structure is constrained. Basically, there are two different sets of constraints under which thermal stresses occur: external constraints and internal constraints. Thermal stresses due to external constraints are readily apparent. The most familiar example is a bar fixed at both ends and subjected to a temperature rise. In this case the bar is at a state of compression except at the fixed ends, which are at a very complex state of stress.

However, thermal stresses due to internal constraints are not so obvious. A structure made of one material may be free to expand and yet have thermal stresses due to a nonuniform temperature distribution. On the other hand, a structure made of more than one material (i.e., a composite structure) may be uniformly heated, have no external constraints, and still have thermal stresses due to different coefficients of expansion and mechanical properties. In both of these cases the constraints occur within the structure.

An electronics package assembly is a typical example of a composite structure that undergoes thermal loadings. It consists of at least two different materials, and is subjected to nonuniform temperature distributions. Due to the geometry, material construction, and thermal expansion mismatch of different parts of the package, thermal stresses can occur inside the packaging system while it is being manufactured and while it is being used.

The determination of thermal stresses in electronic packaging is not an easy task. Closed-form and semi-closed-form solutions for very simple geometries and temperature loadings such as those given by refs. 1–10 are very useful, but are very limited in applications and difficult to obtain. The finite element method[11–20] is one of the best candidates for obtaining numerical results for the thermal stresses and strains in electronics packages. However, as with any popular method, many of the finite element analyses performed are not properly executed due to a limited understanding of the equations of thermomechanics for electronics packaging.

In the present chapter, the basic and governing equations of thermomechanics for electronics packaging are briefly mentioned and their assumptions and limitations are also highlighted. Linear thermoelasticity will be presented first. Nonlinear effects such as large deformations, hyperelasticity, plasticity, viscoelasticity, viscoplasticity, and creep will follow immediately.

## 1.2 FUNDAMENTAL EQUATIONS OF THERMOELASTICITY FOR ELECTRONICS PACKAGING

### 1.2.1 Assumptions

Let us consider a linear elastic package subjected to heating and external forces. We assume that the package is stress-free at a uniform temperature $T_0$ when all external forces are removed. The stress-free state will be referred to as the *reference state*, and the temperature $T_0$ as the *reference temperature*. Furthermore, the velocity and displacement of every particle of the package in the instantaneous state from its position in the reference state will be assumed to be small (i.e., infinitesimal strains, and the material derivative is the same as the partial derivative with respect to time).

### 1.2.2 Fundamental Equations of Thermoelasticity

Based on the foregoing assumptions the fundamental equations of thermoelasticity for the package are:[21–30]

*Infinitesimal strain–displacement relations*

$$\epsilon_{ij} = \frac{1}{2}\left(\frac{\partial u_i}{\partial x_j} + \frac{\partial u_j}{\partial x_i}\right) \tag{1-1}$$

*Constitutive equation (Duhamel–Neumann law)*

$$\sigma_{ij} = C_{ijkl}\epsilon_{kl} - \beta_{ij}(T - T_0) \tag{1-2}$$

*Conservation of mass (continuity equation)*

$$\frac{\partial \rho}{\partial t} + \frac{\partial \rho v_i}{\partial x_i} = 0 \tag{1-3}$$

*Conservation of momentum (Newton's law)*

$$\rho\frac{\partial v_i}{\partial t} = \frac{\partial \sigma_{ij}}{\partial x_j} + X_i \tag{1-4}$$

*Conservation of energy*

$$\frac{\partial \varepsilon}{\partial t} = T\frac{\partial \varphi}{\partial t} + \frac{1}{2\rho}\sigma_{ij}\left(\frac{\partial v_i}{\partial x_j} + \frac{\partial v_j}{\partial x_i}\right) \tag{1-5}$$

*Rate of change of entropy*

$$\rho\frac{\partial \varphi}{\partial t} = -\frac{1}{T}\frac{\partial h_i}{\partial x_i} + \frac{W}{T} \tag{1-6}$$

*Heat conduction (Fourier's law)*

$$h_i = -k_{ij}\frac{\partial T}{\partial x_j} \tag{1-7}$$

*Definition of specific heat (if $\partial \epsilon_{ij}/\partial t = 0$)*

$$-\frac{\partial h_i}{\partial x_i} = \rho C_v \frac{\partial T}{\partial t} \tag{1-8}$$

where $T$ = instantaneous absolute temperature, $x_i$ = space coordinates, $u_i$ = components of the displacement vector, $C_v$ = heat capacity per unit mass, $T_0$ = reference temperature, $X_i$ = body force components per unit volume, $\beta_{ij}$ = thermal moduli, $\sigma_{ij}$ = components of the stress tensor, $\epsilon_{ij}$ = components of strain tensor, $k_{ij}$ = heat conduction coefficients, $C_{ijkl}$ = elastic moduli, $\rho$ = mass density, $W$ = heat generation per unit time per unit volume, $\varepsilon$ = internal energy per unit mass, $\varphi$ = entropy per unit mass, $h_i$ = components of the heat flux vector, and $v_i$ = components of the velocity vector. In the present chapter, all the indices range over 1, 2, 3 and the summation convention for repeated indices is used.

## 1.3  GOVERNING EQUATIONS OF THERMOELASTICITY FOR ELECTRONICS PACKAGING

### 1.3.1  Coupled Thermoelasticity

By combining Eqs. (1-1)–(1-8), we have the following *coupled equations of thermoelasticity* for the package:

$$\frac{\partial}{\partial x_i}\left(k_{ij}\frac{\partial T}{\partial x_j}\right) = \rho C_v \frac{\partial T}{\partial t} + \tfrac{1}{2}T\beta_{ij}\frac{\partial}{\partial t}\left(\frac{\partial u_i}{\partial x_j} + \frac{\partial u_j}{\partial x_i}\right) - W \qquad (1\text{-}9)$$

$$\frac{1}{2}\frac{\partial}{\partial x_j}\left[C_{ijkl}\left(\frac{\partial u_k}{\partial x_l} + \frac{\partial u_l}{\partial x_k}\right)\right] = -X_i + \rho\frac{\partial^2 u_i}{\partial t^2} + \frac{\partial \beta_{ij}T}{\partial x_j} \qquad (1\text{-}10)$$

It can be seen that the heat conduction equation, Eq. (1-9) contains, besides the temperature $T$, the rates of strain

$$\left(\frac{\partial}{\partial t}\left(\frac{\partial u_i}{\partial x_j} + \frac{\partial u_j}{\partial x_i}\right)\right)$$

whereas the equation of motion, Eq. (1-10) contains, besides the displacement components $u_i$, the temperature increase $T$.

### 1.3.2  Coupled-Quasi-Static Thermoelasticity

In principle, every nonstationary problem of thermoelasticity is a dynamic problem. The inertia forces must be taken into account if the temperature field undergoes a sudden change with time, e.g., in cases of sudden cooling or heating of a structure. For small variations of temperature with time, however, the inertia terms may be neglected in the equation of motion, Eq. (1-10), i.e.,

$$\rho\frac{\partial^2 u_i}{\partial t^2} = 0 \qquad (1\text{-}11)$$

Then we have the *coupled, quasi-static thermoelasticity* problem

$$\frac{\partial}{\partial x_i}\left(k_{ij}\frac{\partial T}{\partial x_j}\right) = \rho C_v \frac{\partial T}{\partial t} + \tfrac{1}{2}T\beta_{ij}\frac{\partial}{\partial t}\left(\frac{\partial u_i}{\partial x_j} + \frac{\partial u_j}{\partial x_i}\right) - W \qquad (1\text{-}12)$$

$$\frac{1}{2}\frac{\partial}{\partial x_j}\left[C_{ijkl}\left(\frac{\partial u_k}{\partial x_l} + \frac{\partial u_l}{\partial x_k}\right)\right] = -X_i + \frac{\partial}{\partial x_j}(\beta_{ij}T) \qquad (1\text{-}13)$$

An example of Eq. (1-11) is as follows. If the temperature rise from 0 to 100°C of a 1 in. (25.4 mm) long package (with coefficient of thermal expansion $= 10^{-6}/°C$) is achieved in a time interval of 0.1 sec, then the acceleration is of the order $10^{-4}/(0.1)^2 = 0.01$ in./sec$^2$ (0.254 mm/sec$^2$). The change of stress due to this acceleration may be estimated from $\Delta\sigma_{xx} \simeq \Delta x\rho\,(d^2u/dt^2)$. If the specific gravity of the material is 10 and the material is 1 in. (25.4 mm) thick, we have

$$\Delta\sigma_{xx} = \tfrac{1}{12}\,(10)(62.4)\,\frac{1}{32.2}\,\frac{0.01}{12}\,\frac{1}{144} \simeq 0.00001\ \text{lb/in}^2\ (0.0689\ \text{Pa}).$$

This stress is negligible in most electronics packaging problems in which the magnitude of the stresses concerned are of the order of the strength (yielding stress or ultimate stress) and fracture toughness of the packaging material.

### 1.3.3 Uncoupled-Quasi-Static Thermoelasticity

Boundary-value problems involving Eqs. (1-12) and (1-13) are rather complex and difficult to solve. Fortunately, in most engineering applications it is possible to neglect the mechanical coupling term in Eq. (1-12) without significant error, i.e.,

$$\tfrac{1}{2}\,T\beta_{ij}\,\frac{\partial}{\partial t}\left(\frac{\partial u_i}{\partial x_j} + \frac{\partial u_j}{\partial x_i}\right) = 0 \tag{1-14}$$

Consequently, Eqs. (1-12) and (1-13) become

$$\frac{\partial}{\partial x_i}\left(k_{ij}\,\frac{\partial T}{\partial x_j}\right) = \rho C_v\,\frac{\partial T}{\partial t} - W \tag{1-15}$$

$$\frac{1}{2}\frac{\partial}{\partial x_j}\left[C_{ijkl}\left(\frac{\partial u_k}{\partial x_l} + \frac{\partial u_l}{\partial x_k}\right)\right] = -X_i + \frac{\partial}{\partial x_j}\,(\beta_{ij}T) \tag{1-16}$$

Equations (1-15) and (1-16) are called the equations of *uncoupled, quasi-static thermoelasticity* or the *theory of thermal stresses* for electronics packaging.

The physical meaning of Eq. (1-14) is that the interaction between strain and temperature is ignored and the effects of elasticity (change in dimensions of the package) on the temperature distribution are negligible. For example, the change in dimension of a package is of the order of the product of the linear dimension of the package $L$, the temperature rise $T$, and the coefficient of thermal expansion $\alpha$. If $L = 0.001$ in. (0.25 mm), $T - T_0 = 100$°C, and

$\alpha = 10^{-6}/°C$, the change in dimension is $10^{-7}$ in. (0.000 0025 mm), which is negligible for electronics packaging problems of heat conduction.

### 1.3.4 Theory of Isotropic Thermal Stresses

Fundamentally, most electronics packaging materials are anisotropic. For practical applications, however, it is possible to gain some insights into electronics packaging problems by assuming the materials to be isotropic. In that case, the governing equations of the *theory of isotropic thermal stresses* for electronic packaging are

$$k\delta_{ij}\frac{\partial}{\partial x_i}\left(\frac{\partial T}{\partial x_j}\right) = \rho C_v \frac{\partial T}{\partial t} - W \tag{1-17}$$

$$Gu_{i,\mu\mu} + (\lambda + G)u_{\mu,\mu i} = -X_i + \beta\frac{\partial T}{\partial x_i} \tag{1-18}$$

where

$$\sigma_{ij} = \lambda\epsilon_{\mu\mu}\delta_{ij} + 2G\epsilon_{ij} - \beta\delta_{ij}(T - T_0) \tag{1-19}$$

$$\beta = \frac{\alpha E}{1 - 2v} \tag{1-20}$$

$$\lambda = \frac{Ev}{(1 + v)(1 - 2v)} \tag{1-21}$$

$$G = \frac{E}{2(1 + v)} \tag{1-22}$$

and where $\alpha$ = thermal coefficient of linear expansion, $k$ = heat conductivity, $E$ = Young's modulus, $v$ = Poisson's ratio, $\lambda$ = Lame's constant, $G$ = shear modulus, and $\delta_{ij}$ = Kronecker delta.

For the special case of steady heat flow, i.e., $\partial T/\partial t = 0$, we have

$$\delta_{ij}\frac{\partial}{\partial x_i}\left(\frac{\partial T}{\partial x_j}\right) = -\frac{W}{k} \tag{1-23}$$

$$Gu_{i,\mu\mu} + (\lambda + G)u_{\mu,\mu i} = -X_i + \beta\frac{\partial T}{\partial x_i} \tag{1-24}$$

These are the equations for *isotropic thermal stresses with steady heat flow.* If there is no heat source in the packaging material, then $W = 0$ in the foregoing equations.

For engineering applications, Eqs. (1-17), (1-18), and (1-19) are rewritten as Eqs. (1-25), (1-26), and (1-27), respectively.

$$\nabla^2 T = \frac{\rho C_v}{k} \frac{\partial T}{\partial t} - \frac{W}{k} \tag{1-25}$$

$$\frac{\partial^2 u}{\partial x^2} + \frac{\partial^2 v}{\partial x \, \partial y} + \frac{\partial^2 w}{\partial x \, \partial z} + (1 - 2v)\nabla^2 u + \frac{X_x}{(\lambda + G)} = 2(1 + v)\alpha \frac{\partial T}{\partial x} \tag{1-26a}$$

$$\frac{\partial^2 v}{\partial y^2} + \frac{\partial^2 u}{\partial x \, \partial y} + \frac{\partial^2 w}{\partial y \, \partial z} + (1 - 2v)\nabla^2 v + \frac{X_y}{(\lambda + G)} = 2(1 + v)\alpha \frac{\partial T}{\partial y} \tag{1-26b}$$

$$\frac{\partial^2 w}{\partial z^2} + \frac{\partial^2 u}{\partial x \, \partial z} + \frac{\partial^2 v}{\partial y \, \partial z} + (1 - 2v)\nabla^2 w + \frac{X_z}{(\lambda + G)} = 2(1 + v)\alpha \frac{\partial T}{\partial z} \tag{1-26c}$$

$$\sigma_x = \frac{\lambda}{v}\left[(1 - v)\frac{\partial u}{\partial x} + v\left(\frac{\partial v}{\partial y} + \frac{\partial w}{\partial z}\right)\right] - \beta(T - T_0) \tag{1-27a}$$

$$\sigma_y = \frac{\lambda}{v}\left[(1 - v)\frac{\partial v}{\partial y} + v\left(\frac{\partial w}{\partial z} + \frac{\partial u}{\partial x}\right)\right] - \beta(T - T_0) \tag{1-27b}$$

$$\sigma_z = \frac{\lambda}{v}\left[(1 - v)\frac{\partial w}{\partial z} + v\left(\frac{\partial u}{\partial x} + \frac{\partial v}{\partial y}\right)\right] - \beta(T - T_0) \tag{1-27c}$$

$$\tau_{xy} = G\left(\frac{\partial u}{\partial y} + \frac{\partial v}{\partial x}\right) \tag{1-27d}$$

$$\tau_{yz} = G\left(\frac{\partial v}{\partial z} + \frac{\partial w}{\partial y}\right) \tag{1-27e}$$

$$\tau_{zx} = G\left(\frac{\partial w}{\partial x} + \frac{\partial u}{\partial z}\right) \tag{1-27f}$$

where

$$\nabla^2 = \frac{\partial^2}{\partial x^2} + \frac{\partial^2}{\partial y^2} + \frac{\partial^2}{\partial z^2} \tag{1-28}$$

and in Eqs. (1-27$a$–$f$), the following strain–displacement relations have been substituted.

$$\epsilon_x = \frac{\partial u}{\partial x} \qquad (1\text{-}29a)$$

$$\epsilon_y = \frac{\partial v}{\partial y} \qquad (1\text{-}29b)$$

$$\epsilon_z = \frac{\partial w}{\partial z} \qquad (1\text{-}29c)$$

$$\gamma_{xy} = \frac{\partial u}{\partial y} + \frac{\partial v}{\partial x} \qquad (1\text{-}29d)$$

$$\gamma_{yz} = \frac{\partial v}{\partial z} + \frac{\partial w}{\partial y} \qquad (1\text{-}29e)$$

$$\gamma_{zx} = \frac{\partial w}{\partial x} + \frac{\partial u}{\partial z} \qquad (1\text{-}29f)$$

In Eqs. (1-25)–(1-29), $u$, $v$, and $w$ are the displacement components in the $x$-, $y$-, and $z$-directions, respectively: $X_x$, $X_y$, and $X_z$ are the body force components in the $x$-, $y$-, and $z$-directions, respectively; $\sigma_x$ is the normal stress acting in the $x$-direction; $\sigma_y$ is the normal stress acting in the $y$-direction; $\sigma_z$ is the normal stress acting in the $z$-direction; $\tau_{xy}$ is the shear stress acting in the $y$-direction of the plane normal to the $x$-axis; $\tau_{yz}$ is the shear stress acting in the $z$-direction of the plane normal to the $y$-axis; $\tau_{zx}$ is the shear stress acting in the $x$-direction of the plane normal to the $z$-axis; $\epsilon_x$ is the normal strain acting in the $x$-direction; $\epsilon_y$ is the normal strain acting in the $y$-direction; $\epsilon_z$ is the normal strain acting in the $z$-direction; $\gamma_{xy}$ is the shear strain acting in the $y$-direction of the plane normal to the $x$-axis; $\gamma_{yz}$ is the shear strain acting in the $z$-direction of the plane normal to the $y$-axis; $\gamma_{zx}$ is the shear strain acting in the $x$-direction of the plane normal to the $z$-axis.

### 1.3.5 Temperature-Dependent Strain Energy Density

Finally, the strain energy density for an isotropic linear elastic electronic packaging material is given by

$$U = \tfrac{1}{2}\lambda(\epsilon_x + \epsilon_y + \epsilon_z)^2 + G[\epsilon_x^2 + \epsilon_y^2 + \epsilon_z^2 + \tfrac{1}{2}(\gamma_{xy}^2 + \gamma_{yz}^2 + \gamma_{zx}^2)]$$
$$- \beta(\epsilon_x + \epsilon_y + \epsilon_z)(T - T_0) + \tfrac{3}{2}\beta\alpha(T - T_0)^2 \tag{1-30}$$

It can be seen that the strain energy density has a temperature-dependent term that is proportional to the volumetric strain $(\epsilon_x + \epsilon_y + \epsilon_z)$ and a term proportional to $(T - T_0)^2$.

## 1.4 BOUNDARY VALUE PROBLEMS FOR ELECTRONICS PACKAGING

Heat transfer and thermal stress in electronics packaging are usually applied in two stages, transient (power on/off) and steady-state (during operation). In both cases, for the theory of isotropic thermal stresses and strains, the temperature distribution $T(x_i, t)$ in the package is calculated by solving the heat conduction equation, Eq. (1-25), with the prescribed initial and boundary conditions. The displacement components $(u, v,$ and $w)$ everywhere inside the package are then determined by solving Eq. (1-26) with the prescribed stress–displacement boundary conditions and with the calculated temperature distribution as an imposed boundary condition. (This tempera-ture distribution is shown mathematically in the right-hand side of Eq. (1-26) and is a known function.) The thermal stresses $(\sigma_x, \sigma_y, \sigma_z, \tau_{xy}, \tau_{yz},$ and $\tau_{zx})$ and strains $(\epsilon_x, \epsilon_y, \epsilon_z, \gamma_{xy}, \gamma_{yz},$ and $\gamma_{zx})$ everywhere inside the package are then calculated by Eqs. (1-27) and (1-29), respectively.

## 1.5 THERMOELASTIC EXAMPLE PROBLEMS FOR ELECTRONICS PACKAGING

### 1.5.1 Chip on a Semi-infinite Substrate

Figure 1-1 shows a semi-infinite elastic substrate subjected to a uniform temperature distribution $(T_u)$ over a rectangular region $(2a, 2b)$ on its surface. This temperature distribution could be generated from, for example, tape automated bonding devices[31–35] or normal powered (functional) cycling with self-heating.

The heat-conduction boundary conditions for this problem are:

At $-a \leqslant x \leqslant a;\ -b \leqslant y \leqslant b;$ and $z = 0$

$$T = T_u \tag{1-31}$$

At $x = y - z \to \infty$

$$T \to 0 \tag{1-32}$$

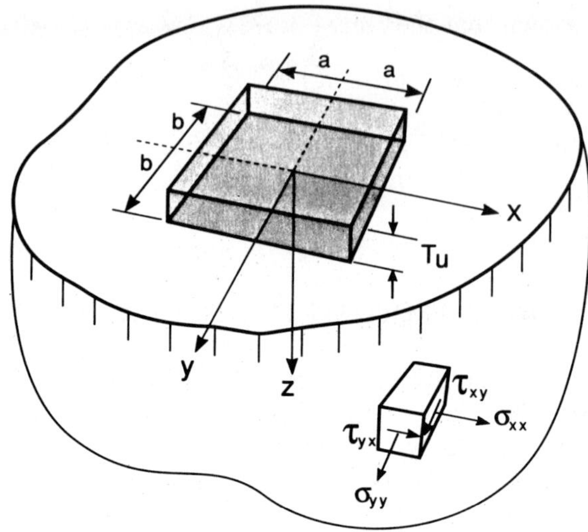


**Figure 1-1**   Thermal stresses in a semi-infinite substrate with a uniform temperature distribution on its surface. Ref. 37.

It can be shown that the temperature distribution in the substrate is given by[36,37]

$$
\begin{aligned}
T = \frac{T_u}{2\pi}\Bigg[ & 2\pi - \cos^{-1}\frac{(a+x)(b-y)}{\sqrt{[(a+x)^2+z^2][(b-y)^2+z^2]}} \\
& - \cos^{-1}\frac{(a+x)(b+y)}{\sqrt{[(a+x)^2+z^2][(b+y)^2+z^2]}} \\
& - \cos^{-1}\frac{(a-x)(b-y)}{\sqrt{[(a-x)^2+z^2][(b-y)^2+z^2]}} \\
& - \cos^{-1}\frac{(a-x)(b+y)}{\sqrt{[(a-x)^2+z^2][(b+y)^2+z^2]}}\Bigg]
\end{aligned}
\tag{1-33}
$$

Equation (1-33) satisfies the boundary conditions, Eqs. (1-31, 1-32), and the heat-conduction equation, Eq. (1-25).

The elasticity boundary conditions for this problem are:

At $z = 0$

$$
\sigma_{zz} = \tau_{zx} = \tau_{zy} = 0
\tag{1-34}
$$

At $x = y = z \to \infty$

$$
\sigma_{xx} = \sigma_{yy} = \sigma_{zz} = \tau_{xy} = \tau_{yz} = \tau_{zx} \to 0
\tag{1-35}
$$

It can also be shown that the stresses in the substrate are given by[36,37]

$$
\begin{aligned}
\sigma_{xx} = \frac{\alpha E T_u}{2\pi} \Bigg[ &\tan^{-1}\frac{(a-x)z}{(b-y)r_1} + \tan^{-1}\frac{(a+x)z}{(b-y)r_2} + \tan^{-1}\frac{(a+x)z}{(b+y)r_3} \\
&+ \tan^{-1}\frac{(a-x)z}{(b+y)r_4} - \tan^{-1}\frac{(a-x)}{(b-y)} - \tan^{-1}\frac{(a+x)}{(b-y)} \\
&- \tan^{-1}\frac{(a+x)}{(b+y)} - \tan^{-1}\frac{(a-x)}{(b+y)} \Bigg]
\end{aligned}
\tag{1-36}
$$

$$
\begin{aligned}
\sigma_{yy} = \frac{\alpha E T_u}{2\pi} \Bigg[ &\tan^{-1}\frac{(b-y)z}{(a-x)r_1} + \tan^{-1}\frac{(b-y)x}{(a+x)r_2} + \tan^{-1}\frac{(b+y)z}{(a+x)r_3} \\
&+ \tan^{-1}\frac{(b+y)z}{(a-x)r_4} - \tan^{-1}\frac{(b-y)}{(a-x)} - \tan^{-1}\frac{(b-y)}{(a+x)} \\
&- \tan^{-1}\frac{(b+y)}{(a+x)} - \tan^{-1}\frac{(b+y)}{(a-x)} \Bigg]
\end{aligned}
\tag{1-37}
$$

and

$$
\tau_{xy} = \frac{\alpha E T_u}{2\pi} \log\left[\frac{(z+r_1)(z+r_3)}{(z+r_2)(z+r_4)}\right]
\tag{1-38}
$$

where

$$
r_1 = \sqrt{(x-a)^2 + (y-b)^2 + z^2}
\tag{1-39a}
$$

$$
r_2 = \sqrt{(x+a)^2 + (y-b)^2 + z^2}
\tag{1-39b}
$$

$$
r_3 = \sqrt{(x+a)^2 + (y+b)^2 + z^2}
\tag{1-39c}
$$

$$
r_4 = \sqrt{(x-a)^2 + (y+b)^2 + z^2}
\tag{1-39d}
$$

and the ranges of $\cos^{-1}$ and $\tan^{-1}$ are $[0, \pi]$ and $[-\pi/2, \pi/2]$, respectively. In Eqs. (1–36)–(1-38), $\alpha$ and $E$ are the coefficient of linear thermal expansion and Young's modulus respectively, of the substrate.

Equations (1-36)–(1-38) satisfy the boundary conditions, Eqs. (1-34, 1-35), the elasticity equation, Eq. (1-26), with $X_x = X_y = X_z = 0$, and the stress–displacement equation, Eq. (1-27). The temperature and stress equations given by, respectively, Eq. (1.33) and Eqs. (1-36)–(1-38), are the exact closed-form solutions for the substrate when its surface is subjected to a uniform temperature distribution over a rectangular region $(2a, 2b)$.

Figure 1-2 shows the temperature distributions in the substrate when its surface is subjected to a square $(2a, 2a)$ uniform temperature distribution

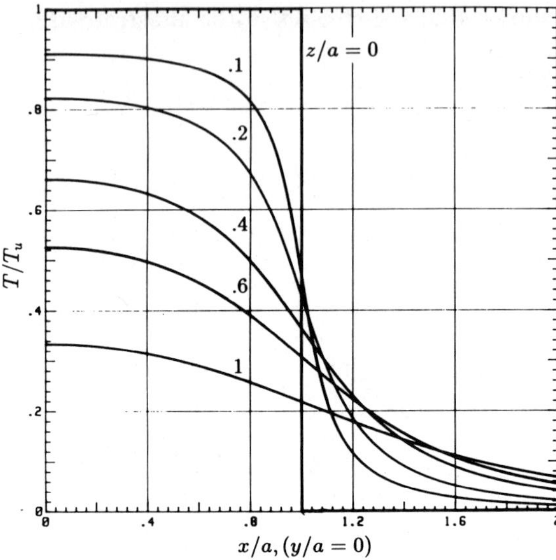


**Figure 1-2**   Temperature distribution along $x/a$ ($y/a = 0$) and through the depth of the substrate. Ref. 37.

with magnitude $T_u$. The plots are along the line $y/a = 0$ with $x/a \geqslant 0$ at various depths from the surface ($z/a = 0, 0.1, 0.2, 0.4, 0.6, 1.0$). It can be seen that the temperature distributions in the substrate ($z/a > 0$) are continuous, even through the edge ($x/a = 1$) of the imposed temperature (Fig. 1.1). Furthermore, the temperature distributions inside the substrate are smaller than those on the surface of the substrate if $x/a \leqslant 1$, but are larger if $x/a \geqslant 1$.

The maximum shear stress ($\tau_{max}$) distributions along $x/a = y/a$ and through the depth of the substrate ($z/a \geqslant 0$) are shown in Fig. 1-3. It can be seen that on the surface of the substrate ($z/a = 0$) and at the corner ($x/a = y/a = 1$) of the imposed temperature, the maximum shear stress tends to infinity. However, it becomes finite once inside the substrate (i.e., $z/a = y/a = 1$, $z/a > 0$). For the plots for other stress components and at other locations of the substrate, see ref. 37.

### 1.5.2  Chip on a Finite Substrate

Figure 1-4 shows a substrate with finite thickness ($2h$). This substrate is subjected to axially symmetrical temperature distributions, $T(r)$, on its upper surface.

The heat-conduction boundary conditions are:

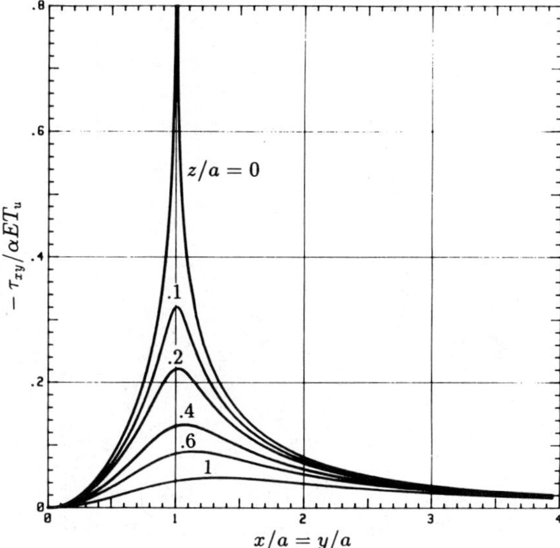



**Figure 1-3**    Shear stress ($T_{xy}$) distribution along $x/a = y/a$, and through the depth of the substrate. Ref. 37.

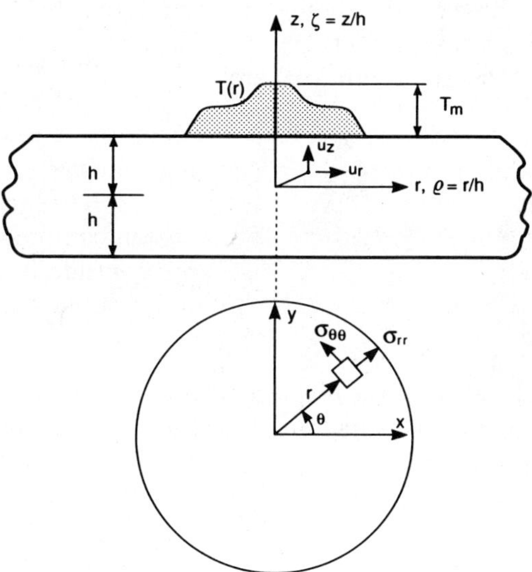



**Figure 1-4**    Thermal displacements and stresses in a substrate (thickness = $2h$) subjected to a temperature $T(r)$ on its upper surface. Ref. 39.

At $z = h$ ($\zeta = 1$)

$$T(\rho, \zeta = 1) = T_m \frac{(g^2 - 4)^2}{8g} \left[ \frac{g - 2}{[\rho2 + (g - 2)^2]^{3/2}} - \frac{g + 2}{[\rho^2 + (g + 2)^2]^{3/2}} \right]$$

(1-40)

where

$$\zeta = \frac{z}{h}$$

(1-41)

$$\rho = \frac{r}{h}$$

(1-42)

and $T_m$ is the temperature at $\rho = 0$. In Eq. (1-40), $g > 2$.

It can be shown that the temperature distribution in the substrate is given by[38,39]

$$T(\rho, \zeta) = T_m \frac{(g^2 - 4)^2}{8g} \left[ \frac{g - 1 - \zeta}{[\rho^2 + (g - 1 - \zeta)^2]^{3/2}} - \frac{g + 1 + \zeta}{[\rho^2 + (g + 1 + \zeta)^2]^{3/2}} \right]$$

(1-43)

Equation (1-43) satisfies the boundary condition, Eq. (1-40), and the heat-conduction equation, Eq. (1-25).

The elasticity boundary conditions are:

At $z = \pm h$

$$\sigma_{zz} = \tau_{zr} = \tau_{z\theta} = 0$$

(1-44)

At $r \to \infty$

$$\sigma_{zz} = \sigma_{rr} = \sigma_{\theta\theta} = \tau_{rz} = \tau_{z\theta} = \tau_{\theta r} \to 0$$

(1-45)

It can also be shown that the displacement components $(u_r, u_z)$ and stress components $(\sigma_{rr}, \sigma_{\theta\theta})$ in the substrate, Fig. 1-4, are given by[38,39]

$$u_r = \frac{(g^2 - 4)^2 (1 + v) \alpha h T_m}{8g} \left[ \frac{g + 1 + \zeta}{\rho[\rho^2 + (g + 1 + \zeta)^2]^{1/2}} \right.$$

$$\left. - \frac{g - 1 - \zeta}{\rho[\rho^2 + (g - 1 - \zeta)^2]^{1/2}} \right]$$

(1-46)

$$u_z = \frac{(g^2 - 4)^2(1 + v)\alpha h T_m}{8g}\left[\frac{1}{[\rho^2 + (g - 1 - \zeta)^2]^{1/2}}\right.$$

$$\left. + \frac{1}{[\rho^2 + (g + 1 + \zeta)^2]^{1/2}}\right] \quad (1\text{-}47)$$

$$\sigma_{\theta\theta} = \frac{(g^2 - 4)^2 E\alpha T_m}{8g}\left[\frac{g + 1 + \zeta}{[\rho^2 + (g + 1 + \zeta)^2]^{3/2}} - \frac{g - 1 - \zeta}{[\rho^2 + (g - 1 - \zeta)^2]^{3/2}}\right.$$

$$\left. + \frac{g + 1 + \zeta}{\rho^2[\rho^2 + (g + 1 + \zeta)^2]^{1/2}} - \frac{g - 1 - \zeta}{\rho^2[\rho^2 + (g - 1 - \zeta)^2]^{1/2}}\right] \quad (1\text{-}48)$$

$$\sigma_{rr} = \frac{(g^2 - 4)^2 E\alpha T_m}{8g}\left[- \frac{g + 1 + \zeta}{\rho^2[\rho^2 + (g + 1 + \zeta)^2]^{1/2}}\right.$$

$$\left. + \frac{g - 1 - \zeta}{\rho^2[\rho^2 + (g - 1 - \zeta)^2]^{1/2}}\right] \quad (1\text{-}49)$$

Equations (1-46)–(1-49) satisfy the boundary conditions, Eqs. (1-44, 1-45), the elasticity equation, Eq. (1-26), with $X_x = X_y = X_z = 0$, and the stress–displacement equation, Eq. (1-27).

Figure 1-5 shows five temperature distributions on the upper surface of the substrate with $g = 2.1, 2.3, 2.5, 2.7,$ and $2.9$, respectively. Figures 1-6 and 1-7

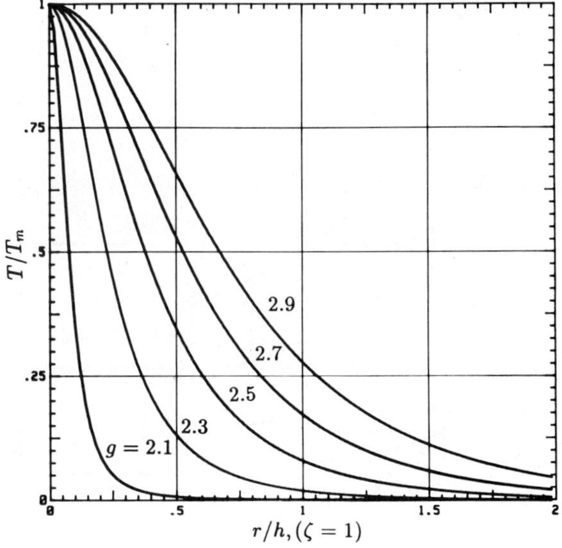


**Figure 1-5**    Temperature distributions $T(r)$ on the upper surface of the substrate. Ref. 39.

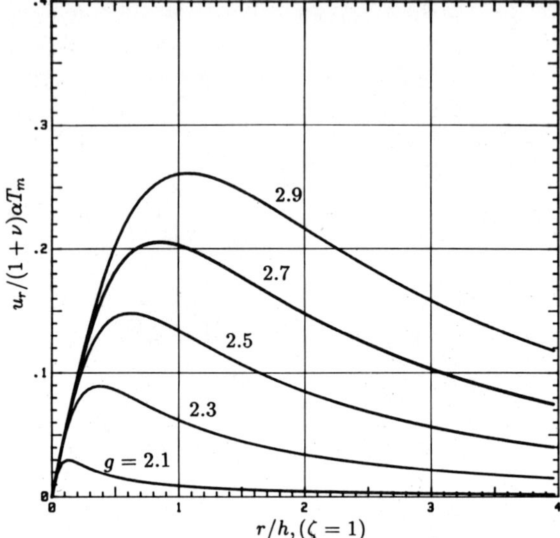

**Figure 1-6**    $u_r$-Displacement on the substrate. Ref. 39.

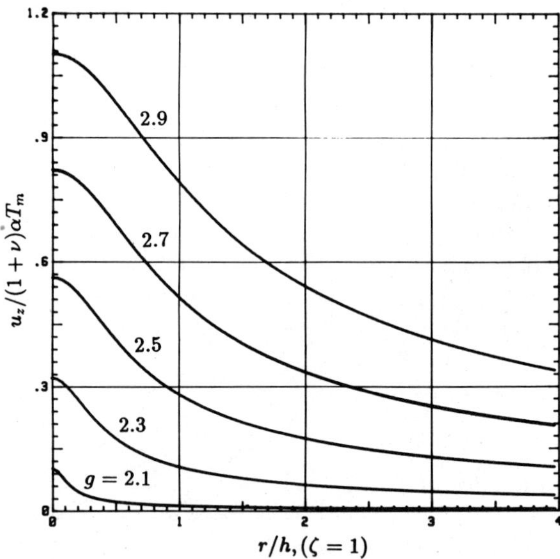

**Figure 1-7**    $u_z$-Displacement on the substrate. Ref. 39.

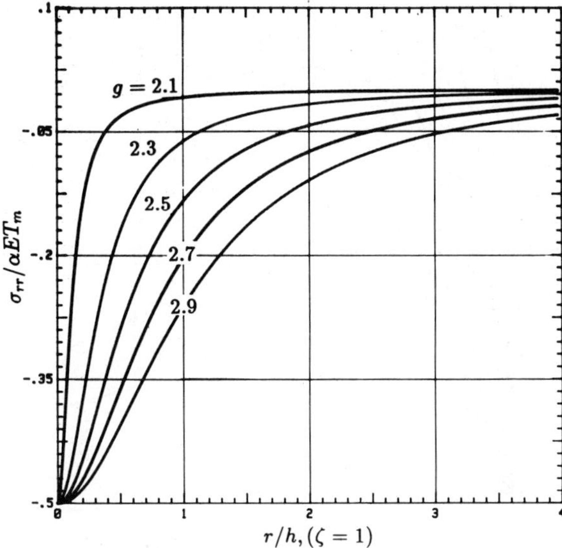



**Figure 1-8**    Radial normal stress on the substrate. Ref. 39.

show the displacement components in the radial ($u_r$) and vertical ($u_z$) directions on the upper surface ($\zeta = 1$), respectively. It can be seen that there is an optimal point of $u_r(\rho)$, and the maximum $u_z$ is at the location of $T_m$ ($\rho = 0$, $\zeta = 1$).

The normal stresses in the radial direction ($\sigma_{rr}$) and in the circumferential direction ($\sigma_{\theta\theta}$) on the surface of the substrate are shown in Figs. 1-8 and 1-9, respectively. It can be seen that $\sigma_{rr}$ is in compression with the absolute maximum value at $\rho = 0$, $\zeta = 1$. The absolute maximum value of the $\sigma_{\theta\theta}$ also occurs at $\rho = 0$, $\zeta = 1$; however, it changes from compression to tension as $\rho$ becomes larger.

The maximum shear stress ($\tau_{\max}$) distributions on the upper surface ($\zeta = 1$) of the substrate for each temperature shown in Fig. 1-5 are shown in Fig. 1-10. It can be seen that, for all the values of $g$, the maximum shear stress is zero at $\rho = 0$ and $\zeta = 1$ (i.e., at the location of the maximum imposed temperature). These maximum shear stresses increase rapidly as soon as they are away from $\rho = 0$ and $\zeta = 1$ and then decrease as $\rho$ increases. For the plots of other stress components and at other locations of the substrate, see ref. 39.

## 1.6 ANALYSIS OF STRESS

### 1.6.1 Three-dimensional Stress State

For a general state of stress at any point in a structure there exist three mutually perpendicular planes at that point on which the shearing stresses

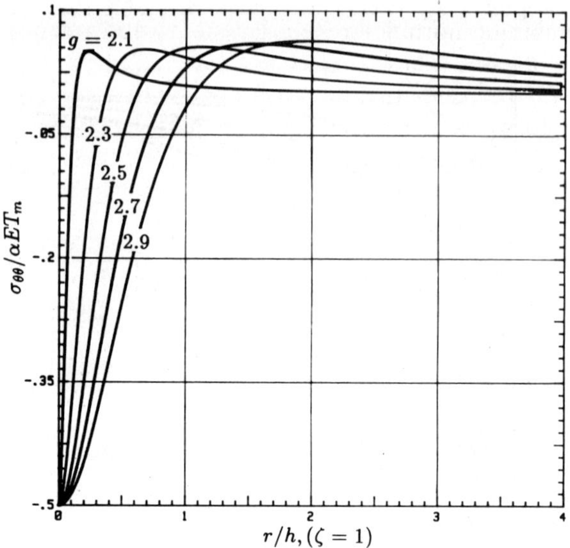


**Figure 1-9**  Circumferential normal stress on the substrate. Ref. 39.

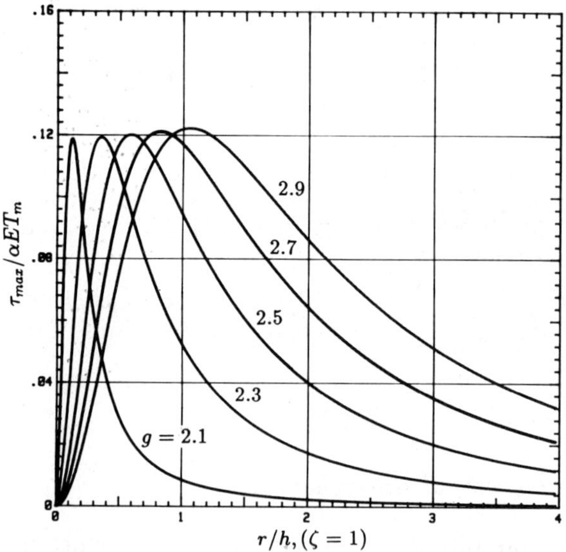


**Figure 1-10**  Maximum shear stress on the substrate. Ref. 39.

vanish. The remaining normal stress components on these three planes are called principal stresses. These three planes are called principal planes and the three mutually perpendicular axes that are normal to the three planes are called principal axes, i.e., the principal axes coincide with the three principal stress directions. Consequently, the principal stresses are, by definition, the stresses that act perpendicularly to the principal planes. These principal stresses can be determined by solving[40]

$$\sigma^3 - I_1\sigma^2 + I_2\sigma - I_3 = 0 \tag{1-50}$$

where

$$I_1 = \sigma_x + \sigma_y + \sigma_z \tag{1-51a}$$

$$I_2 = \sigma_x\sigma_y + \sigma_y\sigma_z + \sigma_z\sigma_x - \tau_{xy}^2 - \tau_{yz}^2 - \tau_{zx}^2 \tag{1-51b}$$

$$I_3 = \sigma_x\sigma_y\sigma_z + 2\tau_{xy}\tau_{yz}\tau_{zx} - \sigma_x\tau_{yz}^2 - \sigma_y\tau_{zx}^2 - \sigma_z\tau_{xy}^2 \tag{1-51c}$$

The three roots of Eq. (1-50) will give the three principal stresses $(\sigma_1, \sigma_2, \sigma_3)$ of the given stress field $(\sigma_x, \sigma_y, \sigma_z, \tau_{xy} = \tau_{yx}, \tau_{yz} = \tau_{zy},$ and $\tau_{zx} = \tau_{xz})$. It can be shown[40] that $\sigma_1, \sigma_2,$ and $\sigma_3$ are not only orthogonal but also real.

The principal direction of the principal stress $\sigma_i$ $(i = 1, 2, 3)$ can be determined by the following simultaneous equations:[41]

$$(\sigma_i - \sigma_x)l_i - \tau_{xy}m_i - \tau_{zx}n_i = 0 \tag{1-52a}$$

$$-\tau_{xy}l_i + (\sigma_i - \sigma_y)m_i - \tau_{yz}n_i = 0 \tag{1-52b}$$

$$-\tau_{zx}l_i - \tau_{yz}m_i + (\sigma_i - \sigma_z)n_i = 0 \tag{1-52c}$$

$$l_i^2 + m_i^2 + n_i^2 = 1 \tag{1-52d}$$

The solutions $(l_i, m_i, n_i)$ of these equations will give the direction of the principal stress $\sigma_i$. In Eqs. (1-52a–d), $l_i, m_i,$ and $n_i$ are the direction cosines of a unit normal vector coinciding with the principal axis of $\sigma_i$.

For example, let the state of stress at a point be given[41] by $\sigma_x = 20$ MPa, $\sigma_y = 40$ MPa, $\sigma_z = -20$ MPa, $\tau_{xy} = -40$ MPa, $\tau_{yz} = 20$ MPa, and $\tau_{zx} = -60$ MPa. Determine the three principal stresses and the directions associated with the three principal stresses. In this case, from Eqs. (1-50, 1-51), we have

$$\sigma^3 - 40\sigma^2 - 6000\sigma + 40\,000 = 0 \tag{1-53}$$

The three roots of this equation are the three principal stresses, $\sigma_1 = 97.39$ MPa, $\sigma_2 = 6.44$ MPa, and $\sigma_3 = -63.83$ MPa.

The direction cosines for $\sigma_1 = 97.39$ MPa can be determined by Eq. (1-52) as follows:

$$(20 - 97.39)l_1 - 40m_1 - 60n_1 = 0 \tag{1-54a}$$

$$-40l_1 + (40 - 97.39)m_1 + 20n_1 = 0 \tag{1-54b}$$

$$-60l_1 + 20m_1 + (-20 - 97.39)n_1 = 0 \tag{1-54c}$$

$$l_1^2 + m_1^2 + n_1^2 = 1 \tag{1-54d}$$

It can be shown that the direction cosines for the principal stress $\sigma_1$ are $l_1 = 0.6574$, $m_1 = -0.6116$, and $n_1 = -0.4402$. The same procedure can be used to determine the principal directions of principal stresses $\sigma_2$ ($l_2 = 0.4488$, $m_2 = 0.7871$, $n_2 = -0.4232$) and $\sigma_3$ ($l_3 = 0.6053$, $m_3 = 0.0807$, $n_3 = 0.7919$).

The stress components with respect to any axes rotated through an angle with respect to the $xyz$-axes are not presented here. However, they can be found in ref. 41.


### 1.6.2 Two-dimensional Stress State

For a two-dimensional plane stress state ($\sigma_x$, $\sigma_y$, and $\tau_{xy}$) in the $xy$-axes, the stress components $\sigma_{x'}$, $\sigma_{y'}$, $\tau_{x'y'}$ with respect to new $x'y'$-axes rotated counterclockwise through an angle $\theta$ with respect to the $xy$-axes are given by

$$\sigma_{x'} = \frac{\sigma_x + \sigma_y}{2} + \frac{\sigma_x - \sigma_y}{2} \cos 2\theta + \tau_{xy} \sin 2\theta \tag{1-55}$$

$$\sigma_{y'} = \frac{\sigma_x + \sigma_y}{2} - \frac{\sigma_x - \sigma_y}{2} \cos 2\theta - \tau_{xy} \sin 2\theta \tag{1-56}$$

$$\tau_{x'y'} = -\frac{\sigma_x - \sigma_y}{2} \sin 2\theta + \tau_{xy} \cos 2\theta \tag{1-57}$$

The principal stresses are given by

$$\sigma_1 = \frac{\sigma_x + \sigma_y}{2} + \sqrt{\tfrac{1}{4}(\sigma_x - \sigma_y)^2 + \tau_{xy}^2} \tag{1-58}$$

$$\sigma_2 = \frac{\sigma_x + \sigma_y}{2} - \sqrt{\tfrac{1}{4}(\sigma_x - \sigma_y)^2 + \tau_{xy}^2} \tag{1-59}$$

and the principal directions of the principal stresses are given by

$$\theta_p = \tfrac{1}{2} \tan^{-1} \frac{2\tau_{xy}}{\sigma_x - \sigma_y} \tag{1-60}$$

where the angle $\theta_p$ is measured counterclockwise from the positive $x$-axis. There are two values of $\theta_p$ and they are 90° apart, one between 0 and 90° and the other between 90° and 180°. The maximum shear stress $\tau_{max}$ and the algebraically minimum shear stress $\tau_{min}$ are given by

$$\tau_{max} = \sqrt{\tfrac{1}{4}(\sigma_x - \sigma_y)^2 + \tau_{xy}^2} \tag{1-61a}$$

$$\tau_{min} = -\tau_{max} \tag{1-61b}$$

and the directions of the planes on which they act are given by

$$\theta_s = \tfrac{1}{2} \tan^{-1} \frac{\sigma_y - \sigma_x}{2\tau_{xy}} \tag{1-61c}$$

where the angle $\theta_s$ is measured counterclockwise from the positive $x$-axis. There are two values of $\theta_s$ and they are 90° apart, one between 0 and 90° and the other between 90° and 180°. It can be shown that the planes of maximum shear stress occur at 45° to the principal planes.

For example,[42] an element in plane stress is subjected to stresses $\sigma_x = 12\,300$ psi (85 MPa), $\sigma_y = -4200$ psi ($-29$ MPa), and $\tau_{xy} = -4700$ psi ($-32$ MPa). Determine the principal stresses and maximum shear stress and the directions associated with these stresses.

By means of Eqs. (1-58, 1-59), we have $\sigma_1 = 13\,540$ psi (93 MPa), and $\sigma_2 = -5440$ psi ($-37$ MPa). The principal directions are, by means of Eq. (1-60), $\theta_p = 75.2°$ and $\theta_p = 165.2°$. To determine which value of $\theta_p$ is associated with each principal stress, substitute $\theta_p = 75.2°$ into Eq. (1-55) and solve for the stress; we have $\sigma_{x'} = -5440$ psi. This result shows that the principal stress $\sigma_2 = -5440$ psi is at the angle $\theta_{p2} = 75.2°$. Similarly, the principal stress $\sigma_1 = 13\,540$ psi is at the angle $\theta_{p1} = 165.2°$. Also, the maximum shear stresses $\tau_{xy}$ can be determined by Eq. (1-61) and are equal to $\pm 9490$ psi ($\pm 65$ MPa). The angle to the plane having the positive maximum shear stress is 120.2° and the angle to the plane having the algebraically minimum shear stress is 30.2°.

## 1.7 ANALYSIS OF STRAIN

### 1.7.1 Three-dimensional Strain State

Just as for the general state of stress, for a general state of strain at any point in a structure there exist three mutually perpendicular planes at that point on which the shearing strains vanish. The remaining normal strain components on these three planes are called principal strains. These three planes are called principal planes and the three mutually perpendicular axes that are normal to the three planes are called principal axes, i.e., the principal axes coincide with the three principal strain directions. Consequently, the principal strains are, by definition, the strains that act perpendicularly to the principal planes. These principal strains can be determined by solving[40]

$$\epsilon^3 - J_1\epsilon^2 + J_2\epsilon - J_3 = 0 \tag{1-62}$$

where

$$J_1 = \epsilon_x + \epsilon_y + \epsilon_z \tag{1-63a}$$

$$J_2 = \epsilon_x\epsilon_y + \epsilon_y\epsilon_z + \epsilon_z\epsilon_x - \tfrac{1}{4}(\gamma_{xy}^2 + \gamma_{yz}^2 + \gamma_{zx}^2) \tag{1-63b}$$

$$J_3 = \epsilon_x\epsilon_y\epsilon_z + \tfrac{1}{4}(\gamma_{xy}\gamma_{yz}\gamma_{zx} - \epsilon_x\gamma_{yz}^2 - \epsilon_y\gamma_{zx}^2 - \epsilon_z\gamma_{xy}^2) \tag{1-63c}$$

The three roots of Eq. (1-62) will give the three principal strains $(\epsilon_1, \epsilon_2, \epsilon_3)$ of the given strain field $(\epsilon_x, \epsilon_y, \epsilon_z, \gamma_{xy} = \gamma_{yx}, \gamma_{yz} = \gamma_{zy}, \text{ and } \gamma_{zx} = \gamma_{xz})$. It can be shown[40] that $\epsilon_1, \epsilon_2,$ and $\epsilon_3$ are not only orthogonal but also real.

The principal direction of the principal strain $\epsilon_i$ $(i = 1, 2, 3)$ can be determined by the following simultaneous equations:[41]

$$2(\epsilon_i - \epsilon_x)l_i - \gamma_{xy}m_i - \gamma_{zx}n_i = 0 \tag{1-64a}$$

$$-\gamma_{xy}l_i + 2(\epsilon_i - \epsilon_y)m_i - \gamma_{yz}n_i = 0 \tag{1-64b}$$

$$-\gamma_{zx}l_i - \gamma_{yz}m_i + 2(\epsilon_i - \epsilon_z)n_i = 0 \tag{1-64c}$$

$$l_i^2 + m_i^2 + n_i^2 = 1 \tag{1-64d}$$

The solutions $(l_i, m_i, n_i)$ of these equations will give the direction of the principal strain $\epsilon_i$. In Eqs. (1-64a–d), $l_i, m_i,$ and $n_i$ are the direction cosines of a unit normal vector coinciding with the principal axis of $\epsilon_i$.

The strain–displacement equations, Eqs. (1-29a–f), have six equations for three unknown displacement component functions $u, v, w$. They will not have a single-valued solution in general if the strain component functions $\epsilon_x, \epsilon_y, \epsilon_z, \gamma_{xy}, \gamma_{yz},$ and $\gamma_{zx}$ are arbitrarily assigned. In order to have a single-valued solution for $u, v, w$, these strain components have to satisfy the following *equations of compatibility:*[43]

$$2\frac{\partial^2 \epsilon_x}{\partial y\,\partial z} = \frac{\partial}{\partial x}\left(-\frac{\partial \gamma_{yz}}{\partial x} + \frac{\partial \gamma_{xz}}{\partial y} + \frac{\partial \gamma_{xy}}{\partial z}\right) \tag{1-65a}$$

$$2\frac{\partial^2 \epsilon_y}{\partial x\,\partial z} = \frac{\partial}{\partial y}\left(\frac{\partial \gamma_{yz}}{\partial x} - \frac{\partial \gamma_{xz}}{\partial y} + \frac{\partial \gamma_{xy}}{\partial z}\right) \tag{1-65b}$$

$$2\frac{\partial^2 \epsilon_z}{\partial x\,\partial y} = \frac{\partial}{\partial z}\left(\frac{\partial \gamma_{yz}}{\partial x} + \frac{\partial \gamma_{xz}}{\partial y} - \frac{\partial \gamma_{xy}}{\partial z}\right) \tag{1-65c}$$

$$\frac{\partial^2 \epsilon_x}{\partial y^2} + \frac{\partial^2 \epsilon_y}{\partial x^2} = \frac{\partial^2 \gamma_{xy}}{\partial x\,\partial y} \tag{1-65d}$$

$$\frac{\partial^2 \epsilon_y}{\partial z^2} + \frac{\partial^2 \epsilon_z}{\partial y^2} = \frac{\partial^2 \gamma_{yz}}{\partial y\,\partial z} \tag{1-65e}$$

$$\frac{\partial^2 \epsilon_z}{\partial x^2} + \frac{\partial^2 \epsilon_x}{\partial z^2} = \frac{\partial^2 \gamma_{xz}}{\partial x\,\partial z} \tag{1-65f}$$

For example, the following strain components

$$\epsilon_x = 2x^2 + 3y^2 + z + 1 \tag{1-66a}$$

$$\epsilon_y = x^2 + 2y^2 + 3z + 2 \tag{1-66b}$$

$$\epsilon_z = 3x + 2y + z^2 + 1 \tag{1-66c}$$

$$\gamma_{xy} = 8xy \tag{1-66d}$$

$$\gamma_{yz} = 0 \tag{1-66e}$$

$$\gamma_{zx} = 0 \tag{1-66f}$$

satisfy Eqs. (1-64*a–f*) and are compatible. On the other hand, the following strain components

$$\epsilon_x = 3y^2 + xy \tag{1-67a}$$

$$\epsilon_y = 2y + 4z + 3 \tag{1-67b}$$

$$\epsilon_z = 2xy + 3yz + 3zx + 2 \tag{1-67c}$$

$$\gamma_{xy} = 6xy \tag{1-67d}$$

$$\gamma_{yz} = 2x \tag{1-67e}$$

$$\gamma_{zx} = 2y \tag{1-67f}$$

do not satisfy Eqs. (1-64*a–f*) and are not compatible.

### 1.7.2 Two-dimensional Strain State

For a two-dimensional plane strain state $(\epsilon_x, \epsilon_y, \gamma_{xy})$ in the $xy$-plane, the strain components $\epsilon_{x'}$, $\epsilon_{y'}$, $\gamma_{x'y'}$ with respect to new $x'y'$-axes rotated counterclockwise through an angle $\theta$ with respect to the $xy$-axes are given by

$$\epsilon_{x'} = \frac{\epsilon_x + \epsilon_y}{2} + \frac{\epsilon_x - \epsilon_y}{2} \cos 2\theta + \tfrac{1}{2}\gamma_{xy} \sin 2\theta \qquad (1\text{-}68a)$$

$$\epsilon_{y'} = \frac{\epsilon_x + \epsilon_y}{2} - \frac{\epsilon_x - \epsilon_y}{2} \cos 2\theta - \tfrac{1}{2}\gamma_{xy} \sin 2\theta \qquad (1\text{-}68b)$$

$$\gamma_{x'y'} = -\frac{\epsilon_x - \epsilon_y}{2} \sin 2\theta + \tfrac{1}{2}\gamma_{xy} \cos 2\theta \qquad (1\text{-}68c)$$

The principal strains are given by

$$\epsilon_1 = \frac{\epsilon_x + \epsilon_y}{2} + \tfrac{1}{2}\sqrt{(\epsilon_x - \epsilon_y)^2 + \gamma_{xy}^2} \qquad (1\text{-}69)$$

$$\epsilon_2 = \frac{\epsilon_x + \epsilon_y}{2} - \tfrac{1}{2}\sqrt{(\epsilon_x - \epsilon_y)^2 + \gamma_{xy}^2} \qquad (1\text{-}70)$$

and the principal directions of the principal strains are given by

$$\theta_p = \tfrac{1}{2} \tan^{-1} \frac{\gamma_{xy}}{\epsilon_x - \epsilon_y} \qquad (1\text{-}71)$$

where the angle $\theta_p$ is measured counterclockwise from the positive $x$-axis and has two values differing by $90°$. For isotropic materials, the principal directions of strain coincide with those of stress. The maximum shear strain and the algebraically minimum shear strain are given by

$$\gamma_{\max} = \sqrt{(\epsilon_x - \epsilon_y)^2 + \gamma_{xy}^2} \qquad (1\text{-}72a)$$

$$\gamma_{\min} = -\gamma_{\max} \qquad (1\text{-}72b)$$

and the directions of the planes on which they act occur at $\theta_p \pm \pi/4$.

For example,[42] an element is at a state of plane strain with $\epsilon_x = 340 \times 10^{-6}$, $\epsilon_y = 110 \times 10^{-6}$, $\gamma_{xy} = 180 \times 10^{-6}$. Determine the strains at an element rotated through an angle $\theta = 30°$ from the positive $x$-axis. Also determine the principal strains and maximum shear strains and the directions associated with these strains.

By means of Eqs. (1-68$a$–$c$), we have $\epsilon_{x'} = 360 \times 10^{-6}$, $\epsilon_{y'} = 90 \times 10^{-6}$, and $\gamma_{x'y'} = -110 \times 10^{-6}$. By means of Eqs. (1-69)–(1-71), we have $\epsilon_1 = 370 \times 10^{-6}$, $\theta_{p1} = 19°$, and $\epsilon_2 = 80 \times 10^{-6}$, $\theta_{p2} = 109°$. Also, by means of Eq. (1-72), we have that the angle to the plane having the positive maximum shear strain ($\gamma_{max} = 290 \times 10^{-6}$) is 154.2° and the angle to the plane having the algebraically minimum shear strain ($\gamma_{min} = -290 \times 10^{-6}$) is 64°.

## 1.8 GEOMETRIC NONLINEARITY

When an electronics packaging system is under very large deformations, e.g., if the maximum displacement of a printed circuit board (PCB) is larger than the thickness of the PCB, the strain–displacement relations, Eqs. (1-29$a$–$f$), have to be modified in order to account for the excessive deformations. There are at least two popular ways to define the strain tensor for large deformations and they are briefly mentioned in this section.

### 1.8.1 Strain Components in Lagrangian Coordinates

The most popular strain tensor for large deformations was introduced by Green and St. Venant (Green's strain tensor) and is often referred to as a strain tensor in Lagrangian coordinates (see below).

$$\epsilon_x = \frac{\partial u}{\partial x} + \frac{1}{2}\left[\left(\frac{\partial u}{\partial x}\right)^2 + \left(\frac{\partial v}{\partial x}\right)^2 + \left(\frac{\partial w}{\partial x}\right)^2\right] \tag{1-73a}$$

$$\epsilon_y = \frac{\partial v}{\partial y} + \frac{1}{2}\left[\left(\frac{\partial u}{\partial y}\right)^2 + \left(\frac{\partial v}{\partial y}\right)^2 + \left(\frac{\partial w}{\partial y}\right)^2\right] \tag{1-73b}$$

$$\epsilon_z = \frac{\partial w}{\partial z} + \frac{1}{2}\left[\left(\frac{\partial u}{\partial z}\right)^2 + \left(\frac{\partial v}{\partial z}\right)^2 + \left(\frac{\partial w}{\partial z}\right)^2\right] \tag{1-73c}$$

$$\gamma_{xy} = \frac{\partial u}{\partial y} + \frac{\partial v}{\partial x} + \frac{\partial u}{\partial x}\frac{\partial u}{\partial y} + \frac{\partial v}{\partial x}\frac{\partial v}{\partial y} + \frac{\partial w}{\partial x}\frac{\partial w}{\partial y} \tag{1-73d}$$

$$\gamma_{yz} = \frac{\partial v}{\partial z} + \frac{\partial w}{\partial y} + \frac{\partial u}{\partial y}\frac{\partial u}{\partial z} + \frac{\partial v}{\partial y}\frac{\partial v}{\partial z} + \frac{\partial w}{\partial y}\frac{\partial w}{\partial z} \tag{1-73e}$$

$$\gamma_{zx} = \frac{\partial w}{\partial x} + \frac{\partial u}{\partial z} + \frac{\partial u}{\partial z}\frac{\partial u}{\partial x} + \frac{\partial v}{\partial z}\frac{\partial v}{\partial x} + \frac{\partial w}{\partial z}\frac{\partial w}{\partial x} \tag{1-73f}$$

It should be pointed out that the displacement components ($u$, $v$, $w$) in Eqs. (1-73$a$–$f$) are considered as functions of ($x$, $y$, $z$), the positions of points in the structure in the *unstrained configuration*.

### 1.8.2 Strain Components in Eulerian Coordinates

The other popular way to define the strain tensor for larger deformations was introduced by Cauchy for infinitesimal strains and by Almansi and Hamel for finite strains (Almansi's strain tensor) and is often referred to as a strain tensor in Eulerian coordinates (see below).

$$\epsilon_x = \frac{\partial u}{\partial \bar{x}} - \frac{1}{2}\left[\left(\frac{\partial u}{\partial \bar{x}}\right)^2 + \left(\frac{\partial v}{\partial \bar{x}}\right)^2 + \left(\frac{\partial w}{\partial \bar{x}}\right)^2\right] \tag{1-74a}$$

$$\epsilon_y = \frac{\partial v}{\partial \bar{y}} - \frac{1}{2}\left[\left(\frac{\partial u}{\partial \bar{y}}\right)^2 + \left(\frac{\partial v}{\partial \bar{y}}\right)^2 + \left(\frac{\partial w}{\partial \bar{y}}\right)^2\right] \tag{1-74b}$$

$$\epsilon_z = \frac{\partial w}{\partial \bar{z}} - \frac{1}{2}\left[\left(\frac{\partial u}{\partial \bar{z}}\right)^2 + \left(\frac{\partial v}{\partial \bar{z}}\right)^2 + \left(\frac{\partial w}{\partial \bar{z}}\right)^2\right] \tag{1-74c}$$

$$\gamma_{xy} = \frac{\partial u}{\partial \bar{y}} + \frac{\partial v}{\partial \bar{x}} - \frac{\partial u}{\partial \bar{x}}\frac{\partial u}{\partial \bar{y}} - \frac{\partial v}{\partial \bar{x}}\frac{\partial v}{\partial \bar{y}} - \frac{\partial w}{\partial \bar{x}}\frac{\partial w}{\partial \bar{y}} \tag{1-74d}$$

$$\gamma_{yz} = \frac{\partial v}{\partial \bar{z}} + \frac{\partial w}{\partial \bar{y}} - \frac{\partial u}{\partial \bar{y}}\frac{\partial u}{\partial \bar{z}} - \frac{\partial v}{\partial \bar{y}}\frac{\partial v}{\partial \bar{z}} - \frac{\partial w}{\partial \bar{y}}\frac{\partial w}{\partial \bar{z}} \tag{1-74e}$$

$$\gamma_{zx} = \frac{\partial w}{\partial \bar{x}} + \frac{\partial u}{\partial \bar{z}} - \frac{\partial u}{\partial \bar{z}}\frac{\partial u}{\partial \bar{x}} - \frac{\partial v}{\partial \bar{z}}\frac{\partial v}{\partial \bar{x}} - \frac{\partial w}{\partial \bar{z}}\frac{\partial w}{\partial \bar{x}} \tag{1-74f}$$

It should be pointed out that the displacement components $(u, v, w)$ in Eqs. (1-74a–f) are considered as functions of $(\bar{x}, \bar{y}, \bar{z})$, the positions of points in the structure in the *strained configuration*. For infinitesimal deformations, the quadratic terms in Eqs. (1-73a–f) and Eqs. (1-74a–f) are neglected. Then the distinction between the Lagrangian and Eulerian strain tensor disappears and Eqs. (1-73a–f) and Eqs. (1-74a–f) reduce to Eqs. (1-29a–f).

### 1.8.3 Large-Deflection Example Problem for Electronics Packaging

Closed-form solutions for problems involving Eq. (1-73) or Eq. (1-74) are very difficult to obtain. Results of this class of problem are usually generated by either finite element or numerical integration methods. In this chapter, however, we will present a closed-form solution for a very simple electronics packaging application.

Figure 1-11 shows a flexible curved optical glass fiber subjected to two diametrically opposite forces $(F)$. The deflections at point B and the bending moment at points A and B are of interest. The free body diagram of a portion

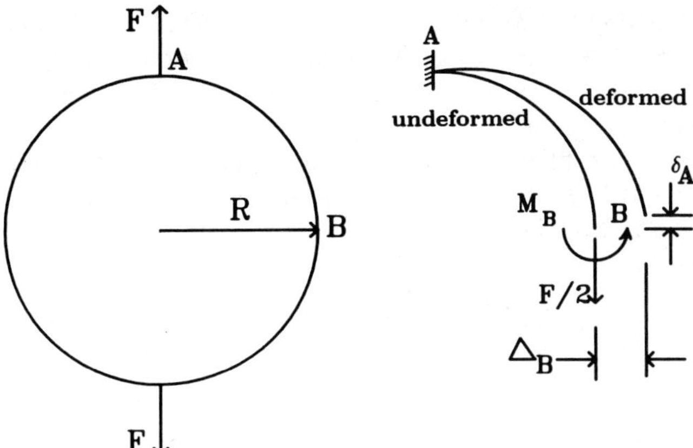

**Figure 1-11**   Circular ring subjected to two diametrically opposite forces. Ref. 49.

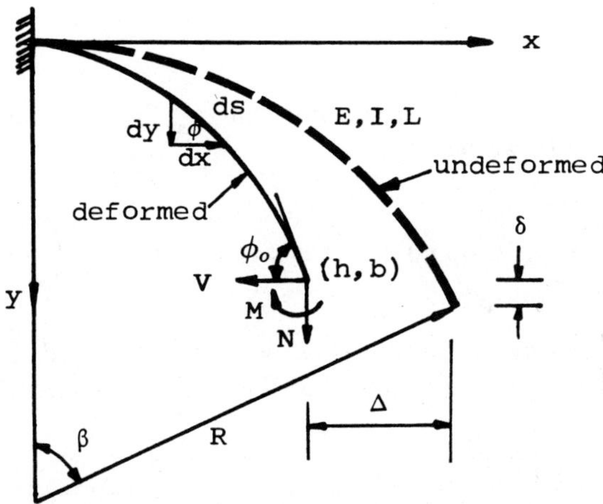

**Figure 1-12**   Large deflection of curved bar under combined loads. Ref. 49.

(defined by the angle $\beta$) of the curved fiber is shown in Fig. 1-12. It can be seen that the fiber is fixed (clamped) at one end and subjected to a horizontal force, a vertical force, and a bending moment at the other end. It can be shown[44] that the differential equation of the deflected curved fiber shown in Fig. 1-11 is

$$\frac{d\phi}{ds} = \frac{1}{R} + \frac{N}{D}(h - x) + \frac{V}{D}(b - y) + \frac{M}{D} \tag{1-75}$$

where $\phi$ = the slope at any point along the deflected curved fiber, $s$ = arc length coordinate, $R$ = radius of the undeformed curved fiber, $N$ = vertical concentrated load, $V$ = horizontal concentrated load, $M$ = bending moment, $D = EI$ = flexural rigidity, $E$ = Young's modulus, $I$ = second moment of the area of the curved fiber, $(h, b)$ = the coordinates of the free end of the deformed curved fiber, and $(x, y)$ = the Cartesian coordinates.

Differentiating Eq. (1-75) with respect to $s$, we have

$$\frac{d^2\phi}{ds^2} = -\frac{1}{D}(N\cos\phi + V\sin\phi) \tag{1-76}$$

where

$$\sin\phi = \frac{dy}{ds} \tag{1-77}$$

$$\cos\phi = \frac{dx}{ds} \tag{1-78}$$

Hence, after integrating and noting that at the free end $\phi = \phi_0$ and $d\phi/ds = 1/R + M/D$, we have

$$\sqrt{\frac{D}{2}}\frac{d\phi}{ds} = \sqrt{N(\sin\phi_0 - \sin\phi) + V(\cos\phi - \cos\phi_0) + \frac{D}{2}\left(\frac{1}{R} + \frac{M}{D}\right)^2} \tag{1-79}$$

where the positive sign of the root is chosen to satisfy $d\phi/ds \geq 0$. Separating variables of Eq. (1-79), and noting that the beam is inextensible, we have

$$\sqrt{2}\int_0^{L=\beta R} ds = \int_0^{\phi_0} \frac{R\,d\phi}{\sqrt{\gamma(\xi\cos\phi - \sin\phi + \lambda)}} \tag{1-80}$$

where

$$\xi = \frac{V}{N} \tag{1-81}$$

$$\gamma = \frac{NR^2}{D} \tag{1-82}$$

$$\eta = \frac{MR}{D} \tag{1-83}$$

and

$$\lambda = \sin \phi_0 - \xi \cos \phi_0 + \frac{1}{2\gamma} (1 + \eta)^2 \tag{1-84}$$

Equation (1-80) can be integrated in closed form and is given (where $\beta$ is the angle of the undeformed fiber and is shown in Fig. 1-12) by

$$\beta = V_1 [F(x_2, \kappa) - F(x_1, \kappa)] \qquad \text{for } \kappa^2 < 1 \tag{1-85}$$

and

$$\beta = V_2 [F(z_2, r) - F(z_1, r)] \qquad \text{for } \kappa^2 \geqslant 1 \tag{1-86}$$

where

$$\kappa^2 = \frac{2\sqrt{1 + \xi^2}}{(\lambda\gamma/|\gamma|) + \sqrt{1 + \xi^2}} \tag{1-87}$$

$$r = \frac{1}{\kappa} \tag{1-88}$$

$$x_1 = \tfrac{1}{2} \tan^{-1}(1/\xi) \tag{1-89}$$

$$x_2 = \tfrac{1}{2} [\phi_0 + \tan^{-1}(1/\xi)] \tag{1-90}$$

$$z_1 = \sin^{-1}(\kappa \sin(x_1)) \tag{1-91}$$

$$z_2 = \sin^{-1}(\kappa \sin(x_2)) \tag{1-92}$$

$$V_1 = \sqrt{\frac{2}{\lambda\gamma + |\gamma|\sqrt{1 + \xi^2}}} \tag{1-93}$$

$$V_2 = \sqrt{\frac{1}{|\gamma|\sqrt{1 + \xi^2}}} \tag{1-94}$$

and $F(\,,\,)$ is the elliptic integral of the first kind.

The vertical ($\delta$) and horizontal ($\Delta$) deflections at the free end of the curved fiber can be obtained from Eqs. (1-77)–(1-79) and are given by the following.

For $\kappa^2 < 1$

$$\frac{\delta}{R} = \nabla_3\left(\frac{2 - \kappa^2}{2}[F(x_2, \kappa) - F(x_1, \kappa)] - E(x_2, \kappa) + E(x_1, \kappa) - \xi\kappa_2 + \xi\kappa_1\right)$$
$$- 1 + \cos\beta \tag{1-95}$$

$$\frac{\Delta}{R} = \nabla_3\left(\xi[E(x_2, \kappa) - E(x_1, \kappa)] - \frac{\xi(2 - \kappa^2)}{2}[F(x_2, \kappa) - F(x_1, \kappa)] - \kappa_2 + \kappa_1\right)$$
$$- \sin\beta \tag{1-96}$$

For $\kappa^2 \geqslant 1$

$$\frac{\delta}{R} = \nabla_4\left(\tfrac{1}{2}[F(z_2, r) - F(z_1, r)] - E(z_2, r) + E(z_1, r) - \frac{\xi}{\kappa}(\kappa_2 - \kappa_1)\right)$$
$$- 1 + \cos\beta \tag{1-97}$$

$$\frac{\Delta}{R} = \nabla_4\left(\xi[E(z_2, r) - E(z_1, r)] - \frac{\xi}{2}[F(z_2, r) - F(z_1, r)] - \frac{1}{\kappa}(\kappa_2 - \kappa_1)\right)$$
$$- \sin\beta \tag{1-98}$$

where

$$\kappa_1 = \sqrt{1 - \kappa^2\sin^2(x_1)} \tag{1-99}$$

$$\kappa_2 = \sqrt{1 - \kappa^2\sin^2(x_2)} \tag{1-100}$$

$$\nabla_3 = \frac{\gamma}{|\gamma|}\sqrt{\frac{2[1 + (\gamma\lambda/|\gamma|\sqrt{1 + \xi^2})]}{|\gamma|(1 + \xi^2)\sqrt{1 + \xi^2}}} \tag{1-101}$$

$$\nabla_4 = \frac{2\gamma/|\gamma|}{\sqrt{|\gamma|(1 + \xi^2)\sqrt{1 + \xi^2}}} \tag{1-102}$$

and $E(\,,)$ is the elliptic integral of the second kind.

Figures 1-13$a$ and 1-13$b$ show the plots of $\delta/R$ and $\Delta/R$ versus $\beta$ for $\xi = \lambda = \eta = 0.4$. Also shown are the results obtained from linear theory.[41] It can be seen that they are very different. The reliability of the present closed-form solutions has been partially verified by checking five degenerate cases with published results.[45–48] For other values of $\xi$, $\lambda$, and $\eta$, see ref. 44.

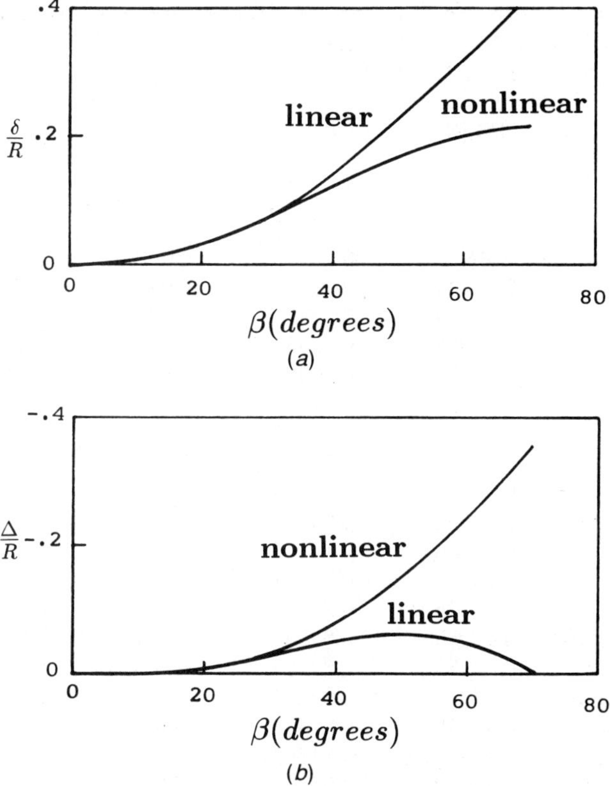


**Figure 1-13**   (a) Vertical ($\delta/R$) and (b) horizontal ($\Delta/R$) deflections for $\xi = \lambda = \eta = 0.4$. Ref. 49.

Equations (1-85–1-86, 1-95–1-98) are the closed-form solutions for a cantilever circular rod of any length subjected to combined loads at the free end. For our present problem, in Eqs. (1-85–1-86, 1-95–1-98), $\beta = \phi_0 = 90°$, $N = F/2$, $V = 0$, and $M = -M_B$ and the results are shown in Table 1-1.


## 1.9 MATERIAL NONLINEARITY

Fundamentally, most electronics packaging materials are nonlinear, although for practical applications it is sometimes possible to gain some insight into electronics packaging problems by assuming the materials to be linear. In order to have an optimal design and perform failure analysis of electronics packaging components and systems, however, nonlinear behavior of materials has to be considered. Furthermore, with the help of high-speed computers, engineers nowadays purposely design packaging components and

**Table 1-1**  Bending Moments $M_A$ and $M_B$ and Maximum Deflections $\delta_A$ and $\Delta_B$ for a Flexible Ring Subjected to Two Diametrically Opposite Forces $F$ (Ref. 49)

| $\dfrac{FR^2}{EI}$ | $\dfrac{\lvert M_B\rvert R}{EI}$ | $\dfrac{\lvert M_A\rvert R}{EI}$ | $\dfrac{\lvert\delta_A\rvert}{R}$ | $\dfrac{\lvert\Delta_B\rvert}{R}$ |
|---|---|---|---|---|
| 0.2 | 0.03559 | 0.06305 | 0.01454 | 0.01355 |
| 0.4 | 0.06970 | 0.12493 | 0.02843 | 0.02687 |
| 0.6 | 0.10241 | 0.18560 | 0.04169 | 0.03994 |
| 0.8 | 0.13375 | 0.24515 | 0.05435 | 0.05276 |
| 1.0 | 0.16379 | 0.30355 | 0.06644 | 0.06532 |
| 1.2 | 0.19257 | 0.36086 | 0.07799 | 0.07761 |
| 1.4 | 0.22015 | 0.41710 | 0.08903 | 0.08964 |
| 1.6 | 0.24659 | 0.47229 | 0.09957 | 0.10140 |
| 1.8 | 0.27193 | 0.52647 | 0.10965 | 0.11290 |
| 2.0 | 0.29623 | 0.57965 | 0.11929 | 0.12412 |
| 2.2 | 0.31952 | 0.63189 | 0.12851 | 0.13508 |
| 2.4 | 0.34186 | 0.68320 | 0.13734 | 0.14579 |
| 2.6 | 0.36329 | 0.73361 | 0.14579 | 0.15623 |
| 2.8 | 0.38385 | 0.78316 | 0.15389 | 0.16642 |
| 3.0 | 0.40359 | 0.83185 | 0.16166 | 0.17637 |
| 3.2 | 0.42253 | 0.87975 | 0.16910 | 0.18607 |
| 3.4 | 0.44072 | 0.92686 | 0.17624 | 0.19554 |
| 3.6 | 0.45819 | 0.97322 | 0.18310 | 0.20478 |
| 3.8 | 0.47498 | 1.01883 | 0.18969 | 0.21379 |
| 4.0 | 0.49112 | 1.06373 | 0.19602 | 0.22258 |
| 4.2 | 0.50663 | 1.10795 | 0.20211 | 0.23116 |
| 4.4 | 0.52156 | 1.15148 | 0.20797 | 0.23952 |
| 4.6 | 0.53591 | 1.19440 | 0.21361 | 0.24769 |
| 4.8 | 0.54972 | 1.23669 | 0.21904 | 0.25566 |
| 5.0 | 0.56302 | 1.27836 | 0.22427 | 0.26345 |

systems to be stressed beyond the elastic limit into the nonlinear region. In this section, nonlinear theories of materials such as hyperelasticity, plasticity, viscoelasticity, viscoplasticity, and creep are briefly discussed.


### 1.9.1 Hyperelasticity

The objective of the theory of hyperelasticity is to offer a mathematical description of the nonlinear mechanical behavior of materials in the *large-strain range*. It is usually applied to solid propellant, elastomer, rubber, and other rubberlike materials. In electronics packaging applications, for example, the rubberlike materials can be used as interface structures between the cap and the substrate of a thermal conduction module.

A hyperelastic material is characterized by the existence of a strain energy $W$, measured per unit volume of the undeformed state, that is a function of the deformation gradient.[50–62] The form of the strain energy as a function of the strain components is restricted by any symmetry properties that the material may possess in the undeformed state. For an isotropic material (no preferred directions) the form of $W$ does not depend on the orientation of the coordinate system used to describe the initial locations of the particles and depends only on the strain components through the three principal extension (stretch) ratios $\lambda_1$, $\lambda_2$, $\lambda_3$ (i.e., ratio of the deformed dimension to the undeformed dimension).

$$W = \sum_{i+j=1}^{N} C_{ij}(J_1 - 3)^i(J_2 - 3)^j + \sum_{i=1}^{N} \frac{1}{D_i}(J - 1 - R)^{2i} \tag{1-103}$$

where

$$J = \lambda_1\lambda_2\lambda_3 \tag{1-104}$$

$$J_1 = J^{-2/3}(\lambda_1^2 + \lambda_2^2 + \lambda_3^2) \tag{1-105}$$

$$J_2 = J^{-4/3}[(\lambda_1\lambda_2)^2 + (\lambda_2\lambda_3)^2 + (\lambda_3\lambda_1)^2] \tag{1-106}$$

$$R = (1 + \alpha\,\Delta T)^3 - 1 \tag{1-107}$$

and $\alpha$ is the thermal coefficient of expansion of the material, $\Delta T$ is the temperature change, $R$ is the nominal volumetric strain associated with $\Delta T$, $J$ is the ratio of current to original volume, $C_{ij}$ and $D_i$ are constants, and $N$ is an integer usually less than 3.

For materials such as elastomer and rubber, which undergo little change in volume at stress levels that cause severe deformations, the material can be treated as incompressible ($J = 1$) and Eq. (1-103) becomes (with $N = 2$)

$$W = C_{01}(J_2 - 3) + C_{10}(J_1 - 3) + C_{11}(J_1 - 3)(J_2 - 3)$$
$$+ C_{02}(J_2 - 3)^2 + C_{20}(J_1 - 3)^2 \tag{1-108}$$

The material constants $C_{01}$, $C_{10}$, $C_{11}$, $C_{02}$, $C_{20}$ can be determined by experiment,[53,54] e.g., the simple tension test, pure shear test, and equibiaxial tension test.

For $N = 1$, Eq. (1-108) becomes ($C_1 = C_{10}$, $C_2 = C_{01}$)

$$W = C_1(J_1 - 3) + C_2(J_2 - 3) \tag{1-109}$$

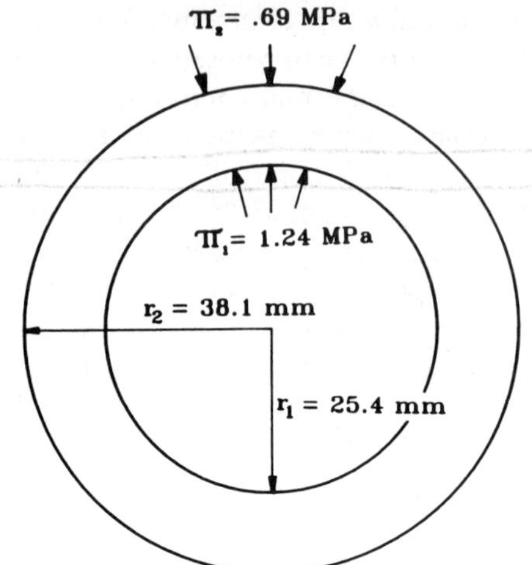


**Figure 1-14**  Hyperelastic thick spherical shell subjected to internal and external pressures. Ref. 62.

which is sometimes referred to as the Mooney–Rivlin law. It has been found to adequately represent rubberlike materials over a moderate range of strain. For strains larger than 450–500 percent, however, experiments on natural rubbers indicate a departure from the Mooney–Rivlin law. In those cases, Eq. (1-108) is necessary.

Figure 1-14 shows a thick-walled spherical shell. The shell material was assumed to be of the Mooney–Rivlin type with material constants $C_1$ and $C_2$ of 0.334 and 0.074 MPa, respectively. The internal and external radii of the shell in the unstrained and strained states are denoted by $r_1$, $r_2$ and $R_1$, $R_2$, respectively, with $r_1 = 25.4$ mm and $r_2 = 38.1$ mm. The sphere was subjected to an internal pressure $\Pi_1$ and an external pressure $\Pi_2$. These pressures were increased at a constant ratio to final values of 1.24 Mpa and 0.69 Mpa, respectively.

It can be shown that the closed-form solutions of the foregoing problem can be represented by the following equations:[53, 54, 62]

$$(R_1^3 - r_1^3)\Pi_1 - (R_2^3 - r_2^3)\Pi_2$$

$$= r_1^3 W_1 - r_2^3 W_2 + 3\left\{ -(C_1 + C_2)(R_2^3 - R_1^3) + C_1 \int_{R_1}^{R_2} \left( Q^4 + \frac{2}{Q^2} \right) R^2 \, dR \right.$$

$$\left. + C_2 \int_{R_1}^{R_2} \left( 2Q^2 + \frac{1}{Q^4} \right) R^2 \, dR \right\} \tag{1-110}$$

where

$$R_2^3 = r_2^3 - r_1^3 + R_1^3 \tag{1-111}$$

$$Q^3 = 1 + \frac{(1 - \lambda_1^3)r_1^3}{R^3} \tag{1-112}$$

$$Q_1 = \frac{r_1}{R_1} = \frac{1}{\lambda_1} \tag{1-113}$$

$$Q_2^3 = 1 + \frac{(r_1^3 - R_1^3)}{R_2^3} \tag{1-114}$$

$$W_1 = C_1\left(Q_1^4 + \frac{2}{Q_1^2} - 3\right) + C_2\left(2Q_1^2 + \frac{1}{Q_1^4} - 3\right) \tag{1-115}$$

$$W_2 = C_1\left(Q_2^4 + \frac{2}{Q_2^2} - 3\right) + C_2\left(2Q_2^2 + \frac{1}{Q_2^4} - 3\right) \tag{1-116}$$

and

$$\sigma_{\text{radial}} = C_1(Q^4 + 4Q) + 2C_2\left(Q^2 - \frac{2}{Q}\right) - C_1\left(\frac{1}{\lambda_1^4} + \frac{4}{\lambda_1}\right) - 2C_2\left(\frac{1}{\lambda_1^2} - 2\lambda_1\right)$$
$$- \Pi_1 \tag{1-117}$$

$$\sigma_{\text{hoop}} = \sigma_{\text{merid}} = C_1\left(\frac{2}{Q^2} + 4Q - Q^4\right) + 2C_2\left(\frac{1}{Q^4} - \frac{2}{Q}\right) - C_1\left(\frac{1}{\lambda_1^4} + \frac{4}{\lambda_1}\right)$$
$$- 2C_2\left(\frac{1}{\lambda_1^2} - 2\lambda_1\right) - \Pi_1 \tag{1-118}$$

The above equations for the shell displacements and stresses can be solved numerically as follows:

1. Substitute Eqs. (1-111)–(1-116) into Eq. (1-110) with $r_1 = 25.4$ mm, $r_2 = 38.1$ mm, $C_1 = 0.334$ MPa, and $C_2 = 0.074$ MPa. This yields an equation where $R_1$ is the only unknown if $\Pi_1$ and $\Pi_2$ are specified.
2. For a given $\Pi_1$ and $\Pi_2$, search for an $R_1$ (the internal radius of the shell in the strained state) that satisfies Eq. (1-110). The integrals in Eq. (1-110) were carried out using Simpson's 1/3 Rule with $N = 100$.

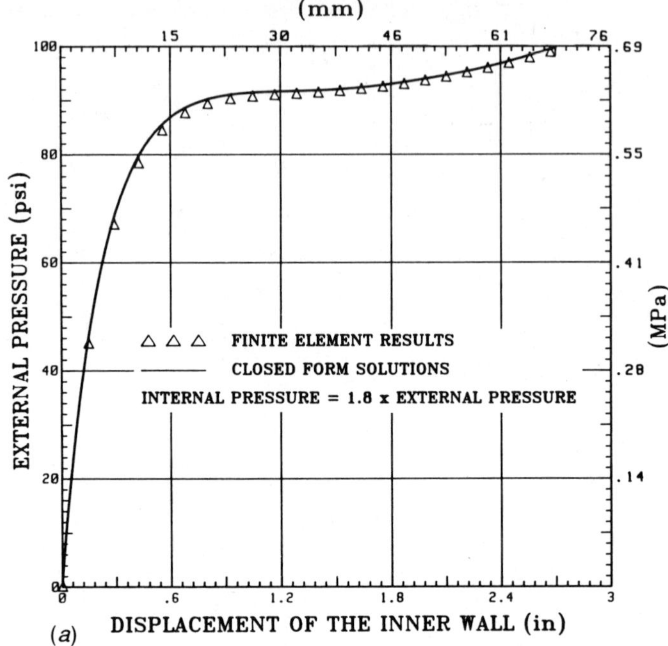


**Figure 1-15**   (a) Load–deflection curve. Ref. 62. (*Continued*)

3. Once $R_1$ has been found, then $R_2$ (the external radius of the shell in the strained state) can be calculated from Eq. (1-111) and the nonzero physical components of stresses can be obtained from Eqs. (1-117, 1-118).

The load–deflection curve of the sphere is shown in Fig. 1-15a. It can be seen that when the applied external pressure was larger than 0.61 MPa, the deflection of the sphere was very sensitive to pressure changes (large changes in displacements for small changes in the applied pressure). Figure 1-15b shows the hoop stress (which is the same as the meridian stress) acting through the deformed wall thickness when the internal and external pressures were equal to 1.24 and 0.69 MPa, respectively. It can be seen that the maximum hoop (or meridian) stress occurred at the inner surface of the sphere and decreased to a minimum at the outer surface. The radial stress acting through the deformed wall thickness is shown in Fig. 1-15c. As expected, all the stresses were compressive and the magnitudes were negligibly small compared with the hoop (or meridian) stresses when the wall thickness became very small. Finite element results for the displacements and stresses were also obtained and are shown in Figs. 1-15a–c by triangles.[62] It can be

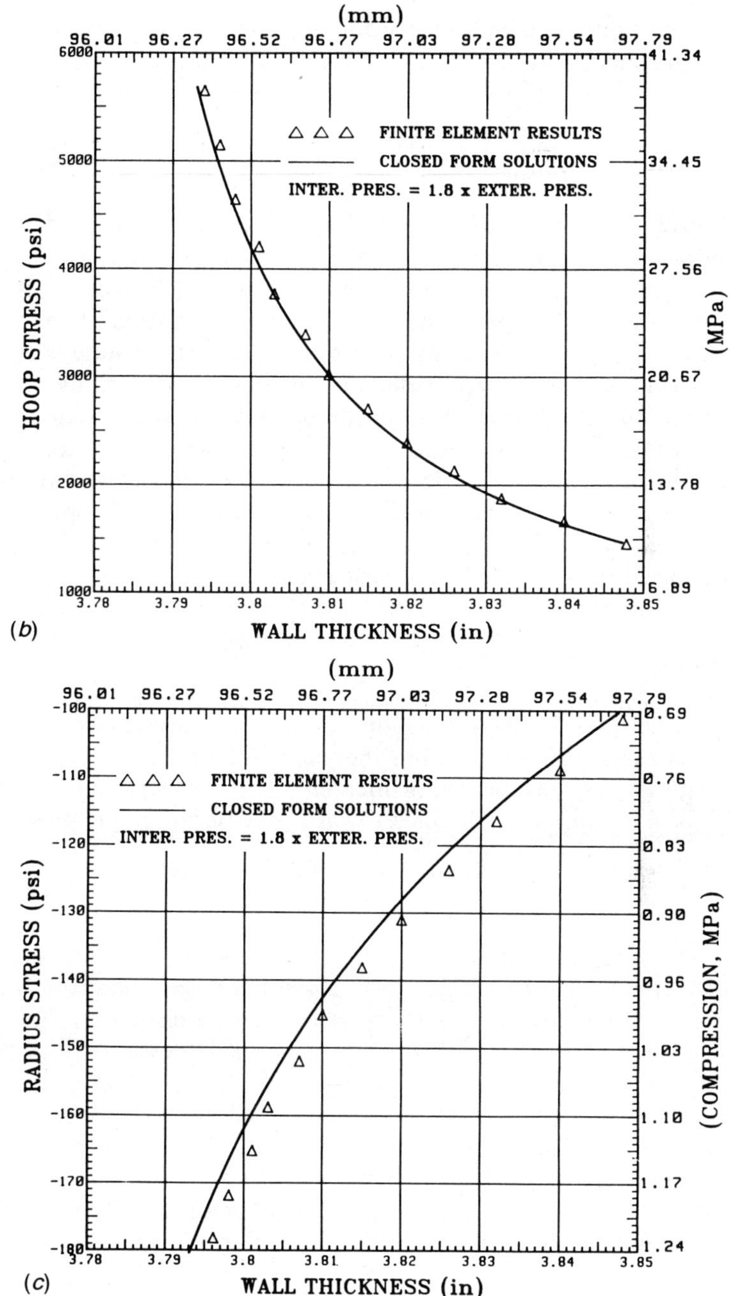



**Figure 1-15** (*continued*)  (*b*) Hoop stress through the wall thickness at external pressure = 0.69 MPa. (*c*) Radius stress through the wall thickness at external pressure = 0.69 MPa. Ref. 62.

seen that the finite element results and the closed-form solutions are in excellent agreement.


### 1.9.2 Plasticity

The objective of the theory of plasticity is to offer a mathematical description of the nonlinear mechanical behavior of materials in the *plastic (beyond elastic) range.* Unlike elastic deformation, plastic deformation is not a reversible process, and depends not only upon the initial and final states of loading but also upon the loading path by which the final state is achieved. Several aspects of real material behavior, such as the Bauschinger effect (a specimen initially stressed in tension often yields at a much reduced stress when restressed in compression, i.e., the yield stresses in tension and compression are not the same, Fig. 1-16), cyclic hardening, plastic anisotropy, elastic hysteresis, etc., can be modeled (at different levels of sophistication) by the theories of plasticity.[63–70] In this section, some well-established and most often used theories are briefly discussed.


**Yield Surface**

The yield surface is defined as the surface in stress space with stress components as coordinates. Within the yield surface the stress vector may change without any plastic strain increment; stress increments beginning from points in the surface, if directed toward the exterior, imply plastic strain increments.

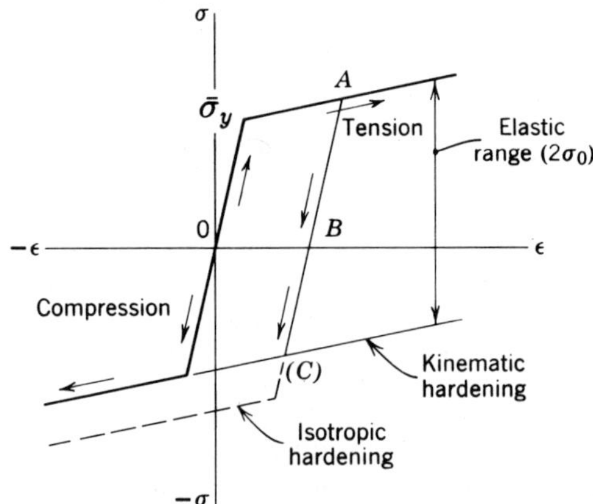


**Figure 1-16**   A stress–strain curve with Bauschinger effect $\sigma_0 = \bar{\sigma}_y$.

## Initial Yield Surface

For isotropic plasticity, the initial yield surface must be independent of the orientation of the reference axes. By choosing the axes of the principal stresses $(\sigma_1, \sigma_2, \sigma_3)$ as the reference axes, the initial yield surface may be expressed in terms of the principal stresses and represented by a surface in a stress space with $\sigma_1, \sigma_2, \sigma_3$ as coordinate axes. Thus the initial yield function may appear as

$$f(\sigma_1, \sigma_2, \sigma_3) = 0 \tag{1-119}$$

Furthermore, experiment indicates that the hydrostatic pressure has very small effect on the plastic deformation. Hence the initial yield condition may be expressed in terms of the deviatoric stress invariants in the form

$$f(\bar{I}_2, \bar{I}_3) = 0 \tag{1-120}$$

where

$$\bar{I}_2 = \tfrac{1}{2} S_{ij} S_{ij} \tag{1-121}$$

$$\bar{I}_3 = \tfrac{1}{3} S_{ij} S_{jk} S_{ki} \tag{1-122}$$

$$S_{ij} = \sigma_{ij} - \tfrac{1}{3} \sigma_{\beta\beta} \delta_{ij} \tag{1-123}$$

$\bar{I}_2$ and $\bar{I}_3$ are the deviatoric stress invariants and $S_{ij}$ is the deviatoric stress tensor. Two simple yield surfaces (conditions) for the initial yield of isotropic materials with isotropic hardening (i.e., the tensile and compressive yield stresses are equal at all times) that have provided highly useful descriptions of many real materials are briefly discussed in the following.

*Von Mises yield condition (distortion energy theory)* is based on the assumption that yielding occurs when the second deviatoric stress invariant attains a prescribed value $k$.

$$f = \bar{I}_2 - k^2 = \tfrac{1}{2} S_{ij} S_{ij} - k^2 = 0 \tag{1-124a}$$

or

$$f = \tfrac{1}{6}[(\sigma_1 - \sigma_2)^2 + (\sigma_2 - \sigma_3)^2 + (\sigma_3 - \sigma_1)^2] - k^2 = 0 \tag{1-124b}$$

or

$$f = \tfrac{1}{6}[(\sigma_x - \sigma_y)^2 + (\sigma_y - \sigma_z)^2 + (\sigma_z - \sigma_x)^2 + 6(\tau_{xy}^2 + \tau_{yz}^2 + \tau_{zx}^2)] - k^2 = 0$$

$$\tag{1-124c}$$

where $k$ may be a function of plastic strain for strain-hardening materials and the relation of $k$ to test data follows $k = \tau_y$ ($\tau_y$ is the yield stress in pure shear) or $k = \bar{\sigma}_y/\sqrt{3}$ ($\bar{\sigma}_y$ is the yield stress in uniaxial tension).

For example, the stress state at a point in a packaging component is $\sigma_x = 200\ \text{MPa}$, $\sigma_y = 100\ \text{MPa}$, $\sigma_z = -50\ \text{MPa}$, $\tau_{xy} = 30\ \text{MPa}$, $\tau_{yz} = \tau_{zx} = 0$. It is interesting to know if the component at that point exhibits yielding. The material of the component has a uniaxial yield stress $\bar{\sigma}_y = 500\ \text{MPa}$. From Eq. (1-124$c$) we have

$$f = \tfrac{1}{6}[100^2 + 150^2 + 250^2 + 6(30)^2] - \tfrac{1}{3}(500)^2 = 50\,200 - 250\,000 < 0$$

$$(1\text{-}124d)$$

Thus, the material at that point of the component will not yield.

***Tresca yield condition (maximum shear theory)*** is based on the assumption that yield occurs when the maximum shear stress reaches a limiting value $k$.

$$f = 4\bar{I}_2^3 - 27\bar{I}_3^2 - 36k^2\bar{I}_2^2 + 96k^4\bar{I}_2 - 64k^6 = 0 \qquad (1\text{-}125a)$$

or

$$f = [(\sigma_1 - \sigma_2)^2 - 4k^2][(\sigma_2 - \sigma_3)^2 - 4k^2][(\sigma_3 - \sigma_1)^2 - 4k^2] = 0 \quad (1\text{-}125b)$$

where $k = \tau_y$ or $k = \bar{\sigma}_y/2$. Experimental data appear to favor the use of the von Mises yield condition for most of the materials.

## Subsequent Yield Surface

Continued loading beyond the initial yield surface leads to plastic deformation, which may be accompanied by changes in both size and shape of the yield surface. For perfect plasticity the yield surface does not change during plastic deformation and the initial yield surface remains valid. For isotropic hardening, however, the size of the yield surface increases but the shape remains the same during loading, Fig. 1-17. To take such changes into account it is necessary to modify the initial yield surface and to define the subsequent yield surface, also known as the loading surface. A general form for the loading surface is given[63-70] by

$$f(\sigma_{ij}, \epsilon_{ij}^p, k) = 0 \qquad (1\text{-}126)$$

which depends not only upon the stresses $\sigma_{ij}$ but also upon the plastic strain $\epsilon_{ij}^p$ and the work-hardening characteristics represented by the parameter $k$.

Differentiating $f = 0$ by the chain rule of calculus, we have

$$df = \frac{\partial f}{\partial \sigma_{ij}} d\sigma_{ij} + \frac{\partial f}{\partial \epsilon_{ij}^p} d\epsilon_{ij}^p + \frac{\partial f}{\partial k} dk \qquad (1\text{-}127)$$

where $df$, $d\sigma_{ij}$, $d\epsilon_{ij}^p$, and $dk$ represent time differentials. If $f = 0$ and $df < 0$, a condition leading to an elastic state is implied, and it must follow that $d\epsilon_{ij}^p = dk = 0$. Thus

$$f = 0 \qquad \frac{\partial f}{\partial \sigma_{ij}} d\sigma_{ij} < 0 \qquad (1\text{-}128a)$$

is defined as *unloading*;

$$f = 0 \qquad \frac{\partial f}{\partial \sigma_{ij}} d\sigma_{ij} = 0 \qquad (1\text{-}128b)$$

is defined as *neutral loading*, since it implies that the stress-point remains on the initial yield surface; and

$$f = 0 \qquad \frac{\partial f}{\partial \sigma_{ij}} d\sigma_{ij} > 0 \qquad (1\text{-}128c)$$

is defined as *loading*, since it implies that the stress-point is moving outward from the current yield surface. For perfectly plastic materials, plastic flow occurs for $f = 0$, $(\partial f/\partial \sigma_{ij})d\sigma_{ij} = 0$, and the case $f = 0$, $(\partial f/\partial \sigma_{ij})d\sigma_{ij} > 0$ does not exist.

**Constitutive Equation: Incremental Theory**

A general equation for determining the plastic stress–strain relation for any yield condition was proposed by Drucker.[63–70] Based on his definition of work-hardening materials ($d\sigma_{ij}\, d\epsilon_{ij} > 0$ upon loading; $d\sigma_{ij}\, d\epsilon_{ij}^p \geq 0$ on completing a cycle), he stated that the plastic strain increment vector must be normal to the yield or loading surface at a smooth point on that surface, Fig. 1-17, and must lie between adjacent normals at a corner point, i.e.,

$$d\epsilon_{ij}^p = d\lambda \frac{\partial f}{\partial \sigma_{ij}} \qquad (1\text{-}129)$$

Equation (1-129) is called the normality principle of plasticity and can be applied to plastic anisotropic hardening; $d\lambda$ is a function that may depend on stress, strain, and strain history.

For isotropic hardening, the von Mises yield condition is given by

$$f = \tfrac{1}{2}S_{ij}S_{ij} - \tfrac{1}{3}\bar{\sigma}_y^2 \tag{1-130}$$

Substituting Eq. (1-130) into Drucker's equation (1-129), we have

$$d\epsilon_{ij}^p = d\lambda\, S_{ij} \tag{1-131a}$$

or

$$d\epsilon_{ij}^p\, d_{ij}^p = (d\lambda)^2 S_{ij}S_{ij} \tag{1-131b}$$

Defining the effective (equivalent) stress

$$\bar{\sigma} = \sqrt{(3/2)S_{ij}S_{ij}} \tag{1-132a}$$

$$\bar{\sigma} = (\sqrt{2}/2)\sqrt{(\sigma_1 - \sigma_2)^2 + (\sigma_2 - \sigma_3)^2 + (\sigma_3 - \sigma_1)^2} \tag{1-132b}$$

$$\bar{\sigma} = (\sqrt{2}/2)\sqrt{(\sigma_x - \sigma_y)^2 + (\sigma_y - \sigma_z)^2 + (\sigma_z - \sigma_x)^2 + 6(\tau_{xy}^2 + \tau_{yz}^2 + \tau_{zx}^2)} \tag{1-132c}$$

and the effective (equivalent) plastic incremental strain

$$d\bar{\epsilon}_p = \sqrt{(2/3)\, d\epsilon_{ij}^p\, d\epsilon_{ij}^p} \tag{1-133a}$$

$$d\bar{\epsilon}_p = (\sqrt{2}/3)\sqrt{(d\epsilon_1^p - d\epsilon_2^p)^2 + (d\epsilon_2^p - d\epsilon_3^p)^2 + (d\epsilon_3^p - d\epsilon_1^p)^2} \tag{1-133b}$$

$$d\bar{\epsilon}_p = (\sqrt{2}/3)\sqrt{(d\epsilon_x^p - d\epsilon_y^p)^2 + (d\epsilon_y^p - d\epsilon_z^p)^2 + (d\epsilon_z^p - d\epsilon_x^p)^2 + d\gamma} \tag{1-133c}$$

where

$$d\gamma = \tfrac{3}{2}(d\gamma_{xy}^{p\,2} + d\gamma_{yz}^{p\,2} + d\gamma_{zx}^{p\,2}) \tag{1-133d}$$

then Eq. (1-131b) becomes

$$d\lambda = \frac{3}{2}\frac{d\bar{\epsilon}_p}{\bar{\sigma}} \tag{1-134}$$

Suppose there exists a universal stress–strain curve (which coincides with a uniaxial true stress vs. true plastic strain curve for the material)

$$\bar{\sigma} = H\left(\int d\bar{\epsilon}_p\right) \tag{1-135}$$

expressing an equivalent stress $\bar{\sigma}$ as a function $H$ of an equivalent plastic strain increment $d\bar{\epsilon}_p$ integrated over the strain history. Then the slope $(H')$ of the equivalent stress $\bar{\sigma}$ vs. equivalent plastic strain $\int d\bar{\epsilon}_p$ curve is given by

$$H' = \frac{d\bar{\sigma}}{d\bar{\epsilon}_p} \tag{1-136}$$

Substituting Eq. (1-136) into Eq. (1-134), then Eq. (1-131$a$), we have

$$d\lambda = \frac{3}{2}\frac{d\bar{\sigma}}{\bar{\sigma}H'} \tag{1-137$a$}$$

and

$$d\epsilon_{ij}^p = \frac{3d\bar{\sigma}}{2\bar{\sigma}H'}S_{ij} \tag{1-137$b$}$$

Equation (1-137$b$) is referred to as the Levy–Mises equation and applied to problems of mostly plastic deformation (the elastic deformation is very small and is neglected). However, in most of electronics packaging problems the elastic strains cannot be neglected. In that case the incremental total strain is given by

$$d\epsilon_{ij} = d\epsilon_{ij}^e + d\epsilon_{ij}^p \tag{1-138$a$}$$

where $d\epsilon_{ij}^e$ is the incremental elastic strain tensor and can be derived from Eq. (1-19) with $\beta = 0$, and $d\epsilon_{ij}^p$ is given by Eq. (1-137$b$). Thus Eq. (1-138$a$) becomes

$$d\epsilon_{ij} = \frac{dS_{ij}}{2G} + (1 - 2v)\delta_{ij}\frac{d\sigma_{ij}}{3E} + \frac{3d\bar{\sigma}}{2\bar{\sigma}H'}S_{ij} \tag{1-138$b$}$$

Equation (1-138$b$) is called the Prandtl–Reuss equation.

Equation (1-138) can only be applied for isotropic hardening materials. For kinematic hardening materials that obey the von Mises yield condition for the initial yield surface, Prager's loading function corresponding to a translation of this surface is expressed as

$$f = \tfrac{1}{2}(S_{ij} - \alpha_{ij})(S_{ij} - \alpha_{ij}) - \frac{\bar{\sigma}_y^2}{3} = 0 \tag{1-139$a$}$$

where $\alpha_{ij}$ represents the translation of the center of the initial yield surface in the stress space, see Fig. 1-17. For linear hardening, we have

$$\alpha_{ij} = c\epsilon_{ij}^{p} \tag{1-139b}$$

and Eq. (1-139a) becomes

$$f = \tfrac{1}{2}(S_{ij} - c\epsilon_{ij}^{p})(S_{ij} - c\epsilon_{ij}^{p}) - \frac{\bar{\sigma}_{y}^{2}}{3} = 0 \tag{1-139c}$$

where $c$ is a constant and $\epsilon_{ij}^{p}$ is a plastic strain tensor. From Drucker's normality principle of plasticity, Eq. (1-129), we have

$$d\epsilon_{ij}^{p} = \frac{3}{2}\frac{(S_{mn} - c\epsilon_{mn}^{p})\,dS_{mn}}{c\bar{\sigma}_{y}^{2}}(S_{ij} - c\epsilon_{ij}^{p}) \tag{1-140a}$$

The material constant $c$ can be determined from the uniaxial test similarly to the isotropic hardening case and is equal to two-thirds of the slope $(H')$ of the uniaxial stress–plastic strain diagram. Then we have Prager's constitutive equation for kinematic hardening materials:

$$d\epsilon_{ij}^{p} = \frac{9}{4}\frac{[S_{mn} - (2/3)H'\epsilon_{mn}^{p}]\,dS_{mn}}{H'\bar{\sigma}_{y}^{2}}(S_{ij} - \tfrac{2}{3}H'\epsilon_{ij}^{p}) \tag{1-140b}$$

For elastoplastic solids with kinematic hardening, we have

$$d\epsilon_{ij} = \frac{dS_{ij}}{2G} + (1 - 2v)\delta_{ij}\frac{d\sigma_{ij}}{3E} + \frac{9}{4}\frac{[S_{mn} - (2/3)H'\epsilon_{mn}^{p}]\,dS_{mn}}{H'\bar{\sigma}_{y}^{2}}(S_{ij} - \tfrac{2}{3}H'\epsilon_{ij}^{p}) \tag{1-140c}$$

Equation (1-140c) is sometimes called the Reuss–Prager equation. The qualitative difference between isotropic and kinematic hardening can be illustrated for the same loading path in the $\pi$-plane $(\sigma_{1} + \sigma_{2} + \sigma_{3} = 0)$ in the stress space, Fig. 1-17. It can be seen that, for kinematic hardening, the initial yield surface translates in the $\pi$-plane without rotation and without change in size (the centre of the yield surfaces $f_{0}$ moves from 0 to $0_{1}$ for the subsequent yield surface $f_{1}$, etc.) On the other hand, isotropic hardening assumes a uniform expansion of the initial yield surface in the $\pi$-plane. Also, the resultant of the incremental plastic strain vectors predicted by isotropic hardening lags farther behind the resultant stress vector than the strain resultant predicted by Prager's kinematic hardening. For many other yielding and loading functions including the Bauschinger effect, see refs. 63–70.

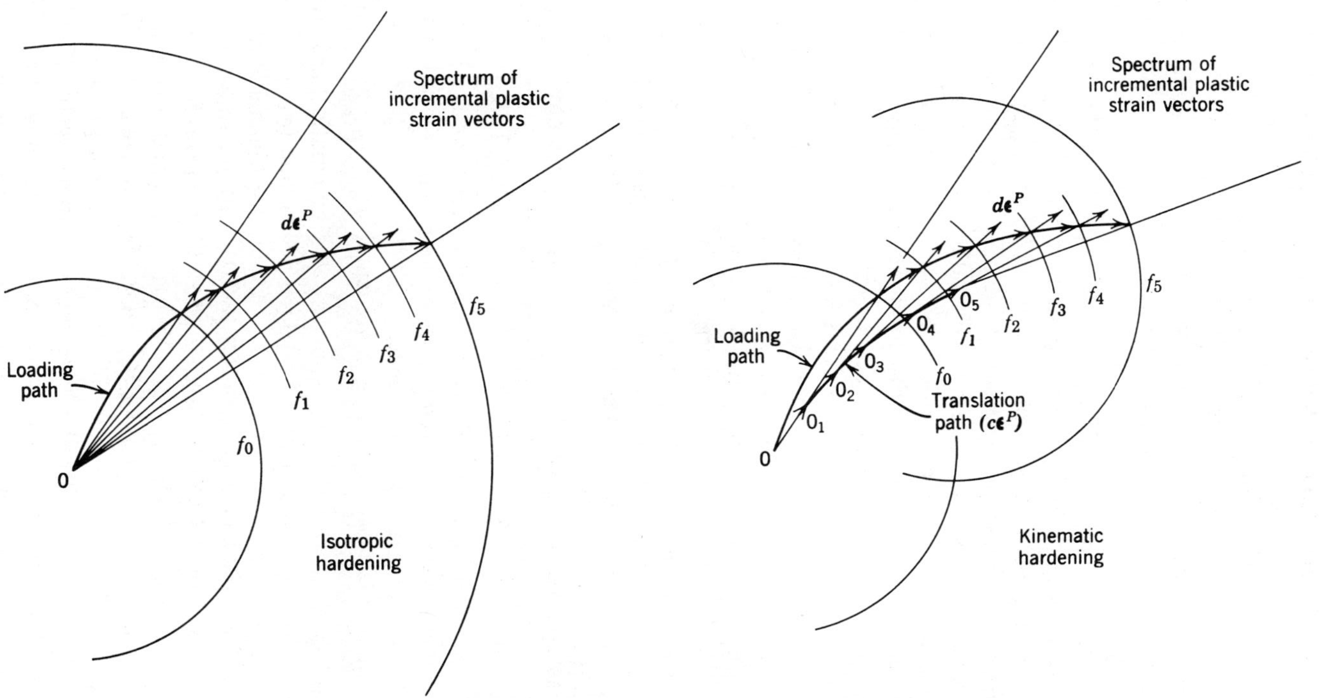


**Figure 1-17**   Isotropic hardening vs kinematic hardening.

### Elastoplastic Finite Element Method

The thermal stresses and strains in electronics packaging components and systems with materials governed by Eq. (1-138) are very difficult to obtain. The finite element method can be one of the best candidates for obtaining numerical results for the thermal stresses and deformations in electronics packages and interconnects.

The basic concept of the finite element method is that a boundary-value problem can be decomposed into a finite number of regions (elements). For each element, trial function approximations of displacement components are used in conjunction with variational principles or Galerkin's approximation and matrix methods to transform the boundary-value problem into a system of simultaneous algebraic equations. Since the method may be applied to individual discrete elements of the continuum, each element may be given distinct physical and material properties, thus achieving very general descriptions of a continuum as a whole. This feature of the finite element method is very attractive to practicing analysts dealing with composite structures such as electronic components and systems.

In formulating the nonlinear finite element method by the principle of incremental minimum potential energy, the incremental displacement field is represented by interpolation functions together with incremental generalized displacements at a finite number of nodal points in each element. In matrix form the assumed incremental displacement may be written as (Fig. 1-18)

$$[\Delta u] = [A][\Delta s] \qquad (1\text{-}141)$$

where $[\Delta u]$ is an incremental displacement column matrix, $[\Delta s]$ is a nodal incremental displacement column matrix, and $[A]$ is a shape function matrix. In Fig. 1-18, for the sake of simplicity, all the $\Delta s$ in front of $u, s, \ldots$ have been omitted. The corresponding incremental strain column matrix is

$$[\Delta \epsilon] = [L][\Delta u] \qquad (1\text{-}142)$$

where $[L]$ is the differential operator matrix.

The corresponding incremental stress column matrix is

$$[\Delta \sigma] = [D][\Delta \epsilon] - [D][\Delta \epsilon_0] + [\sigma_0] \qquad (1\text{-}143)$$

where $[\Delta \epsilon_0]$ and $[\sigma_0]$ are the incremental column matrix of initial strains and the column matrix of initial stresses, respectively, and $[D]$ is a material matrix that depends on the current state of stress and hardening of the material. If the material follows the Prandtl–Reuss theory and the von Mises

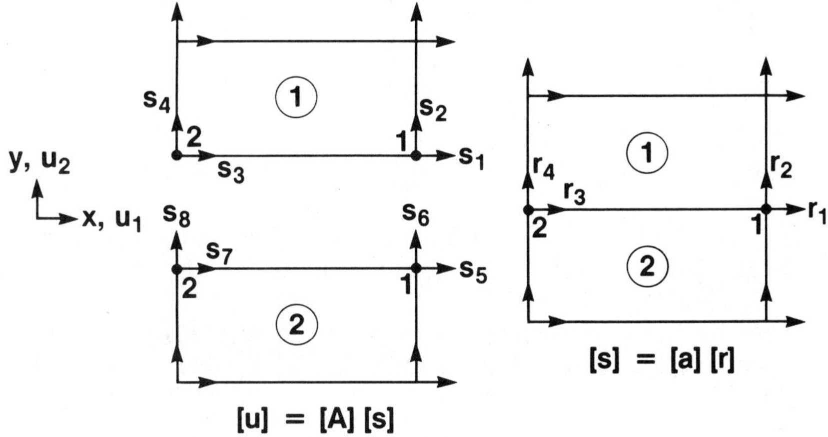


Where  [u] = Displacement Column Matrix
       [s] = Nodal Displacement Column Matrix
       [A] = Shape Function Matrix
       [a] = Compatibility Transformation Matrix
       [r] = Global Displacement Column Matrix

**Figure 1-18**  Nodal points, nodal displacements, and global nodal displacements. Ref. 17.

yield criterion, then the material matrix $[D]$ has the form

$$[D] = [D]_e + [D]_p \tag{1-144}$$

where

$$[D]_e = 2G \begin{bmatrix} \eta & \mu & \mu & 0 & 0 & 0 \\ \mu & \eta & \mu & 0 & 0 & 0 \\ \mu & \mu & \eta & 0 & 0 & 0 \\ 0 & 0 & 0 & \frac{1}{2} & 0 & 0 \\ 0 & 0 & 0 & 0 & \frac{1}{2} & 0 \\ 0 & 0 & 0 & 0 & 0 & \frac{1}{2} \end{bmatrix} \tag{1-145}$$

$$[D]_p = \frac{-9G^2}{\bar{\sigma}^2(3G + H')} \begin{bmatrix} S_x S_x & S_x S_y & S_x S_z & S_x S_{xy} & S_x S_{yz} & S_x S_{zx} \\ S_x S_y & S_y S_y & S_y S_z & S_y S_{xy} & S_y S_{yz} & S_y S_{zx} \\ S_x S_z & S_y S_z & S_z S_z & S_z S_{xy} & S_z S_{yz} & S_z S_{zx} \\ S_x S_{xy} & S_y S_{xy} & S_z S_{xy} & S_{xy} S_{xy} & S_{xy} S_{yz} & S_{xy} S_{zx} \\ S_x S_{yz} & S_y S_{yz} & S_z S_{yz} & S_{xy} S_{yz} & S_{yz} S_{yz} & S_{yz} S_{zx} \\ S_x S_{zx} & S_y S_{zx} & S_z S_{zx} & S_{xy} S_{zx} & S_{yz} S_{zx} & S_{zx} S_{zx} \end{bmatrix} \tag{1-146}$$

$$S_x = \frac{2\sigma_x - \sigma_y - \sigma_z}{3} \qquad S_y = \frac{2\sigma_y - \sigma_x - \sigma_z}{3} \qquad S_z = \frac{2\sigma_z - \sigma_y - \sigma_x}{3}$$

$$S_{xy} = \tau_{xy} \qquad S_{yz} = \tau_{yz} \qquad S_{zx} = \tau_{zx} \tag{1-147}$$

$$\eta = \frac{1-v}{1-2v} \qquad \mu = \frac{v}{1-2v}$$

and $G$ is given by Eq. (1-22), $\bar{\sigma}$ by Eq. (1-132), $d\bar{\epsilon}_p$ by Eq. (1-133), $H'$ by Eq. (1-136), and $[D]_e$ is the elastic material matrix, $[D]_p$ is the plastic material matrix, and $v$ is Poisson's ratio.

The total incremental potential energy functional ($\Delta V_p$) for any elasto-plastic solid divided into a finite number ($n$) of discrete elements ($V_n$) is given by

$$\Delta V_p = \sum_n \left( -\int_{S_n} [\Delta u]^T [\Delta T] \, dS - \int_{V_n} [\Delta u]^T [\Delta F] \, dV \right)$$

$$+ \tfrac{1}{2} \sum_n \int_{V_n} [\Delta\sigma]^T [\Delta\epsilon] \, dV \tag{1-148}$$

where $[\Delta T]$ is the incremental traction column matrix acting on a surface, $S_n$, and $[\Delta F]$ is the incremental body force column matrix.

Substituting Eqs. (1-141)–(1-143) into Eq. (1-148) leads to

$$\Delta V_p = \sum_n \left( -[\Delta s]^T [\Delta S] + \tfrac{1}{2} [\Delta s]^T [k][\Delta s] \right) \tag{1-149}$$

where

$$[k] = \int_{V_n} ([L][A])^T [D]([L][A]) \, dV \tag{1-150}$$

and

$$[\Delta S] = \int_{S_n} [A]^T [\Delta T] \, dS + \int_{V_n} [A]^T [\Delta F] \, dV + \int_{V_n} ([L][A])^T [D][\Delta\epsilon_0] \, dV$$

$$- \int_{V_n} ([L][A])^T [\sigma_0] \, dV \tag{1-151}$$

are the element stiffness matrix and the incremental nodal force column matrix, respectively.

The incremental nodal displacements $[\Delta s]$ for different elements are not completely independent; a transformation is needed to relate the element incremental nodal displacements to the independent incremental generalized global displacements. Then the compatibility equations of the assembled structure are written in matrix form as (Fig. 1-18)

$$[\Delta s] = [a][\Delta r] \qquad (1\text{-}152)$$

where $[a]$ is the compatibility transformation matrix and $[\Delta r]$ is the incremental global displacement column matrix.

Substituting Eq. (1-152) into Eq. (1-149), yields

$$\Delta V_p = -[\Delta r]^{\mathrm{T}}[\Delta R] + \tfrac{1}{2}[\Delta r]^{\mathrm{T}}[K][\Delta r] \qquad (1\text{-}153)$$

where

$$[K] = \sum_n [a]^{\mathrm{T}}[k][a] \qquad (1\text{-}154)$$

and

$$[\Delta R] = \sum_n [a]^{\mathrm{T}}[\Delta S] \qquad (1\text{-}155)$$

are the global stiffness matrix and the incremental global force column matrix, respectively.

The principle of minimum incremental potential energy requires that $\delta \Delta V_p = 0$, i.e.,

$$\frac{\partial \Delta V_p}{\partial [\Delta r]} \delta[\Delta r] = (-[\Delta R] + [K][\Delta r]) \, \delta[\Delta r] = 0 \qquad (1\text{-}156)$$

For arbitrary $\delta[\Delta r]$,

$$[K][\Delta r] = [\Delta R] \qquad (1\text{-}157)$$

In view of Eqs. (1-150) and (1-154) and the definition of $[D]$, it can be seen that the global stiffness matrix $[K]$ for elastoplastic problems is itself a function of current state of stress. The incremental equilibrium equation, Eq. (1-157), is therefore nonlinear and cannot be solved by a simple matrix inversion. One of the popular techniques for solving Eq. (1-157) is combining the method of incremental loading and the modified Newton–Raphson method (for examples, see refs. 15–20) to generate the complete nonlinear response by a sequence of piecewise linear steps.

Figure 1-19 shows a 3-D finite element model of one of the corner leads

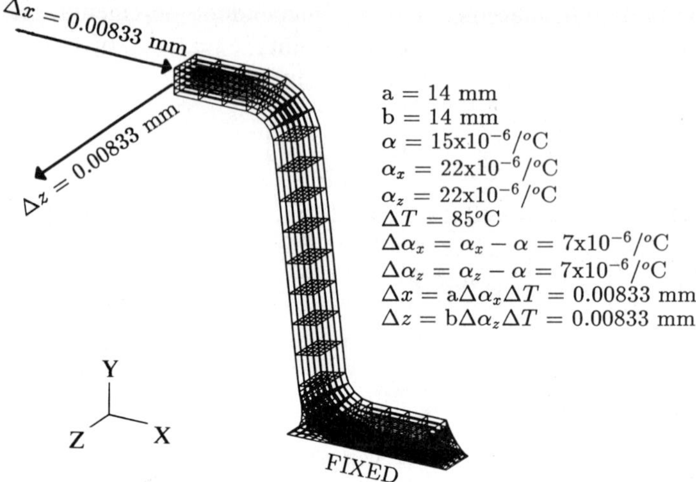


**Figure 1-19**　3-D finite element model of a 256 QFP corner lead and solder joint. Ref. 71.

and solder joints of a 256-pin fine-pitch plastic quad flat pack (QFP) surface mount assembly (Fig. 1-20). The imposed displacement boundary conditions were: (1) at the bottom end of the solder joint all points cannot move in the $x$-, $y$-, and $z$-directions and (2) at the uppr end of the lead all points moved 0.008 33 mm in the $x$-direction and 0.008 33 mm in the $z$-direction. These values were determined by the global thermal expansion mismatch between the plastic QFP and the FR-4 printed circuit board (PCB). The whole system is subjected to a temperature rise of 85°C.[71]

Due to the low yield stress and high ductility of the 63%wtSn/37%wtPb solder, it is assumed to be an elastoplastic material. The Young's modulus of the solder is $1.5 \times 10^6$ psi (10 000 MN/m²) and Poisson's ratio is 0.4. The thermal coefficient of linear expansion of the solder is $21 \times 10^{-6}/°C$. It should be emphasized that the mechanical properties of solders are strongly temperature-, frequency-, rate-, and time-dependent.[72] These variables have to be included in the analysis to produce an accurate estimation of the strains, and consequently the number of cycles-to-failure of the solder joint. However, for the sake of simplicity and due to the lack of a solder constitutive equation that includes all the important variables, all the material properties are assumed to be constant. Furthermore, in order to have a conservative estimate of the fatigue life (based on the Coffin–Manson law[72] and plastic strain calculation), the stress–strain curve at 120°C (yield stress = 1200 psi or 8.3 MN/m²; yield strain = 0.0008; strain-hardening parameter = 0.1) has been used (Fig. 1-21).

(a)

## THREE MAJOR SOURCES TO PRODUCE THERMAL STRESS IN THE 256-PIN QFP SOLDER JOINT

1. Global thermal coefficient of expansion (TCE) mismatch between the QFP and PCB

2. Local TCE mismatch between the lead and solder

3. Lead stiffness

$$a = 14 \text{ mm}$$
$$b = 14 \text{ mm}$$
$$\alpha = 15 \times 10^{-6}/°C$$
$$\alpha_x = 22 \times 10^{-6}/°C$$
$$\alpha_z = 22 \times 10^{-6}/°C$$
$$\Delta T = 85°C$$

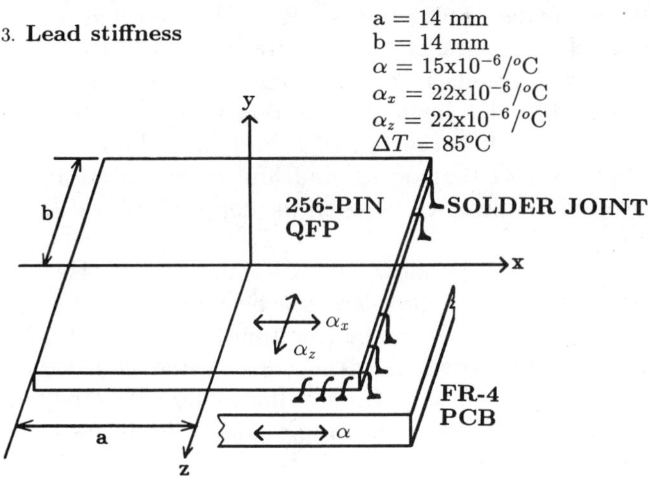



(b)

**Figure 1-20**    (a) Cross-section of the 256 QFP solder joint and lead. (b) 256 QFP surface mount assembly. Ref. 71.

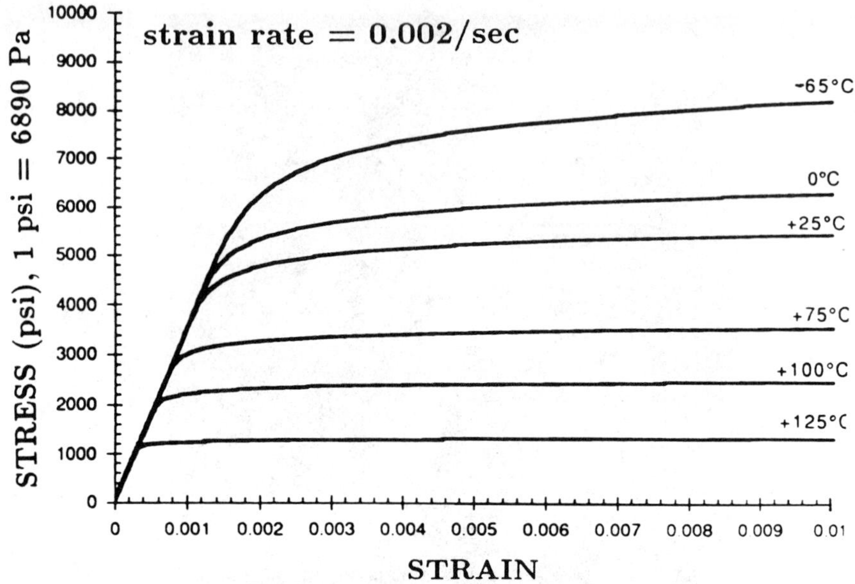


**Figure 1-21**   True stress–strain curves for the 63Sn/37Pb solder. Ref. 71.

Figure 1-22 shows the whole-field displacements (deflections) of the QFP assembly under heating. The dotted lines are for the original mesh and the solid lines are for the displaced mesh. It can be seen that the gull-wing lead not only expands in the $x$- and $y$-directions but also expands in the $z$-direction, which results in a very complex state of stresses and strains in the lead and solder joint. This is due to (1) the horizontal thermal expansion mismatch between the QFP body and the PCB, (2) the vertical thermal expansion of the copper lead and solder joint, and (3) the local thermal expansion mismatch between the copper lead and the Sn/Pb solder joint.

Figures 1-23 and 1-24 show the von Mises (equivalent) stress, $\bar{\sigma}$, Eq. (1-132), and cumulative equivalent plastic strain, $\bar{\epsilon}_p = \int d\bar{\epsilon}_p$, Eq. (1-133), acting at the solder joint. It can be seen that both $\bar{\sigma}$ and $\bar{\epsilon}_p$ are not uniform. The maximum von Mises stress occurs at the edge (in the negative $z$-direction) near the middle portion of the solder joint, and is concentrated in a very small area and decreases rapidly away from this location.

The maximum $\bar{\epsilon}_p$ (0.0083) is concentrated at a very small area at the edge in the negative $z$-direction near the center portion of the solder joint. Just as in the case of stress this maximum value decreases rapidly to a much smaller value a short distance away from the center portion.

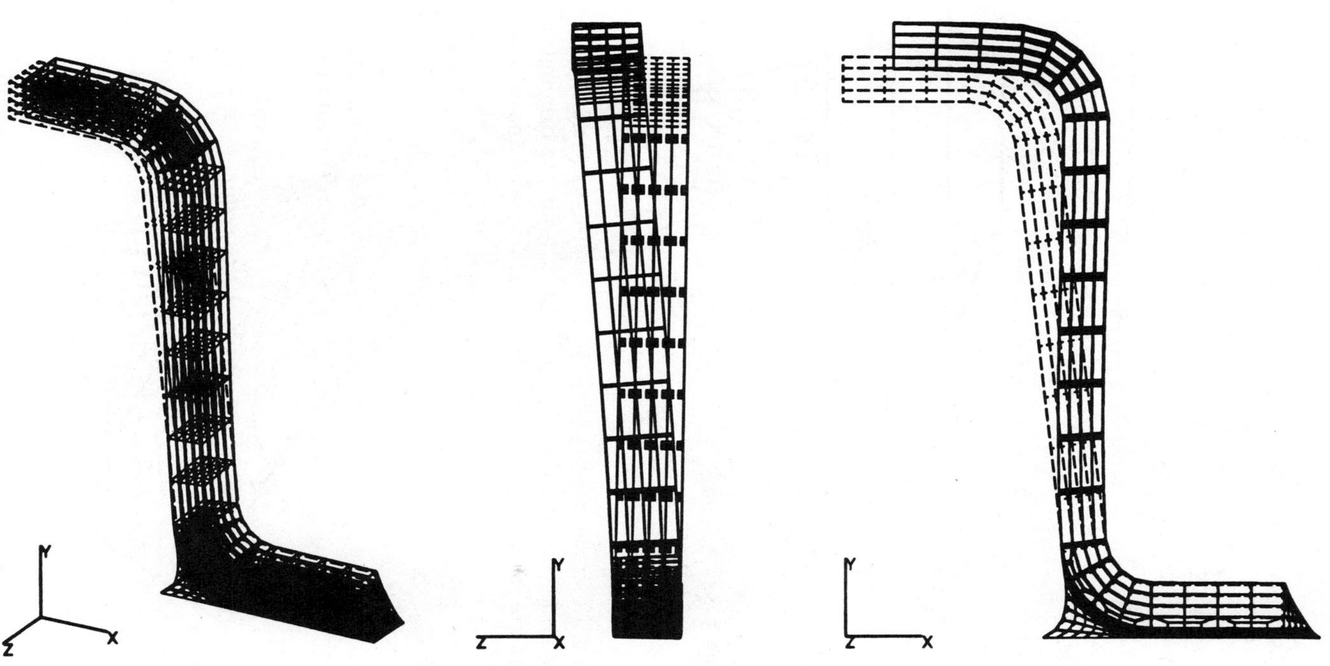

**Figure 1-22**  Deformation (solid line) of the 256-pin QFP corner lead and solder joint ($\Delta T = 85°C$). Ref. 71.

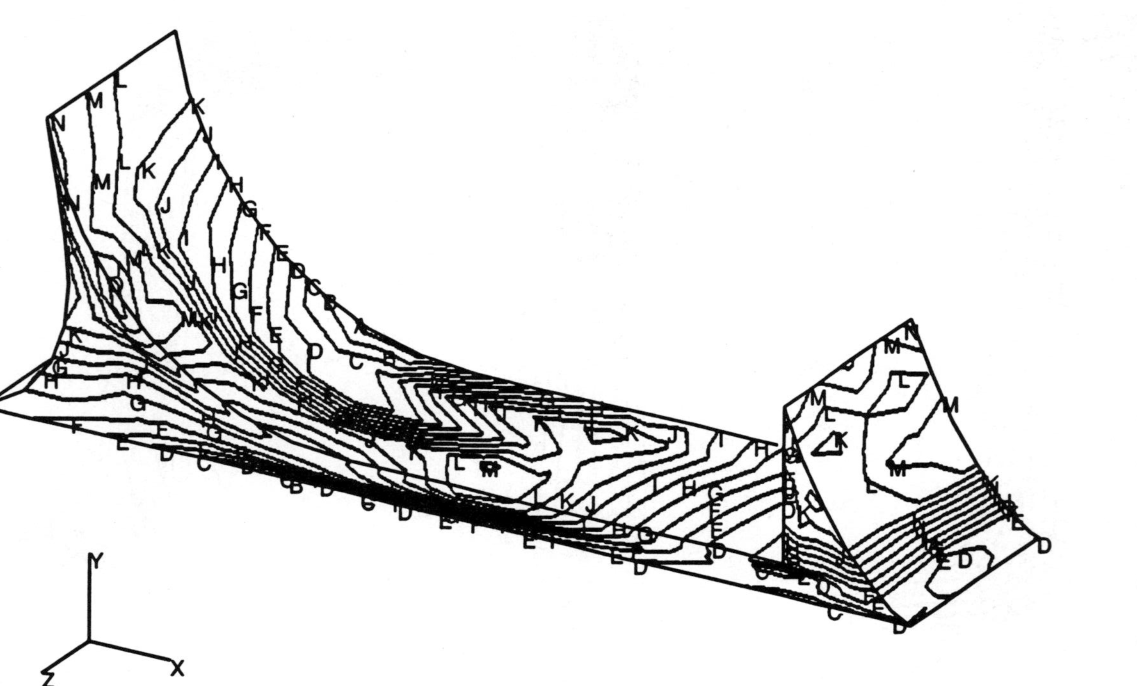


**Figure 1-23**  Von Mises stress (MPa) in the solder joint. Ref. 71.

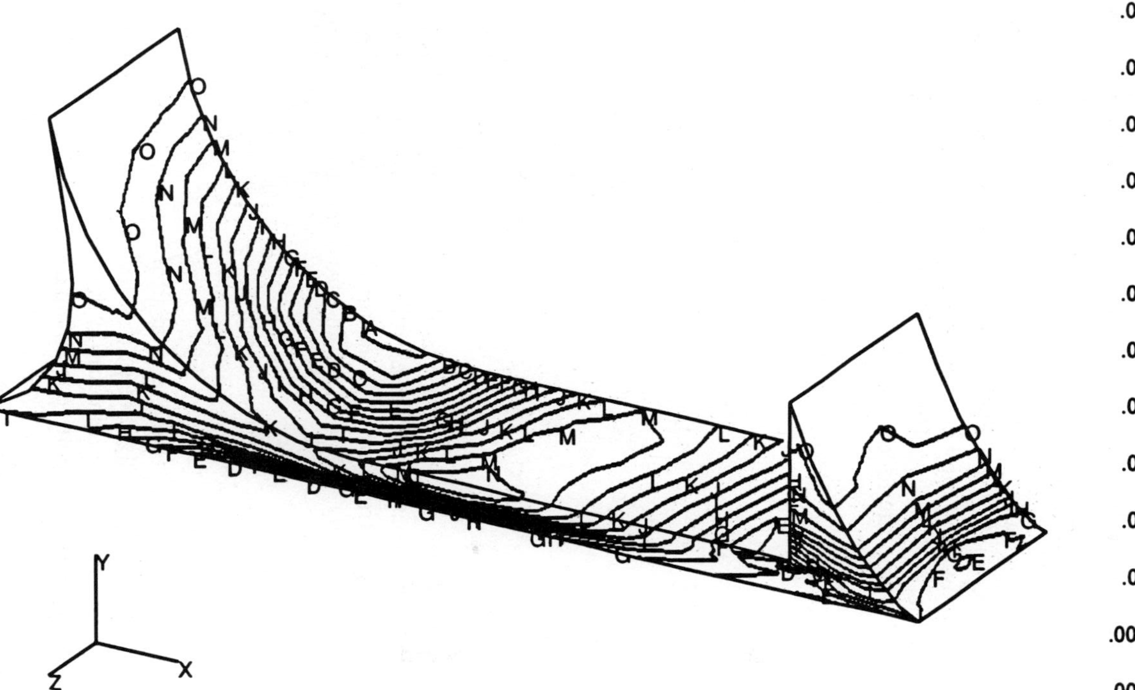


**Figure 1-24**  Cumulative equivalent plastic strain in the solder joint. Ref. 71.

### Constitutive Equation: Total Strain Theory

A very simple nonlinear stress–strain relation was provided by Hencky.[63–70] He proposed a one-to-one correspondence between the stress and strain. Thus, the total plastic strain components are taken to be proportional to the corresponding deviatoric stress components, i.e.,

$$\epsilon_{ij}^{p} = \Lambda S_{ij} \tag{1-158}$$

where $\epsilon_{ij}^{p}$ is the total plastic strain and $\Lambda$ is positive during loading and zero during unloading and may depend on stress and strain. For von Mises yield condition, Eq. (1-124), we use the same definition of equivalent stress, Eq. (1-132) and define the following equivalent total plastic strain:

$$\bar{\epsilon}_p = \sqrt{\tfrac{2}{3}\epsilon_{ij}^{p}\epsilon_{ij}^{p}} \tag{1-159}$$

Then, we have

$$\Lambda = \frac{3}{2}\frac{\bar{\epsilon}_p}{\bar{\sigma}} \tag{1-160}$$

and

$$\epsilon_{ij}^{p} = \frac{3}{2}\frac{\bar{\epsilon}_p}{\bar{\sigma}}S_{ij} = \frac{3}{2}\left(\frac{1}{S} - \frac{1}{E}\right)S_{ij} \tag{1-161}$$

where $E$ and $S$ are, respectively, the Young's modulus and secant modulus of the uniaxial true stress vs. true strain curve. Once again, the existence of a universal stress–strain curve is assumed and coincides with the uniaxial tension curve. Finally, the total strain–stress relation is given by

$$\epsilon_{ij} = \frac{S_{ij}}{2G} + (1 + 2v)\delta_{ij}\frac{\sigma_{ij}}{3E} + \frac{3}{2}\left(\frac{1}{S} - \frac{1}{E}\right)S_{ij} \tag{1-162}$$

It should be pointed out that the total strain theory (deformation plasticity) is not acceptable physically, except that for the case of proportional loading (a sufficient condition) the total strain theory coincides with the incremental theory. Because of the relative mathematical simplicity of the total strain theory, however, it has been used in many cases.[63–70, 73, 74]

Figure 1-25 shows a thin-walled circular cylindrical specimen subjected to the simultaneous action of bending $(M)$ and twisting $(T)$ moments. The cylinder material is made of solder and has the average uniaxial stress–strain

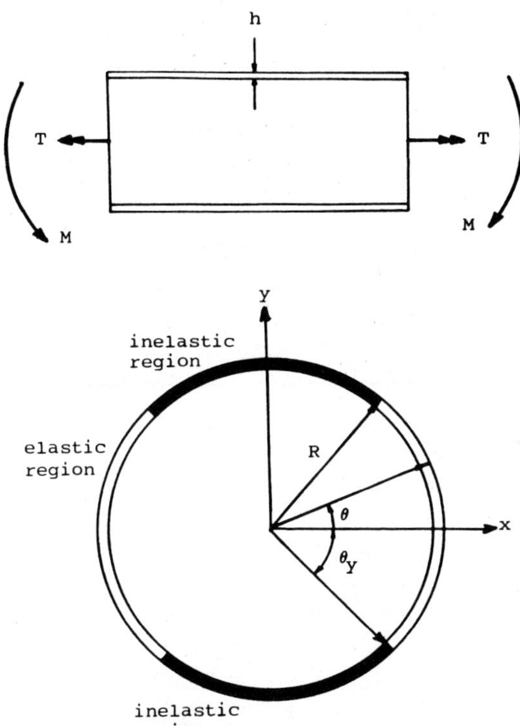


**Figure 1-25**   Thin-walled circular cylinder subjected to bending and twisting. Ref. 73.

diagram shown in Fig. 1-26. This kind of specimen is often used to determine the isothermal fatigue behavior of solders.[75]

It can be shown that the closed-form solutions for the interaction between the bending and twisting moments are[73]

$$\frac{M}{M_0} = \frac{\sqrt{1-\beta^2}}{\pi}\left\{\alpha\lambda\pi + \frac{4(1-\lambda)}{k^2}\left[E(\mu,k) - (1-k^2)F(\mu,k)\right] + 2\alpha(1-\lambda)\right.$$

$$\left. \times\left[\sin^{-1}\sqrt{\frac{1-\alpha^2\beta^2}{\alpha^2(1-\beta^2)}} - \frac{\sqrt{(1-\alpha^2\beta^2)(\alpha^2-1)}}{\alpha^2(1-\beta^2)}\right]\right\} \quad (1\text{-}163)$$

$$\frac{T}{T_0} = \frac{\beta}{\pi}\left[\alpha\lambda\pi + 2\alpha(1-\lambda)\sin^{-1}\sqrt{\frac{1-\alpha^2\beta^2}{\alpha^2(1-\beta^2)}} + 2(1-\lambda)F(\mu,k)\right] \quad (1\text{-}164)$$

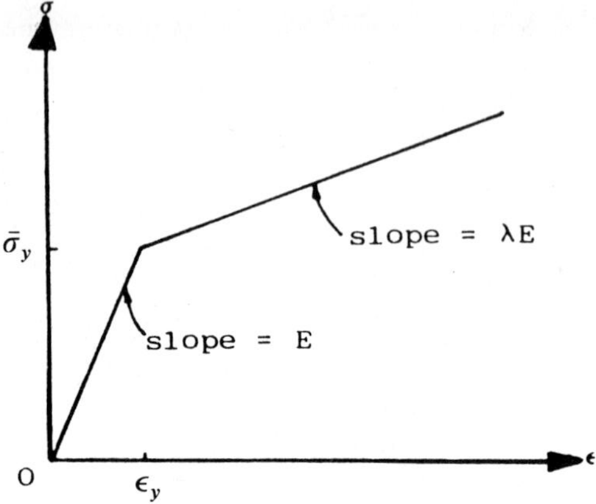


**Figure 1-26**   Average uniaxial stress–strain diagram. Ref. 73.

where

$$\alpha = \frac{1}{\epsilon_y} \sqrt{\tfrac{1}{3}\gamma^2 + \left(\frac{R}{\rho}\right)^2} \geqslant 1 \tag{1-165}$$

$$-1 \leqslant \beta = \frac{\gamma}{\sqrt{3\alpha\epsilon_y}} \leqslant 1 \tag{1-166}$$

$$\theta_y = \sin^{-1}\sqrt{\frac{1 - \alpha^2\beta^2}{\alpha^2(1 - \beta^2)}} \tag{1-167}$$

$$\mu = \frac{\pi}{2} - \sin^{-1}\sqrt{\frac{1 - \alpha^2\beta^2}{\alpha^2(1 - \beta^2)}} \tag{1-168}$$

$$k = \sqrt{1 - \beta^2} \tag{1-169}$$

$$M_0 = \pi R^2 h\bar{\sigma}_y \tag{1-170}$$

$$T_0 = \frac{2\pi R^2 h\bar{\sigma}_y}{\sqrt{3}} \tag{1-171}$$

and $F(\mu, \kappa)$ and $E(\mu, \kappa)$ are the elliptic integrals of the first and second kinds; $\gamma$ is the shear strain due to twisting moment; $E$ is the Young's modulus; $h$ is the wall thickness; $R$ is the middle radius of the thin-walled cylinder; $\rho$ is the radius of curvature; $\bar{\sigma}_y$ and $\epsilon_y$ are, respectively, the yield stress and strain in uniaxial tension; $M_0$ and $T_0$ are, respectively, the elastic bending and twisting moments of a circular thin-walled cylinder; $\alpha$ and $\beta$ are dimensionless parameters introduced for the sake of obtaining closed-form solutions; and $\theta_y$ defines the elastoplastic boundary. It is noted that Eqs. (1-163, 1-164) are valid only in the range $0 \leqslant \theta_y \leqslant \pi/2$, i.e., $0 \leqslant \alpha\beta \leqslant 1$.

When $\theta_y = \pi/2$, the cylinder is subjected to maximum elastic loads that are specified by $\alpha = 1$. In that case, we have

$$\frac{M}{M_0} = \sqrt{1 - \beta^2} \qquad (1\text{-}172)$$

$$\frac{T}{T_0} = \beta \qquad (1\text{-}173)$$

It can be seen that, when $\beta = 1$, we have the maximum elastic pure twisting. One the other hand, when $\beta = 0$, we have the maximum elastic pure bending.

The cylinder becomes completely inelastic when $\theta_y = 0$, i.e., $\alpha\beta = 1$. If $\alpha\beta$ is greater than 1, we have[73]

$$\frac{M}{M_0} = \frac{4}{\pi} \sqrt{1 - \beta^2} \left\{ \frac{\alpha\lambda\pi}{4} + \frac{(1 - \lambda)}{k^2} [E(\pi/2, k) - (1 - k^2)F(\pi/2, k)] \right\} \qquad (1\text{-}174)$$

$$\frac{T}{T_0} = \frac{\beta}{\pi} [\alpha\lambda\pi + 2(1 - \lambda)F(\pi/2, k)] \qquad (1\text{-}175)$$

For $\lambda = 0$, Eqs. (1-174, 1-175) give the fully plastic values for $M$ and $T$:

$$\frac{M}{M_0} = \frac{4}{\pi} \frac{\sqrt{1 - \beta^2}}{k^2} [E(\pi/2, k) - (1 - k^2)F(\pi/2, k)] \qquad (1\text{-}176)$$

$$\frac{T}{T_0} = \frac{2\beta}{\pi} F(\pi/2, k) \qquad (1\text{-}177)$$

For a given set of values of $\alpha$, $\beta$, and $\lambda$, the resultant bending and twisting moments can be evaluated from Eqs. (1-163)–(1-177). However, for a given set of values of bending and twisting moments, in order to determine the stress distribution, Eqs. (1-163)–(1-177) have to be plotted for a wide range of values of $\alpha$, $\beta$, and $\lambda$. Figure 1-27 shows a family of dimensionless bending and twisting moment interaction curves for $\lambda = 0.1$. (In this family of curves, $\alpha = 1, 2, 4, 6, 8, 10$, and $\beta = 0, 0.1, 0.2, 0.3, 0.4, 0.5, 0.6, 0.7, 0.8, 0.9, 1.0$). Thus, for a given set of $M$, $T$, and $\lambda = 0.1$, the values of $\alpha$ and $\beta$ can be read

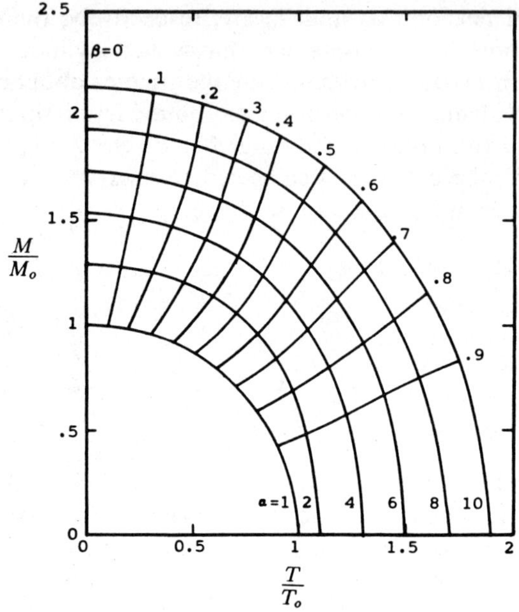


**Figure 1-27**    Dimensionless bending moment and twisting moment interaction curves
($\lambda = 0.1$). Ref. 73.

from Fig. 1-27 and the normal ($\sigma_z$) and shear ($\tau_{\theta z}$) stress distributions are
given by[73]

$$\sigma_z = \sqrt{1 - \beta^2}\,\bar{\sigma}_y\left[\alpha\lambda + \frac{1 - \lambda}{\sqrt{\beta^2 + (1 - \beta^2)\sin^2\theta}}\right]\sin\theta \qquad (1\text{-}178)$$

$$\tau_{\theta z} = \frac{\beta}{\sqrt{3}}\,\bar{\sigma}_y\left[\alpha\lambda + \frac{1 - \lambda}{\sqrt{\beta^2 + (1 - \beta^2)\sin^2\theta}}\right] \qquad (1\text{-}179)$$

in the plastic portion of the cylinder, and

$$\sigma_z = \alpha\sqrt{1 - \beta^2}\,\bar{\sigma}_y\sin\theta \qquad (1\text{-}180)$$

$$\tau_{\theta z} = \frac{1}{\sqrt{3}}\,\alpha\beta\bar{\sigma}_y \qquad (1\text{-}181)$$

in the elastic portion of the cylinder. For more families of dimensionless
interaction curves with other values of $\lambda$, see ref. 73.

### 1.9.3  Viscoelasticity

Viscoelastic materials (e.g., polymers, silicon) display a pronounced influence of the *rate of straining or stressing* and possess *time-dependent* material properties. Under a constant loading in time, viscoelastic materials may exhibit an initial elastic strain followed by *creep* where the materials continue to strain at a rate that depends on the magnitude of the applied load. Upon removal of the load there may be a partial instantaneous elastic recovery followed by a decrease of strain until the strain recovery is complete (delayed elasticity). On the other hand, if viscoelastic materials are subjected to a constant deformation in time, the materials will relax in the sense that the stress decreases with time (*stress relaxation*) from its initial value. Consequently, for viscoelastic materials, the current stress is a function of the entire history of deformation, and conversely, the current strain is a function of the entire history of loading. For linear viscoelastic materials (i.e., obeying a linear heredity law), the stress–strain law of *relaxation type* viscoelastic materials is given by[40]

$$\sigma_{ij}(x, t) = \int_{\infty}^{t} G_{ijkl}(x, t - \tau) \frac{\partial \epsilon_{kl}}{\partial \tau}(x, \tau)\, d\tau \qquad (1\text{-}182)$$

and the stress–strain law of *creep type* viscoelastic materials is given by[40]

$$\epsilon_{ij}(x, t) = \int_{\infty}^{t} J_{ijkl}(x, t - \tau) \frac{\partial \sigma_{kl}}{\partial \tau}(x, \tau)\, d\tau \qquad (1\text{-}183)$$

where $G_{ijkl}$ and $J_{ijkl}$ are fourth-order tensors and are called the *tensorial relaxation function* and *tensorial creep function*, respectively.

By means of simple models built of springs (to produce instantaneous deformations in response to the applied load) and dashpots (to produce velocity response to the applied load at any instant), the stress–strain relaxations, Eqs. (1-182, 1-183), may be put into the form of differential equations. It can be shown[40] that the Laplace or Fourier transformation equations of the linear viscoelasticity equations (equilibrium, strain–displacement, stress–strain, and boundary condition) have the same form as the corresponding equations of linear, isotropic elasticity (the *correspondence principle*). If the elasticity problem can be solved in closed form, the viscoelasticity solution can be obtained by an inverse transformation.

For example, Fig. 1-28 shows a half-space viscoelastic (Voigt-type) structure with bulk modulus $K = E/3(1 - 2v)$ and shear modulus $G = E/2(1 + v)$ subjected to a concentrated load $Z(t) = Z_0 \mathbf{1}(t)$ ($Z_0$ a constant, $\mathbf{1}(t)$ a unit-step function) acting normal to the free surface. $E$ is the Young's

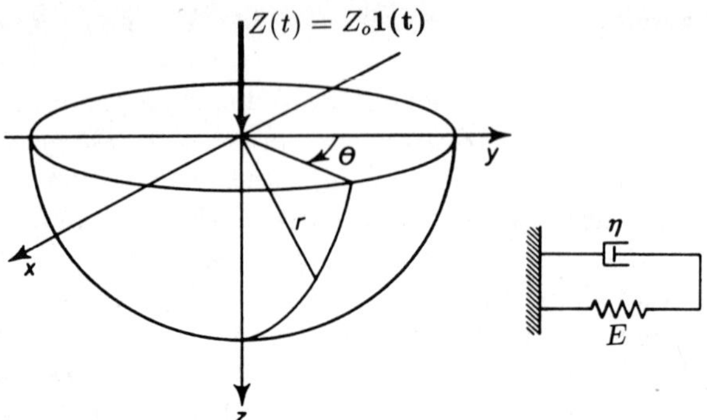


**Figure 1-28**   A concentrated load acts to the free surface of a half-space of Voigt material. Ref. 40.

modulus, $\eta$ is the coefficient of viscosity, and $v$ is the Poisson's ratio of the structure. It can be shown[40] that the radial stress component is given by

$$\sigma_{rr} = \frac{Z_0}{2\pi}\left(\frac{3}{3K+G}\left[G + 3K\exp\left(-\frac{3K+G}{\eta}t\right)\right]\left[\frac{1}{r^2} - \frac{z}{r^2}\frac{1}{(r^2+z^2)^{1/2}}\right]\right.$$

$$\left. - 3r^2\frac{z}{(r^2+z^2)^{5/2}}\right)1(t) \tag{1-184}$$

### 1.9.4 Viscoplasticity

Viscoplasticity is a mathematical model for *rate-sensitive plastic* materials. Unlike viscoelastic solids that have some rate dependence but still possess a well-defined natural state to which they eventually return when the applied loads are removed, a viscoplastic solid has some rate dependence but never returns to its natural state. A viscoplastic material differs from a fluid in that it can sustain a shear stress even when it is at rest. A stressed viscoplastic solid will not deform and flow until the magnitude of the stress reaches the critical value specified by the yield function of the solid. Thus, the responses of viscoplastic materials exhibit *strain rate-dependent permanent deformation* and the viscoplastic strain rate tensor $\dot{\epsilon}_{ij}^{vp}$ may be defined by

$$2\eta\dot{\epsilon}_{ij}^{vp} = 2k\langle F\rangle\frac{\partial F}{\partial\sigma_{ij}} \tag{1-185}$$

where

$$\langle F \rangle = \begin{cases} 0 & \text{if } F < 0 \\ F & \text{if } F \geqslant 0 \end{cases} \tag{1-186}$$

$$F = \frac{\sqrt{S_{ij}S_{ij}/2}}{k} - 1 \tag{1-187}$$

and $\eta$ is the coefficient of viscosity, $S_{ij}$ is defined in Eq. (1-129), and $k = \tau_y$ is the yield stress in pure shear. Equation (1-185) is called the Hohenemser–Prager–Bingham equation.[76] It should be noted that in contrast to rate-independent plasticity, in viscoplasticity, because of the rate effects, the stress point may be outside of the limiting quasi-static (or equilibrium) yield surface.

### 1.9.5 Creep

Creep is a mathematical model for *rate-sensitive elastoplastic* materials operating at *elevated temperature*. Creep strain may be broadly defined as elastoplastic time-dependent deformation under constant load at "high" temperature. Since different materials have different melting temperatures, it is convenient to define a homologous temperature (the ratio of the test or use temperature to the melting temperature on an absolute temperature scale). In general, creep becomes of engineering significance at a homologous temperature greater than 0.5. For some electronics packaging materials such as solder, creep deformation becomes important even at room temperature.

**The Creep Curve**

Figures 1-29 and 1-30 show a typical creep curve that can be obtained from "long-time" uniaxial test at constant load and constant temperature.[77–100] The slope of this curve ($\dot{\epsilon} = d\epsilon/dt$, $\epsilon$ = strain, $t$ = time) is referred to as the creep rate ($\dot{\epsilon}$). At the first stage, the creep rate is undefined and the initial creep strain consists of either entirely elastic strain or partially elastic strain and partially plastic strain. During the second stage, the creep rate decreases with time because the effect of strain hardening is greater than that of annealing (recovery). These two effects are in equilibrium (balance) during the third stage and the creep rate reaches essentially a steady state and changes very little with time. However, during the fourth stage, the creep rate increases rapidly with time until fracture occurs at point D. This is because the reduced cross-sectional area (either due to necking or to internal void formation) causes an increase in stress.

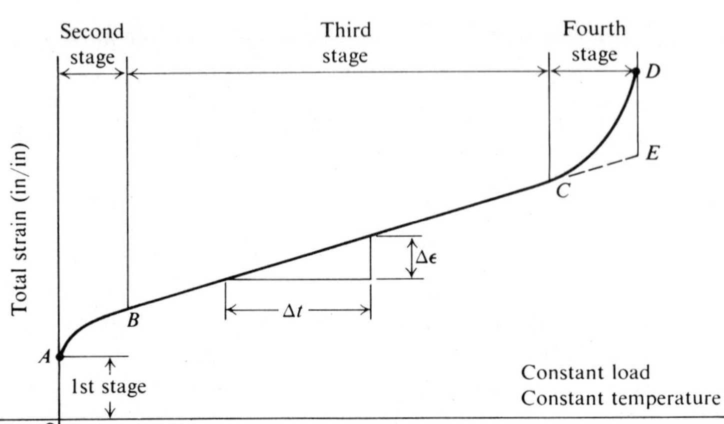


**Figure 1-29**    A typical creep curve.

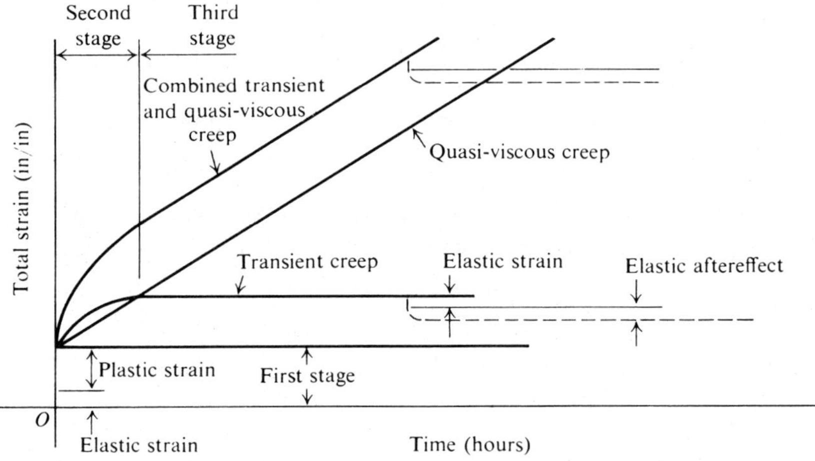


**Figure 1-30**    Andrade's analysis of the creep curve.

Andrade's pioneering work on pure metals has had considerable influence on the thinking about the mechanisms of the second and third stages of creep.[77–100] He considered the second and third stages of creep as being the superposition of transient creep (with a creep rate decreasing with time) and quasi-viscous creep (with a constant creep rate) processes that occur immediately after the sudden strain (first stage of creep) that results from applying the load, Figs. 1-29 and 1-30.

The second stage of creep is called primary creep. During this period, the primary creep is dominated by transient creep. For low temperatures and stresses, as in the creep of lead at room temperature, primary creep is the predominant creep process. The third stage of creep is called secondary creep or steady-state creep. The average value of the creep rate during secondary creep is called the minimum creep rate. Log–log plots of stress vs. minimum creep rate for various temperatures are essentially straight lines for many metals. The fourth stage of creep is called tertiary creep, which is often associated with microstructural changes such as coarsening of precipitate particles, recrystallization, or diffusional changes in the phases that are present.

The creep curves shown in Figs. 1-29 and 1-30 can be plotted in various ways with certain objectives in mind. For mechanics analysts the creep curves are usually plotted in the following two ways.

Figure 1-31$a$ shows a family of creep curves tested at constant temperature and various stress conditions ($\sigma_4 > \sigma_3 > \sigma_2 > \sigma_1$). The angles of the minimum creep rate at various creep curves are $\Psi_4 > \Psi_3 > \Psi_2 > \Psi_1$. Figure 1-31$b$ shows the corresponding stress vs. minimum creep rate curve at constant temperature. This curve can be used as the constitutive relation for steady-state creep analysis of structures at "high" temperature.[101]

Another useful way to plot the creep curves (Fig. 1-32$a$) generated at constant temperature and various stress conditions ($\sigma_1 > \sigma_2 > \sigma_3 > \sigma_4$) is shown in Fig. 1.32$b$. These curves relating the stress and creep strain at selected times for a contant temperature are called isochronous stress–strain curves and can be used as the constitutive relations for creep strain analysis of structures.[101]

## Empirical Creep Equations

In this section some of the commonest empirical uniaxial constitutive equations for describing the second (primary) and third (secondary) stages of

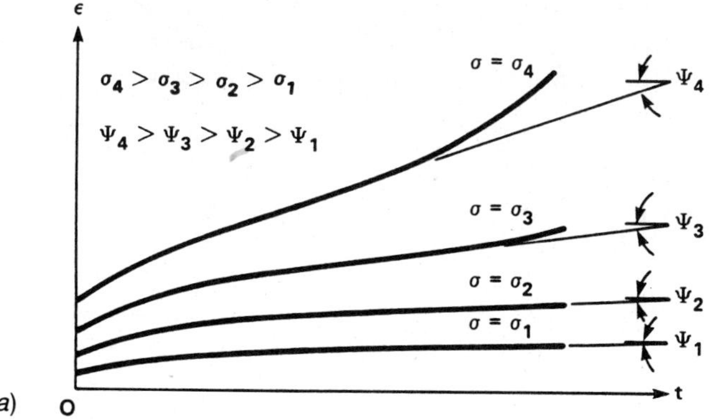


**Figure 1-31**    (a) Constant-stress creep–time curves. Ref. 101. (*Continued*)

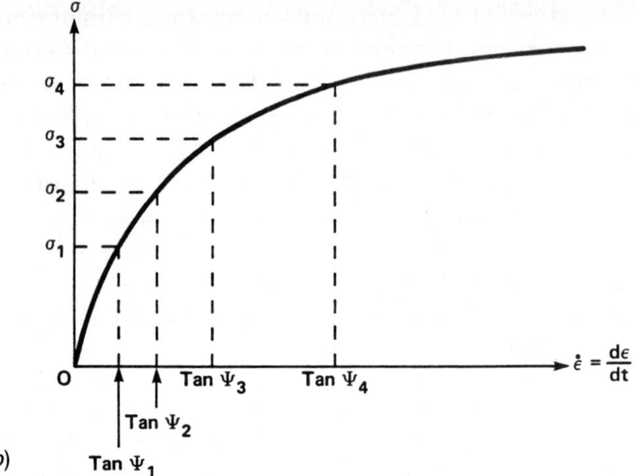



**Figure 1-31** (*continued*)   (*b*) Stress–strain rate diagram. Ref. 101.

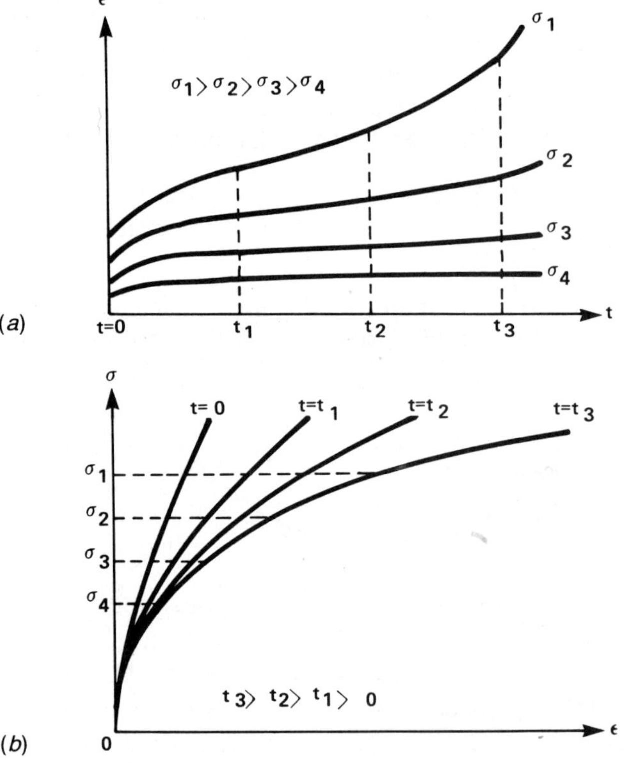



**Figure 1-32**   (*a*) Constant-stress creep–time curves. (*b*) Isochronous stress–strain diagram. Ref. 101.

creep (Fig. 1-29 to 1-32) are presented. In the broadest sense, the creep strain ($\epsilon_c$) is a function of applied load (or stress, $\sigma$), time ($t$), and temperature ($T$), i.e.,

$$\epsilon_c = p(\sigma, t, T) \qquad (1\text{-}188a)$$

which is usually assumed to be separable into

$$\epsilon_c = f(\sigma)g(t)h(T) \qquad (1\text{-}188b)$$

In Eq. (1-188$b$), the stress dependence term has been proposed by Norton, Prandtl, Dorn, and Garofalo[77–100] as

$$f(\sigma) = \begin{cases} A\sigma^n & \text{Norton} \\ B(\sigma - F)^n & \text{Friction stress} \\ C\sinh(\alpha\sigma) & \text{Prandtl} \\ D\exp(\beta\sigma) & \text{Dorn} \\ E[\sinh(\gamma\sigma)]^n & \text{Garofalo} \end{cases} \qquad (1\text{-}189)$$

The time dependence term has been proposed by Bailey, Andrade, Graham, and Walls[77–100]

$$g(t) = \begin{cases} t & \text{Secondary creep} \\ \eta t^m & \text{Bailey} \\ (1 + \eta t^{1/3})e^{kt} & \text{Andrade} \\ \sum \eta_i t^{m_i} & \text{Graham and Walles} \end{cases} \qquad (1\text{-}190)$$

The temperature dependence term in Eq. (1-188$b$) is usually associated with the Arrhenius law and has the form:

$$h(T) = G\exp(-\Delta H/RT) \qquad (1\text{-}191)$$

In Eqs. (1-189)–(1-191), $\Delta H$ is the activation energy, $R$ is Boltzmann's constant, $T$ is the absolute temperature, $t$ is time, and all the remaining symbols other than $\sigma$ are material constants that can be determined by fitting the creep curves from experiment.

It can be seen from Eqs. (1-188) to (1-191) that there are many ways to write the creep equation. One of the simplest creep equations, for example, can be obtained using Norton's term in Eq. (1-189) and Bailey's term in Eq.

(1-190):

$$\epsilon_c = At^m\sigma^n \exp(-\Delta H/RT) \tag{1-192}$$

Equation (1-192) is called the Bailey–Norton creep equation and for isothermal cases it becomes

$$\epsilon_c = Bt^m\sigma^n \tag{1-193}$$

The material constants can be determined by fitting the creep curves (such as those in Fig. 1-32$b$) from experiment. The creep rate of Eq. (1-193) can be determined as

$$\frac{d\epsilon_c}{dt} = \dot{\epsilon}_c = mBt^{m-1}\sigma^n \tag{1-194a}$$

In view of Eq. (1-193), Eq. (1-194$a$) can be written in a form independent of time:

$$\dot{\epsilon}_c = mB^{1/m}\sigma^{n/m}\epsilon_c^{(m-1)/m} \tag{1-194b}$$

Equation (1-194$a$) is called *time hardening* and Eq. (1-194$b$) is called *strain hardening*. Equations (1-193), (1-194$a$), and (1-194$b$) are valid for constant stress; however, when integrated for variable stress histories from Eqs. (1-194$a$) and (1-194$b$), they do not usually give the same results. In general, strain hardening, Eq. (1-194$b$), leads to better agreement with experimental results for variable stresses.

For steady-state creep (the third stage of creep, Figs. 1-29 to 1-32), the creep strain rate ($\dot{\epsilon}_c$) is independent of time and may be written as a function of stress and temperature:

$$\dot{\epsilon}_c = q(\sigma, T) = f(\sigma)h(T) \tag{1-195}$$

For isothermal conditions, we have

$$\dot{\epsilon}_c = f(\sigma) \tag{1-196}$$

where $f(\sigma)$ is given by Eq. (1-189) and the material constants can be determined by the creep curves (such as Fig. 1-31$b$) from experiment. It is noted that for most metals the steady-state creep strains are associated largely with plastic deformation. Also, the concepts of equivalent stress, equivalent strain, the existence of a universal strain–strain curve, Drucker's

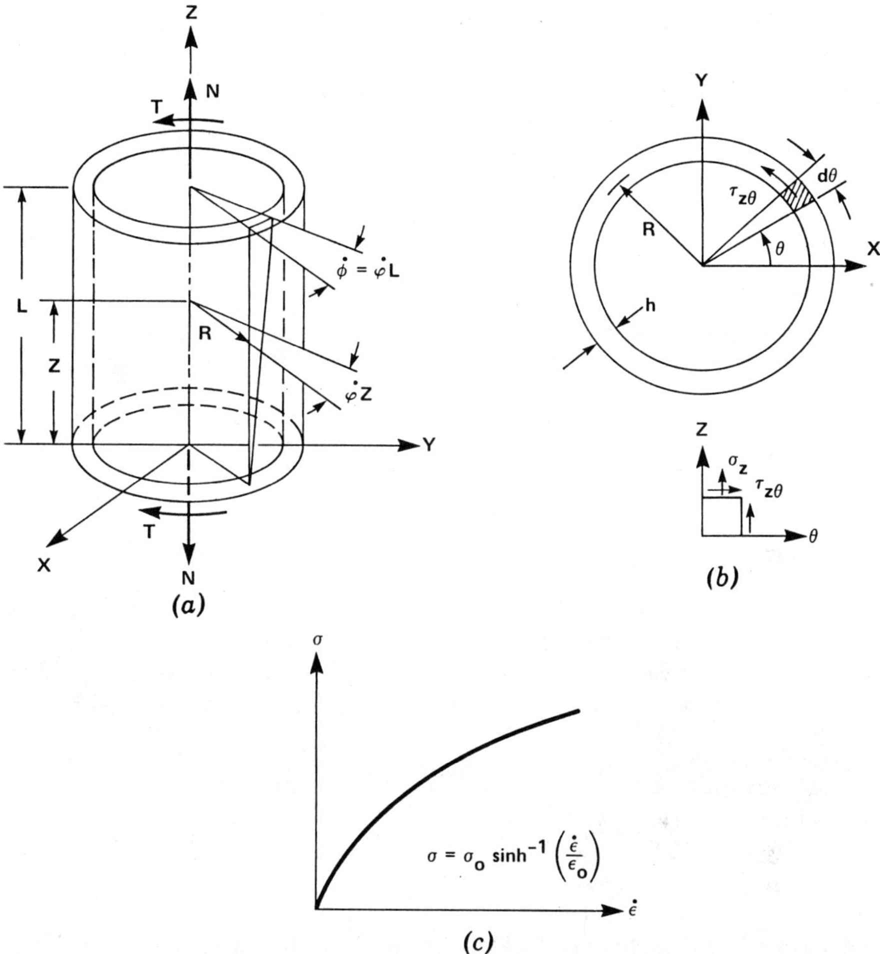



**Figure 1-33**   Torsion of thin-walled circular cylinder in the presence of axial force. Ref. 100.

flow rule, incremental strain, total strain, etc., for the theory of plasticity can be applied to the theory of creep.

For example, Figs. 1-33*a* and 1-33*b* show a thin-walled circular cylinder subjected to a twisting moment $(T)$ and an axial force $(N)$. This kind of specimen is frequently used for the creep test of solder materials under combined loads.[102] In this case, the stress–strain rate relation is assumed to obey Prandtl's equation, Eq. (1-189), and is shown in Fig. 1-33*c*. In Fig. 1-33, $R$ is the radius of the cylinder, $h$ is the thickness, $L$ is the length, $\dot{\varphi}$ is the twist rate per unit length, $\dot{\phi}$ is the total angle of twist rate, and $\sigma_0$ and $\epsilon_0$ are material constants of the solder.

By means of total creep strain theory, it can be shown that the interaction equations are[100]

$$N = 2\pi Rh\sigma_0\beta \sinh^{-1} K \qquad (1\text{-}197)$$

and

$$T = \frac{2\pi R^2 h\sigma_0}{\sqrt{3}} \sqrt{1 - \beta^2}\, \sinh^{-1} K \qquad (1\text{-}198)$$

and the normal stress $(\sigma_z)$ and shear stress $(\tau_{\theta z})$ are, respectively,

$$\sigma_z = \beta\sigma_0 \sinh^{-1} K \qquad (1\text{-}199)$$

and

$$\tau_{\theta z} = \frac{\sigma_0}{\sqrt{3}} \sqrt{1 - \beta^2}\, \sinh^{-1} K \qquad (1\text{-}200)$$

Equations (1-197) and (1-198) are plotted in Fig. 1-34 for a wide range of values of $\beta$ and $K$ that were introduced for the sake of convenience. Thus, for a given set of values of $T$ and $N$, the values of $\beta$ and $K$ can be read from Fig. 1-34 and the stresses in the solder cylinder can be determined by Eqs. (1-199) and (1-200).

The twisting moment vs. twist rate per unit length $(\dot{\phi})$ is given by[100]

$$T = \frac{2\pi R^2 h\sigma_0}{\sqrt{3}} \sqrt{1 - \beta^2}\, \sinh^{-1}\left(\frac{1}{\sqrt{1 - \beta^2}} \frac{R\dot{\phi}}{\sqrt{3\epsilon_0}}\right) \qquad (1\text{-}201)$$

which is plotted in Fig. 1-35. Thus, for a given $T$ and $N$, we can read the value of $\beta$ from Fig. 1-34 and then read the value of $\dot{\phi}$ from Fig. 1-35 with $\beta$ and $T$ known.


## 1.10 SUMMARY AND RECOMMENDATIONS

The fundamental and governing equations of thermomechanics for electronics packaging have been briefly compiled in this chapter. The limitations and assumptions of these equations for electronics packaging applications have also been highlighted. Furthermore, closed-form and numerical solutions for some electronics packaging problems have been provided.

Because of the geometry and material construction of most electronics packages, exact solutions of boundary-value problems involving Eqs. (1-25,

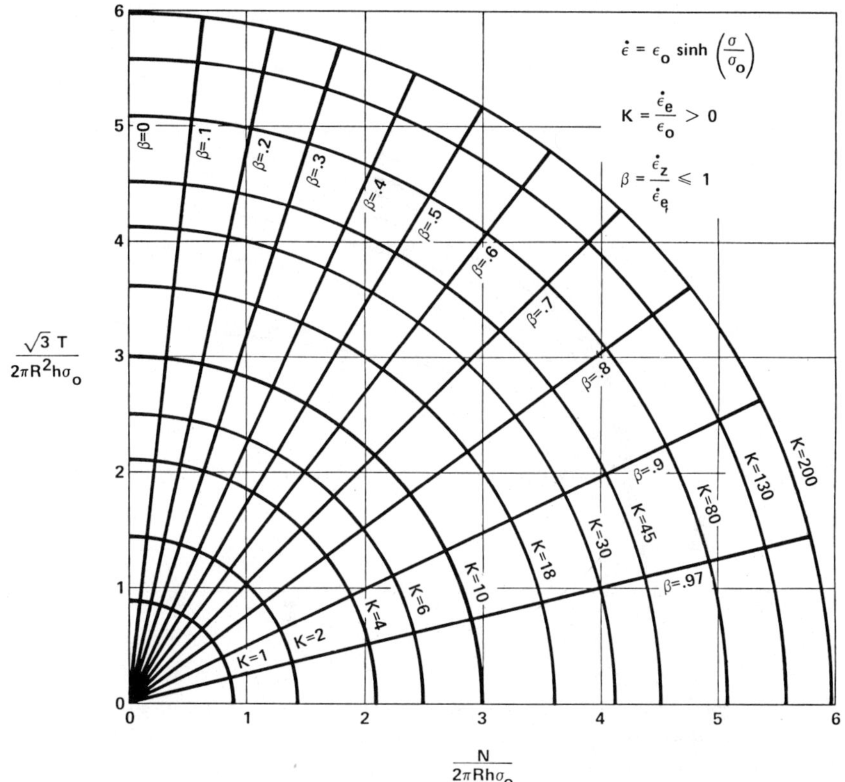



**Figure 1-34**   Dimensionless twisting moment and axial force interaction curves. Ref. 100.

1-26, 1-73, 1-74, 1-108, 1-138, 1-140, 1-162, 1-182, 1-183, 1-185, 1-189–1-191) are very difficult to obtain. Therefore, numerical schemes, of which the finite element method is one of the best candidates, are required.

With the advance of computers and the rapid development of commercially available finite element codes, boundary-value problems with coupling/ nonlinearity/anisotropicity/material property temperature-dependent effects can be solved. However, it should be pointed out that no boundary-value problem can completely describe the complex world around us. Every boundary-value problem aims at a certain class of phenomena, formulates their essential features, and disregards what is of minor importance. The finite element method is merely a technique for obtaining approximate solutions of boundary-value problems. Thus the basic assumptions, governing equations, and special features of the boundary-value problem should be borne in mind while constructing and analyzing a finite element model of the problem. Verify the finite element results by actual tests, if possible.

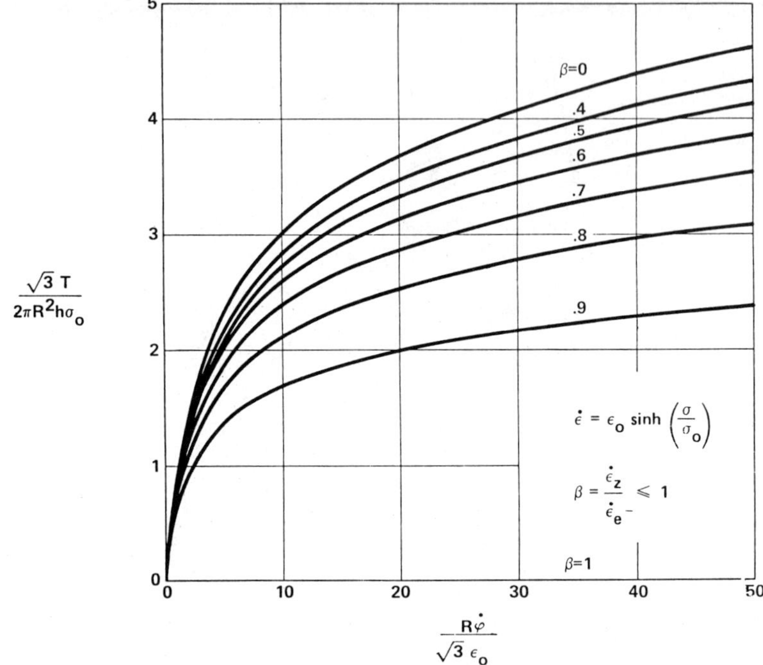


**Figure 1-35**  Dimensionless twisting moment–twist rate per unit length relations. Ref. 100.

If not, then verify the analysis procedures and the accuracy of the finite element codes by solving similar types of problems that have experimental results or closed-form solutions. In fact, this last step is vital and should be done first.


## ACKNOWLEDGMENTS

The author thanks Drs. Albert Jeans, Teresa Lau, and Steve Erasmus for their contributions to this chapter.


## REFERENCES


1. Chen, W. T., and C. W. Nelson, "Thermal Stresses in Bonded Joints," *IBM J. Research and Development*, **23**(2), 1979, pp. 179–187.
2. Suhir, S., "Interfacial Stresses in Bimetal Thermostats," *ASME J. Applied Mechanics*, **55**, 1989, pp. 595–600.
3. Durelli, A. J., and C. H. Tsao, "Determination of Thermal Stresses in Three-Ply Laminates," *J. Applied Mechanics, Transactions of ASME*, **77**, 1955, pp. 190–192.

4. Evans, A. G., and J. W. Hutchinson, "On the Mechanics of Delamination and Spalling in Compressed Films," *Int. J. Solids and Structures*, **20**, 1984, 455–466.

5. Hu, S. M., "Film-Edge Induced Edge Stress in Substrates," *J. Applied Physics*, **50**(7), 1979, pp. 4661–4666.

6. Isomae, S., 'Stress Distributions in Silicon Crystal Substrates With Thin Films," *J. Applied Physics*, **52**(4), 1981, pp. 2782–2791.

7. Eischen, J. W., C. Chung, and J. H. Kim, "Realistic Modeling of Edge Effect Stresses in Bimaterial Elements," *J. Electronic Packaging, Trans. ASME*, **112**, March 1990, pp. 16–23.

8. Roll, K., "Analysis of Stress and Strain Distribution in Thin Films and Substrates," *J. Applied Physics*, **47**(7), 1976, pp. 3224–3229.

9. Engel, P. A., "Structural Analysis for Circuit Card Systems Subjected to Bending," *J. Electronic Packaging, Trans. ASME*, **112**, March 1990, pp. 2–10.

10. Kuo, A., "Thermal Stresses at the Edge of a Bimetallic Thermostat," *J. Applied Mechanics, Trans. ASME*, **56**, 1989, pp. 585–589.

11. Silvester, P., and R. L. Ferrari, *Finite Element for Electrical Engineers*, Cambridge University Press, Cambridge, UK, 1983.

12. Strang, G., and G. J. Fix, *An Analysis of the Finite Element Method*, Prentice-Hall, Englewood Cliffs, NJ, 1973.

13. Bathe, K. J., *Finite Element Procedures in Engineering Analysis*, Prentice-Hall, Englewood Cliffs, NJ, 1973.

14. Weaver, W., and P. R. Johnston, *Finite Elements for Structural Analysis*, Prentice-Hall, Englewood Cliffs, NJ, 1984.

15. Oden, J. T., *Finite Elements of Nonlinear Continua*, McGraw-Hill, New York, 1973.

16. Tong, P., and J. N. Rossettos, *Finite-Element Method: Basic Technique and Implementation*, MIT Press, Cambridge, MA, 1977.

17. Lau, J. H., "A Note on the Calculation of Thermal Stresses in Electronic Packaging by Finite Element Methods," *J. Electronic Packaging, Trans. ASME*, **111**, Dec. 1989, pp. 313–320.

18. Kikuchi, N., *Finite Element Methods in Mechanics*, Cambridge University Press, Cambridge, UK, 1986.

19. Hughes, T., *The Finite Element Method*, Prentice-Hall, Englewood Cliffs, NJ, 1987.

20. Zienkiewicz, O. C., *The Finite Element Method in Engineering Science*, McGraw-Hill, London, 1971.

21. Gatewood, B. E., *Thermal Stresses*, McGraw-Hill, New York, 1957.

22. Boley, B. A., and J. H. Weiner, *Theory of Thermal Stresses*, Wiley, New York, 1960.

23. Nowacke, W., *Dynamic Problems of Thermoelasticity*, Noordhoff International Publishing, Leyden, 1975.

24. Lau, J. H., "Thermoelastic Problems for Electronic Packaging," *J. Hybrid Circuits*, **25**, May 1991, pp. 11–15.

25. Nowacke, W., *Thermoelasticity*, 2d edn., Pergamon Press, New York, 1986.

26. Kupradze, V. D., *Three-Dimensional Problems of the Mathematical Theory of Elasticity and Thermoelasticity*, North-Holland, New York, 1976.

27. Nowinski, J. L., *Theory of Thermoelasticity with Applications*, Sijthoff & Noordhoff, Sephin AAN Rijn, 1978.
28. Kovalenko, A. D., *Thermoelasticity*, Wolters-Norrdhoff, Groningen, 1969.
29. Parkus, H., *Thermoelasticity*, Blaisdell, London, 1968.
30. Boley, B. A., *Thermoinelasticity*, Springer-Verlag, New York, 1970.
31. Lau, J. H., S. J. Erasmus, and D. W. Rice, "Overview of Tape Automated Bonding Technology," *Electronic Materials Handbook*, Vol. 1: *Packaging*, ASM International, November 1989.
32. Lau, J. H., D. W. Rice, and G. Harkins, "Thermal Stress Analysis of TAB Packagings and Interconnections," *IEEE Trans. Components, Hybrids, and Manufacturing Technology*, **13**(1), March 1990, pp. 183–188.
33. Lau, J. H., S. J. Erasmus, and D. W. Rice, "An Introduction to Tape Automated Bonding Technology," in *Electronics Packaging Forum*, ed. J. E. Morris, Van Nostrand Reinhold, New York, 1991, pp. 1–83.
34. Lau, J. H., S. J. Erasmus, and D. W. Rice, "Overview of Tape Automated Bonding Technology," *Circuit World*, **16**(2), 1990, pp. 5–24.
35. Lau, J. H., *Handbook of Tape Automated Bonding*, Van Nostrand Reinhold, New York, 1992.
36. Sternberg, E., and E. L. McDowell, "On The Steady-State Thermoelastic Problem For The Half-Space," *Quarterly J. Applied Mathematics*, **XIV**(4), 1957, pp. 381–398.
37. Lau, J. H., "Thermoelastic Solutions for a Semi-Infinite Substrate with an Electronic Device," *J. Electronic Packaging, Trans. ASME*, **114**, September 1992, pp. 353–358.
38. Sneddon, I. N., and F. J. Lockett, "On The Steady-State Thermoelastic Problem For The Half-Space and The Thick Plate," *Quarterly J. Applied Mathematics*, **XVIII**(2), 1960, pp. 145–153.
39. Lau, J. H., "Thermoelastic Solutions for a Finite Substrate with an Electronic Device," *J. Electronic Packaging, Trans. ASME*, **113**, March 1991, pp. 84–88.
40. Fung, Y. C., *Foundations of Solid Mechanics*, Prentice-Hall, Englewood Cliffs, NJ, 1965.
41. Boresi, A. P., and O. M. Sidebottom, *Advanced Mechanics of Materials*, Wiley, New York, 1984.
42. Gere, J. M., and S. P. Timoshenko, *Mechanics of Materials*, 2d edn., PWS Engineering, Boston, MA, 1984.
43. Timoshenko, S. P., and J. N. Goodier, *Theory of Elasticity*, McGraw-Hill, New York, 1970.
44. Lau, J. H., "Closed-Form Solutions for The Large Deflection of Curved Optical Glass Fibers Under Combined Loads," to be published in *J. Electronic Packaging*, **115**, June 1993.
45. Suhir, E., "Effect of the Nonlinear Behavior of the Material on Two-Point Bending of Optical Glass Fibers," *J. Electronic Packaging*, **114**, June 1992, pp. 246–250.
46. Pan, H. H., "Non-linear Deformation of Flexible Ring," *Quarterly J. Mechanics and Applied Mathematics*, **XV**, 1962, pp. 402–412.
47. Lau, J. H., "Large Deflection of Cantilever Beam," *J. Engineering Mechanics*, **107**, February 1981, pp. 259–264.

48. Lau, J. H., "Large Deflections of Beam With Combined Loads," *J. Engineering Mechanics*, **108**, February 1982, pp. 180–185.

49. Conway, H. D., "The Nonlinear Bending of Thin Circular Rods," *J. Applied Mechanics*, **23**, March 1956, pp. 7–10.

50. *ABAQUS Theory Manual*, Hibbitt, Karlsson and Sorensen, Inc., Providence, 1987.

51. *ABAQUS User's Manual*, Hibbitt, Karlsson and Sorensen, Inc., Providence, 1987.

52. Alexander, H., "A Constitutive Relation for Rubber-Like Materials," *Int. J. Engineering Science*, **6**, 1968, pp. 549–563.

53. Green, A. E., and J. E. Adkins, *Large Elastic Deformations*, Oxford University Press, London, 1960.

54. Green, A. E., and W. Zerna, *Theoretical Elasticity*, 2d edn., Oxford University Press, London, 1968.

55. Hart-Smith, L. J., and J. D. C. Crisp, "Large Elastic Deformations of Thin Rubber Membranes," *Int. J. Engineering Science*, **5**, 1967, pp. 1–24.

56. Mooney, M., "Theory of Large Elastic Deformation," *J. Applied Physics*, **11**, 1940, pp. 582–592.

57. Ogden, R. W., "Elastic Deformations of Rubberlike Solids," in *Mechanics of Solids*, ed. H. G. Hopkins and M. J. Sewell, Pergamon Press, Oxford, 1982, pp. 499–537.

58. Shield, R. T., "Equilibrium Solutions in Finite Elasticity," *ASME J. Applied Mechanics*, **50**, 1983, pp. 1171–1180.

59. Smith, G. F., and R. S. Rivlin, "The Strain-Energy Function for Anisotropic Elastic Materials," *Trans. American Mathematical Society*, **88**, 1958, pp. 175–193.

60. Treloar, L. R. G., *The Physics of Rubber Elasticity*, 2d edn., Oxford University Press, London, 1958.

61. Treloar, L. R. G., "The Elasticity and Related Properties of Rubbers," *Reports on Progress in Physics*, **36**, 1973, pp. 755–826.

62. Lau, J. H., and A. H. Jeans, "Nonlinear Analysis of Elastomeric Keyboard Domes," *J. Applied Mechanics, Trans. ASME*, **56**, December 1989, pp. 751–755.

63. Polukhin, P., S. Gorelik, and V. Vorontsov, *Physical Principles of Plastic Deformation*, Mir, Moscow, 1983.

64. Atkins, A. G., and Y. W. Mai, *Elastic and Plastic Fractures*, Horwood, Chichester, 1985.

65. Hill, R., *The Mathematical Theory of Plasticity*, Clarendon Press, Oxford, 1983.

66. Thomas, T. Y., *Plastic Flow and Fracture in Solids*, Academic Press, New York, 1961.

67. Johnson, W., and P. B. Meller, *Engineering Plasticity*, Horwood, Chichester, 1983.

68. Washizu, K., *Variational Methods in Elasticity and Plasticity*, Pergamon Press, Oxford, 1982.

69. Gopinathan, V., *Plasticity Theory and Its Application in Metal Forming*, Wiley, New York, 1982.

70. Mendelson, A., *Plasticity, Theory and Application*, Krieger, Malabar, FL, 1983.

71. Lau, J. H., D. Rice, and S. Erasmus, "Thermal Fatigue Life of 256-Pin, 0.4 mm Pitch Plastic Quad Flat Pack (QFP) solder Joints," *Proc. 1st ASME/JSME Electronic Packaging Conference*, April 1992, pp. 855–863.

72. Lau, J. H., *Solder Joint Reliability: Theory and Applications*, Van Nostrand Reinhold, New York, 1991.

73. Lau, J. H., and T. T. Lau, "Bending and Twisting of Pipes with Strain Hardening," *J. Pressure Vessel Technology, Trans. ASME*, **106**, May 1984, pp. 188–195.

74. Lau, J. H., and C. K. Hu, "Nonlinear Stress Analysis of Curved Bars," *Proc. 5th ASCE Engineering Mechanics Conferences*, 1984, pp. 917–920.

75. Cortez, R., E. Cutiongco, M. Fine, and D. Jeannotte, "Correlation of Uniaxial Tension–Tension, Torsion, and Multiaxial Tension–Torsion Fatigue Failure in a 63Sn–37Pb Solder Alloy," *Proc. 42nd IEEE Electronic Components and Technology Conference*, May 1992, pp. 354–359.

76. Malvern, L. E., *Introduction to the Mechanics of a Continuous Medium*, Prentice-Hall, Englewood Cliffs, NJ, 1969.

77. Kennedy, A. J., *Processes of Creep and Fatigue in Metals*, Oliver & Boyd, New York, 1962.

78. Gittus, J., *Viscoelasticity and Creep Fracture in Solids*, John Wiley–Halsted Press, New York, 1975.

79. Evans, H. E., *Mechanics of Creep Fracture*, Elsevier Applied Science, New York, 1984.

80. Evans, H. E., and B. Wilshire, *Creep of Metals and Alloys*, The Institute of Metals, London, 1985.

81. Conway, S. B., *Numerical Methods for Creep and Rupture*, Gordon and Breach, New York, 1967.

82. Garofalo, F., *Fundamentals of Creep and Creep-Rupture in Metals*, The Macmillan Company, New York, 1965.

83. Clauss, F. J., *Engineer's Guide to High Temperature Materials*, Addison-Wesley, Reading, MA, 1969.

84. Wilshire, B., and R. J. Owen, *Recent Advances in Creep and Fracture of Engineering Materials and Structures*, Pineridge Press, Swansea, UK, 1982.

85. Lubahn, J. D., and R. P. Felgar, *Plasticity and Creep of Metals*, Wiley, New York, 1961.

86. Sully, A. H., *Metallic Creep and Creep Resistant Alloys*, Interscience, New York, 1949.

87. Ponter, A. R. S., and F. A. Leckie, "Constitutive Relationships for the Time Dependent Deformation of Metals," *J. Engineering Materials and Technology, Trans. ASME*, **98**, 1976, pp. 47–51.

88. Miller, A. K., "An Inelastic Constitutive Model for Monotonic, Cyclic, and Creep Deformation," *J. Engineering Materials and Technology, Trans. ASME*, **98**, 1976, pp. 97–105.

89. Hart, E. W., "Constitutive Relations for the Non-elastic Deformation of Metals," *J. Engineering Materials and Technology, Trans. ASME*, **98**, 1976, pp. 193–202.

90. Perzyna, P., "The Constitutive Equations for Rate Sensitive Plastic Materials," *Quarterly J. Mechanics and Applied Mathematics*, **XX**, 1963, 321–332.

91. Lau, J. H., and G. K. Listvinsky, "Bending and Twisting of Internally Pressurized Thin-Walled Cylinder With Creep," *J. Applied Mechanics, Trans. ASME*, **48**, June 1981, pp. 439–441.

92. Lau, J. H., "Bending of Circular Cylinder with Creep," *J. Engineering Mechanics Division, Proc. ASCE,* **107**, 1981, pp. 265–270.

93. Lau, J. H., "Bending and Twisting of Pipe with Creep," *Int. J. Nuclear Engineering and Design,* June 1981, pp. 367–374.

94. Lau, J. H., and T. T. Lau, "Creep of Pipes Under Axial Force and Bending Moment," *J. Engineering Mechanics Division, Proc. ASCE,* **108**, 1982, pp. 190–195.

95. Lau, J. H., and T. T. Lau, "Deformation of Elbows With Creep," *Proc. 4th ASME National Congress on Pressure Vessel and Piping Technology,* June 1983.

96. Lau, J. H., S. S. Jung, and T. T. Lau, "Creep of Thin-Wall Cylinder Under Axial Force, Bending, and Twisting Moments," *J. Engineering for Power, Trans. ASME,* **106**, 1984, pp. 79–83.

97. Lau, J. H., and C. K. Hu, "Creep of Thin-Wall Cylinder Under Combined Loads," *Proc. 5th ASCE Engineering Mechanics Conference,* 1984, pp. 917–920.

98. Lau, J. H., and T. T. Lau, "Bending and Twisting of Pipes With Creep," *J. Pressure Vessel Technology, Trans. ASME,* **106**, May 1984, pp. 188–195.

99. Lau, J. H., and C. K. Hu, "Deformation of Curved Bars With Creep," *J. Engineering for Power, Trans. ASME,* **107**, 1985, pp. 225–230.

100. Lau, J. H., and T. T. Lau, "Creep of Pipes Under Axial Force and Twisting Moment," *J. Engineering Mechanics, Proc. ASCE,* **108**, 1982, pp. 174–179.

101. Lau, J. H., "Creep of Solder Interconnects Under Combined Loads," to be published in *Proc. 43rd IEEE Electronic Components and Technology Conference,* June 1993.

102. Schroeder, S. A., and M. R. Mitchell, "Torsional Creep Behavior of 63Sn–37Pb Solder," *Proc. 1st ASME/JSME Electronic Packaging Conference,* April 1992, pp. 649–653.

# 2

# Thermal Expansivity and Thermal Stress in Multilayered Structures

*Peter M. Hall*

## 2.1 INTRODUCTION

Thermal stress is one of the most serious of reliability problems for microelectronic circuits. This is aggravated because the structures are mechanically complex, involving many different materials, each with its own set of elastic coefficients, expansivities, and stress limitations. Many of these effects require long service times or many temperature cycles to cause failures or even measurable changes, so testing new designs is an extremely time-consuming and costly affair. Many designers are resorting to numerical stress analysis, such as finite element studies, and these have helped greatly. There are good reasons, however, to perform analytical calculations of these structures as well, both to verify the finite element studies and to provide the designer with a feeling of how the various controllable parameters will affect the reliability of the product. Also, there are some situations where finite element analysis is not available, or requires unavailable expertise, or may even be too expensive to be practical. In these cases, sometimes a quick analysis of a simplified structure will provide some immediate advice, and perhaps avoid, or at least minimize, thermal stress problems.

The chapter presents a methodology for calculating thermal stresses and strains in layered structures with any number of layers. The structure is shown schematically in Fig. 2-1. This configuration is especially important today, as it approximates the structure of many multichip module designs, as well as thin- and thick-film circuits based on ceramic substrates. Such structures are becoming increasingly popular; for example, the 1991

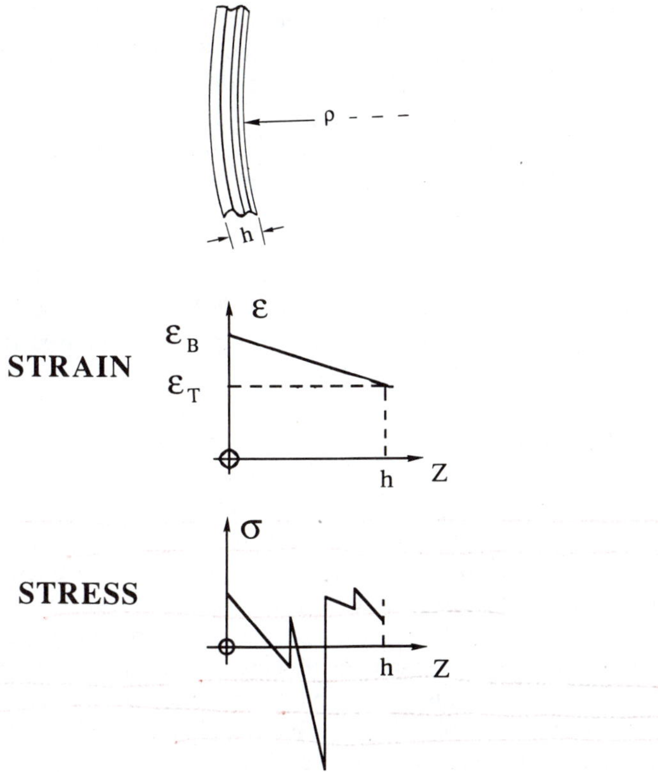



**Figure 2-1**   Multilayer stack of $N$ different materials, showing linear strain and sawtooth stress functions.

International Symposium on Microelectronics sponsored by ISHM (the International Society for Hybrid Microelectronics) had eleven papers on recent developments in multichip modules. And other electronic conferences such as the Electronic Components and Technology Conference, sponsored by the Institute of Electrical and Electronics Engineers and the Electronic Industries Association, have had similar interest.

The present analysis is basically a generalization of the "bimetallic strip" solution. Many such calculations have been done already,[1-7] but many of them are limited to two or three layers, or special thickness ratios, and some[1,3,4] are applicable to "strips," or to cylindrical bending, but not to "plates" for which the length and width are similar in magnitude. In addition, when more than three layers are present, the calculations can become extensive. A concise, but very good, review of early work is given by Suhir.[3] Recently several relevant papers have appeared, mostly dealing with shear stresses and "edge effects." Many of these offer finite element analysis calculations for certain specific geometries, and most are difficult to apply to one's own situation for the primary effects in the middle of the structure. Some of the better recent papers are refs. 8–14.

The present chapter presents a simplified method of calculating and presenting the mechanics of such structures, using a spreadsheet format that is available now to anyone with a PC or Macintosh computer having spreadsheet software such as LOTUS 1-2-3® or Microsoft WORKS®.

This analysis has been presented before,[15] but not in this form, and that paper contains some errors of a factor of 2, caused by this author. Of course, errata have been published,[16] but it is felt that a complete and corrected presentation would help alleviate any problems caused thereby.


## 2.2 ANALYSIS

The analysis is for uniform layers and applies to the case where the total thickness is much less than the in-plane dimensions. It also assumes that the total deflection (out-of-plane) is less than the total thickness of the layered structure. It does not take into account "end effects," assuming that the curvature of the plate is the same everywhere. Thus the entire analysis is independent of the in-plane dimensions. It assumes that all materials are isotropic in the plane of the plate, that they are in their elastic range, and that the stresses and strains are axisymmetric. The axisymmetric assumption is discussed further in a later section.

We begin with the general equations of mechanical equilibrium within a solid body. We start with Cartesian coordinates, and include thermal stress and strain, as in standard texts such as Timoshenko and Goodier.[17] The plate is assumed to be in the $x$–$y$ plane. Following convention, we use the term "strain" and the symbol $\varepsilon$ to include the thermal expansion part of the strain as well as the other term, denoted "stress-related" strain.

$$\varepsilon_x = \alpha\, \Delta T + [\sigma_x - v(\sigma_y + \sigma_z)]/E \qquad (2\text{-}1)$$

$$\varepsilon_y = \alpha\, \Delta T + [\sigma_y - v(\sigma_x + \sigma_z)]/E \qquad (2\text{-}2)$$

$$\varepsilon_z = \alpha\, \Delta T + [\sigma_z - v(\sigma_x + \sigma_y)]/E \qquad (2\text{-}3)$$

where $\sigma$ is the stress, $E$ is the elastic modulus, $\alpha$ is the coefficient of linear thermal expansion, $v$ is Poisson's ratio and $\Delta T$ is the temperature change from the temperature where the stresses are zero. (This may be the assembly temperature, or, if relaxation has occurred, it may be near room temperature.) Now if the plate is in the $x$–$y$ plane, we have $\sigma_z$ equal to zero and this

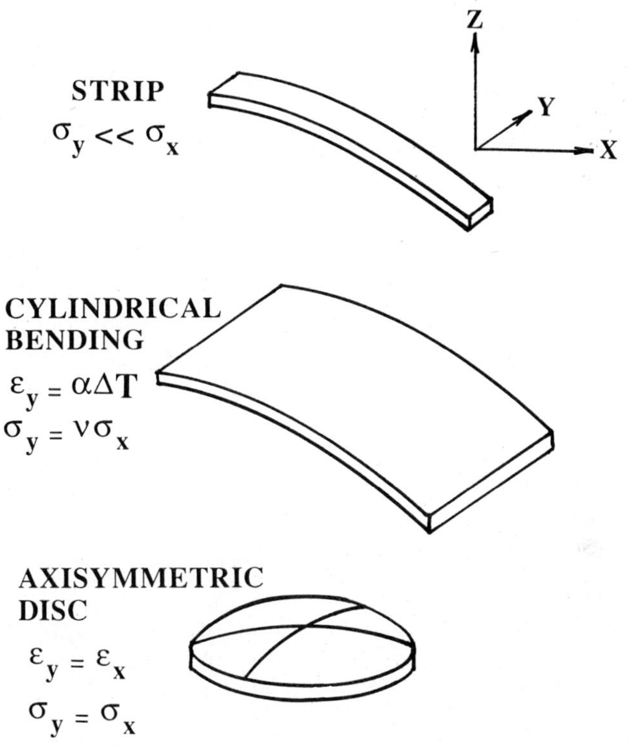


**Figure 2-2**   Three possible models for multilayered structures. This work uses the axisymmetric model.

gives us

$$\varepsilon_x = \alpha\, \Delta T + (\sigma_x - \nu\sigma_y)/E \tag{2-4}$$

$$\varepsilon_y = \alpha\, \Delta T + (\sigma_y - \nu\sigma_x)/E \tag{2-5}$$

$$\varepsilon_z = \alpha\, \Delta T - \nu(\sigma_x + \sigma_y)/E \tag{2-6}$$

There are three special cases of some interest here as shown in Fig. 2-2.

First, the "strips." A strip is much thinner in the $z$-direction than in the $x$ and $y$ directions, but its $x$-dimension is much larger than its $y$-dimension. In this case, we can assume that $\sigma_y$ is much less than $\sigma_x$. This leads to

$$\varepsilon_x = \alpha\, \Delta T + \frac{\sigma_x}{E} \tag{2-7}$$

This is the equation used for most of the bimetallic strip analyses, including the classic by Timoshenko.[1]

Second, if the bending is all in one direction (cylindrical bending), then the only strain in the $y$-direction is thermal strain, and we can see that $\sigma_y$ equals $v\sigma_x$. We can therefore substitute this into the first equation, giving

$$\varepsilon_x = \alpha\,\Delta T + \sigma_x(1 - v^2)/E \tag{2-8}$$

This is consistent with setting the "flexural rigidity" $= Eh^3/[(12(1 - v^2)]$. As Timoshenko[18] shows, this is appropriate for cylindrical bending.

The third case is the one we will use. It is for the axisymmetric situation, where by symmetry $\varepsilon_y$ equals $\varepsilon_x$ and $\sigma_y$ equals $\sigma_x$. In this case, we have

$$\varepsilon_x = \alpha\,\Delta T + \sigma_x(1 - v)/E \tag{2-9}$$

If Poisson's ratio, $v$, is 0.3, then numerically, the stress-related term in Eq. (2-8) is 30% more than in Eq. (2-9), and the corresponding term in Eq. (2-7) is 43% more than in Eq. (2-9).

Now since we are assuming an axisymmetric situation, we can eliminate the subscript $(x)$ from here on. Next we assume that the strain is a linear function of $z$, where $z$ is the dimension of the thickness of the plate. This can be written

$$\varepsilon = \varepsilon_B + \frac{z}{h}(\varepsilon_T - \varepsilon_B) \tag{2-10}$$

where $h$ is the total thickness of the structure, and the subscript $B$ represents the bottom, whereas $T$ represents the top of the entire plate. Within any given layer, then, the stress $(\sigma)$ is also a linear function of $z$. At each interface, however, $\sigma$ may be discontinuous, with the resulting "sawtooth" type of stress plot as a function of $z$ as indicated in Fig. 2-1.

We consider now a plate made up of $N$ layers, and let each of the layers have its own elastic modulus $(E)$, expansivity $(\alpha)$, and Poisson's ratio $(v)$. We assume the temperature is the same for all the layers. We denote the $i$th layer with the subscript $i$. Then the stress within the $i$th layer is given by combining Eqs. (2-9) and (2-10). The result is

$$\sigma = \frac{E_i}{1 - v_i}\left(\varepsilon_B + \frac{z}{h}(\varepsilon_T - \varepsilon_B) - \alpha_i\,\Delta T\right) \tag{2-11}$$

Recalling now that we are dealing with an axisymmetric case, we set the total radial force per unit perimeter in the $i$th layer equal to

$$P_i = \int_{z_{i-1}}^{z_i} \sigma \, dz \qquad (2\text{-}12)$$

where $z_i$ is the $z$-location of the top of the $i$th layer, and $z_0$ is zero. This makes $z_N$ equal to $h$. Next we substitute from Eq. (2-11) for the integrand, and integrate, giving a quadratic equation in $z_i$ and $z_{i-1}$:

$$P_i = \frac{E_i}{1 - v_i} \left( (\varepsilon_B - \alpha_i \, \Delta T)(z_i - z_{i-1}) + \frac{(\varepsilon_T - \varepsilon_B)(z_i^2 - z_{i-1}^2)}{2h} \right) \qquad (2\text{-}13)$$

We can also calculate the moment per unit perimeter in the $i$th layer, taking moments about the location $z = 0$:

$$M_i = \int_{z_{i-1}}^{z_i} z\sigma \, dz \qquad (2\text{-}14)$$

As before, we substitute for the stress from Eq. (2-11), finding now a cubic equation in $z_i$ and $z_{i-1}$:

$$M_i = \frac{E_i}{1 - v_i} \left( \frac{(\varepsilon_B - \alpha_i \, \Delta T)(z_i^2 - z_{i-1}^2)}{2} + \frac{(\varepsilon_T - \varepsilon_B)(z_i^3 - z_{i-1}^3)}{3h} \right) \qquad (2\text{-}15)$$

We are assuming here no external forces or moments, so the system must be "self-equilibrating." This means the sum of all the loads (forces and moments) at any perimeter must be zero. Thus,

$$\sum_{i=1}^{N} P_i = 0 \qquad (2\text{-}16)$$

and

$$\sum_{i=1}^{N} M_i = 0 \qquad (2\text{-}17)$$

From these equations, we will calculate the top and bottom strains, from which we can find the strains and stresses anywhere in the structure from Eqs. (2-10) and (2-11). These summations become rather cumbersome, so we define some factors to express the result in simpler equations. Both summations result in linear equations for $\varepsilon_T$ and $\varepsilon_B$ with a constant term involving $\Delta T$. From the force summation, we have

$$\varepsilon_B[3h(A_1 - B_1) + B_2 - A_2] + \varepsilon_T(A_2 - B_2) + 2h\,\Delta T(D_1 - C_1) = 0 \quad (2\text{-}18)$$

And from the moment summation we have

$$\varepsilon_B[3h(A_2 - B_2) + 2B_3 - 2A_3] + 2\varepsilon_T(A_3 - B_3) + 3h\,\Delta T(D_2 - C_2) = 0$$

$$(2\text{-}19)$$

(This equation is corrected from that in the original paper.)
   The $A$, $B$, $C$, and $D$ summations are defined as

$$A_k \equiv \sum \left(\frac{E_i}{1 - v_i}\right) z_i^k \qquad \text{for } k = 1, 2, 3 \qquad (2\text{-}20)$$

$$B_k \equiv \sum \left(\frac{E_i}{1 - v_i}\right) z_{i-1}^k \qquad \text{for } k = 1, 2, 3 \qquad (2\text{-}21)$$

$$C_k \equiv \sum \left(\frac{E_i}{1 - v_i}\right) \alpha_i z_i^k \qquad \text{for } k = 1, 2 \qquad (2\text{-}22)$$

and

$$D_k \equiv \sum \left(\frac{E_i}{1 - v_i}\right) \alpha_i z_{i-1}^k \qquad \text{for } k = 1, 2 \qquad (2\text{-}23)$$

where all the summations are from $i = 1$ to $i = N$, as before.
   The solution of these two equations in $\varepsilon_T$ and $\varepsilon_B$ is now straightforward. The result is

$$\varepsilon_B = \frac{h\,\Delta T}{F} [4(D_1 - C_1)(A_3 - B_3) - 3(D_2 - C_2)(A_2 - B_2)] \quad (2\text{-}24)$$

and

$$\varepsilon_T = \frac{h\,\Delta T}{F} [3(D_2 - C_2)(2hA_1 - 2hB_1 + B_2 - A_2) - 2(D_1 - C_1)$$

$$\times (3hA_2 - 3hB_2 + 2B_3 - 2A_3)] \qquad (2\text{-}25)$$

where

$$F \equiv -2(A_3 - B_3)(2hA_1 - 2hB_1 + B_2 - A_2)$$
$$+ (A_2B_2)(3hA_2 - 3hB_2 + 2B_3 - 2A_3) \qquad (2\text{-}26)$$

We are now in a position to calculate the stress and strain at any point in any of the $N$ layers using Eqs. (2-10) and (2-11). Also the radius of curvature of the bending is given by $\rho$ where

$$\frac{1}{\rho} = \frac{\varepsilon_T - \varepsilon_B}{h} \qquad (2\text{-}27)$$

and the deflection at the center of a disk of radius $L$ is approximately $\delta$ where

$$\delta = \frac{L^2}{2\rho} \qquad (2\text{-}28)$$

This analysis is still cumbersome to evaluate on a pocket calculator, but it can be set up conveniently on a spreadsheet such that new parameters, dimensions, or layers can easily be added, and all new results are instantly displayed.

## 2.3 SPREADSHEET CALCULATION OF STRESS IN $N$ LAYERS

We have prepared a spreadsheet calculation (Figs. 2-3 and 2-4) to perform these equations. The equations are entered into the appropriate cells of the spreadsheet once, using the "fill right" and "fill down" features. The spreadsheet software automatically increases cell reference locations as you fill right or down. In a few cases, we need to inhibit this increase by inserting "$" before the cell reference. This keeps the column letter the same as you fill right. Thus only the encircled cells have to be keyed in manually. This eliminates most of the entering of equations. We entered these the first time in less than two hours. As the equations are entered, each step is immediately verifiable by the numerical results of the sample provided in Fig. 2-4, so that typographical errors are readily found and corrected as they are made. The spreadsheet is then saved and used for any new set of input variables.

This work was done using Microsoft WORKS® on a Macintosh SE®, but it could be done just as well with any other spreadsheet. The best part of it is that once set up, the input parameters can be changed, with results instantly available.

Figure 2-3 shows the spreadsheet with equations, and Fig. 2-4 shows it

|    | A | B | C | D |
|----|---|---|---|---|
| 1 | | INPUT: | bot. layer | 2nd |
| 2 | Total Thickness | Material | CERAMIC | ADHESIVE |
| 3 | =Sum(C3:L3) | Thick (in.) | .021 | .006 |
| 4 | | Alpha(ppm/degC) | 6 | 80 |
| 5 | | E(Mpsi) | 15 | .1 |
| 6 | | Nu | .23 | .3 |
| 7 | | Delta T | -180 | |
| 8 | | | | |
| 9 | OVERALL: | | | |
| 10 | TCE (Bottom) | TCE (Midpoint) | TCE (Top) | Curvature |
| 11 | ppm/deg C | ppm/deg C | ppm/deg C | 1/ inch |
| 12 | =C24/C7 | =.5*(A12+C12) | =C23/C7 | =H12/10^6 |
| 13 | STRESSPLOT DATA: | | | |
| 14 | =C21 | =C20 | =D21 | =D20 |
| 15 | 0 | =C26 | =C26 | =D26 |
| 16 | eps layer top | ppm | =$C24+C26*$H12 | =$C24+D26*$H12 |
| 17 | eps layer bot. | ppm | =$C24+B26*$H12 | =$C24+C26*$H12 |
| 18 | eps Mech top | ppm | =C16-C4*$C7 | =D16-D4*$C7 |
| 19 | eps Mech bot. | ppm | =C17-C4*$C7 | =D17-D4*$C7 |
| 20 | Stress top | psi | =C18*C27 | =D18*D27 |
| 21 | Stress bottom | psi | =C19*C27 | =D19*D27 |
| 22 | CALCULATIONS: | | | |
| 23 | eps. overall top | ppm | =2*A3*C7*D23/D25 | =E23-F23 |
| 24 | eps. overall bot | ppm | =2*A3*C7*D24/D25 | =E24-F24 |
| 25 | intermed. calcs. | | | =E25+F25 |
| 26 | | Z sub i | =C3 | =C26+D3 |
| 27 | Sums | E/(1-Nu) | =C5/(1-C6) | =D5/(1-D6) |
| 28 | =Sum(C28:L28) | Asub1 | =C26*C27 | =D26*D27 |
| 29 | =Sum(C29:L29) | Asub2 | =C26*C28 | =D26*D28 |
| 30 | =Sum(C30:L30) | Asub3 | =C26*C29 | =D26*D29 |
| 31 | =Sum(C31:L31) | Bsub1 | 0 | =C26*D27 |
| 32 | =Sum(C32:L32) | Bsub2 | 0 | =C26*D31 |
| 33 | =Sum(C33:L33) | Bsub3 | 0 | =C26*D32 |
| 34 | =Sum(C34:L34) | Csub1 | =C4*C28 | =D4*D28 |
| 35 | =Sum(C35:L35) | Csub2 | =C26*C34 | =D26*D34 |
| 36 | =Sum(C36:L36) | Dsub1 | 0 | =D4*D31 |
| 37 | =Sum(C37:L37) | Dsub2 | 0 | =C26*D36 |

**Figure 2-3**  Spreadsheet with equations used to calculate stresses. Only the encircled cells need be entered manually. (*Continued*)

| E | F | G | H | |
|---|---|---|---|---|
| 3rd | 4th | 5th | 6th | ••• |
| ALUMINUM | ENCAPS. | ••• | | |
| .030 | .001 | ••• | | |
| 23 | 40 | ••• | | |
| 10 | 1 | ••• | | |
| .32 | .3 | ••• | | |
| Curv/degC | Rad Curv. | Deflect. | Curv. | |
| 1/(in. deg C) | inch | inch | 10^6/in. | |
| =D12/C7 | =1/D12 | =.5*D12 | =(C23-C24)/A3 | |
| =E21 | =E20 | =F21 | =F20 | ••• |
| =D26 | =E26 | =E26 | =F26 | ••• |
| =$C24+E26*$H12 | =$C24+F26*$H12 | =$C24+G26*$H12 | =$C24+H26*$H12 | ••• |
| =$C24+D26*$H12 | =$C24+E26*$H12 | =$C24+F26*$H12 | =$C24+G26*$H12 | ••• |
| =E16-E4*$C7 | =F16-F4*$C7 | =G16-G4*$C7 | =H16-H4*$C7 | ••• |
| =E17-E4*$C7 | =F17-F4*$C7 | =G17-G4*$C7 | =H17-H4*$C7 | ••• |
| =E18*E27 | =F18*F27 | =G18*G27 | =H18*H27 | ••• |
| =E19*E27 | =F19*F27 | =G19*G27 | =H19*H27 | ••• |
| =H24*(G24-2*G23) | =H23*(G25-3*H25) | =A3*(A31-A28) | =A36-A34 | |
| =-H23*G25 | =-H24*G24 | =A32-A29 | =1.5*(A37-A35) | |
| =G25*(G24-2*G23) | =-G24*(G25-3*H25) | =2*(A33-A30) | =A3*G24 | |
| =D26+E3 | =E26+F3 | =F26+G3 | =G26+H3 | ••• |
| =E5/(1-E6) | =F5/(1-F6) | =G5/(1-G6) | =H5/(1-H6) | ••• |
| =E26*E27 | =F26*F27 | =G26*G27 | =H26*H27 | ••• |
| =E26*E28 | =F26*F28 | =G26*G28 | =H26*H28 | ••• |
| =E26*E29 | =F26*F29 | =G26*G29 | =H26*H29 | ••• |
| =D26*E27 | =E26*F27 | =F26*G27 | =G26*H27 | ••• |
| =D26*E31 | =E26*F31 | =F26*G31 | =G26*H31 | ••• |
| =D26*E32 | =E26*F32 | =F26*G32 | =G26*H32 | ••• |
| =E4*E28 | =F4*F28 | =G4*G28 | =H4*H28 | ••• |
| =E26*E34 | =F26*F34 | =G26*G34 | =H26*H34 | ••• |
| =E4*E31 | =F4*F31 | =G4*G31 | =H4*H31 | ••• |
| =D26*E36 | =E26*F36 | =F26*G36 | =G26*H36 | ••• |

**Figure 2-3** (*continued*)

|    | A | B | C |
|----|---|---|---|
| 1  |   | INPUT: | bot. layer |
| 2  | Total Thickness | Material | CERAMIC |
| 3  | 0.058 | Thick (in.) | 0.021 |
| 4  |   | Alpha(ppm/degC) | 6 |
| 5  |   | E(Mpsi) | 15 |
| 6  |   | Nu | 0.23 |
| 7  |   | Delta T | -180 |
| 8  |   |   |   |
| 9  | OVERALL: |   |   |
| 10 | TCE  (Bottom) | TCE (Midpoint) | TCE (Top) |
| 11 | ppm/deg C | ppm/deg C | ppm/deg C |
| 12 | 3.08 | 15.86 | 28.64 |
| 13 | STRESSPLOT DATA: |   |   |
| 14 | 10245 | -22203 | 1740 |
| 15 | 0 | 0.021 | 0.021 |
| 16 | eps layer top | ppm | -2219.7 |
| 17 | eps layer bot. | ppm | -554.1 |
| 18 | eps Mech top | ppm | -1139.7 |
| 19 | eps Mech bot. | ppm | 525.9 |
| 20 | Stress top | psi | -22202.6 |
| 21 | Stress bottom | psi | 10244.5 |
| 22 | CALCULATIONS: |   |   |
| 23 | eps. overall top | ppm | -5154.40 |
| 24 | eps. overall bot | ppm | -554.11 |
| 25 | intermed. calcs. |   |   |
| 26 |   | Z sub i | 0.021 |
| 27 | Sums | E/(1-Nu) | 19.48052 |
| 28 | 1.33404 | Asub1 | 0.409091 |
| 29 | 0.06128 | Asub2 | 0.00859 |
| 30 | 0.00319 | Asub3 | 1.8041E-04 |
| 31 | 0.48149 | Bsub1 | 0 |
| 32 | 0.01543 | Bsub2 | 0 |
| 33 | 0.00056 | Bsub3 | 0 |
| 34 | 25.35681 | Csub1 | 2.45455 |
| 35 | 1.35103 | Csub2 | 0.05155 |
| 36 | 12.62950 | Dsub1 | 0.00000 |
| 37 | 0.43727 | Dsub2 | 0.00000 |

**Figure 2-4**   Same spreadsheet as Fig. 2-3, but with values displayed. Example is for a four-layer system. (*Continued*)

| D | E | F | G | H |
|---|---|---|---|---|
| 2nd | 3rd | 4th | 5th | 6th |
| ADHESIVE | ALUMINUM | ENCAPS. | ••• | |
| 0.006 | 0.03 | 0.001 | ••• | |
| 80 | 23 | 40 | ••• | |
| 0.1 | 10 | 1 | ••• | |
| 0.3 | 0.32 | 0.3 | ••• | |

| Curvature | Curv/degC | Rad Curv. | Deflect. | Curv. | |
|---|---|---|---|---|---|
| 1/ inch | 1/(in. deg C) | inch | inch | 10^6/in. | |
| -0.079315 | 4.4064E-04 | -12.608 | -0.0397 | -79315 | |
| 1672 | 21241 | -13751 | 3036 | 2922 | ••• |
| 0.027 | 0.027 | 0.057 | 0.057 | 0.058 | ••• |
| -2695.6 | -5075.1 | -5154.4 | -5154.4 | -5154.4 | ••• |
| -2219.7 | -2695.6 | -5075.1 | -5154.4 | -5154.4 | ••• |
| 11704.4 | -935.1 | 2045.6 | -5154.4 | -5154.4 | ••• |
| 12180.3 | 1444.4 | 2124.9 | -5154.4 | -5154.4 | ••• |
| 1672.1 | -13751.2 | 2922.3 | 0.0 | 0.0 | ••• |
| 1740.0 | 21240.8 | 3035.6 | 0.0 | 0.0 | ••• |
| -0.0381 | -0.072700 | -0.0346 | -0.0494 | -12.7273 | |
| -0.0041 | -0.066947 | -0.0629 | -0.0459 | -1.3706 | |
| -1.543E-04 | -2.790E-04 | 1.247E-04 | -0.0053 | -0.0027 | |
| 0.027 | 0.057 | 0.058 | 0.058 | 0.058 | ••• |
| 0.14286 | 14.70588 | 1.42857 | 0 | 0 | ••• |
| 0.003857 | 0.838235 | 0.082857 | 0 | 0 | ••• |
| 0.00010 | 0.04778 | 0.00481 | 0 | 0 | ••• |
| 2.8119E-06 | 2.7234E-03 | 2.7873E-04 | 0 | 0 | ••• |
| 0.00300 | 0.39706 | 0.08143 | 0 | 0 | ••• |
| 0.000063 | 0.010721 | 0.004641 | 0 | 0 | ••• |
| 1.3230E-06 | 2.8946E-04 | 2.6456E-04 | 0 | 0 | ••• |
| 0.30857 | 19.27941 | 3.31429 | 0 | 0 | ••• |
| 0.00833 | 1.09893 | 0.19223 | 0 | 0 | ••• |
| 0.24000 | 9.13235 | 3.25714 | 0 | 0 | ••• |
| 0.00504 | 0.24657 | 0.18566 | 0 | 0 | ••• |

**Figure 2-4** (*continued*)

with evaluated numbers. It is set up for ten layers, but only six are shown. It can be easily extended to any number. And it need not be changed to analyze a plate with fewer layers than that for which it was set up. Our example uses four layers.

## Input

The input data are near the top, in rows 2 through 6 and columns C through F (or more if required). Here the user inserts the names and required data for each of the layers. The Poisson's ratio is sometimes not well known, and if that is the case, a value of 0.3 is often reasonable. The temperature change is entered in cell C7. This can be positive or negative. In our case, it was negative, since the temperature of curing of the adhesive was 205°C. The units, of course, can be any consistent set of units. The rest of the spreadsheet is calculated by the computer, including all the intermediate calculations and final results.

## Calculations

The results of the calculations are in the next section, rows 9 through 21. The intermediate calculations are located in rows 22 through 37. It is arranged this way to simplify printing of the input and final results without printing the intermediate results if so desired. Logically, the calculations in rows 26 and 27 are first, and it is recommended that the equations be entered in the logical order rather than from top to bottom. That way, numerical verification of each step is immediately available. In row 26 we calculate the $z$-location of each interface, and in row 27 the modified elastic modulus for each layer. Then in rows 28 through 37 we calculate the summations required for the $A$, $B$, $C$, and $D$ summations. These are used in rows 23 through 25 to calculate the strain at the top and bottom of the stack that constitutes the plate. Rows 23 through 25 proceed logically from right to left, such that columns G and H are calculated first, then columns E and F, and finally D and C. The results from C23 and C24 are then used to calculate the overall results in row 12 and the results for each layer in rows 16 through 21. In row 12 we have available the temperature coefficient of expansion of the top and bottom surface, as well as that of the midpoint, which is midway between them. We also list the curvature, the radius of curvature, and the deflection in inches of a 2-inch-diameter disk of the stack. This gives a more intuitive feeling for the bowing than does the radius of curvature. Column H12 is included because it is helpful in calculating rows 16 and 17.

### Results for Each Layer

For each layer, we have included calculations of both the total strain (rows 16 and 17) and the "mechanical" strain (rows 18 and 19), which excludes the thermal expansivity part of the strain and is proportional to the stress. In each case, a calculation is made at the top and bottom of each layer, and linear interpolation can be made for any point within any layer. Of greatest interest, however, are the extremes, as given.

It is often convenient to have available a set of data points for a plot of stress vs. $z$. The required data are available in rows 20 and 21, but they have been repeated in rows 14 and 15 so as to be copied directly into a plot program. We used Cricket Graph to obtain the plot on Fig. 2-5. Note that because of the need to duplicate the $z$-location values, it is not a good idea to use fill right for rows 14 and 15. They are very short entries, however, and thus do not require much time.

It is suggested that anybody who wants to use this calculation with a spreadsheet, or in any other way, should verify it using the numerical example given here. That will probably detect any errors that may arise in programming or typing.

This routine has been helpful to us in calculating stresses in simple layered structures, and it is offered in the hope that others will also find it useful. It is, of course, highly idealized, assuming uniformity of the curvature, and elasticity and isotropy of the materials.

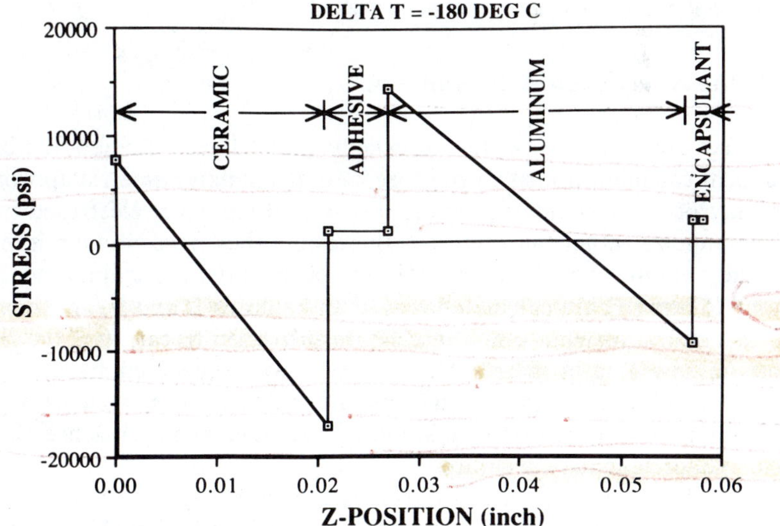


**Figure 2-5**   Stress in a four-layer system, as calculated from Figs. 2-3 and 2-4.

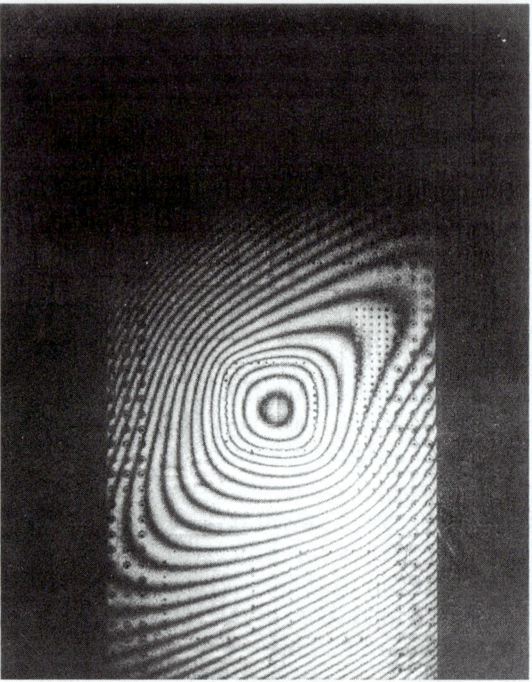

**Figure 2-6** Holographic interferogram for a 2.2 inch × 3.8 inch printed circuit board.[19] A square ceramic chip carrier mounted in the central four fringes is dissipating 0.28 watts. The interference fringes constitute a contour plot of the deflections caused by the heat. The resulting diagonal symmetry was not expected.

### 2.3.1 The Axisymmetric Assumption

One effect that may not be well appreciated, and is generally ignored by any analytical calculation for this purpose, is that the axisymmetric approximation may produce significant errors. For one thing, most electronic circuits are rectangular rather than circular. But even if they were circular, it may be that the bending would not be axisymmetric but more cylindrical, like a "taco." If bending starts along any one direction for any reason, the board becomes much stiffer against bending in other directions, and bending in those directions is inhibited. This effect is not considered by any of the "axisymmetric analyses, including this one, but it may be significant. From our holographic interferometry studies[19] we have some evidence that the "taco" shape is indeed preferred.

One such holographic interferogram is shown in Fig. 2-6. It was taken on a 3.8 × 2.2 inch printed wiring board with a 0.65 × 0.65 inch ceramic chip carrier surface mounted to it in the center (occupying approximately the middle four fringes on the figure). There was a power dissipation in the carrier of 0.28 watts. The circuit seemed initially to have a truly rectangular symmetry and all of the fibers of the fabrics were in either the $X$ or the $Y$ direction. The interferogram displays contour curves of the deflection relative to the unheated surface. Upon heating, the two diagonals showed markedly different curvature. Clearly the axisymmetric analysis would not predict this type of bending, and would give major errors, except, perhaps near the center of the figure. The deflections were all proportional to the power applied, but the board seemed to prefer to keep one diagonal almost straight, while the other had serious bending. It may be that the bending would also have been stable with a bend on the other diagonal if we had influenced it to start that way. There may also have been some unintentional asymmetry in the structure. This is an effect that requires further study, and it is suggested as an experimental project for anyone who is interested.

## 2.4  CONCLUSION

We have presented here a method of estimating the bending and stresses caused by temperature changes in multilayered structures with many layers. It is intended to provide insight into what happens inside such structures, which are common in microelectronic packages. In real life, of course, such idealized geometries do not exist, and temperature gradients, nonlinear, and anisotropic materials may spoil even the best of finite element analyses. Nevertheless, it seems that such simple calculations can provide first-order guidance in designing packages.

## ACKNOWLEDGMENTS

I am indebted to Frank Howland for working out this analysis in the first place, and for introducing me to the mysteries of mechanical engineering. I am also indebted to Dixon Dudderar, who did the holography.

## REFERENCES


1. Timoshenko, S., "Analysis of Bi-Metal Thermostats," *Journal of the Optical Society of America*, **11**, 1925, pp. 233–255.
2. McClintock, F. A., and A. S. Argon, *Mechanical Behavior of Materials*, Addison-Wesley, Reading, MA, 1966, p. 351.
3. Suhir, E., "Stresses in Bi-Metal Thermostats," *ASME Journal of Applied Mechanics*, **53**, 1986, pp. 657–660.

4. Suhir, E., "Die Attachment Design and its influence on Thermal Stresses in the Die and its Attachment," *Electronic Components Conference*, pp. 508–517.
5. Boley, B. A., and J. H. Weiner, *Theory of Thermal Stresses*, Wiley, New York, 1960, p. 429.
6. Durelli, A. J., and C. H. Tsao, "Determination of Thermal Stresses in Three-ply Laminates," *ASME Journal of Applied Mechanics*, **77**, 1955, pp. 190–192.
7. Goldberg, J. E., "Axisymmetric Flexural Temperature Stresses in Circular Plates," *ASME Journal of Applied Mechanics*, **20**, 1953, pp. 257–260.
8. Morgan, H. S., "Thermal Stresses in Layered Electrical Assemblies Bonded with Solder," *ASME Journal of Electronic Packaging*, **113**, 1991, pp. 350–354.
9. Yin, W., "Thermal Stresses in Free-Edge Effects in Laminated Beams: A Variational Approach Using Stress Functions," *ASME Journal of Electronic Packaging*, **113**, 1991, pp. 68–75.
10. Daniel, I. M., T.-M. Wang, and J. T. Gotro, "Thermomechanical Behavior of Multilayered Structures in Microelectronics," *ASME Journal of Electronic Packaging*, **112**, 1990, pp. 11–15.
11. Eischen, J. W., C. Chung, and J. H. Kim, "Realistic Modeling of Edge Effect Stresses in Bimaterial Elements," *ASME Journal of Electronic Packaging*, **112**, 1990, pp. 16–23.
12. Glaser, J. C., "Thermal Stresses in Compliantly Joined Materials," *ASME Journal of Electronic Packaging*, **112**, 1990, pp. 24–29.
13. Erdogan, F., and P. F. Joseph, "Mechanical Modeling of Multilayered Films on an Elastic Substrate," *ASME Journal of Electronic Packaging*, **112**, 1990, pp. 309–332.
14. Pan, T., and Y. Pao, "Deformation in Multilayer Stacked Assemblies," *ASME Journal of Electronic Packaging*, **112**, 1990, pp. 30–34.
15. Hall, P. M., F. L. Howland, Y. S. Kim, and L. H. Herring, "Strains in Aluminum-Adhesive-Ceramic Tri-layers," *ASME Journal of Electronic Packaging*, **112**, 1990, 288–302.
16. Hall, P. M., F. L. Howland, Y. S. Kim, and L. H. Herring, "Errata", *ASME Journal of Electronic Packaging*, **113**, 1991, p. 26.
17. Timoshenko, S., and J. N. Goodier, *Theory of Elasticity*, 2nd edn., McGraw-Hill, New York, 1951, p. 421.
18. Timoshenko, S., *Theory of Plates and Shells*, McGraw-Hill, New York, 1940, p. 3.
19. Hall, P. M., T. D. Dudderar, and J. F. Argyle, "Thermal Deformations Observed in Leadless Ceramic Chip Carriers Surface Mounted to Printed Wiring Boards," *IEEE Trans. Components, Hybrids, and Manufacturing Technology*, **CHMT-6**, 1983, pp. 544–552.

# 3

# Thermal Stresses in Anisotropic Multilayered Structures

*Wan-Lee Yin*

## 3.1 INTRODUCTION

Bimetal thermostats and multilayered beams and laminates are often subjected to severe interfacial stresses under mechanical and temperature loads. According to the predictions of the linear theory of elasticity, mismatches in the elastic properties of two adjacent layers generally result in a stress singularity at the intersection of their interface with a free edge.[1–4] The intense but localized peeling and shearing action near the free edge may initiate interfacial fracture and may lead to failure of the component. Such adverse conditions can sometimes be avoided or rendered less harmful by rearranging the layers or by altering certain geometrical parameters. This requires, however, efficient and accurate analytical methods for determining the interlaminar stresses across all interfaces of a layered beam or laminate under various mechanical and thermal loads.

Because severe interlaminar stress is localized to narrow regions near a free edge, in which the stress fields depend crucially on the disparate elastic moduli and thermal expansion coefficients of the adjacent layers, a theoretical or numerical analysis capable of predicting the interlaminar stresses with reasonable accuracy must take into account various geometrical and material parameters of a multilayered structure. Consequently, accurate finite element solutions of layered beams and plates under mechanical and thermal loads usually require elaborate modeling and a large number of degrees of

95

freedom.* Repeated execution of refined finite element analysis for various loads and for different combinations of problem parameters may be impractical in applications related to the design and optimization of a layered structure. Furthermore, conventional finite element solutions yield interlaminar shear stresses that fail to satisfy the traction-free boundary conditions at the free edge.

On the other hand, a simplified analysis that treats the two groups of layers below and above a particular interface as classical or high-order beams or laminates immediately loses the distinctive elastic properties of the layers adjacent to the interface—properties that determine the eigenvalues and eigenfunctions associated with the stress singularity of the elasticity solution. While such a simplified analysis may sometimes be used to estimate the significance of the interlaminar action, it cannot be relied upon to provide consistently accurate solutions of interlaminar stresses.†

The variational method of analysis presented in this chapter is intended to provide an efficient yet accurate method for the steady-state thermal stress analysis of a layered beam or plate. Since our objective is to obtain reasonably accurate solutions for the interlaminar stresses, it is natural to adopt a stress formulation, rather than a displacement formulation, in the variational approach. The stress components in each layer will be represented by the derivatives of stress functions. This implies that the equilibrium equations are *exactly* satisfied by the variational solution in all regions of the layered beam or plate, including the regions of high stress gradient near a free edge. Furthermore, traction-free boundary conditions and interfacial continuity of interlaminar stresses are strictly imposed, while the compatibility of strain and the interfacial continuity of displacements are enforced in the sense of weighted integrals through the use of the principle of complementary virtual work.

The main problem considered in this article is that of a layered beam of finite length, or a rectangular laminate, composed of a number of layers of homogeneous, anisotropic materials and subjected to a temperature increment depending only on the thickness coordinate. While the constituent

---

* While there is a large body of finite element solutions of interlaminar stress problems under mechanical loading, one finds relatively few works (based on the finite element method) concerning free-edge interlaminar stresses induced by steady-state thermal loads. Among these works we should mention the papers of Wang and Crossman,[5] Stango and Wang,[6] Cho et al.,[7] Glaser,[8] and Kuo and Chen.[9] Furthermore, Hess[10] investigated the problem for a two-layer laminate using eigenfunction expansions, and Kuo[11] presented an approximate elasticity solution of a two-layer beam by determining the stress singularities and using the Fourier transform method.

† Various approximate analyses of the interlaminar thermal stresses in laminated beams composed of isotropic layers were given by Grimado,[12] Chen and Nelson,[13] Suhir (see his recent book),[14] and others.

layers in electronic packaging components are often isotropic or transversely isotropic, composite laminates used in aerospace structures are intentionally made of strongly anisotropic layers with various orientation angles. Compared to several recent numerical and approximate studies of the same subject, the scope of the present analysis is considerably broader because it encompasses general elastic and thermal anisotropy of the constituent layers.*

The thermoelastic constitutive equations of an anisotropic layer are introduced in Section 3-2. In Section 3-3, we show that the thermal stress problem for the finite beam or rectangular laminate may be decomposed into (i) a trivial problem for an *infinite* but otherwise identical beam or laminate under the same temperature load, and (ii) a complementary, purely mechanical problem in which the finite beam or rectangular laminate is subjected only to boundary tractions equal but opposite to the in-plane stresses of the first (trivial) problem. After introducing the stress functions and the displacement expressions in Sections 3-4 and 3-5, we apply the complementary virtual work principle to the layered structure, and obtain a variational equation for the complementary problem (Section 3-6). The stress functions in each layer are approximated by polynomial functions of the thickness coordinate. The differential equations governing the coefficients of the polynomials and the associated boundary conditions are derived, respectively, from the variational equation and from the trivial thermal stress solution of the infinite beam or infinite laminate (Sections 3-7 and 3-8). These differential equations and boundary conditions define an eigenvalue problem. A procedure for constructing the final solution of the complementary problem from the solution of the eigenvalue problem is described in Sections 3-9 and 3-10. In Section 3-11, we point out the simplification resulting from the case of isotropic or specially orthotropic laminates. The case of laminated beams is considered in Section 3-13.

Although the present analysis method was developed for a straight free edge, it may be applicable in an approximate sense to a curved free edge provided that the radius of curvature of the free edge is an order of magnitude greater than the thickness of the laminate. This is because the characteristic length of decay of the interlaminar stresses is usually comparable to the thickness of the laminate. Hence, within the narrow boundary region where the interlaminar stresses are significant, the curved boundary may be approximated by a straight edge (see Section 3-12).

---

* For the special case of a laminated beam composed of two *isotropic* layers and a thin adhesive middle layer, a variational analysis based on the complementary energy principle was given by Chen et al.[15] These authors did not introduce stress functions but instead relied on integration of the equilibrium equations. The present general method of analysis, when properly modified to account for the lower-order approximation in the adhesive layer, yields essentially the same eigenvalue problem and variational solutions as in their study.

The accuracy of the present method of analysis depends naturally on the degree of the polynomial approximation for the stress functions. The approximation presented in Section 3-7 generally yields reasonably accurate results for the interlaminar stresses, except in exceedingly short segments of the interfaces adjacent to the free edge. Higher-order polynomial approximations may be used (especially for the stress function in a relatively thick layer) to improve the solution near the free edge. However, in a very thin layer it may be desirable to use lower-order polynomial approximations to avoid numerical difficulties associated with the occurrence of unusually large eigenvalues (see Section 3-14).

It should be mentioned that the pointwise values of the interlaminar stresses near a stress singularity of the elasticity solution have dubious physical significance because, in a real material, excessively large interlaminar stresses are invariably modified by nonlinear and inelastic effects. It is pointed out that, at any point on the interface, the values of one stress function, $\Psi$, and of the first derivatives of the other function, $F$, are equal to the integrals of the interlaminar stresses over an end segment of the interface (Section 3-15). Hence the maximum values of these functions indicate the total peeling and shearing action acting across the end segment. They are useful measures of the criticality of the interlaminar stresses. The integrals also yield directly the average values of the interlaminar stresses over short end intervals, and these averages may be used as parameters in a failure criterion associated with interlaminar fracture. Furthermore, the integrals may be calculated accurately from the variational solutions based on lower-order polynomial approximations of the stress functions, even though the lower-order solutions do not yield accurate results of the interlaminar stresses in regions extremely close to a free edge. Hence the proposed measures of the criticality of interlaminar stresses have the additional advantage that their values may be calculated accurately without excessive computational effort.

The present method of solution is applied to three-layer isotropic beams and anisotropic laminates, and the results for the interlaminar stresses are shown in Section 3-16. The symmetric matrices characterizing the eigenvalue problem are determined by using the symbolic algebraic program MACSYMA, and a transcript of the MACSYMA session is shown in the Appendix.

## 3.2 ELASTICITY OF AN ORTHOTROPIC LAYER REFERRED TO THE GLOBAL COORDINATES OF THE LAMINATE

Consider a multilayered laminate, consisting of $n + 1$ homogeneous, orthotropic elastic layers with different orientation angles and possibly made of

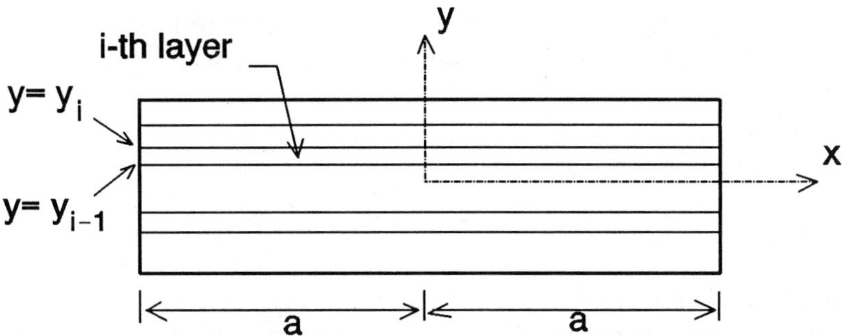


**Figure 3-1**   A multilayer laminate.

different materials. The layers are separated by $n$ parallel interfaces at $y = y_i$ ($i = 1, 2, \ldots, n$), where the thickness coordinate $y$ is defined in such a way that the coordinate plane $y = 0$ coincides with the middle plane of the laminate. The layers and their interfaces will be numbered from the bottom up. Thus the $i$th layer is bounded below by the $(i - 1)$th interface and bounded above by the $i$th interface (Fig. 3-1). This layer has thickness $h_i = y_i - y_{i-1}$, and is composed of an elastic material with the extensional elastic moduli $E_1^{(i)}$, $E_2^{(i)}$ and $E_3^{(i)}$ along the three axes of orthotropy. Other material moduli needed to characterize the elasticity of the layer are $v_{23}^{(i)}$, $v_{31}^{(i)}$, $v_{12}^{(i)}$, $G_{23}^{(i)}$, $G_{31}^{(i)}$ and $G_{12}^{(i)}$. It will be assumed in the following analysis that the elastic orthotropy axes coincide with the three symmetry axes of thermal expansion (the cases when this assumption is violated may be analyzed with only minor changes in the formulation). If this assumption is valid, and if the coefficients of thermal expansion along the three orthotropic directions are $\alpha_1^{(i)}$, $\alpha_2^{(i)}$ and $\alpha_3^{(i)}$, then the thermoelastic constitutive equation of the layer has the form (generalizing eq. (1) of ref. 16 to include the thermal strain)

$$
\begin{Bmatrix} \varepsilon_1 \\ \varepsilon_2 \\ \varepsilon_3 \\ \varepsilon_4 \\ \varepsilon_5 \\ \varepsilon_6 \end{Bmatrix}
=
\begin{bmatrix}
S_{11}^{(i)} & S_{12}^{(i)} & S_{13}^{(i)} & 0 & 0 & 0 \\
S_{21}^{(i)} & S_{22}^{(i)} & S_{23}^{(i)} & 0 & 0 & 0 \\
S_{31}^{(i)} & S_{32}^{(i)} & S_{33}^{(i)} & 0 & 0 & 0 \\
0 & 0 & 0 & S_{44}^{(i)}/2 & 0 & 0 \\
0 & 0 & 0 & 0 & S_{55}^{(i)}/2 & 0 \\
0 & 0 & 0 & 0 & 0 & S_{66}^{(i)}/2
\end{bmatrix}
\begin{Bmatrix} \sigma_1 \\ \sigma_2 \\ \sigma_3 \\ \tau_{23} \\ \tau_{12} \\ \tau_{13} \end{Bmatrix}
+
\begin{Bmatrix} \alpha_1^{(i)} T \\ \alpha_2^{(i)} T \\ \alpha_3^{(i)} T \\ 0 \\ 0 \\ 0 \end{Bmatrix}
$$

$$(3\text{-}1)$$

where $T$ is the temperature field (relative to the stress-free temperature), $\varepsilon_1$, $\varepsilon_2$, and $\varepsilon_3$ are the extensional strains along the three orthotropic axes, and where $\varepsilon_4 = \gamma_{23}/2$, $\varepsilon_5 = \gamma_{13}/2$ and $\varepsilon_6 = \gamma_{12}/2$ are the shearing components of the infinitesimal strain tensor. The compliance coefficients $S_{mn}^{(i)}$ are given by

$$S_{11}^{(i)} = 1/E_1^{(i)}, \qquad S_{22}^{(i)} = 1/E_2^{(i)}, \qquad S_{33}^{(i)} = 1/E_3^{(i)}$$

$$S_{12}^{(i)} = -v_{12}^{(i)}/E_1^{(i)}, \qquad S_{13}^{(i)} = -v_{13}^{(i)}/E_1^{(i)}, \qquad S_{23}^{(i)} = -v_{23}^{(i)}/E_2^{(i)} \qquad (3\text{-}2)$$

$$S_{44}^{(i)} = 1/G_{23}^{(i)}, \qquad S_{55}^{(i)} = 1/G_{13}^{(i)}, \qquad S_{66}^{(i)} = 1/G_{12}^{(i)}$$

The thickness direction of the laminate (i.e., the $y$-axis) is an orthotropic direction and this direction will be taken as the 3-direction in every layer. In the tangential plane of the $i$th layer, the orthotropy direction 1 is oriented at a constant angle $\theta_i$ with respect to the $z$-direction (the direction parallel to the free edges). Transformation of the stress components from the $(z, x, y)$ coordinate system to the orthotropic material system $(1, 2, 3)$ follows the rule

$$\begin{Bmatrix} \sigma_1 \\ \sigma_2 \\ \sigma_3 \\ \tau_{23} \\ \tau_{13} \\ \tau_{12} \end{Bmatrix} = \begin{bmatrix} \cos^2 \theta_i & \sin^2 \theta_i & 0 & 0 & 0 & \sin 2\theta_i \\ \sin^2 \theta_i & \cos^2 \theta_i & 0 & 0 & 0 & -\sin 2\theta_i \\ 0 & 0 & 1 & 0 & 0 & 0 \\ 0 & 0 & 0 & \cos \theta_i & -\sin \theta_i & 0 \\ 0 & 0 & 0 & \sin \theta_i & \cos \theta_i & 0 \\ -\tfrac{1}{2}\sin 2\theta_i & \tfrac{1}{2}\sin 2\theta_i & 0 & 0 & 0 & \cos 2\theta_i \end{bmatrix} \begin{Bmatrix} \sigma_z \\ \sigma_x \\ \sigma_y \\ \tau_{xy} \\ \tau_{yz} \\ \tau_{xz} \end{Bmatrix}$$

$$(3\text{-}3)$$

Both the tensorial components of the total strain and those of the thermal expansional strain transform in exactly the same manner. Hence the components of the mechanical strain (i.e., the total strain minus the thermal strain) also transform in the same way. For simplicity we denote the matrix of transformation in the last equation by $\mathbf{H}(\theta_i)$ and the matrix of Eq. (3-1) by $\mathbf{S}^{(i)}$. Then $\mathbf{H}(-\theta_i)$ is the inverse matrix of $\mathbf{H}(\theta_i)$. It follows that the mechanical strain $\varepsilon_{ij}^*$ is related to the stress in the $i$th layer by the constitutive equation

$$\begin{Bmatrix} \varepsilon_z^* \\ \varepsilon_x^* \\ \varepsilon_y^* \\ \gamma_{xy}^*/2 \\ \gamma_{yz}^*/2 \\ \gamma_{xz}^*/2 \end{Bmatrix} = \mathbf{H}(-\theta_i)\mathbf{S}^{(i)}\mathbf{H}(\theta_i) \begin{Bmatrix} \sigma_z \\ \sigma_x \\ \sigma_y \\ \tau_{xy} \\ \tau_{yz} \\ \tau_{xz} \end{Bmatrix} \qquad (3\text{-}4)$$

where the matrix $\mathbf{H}(-\theta_i)\mathbf{S}^{(i)}\mathbf{H}(\theta_i)$ has the form

$$
\begin{bmatrix}
a_{11}^{(i)} & a_{12}^{(i)} & a_{13}^{(i)} & 0 & 0 & a_{16}^{(i)} \\
a_{12}^{(i)} & a_{22}^{(i)} & a_{23}^{(i)} & 0 & 0 & a_{26}^{(i)} \\
a_{13}^{(i)} & a_{23}^{(i)} & a_{33}^{(i)} & 0 & 0 & a_{36}^{(i)} \\
0 & 0 & 0 & a_{44}^{(i)}/2 & a_{45}^{(i)}/2 & 0 \\
0 & 0 & 0 & a_{45}^{(i)}/2 & a_{55}^{(i)}/2 & 0 \\
a_{16}^{(i)}/2 & a_{26}^{(i)}/2 & a_{36}^{(i)}/2 & 0 & 0 & a_{66}^{(i)}/2
\end{bmatrix}
$$

From the first row of Eq. (3-4) one obtains

$$
\sigma_z = (\varepsilon_z^* - a_{12}^{(i)}\sigma_x - a_{13}^{(i)}\sigma_y - a_{16}^{(i)}\tau_{xz})/a_{11}^{(i)} \tag{3-5}
$$

Substituting the expression into the right-hand side of Eq. (3-3), and replacing the first row of Eq. (3-3) by Eq. (3-5), one has

$$
\begin{Bmatrix}
\sigma_z \\ \varepsilon_x^* \\ \varepsilon_y^* \\ \gamma_{xy}^* \\ \gamma_{yz}^* \\ \gamma_{xz}^*
\end{Bmatrix}
=
\begin{bmatrix}
1/a_{11}^{(i)} & -a_{12}^{(i)}/a_{11}^{(i)} & -a_{13}^{(i)}/a_{11}^{(i)} & 0 & 0 & -a_{16}^{(i)}/a_{11}^{(i)} \\
a_{12}^{(i)}/a_{11}^{(i)} & \beta_{22}^{(i)} & \beta_{23}^{(i)} & 0 & 0 & \beta_{26}^{(i)} \\
a_{13}^{(i)}/a_{11}^{(i)} & \beta_{23}^{(i)} & \beta_{33}^{(i)} & 0 & 0 & \beta_{36}^{(i)} \\
0 & 0 & 0 & \beta_{44}^{(i)} & \beta_{45}^{(i)} & 0 \\
0 & 0 & 0 & \beta_{45}^{(i)} & \beta_{55}^{(i)} & 0 \\
a_{16}^{(i)}/a_{11}^{(i)} & \beta_{26}^{(i)} & \beta_{36}^{(i)} & 0 & 0 & \beta_{66}^{(i)}
\end{bmatrix}
\begin{Bmatrix}
\varepsilon_z^* \\ \sigma_x \\ \sigma_y \\ \tau_{xy} \\ \tau_{yz} \\ \tau_{xz}
\end{Bmatrix}
$$

$$\tag{3-6}$$

where the coefficients

$$
\beta_{mn}^{(i)} = a_{mn}^{(i)} - \frac{a_{1m}^{(i)}a_{1n}^{(i)}}{a_{11}^{(i)}} \qquad \text{for } m, n \neq 1
$$

were introduced by Lekhnitskii.[17]


## 3.3 THERMAL STRESS PROBLEM OF A RECTANGULAR LAMINATE

We assume that the multilayered laminate considered in the preceding section has free edges $x = \pm a$ and $z = \pm b$ (where $a$ and $b$ are significantly greater than $t$, the thickness of the laminate) and is subjected to a temperature load $T(y)$ that varies linearly with respect to the thickness coordinate in each layer. Then, in the $i$th layer, we have

$$T(y) = T_i^0 + yT_i^1 \tag{3-7}$$

where $T_i^0$ and $T_i^1$ are constants related in the following manner to ensure continuity of the temperature across the interfaces of the layers:

$$T_i^0 + y_i T_i^1 = T_{i+1}^0 + y_i T_{i+1}^1$$

The temperature load produces interlaminar thermal stresses across the interfaces. Generally, significant interlaminar stress is confined to narrow regions along the free edges of the laminate (including the four corner regions) and decays rapidly away from the free edges. On the various cross-sections parallel to the coordinate plane $z = 0$, the thermal stress has almost an identical pattern (i.e., nearly independent of the coordinate $z$), provided that $z$ is not close to $\pm b$. Our problem is to obtain an accurate approximate solution to the thermal stress distribution in the region away from $z = \pm b$. Notice that, in the region near $z = \pm b$ but away from the corners of the laminate, the interlaminar stresses may be determined by solving a similar problem in which the coordinates $x$ and $z$ are interchanged.

In the $i$th layer ($i = 1, 2, \ldots, n + 1$), we define

$$h_i = y_i - y_{i-1} \qquad g_i = (y_i^2 - y_{i-1}^2)/2, \qquad f_i = (y_i^3 - y_{i-1}^3)/3 \tag{3-8}$$

We now consider an *infinite* laminate ($-\infty < z < \infty$, $-\infty < x < \infty$), having the same ply layup, and subjected to the same temperature load $T(y)$. If this infinite laminate is free from mechanical loads, and has vanishing resultant forces and moments, then the stress in the laminate depends only on the thickness coordinate and, furthermore,

$$\sigma_y = \tau_{xy} = \tau_{yz} = 0 \tag{3-9}$$

in the layers of the laminate. The tangential strains in the infinite laminate are determined by the middle-plane strains $\varepsilon_z^0$, $\varepsilon_x^0$, $\gamma_{xz}^0$ and the curvatures $\kappa_z$, $\kappa_x$, $\kappa_{xz}$ according to the expressions

$$\varepsilon_z = \varepsilon_z^0 - y\kappa_z, \qquad \varepsilon_x = \varepsilon_x^0 - y\kappa_x, \qquad \gamma_{xz} = \gamma_{xz}^0 - y\kappa_{xz} \tag{3-10}$$

These tangential components of the total strain include the thermal part and the mechanical part. Substituting Eq. (3-9) into Eq. (3-4), we obtain the following expressions for the tangential components of the mechanical strain in the $i$th layer:

$$\begin{Bmatrix} \varepsilon_z^* \\ \varepsilon_x^* \\ \gamma_{xz}^* \end{Bmatrix} = \begin{bmatrix} a_{11}^{(i)} & a_{12}^{(i)} & a_{16}^{(i)} \\ a_{12}^{(i)} & a_{22}^{(i)} & a_{26}^{(i)} \\ a_{16}^{(i)} & a_{26}^{(i)} & a_{66}^{(i)} \end{bmatrix} \begin{Bmatrix} \sigma_z \\ \sigma_x \\ \tau_{xz} \end{Bmatrix} \tag{3-11}$$

The corresponding components of the thermal strain are

$$\varepsilon_z - \varepsilon_z^* = \alpha_z^{(i)} T(y), \qquad \varepsilon_x - \varepsilon_x^* = \alpha_x^{(i)} T(y), \qquad \gamma_{xz} - \gamma_{xz}^* = \alpha_{xz}^{(i)} T(y) \tag{3-12}$$

where

$$\begin{aligned} \alpha_z^{(i)} &= \alpha_1^{(i)} \cos^2 \theta_i + \alpha_2^{(i)} \sin^2 \theta_i \\ \alpha_x^{(i)} &= \alpha_1^{(i)} \sin^2 \theta_i + \alpha_2^{(i)} \cos^2 \theta_i \\ \alpha_{xz}^{(i)} &= (\alpha_1^{(i)} + \alpha_2^{(i)}) \sin 2\theta_i \end{aligned} \tag{3-13}$$

Inverting Eq. (3-11) and substituting Eqs. (3-7), (3-10), and (3-12) into the resulting expression, we obtain

$$\begin{Bmatrix} \sigma_z \\ \sigma_x \\ \tau_{xz} \end{Bmatrix} = [Q]^{(i)} \left( \begin{Bmatrix} \varepsilon_z^0 - \alpha_z^{(i)} T_i^0 \\ \varepsilon_x^0 - \alpha_x^{(i)} T_i^0 \\ \gamma_{xz}^0 - \alpha_{xz}^{(i)} T_i^0 \end{Bmatrix} - y \begin{Bmatrix} \kappa_z + \alpha_z^{(i)} T_i^1 \\ \kappa_x + \alpha_x^{(i)} T_i^1 \\ \kappa_{xz} + \alpha_{xz}^{(i)} T_i^1 \end{Bmatrix} \right) \tag{3-14}$$

where $[Q]^{(i)}$ stands for the inverse matrix of the $3 \times 3$ matrix in Eq. (3-11). Integrating the tangential stress components across the thickness of the successive layers and summing the results over all layers, one obtains the stress and moment resultants. These stress and moment resultants vanish in the infinite laminate, i.e.,

$$N_z = \int \sigma_z \, dy = 0, \qquad N_x = \int \sigma_x \, dy = 0, \qquad N_{xz} = \int \tau_{xz} \, dy = 0$$

$$M_z = -\int y\sigma_z \, dy = 0, \qquad M_x = -\int y\sigma_x \, dy = 0, \qquad M_{xz} = -\int y\tau_{xz} \, dy = 0$$

The preceding six conditions provide the following system of linear equations for the middle-plane strains and the curvatures:

$$
\begin{bmatrix} \mathbf{A} & \mathbf{B} \\ \mathbf{B} & \mathbf{D} \end{bmatrix}
\left\{ \begin{array}{c} \varepsilon_z^0 \\ \varepsilon_x^0 \\ \gamma_{xz}^0 \\ -\kappa_z \\ -\kappa_x \\ -\kappa_{xz} \end{array} \right\}
= \sum_i \left( \begin{bmatrix} h_i[Q]^{(i)} & g_i[Q]^{(i)} \\ g_i[Q]^{(i)} & f_i[Q]^{(i)} \end{bmatrix}
\left\{ \begin{array}{c} \alpha_z^{(i)} T_i^0 \\ \alpha_x^{(i)} T_i^0 \\ \alpha_{xz}^{(i)} T_i^0 \\ \alpha_z^{(i)} T_i^1 \\ \alpha_x^{(i)} T_i^1 \\ \alpha_{xz}^{(i)} T_i^1 \end{array} \right\} \right)
\tag{3-15}
$$

where $\mathbf{A}$, $\mathbf{B}$ and $\mathbf{D}$ denote, respectively, the stiffness matrices of the laminate associated with extension, extension-bending coupling, and bending (see, for example, ref. 18, p. 154). They are given by

$$
\mathbf{A} = \sum h_i[Q]^{(i)}, \qquad \mathbf{B} = \sum g_i[Q]^{(i)}, \qquad \mathbf{D} = \sum f_i[Q]^{(i)}
\tag{3-16}
$$

When the middle-plane strains and curvatures as determined by Eq. (3-15) are substituted into Eq. (3-14), we obtain the stresses in the successive layers of the infinite laminate. These stresses vary only in the thickness direction, i.e., they are independent of the coordinates $x$ and $z$. Along the sections $x = \pm a$ of the infinite laminate, the tangential stresses $\sigma_x$ and $\tau_{xz}$ are the boundary tractions of the laminated strip $-a \leqslant x \leqslant a$.

We now consider a purely mechanical problem (i.e., without an accompanying temperature load) for a $z$-independent stress field $\sigma_{ij}(x, y)$ satisfying the following conditions: (i) $\sigma_x$ and $\tau_{xz}$ along the edges $x = \pm a$ are equal and opposite to those given by the second and third rows of Eq. (3-14); $\tau_{xy} = 0$ along the free edges; (ii) $\sigma_y = \tau_{xy} = \tau_{yz} = 0$ on $y = \pm t/2$; (iii) the cross-sections of the strip are subjected to vanishing total axial force, total bending moment and twisting moment:

$$
\int_{-a}^{a} N_z \, dx = 0, \qquad \int_{-a}^{a} M_z \, dx = \int_{-a}^{a} M_{xz} \, dx = 0
\tag{3-17}
$$

The stress solution of this purely mechanical problem depends only on the coordinates $x$ and $y$. When this solution is superimposed upon the trivial thermal stress solution described by Eqs. (3-9) and (3-14), the result is a stress field in an infinitely long laminated strip with free edges along $x = \pm a$, subjected to the temperature load of Eq. (3-7), and characterized by the conditions of Eq. (3-17). This stress field, which also depends only on the coordinates $x$ and $y$, closely approximates the solution of the original thermoelastic problem for the *finite* rectangular laminate (bounded by the edges $x = \pm a$ and $z = \pm b$) under the same temperature load, except in the boundary regions close to the end sections $z = \pm b$. Consequently, the original thermoelastic problem for the finite rectangular laminate has been reduced to a purely *mechanical* problem for an infinite strip of width $2a$. The cross-sections of this infinite strip are subjected to a vanishing total axial force and vanishing total bending and twisting moments (see Eq. (3-17)), and the edges of the strip are subjected to tractions equal and *opposite* to those given by the second and third rows of Eq. (3-14). That is, on $x = \pm a$,

$$\tau_{xy} = 0 \tag{3-18a}$$

$$\begin{Bmatrix} \sigma_x \\ \tau_{xz} \end{Bmatrix} = \begin{bmatrix} Q_{12}^{(i)} & Q_{22}^{(i)} & Q_{26}^{(i)} \\ Q_{16}^{(i)} & Q_{26}^{(i)} & Q_{66}^{(i)} \end{bmatrix} \left( \begin{Bmatrix} -\varepsilon_z^0 + \alpha_z^{(i)} T_i^0 \\ -\varepsilon_x^0 + \alpha_x^{(i)} T_i^0 \\ -\gamma_{xz}^0 + \alpha_{xz}^{(i)} T_i^0 \end{Bmatrix} + y \begin{Bmatrix} \kappa_z + \alpha_z^{(i)} T_i^1 \\ \kappa_x + \alpha_x^{(i)} T_i^1 \\ \kappa_{xz} + \alpha_{xz}^{(i)} T_i^1 \end{Bmatrix} \right) \tag{3-18b}$$

The reduction to a purely mechanical problem is achieved by eliminating (from the solution of the original thermal expansion problem for the finite, rectangular laminate) the thermal expansion solution for an *infinite* laminate subjected to the same temperature load. In the following, the reduced mechanical problem (henceforth referred to as the *complementary problem*) will be solved approximately by a variational approach using stress functions.

## 3.4 STRESS FUNCTIONS: INTERFACE AND BOUNDARY CONDITIONS

A stress field in the $i$th layer depending only on the coordinate variables $x$ and $y$ and satisfying the differential equations of equilibrium may be expressed in terms of a pair of stress functions $F^{(i)}(x, y)$ and $\Psi^{(i)}(x, y)$ in the following manner:

$$\sigma_x^{(i)} = F^{(i)},_{yy}, \qquad \sigma_y^{(i)} = F^{(i)},_{xx}, \qquad \tau_{xy}^{(i)} = -F^{(i)},_{xy}$$

$$\tau_{xz}^{(i)} = \Psi^{(i)},_y, \qquad \tau_{yz}^{(i)} = -\Psi^{(i)},_x \tag{3-19}$$

where the subscripts following the commas indicate partial differentiation. Obviously, an arbitrary bilinear function of $x$ and $y$ and an arbitrary constant may be added to the first and the second stress functions, respectively, without affecting the resulting stress field in the layer.

The continuity of the normal and shearing stresses across the $i$th interface requires that, along that interface,

$$(F^{(i+1)} - F^{(i)})_{,xx} = 0, \qquad (F^{(i+1)}_{,y} - F^{(i)}_{,y})_{,x} = 0, \qquad (\Psi^{(i+1)} - \Psi^{(i)})_{,x} = 0$$

Hence we may appropriately choose an additive bilinear function and an additive constant for the stress functions in the $(i + 1)$th layer so that, along the interface $y = y_i$,

$$F^{(i+1)} = F^{(i)}, \qquad F^{(i+1)}_{,y} = F^{(i)}_{,y}, \qquad \Psi^{(i+1)} = \Psi^{(i)} \qquad (3\text{-}20)$$

Furthermore, without loss of generality the stres functions in the bottom layer may be chosen in such a way that, over the lower surface of the laminate, $y = -t/2$, one has

$$F^{(1)} = 0, \qquad F^{(1)}_{,y} = 0, \qquad \Psi^{(1)} = 0 \qquad (3\text{-}21a)$$

By integrating $\sigma_x$ and $\tau_{xz}$ through the thickness of all layers, and making use of the continuity of $F_{,y}$ and $\Psi$ across the interfaces, one finds that the stress resultants $N_x$ and $N_{xz}$ are equal to the values of $F_{,y}$ and $\Psi$, respectively, on the upper surface of the laminate. Let the values be denoted by $G^*$ and $\Psi^*$, i.e.,

$$N_x = \sum \int \sigma_x^{(i)}\, dy = \sum \int F^{(i)}_{,yy}\, dy = \sum F_{,y}\Big|_{y=y_{i-1}}^{y=y_i} = F_{,y}\Big|_{y=-t/2}^{y=t/2} = G^*$$

$$N_{xz} = \sum \int \tau_{xz}^{(i)}\, dy = \sum \int \Psi^{(i)}_{,y}\, dy = \Psi^*$$

Now the stress moment $M_x$ is given by

$$M_x = -\sum \int y\sigma_x^{(i)}\, dy = -\sum \int yF_{,yy}\, dy = -\sum \int \{(yF_{,y})_{,y} - F_{,y}\}\, dy$$

$$= F^* - (t/2)G^*$$

where $F^*$ is the value of the stress function on the top surface. Since $N_x$, $N_{xz}$, and $M_x$ all vanish in the present problem, we have

$$F^* = G^* = \Psi^* = 0 \qquad (3\text{-}21b)$$

Equations (3-20) and (3-21a, b) imply that, for the complementary problem, the stress functions in the layers may be chosen in such a way that $F$, $\Psi$, and the normal derivative of $F$ vanish on the upper and lower laminate surfaces ($y = \pm t/2$) and are continuous across all layer interfaces.

On the two boundary edges of the laminated strip, the boundary values of the stress functions for the complementary problem may be obtained by integrating the boundary tractions $\tau_{xy}$, $\sigma_x$, and $\tau_{xz}$ with respect to $y$, where $\tau_{xy} \equiv 0$ and the other two traction components are given by Eq. (3-18$b$). Integration of $\tau_{xy} \equiv -F,_{xy} \equiv 0$ shows that the normal derivative of $F$ also vanishes on the two edges. Hence

$$F,_{xy} = 0, \qquad F,_x = 0 \qquad \text{(on } x = \pm a) \tag{3-22}$$

## 3.5  GENERALIZED PLANE DEFORMATION OF A LAMINATED STRIP

For a homogeneous anisotropic elastic medium of cylindrical shape with the generators parallel to the $z$-direction, Lekhnitskii investigated the class of infinitesimal deformations for which the stress tensor is independent of the $z$-coordinate. He found that the class consists of deformations whose displacement components have the form (ref. 17, pp. 107–108)

$$w(z, x, y) = (Ax - By + C)z + W(x, y) + \omega_1 y - \omega_2 x + w_0$$

$$u(z, x, y) = -Az^2/2 - \Theta yz + U(x, y) + \omega_2 z - \omega_3 y + u_0 \tag{3-23}$$

$$v(z, x, y) = Bz^2/2 + \Theta xz + V(x, y) + \omega_3 x - \omega_1 z + v_0$$

For the complementary problem considered here, each layer of the laminated strip undergoes a generalized plane deformation, as described by Eq. (3-23). In order to satisfy the interfacial continuity of the displacements, either exactly or in an averaged sense, it is necessary that the set of constants $A$, $B$, $C$, and $\Theta$ be the same for all layers.

Along the edges of the laminated strip, the in-plane normal and shearing stress resultants $N_x$ and $N_{xz}$ vanish, and so does the moment resultant $M_x$. The nonvanishing stress and moment resultants are $N_z$, $M_z$, and $M_{zx}$, and their corresponding kinematical variables are the axial extensional strain $\varepsilon_z^0 = C$, the bending curvature $\kappa_z = B$, and the twisting curvature $\kappa_{zx} = \Theta$. Bending with respect to the $y$-axis introduces large bending strains near the free edges, and this usually does not happen unless the thickness and width of the laminated strip ($t$ and $2a$) are comparable in magnitude. Consequently, we shall set $A = 0$ in Eq. (3-23). Hence the external loading of the complementary problem for the laminated strip may be characterized by three parameters, $B$, $C$, and $\Theta$ of Eq. (3-23) and the traction boundary conditions along the free edges $x = \pm a$, Eqs. (3-18$a$, $b$). The three parameters $B$, $C$, and $\Theta$ are determined by the three conditions of Eq. (3-17).

## 3.6 THE PRINCIPLE OF COMPLEMENTARY VIRTUAL WORK

An *admissible variation* in the stress field satisfies (i) the equilibrium equations

$$\delta\sigma_{kl,l}^{(i)} = 0$$

in the $i$th layer, (ii) the homogeneous boundary conditions $\delta\sigma_x = \delta\tau_{xy} = \delta\tau_{xz} = 0$ on $x = \pm a$ and $\delta\sigma_y = \delta\tau_{xy} = \delta\tau_{yz} = 0$ on $y = \pm t/2$, and (iii) the continuity of $\delta F^{(i)}$, $\delta F^{(i)}_{,y}$ and $\delta\Psi^{(i)}$ across all interfaces. Since, for the complementary problem, the stresses in all layers are independent of the axial coordinate $z$, it is sufficient to consider only those variations of the stresses that are also independent of $z$. Furthermore, one need only consider a segment of the laminated strip of unit length along the axial direction, $0 \leqslant z \leqslant 1$. For this segment of the laminated strip, one has

$$0 = \iiint u_i \,\delta\sigma_{ij,j}\, dx\, dy\, dz = \iiint \{(u_i\,\delta\sigma_{ij})_{,j} - u_{i,j}\,\delta\sigma_{ij}\}\, dx\, dy\, dz$$

$$= \iint u_i\,\delta\sigma_{ij}n_j\, dA - \frac{1}{2}\iiint (u_{i,j} + u_{j,i})\,\delta\sigma_{ij}\, dx\, dy\, dz \qquad (3\text{-}24)$$

where the surface integral extends over the entire boundary surface of the segment. By summing Eq. (3-24) over all layers, and making use of the traction boundary conditions and the interfacial continuity of the inter-laminar stresses, one obtains

$$\sum \iint (\Delta u\,\delta\tau_{xz} + \Delta v\,\delta\tau_{yz} + \Delta w\,\delta\sigma_z)\, dx\, dy$$

$$- \sum \int ([u]_i\,\delta\tau_{xy} + [v]_i\,\delta\sigma_y + [w]_i\,\delta\tau_{yz})\, dx = \sum \iint \varepsilon_{ij}\,\delta\sigma_{ij}\, dx\, dy \qquad (3\text{-}25)$$

where the double integrals are summed over all layers and the single integrals are summed over all interfaces $y = y_i$. The bracket symbol, $[\ ]_i$, denotes the jump of the quantity inside the bracket across the $i$th interface. $\Delta u$, $\Delta v$ and $\Delta w$ stand for the differences of the displacements at $z = 1$ and $z = 0$ and they may be obtained from Eq. (3-23) with $A = 0$. The resulting expressions are

$$\Delta u = -\Theta y + \omega_2, \qquad \Delta v = \frac{B}{2} + \Theta x - \omega_1, \qquad \Delta w = C - By$$

Equations (3-18) and (3-5) yield, in each layer,

$$\delta\tau_{xz} = \delta\Psi,_y, \qquad \delta\tau_{yz} = -\delta\Psi,_x$$

$$\delta\sigma_z = -(a_{12}\,\delta F,_{yy} + a_{13}\,F,_{xx} + a_{16}\,\delta\Psi,_y)/a_{11}$$

(3-26)

Hence the double integrals on the left-hand side of Eq. (3-25) yield, after integration by parts,

$$-\sum (Ba_{16}^{(i)}/a_{11}^{(i)} - 2\Theta) \iint \delta\Psi\, dx\, dy + B\sum [a_{12}/a_{11}]_i\,\delta F_i\, dx$$

$$+ \sum (C - By_i) \int ([a_{12}/a_{11}]_i\,\delta G_i + [a_{16}/a_{11}]_i\,\delta\Psi_i\, dx - \sum \int [\omega_2]_i\,\delta\Psi_i\, dx$$

while the single integrals in Eq. (3-25) yield

$$-\sum \int ([u,_x]_i\,\delta G_i + [v,_{xx}]_i\,\delta F_i + [u,_z + w,_x]_i\,\delta\Psi_i)\, dx + \sum \int [\omega_2]_i\,\delta\Psi_i\, dx$$

Here $F_i$, $G_i$, $\Psi_i$ stand for the values of $F$, $F,_y$, and $\Psi$, respectively, on the $i$th interface. Substituting the preceding results into Eq. (3-25) and using the interfacial continuity of the tangential strains $\varepsilon_x$, $\gamma_{xz}$ and the curvature $v,_{xx}$, one obtains

$$\sum \iint \varepsilon_{ij}\,\delta\sigma_{ij}\, dx\, dy + \sum (Ba_{16}^{(i)}/a_{11}^{(i)} - 2\Theta) \iint \delta\Psi\, dx\, dy - \sum (C - By_i)$$

$$\times \int ([a_{12}/a_{11}]_i\,\delta G_i + [a_{16}/a_{11}]_i\,\delta\Psi_i)\, dx - B\sum [a_{12}/a_{11}]_i \int \delta F_i\, dx = 0$$

(3-27a)

The following expression for the first term of the preceding equation may be obtained from Eqs. (3-6) and (3-26):

$$\sum \iint \varepsilon_{ij}\,\delta\sigma_{ij}\, dx\, dy = \sum \iint \{F,_{yy}, F,_{xx}, -F,_{xy}, -\Psi,_x, \Psi,_y\}$$

$$\begin{bmatrix} \beta_{22} & \beta_{23} & 0 & 0 & \beta_{26} \\ \beta_{23} & \beta_{33} & 0 & 0 & \beta_{36} \\ 0 & 0 & \beta_{44} & \beta_{45} & 0 \\ 0 & 0 & \beta_{45} & \beta_{55} & 0 \\ \beta_{26} & \beta_{36} & 0 & 0 & \beta_{66} \end{bmatrix} \begin{Bmatrix} \delta F,_{yy} \\ \delta F,_{xx} \\ -\delta F,_{xy} \\ -\delta\Psi,_x \\ \delta\Psi,_y \end{Bmatrix} dx\, dy \qquad (3\text{-}27b)$$

## 3.7 POLYNOMIAL APPROXIMATIONS OF THE STRESS FUNCTIONS

In a variational method of solution of the complementary problem, the stress functions $F$ and $\Psi$ in all layers are to be determined within a class of admissible stress functions satisfying the traction boundary conditions (Eqs. (3-18) and (3-21a, b)) and the interfacial continuity of $F$, $F_{,y}$, and $\Psi$. The criterion of selection is that Eq. (3-27) should be satisfied for arbitrary admissible variations $\delta F$ and $\delta \Psi$. The criterion ensures, in the weak sense, the compatibility of strain and interfacial continuity of the displacements. If the layers constituting the laminate are relatively thin, one may approximate the stresses in each layer by polynomial functions of the thickness coordinate. Then the class of admissible stress functions consists of polynomial functions of $y$, with the coefficients depending on $x$.

In the $i$th layer, we define the nondimensional thickness coordinate $\eta$ by

$$\eta = \frac{y - y_{i-1}}{y_i - y_{i-1}}$$

In the classical plate theory the in-plane stresses vary linearly in the thickness direction. This feature is approximately valid in each thin layer of a laminated plate, except in regions close to the free edges. Hence the stress function $F(x, y)$ in each layer may be approximated by a polynomial function of degree 3 in the normalized thickness coordinate $\eta$. Now the stress function in the $i$th layer must assume the values $F_i(x)$ and $F_{i-1}(x)$, respectively, on the interfaces $y = y_i$ and $y = y_{i-1}$, and the $y$-derivative of the stress function must assume the values $G_i(x)$ and $G_{i-1}(x)$ on these interfaces. Hence the cubic polynomial approximation of the stress function in the $i$th layer must have the following expression:

$$F^{(i)}(x, \eta) = (1 - 3\eta^2 + 2\eta^3)F_{i-1}(x) + (\eta - 2\eta^2 + \eta^3)h_i G_{i-1}(x)$$
$$+ (3\eta^2 - 2\eta^3)F_i(x) + (\eta^3 - \eta^2)h_i G_i(x) \qquad (i = 2, 3, \ldots, n)$$

$$(3\text{-}28a)$$

The cubic polynomial approximations of the stress functions in the bottom and top layers are given, respectively, by

$$F^{(1)}(x, \eta) = (3\eta^2 - 2\eta^3)F_1(x) + (\eta^3 - \eta^2)h_1 G_1(x) \qquad (3\text{-}28b)$$

and

$$F^{(n+1)}(x, \eta) = (1 - 3\eta^2 + 2\eta^3)F_n(x) + (\eta - 2\eta^2 + \eta^3)h_{n+1}G_n(x) \qquad (3\text{-}28c)$$

The stress functions given by Eqs. (3-28a, b, c) yield an interlaminar peeling stress $\sigma_y$ having a cubic dependence on the thickness coordinate in each layer, an interlaminar shearing stress $\tau_{xy}$ depending quadratically on $\eta$, and an in-plane stress $\sigma_x$ depending linearly on $\eta$. For the sake of consistency the interlaminar shearing stress $\tau_{yz}$ and the in-plane shearing stress $\tau_{xz}$ will be approximated, respectively, by quadratic and linear functions of $\eta$. Then, according to Eq. (3-19), the stress function $\Psi$ depends quadratically on the thickness coordinate. It follows that

$$\Psi^{(i)}(x, \eta) = (1 - \eta^2)\Psi_{i-1}(x) + \eta^2\Psi_i(x) + (\eta - \eta^2)h_i H_{i-1}(x)$$

$$(i = 2, \ldots, n) \quad (3\text{-}29a)$$

$$\Psi^{(1)}(x, \eta) = \eta^2\Psi_1(x) + (\eta - \eta^2)h_1 H_0(x) \qquad (3\text{-}29b)$$

$$\Psi^{(n+1)}(x, \eta) = (1 - \eta^2)\Psi_n(x) + (\eta - \eta^2)h_{n+1}H_n(x) \qquad (3\text{-}29c)$$

where $\Psi_i(x)$ is the value of the stress function $\Psi$ on the $i$th interface and $H_i(x)$ is the value of $\Psi,_y$ on the *upper* side of the same interface (notice that, although $\Psi$ is continuous across each interace, $\Psi,_y = \tau_{xz}$ is generally discontinuous). Equations (3-28) and (3-29) assume in effect that the stress functions in all layers are completely determined by $3n$ functions $F_i(x)$, $G_i(x)$, $\Psi_i(x)$ $(i = 1, 2, \ldots, n)$ and $n + 1$ functions $H_i(x)$ $(i = 0, 1, \ldots, n)$, where $H_0(x)$ is the value of $\Psi,_y$ on the bottom surface $y = -t/2$.

The traction boundary conditions of the complementary problem, i.e., Eqs. (3-18) and (3-21a, b), may be integrated with respect to the thickness coordinate. This yields the following recurrence relations for the boundary values of the functions $F_i$, $G_i$, $\Psi_i$, and $H_i$ at $x = \pm a$:

$$\begin{Bmatrix} G_i \\ \Psi_i \end{Bmatrix} = \begin{Bmatrix} G_{i-1} \\ \Psi_{i-1} \end{Bmatrix} + \begin{bmatrix} Q_{12}^{(i)} & Q_{22}^{(i)} & Q_{26}^{(i)} \\ Q_{16}^{(i)} & Q_{26}^{(i)} & Q_{66}^{(i)} \end{bmatrix}$$

$$\cdot \left( -h_i \begin{Bmatrix} \varepsilon_z^0 - \alpha_z^{(i)}T_i^0 \\ \varepsilon_x^0 - \alpha_x^{(i)}T_i^0 \\ \gamma_{xz}^0 - \alpha_{xz}^{(i)}T_i^0 \end{Bmatrix} + g_i \begin{Bmatrix} \kappa_z + \alpha_z^{(i)}T_i^1 \\ \kappa_x + \alpha_x^{(i)}T_i^1 \\ \kappa_{xz} + \alpha_{xz}^{(i)}T_i^1 \end{Bmatrix} \right) \quad (i = 1, 2, \ldots, n)$$

$$F_i = F_{i-1} + h_i G_{i-1} - \frac{h_i^2}{2}\{Q_{12}^{(i)}(\varepsilon_z^0 - \alpha_z^{(i)}T_i^0) + Q_{22}^{(i)}(\varepsilon_x^0 - \alpha_x^{(i)}T_i^0)$$

$$+ Q_{26}^{(i)}(\gamma_{xz}^0 - \alpha_{xz}^{(i)}T_i^0)\} + \left\{ f_i - \frac{(g_i - h_i^2/2)^2}{2h_i} \right\}\{Q_{12}^{(i)}(\kappa_z + \alpha_z^{(i)}T_i^1)$$

$$+ Q_{22}^{(i)}(\kappa_x + \alpha_x^{(i)}T_i^1) + Q_{66}^{(i)}(\kappa_{xz} + \alpha_{xz}^{(i)}T_i^1)\} \qquad (i = 1, 2, \ldots, n)$$

$$H_{i-1} = -Q_{16}^{(i)}(\varepsilon_z^0 - \alpha_z^{(i)}T_i^0) - Q_{26}^{(i)}(\varepsilon_x^0 - \alpha_x^{(i)}T_i^0) - Q_{66}^{(i)}(\gamma_{xz}^0 - \alpha_{xz}^{(i)}T_i^0)$$

$$+ y_{i-1}\{Q_{16}^{(i)}(\kappa_z + \alpha_z^{(i)}T_i^1) + Q_{26}^{(i)}(\kappa_x + \alpha_x^{(i)}T_i^1) + Q_{66}^{(i)}(\kappa_{xz} + \alpha_{xz}^{(i)}T_i^1)\}$$

$$(i = 1, 2, \ldots, n + 1) \quad (3\text{-}30)$$

Furthermore, Eq. (3-22) yields additional conditions for the first derivatives of $F_i$ and $G_i$:

$$F_i'(\pm a) = G_i'(\pm a) = 0 \qquad (3\text{-}31)$$

where the primes indicate differentiation with respect to $x$.

## 3.8 DIFFERENTIAL EQUATIONS AND BOUNDARY CONDITIONS FOR THE COEFFICIENT FUNCTIONS

Let the set of $4n + 1$ functions $F_i$, $G_i$, $\Psi_i$, and $H_i$ be arranged as the components of a column vector $\{X\}$ according to

$$X_i = F_i(x), \qquad X_{i+n} = G_i(x), \qquad X_{i+2n} = \Psi_i(x) \qquad (i = 1, 2, \ldots, n)$$

$$X_{i+3n+1} = H_i(x) \qquad (i = 0, 1, \ldots, n)$$

Then, by substituting Eqs. (3-28) and (3-29) into Eq. (3-27), performing the integration with respect to $\eta$, and integrating by parts with respect to $x$, we obtain

$$\{\delta X\}^t\left[\left(\mathbf{W}\frac{d^4}{dx^4} + \mathbf{V}\frac{d^2}{dx^2} + \mathbf{U}\right)\{X\} - \{b\}\right] = 0 \qquad (3\text{-}32)$$

where $\{\delta X\}^t$, a row vector, is the variation of the transpose of $\{X\}$ and the column vector $\{b\}$ has the components

$$b_i = B\left[\frac{a_{12}}{a_{11}}\right]_i, \qquad b_{i+n} = (C - By_i)\left[\frac{a_{12}}{a_{11}}\right]_i$$

$$b_{i+2n} = (C - By_i)\left[\frac{a_{16}}{a_{11}}\right]_i + 2h_{i+1}\frac{2\Theta - Ba_{16}^{(i+1)}/a_{11}^{(i+1)}}{3}$$

$$+ \frac{h_i(2\Theta - Ba_{16}^{(i)}/a_{11}^{(i)})}{3} \qquad (i = 1, 2, \ldots, n) \qquad (3\text{-}33)$$

$$b_{i+3n-1} = \frac{(2\Theta - Ba_{16}^{(i+1)}/a_{11}^{(i+1)})h_{i+1}^2}{6} \qquad (i = 0, 1, \ldots, n)$$

As mentioned previously, $[a_{12}/a_{11}]_i$ denotes the jump of $a_{12}/a_{11}$ across the $i$th interface.

The $(4n + 1) \times (4n + 1)$ square matrices $\mathbf{U}$, $\mathbf{V}$, and $\mathbf{W}$ of Eq. (3-32) are constant real symmetric matrices. They are completely determined by the integral expressions in Eq. (3-27$b$), i.e., by the first variation of the strain energy of the layers. It is clear from Eq. (3-27$b$) that the fourth-order derivatives in Eq. (3-32) arise only from the integrals $\sum \iint \beta_{33}^{(i)} F^{(i)},_{xx}$ $\delta F^{(i)},_{xx} \, dx \, dy$, i.e., they involve only the stress functions $F^{(i)}$ but not $\Psi^{(i)}$. Hence, by virtue of the representation of Eq. (3-28), the coefficient matrix $\mathbf{W}$ of the operator $d^4/dx^4$ has nonvanishing elements only in a $2n \times 2n$ square submatrix in the upper left corner (i.e., in the rows and columns corresponding to the unknown functions $F_i$ and $G_i$, $i = 1, 2, \ldots, n$).

The elements of the three matrices have lengthy and complex expressions and it is impractical to obtain these expressions without using symbolic algebraic programs. In the present analysis, the symbolic algebraic program MACSYMA[19] is used to generate the matrices $\mathbf{U}$, $\mathbf{V}$, and $\mathbf{W}$ from Eqs. (3-27$b$), (3-28), and (3-29), and to obtain the compliance coefficients $\beta_{ij}^{(i)}$ of each layer from Eqs. (3-1), (3-4), and (3-6). (See Section 3-16 and the Appendix for the $\mathbf{U}$, $\mathbf{V}$, and $\mathbf{W}$ matrices of a three-layer laminate.)

Equation (3-32) is satisfied for arbitrary admissible variations $\delta H_0$, $\delta F_1$, $\delta G_1$, $\delta \Psi_1, \ldots$, and $\delta H_n$ if and only if

$$\left( \mathbf{W} \frac{d^4}{dx^4} + \mathbf{V} \frac{d^2}{dx^2} + \mathbf{U} \right)\{X\} = \{b\} \tag{3-34}$$

This is the Euler–Lagrange equation associated with the complementary virtual work principle. Notice that according to Eq. (3-33), the right-hand side of Eq. (3-34) contains the parameters $B$, $C$, and $\Theta$. For any set of parameter values, Eq. (3-34) and the boundary conditions of Eqs. (3-30) and (3-31) completely determine the $4n + 1$ functions in $\{X\}$, which in turn determine the stress functions and the stess field through Eqs. (3-28), (3-29), (3-19), and (3-5), with $\varepsilon_z^* = C$ in Eq. (3-5). One must, however, make an appropriate choice of the parameters so as to satisfy Eq. (3-17). Only the stress field associated with this unique set of parameters is the solution of the complementary problem.

## 3.9 DETERMINATION OF THE DEFORMATON PARAMETERS $B$, $C$, AND $\Theta$

The search for the unique set of parameters satisfying Eq. (3-17) is not difficult because the problem is linear. We decompose the complementary problem into four subproblems. For the first three subproblems the boundary conditions of Eqs. (3-30) and (3-31) are replaced by homogeneous boundary conditions, i.e.,

$$F_i(\pm a) = F'_i(\pm a) = G_i(\pm a) = G'_i(\pm a) = \Psi_i(\pm a) = H_i(\pm a) = 0$$

$$H_0(\pm a) = 0 \qquad\qquad\qquad\qquad (i = 1, 2, \dots, n) \quad (3\text{-}35)$$

The right-hand side of Eq. (3-34) for the first, second, and third subproblems corresponds, respectively, to the cases of unit axial strain ($C = 1$), unit bending ($B = 1$), and unit twisting deformation ($\Theta = 1$). Solutions for these three subproblems may be obtained according to the procedure described elsewhere.[16] The fourth subproblem is characterized by the boundary conditions of the complementary problem, Eqs. (3-30) and (3-31), and the *homogeneous* differential equation associated with Eq. (3-34), i.e., the equation obtained by setting the right-hand side of Eq. (3-34) to be the zero vector. The solution of each subproblem yields a total axial force, a total bending moment, and a total twisting moment. The solutions of the first three subproblems are multiplied by undetermined coefficients and combined with the solution of the fourth subproblem. Applying the three conditions of Eq. (3-17) to the combined solution, one obtains a system of three linear algebraic equations for the three undetermined coefficients in the combination.

In practice, the total axial force, bending moment, and twisting moment associated with the fourth subproblem are all extremely small if the width $2a$ of the laminated strip is significantly greater than the thickness $t$. This is because the edge tractions associated with the inhomogeneous boundary conditions of Eq. (3-30) are a self-equilibrating system (since the edge tractions are equal and opposite to the in-plane stresses in an infinite laminate that is subjected to vanishing stress and moment resultants). The self-equilibrating edge tractions produce significant internal stresses only in a narrow region in the vicinity of the edge. The stresses decay rapidly away from the edge so that, for the solution of the fourth subproblem, the interior region of the strip is nearly stress-free and, consequently, the total axial force, bending moment, and twisting moment over the cross-sections of the strip are all small. Now the solutions of the first three subproblems are needed only to offset this system of axial force and bending and twisting moments so as to ensure the satisfaction of Eq. (3-17). The contributions of these three component solutions to the solution of the complementary problem are small since the resultant axial force, bending, and twisting moments are small. Hence the interlaminar stresses near the free edge are usually predicted with reasonable accuracy by the solution of the fourth subproblem alone. The method of solving this subproblem is described in the next section.


## 3.10 SOLUTION OF THE EIGENVALUE PROBLEM

The general solutions of the homogeneous differential equation associated with Eq. (3-34) are combinations of the eigenfunctions of the form $\{c\}_k \exp(\pm\lambda_k x)$, where $\{c\}_k$ is a constant vector with real or complex components depending on whether $\lambda_k$ is a real or complex eigenvalue. The vector $\{c\}_k$ may be normalized in an appropriate way to rid it of an undetermined multiplicative factor. The eigenvalues $\lambda_k$ are the roots of the characteristic equation

$$\text{Determinant}(\mathbf{W}\lambda^4 + \mathbf{V}\lambda^2 + \mathbf{U}) = 0 \qquad (3\text{-}36)$$

Since the nonvanishing elements of $\mathbf{W}$ are limited to a $2n \times 2n$ submatrix, the characteristic equation is a polynomial equation of order $2 \times 2n + 2n + 1 = 6n + 1$ in the variable $\lambda^2$. Hence there are $6n + 1$ pairs of (real and complex) eigenvalues, with each pair consisting of eigenvalues that differ only in algebraic sign. Considered as an equation for $\lambda^2$, the characteristic equation yields no solutions $\lambda^2$ with negative real parts. Hence there are no purely imaginary eigenvalues. Notice that the total number of eigenvalues, $12n + 2$, is equal to the number of boundary conditions in Eqs. (3-30) and (3-31).

A particular combination of eigenfunctions is determined by choosing the coefficients in such a way that all boundary conditions of Eq. (3-30) and (3-31) are satisfied. This provides the solution of the fourth subproblem and, as mentioned in the preceding section, it is often a close approximate solution to the complementary problem for the laminated strip. If the width of the strip, $2a$, is sufficiently large compared to the thickness, then the eigenfunctions associated with the eigenvalues having negative (positive) real parts make negligible contribution to the solution in the right (left) half of the strip, since such eigenfunctions decay rapidly away from the left (right) free edge. In other words, the solutions for the left and right parts of the strip are approximately uncoupled.

Equation (3-36) is not in the usual form of a characteristic equation associated with a single real square matrix. However, it can be reduced to an equation of the latter type through algebraic manipulations, so that commercial and public-domain computer programs can be readily used to obtain all real and complex roots. Let $\mathbf{U} = \mathbf{P\Lambda Q}$ be the singular-value decomposition of the real symmetric matrix $\mathbf{U}$, where $\mathbf{\Lambda}$ is a diagonal matrix and $\mathbf{P}$ and $\mathbf{Q}$ are, respectively, column-orthogonal and row-orthogonal matrices (ref. 20, p. 665). Let $\mathbf{W}^*$ be the $2n \times (4n + 1)$ submatrix composed of the first $2n$ rows of $\mathbf{W}$ (notice that the elements left out are all zero elements). Let $(\mathbf{P}^{-1})^*$ be the $(4n + 1) \times 2n$ rectangular matrix obtained by eliminating the last $2n + 1$ columns of the matrix $\mathbf{P}^{-1}$, and let $\mathbf{I}_{2n}$ denote the $2n \times 2n$ identity matrix. We define $\mathbf{M}$ as the matrix product of two $(6n + 1) \times (6n + 1)$ real matrices in the following manner:

$$\mathbf{M} = \begin{bmatrix} \mathbf{\Lambda}^{-1} & \mathbf{0} \\ \mathbf{0} & \mathbf{I}_{2n} \end{bmatrix} \begin{bmatrix} \mathbf{P}^{-1}\mathbf{V}\mathbf{Q}^{-1} & (\mathbf{P}^{-1})^* \\ -\mathbf{W}^*\mathbf{Q}^{-1} & \mathbf{0} \end{bmatrix} \tag{3-37}$$

It can be shown that the square roots of the eigenvalues of $\mathbf{M}$ are the *reciprocals* of the roots of the characteristic equation (3-36). Once a real or complex eigenvalue $\lambda_k$ has been computed, the corresponding eigenvector $\{c\}_k$ may be obtained by solving the equation

$$(\mathbf{W}\lambda_k^4 + \mathbf{V}\lambda_k^2 + \mathbf{U})\{c\}_k = \{0\} \tag{3-38}$$

As mentioned previously, the boundary conditions of Eqs. (3-30) and (3-31) determine the coefficients of that linear combination of eigenfunctions that represents the solution of the fourth subproblem.

When the functions $F_i$, $G_i$, $\Psi_i$, and $H_i$ are determined by taking the proper combination of eigenfunctions, the stress functions in the various layers may be obtained from Eqs. (3-28) and (3-29). Equation (3-19) then yields the interlaminar and in-plane stresses. Since differentiation of the stress functions is performed *analytically* on exponential and polynomial functions, there is no degradation of accuracy as usually happens in postprocessing finite element and other numerical solutions.

## 3.11 ISOTROPIC AND SPECIALLY ORTHOTROPIC LAMINATES

In the case of a specially orthotropic laminate, where the material axes of all layers are parallel or perpendicular to the free edges, the solution space of the eigenvalue problem decomposes orthogonally into subspaces of dimensions $8n$ and $4n + 2$. The two subspaces are associated, respectively, with the two stress functions $F$ and $\Psi$. If the temperature load depends only on the thickness coordinate $y$, then the second stress function vanishes identically and $\tau_{xz}$ and $\tau_{yz}$ vanish in the entire laminate. The solutions of such problems require the stress function $F$ only.

Since $\theta_i = 0$ for each $i$, the stiffness matrix $[Q]^{(i)}$ in Eqs. (3-14)–(3-16) reduces to

$$[Q]^{(i)} = (1 - v_{12}^{(i)}v_{21}^{(i)})^{-1} \begin{bmatrix} E_1^{(i)} & v_{12}^{(i)}E_2^{(i)} & 0 \\ v_{12}^{(i)}E_2^{(i)} & E_2^{(i)} & 0 \\ 0 & 0 & (1 - v_{12}^{(i)}v_{21}^{(i)})G_{12}^{(i)} \end{bmatrix}$$

The shearing stress $\tau_{xz}$ and the shearing strain $\gamma_{xz}$ vanish identically as do $\tau_{yz}$ and $\gamma_{yz}$. Hence the twisting parameter $\Theta$ of Eq. (3-23) also vanishes. The variational equation (3-27b) reduces to

$$\sum \iint \varepsilon_{ij}\, \delta\sigma_{ij}\, dx\, dy =$$

$$\sum \iint \{F^{(i)},_{yy},\ F^{(i)},_{xx},\ -F^{(i)},_{xy}\} \begin{bmatrix} \beta_{22}^{(i)} & \beta_{23}^{(i)} & 0 \\ \beta_{23}^{(i)} & \beta_{33}^{(i)} & 0 \\ 0 & 0 & \beta_{44}^{(i)} \end{bmatrix} \begin{Bmatrix} \delta F,_{yy} \\ \delta F,_{xx} \\ -\delta F,_{xy} \end{Bmatrix} dx\, dy \qquad (3\text{-}39a)$$

where

$$\beta_{22}^{(i)} = (1 - v_{12}v_{21})/E_2, \qquad\qquad \beta_{33}^{(i)} = (1 - v_{13}v_{31})/E_3,$$
$$\beta_{23}^{(i)} = -(v_{23} + v_{21}v_{13})/E_2, \qquad \beta_{44}^{(i)} = 1/G_{23}^{(i)} \qquad\qquad (3\text{-}39b)$$

and the material directions 2 and 3 coincide, respectively, with the $x$- and $y$-directions.

In the absence of the stress function $\Psi$, the dimension of the vectors $\{x\}$ and $\{b\}$ in the governing differential equation, Eq. (3-34), is reduced from $4n + 1$ to $2n$, while the total number of eigenfunctions and the number of boundary conditions (Eqs. (3-30) and (3-31)) are each reduced from $12n + 2$ to $8n$.


## 3.12 THERMAL STRESS IN THE VICINITY OF A CURVED FREE EDGE

If the in-plane dimension of the laminate is at least an order of magnitude greater than the thickness of the laminate, then in the vicinity of any portion of the free edge only the (real and complex) eigenfunctions which decay rapidly away from the free edge contribute significantly to the local stress. The preceding analysis method may be directly applied to obtain accurate approximate solutions of the interlaminar thermal stresses near a straight free edge, except in the corner region near the intersection of two straight free edges. If a portion of the free edge has a curved geometry and if the local radius of curvature is an order of magnitude greater than the laminate thickness (usually, this is true around holes and cutouts of a composite laminate in aerospace applications but not true near small holes in electronics packaging components), then the preceding analysis method may still be applied to any short segment of the curved free edge, which is approximated by and treated as a *straight* segment. In this case, however, the traction boundary conditions of the complementary problem (Eq. (3-30)) generally vary along the curved free edge. Hence the local values of the traction data will be used in each short segment of the edge. Furthermore, for laminates composed of anisotropic layers, the governing differential equation (Eq. (3-34)) also varies with the segment, because the continuous change of the tangent direction along the free edge results in a corresponding change in the angle $\theta_i$ for each layer. The actual three-dimensional stress distribution near the curved free edge is effectively replaced by a local two-dimensional stress field. This approximation assumes that the characteristic length of decay of the interlaminar stresses, which is usually comparable to the laminate thickness, should be small compared to the local radius of curvature of the free edge.

## 3.13 LAYERED BEAMS

For layered beams the axial length $2a$ (along the $x$-direction) is significantly greater than both the thickness $t$ and the width $w$. The latter two dimensions (along the $y$- and $z$-directions, respectively) may or may not be comparable. In either case it is assumed that the stress components $\tau_{xz}$, $\tau_{yz}$, and $\sigma_z$ are negligibly small. Consequently, the remaining stress components may be expressed, according to the first three equations of Eq. (3-19), in terms of a single stress function $F^{(i)}$ in each layer. Again, we assume that the thickness direction is an orthotropy direction in each layer. With $\sigma_z = \tau_{xz} = 0$, Eq. (3-5) merely yields the mechanical strain $\varepsilon_z^*$ in terms of the axial stress $\sigma_x$ and the normal stress in the thickness direction, $\sigma_y$. The relevant constitutive relations are to be obtained from Eq. (3-4) rather than Eq. (3-6). We have

$$
\left\{ \begin{array}{c} \varepsilon_x^* \\ \varepsilon_y^* \\ \gamma_{xy}^* \end{array} \right\} = \left[ \begin{array}{ccc} a_{22}^{(i)} & a_{23}^{(i)} & 0 \\ a_{23}^{(i)} & a_{33}^{(i)} & 0 \\ 0 & 0 & a_{44}^{(i)} \end{array} \right] \left\{ \begin{array}{c} \sigma_x \\ \sigma_y \\ \tau_{xy} \end{array} \right\} \tag{3-40}
$$

All terms on the left-hand side of Eq. (3-27a), except the first term, are derived from the first group of integrals in Eq. (3-25). But these integrals make zero contribution since $\delta\tau_{xz} = \delta\tau_{yz} = \delta\sigma_z = 0$. Consequently, for a laminated beam, Eq. (3-27a) reduces to $\sum \iint \varepsilon_{ij}\,\delta\sigma_{ij}\,dx\,dy = 0$. Using Eq.

(3-40), we obtain

$$\sum \iint \{F^{(i)},_{yy}, F^{(i)},_{xx}, -F^{(i)},_{xy}\} \begin{bmatrix} a_{22}^{(i)} & a_{23}^{(i)} & 0 \\ a_{23}^{(i)} & a_{33}^{(i)} & 0 \\ 0 & 0 & a_{44}^{(i)} \end{bmatrix} \begin{Bmatrix} \delta F,_{yy} \\ \delta F,_{xx} \\ -\delta F,_{xy} \end{Bmatrix} dx\, dy = 0$$

$$(3\text{-}41)$$

where

$$a_{22}^{(i)} = \frac{\sin^4 \theta_i}{E_1^{(i)}} + \frac{\cos^4 \theta_i}{E_2^{(i)}} - \sin^2 \theta_i \cos^2 \theta_i \left( \frac{2v_{12}^{(i)}}{E_1^{(i)}} + \frac{1}{G_{12}^{(i)}} \right)$$

$$a_{23}^{(i)} = -\frac{v_{31}^{(i)} \sin^2 \theta_i + v_{32}^{(i)} \cos^2 \theta_i}{E_3^{(i)}}$$

$$a_{33}^{(i)} = \frac{1}{E_3^{(i)}}, \qquad a_{44}^{(i)} = \frac{\cos^2 \theta_i}{G_{23}^{(i)}} + \frac{\sin^2 \theta_i}{G_{13}^{(i)}}$$

If $\theta_i = 0$ in all layers (the case of a specially orthotropic multilayered beam), then the preceding compliance coefficients reduce to $a_{22}^{(i)} = 1/E_2^{(i)}$, $a_{23}^{(i)} = -v_{32}^{(i)}/E_3^{(i)}$, $a_{33}^{(i)} = 1/E_3^{(i)}$, and $a_{44}^{(i)} = 1/G_{23}^{(i)}$. In particular, for a laminated beam composed of several *isotropic* layers with the elastic moduli $E^{(i)}$ and $v^{(i)}$, the variational equation reduces to

$$\delta \Omega = 0 \qquad (3\text{-}42a)$$

where the complementary energy functional $\Omega$ has the expression

$$\Omega = \sum \iint (1/2E^{(i)})\{(F^{(i)},_{yy})^2 + (F^{(i)},_{xx})^2 - 2v^{(i)}F^{(i)},_{yy}F^{(i)},_{xx}$$

$$+ 2(1 + v^{(i)})(F^{(i)},_{xy})^2\}\, dx\, dy \qquad (3\text{-}42b)$$

Notice that there are two major differences between the governing equations for laminated plates and those for layered beams. First, the deformation parameters $B$, $C$, and $\Theta$ occur in Eq. (3-27a) for a laminated plate but not in the corresponding equation for a layered beam. Secondly, in Eq. (3-41) for a layered beam, the compliance coefficients are $a_{ij}^{(i)}$ rather than $\beta_{ij}^{(i)}$. The complementary problem for a layered beam may be solved directly, without separating into four subproblems as in the case of a layered plate.

## 3.14 REFINEMENT AND REGRESSION OF THE POLYNOMIAL APPROXIMATION

Stress functions that are cubic functions of the thickness coordinate in each layer (Eq. (3-28)) yield an axial stress that has a linear variation across the thickness of the layer. Such stress functions, therefore, provide the lowest-order polynomial approximation needed to account for the effect of bending of the layer. Numerical results show that even this lowest-order variational solution yields reasonably accurate results for the interlaminar stresses, except in narrow regions adjacent to the free edge. More accurate solutions for the interfacial stresses may be obtained, in the present approach, by using higher-order polynomial expansions of the stress functions $F^{(i)}$ and $\Psi^{(i)}$, particularly for a very thick layer among several layers of uneven thicknesses. For example, the polynomial expansion of Eq. (3-28$a$) may be expanded to one of the fifth order by adding the following terms:[21]

$$\eta^2(1 - \eta)^2 P_i(x) + \eta^2(1 - \eta)^3 Q_i(x)$$

Consistently with the improved approximation for the stress function $F^{(i)}$, Eq. (3-29$a$) requires the inclusion of a cubic term, $\eta^2(1 - \eta)R_i(x)$, in the expansion of $\Psi^{(i)}$. After the addition of these terms, $F_i(x)$, $G_i(x)$, and $\Psi_i(x)$ are still equal, respectively, to the values of $F$, $F_{,y}$, and $\Psi$ on the $i$th interface, while $H_{i-1}(x)$ continues to be the value of $\Psi_{,y}$ on the upper side of the $(i - 1)$th interface. It is now possible to deal with a temperature load that has a cubic variation across the thickness of the $i$th layer. In the complementary problem associated with the new temperature load, the two edges of the strip, $x = \pm a$, are subjected to boundary tractions that also have a cubic variation across the thickness of the layer. These boundary tractions provide appropriate boundary conditions, including those for the additional unknown functions $P_i$, $Q_i$, and $R_i$. The dimension of the matrices $\mathbf{W}$, $\mathbf{V}$, and $\mathbf{U}$ and the number of eigenvalues associated wtih Eq. (3-34) are increased. Because of the corresponding increase in the number of eigenfunctions, the stress field near the free edge may be better approximated.

Numerical results have been obtained for the variational solutions of two-layer and three-layer beams composed of different isotropic layers and subjected to a uniform temperature load.[22] Cubic, quartic, and quintic polynomial approximations of the stress function $F$ were used successively to obtain increasingly refined solutions.[21] Variational solutions based on the fifth-order polynomial stress function for a three-layer beam are found to be in excellent agreement with available finite-element solutions using an exceedingly large number of degrees of freedom. The solutions based on lower-order polynomial approximations yield good results for the interlaminar stresses except in an immediate neighborhood of the free edge.

For a two-layer bimetal thermostat, the variational solutions of increasing orders converge rapidly, in the sense appropriate to variational calculus, to an approximate elasticity solution that was obtained by using the Fourier transform method and the knowledge of the stress singularity associated with two joined quarter-spaces.[11] The comparisons confirm the efficacy and accuracy of the present method of analysis.

If the thickness of the $i$th layer in a laminated beam or plate is much smaller than those of the adjacent layers, then in this layer the boundary region with steep stress gradient may be especially narrow. This situation is generally associated with the occurrence of one or more unusually large eigenvalues, which may introduce significant computational errors in determining the appropriate combination of eigenfunctions. These unusually large eigenvalues and their associated eigenfunctions have little practical significance because they make insignificant contributions to the stress except in an extremely small region adjacent to the free edge. In this small region the highly localized stress predicted by the linear theory of elasticity is invariably modified by the inelastic behavior of the material, including the effect of plasticity. Such numerically large eigenvalues and their associated eigenfunctions may be excluded, without appreciably affecting the computed stresses in the surrounding layers, by using a lower-order polynomial expansion for the stress function in the layer. This is often advisable when the laminate contains a relatively soft interleaf or a thin adhesive layer.

## 3.15 MEASURES OF THE CRITICALITY OF THE INTERLAMINAR STRESSES

The functions $F_i(x)$, $\Psi_i(x)$, $G_i(x)$, and $H_i(x)$ in Eqs. (3-28) and (3-29) are the values of the stress functions and their $y$-derivatives along the $i$th interface. They are related to the resultant forces and moments of the normal and shearing interlaminar stresses over interfacial intervals of varying lengths adjacent to the free edge. Let the origin of the axial coordinate $x$ be shifted to the left free edge of the strip. Then, over the interval $[0, x]$ of the $i$th interface in the new coordinate system, we have

$$\int \sigma_y \, dx = \int F_i'' \, dx = F_i'(x) \tag{3-43a}$$

$$\int \tau_{xy} \, dx = -\int G_i' \, dx = -G_i(x) \tag{3-43b}$$

$$\int \tau_{yz} \, dx = -\int \Psi_i' \, dx = -\Psi_i(x) \tag{3-43c}$$

and

$$\int x\sigma_y\,dx = xF'_i(x) - F_i(x) \tag{3-43d}$$

In particular, the maximum absolute values of $G_i$ and $\Psi_i$, reached at the centerline of the strip, are equal to the resultant forces of the interlaminar shearing stresses $\tau_{xy}$ and $\tau_{yz}$, respectively, over one half of the $i$th interface. Generally, $G_i$ and $\Psi_i$ increase or decrease monotonically from zero to their maximum or minimum values at the centerline of the strip. If $G_i$ or $\Psi_i$ attains a large value in a very short distance from the free edge, then an intense interlaminar shearing stress (of mode II or III, respectively) occurs in a very short interval from the free edge. This situation is susceptible to delamination failure under the shear fracture mode.

The behavior of the function $F'_i(x)$ has a similar implication for the interlaminar normal stress. This function generally increases from zero at the free edge to a maximum value in a short distance from the free edge. It then decays slowly to zero at the midpoint of the beam. If $F'_i$ increases from zero to a large maximum value in a very short interval from the free edge, then an intense *tensile* interlaminar normal stress acts in that short end interval, and the resultant tensile force is balanced by an equal compressive force distributed over the remaining portion of the left half of the interface. This situation is susceptible to delamination failure under the peeling fracture mode.

The maximum values of the functions $G_i$, $\Psi_i$, and $F'_i$ and the characteristic lengths of their regions of steep gradient are meaningful measures of the criticality of the interlaminar stresses in the $i$th interface near the free edge. There are good reasons to suggest the use of these maximum values and characteristic lengths as parameters in the interlaminar fracture criteria for the peeling and shearing modes of failure. These parameters are readily given by a solution based on the stress formulation. Being global measures, their values are less affected by the crudeness or refinement of the approximate analysis than the pointwise values of the interlaminar stresses are. That is, a relatively simple analysis (involving low-order polynomial approximations of the stress functions) may provide unreliable results for the detailed interlaminar stress distributions along the interfaces but may still yield accurate results for the maximum values of $G_i$, $\Psi_i$, and $F'_i$, as well as the characteristic lengths of their regions of steep gradient. The validity of this statement has been supported by a comparison of the variational solutions of a two-layer beam based on cubic, quartic and quintic polynomial approximations of the stress functions[21] (see also the second example in the following section).

## 3.16 EXAMPLES: THREE-LAYER ANISOTROPIC LAMINATES AND ISOTROPIC BEAMS

A transcript of the MACSYMA session (edited to eliminate inessential information) for obtaining the matrices $\mathbf{U}$, $\mathbf{V}$, and $\mathbf{W}$ of Eq. (3-34) for a laminate consisting of three anisotropic layers is given in the Appendix 3A. In the case of a layered beam with three isotropic layers, the stress function $\Psi^{(i)}$ vanishes in each layer and the $9 \times 9$ symmetric matrices $\mathbf{U}$, $\mathbf{V}$, and $\mathbf{W}$ reduce to $4 \times 4$ matrices $\mathbf{UU}$, $\mathbf{VV}$, and $\mathbf{WW}$, which are also obtained. In this derivation, FL, FM, FU, SL, SM, and SU are, respectively, representations of the cubic or quadratic polynomial expansions of the stress functions $F^{(1)}$, $F^{(2)}$, $F^{(3)}$, $\Psi^{(1)}$, $\Psi^{(2)}$, and $\Psi^{(3)}$ associated with the lower, the middle, and the upper layer. The elements of the row vectors in these representations are the polynomials appearing in Eqs. (3-28) and (3-29), arranged in the order of their associated coefficient functions $F_1(x)$, $F_2(x)$, $G_1(x)$, $G_2(x)$, $\Psi_1(x)$, $\Psi_2(x)$, $H_0(x)$, $H_1(x)$, and $H_2(x)$. The matrices BL, BM, and BU in the statements (C26)–(D28) are the $[\beta_{ij}]$ matrices of Eq. (3-27$b$) for the three layers, where H1, H2, and H3 denote the thicknesses. Notice that only the $4 \times 4$ submatrix in the upper left corner of the matrix $\mathbf{W}$ is shown, because all the remaining elements of $\mathbf{W}$ vanish. According to Eq. (3-41), the elements of the matrices BL, BM, and BU for a beam with three isotropic layers are given by the statements (C48)–(C74) in terms of the Young's moduli and the Poisson ratios of the layers. The resulting matrices $\mathbf{UU}$, $\mathbf{VV}$, and $\mathbf{WW}$ are identical to the matrices $\mathbf{C}$, $\mathbf{B}$, and $\mathbf{A}$, respectively, shown in the Appendix of ref. 22.

Glaser[8] presented a finite-element solution of the thermal stresses in a three-layer beam of length $2a = 30.48$ mm subjected to a uniform temperature increment of 240°C. The thicknesses, elastic moduli, and thermal expansion coefficients of the successive layers are given as follows:

$$h_1 = 2.032 \text{ mm}, \quad E_1 = 68.95 \text{ GPa}, \quad v_1 = 0.33, \quad \alpha_1 = 23.6 \times 10^{-6}/°C$$

$$h_2 = 0.0508 \text{ mm}, \quad E_2 = 13.0 \text{ GPa}, \quad v_2 = 0.30, \quad \alpha_2 = 11.7 \times 10^{-6}/°C$$

$$h_3 = 0.508 \text{ mm}, \quad E_3 = 120.66 \text{ GPa}, \quad v_2 = 0.28, \quad \alpha_2 = 3.2 \times 10^{-6}/°C$$

For this problem, two variational solutions are obtained by the present method of analysis using, respectively, cubic and quintic polynomial approximations of the stress function in the thickness direction. The results for the interlaminar and in-plane stresses are shown in Figs. 3-2 to 3-5. In these figures the axial coordinate $x$ originates from the center of the beam and is normalized with respect to the half-length $a = 15.24$ mm. The solid and broken curves in the figures refer, respectively, to the solutions based on the fifth- and the third-degree polynomial expansions of the stress function. The higher-order polynomial solution is in better agreement with Glaser's

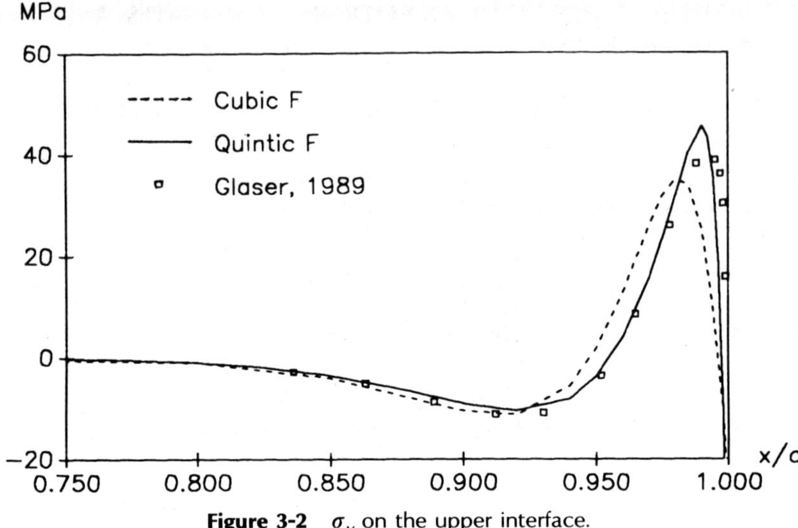

**Figure 3-2**   $\sigma_y$ on the upper interface.

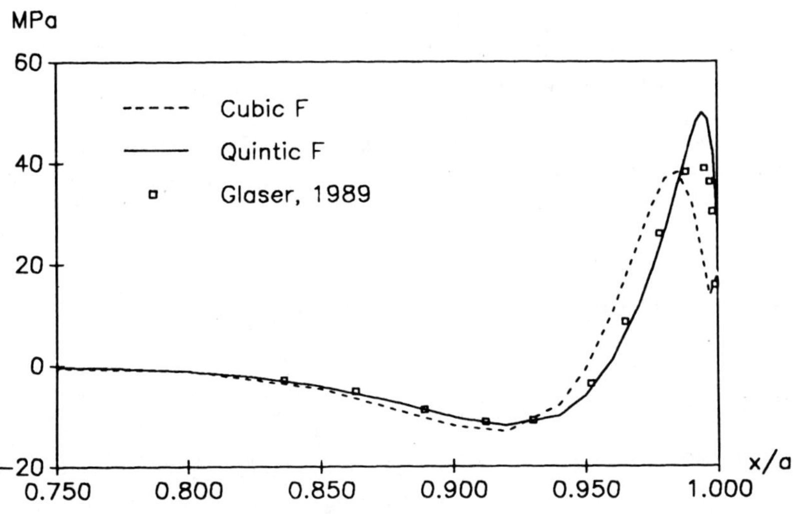

**Figure 3-3**   $\sigma_y$ on the lower interface.

finite element solution (shown by the scattered open squares in the figures) than the lower-order solution. However, significant discrepancies between the two variational solutions are confined to a very short end interval of the interface with a length smaller than the thicknesses of the adjacent layers.

As the next example, we consider a specially orthotropic laminate composed of four identical *orthotropic* layers with orientation angles 0°, 90°,

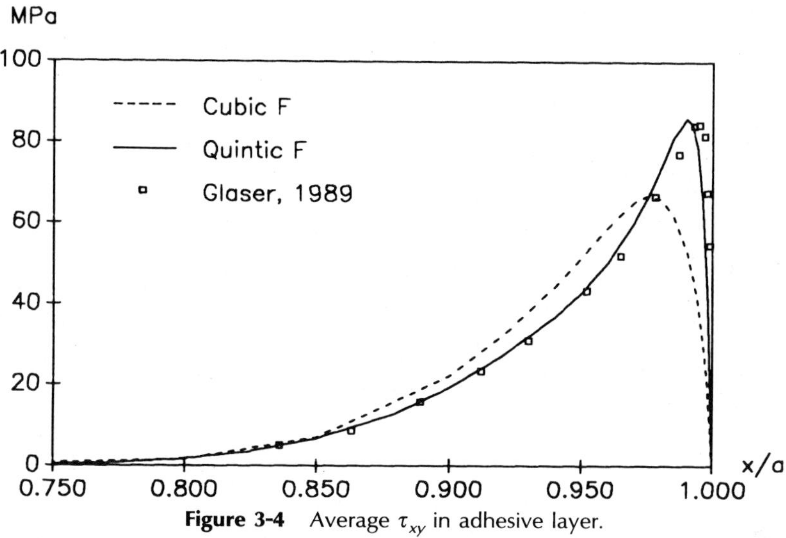



**Figure 3-4**  Average $\tau_{xy}$ in adhesive layer.

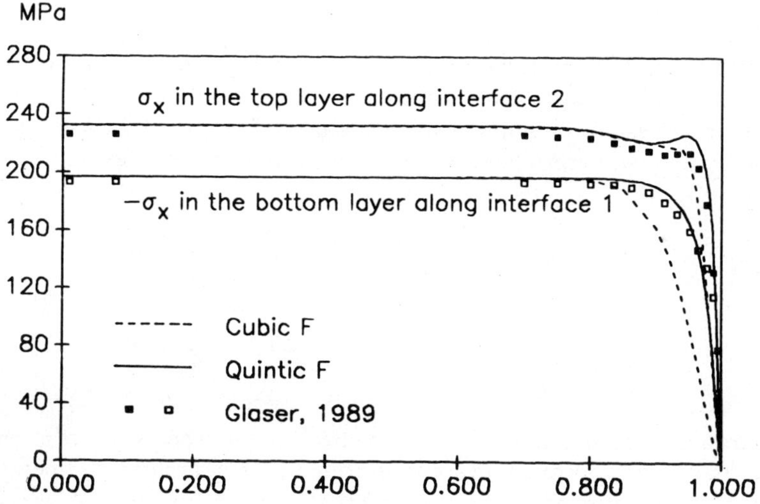



**Figure 3-5**  $\sigma_x$ in the top layer along interface 2 and the bottom layer along interface 1.

90°, and 0° relative to the $x$-axis. This laminate may also be considered as a three-layer laminate because the two 90° layers actually form a single layer of double thickness. Since the laminate is symmetric with respect to the middle plane, the bending–extension coupling matrix **B** (see Eq. (3-16)) vanishes so that a temperature field symmetric with respect to the middle plane will cause only extensional deformation but no bending or twisting

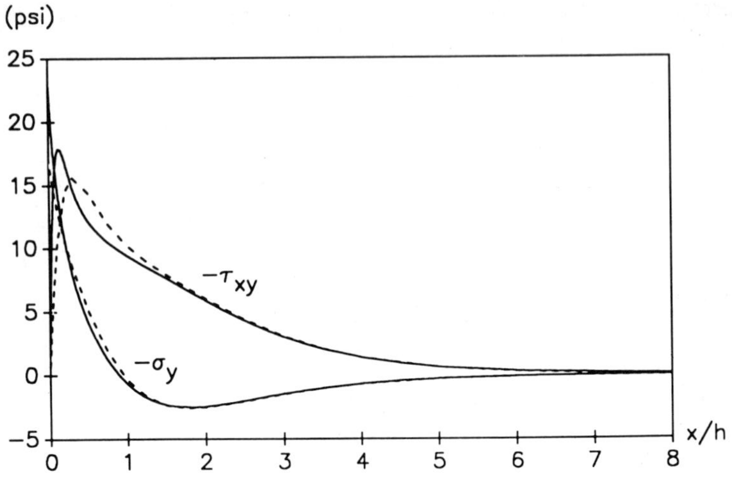



**Figure 3-6**  $[0/90]_s$ laminate, 1°F uniform temperature increment: $\tau_{xy}$ and $\sigma_y$.

deformation. Solutions of the interlaminar thermal stresses in the four-layer laminate due to a uniform temperature increment were obtained by Wang and Crossman[5] using the finite element method. The material properties of the orthotropic layer considered by these authors were as follows:

$$E_1 = 20 \times 10^6 \text{ psi,} \qquad E_2 = E_3 = 2.1 \times 10^6 \text{ psi}$$

$$\nu_{12} = \nu_{13} = \nu_{23} = 0.21, \qquad G_{12} = G_{13} = G_{23} = 0.85 \times 10^6 \text{ psi}$$

$$\alpha_1 = 0.2 \times 10^{-6}/°F, \qquad \alpha_2 = \alpha_3 = 16 \times 10^{-6}/°F$$

The present variational method is applied to this specially orthotropic laminate to obtain approximate solutions due to a 1°F uniform temperature increment. As mentioned in Section 3-11, the stress function $\Psi$ vanishes identically. The results for the interlaminar stresses on the 0°/90° interface are shown in Fig. 3-6, where the coordinate origin $x = 0$ is located at the free edge and where the solid and broken curves in the figure indicate, respectively, the solutions based on quintic and cubic polynomial expansions of the stress function $F$. The two solutions are close except in a narrow region (of width comparable to the thickness $h$ of a single layer) adjacent to the free edge.

When the orientation angles of the interior and exterior layers of the four-layer laminate are interchanged, one obtains the interlaminar stresses

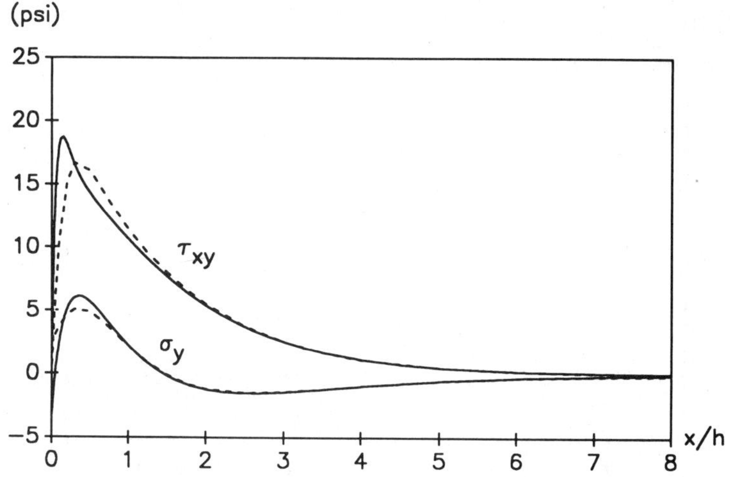

**Figure 3-7**   $[90/0]_s$ laminate, 1°F uniform temperature increment: $\tau_{xy}$ and $\sigma_y$.

as shown in Fig. 3-7 for a $[90/0]_s$ laminate.* The interlaminar shearing stress is almost identical to that shown in the previous figure, except for a sign reversal. However, the behavior of the normal stress $\sigma_y$ in the boundary region is significantly different between the two laminates.

While the figures show significant differences of the interlaminar stresses in an immediate neighborhood of the free edge between the two solutions based on polynomial expansions of different degrees, these solutions yield close results for the resultant peeling and shearing forces over end intervals of any given length, as given by Eqs. (3-43a, b). The function $-G_3(x)$ along the highest interface in the $[90/0]_s$ laminate is shown in Fig. 3-8, where the solid and broken curves denote, respectively, the solutions based on cubic and quintic polynomial expansions of the stress function. The corresponding results for the $[0°/90°]_s$ laminate are nearly identical except for a reversal of the algebraic sign.

## 3.17 CONCLUDING REMARKS

A systematic approximate method has been developed for determining the interlaminar stresses under temperature loads in a layered beam or laminate

---

* This notation is commonly used in the literature of composite structures to denote a symmetric laminate with the stacking sequence 90°/0°/0°/90°. Notice that, in case of square laminates bounded by free edges $x = \pm a$ and $y = \pm a$, the interlaminar stresses $\{\sigma_y, \tau_{zy}\}$ along the $z$-axis of the $[0/90]_s$ laminate are identical to he interlaminar stresses $\{\sigma_y, \tau_{xy}\}$ along the $x$-axis of the $[90/0]_s$ laminate.

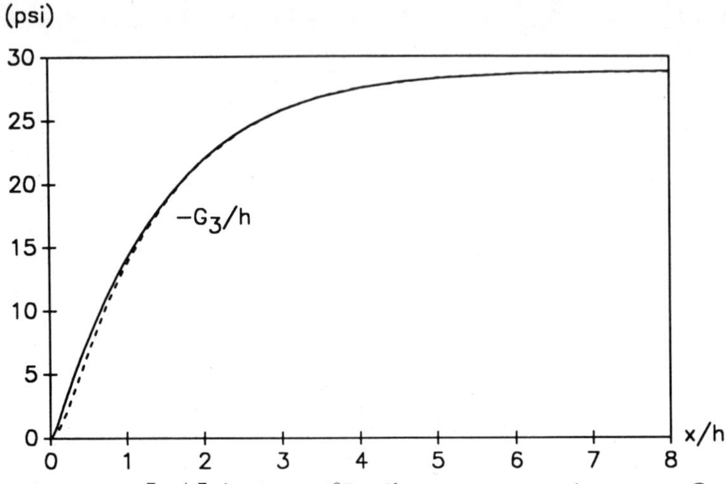



**Figure 3-8**  [90/0]$_s$ laminate, 1°F uniform temperature increment: $G_3$.

composed of isotropic or anisotropic layers. The method is based on the complementary virtual work principle and formulated in terms of Lekhnitskii's stress functions for the various layers. The class of admissible stress fields in the present analysis ensures *exact* satisfaction of the equilibrium equations in the entire region of the laminate, all tractions boundary conditions, and the continuity of interlaminar stresses across all interfaces. The compatibility conditions for the strain and the interfacial continuity of the displacements are enforced in an averaged sense by extremizing the complementary energy functional with respect to the class of admissible stress functions. Consequently, the resulting variational solution yields a more accurate stress field, especially in regions near the free edge, compared to the variational solutions and other numerical solutions based on a displacement formulation.

Besides the constitutive assumption of linear elasticity, the only approximation used in the present analysis is that the stress functions in each layer are polynomial functions of the thickness coordinate. No a priori assumption is made concerning the dependence of the stress functions on the coordinate parallel to the interfaces. Rather, it follows as a consequence of the principle of complementary virtual work that the form of dependence is governed by a system of linear ordinary differential equations with constant coefficients. The coefficients are determined by the thicknesses, orientation angles, and the thermal and elastic properties of the layers, while the constant terms in the differential equations depend on the thermal and mechanical loads. For a layered beam or laminate composed of only a relatively small number of layers, numerical solution of the eigenvalue problem associated with the differential equation may be easily implemented. If the beam or laminate contains a large number of layers, then a substructure approach should be used to reduce the dimension of the eigenvalue problem. In the substructure approach, two interior layers adjacent to a particular interface are modeled as elastic media, while all remaining layers are grouped into an upper and a lower sublaminate. This sublaminate/layer modeling and analysis takes into account the combined stiffness matrices of the remote layers as well as the distinct elastic and thermal properties of the two adjacent layers. The latter properties directly affect the nature of the interlaminar stresses across the interface of the layers. The sublaminate/layer approach has been developed recently for both mechanical[23] and thermal loads.[24]

In the $i$th layer, the first-order partial derivatives of the stress function $F$ and the value of the stress function $\Psi$ provide useful measures of the criticality of the interlaminar stresses near the free edge. These values are less affected by the degree of refinement of the approximate analysis method than the detailed distributions of the interlaminar stresses are. They may be used as parameters in the fracture criteria for the initiation of delamination failure.

FORTRAN computer programs have been developed to implement the analysis of interlaminar thermal stresses using the present method for three-layer and symmetric four-layer laminates and beams. Furthermore, computer programs have also been written for computing the interlaminar stresses in multilayer laminates by the sublaminate/layer approach. These programs are extremely easy to use, because of the simplicity of data input. The data files include a few lines of numbers referring to the thicknesses, the orientation angles, the anisotropic elastic moduli and thermal expansion coefficients of successive layers, as well as the temperature and temperature gradient in each layer. Computing time for executing each program on a 486 personal computer is of the order of one second. Hence the method provides a reliable, efficient, and extremely inexpensive analysis tool for the practical task of assessing interlaminar thermal stresses in multilayered structures.

## ACKNOWLEDGMENTS

This article is based partially on research work performed under a U.S. Army Research Office contract to Georgia Institute of Technology and partially on work supported by a contract from the NASA Langley Research Center to Lockheed Aeronautical Systems Company. The author appreciates the policy of *ASME Journal of Electronic Packaging*, which allows the use of certain figures previously published in the journal.

## NOMENCLATURE

| | |
|---|---|
| $y_i$ | Position of the $i$th interface |
| $h_i = y_i - y_{i-1}$ | Thickness of the $i$th layer |
| $g_i = (y_i^2 - y_{i-1}^2)/2$ | |
| $f_i = (y_i^3 - y_{i-1}^3)/3$ | |
| $E_1^{(i)}, E_2^{(i)}, E_3^{(i)}$ | Extensional elastic moduli of the $i$th layer along the three orthotropy axes of elasticity |
| $v_{23}^{(i)}, v_{31}^{(i)}, v_{12}^{(i)}$ | Poisson ratios of the $i$th layer |
| $G_{23}^{(i)}, G_{31}^{(i)}, G_{12}^{(i)}$ | Shear moduli |
| $S_{mn}^{(i)}$ (or $\mathbf{S}^{(i)}$) | Elastic compliance matrix of the $i$th layer referred to the orthotropy axes |
| $\alpha_k^{(i)}$ | Thermal expansion coefficients of the $i$th layer along the $k$th orthotropy axes |
| $T$ | Temperature field referred to the stress-free temperature |
| $T_i^1$ | (Constant) Temperature gradient in the $i$th layer |
| $T_i^0$ | Temperature on the $i$th interface minus $y_i T_i^1$ |
| $\sigma_{ij}$ | Stresses |
| $\varepsilon_{ij}$ | Strains |
| $\varepsilon_{ij}^*$ | Mechanical strains |
| $u, v, w$ | Displacements along the $x$, $y$ and $z$ coordinate directions |
| $\theta_i$ | Orientation angle of the $i$th layer with respect to the free edge ($x$-axis) |
| $a_{mn}^{(i)}$ | Elastic compliance matrix of the $i$th layer referred to $(x, y, z)$ coordinates |
| $\beta_{mn}^{(i)} = a_{mn}^{(i)} - a_{1m}^{(i)} a_{1n}^{(i)} / a_{11}^{(i)}$ | for $m, n \neq 1$ |
| $\mathbf{H}(\theta_i)$ | Transformation matrix of the stress components from the $(x, y, z)$ coordinates to the material coordinates of the $i$th layer (Eq. (3-3)) |
| $t$ | Thickness of the multilayer beam or laminate |
| $2a$ | Length of the multilayer beam or width of the multilayer laminate |
| $\varepsilon_z^0, \varepsilon_x^0, \gamma_{xz}^0$ | Middle plane strains |
| $\kappa_z, \kappa_x, \kappa_{xz}$ | Curvatures of the deformed middle plane |
| $C, B, \Theta$ | Constant values of $\varepsilon_z^0$, $\kappa_z$ and $\kappa_{xz}$, respectively, in a generalized plane deformation (Eq. (3-23)) |
| $N_z, N_x, N_{xz}$ | Stress resultants (membrane forces) |
| $M_z, M_x, M_{xz}$ | Stress moments |
| $[Q]^{(i)}$ | In-plane stiffness of the $i$th layer (Eq. (3-14)) |
| $\mathbf{A} = \sum h_i [Q]^{(i)}$ | Extensional stiffness matrix of the laminate |
| $\mathbf{B} = \sum g_i [Q]^{(i)}$ | Bending stiffness matrix |
| $\mathbf{D} = \sum f_i [Q]^{(i)}$ | Extension–bending coupling matrix |
| $\eta$ | Normalized thickness coordinate ($\eta = (y - y_{i-1})/(y_i - y_{i-1})$ in the $i$th layer) |
| $F^{(i)}(x, y), \Psi^{(i)}(x, y)$ | Stress functions in the $i$th layer |
| $F_i(x), G_i(x), \Psi_i(x)$ | Value of $F$, $\partial F/\partial y$, and $\Psi$, respectively, on the $i$th interface |

| | |
|---|---|
| $H_i(x)$ | Value of $\partial F/\partial y$ on the upper side of the $i$th interface |
| $\{X\}$ | Vector of dimension $4n + 1$ with the components $F_i(x)$, $G_i(x)$, $\Psi_i(x)$, and $H_i(x)$ |
| $\{b\}$ | Vector of dimension $4n + 1$ defined by Eq. (3-33) |
| $\mathbf{W}, \mathbf{V}, \mathbf{U}$ | Constant symmetric matrices in the Euler–Lagrange equation associated with the variational problem (Eqs. (3-32) and (3-34)) |
| $\lambda$ | Eigenvalues of the characteristic equation (Eq. (3-36)) |
| $\{c\}_k$ | $k$th (real or complex) eigenvector associated with the $k$th eigenvalue $\lambda_k$ |

## Appendix 3A

A MACSYMA Session for Deriving the $\mathbf{U}$, $\mathbf{V}$, and $\mathbf{W}$ Matrices of

  (a) A laminate with three anisotropic layers

  (b) A beam with three isotropic layers (results shown by $\mathbf{UU}$, $\mathbf{VV}$, $\mathbf{WW}$)

```
(D1)                    DUA0:[MACSYMA_412]F050492.TXT;1
(C2)   FL:MATRIX([3*Y^2-2*Y^3,0,H1*(Y^3-Y^2),0,0,0,0,0,0])$
(C3)   FM:MATRIX([1-3*Y^2+2*Y^3,3*Y^2-2*Y^3,H2*(Y-2*Y^2+Y^3),
       H2*(Y^3-Y^2),0,0,0,0,0])$
(C4)   FU:MATRIX([0,1-3*Y^2+2*Y^3,0,H3*(Y-2*Y^2+Y^3),0,0,0,0,0])$
(C5)   SL:MATRIX([0,0,0,0,Y^2,0,H1*(Y-Y^2),0,0])$
(C6)   SM:MATRIX([0,0,0,0,1-Y^2,Y^2,0,H2*(Y-Y^2),0])$
(C7)   SU:MATRIX([0,0,0,0,0,1-Y^2,0,0,H3*(Y-Y^2)])$
(C8)   STLX:EXPAND(DIFF(FL,Y,2)/H1^2)$
(C9)   STMX:EXPAND(DIFF(FM,Y,2)/H2^2)$
(C10)  STUX:EXPAND(DIFF(FU,Y,2)/H3^2)$
(C11)  STLY:EXPAND(D^2*FL)$
(C12)  STMY:EXPAND(D^2*FM)$
(C13)  STUY:EXPAND(D^2*FU)$
(C14)  STLXY:EXPAND(-DIFF(FL,Y)*D/H1)$
(C15)  STMXY:EXPAND(-DIFF(FM,Y)*D/H2)$
(C16)  STUXY:EXPAND(-DIFF(FU,Y)*D/H3)$
(C17)  STLYZ:EXPAND(-D*SL)$
(C18)  STMYZ:EXPAND(-D*SM)$
(C19)  STUYZ:EXPAND(-D*SU)$
(C20)  STLXZ:EXPAND(DIFF(SL,Y)/H1)$
(C21)  STMXZ:EXPAND(DIFF(SM,Y)/H2)$
(C22)  STUXZ:EXPAND(DIFF(SU,Y)/H3)$
(C23)  STL:ADDROW(STLX,STLY,STLXY,STLYZ,STLXZ)$
(C24)  STM:ADDROW(STMX,STMY,STMXY,STMYZ,STMXZ)$
(C25)  STU:ADDROW(STUX,STUY,STUXY,STUYZ,STUXZ)$
(C26)  BL:ENTERMATRIX(5,5);
                       [ BL22   BL23   0      0      BL26 ]
                       [ BL23   BL33   0      0      BL36 ]
(D26)                  [ 0      0      BL44   BL45   0    ]
                       [ 0      0      BL45   BL55   0    ]
                       [ BL26   BL36   0      0      BL66 ]
(C27)  BM:ENTERMATRIX(5,5);
                       [ BM22   BM23   0      0      BM26 ]
                       [ BM23   BM33   0      0      BM36 ]
```

```
(D27)                              [   0       0      BM44    BM45     0    ]
                                   [   0       0      BM45    BM55     0    ]
                                   [ BM26    BM36      0       0     BM66   ]
(C28) BU:ENTERMATRIX(5,5);
                                   [ BU22    BU23      0       0     BU26   ]
                                   [ BU23    BU33      0       0     BU36   ]
(D28)                              [   0       0      BU44    BU45     0    ]
                                   [   0       0      BU45    BU55     0    ]
                                   [ BU26    BU36      0       0     BU66   ]
(C29) TRANSPOSE(SUBST(F,D,STL)) . BL$
(C30) EXPAND(H1*%  .  STL)$
(C31) INTEGRATE(%,Y)$
(C32) ENGL:EV(%,Y:1,F:-D)$
(C33) TRANSPOSE(SUBST(F,D,STM)) . BM$
(C34) EXPAND(H2*%  .  STM)$
(C35) INTEGRATE(%,Y)$
(C36) ENGM:EV(%,Y:1,F:-D)$
(C37) TRANSPOSE(SUBST(F,D,STU)) . BU$
(C38) EXPAND(H3*%  .  STU)$
(C39) INTEGRATE(%,Y)$
(C40) ENGU:EV(%,Y:1,F:-D)$
(C41) ENG:ENGL+ENGM+ENGU$
(C42) U:EXPAND(EV(ENG,D:0)/2)$
(C43) V:EXPAND(COEFF(ENG,D,2)/2)$
(C44) W:EXPAND(COEFF(ENG,D,4)/2)$
(C45) FORTRAN(U);
      U(1,1) = 6*BM22/H2**3+6*BL22/H1**3
      U(1,2) = -6*BM22/H2**3
      U(1,3) = 3*BM22/H2**2-3*BL22/H1**2
      U(1,4) = 3*BM22/H2**2
      U(1,5) = -BM26/H2**2-BL26/H1**2
      U(1,6) = BM26/H2**2
      U(1,7) = BL26/H1
      U(1,8) = -BM26/H2
      U(1,9) = 0
      U(2,1) = -6*BM22/H2**3
      U(2,2) = 6*BU22/H3**3+6*BM22/H2**3
      U(2,3) = -3*BM22/H2**2
      U(2,4) = 3*BU22/H3**2-3*BM22/H2**2
      U(2,5) = BM26/H2**2
      U(2,6) = -BU26/H3**2-BM26/H2**2
      U(2,7) = 0
      U(2,8) = BM26/H2
      U(2,9) = -BU26/H3
      U(3,1) = 3*BM22/H2**2-3*BL22/H1**2
      U(3,2) = -3*BM22/H2**2
      U(3,3) = 2*BM22/H2+2*BL22/H1
      U(3,4) = BM22/H2
      U(3,5) = BL26/H1
      U(3,6) = 0
      U(3,7) = -BL26/2.0
      U(3,8) = -BM26/2.0
      U(3,9) = 0
      U(4,1) = 3*BM22/H2**2
      U(4,2) = 3*BU22/H3**2-3*BM22/H2**2
      U(4,3) = BM22/H2
      U(4,4) = 2*BU22/H3+2*BM22/H2
      U(4,5) = -BM26/H2
      U(4,6) = BM26/H2
      U(4,7) = 0
      U(4,8) = -BM26/2.0
```

```
       U(4,9)  =  -BU26/2.0
       U(5,1)  =  -BM26/H2**2-BL26/H1**2
       U(5,2)  =  BM26/H2**2
       U(5,3)  =  BL26/H1
       U(5,4)  =  -BM26/H2
       U(5,5)  =  2.0*BM66/(3.0*H2)+2.0*BL66/(3.0*H1)
       U(5,6)  =  (-2.0*BM66)/(3.0*H2)
       U(5,7)  =  -BL66/6.0
       U(5,8)  =  BM66/6.0
       U(5,9)  =  0
       U(6,1)  =  BM26/H2**2
       U(6,2)  =  -BU26/H3**2-BM26/H2**2
       U(6,3)  =  0
       U(6,4)  =  BM26/H2
       U(6,5)  =  (-2.0*BM66)/(3.0*H2)
       U(6,6)  =  2.0*BU66/(3.0*H3)+2.0*BM66/(3.0*H2)
       U(6,7)  =  0
       U(6,8)  =  -BM66/6.0
       U(6,9)  =  BU66/6.0
       U(7,1)  =  BL26/H1
       U(7,2)  =  0
       U(7,3)  =  -BL26/2.0
       U(7,4)  =  0
       U(7,5)  =  -BL66/6.0
       U(7,6)  =  0
       U(7,7)  =  BL66*H1/6.0
       U(7,8)  =  0
       U(7,9)  =  0
       U(8,1)  =  -BM26/H2
       U(8,2)  =  BM26/H2
       U(8,3)  =  -BM26/2.0
       U(8,4)  =  -BM26/2.0
       U(8,5)  =  BM66/6.0
       U(8,6)  =  -BM66/6.0
       U(8,7)  =  0
       U(8,8)  =  BM66*H2/6.0
       U(8,9)  =  0
       U(9,1)  =  0
       U(9,2)  =  -BU26/H3
       U(9,3)  =  0
       U(9,4)  =  -BU26/2.0
       U(9,5)  =  0
       U(9,6)  =  BU66/6.0
       U(9,7)  =  0
       U(9,8)  =  0
       U(9,9)  =  BU66*H3/6.0
(D45)                                    DONE
(C46)  FORTRAN(V);
       V(1,1)  =  (-3.0*BM44)/(5.0*H2)+(-6.0*BM23)/(5.0*H2)
     1    +(-3.0*BL44)/(5.0*H1)+(-6.0*BL23)/(5.0*H1)
       V(1,2)  =  3.0*BM44/(5.0*H2)+6.0*BM23/(5.0*H2)
       V(1,3)  =  -BM44/20.0+(-3.0*BM23)/5.0+BL44/20.0+3.0*BL23/5.0
       V(1,4)  =  -BM44/20.0-BM23/10.0
       V(1,5)  =  7.0*BM45/20.0+(-3.0*BM36)/20.0
     1    +(-3.0*BL45)/20.0+7.0*BL36/20.0
       V(1,6)  =  3.0*BM45/20.0+3.0*BM36/20.0
       V(1,7)  =  -BL45*H1/10.0-BL36*H1/10.0
       V(1,8)  =  BM45*H2/10.0+BM36*H2/10.0
       V(1,9)  =  0
       V(2,1)  =  3.0*BM44/(5.0*H2)+6.0*BM23/(5.0*H2)
       V(2,2)  =  (-3.0*BU44)/(5.0*H3)+(-6.0*BU23)/(5.0*H3)
     1    +(-3.0*BM44)/(5.0*H2)+(-6.0*BM23)/(5.0*H2)
```

```
  V(2,3)  =  BM44/20.0+BM23/10.0
  V(2,4)  =  -BU44/20.0+(-3.0*BU23)/5.0+BM44/20.0+3.0*BM23/5.0
  V(2,5)  =  (-7.0*BM45)/20.0+(-7.0*BM36)/20.0
  V(2,6)  =  7.0*BU45/20.0+(-3.0*BU36)/20.0
1    +(-3.0*BM45)/20.0+7.0*BM36/20.0
  V(2,7)  =  0
  V(2,8)  =  -BM45*H2/10.0-BM36*H2/10.0
  V(2,9)  =  BU45*H3/10.0+BU36*H3/10.0
  V(3,1)  =  -BM44/20.0+(-3.0*BM23)/5.0+BL44/20.0+3.0*BL23/5.0
  V(3,2)  =  BM44/20.0+BM23/10.0
  V(3,3)  =  -BM44*H2/15.0+(-2.0*BM23*H2)/15.0
1    -BL44*H1/15.0+(-2.0*BL23*H1)/15.0
  V(3,4)  =  BM44*H2/60.0+BM23*H2/30.0
  V(3,5)  =  -BM45*H2/30.0-BM36*H2/30.0-BL45*H1/20.0-BL36*H1/20.0
  V(3,6)  =  BM45*H2/30.0+BM36*H2/30.0
  V(3,7)  =  BL45*H1**2/120.0+BL36*H1**2/120.0
  V(3,8)  =  BM45*H2**2/120.0+BM36*H2**2/120.0
  V(3,9)  =  0
  V(4,1)  =  -BM44/20.0-BM23/10.0
  V(4,2)  =  -BU44/20.0+(-3.0*BU23)/5.0+BM44/20.0+3.0*BM23/5.0
  V(4,3)  =  BM44*H2/60.0+BM23*H2/30.0
  V(4,4)  =  -BU44*H3/15.0+(-2.0*BU23*H3)/15.0
1    -BM44*H2/15.0+(-2.0*BM23*H2)/15.0
  V(4,5)  =  BM45*H2/20.0+BM36*H2/20.0
  V(4,6)  =  -BU45*H3/30.0-BU36*H3/30.0-BM45*H2/20.0-BM36*H2/20.0
  V(4,7)  =  0
  V(4,8)  =  BM45*H2**2/120.0+BM36*H2**2/120.0
  V(4,9)  =  BU45*H3**2/120.0+BU36*H3**2/120.0
  V(5,1)  =  7.0*BM45/20.0+(-3.0*BM36)/20.0
1    +(-3.0*BL45)/20.0+7.0*BL36/20.0
  V(5,2)  =  (-7.0*BM45)/20.0+(-7.0*BM36)/20.0
  V(5,3)  =  -BM45*H2/30.0-BM36*H2/30.0-BL45*H1/20.0-BL36*H1/20.0
  V(5,4)  =  BM45*H2/20.0+BM36*H2/20.0
  V(5,5)  =  (-4.0*BM55*H2)/15.0-BL55*H1/10.0
  V(5,6)  =  -BM55*H2/15.0
  V(5,7)  =  -BL55*H1**2/40.0
  V(5,8)  =  (-7.0*BM55*H2**2)/120.0
  V(5,9)  =  0
  V(6,1)  =  3.0*BM45/20.0+3.0*BM36/20.0
  V(6,2)  =  7.0*BU45/20.0+(-3.0*BU36)/20.0+(-3.0*BM45)/20.0
1    +7.0*BM36/20.0
  V(6,3)  =  BM45*H2/30.0+BM36*H2/30.0
  V(6,4)  =  -BU45*H3/30.0-BU36*H3/30.0-BM45*H2/20.0-BM36*H2/20.0
  V(6,5)  =  -BM55*H2/15.0
  V(6,6)  =  (-4.0*BU55*H3)/15.0-BM55*H2/10.0
  V(6,7)  =  0
  V(6,8)  =  -BM55*H2**2/40.0
  V(6,9)  =  (-7.0*BU55*H3**2)/120.0
  V(7,1)  =  -BL45*H1/10.0-BL36*H1/10.0
  V(7,2)  =  0
  V(7,3)  =  BL45*H1**2/120.0+BL36*H1**2/120.0
  V(7,4)  =  0
  V(7,5)  =  -BL55*H1**2/40.0
  V(7,6)  =  0
  V(7,7)  =  -BL55*H1**3/60.0
  V(7,8)  =  0
  V(7,9)  =  0
  V(8,1)  =  BM45*H2/10.0+BM36*H2/10.0
  V(8,2)  =  -BM45*H2/10.0-BM36*H2/10.0
  V(8,3)  =  BM45*H2**2/120.0+BM36*H2**2/120.0
  V(8,4)  =  BM45*H2**2/120.0+BM36*H2**2/120.0
```

```
          V(8,5)  =  (-7.0*BM55*H2**2)/120.0
          V(8,6)  =  -BM55*H2**2/40.0
          V(8,7)  =  0
          V(8,8)  =  -BM55*H2**3/60.0
          V(8,9)  =  0
          V(9,1)  =  0
          V(9,2)  =  BU45*H3/10.0+BU36*H3/10.0
          V(9,3)  =  0
          V(9,4)  =  BU45*H3**2/120.0+BU36*H3**2/120.0
          V(9,5)  =  0
          V(9,6)  =  (-7.0*BU55*H3**2)/120.0
          V(9,7)  =  0
          V(9,8)  =  0
          V(9,9)  =  -BU55*H3**3/60.0
(D46)                                      DONE
(C47)  FORTRAN(SUBMATRIX(5,6,7,8,9,W,5,6,7,8,9));
          W(1,1)  =  13.0*BM33*H2/70.0+13.0*BL33*H1/70.0
          W(1,2)  =  9.0*BM33*H2/140.0
          W(1,3)  =  11.0*BM33*H2**2/420.0+(-11.0*BL33*H1**2)/420.0
          W(1,4)  =  (-13.0*BM33*H2**2)/840.0
          W(2,1)  =  9.0*BM33*H2/140.0
          W(2,2)  =  13.0*BU33*H3/70.0+13.0*BM33*H2/70.0
          W(2,3)  =  13.0*BM33*H2**2/840.0
          W(2,4)  =  11.0*BU33*H3**2/420.0+(-11.0*BM33*H2**2)/420.0
          W(3,1)  =  11.0*BM33*H2**2/420.0+(-11.0*BL33*H1**2)/420.0
          W(3,2)  =  13.0*BM33*H2**2/840.0
          W(3,3)  =  BM33*H2**3/210.0+BL33*H1**3/210.0
          W(3,4)  =  -BM33*H2**3/280.0
          W(4,1)  =  (-13.0*BM33*H2**2)/840.0
          W(4,2)  =  11.0*BU33*H3**2/420.0+(-11.0*BM33*H2**2)/420.0
          W(4,3)  =  -BM33*H2**3/280.0
          W(4,4)  =  BU33*H3**3/210.0+BM33*H2**3/210.0
(D47)                                      DONE
(C48)  BL26:0$
(C49)  BM26:0$
(C50)  BU26:0$
(C51)  BL36:0$
(C52)  BM36:0$
(C53)  BU36:0$
(C54)  BL45:0$
(C55)  BM45:0$
(C56)  BU45:0$
(C57)  BL55:0$
(C58)  BM55:0$
(C59)  BU55:0$
(C60)  BL66:0$
(C61)  BM66:0$
(C62)  BU66:0$
(C63)  BL22:1/E1$
(C64)  BM22:1/E2$
(C65)  BU22:1/E3$
(C66)  BL23:-PS1/E1$
(C67)  BM23:-PS2/E2$
(C68)  BU23:-PS3/E3$
(C69)  BL33:1/E1$
(C70)  BM33:1/E2$
(C71)  BU33:1/E3$
(C72)  BL44:2*(1+PS1)/E1$
(C73)  BM44:2*(1+PS2)/E2$
(C74)  BU44:2*(1+PS3)/E3$
(C75)  UU:EV(SUBMATRIX(5,6,7,8,9,U,5,6,7,8,9))$
```

```
(C76)  VV:EV(SUBMATRIX(5,6,7,8,9,V,5,6,7,8,9))$
(C77)  WW:EV(SUBMATRIX(5,6,7,8,9,W,5,6,7,8,9))$
(C78)  FORTRAN(UU);
       UU(1,1) := 3/(E2*H2**3)+3/(E1*H1**3)
       UU(1,2) = -3/(E2*H2**3)
       UU(1,3) = 3.0/(2.0*E2*H2**2)+(-3.0)/(2.0*E1*H1**2)
       UU(1,4) = 3.0/(2.0*E2*H2**2)
       UU(2,1) = -3/(E2*H2**3)
       UU(2,2) = 3/(E3*H3**3)+3/(E2*H2**3)
       UU(2,3) = (-3.0)/(2.0*E2*H2**2)
       UU(2,4) = 3.0/(2.0*E3*H3**2)+(-3.0)/(2.0*E2*H2**2)
       UU(3,1) = 3.0/(2.0*E2*H2**2)+(-3.0)/(2.0*E1*H1**2)
       UU(3,2) = (-3.0)/(2.0*E2*H2**2)
       UU(3,3) = 1/(E2*H2)+1/(E1*H1)
       UU(3,4) = 1/(E2*H2)/2.0
       UU(4,1) = 3.0/(2.0*E2*H2**2)
       UU(4,2) = 3.0/(2.0*E3*H3**2)+(-3.0)/(2.0*E2*H2**2)
       UU(4,3) = 1/(E2*H2)/2.0
       UU(4,4) = 1/(E3*H3)+1/(E2*H2)
(D78)                                        DONE
(C79)  FORTRAN(VV);
       VV(1,1) = (-3.0)/(5.0*E2*H2)+(-3.0)/(5.0*E1*H1)
       VV(1,2) = 3.0/(5.0*E2*H2)
       VV(1,3) = PS2/E2/4.0-PS1/E1/4.0-1/E2/20.0+1/E1/20.0
       VV(1,4) = -1/E2/20.0
       VV(2,1) = 3.0/(5.0*E2*H2)
       VV(2,2) = (-3.0)/(5.0*E3*H3)+(-3.0)/(5.0*E2*H2)
       VV(2,3) = 1/E2/20.0
       VV(2,4) = PS3/E3/4.0-PS2/E2/4.0-1/E3/20.0+1/E2/20.0
       VV(3,1) = PS2/E2/4.0-PS1/E1/4.0-1/E2/20.0+1/E1/20.0
       VV(3,2) = 1/E2/20.0
       VV(3,3) = -H2/E2/15.0-H1/E1/15.0
       VV(3,4) = H2/E2/60.0
       VV(4,1) = -1/E2/20.0
       VV(4,2) = PS3/E3/4.0-PS2/E2/4.0-1/E3/20.0+1/E2/20.0
       VV(4,3) = H2/E2/60.0
       VV(4,4) = -H3/E3/15.0-H2/E2/15.0
(D79)                                        DONE
(C80)  FORTRAN(WW);
       WW(1,1) = 13.0*H2/(140.0*E2)+13.0*H1/(140.0*E1)
       WW(1,2) = 9.0*H2/(280.0*E2)
       WW(1,3) = 11.0*H2**2/(840.0*E2)+(-11.0*H1**2)/(840.0*E1)
       WW(1,4) = (-13.0*H2**2)/(1680.0*E2)
       WW(2,1) = 9.0*H2/(280.0*E2)
       WW(2,2) = 13.0*H3/(140.0*E3)+13.0*H2/(140.0*E2)
       WW(2,3) = 13.0*H2**2/(1680.0*E2)
       WW(2,4) = 11.0*H3**2/(840.0*E3)+(-11.0*H2**2)/(840.0*E2)
       WW(3,1) = 11.0*H2**2/(840.0*E2)+(-11.0*H1**2)/(840.0*E1)
       WW(3,2) = 13.0*H2**2/(1680.0*E2)
       WW(3,3) := H2**3/E2/420.0+H1**3/E1/420.0
       WW(3,4) = -H2**3/E2/560.0
       WW(4,1) = (-13.0*H2**2)/(1680.0*E2)
       WW(4,2) = 11.0*H3**2/(840.0*E3)+(-11.0*H2**2)/(840.0*E2)
       WW(4,3) = -H2**3/E2/560.0
       WW(4,4) = H3**3/E3/420.0+H2**3/E2/420.0
(D80)                                        DONE
(C81)  VALUES;
(D81)  [FL, FM, FU, SL, SM, SU, STLX, STMX, STUX, STLY, STMY, STUY,
       STLXY, STMXY, STUXY, STLYZ, STMYZ, STUYZ, STLXZ, STMXZ,STUXZ,
       STL, STM, STU, BL, BM, BU, ENGL, ENGM, ENGU, ENG, U, V, W,
       BL26, BM26, BU26, BL36, BM36, BU36, BL45, BM45, BU45, BL55,
```

```
        BM55, BU55, BL66, BM66, BU66, BL22, BM22, BU22, BL23, BM23,
        BU23, BL33, BM33, BU33, BL44, BM44, BU44, UU, VV, WW]
(C82)   SAVE("eng3lyr.mac",FL,FM,FU,SL,SM,SU,STL,STM,STU,BL,BM,BU,
        ENG,U,V,W,UU,VV,WW)$
(C83)   CLOSEFILE();
```

# REFERENCES


1. Bogy, D. B., "Edge-bonded Dissimilar Orthogonal Elastic Wedges Under Normal and Shear Loading," *ASME J. Appl. Mech.*, **35**, 1968, pp. 460–466.
2. Bogy, D. B., "On the Problem of Edge-bonded Elastic Quarter-planes Loaded at the Boundary," *Int. J. Solids Struct.*, **6**, 1970, pp. 1287–1313.
3. Wang, S. S., and I. Choi, "Boundary-layer Effects in Composite Laminate—I. Free-edge Stress Singularities—II. Free-edge Stress Solutions and Basic Characteristics," *ASME J. Appl. Mech.*, **49**, 1982, pp. 541–560.
4. Zwiers, R. I., T. C. T. Ting, and R. L. Spilker, "On the Logarithmic Singularity of Free-edge Stress in Laminated Composites Under Uniform Extension," *ASME J. Appl. Mech.*, **49**, 1982, pp. 561–569.
5. Wang, A. S. D., and F. W. Crossman, "Edge Effects on Thermally Induced Stresses in Composite Laminates," *J. Composite Materials*, **11**, 1977, pp. 300–312.
6. Stango, R. J., and S. S. Wang, "Process-induced Residual Thermal Stresses in Advanced Fiber-reinforced Composite Laminates," *ASME J. Engr. for Power*, **106**, 1984, pp. 48–54.
7. Cho, K. N., A. G. Striz, and C. W. Bert, "Thermal Stress Analysis of Laminate Using Higher-order Theory in Each Layer," *J. Thermal Stresses*, **12**, 1989, pp. 321–332.
8. Glaser, J. C., "Thermal Stresses in Compliantly-Jointed Materials," ASME Winter Annual Meeting, Paper No. 89-WA/EEP-14, San Francisco, CA, December 1989.
9. Kuo, A.-Y., and K.-L. Chen, "Thermal Stresses at the Edge of a Multi-layered Composite Strip," Paper presented in ASME Winter Annual Meeting, December 1992, Atlanta, GA.
10. Hess, M. S., "The End Problem for a Laminated Elastic Strip—II. Differential Expansion Stresses," *J. Composite Materials*, **3**, 1969, pp. 630–641.
11. Kuo, A.-Y., "Thermal Stresses at the Edge of a Bimetallic Thermostat," *ASME J. Appl. Mech.*, **56**, 1989, pp. 585–589.
12. Grimado, P. B., "Interlaminar Thermoelastic Stresses in Layered Beams," *J. Thermal Stresses*, **1**, 1978, pp. 75–86.
13. Chen, W. T., and C. W. Nelson, "Thermal Stress in Bonded Joints," *IBM J. Res. Develop.*, **23**, 1979, pp. 179–188.
14. Suhir, E., *Structural Analysis in Microelectronic and Fiber-Optic Systems*, Van Nostrand Reinhold, New York, 1991.
15. Chen, D., S. Cheng, and T. D. Gerhardt, "Thermal Stresses in Laminated Beams," *J. Thermal Stresses*, **5**, 1982, pp. 67–84.
16. Yin, W.-L., "Free-edge Effects in Laminates Under Extension, Bending and Twisting, Part I: A Stress Function Approach," *Proc. AIAA/ASME/ASCE/AHS/*

*ASC 32nd Structures, Structural Dynamics and Materials Conference*, April, 1991, Baltimore, MD, pp. 985–995.

17. Lekhnitskii, S. G., *Theory of Elasticity of an Anisotropic Elastic Body*, Holden-Day, San Francisco, 1963.

18. Jones, R. M., *Mechanics of Composite Materials*, McGraw-Hill, New York, 1975.

19. *MACSYMA Refernce Manual, Version 13*. Symbolics, Inc., Burlington, MA, 1988.

20. Wolfram, S., *Mathematica: A System for Doing Mathematics by Computer*, 2d edn. Addison-Wesley, Redwood City, CA, 1990.

21. Yin, W.-L., "Refined Variational Solutions of the Interfacial Thermal Stresses in a Laminated Beam," *ASME J. Electronic Packaging*, **114**, 1992, pp. 193–198.

22. Yin, W.-L., "Thermal Stresses and Free-edge Effects in Laminated Beams: A Variational Approach Using Stress Functions," *ASME J. Electronic Packaging*, **113**, 1991, pp. 68–75.

23. Yin, W.-L., "Free-edge Effects in Laminates Under Extension, Bending and Twisting, Part II: Sublaminate/Layer Modeling and Analysis," *Proc. AIAA/ ASME/ASCE/AHS/ASC 33rd Structures, Structural Dynamics and Materials Conference*, April, 1992, Dallas, TX, pp. 48–58.

24. Yin, W.-L., *Free-Edge Effects in Laminates Subjected to Hygrothermal Loading*. Final Report, Contract NAS1-17925, Task Assignment No. 10, Lockheed Aeronautical Systems Co., Marietta, GA, March 1992.

# 4

# Transient Thermal Stresses in Multilayered Devices

*An-Yu Kuo and Kuan-Luen Chen*

## 4.1 INTRODUCTION

Thermal stress in a multilayered strip is an important problem in electronics packaging and has received great attention in recent years. Thermal stress solutions to the multilayered strip problem under steady-state temperature conditions have been discussed in Chapters 1–3 of this book. This chapter focuses only on the transient behaviors of thermal stresses in multilayered devices.

Thermal stresses in multilayered devices have been known to be highly concentrated at free edges of material interfaces, which are usually one of the most critical life-controlling locations. Therefore, more emphasis will be placed on transient thermal stresses near the free edges in this chapter. Only linear elastic materials are discussed in this chapter. Nonlinear transient thermal stresses will not be discussed. Both theoretical as well as application aspects of the transient thermal stress problem will be addressed in this chapter. Since this chapter is not intended to serve as a review paper, neither exhaustive nor extensive references on related subjects will be provided, to make this chapter more concise and focused. The authors wish to express their apology for not citing as many references as they would like to.

Generally speaking, solutions to the thermal stress problem of multi-layered devices can be categorized into four different categories: (1) analytical solutions without any edge effects,[1] (2) analytical solutions with concentrated but nonsingular edge stresses,[2–5] (3) analytical solutions with singular edge stresses,[6–9] and (4) finite element solutions.[10] Thermal stresses at material interfaces have been found to be highly concentrated at the free edge in all but the first solution categories mentioned above. A recent experimental study by Hattori, Sakata, and Murakami[11] has demonstrated that integrity of a multilayered electronic device is a strong function of $K_{ij}$ and $\delta$, where $K_{ij}$ and $\delta$ are, respectively, generalized stress intensity factor and stress singularity order, which are related to the interfacial edge stresses $\sigma_{ij}$ and the radial distance to the edge, $r$, by $\sigma_{ij} = K_{ij}r^{\delta}$. Thus, use of the two singular parameters $K_{ij}$ and $\delta$ as controlling parameters for life and/or reliability evaluations of the multilayered electronic devices appears to be a promising concept.

Steady-state $K_{ij}$ and $\delta$ solutions for a multilayered device have been obtained by many researchers with different methods. For instance, Kuo[6,7] used an analytical alternating method; Blanchard and Ghoniem[9] used a boundary collocation method; and Hattori, Sakata, and Murakami[11] used the finite element method. Nevertheless, an important question that remained unanswered is how the transient (instead of steady-state) thermal loading will affect the edge stress solutions. After a preliminary (of course not complete or exhaustive) literature survey, the authors identified only a handful of papers in the area of transient thermal stresses at the free-edge problem of a multilayered device. Wang and Chou,[12–14] and Wang, Pipes, and Chou[15] have presented transient thermal stresses for laminated composites. In their studies, no stress singularities were included and no through-thickness temperature variations were allowed. In reality, temperatures are likely to vary in the thickness direction and such through-thickness temperature variations can have significant impact on the resulting thermal stress solutions. Also, under a linear elastic assumption, it has been shown by many authors that stresses at the free edge are singular.

Chen and Kuo[8,16] have recently completed their studies on the transient thermal stresses of a multilayered device. In these studies, the conventional finite element method was used in the transient heat transfer analysis while a mixture of the conventional finite element method and a more complicated hybrid finite element formulation were used in the subsequent thermal stress calculations. They have concluded that advantages of using the hybrid finite element method are two-fold: (1) stress singularities at the free edge can be accounted for properly and automatically, and (2) a coarse finite element mesh can be used at the free edge region and, thus, shorter computation time and less disk storage space are required.

The hybrid finite element method presented by Chen and Kuo[8,16] appears to be an efficient and accurate way of solving the transient problem and, thus, will be discussed in detail later in this chapter. In short, a singular element based on the hybrid finite element formulation[17] has been developed and used in conjunction with the conventional four-node or eight-node elements to solve the transient thermal stress problem of multilayered devices. The hybrid finite element method was found to be an efficient and accurate tool.

Section 4.2 describes the methodology and procedure for solving the transient heat transfer part of the problem. Section 4.3 presents derivation of a hybrid functional which will be used later as the bases for the special element formulation. Section 4.4 discusses an asymptotic solution to the edge problem. The asymptotic solution is an important part in the hybrid finite element formulation. Section 4.5 presents a theoretical and numerical formulation of the special hybrid finite element method. Verification of the hybrid finite element method is also covered in this section. Section 4.6 outlines applications of the Green's function approach, which enables one to handle arbitrary thermal transients based on transient responses due to a unit step thermal loading. Section 4.7 discusses transient characteristics of the free edge thermal stresses. Section 4.8 describes applications of the transient thermal stresses in life assessment of multilayered devices. Section 4.9 contains a discussion and summary, and reference papers are listed at the end of the chapter. Those who are interested in understanding the transient thermal behavior of multilayered devices but do not wish to spend too much time going through the mathematics and numerical manipulations can turn directly to Section 4.7 and revisit Sections 4.2 through 4.6 at a later time.

## 4.2 TRANSIENT HEAT TRANSFER SOLUTIONS

For most electronic applications, it is appropriate to assume that the heat transfer and the thermal stress problems are quasi-static and decoupled. The coupling between heat transfer and thermal stress equations is negligible when $\dot{e}/\dot{T}$ is much smaller than

$$\frac{\rho c}{(3\lambda + 2\mu)\alpha T_0} \qquad \text{where} \qquad \dot{e} = \frac{\partial e}{\partial t}, \qquad e = \varepsilon_{xx} + \varepsilon_{yy} + \varepsilon_{zz}$$

where $\lambda$ and $\mu$ are Lamé's elastic constants, $\rho$ is the mass density, $c$ is the specific heat, $\alpha$ is the coefficient of thermal expansion, and $T_0$ is the stress-free temperature of the material. The condition for the decoupled heat transfer and thermal stress analyses can be met easily for most electronics packaging problems. That is, for practical purposes, the transient heat transfer equations and the thermoelasticity equations can be solved separately. The analysis is quasi-static because the thermoelasticity portion of the analysis is static (no inertia effects) while the heat transfer portion of the analysis is transient. The transient heat transfer portion of the analysis will be discussed in this section.

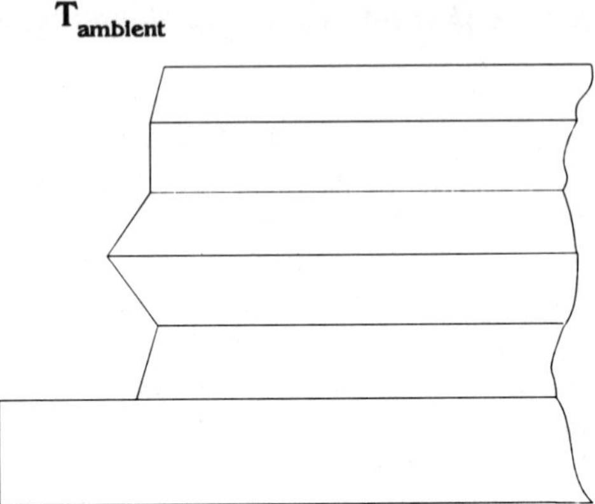


**Figure 4-1**    A multilayered electronic device.

For a multilayered device, such as the one illustrated in Fig. 4-1, the governing equation for the transient heat transfer in each of the constituent layers can be written (assuming no heat sources or sinks in the layers) as

$$\nabla^2 \theta_j = \frac{1}{\mu_j} \frac{\partial \theta_j}{\partial t} \tag{4-1}$$

where $\nabla^2$ is the Laplace operator, $\theta_j$ is the temperature variation of the $j$th layer, $t$ is time, $\mu_j = K_j/\rho_j c_j$ is the thermal diffusivity, $K_j$ is the thermal conductivity, $\rho_j$ is the mass density, and $c_j$ is the specific heat. In the remaining part of this chapter, the subscript $j$ will be dropped when no confusion will be created by doing so. On the boundary of the multilayered device, $\partial A$, one of the following three boundary conditions must be met:

$$\theta(x_i, t) = F(x_i, t) \tag{4-2}$$

$$K \frac{\partial \theta(x_i, t)}{\partial n} = -h[\theta(x_i, t) - \theta_0(x_i, t)] \tag{4-3}$$

$$K \frac{\partial \theta(x_i, t)}{\partial n} = -q(x_i, t) \tag{4-4}$$

where $F(x_i, t)$ and $q(x_i, t)$ are known functions representing prescribed temperature and heat flux, respectively, $\theta_0$ is the prescribed ambient temperature variation, and $n$ is the outer normal of the structure boundary. In addition, it is required that both temperature and heat flux be continuous across the material interfaces.

For a finite-dimensional, multilayered device with an arbitrary geometrical shape, the above transient heat transfer equations can easily be solved by many commercial finite element programs. For two-dimensional problems, as illustrated in Fig. 4-1, the four-node or eight-node isoparametric elements are usually used in the finite element mesh. Since transient heat transfer analyses of the multilayered devices by the finite element methods have been done routinely by most engineers and researchers, details of this transient finite element analysis will be skipped in this chapter.

The temperature distribution, $\theta(x_i, t)$, in the multilayered device resulting from the transient heat transfer analysis will serve as an input to the subsequent thermal stress evaluation, which will be discussed in the following sections.

## 4.3 VARIATIONAL PRINCIPLE FOR THE THERMOELASTICITY PROBLEM

As shown in Fig. 4-2, both the conventional isoparametric elements and the hybrid special elements will be used in the thermal stress analysis. The hybrid element theory has been developed to evaluate the free-edge problem

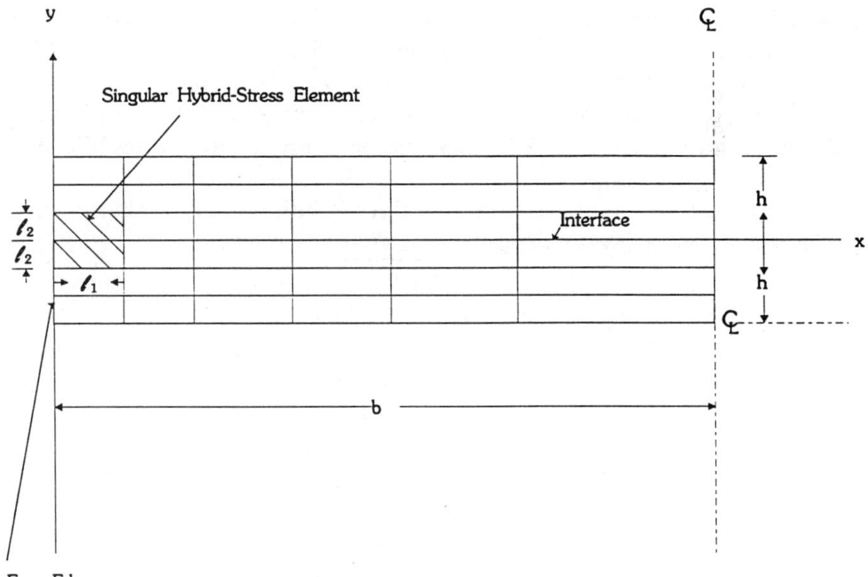


**Figure 4-2**  A special hybrid element surrounded by regular elements.

of composite laminates by Wang and Yuan.[18] In Wang and Yuan's work, the constituent layers of the composite laminate are made of anisotropic materials and the composite strip is subjected to an uniaxial force. Chen and Kuo[8,16] have recently extended the hybrid finite element method to the thermoelasticity problem of multilayered devices. As shown in Fig. 4-2, the special elements are placed at the free-edge locations while conventional elements are used for the rest of the component. Advantages of using the hybrid-singular elements are that a coarse mesh can be used and calculation of the singular parameters $K_{ij}$ and $\delta$ is automatic and very accurate. Again, the two singular parameters are related to the asymptotic stress distribution at the free edge by $\sigma_{ij} = K_{ij} r^{\delta}$.

Both the conventional and the hybrid finite element formulations are based on the first variation of a functional. To derive the functional, consider a plane problem with prescribed boundary traction $\bar{T}_i$ over the boundary $\partial A_\sigma$ and temperature variation $\theta(x_i, t)$. $\theta(x_i, t)$ is defined as the temperature above the stress-free temperature at location $x_i$. In this problem, $\theta(x_i, t)$ are the temperature solutions resulting from the transient heat transfer analysis described in the previous section. One major difference between the hybrid finite element formulation and the conventional finite element formulation is that, in the hybrid finite element theory, displacements and stresses on the element boundary and those within the element can be assumed independently as long as certain variational conditions are met. In contrast, in the conventional finite element theory, displacements on the element boundary as well as displacements within the element are both determined from the element nodal displacements. Throughout this chapter, variables with a tilde "~" above them designate quantities on the element boundary, variables with a bar "‾" above them are designated for prescribed quantities on the device boundary, and variables without any overhead are for quantities inside the element.

The governing equations and boundary conditions of the thermoelasticity problem for each element can be written as follows:

*Within element area, $A_m$ (m is an element number):*

$$\sigma_{ij,j} = 0 \tag{4-5}$$

and

$$\tfrac{1}{2}(u_{i,j} + u_{j,i}) - \varepsilon_{ij}^0 = S_{ijkl}\sigma_{kl} \tag{4-6}$$

where $\sigma_{ij}$ are components of the stress tensor, $u_i$ are components of the displacement vector, $\varepsilon_{ij}^0$ are components of the initial (thermal) strain tensor, and $S_{ijkl}$ are components of the compliance tensor.

*On the boundary of the element, $\partial A_m$:*

$$\sigma_{ij} n_j = T_i \tag{4-7}$$

and

$$u_i = \tilde{u}_i \tag{4-8}$$

where $n_j$ are boundary unit normal vectors, $T_i$ are tractions, and $\tilde{u}_i$ are boundary displacements defined over the element boundary $\partial A_m$.

*On the interelement portion of the boundary, $\partial A_{mj}$:*

$$T_i^+ + T_i^- = 0 \tag{4-9}$$

where $\partial A_{mj}$ is the common boundary between two adjacent elements, and $T_i^+$ and $T_i^-$ are interelement boundary tractions of the two adjacent elements.

*On the prescribed traction portion, $\partial A_{m\sigma} = \partial A_m \cap \partial A_\sigma$:*

$$T_i = \bar{T}_i \tag{4-10}$$

Based on Eqs. (4-5) through (4-10), a weak variational form of the thermoelasticity can be defined as follows:

$$\sum_{m=1}^{M} \left[ \int_{A_m} \{ -\sigma_{ij,j}\, \delta u_i + (\tfrac{1}{2}(u_{i,j} + u_{j,i}) - \varepsilon_{ij}^0 - S_{ijkl}\sigma_{kl})\, \delta\sigma_{ij} \}\, dA \right.$$

$$+ \int_{\partial A_m} \{ (\sigma_{ij}n_j - T_i)\, \delta u_i + (\tilde{u}_i - u_i)\, \delta T_i + T_i\, \delta\tilde{u}_i \}\, dS$$

$$\left. - \int_{\partial A_{m\sigma}} \bar{T}_i\, \delta\tilde{u}_i\, dS \right] = 0 \tag{4-11}$$

where $M$ is the total number of elements in the finite element model.

In principle, all of the quantities in Eq. (4-11) can be assumed independently. However, in the hybrid stress finite element theory, they have been chosen in such a fashion that some of the Euler equations, namely Eqs. (4-5) through (4-10), are satisfied exactly while the rest of them are satisfied variationally.[17] In this chapter, it is assumed that no external mechanical loads are present, i.e., $T_i = 0$ on $\partial A_{m\sigma}$. One can choose $\sigma_{ij}$ and $u_i$ such that Eqs. (4-5), (4-6), (4-7), and (4-10) are satisfied exactly in the hybrid singular element and Eqs. (4-6), (4-7), and (4-8) are satisfied exactly in the regular isoparametric elements. The weak variational form, Eq. (4-11), can then be reduced to a simpler form:

$$\sum_{m=1}^{M_s} \left[ \int_{\partial A_m} \{(\tilde{u}_i - u_i)\, \delta T_i + T_i\, \delta\tilde{u}_i\}\, dS \right]$$

$$+ \sum_{m=1}^{M_r} \left[ \int_{A_m} (-\sigma_{ij,\,j}\, \delta u_i)\, dA + \int_{\partial A_m} T_i\, \delta\tilde{u}\, dS \right] = 0 \quad (4\text{-}12)$$

where $M_s$ is the total number of singular elements, $M_r$ is the total number of regular elements, and $M_s + M_r = M$. The reduced weak variational form, Eq. (4-12), can be further transformed into an alternative symmetric form as follows:

$$\sum_{m=1}^{M_s} \left[ \int_{\partial A_m} \{(T_i\, \delta\tilde{u}_i + \tilde{u}_i\, \delta T_i) - \tfrac{1}{2}(T_i\, \delta u_i + u_i\, \delta T_i)\}\, dS - \tfrac{1}{2}\int_{A_m} \varepsilon_{ij}^0\, \delta\sigma_{ij}\, dA \right]$$

$$+ \sum_{m=1}^{M_r} \left[ \int_{A_m} (u_{k,\,l} S_{ijkl}^{-1}\, \delta u_{k,\,l} - \varepsilon_{kl}^0 S_{ijkl}^{-1}\, \delta u_{i,\,j})\, dA \right] = 0 \quad (4\text{-}13)$$

This transformation is necessary because it is more convenient and efficient to deal with a symmetric functional in the finite element formulation. From Eq. (4-13), a corresponding functional, $\pi$, can be derived

$$\pi = \sum_{m=1}^{M_s} \left[ \int_{\partial A_m} (T_i\tilde{u}_i - \tfrac{1}{2}T_i u_i)\, dS - \tfrac{1}{2}\int_{A_m} \varepsilon_{ij}^0 \sigma_{ij}\, dA \right]$$

$$+ \sum_{m=1}^{M_r} \left[ \int_{A_m} (\tfrac{1}{2}u_{i,\,j} S_{ijkl}^{-1} u_{k,\,l} - \varepsilon_{kl}^0 S_{ijkl}^{-1} u_{i,\,j})\, dA \right] \quad (4\text{-}14)$$

This hybrid functional will be used later to derive the stiffness matrix and force vectors for the finite element analysis.


## 4.4 ASYMPTOTIC THERMAL STRESS DISTRIBUTION NEAR FREE EDGE

According to the hybrid finite element formulation, it is required that asymptotic solutions that satisfy Eqs. (4-5), (4-6), (4-7), and (4-10) be used in the functional $\pi$. Thus, the next step of the hybrid element formulation is to derive the asymptotic solutions for the problem of a bimaterial semi-infinite plane subjected to thermal loading, as shown in Fig. 4-3.

For convenience, the two-dimensional thermoelasticity problem is evaluated with a complex variable, $z$,

$$z = x + iy \tag{4-15}$$

where $x$ and $y$ are Cartesian coordinates of the semi-infinite plane as shown in Fig. 4-3 and $i$ is the imaginary number. The stresses and displacements in an isotropic, bimaterial plane can be expressed in terms of Muskhelishvili's[19]

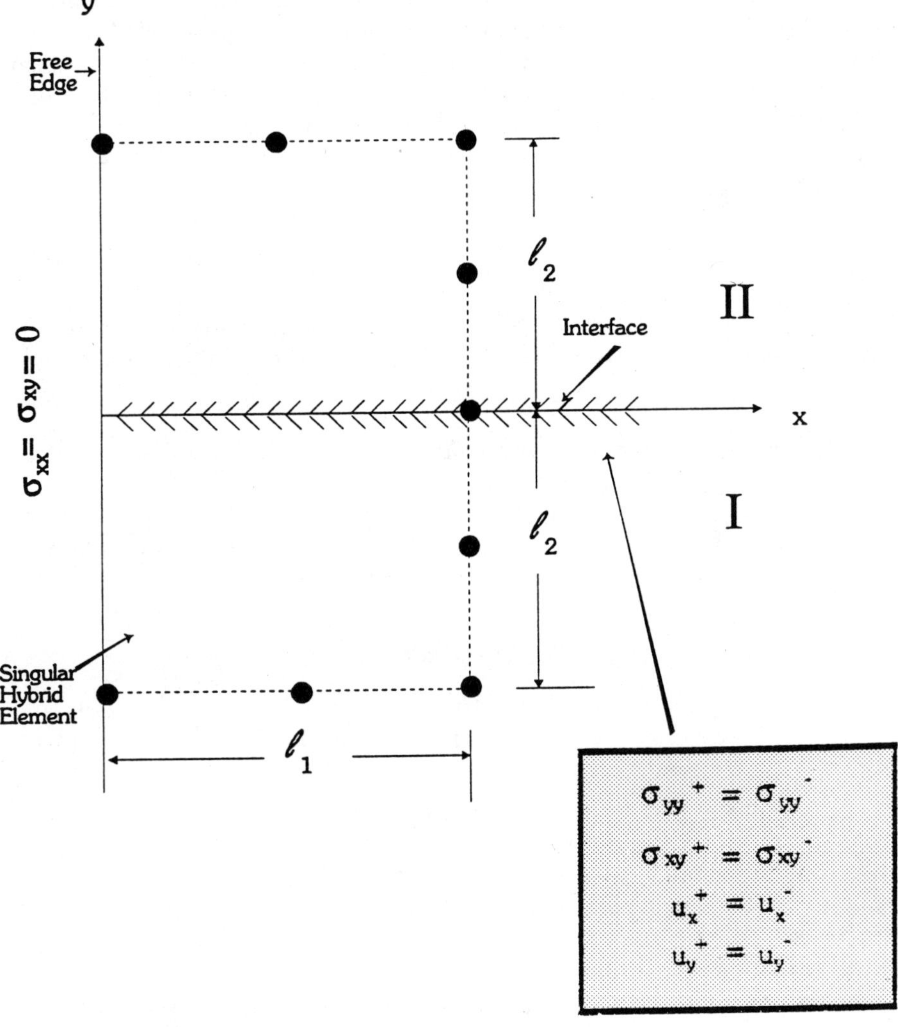


**Figure 4-3**   Geometric dimensions and boundary conditions of hybrid element.

stress functions as[20]

$$\sigma_{xx} + \sigma_{yy} = 4G^I[\varphi_I'(z) + \overline{\varphi_I'(z)} - \lambda^I\Theta]$$

$$\sigma_{xx} - \sigma_{yy} + 2i\sigma_{xy} = 2G^I[\phi_{,xx}^I - \phi_{,yy}^I + 2i\phi_{,xy}^I]$$
$$- 4G^I[\overline{z\varphi_I''(z)} + \overline{\chi_I(z)''}]$$

$$u_x + iu_y = \phi_{,x}^I + i\phi_{,y}^I + \kappa^I\varphi_I(z) - \overline{z\varphi_I'(z)} - \overline{\chi_I'(z)}$$

(4-16)

in which $G^I = E^I/2(1 + v^I)$; $\kappa^I = 3 - 4v^I$ for plane strain problems or $\kappa^I = (3 - v^I)/(1 + v^I)$ for plane stress problems, $\lambda^I = \alpha^I(1 + v^I)/2(1 - v^I)$, $\theta = \theta(x_i, t)$ is temperature distribution, $G^I$ and $v^I$ are, respectively, shear modulus and Poisson's ratio of material $I$ ($I = 1, 2$), ( )$'$ denotes differentiation with respect to $z$, and $\overline{(\ )}$ is the complex conjugate. Both $\varphi_I$ and $\chi_I$ are analytical functions in the region occupied by material $I$. In Eq. (4-16), $\phi^I$ is a particular solution of

$$\nabla^2\phi^2 = 2\lambda^I\Theta$$

(4-17)

In addition to the particular solution, it is also necessary to include the homogeneous solution to satisfy the boundary and continuity conditions.


## 4.4.1 Homogeneous Asymptotic Solution

The homogeneous asymptotic solution can be obtained by an eigenfunction expansion technique. As a typical procedure in the eigenfunction expansion technique, the stress functions $\varphi_I$ and $\chi_I$ can be expanded as

$$\varphi_I'(z) = \sum_{i=1}^{\infty} (A_I^i z^{\delta_i} + C_I^i z^{\bar{\delta}_i})$$

$$\chi_I''(z) = \sum_{i=1}^{\infty} (B_I^i z^{\delta_i} + D_I^i z^{\bar{\delta}_i})$$

(4-18)

$$(C_I^1 = D_I^1 = 0)$$

where $\delta_1$ is a real eigenvalue and $\delta_i$ ($i = 2, 3, \ldots$) are complex eigenvalues, and $A_I^i$, $B_I^i$, $C_I^i$, and $D_I^i$ are complex eigencoefficients. The complex variables, $\delta_i$, $A_I^i$, $B_I^i$, $C_I^i$, and $D_I^i$, will be determined based on an eigenvalue problem to be discussed in the remainder of the section.

From Eq. (4-18) the homogeneous solutions for the stresses, $\sigma_{ij}^h$, and displacements, $u_i^h$, have the following forms:

$$\sigma_{xx}^h + i\sigma_{xy}^h = G^I \sum_{i=1}^{\infty} [2A_I^i z^{\delta_i} + 2\bar{A}_I^i \bar{z}^{\bar{\delta}_i} - 2z\bar{A}_I^i \bar{\delta}_i \bar{z}^{\bar{\delta}_i - 1} - 2\bar{B}_I^i \bar{z}^{\delta_i}$$

$$+ 2C_I^i z^{\bar{\delta}_i} + 2\bar{C}_I^i \bar{z}^{\delta_i} - 2z\bar{C}_I^i \delta_i \bar{Z}^{\delta_i - 1} - 2\bar{D}_I^i \bar{z}^{\delta_i}]$$

$$\sigma_{yy}^h - i\sigma_{xy}^h = G^I \sum_{i=1}^{\infty} [2A_I^i z^{\delta_i} + 2\bar{A}_I^i \bar{z}^{\bar{\delta}_i} + 2z\bar{A}_I^i \bar{\delta}_i \bar{z}^{\bar{\delta}_i - 1} + 2\bar{B}_I^i \bar{z}^{\bar{\delta}_i}$$

$$+ 2C_I^i z^{\delta_i} + 2\bar{C}_I^i \bar{z}^{\delta_i} + 2z\bar{C}_I^i \delta_i \bar{z}^{\delta_i - 1} + 2\bar{D}_I^i \bar{z}^{\delta_i}] \quad (4\text{-}19)$$

$$u_x^h + iu_y^h = \sum_{i=1}^{\infty} \left[ \frac{\kappa^I}{\delta_i + 1} A_I^i z^{\delta_i + 1} - z\bar{A}_I^i \bar{z}^{\bar{\delta}_i} - \frac{1}{\bar{\delta}_i + 1} \bar{B}_I^i \bar{z}^{\bar{\delta}_i + 1} \right.$$

$$\left. + \frac{\kappa^I}{\bar{\delta}_i + 1} C_I^i z^{\bar{\delta}_i + 1} - z\bar{C}_I^i \bar{z}^{\delta_i} - \frac{1}{\delta_i + 1} \bar{D}_I^i \bar{z}^{\delta_i + 1} \right]$$

$$(C_I^1 = D_I^1 = 0)$$

in which variables with a superscript $h$ represent the homogeneous solutions. The homogeneous solutions for stress and displacements are required to satisfy the traction-free boundary conditions along the free edge and the continuity conditions along the bimaterial interface. As illustrated in Fig. 4-2, the near-field boundary conditions along the free edge, $x = 0$, read

$$\sigma_{xx}^h + i\sigma_{xy}^h = 0 \qquad \begin{cases} z = -iy & \text{for } I = 1 \\ z = iy & \text{for } I = 2 \end{cases} \qquad (4\text{-}20)$$

The continuity conditions along the interface $y = 0$ require

$$[\sigma_{yy}^h - i\sigma_{xy}^h, u_x^h + iu_y^h] = 0 \qquad (4\text{-}21)$$

where the bracket represents the jump of the associated quantity across the interface, e.g.,

$$[f(r, 0)] = f(r, 0^+) - f(r, 0^-) \qquad (4\text{-}22)$$

For each $\delta_i$, the complex equations (4-20) and (4-21) consist of eight complex equations. For complex eigenvalues, $\delta_i$ ($i \geqslant 2$), each of the eight complex equations can be further split into two linear independent real equations according to $r^{\delta_{ir}} \cos[\ln(r)\delta_{ii}]$ and $r^{\delta_{ir}} \sin[\ln(r)\delta_{ii}]$, where $r^2 = z\bar{z}$ and $\delta_i = \delta_{ir} + i\delta_{ii}$. Thus, for complex eigenvalues, the eigenvalue problem consists of 16 real linear equations for 16 unknowns, namely, real and imaginary parts of $A_I^i$, $B_I^i$, $C_I^i$, and $D_I^i$ ($I = 1, 2$). For the real eigenvalue, $\delta_1$, only eight real equations exist because $\sin[\ln(r)\delta_{ii}] = 0$ for $\delta_{ii} = 0$. However, there are only eight unknown complex coefficients, namely, real and imaginary parts of $A_I^i$ and $B_I^i$ ($C_I^i = D_I^i = 0$, $i = 1, 2$). The existence of a nontrivial solution requires the coefficient determinant to vanish:

$$|\Delta(\delta_i)| = 0 \tag{4-23}$$

where $\Delta(\delta_i)$ is a $16 \times 16$ (or $8 \times 8$ for $\delta_1$) matrix inolving $\delta_i$ and $\bar{\delta}_i$ in transcendental form. The eigenproblem defined by Eq. (4-23) can usually be solved by any standard numerical scheme for eigenvalue problems, such as Muller's method.[21] It is worth mentioning that the first eigenvalue, $\delta_1$, which is real and bounded by

$$-1 < \delta_1 < 0 \tag{4-24}$$

characterizes the order of stress singularity at the free edge. In other words, $\delta_1$ is one of the two singular parameters mentioned in the first paragraph of Section 4.3.

Once the eigenvalues $\delta_i$ ($i = 1, 2, 3, \ldots$) are obtained, the eigenvectors can be obtained from the above 16 or 8 real equations. Therefore, the complete states of the free-edge stresses and displacements can be expressed explicitly as the sum of the particular solution and a series of eigenfunctions with undetermined coefficients $\beta_i^{(k)}$. $\beta_i^{(k)}$ ($k = 1$ for real $\delta_i$, $k = 1, 2$ for complex $\delta_i$) as their multipliers. It is found by numerical experiments that there is only *one* eigenvector for the real eigenvalue ($\delta_1$), but there are *two* eigenvectors associated with the complex eigenvalues ($\delta_2, \delta_3, \ldots$). The hybrid finite element theory, which will be discussed in Section 4.5, will be used to determine these coefficients $\beta_i^{(k)}$.


### 4.4.2 Particular Asymptotic Solution

Temperature distributions in a multilayered device resulting from the transient finite element analysis can be expressed as a polynomial within each element. For example, as illustrated in Fig. 4-2, when the regular four-node isoparametric elements are used in the transient heat transfer evaluation, temperature at any points, $(x, y)$, inside the hybrid element can be expressed (note that there is *no* singularity in temperature field) as

$$\theta = a_1^I + a_2^I x + a_3^I y + a_4^I xy \tag{4-25}$$

where $a_i^I$ ($I = 1, 2$; $i = 1\text{–}4$) are determined from nodal temperatures. The coefficients $a_i$ are different for the upper and lower parts of the special hybrid element because, as illustrated in Figs. 4-2 and 4-3, the special hybrid element area consists of two regular elements in the transient heat transfer analysis. With this polynomial temperature distribution, the particular solution, $\phi^I$, defined by Eq. (4-17) can be expressed as

$$\phi^I = \lambda^I \left[ \frac{a_1^I}{2} z\bar{z} + \frac{a_2^I}{8} (z^2\bar{z} + \bar{z}^2 z) + \frac{a_3^I}{8i} (\bar{z}z^2 - z\bar{z}^2) + \frac{a_4^I}{24i} (z^3\bar{z} - \bar{z}^3 z) \right] \quad (4\text{-}26)$$

Substituting Eq. (4-26) into Eq. (4-16), one can obtain the particular stress solutions, $\sigma_{ij}^p$, and particular displacement solutions, $u_{ij}^p$, as follows:

$$\sigma_{xx}^p + i\sigma_{xy}^p = G^I \sum_{i=1}^{3} \left[ 2A_I^{pi} z^{\delta_i^p} + 2\bar{A}_I^{pi} \bar{z}^{\bar{\delta}_i^p} - 2z\bar{A}_I^{pi} \bar{\delta}_i^p \bar{z}^{\bar{\delta}_i^p - 1} \right.$$

$$\left. - 2\bar{B}_I^{pi} \bar{z}^{\bar{\delta}_i^p} \right] + G^I \lambda^I (a_2^I z + ia_3^I z + ia_4^I z\bar{z} - 2\Theta)$$

$$\sigma_{yy}^p - i\sigma_{xy}^p = G^I \sum_{i=1}^{3} \left[ 2A_I^{pi} z^{\delta_i^{pi}} + 2\bar{A}_I^{pi} \bar{z}^{\bar{\delta}_i^p} + 2z\bar{A}_I^{pi} \bar{\delta}_i^p \bar{z}^{\bar{\delta}_i^p - 1} \right.$$

$$\left. + 2\bar{B}_I^{pi} \bar{z}^{\bar{\delta}_i^p} \right] - G^I \lambda^I (a_2^I z + ia_3^I z + ia_4^I z\bar{z} + 2\Theta) \quad (4\text{-}27)$$

$$u_x^p + iu_y^p = \sum_{i=1}^{3} \left[ \frac{\kappa^I}{\delta_i^p + 1} A_I^{pi} z^{\delta_i^p + 1} - z\bar{A}_I^{pi} \bar{z}^{\bar{\delta}_i^p} - \frac{1}{\bar{\delta}_i^p + 1} \bar{B}_I^{pi} \bar{z}^{\bar{\delta}_i^p + 1} \right]$$

$$+ \lambda^I \left[ a_1^I z + \frac{a_2^I}{4} (2z\bar{z} + z^2) - \frac{ia_3^I}{4} (z^2 - 2z\bar{z}) \right.$$

$$\left. - \frac{ia_4^I}{12} (z^3 - 3z\bar{z}^2) \right]$$

where

$$\delta_1^p = 0, \qquad \delta_2^p = 1, \qquad \delta_3^p = 2 \quad (4\text{-}28)$$

and $A_I^{pi}$ and $B_I^{pi}$ are determined by requiring that the particular stress solutions and the particular displacement solutions satisfy the traction-free and continuity boundary conditions defined by Eqs. (4-20) and (4-21). The superscripts $p$ are used to identify the particular solutions.


## 4.5 FORMULATION OF HYBRID SINGULAR ELEMENT

### 4.5.1 Formulation

This section will outline the finite element formulation for the special hybrid elements to be placed at the edges of material interfaces. As mentioned in

Section 4.4, formulation of the special hybrid element starts from the hybrid functional $\pi$ defined in Eq. (4-14). Using the stress and displacement expressions described in Section 4.4, one can transform the hybrid functional into a more convenient matrix form

$$\pi = \int_{\partial A_s} \{\mathbf{T}^\mathrm{T}\tilde{\mathbf{u}} - \tfrac{1}{2}\mathbf{T}^\mathrm{T}\mathbf{u}\}\, dS - \frac{1}{2}\int_{A_s} \boldsymbol{\varepsilon}^0 \boldsymbol{\sigma}\, dA \qquad (4\text{-}29)$$

where $A_s$ is area of the singular element, $\partial A_s$ is boundary of $A_s$, $\tilde{\mathbf{u}}$ is element boundary displacements matrix, $\boldsymbol{\varepsilon}^0$ is initial (thermal) strain matrix, $\boldsymbol{\sigma}$, $\mathbf{T}$, and $\mathbf{u}$ are the matrix forms of element stresses, boundary tractions and interior displacements, respectively. $(\ )^\mathrm{T}$ denotes transpose of $(\ )$. Since only a single element is considered here, the upper limit of the first summation in Eq. (4-14) is simply 1 while the second summation in the same equation vanishes altogether. The superscript $e$ and subscript $s$ are used to identify the special element. As discussed in Section 4.4, $\boldsymbol{\sigma}$, $\mathbf{T}$, and $\mathbf{u}$ are summations of the homogeneous solution and the particular solution. These quantities can be expressed in terms of the undetermined coefficient vector $\boldsymbol{\beta}$ as

$$\boldsymbol{\sigma} = \mathbf{P}\boldsymbol{\beta} + \boldsymbol{\sigma}_p$$

$$\mathbf{T} = \mathbf{R}\boldsymbol{\beta} + \mathbf{T}_p \qquad (4\text{-}30)$$

$$\mathbf{u} = \mathbf{U}\boldsymbol{\beta} + \mathbf{u}_p$$

where $\mathbf{P}$, $\mathbf{R}$, and $\mathbf{U}$ are matrices containing the homogeneous solutions of stresses, tractions, and interior displacements, respectively; and $\boldsymbol{\sigma}_p$, $\mathbf{T}_p$, and $\mathbf{u}_p$ are matrices containing the particular solutions.

Compatibility of the special singular element with the adjacent regular elements is achieved by expressing the element boundary displacement $\tilde{\mathbf{u}}$ in terms of element nodal displacements $\mathbf{q}$; namely,

$$\tilde{\mathbf{u}} = \mathbf{L}\mathbf{q} \qquad (4\text{-}31)$$

where $\mathbf{L}$ is the interpolation function matrix defined on $\partial A_s$. The interpolation function is chosen such that when the corresponding nodal displacements of the adjacent elements are matched, $\tilde{\mathbf{u}}$ is the same for the two elements over their common boundaries.

Substitution of Eqs. (4-30) and (4-31) into Eq. (4-29) yields

$$\pi_s^e = \boldsymbol{\beta}^\mathrm{T}\mathbf{G}\mathbf{q} - \tfrac{1}{2}\boldsymbol{\beta}^\mathrm{T}\mathbf{H}\boldsymbol{\beta} + \boldsymbol{\beta}^\mathrm{T}\mathbf{F}_\beta + \mathbf{F}_q^\mathrm{T}\mathbf{q} - \frac{1}{2}\int_{\partial A_s}\mathbf{T}_p^\mathrm{T}\mathbf{u}_p\, dS - \frac{1}{2}\int_{A_s}\boldsymbol{\varepsilon}^0\boldsymbol{\sigma}_p\, dA$$

$$(4\text{-}32)$$

where

$$G = \int_{\partial A_s} \mathbf{R}^\mathrm{T} \mathbf{L} \, dS$$

$$H = \frac{1}{2} \int_{\partial A_s} (\mathbf{R}^\mathrm{T} \mathbf{U} + \mathbf{U}^\mathrm{T} \mathbf{R}) \, dS$$

$$\mathbf{F}_\beta = -\frac{1}{2} \int_{\partial A_s} (\mathbf{R}^\mathrm{T} \mathbf{u}_p + \mathbf{U}^\mathrm{T} \mathbf{T}_p) \, dS - \frac{1}{2} \int_{A_s} \mathbf{P}^\mathrm{T} \varepsilon^0 \, dA \tag{4-33}$$

$$\mathbf{F}_q = \int_{\partial A_s} \mathbf{L}^\mathrm{T} \mathbf{T}_p \, dS$$

Since the $\boldsymbol{\beta}$'s can be assumed independently from those of the surrounding elements, one may take a first variation of the functional $\pi_s^e$ with respect to $\boldsymbol{\beta}$ and obtain

$$\mathbf{H}\boldsymbol{\beta} = \mathbf{G}\mathbf{q} + \mathbf{F}_\beta \tag{4-34}$$

$\mathbf{H}$ is always positive definite because $\beta^\mathrm{T} H \beta$ is a quantity corresponding to strain energy. Therefore, Eq. (4-34) can be rewritten as

$$\boldsymbol{\beta} = \mathbf{H}^{-1}(\mathbf{G}\mathbf{q} + \mathbf{F}_\beta) \tag{4-35}$$

Substituting Eq. (4-35) into Eq. (4-32) and taking another first variation of $\pi_s^e$ with respect to $\mathbf{q}$, one obtains the following finite element equations

$$\mathbf{k}\mathbf{q} = \mathbf{F} \tag{4-36}$$

where

$$\mathbf{K} = \mathbf{G}^\mathrm{T} \mathbf{H}^{-1} \mathbf{G}$$

$$\mathbf{F} = -\mathbf{G}^\mathrm{T} \mathbf{H}^{-1} \mathbf{F}_\beta - \mathbf{F}_q \tag{4-37}$$

are the element stiffness matrix and force vector for the hybrid singular elements, respectively. Once the stiffness matrix and the force vector for the special hybrid element are formulated, they can be assembled together with the other regular elements in the finite element analysis.

### 4.5.2 Verification of the Special Hybrid Element

To check the accuracy and efficiency of the special hybrid element, two steady-state problems were analyzed using the hybrid finite element method. Finite element results were then compared with reference solutions.

**Double-Layer Steady-State Problem**

Figure 4-4 shows the geometry, material properties, and loading condition of the first check problem. The two materials illustrated in this figure are molybdenum and aluminum, respectively, which have been widely used in electronics packaging. The problem is assumed to be under a plane stress condition. Kuo[6] has studied this problem by an "alternating analytical method," by which the final solution is expressed as a series of two analytical solutions.

In the hybrid element formulation described above, eigenfunction truncation, i.e., the number of the $\beta$ terms used in the eigenfunction expansion in Eq. (4-30), needs to be studied first. Figure 4-5 shows the generalized stress intensity factors $K_{yy}$ vs. the number of eigenfunctions included in the numerical calculation. Near the free edge, the stresses have an asymptotic form $\sigma_{ij} = K_{ij}(r/h)^{\delta_1}$ where $r$ is the radial distance from the edge, and $h$ is a characteristic length of the structure, which is the layer thickness (2.54 mm) in this case. A hybrid element with $l_1 = 0.5$ mm and $l_2 = 1$ mm, where $l_1$ and $l_2$ are width and half-height of the element, respectively, was used in the finite element mesh in this convergence study. From Fig. 4-5, it is found that, the finite element solution converges if more than three eigenfunctions are included.

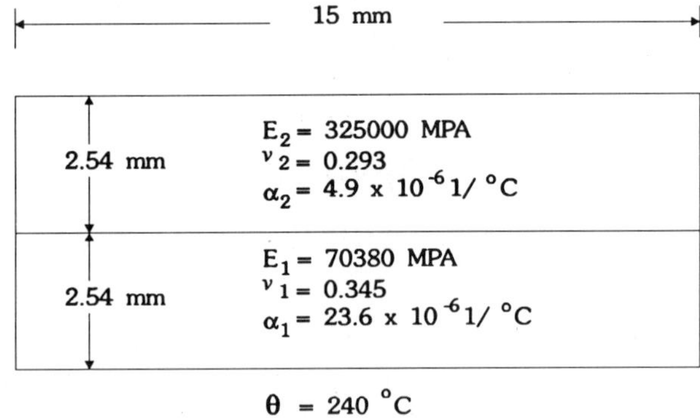


**Figure 4-4**   A two-layer composite strip.

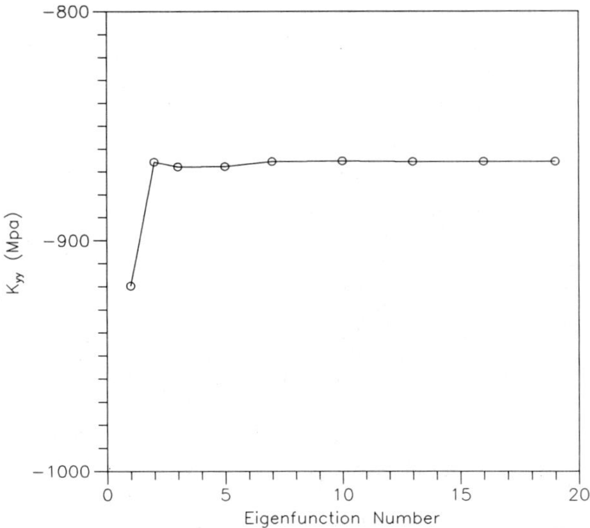

**Figure 4-5**    Effects of eigenfunction truncation on $K_{yy}$.

Effects of size and aspect ratio of the special element on the accuracy and convergence of the hybrid finite element solution have also been examined. Table 4-1 tabulates comparison of the generalized stress intensity factors with reference solutions at various element sizes and aspect ratios.

The results in the table indicate that the finite element solutions are insensitive to the size of the element. However, it appears that the optimal aspect ratio for the special hybrid element is approximately a square shape. In the remainder of the study, 10 eigenfunctions and square shape ($l_1/l_2 = 1$)

**Table 4-1y**

| $l_1$ (mm) | $l_1/l_2$ | $K_{yy}$ (MPa) | $K_{xy}$ (MPa) | $K_{xx}^+$ (MPa) | $K_{xx}^-$ (MPa) | $\delta_1$ |
|---|---|---|---|---|---|---|
| 0.25 | 1.0 | −867.4 | 135.2 | −170.3 | 346.1 | −0.113 |
| 0.5 | 1.0 | −865.9 | 135.0 | −170.0 | 345.7 | −0.113 |
| 1.0 | 1.0 | −863.9 | 134.6 | −169.6 | 344.8 | −0.113 |
| 1.5 | 1.0 | −864.1 | 134.6 | −169.7 | 344.8 | −0.113 |
| 0.5 | 2.0 | −878.9 | 137.0 | −172.6 | 350.7 | −0.113 |
| 0.5 | 0.5 | −847.4 | 132.1 | −166.3 | 338.2 | −0.113 |
| *Kuo's solution* | | −850.0 | 132.0 | −166.2 | 337.5 | −0.113 |

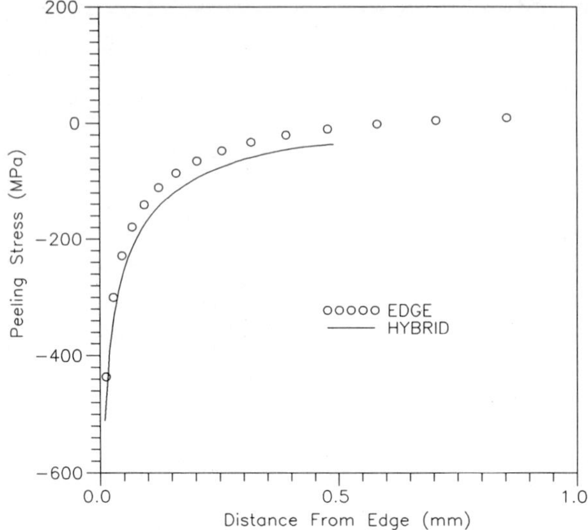


**Figure 4-6**    Peeling stress $(\sigma_{yy})$ of the two-layer problem.

elements are always used unless stated otherwise. Figures 4-6 and 4-7 show peeling and shear stresses along the interface. It is seen from these figures that the hybrid finite element solutions are in good agreement with solutions by a different method, the alternating analytical solutions.[6] It is worth mentioning that the solutions shown in Figs. 4-6 and 4-7 and Table 4-1 are under the plane stress assumption. Both Suhir's[3] and Kuo's[6] original solutions are not correct because they are under a mixture of plane strain and plane stress conditions. The reference solutions depicted in these two figures and the table have been recalculated according to Kuo's formulation[6] under the plane stress assumption.

The finite element mesh for this check problem is illustrated in Fig. 4-8. It is seen that with the hybrid finite element formulation the finite element mesh can be very coarse and yet the numerical solutions are still very accurate. The hybrid finite element method seems to be a very efficient and effective solution scheme for the free-edge problem for multilayered devices. Coarse mesh and accurate results do not usually come together in the conventional finite element method.

### Three-Layer Steady-State Problem

To assess the ability of the hybrid finite element method in handling multiple singular edges, a three-layer device subjected to a uniform temperature variation $(\theta = 240°C)$ is considered as the second check problem. To

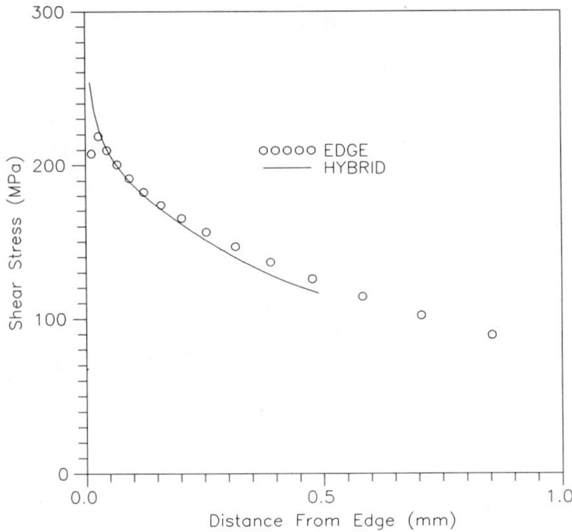



**Figure 4-7**   Shear stress $(\sigma_{xy})$ of the two-layer problem.

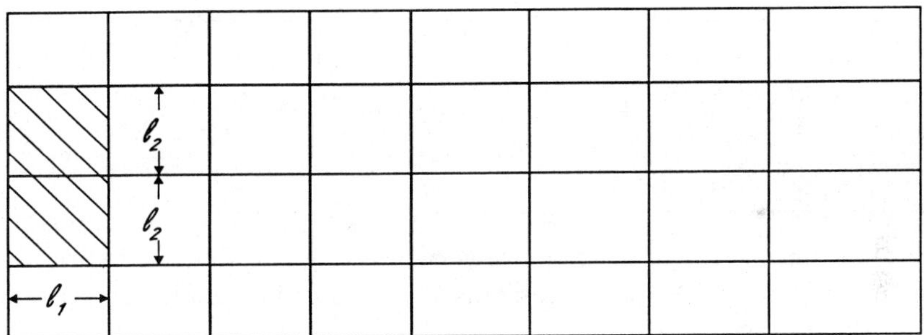



**Figure 4-8**   Hybrid finite element mesh for $l_1 = 1.5$ mm and $l_1/l_2 = 1$.

cover the whole spectrum, the problem is assumed to be under a plane strain condition. The thickness for each of the three layers is 2 mm. Table 4-2 shows material properties of the three layers.

Since there are two material interfaces in this problem, two hybrid elements are required in the finite element modeling, one at the edge between the top and the middle layers and the other at the edge between the middle and the bottom layers. Finite element meshes for the second check problem are very similar to the one shown in Fig. 4-8 except for the addition of two more layers of elements to the model to account for the additional third material layer. Comparisons of the resulting generalized stress intensity factors $K_{ij}$ with a reference solution[7] are summarized in Table 4-3. Again, in this table,

**Table 4-2**

|              | $E$ (MPa) | $\nu$ | $\alpha$ $(10^{-6}/°C)$ |
| ------------ | --------- | ----- | ----------------------- |
| Upper layer  | 20 660    | 0.28  | 3.2                     |
| Middle layer | 13 000    | 0.30  | 11.7                    |
| Lower layer  | 68 950    | 0.33  | 23.6                    |

**Table 4-3**

|                   | $K_{yy}$ (MPa) | $K_{xy}$ (MPa) | $K_{xx}^{+}$ (MPa) | $K_{xx}^{-}$ (MPa) | $\delta_1$ (MPa) |
| ----------------- | -------------- | -------------- | ------------------ | ------------------ | ---------------- |
| *Upper interface* |                |                |                    |                    |                  |
| Hybrid FEM        | $-84.9$        | 15.0           | $-18.2$            | 43.6               | $-0.155$         |
| Alternating       | $-78.7$        | 13.9           | $-16.9$            | 40.4               | $-0.155$         |
| *Lower interface* |                |                |                    |                    |                  |
| Hybrid FEM        | 128.9          | 18.1           | $-46.7$            | 21.8               | $-0.109$         |
| Alternating       | 128.6          | 18.0           | $-46.6$            | 21.8               | $-0.109$         |

the asymptotic stress field near the edge takes the form $\sigma_{ij} = K_{ij}(r/h)^{\delta_1}$. Distributions of the interfacial stresses along the two material interfaces are shown in Figs. 4-9 and 4-10.

From Table 4-3 and Figs. 4-9 and 4-10, it is seen that the solutions from the hybrid finite element method are very accurate and that the hybrid finite element method can handle problems with multiple singular stress points efficiently and accurately.


## 4.6  GREEN'S FUNCTION INTEGRATION METHOD

Thermal response of a structure subjected to a unit step thermal loading is usually called the Green's function $G(x_i, t)$.[22] The Green's function can be a temperature, displacement, stress, stress intensity factor, or any other physical quantity associated with the heat transfer/thermal stress problem. For instance, the Green's function $G(x_i, t)$ could be the transient thermal stress response at location $x_i$ of a bimaterial strip, as illustrated in Fig. 4-4, due to a unit step ambient temperature change at time zero. It has been shown by Kyo et al.[22] that thermal responses due to arbitrary thermal transients can be calculated by a convolution integration of the Green's function. For instance, for the two-layer example shown in Fig. 4-4, the

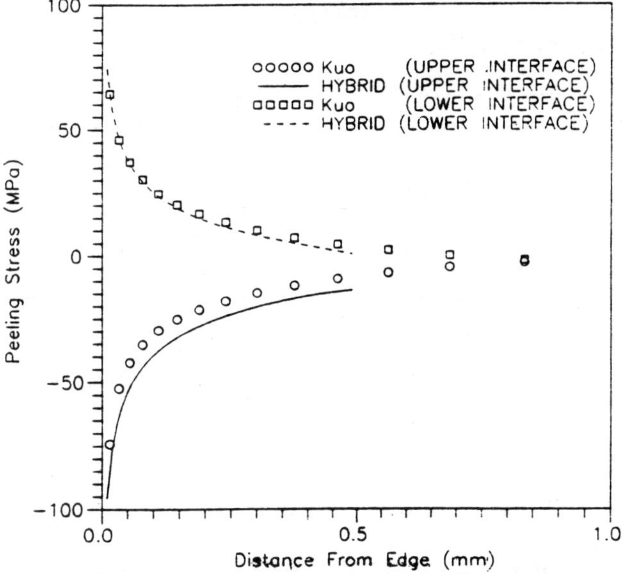


**Figure 4-9**    Peeling stress ($\sigma_{yy}$) of the three-layer problem.

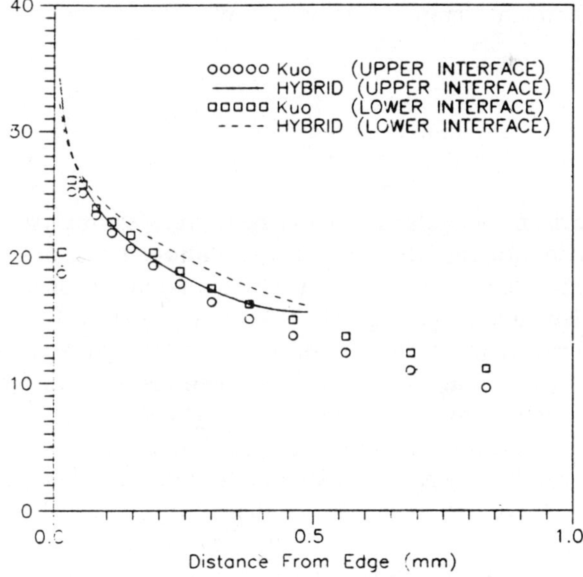


**Figure 4-10**    Shear stress ($\sigma_{xy}$) of the three-layer problem.

**Table 4-4**

| | $\rho$<br>$(10^{-3}\ \text{g/mm}^3)$ | $C_d$<br>$(\text{J/g}^\circ\text{C})$ | $k$<br>$(\text{J/mm-}^\circ\text{C-sec})$ |
|---|---|---|---|
| Upper layer | 0.0027 | 2.763 | 0.1424 |
| Lower layer | 0.0102 | 8.625 | 0.2010 |

transient responses resulting from an arbitrary ambient temperature variation $\theta(t)$ can be obtained by

$$F(x_i, t) = \int_0^t G(x_i, t - \tau)\, d\theta(\tau) \tag{4-38}$$

This technique is also known as the Duhamel integral method in structural dynamic theory.

To demonstrate the Green's function method, the same two-layer device shown in Fig. 4-4 was chosen as the third example. The geometry and material properties have already been shown in Fig. 4-4 and the additional heat transfer properties are shown in Table 4-4. Again, these two materials are molybdenum and aluminum, respectively. The heat transfer coefficient between the ambient and the boundary is $0.002\,8391\ \text{J/mm}^2\text{-sec-}^\circ\text{C}$.

First, the ambient temperature is assumed to undergo a $240^\circ\text{C}$ step change, i.e.,

$$\Theta(t) = \begin{cases} 0^\circ C & t \leqslant 0^- \\ 240^\circ C & t > 0^+ \end{cases} \tag{4-39}$$

Because there is no singularity in the heat transfer analysis, no singular element is needed for the heat transfer calculation. Thus, only regular heat transfer elements were used in the transient temperature calculation. But the two regular elements at the edge of the material interface were replaced by a special hybrid element in the subsequent thermal stress calculation. Temperature histories at three interface locations, (0 mm, 0 mm), (1 mm, 0 mm), and (2 mm, 0 mm) are depicted in Fig. 4-11. Vertical ($y$-direction) displacements at locations (1 mm, 0 mm) and (2 mm, 0 mm) are shown in Fig. 4-12. The resulting generalized stress intensity factor histories are shown in Fig. 4-13. It is seen from this figure that the generalized stress intensity factors reach their peak values at about 1 second and start to decay and change their signs after the peaks. This is because, in the first few seconds, temperature of the upper layer is much higher than that of the lower layer

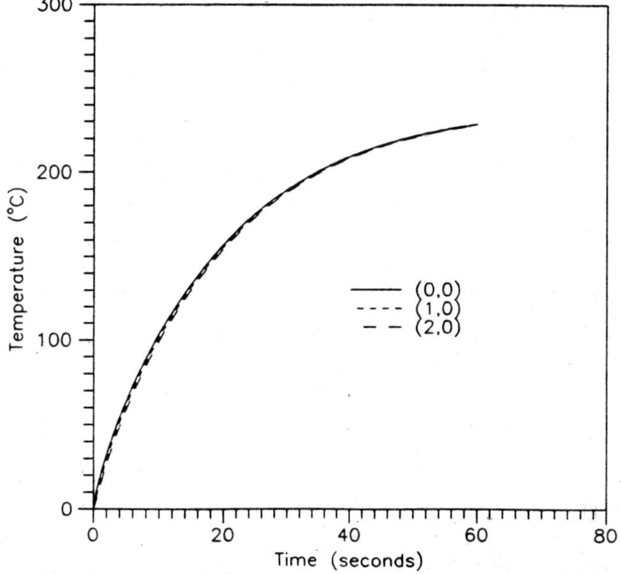


**Figure 4-11**    Temperature history at various locations.

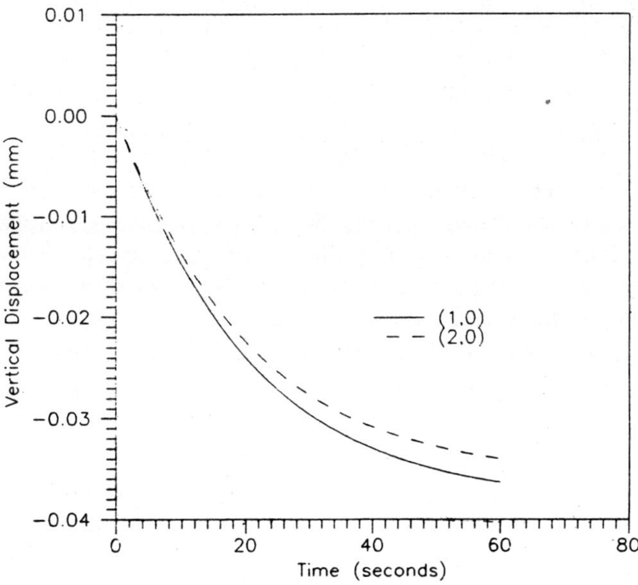


**Figure 4-12**    Displacement history at various locations.

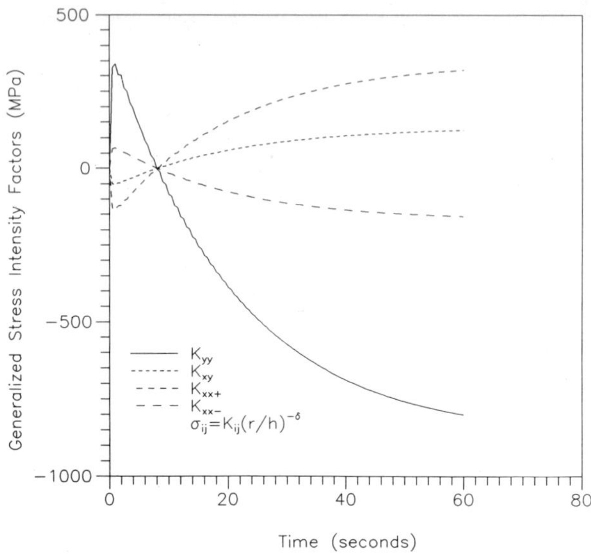



**Figure 4-13**   Histories of generalized stress intensity factors.

due to its higher thermal diffusivity (19.09 mm$^2$/sec versus 0.23 mm$^2$/sec). The higher upper layer temperature would make the upper layer try to expand more than the lower layer. On the other hand, the lower layer has a higher coefficient of thermal expansion (CTE) (4.9 × 10$^{-6}$/°C versus 23.6 × 10$^{-6}$/°C), which will make the lower layer expand more than the upper layer. As illustrated in Fig. 4-13, the first factor (thermal diffusivity) is dominant during the first 8 seconds, while the second factor (CTE) is in control afterwards. The confluence point at 8 seconds represents a zero-stress intensity factor state where the upper and the lower layers expand an almost equal amount. The Green's functions for this problem are simply the transient responses shown in Figs. 4-11 through 4-13 multiplying a scale factor of 1/240. The scale factor 1/240 is necessary because, instead of a unit step ambient temperature change, a 240°C temperature jump was assumed in the finite element calculation. As illustrated in Figs. 4-11 to 4-13, the Green's function responses will usually approach their steady-state values after a decay time, $t_d$. Therefore, the Green's function integration defined in Eq. (4-38) can be further shortened to

$$F(x_i, t) = \int_{t-t_d}^{t} \bar{G}(x_i, t - \tau)\, d\theta(\tau) + G_0(x_i)\theta(t) \tag{4-40}$$

where $\bar{G}(x_i, t) = G(x_i, t) - G_0(x_i)$, and $G_0$ is the steady-state value of the Green's function $G(x_i, t)$. This shortened Green's function integration is

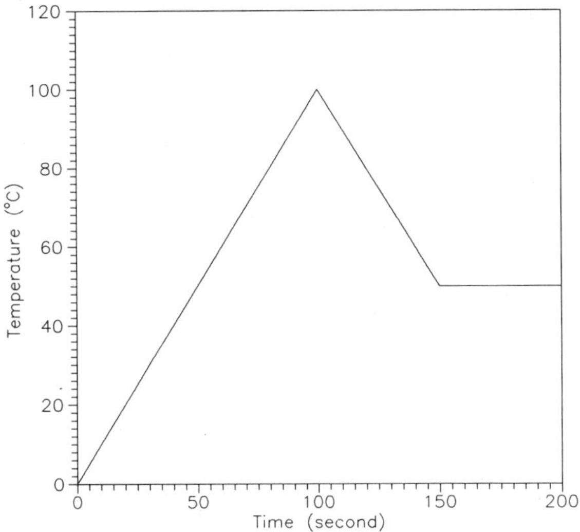

**Figure 4-14**    An arbitrary ambient temperature change.

extremely useful when evaluating a length thermal transient, because, instead of integrating from time 0 to $t$, we need only integrate from $t - t_d$ to $t$.

Next, let us assume that an ambient temperature variation as shown in Fig. 4-14 is the thermal transient of interest. Instead of running another straightforward finite element analysis to account for this particular ambient temperature variation, we can calculate the new transient responses by the Green's function integration. Comparisons of the generalized stress intensity factor solutions via the Green's function integration method and the straightforward finite element analysis are shown in Fig. 4-15. It is seen from this figure that the transient responses given by the two different approaches are essentially identical.

The Green's function integration method is even more attractive when, for the same electronic device, there is more than one thermal transient to be considered in the evaluation. Applying the Green's function method, only one finite element analysis for a unit step thermal transient is required and responses for all the other thermal transients are simply the integrations of the Green's functions.

## 4.7 TRANSIENT BEHAVIORS OF MULTILAYERED DEVICES

Up to now most of the thermal stress evaluations of multilayered devices have been based on steady-state solutions. However, it has been pointed out

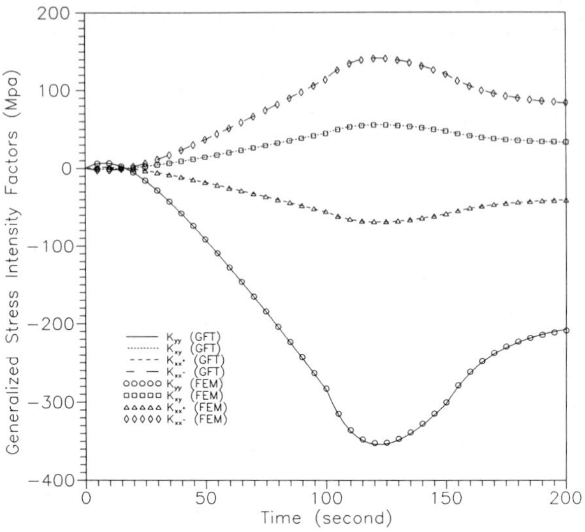



**Figure 4-15**  Comparisons of FEM and Green's function solutions.

by Chen and Kuo[8,16] that estimated stress ranges based on the steady-state solutions may be more than one order of magnitude less than what they should be, if the thermal loads are transient. As illustrated in Fig. 4-16, the authors have found that, for a two-layer strip as shown in Fig. 4-4 under a step ambient temperature change, the generalized stress intensity factors at the edge exhibit three distinctly different patterns in their transient responses. The transient $K$ could (1) increase monotonically to the steady-state value ($K_{ss}/K_{peak} = 1$), (2) increase to reach a peak value of the same sign and decay to the steady-state value ($0 < K_{ss}/K_{peak} < 1$), or (3) reach a peak value of the opposite sign and decay to the steady-state value ($K_{ss}/K_{peak} < 0$). A real example of the third transient pattern has been shown in Fig. 4-13 in one of the earlier example problems. For the two-layer strip, whether its transient behavior is of the first, second, or third pattern depends on the combination of material properties and the heat transfer coefficient. If the combination of material properties and heat transfer coefficient happens to yield a transient behavior of the second or the third transient pattern, the $\Delta K$ range determined based on the steady-state solutions may be off by a huge factor. To further understand this phenomenon, a parametric study has been conducted by Kuo and Chen[8] and their results are shown in Figs. 4-17 and 4-18. In these two figures, contours of the $K_{ss}/K_{peak}$ ratios are plotted on a plane with relative diffusivity and relative coefficient of thermal expansion as the horizontal and vertical axes, respectively. The relative diffusivity is defined as $\mu_{upper}/\mu_{lower}$ while the relative coefficient of thermal expansion is

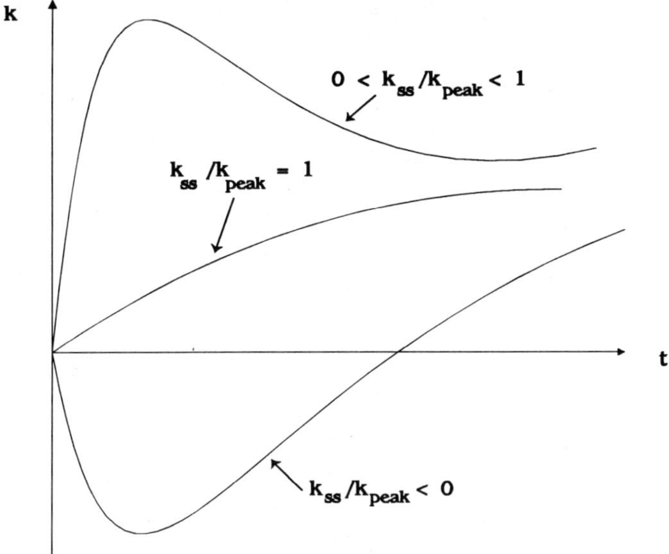


**Figure 4-16**    Three transient thermal stress patterns.

defined as $\alpha_{upper}/\alpha_{lower}$, where $\mu = K/\rho C$ is the thermal diffusivity and $\alpha$ is the coefficient of thermal expansion of the material. The heat transfer coefficient $h$ between the composite layer and the ambient is 0.002 839 1 J/mm$^2$-°C-sec for the contours shown in Fig. 4-17 and 0.011 356 4 J/mm$^2$-°C-sec for the contours shown in Fig. 4-18. On these two contour plots, points corresponding to material combinations of Al/Si, Cu/Al, Al/solder, Cu/solder, and BeO/solder have also been marked for reference purposes. From these two figures, the following conclusions have been drawn by Kuo and Chen.[8]

1. Estimation of $\Delta K$ based on the steady-state solution is valid only for a limited range of material property combinations. In the contour maps shown in Figs. 4-17 and 4-18, this is valid only in the small region bounded by the upper right-hand corner of the map and the contour line corresponding to $K_{ss}/K_{peak} = 0.99$. For most engineering applications, the calculation of $\Delta K$ from the steady-state solutions could be off by a factor of more than 10. This could have significant impact on the structural integrity evaluation of the multilayered device when $\Delta K$ is used in conjunction with design curves like the ones illustrated in Figs. 4-19 and 4-20.

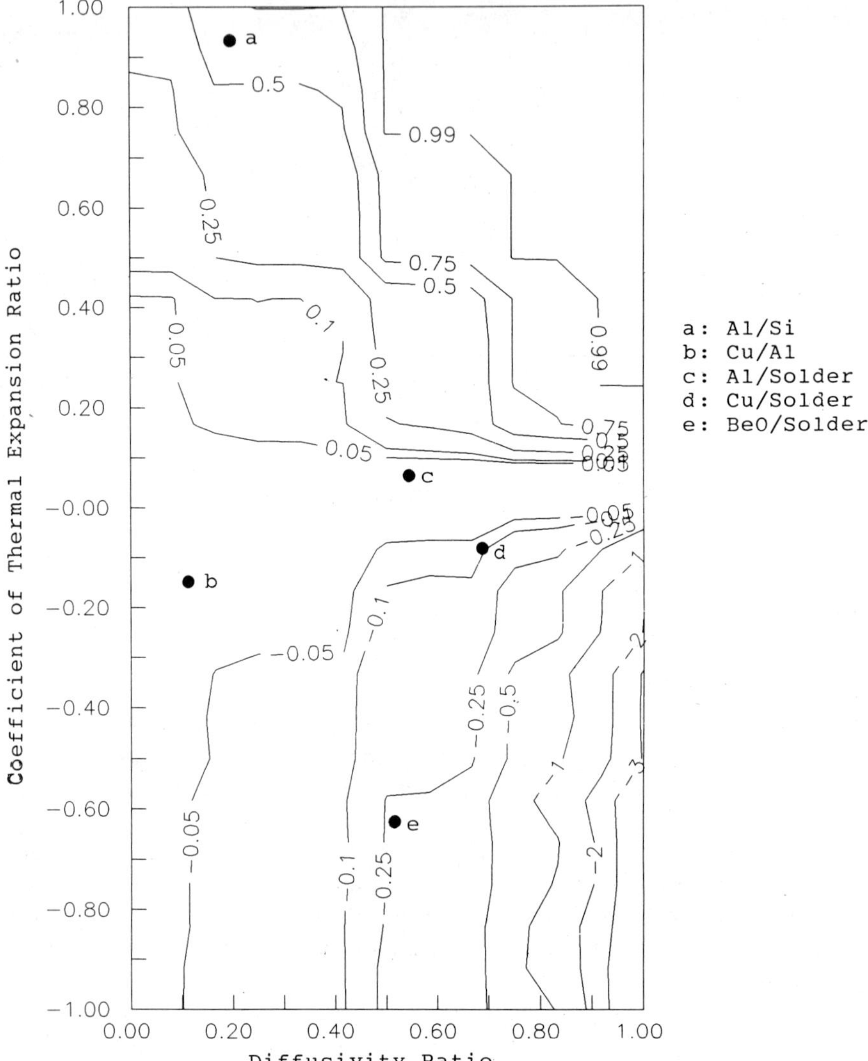


**Figure 4-17**   Transient amplification factors ($h = 0.002\,839\,1$ J/mm$^2$-°C-sec).

2. The transient behavior is more pronounced when the heat transfer coefficient between the device and the ambient is higher. This can be easily deduced from the two contour maps.

3. For many material combinations, the peak thermal stress is actually of the opposite sign to the steady-state stress. Material combinations falling in the lower halves of the maps shown in Figs. 4-17 and 4-18 are in this class of materials.

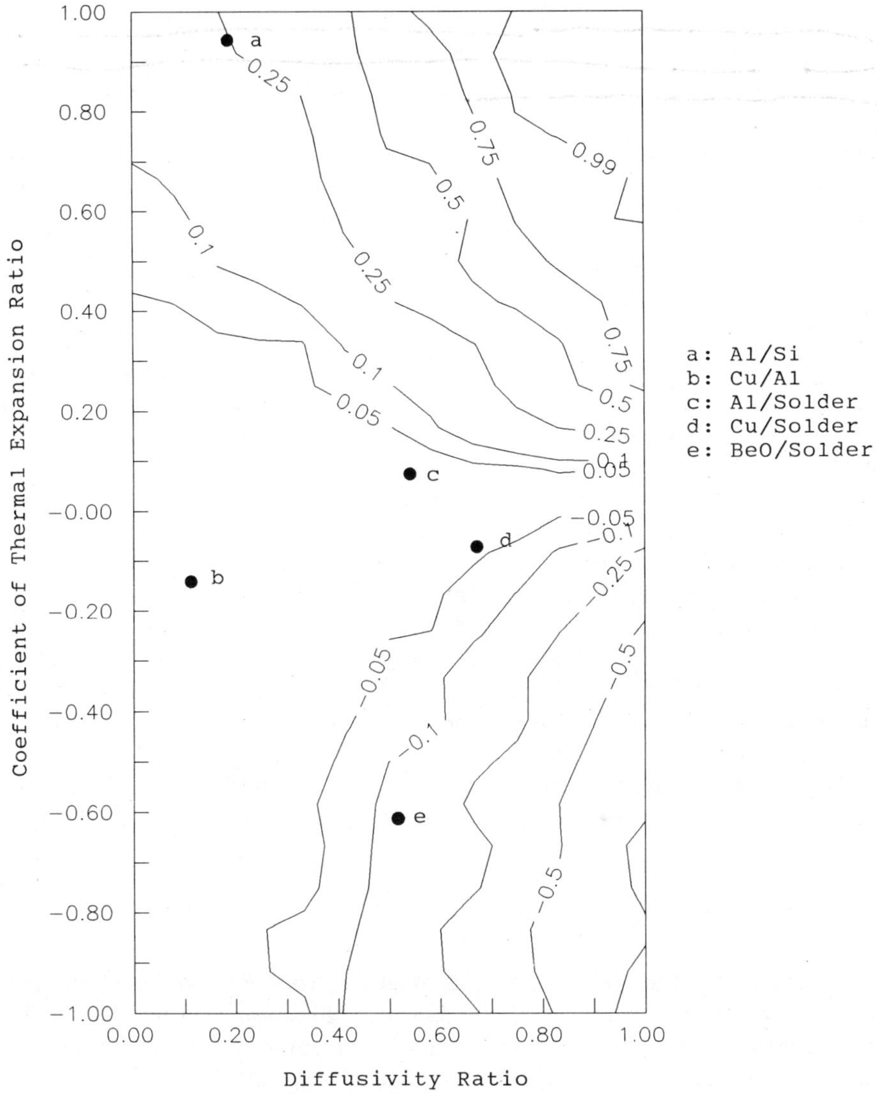


**Figure 4-18** Transient amplification factors ($h = 0.011\,356\,4$ J/mm$^2$-°C-sec).

## 4.8 DESIGN BASED ON TRANSIENT THERMAL STRESSES

A multilayered electronic device is usually expected to endure numerous thermal cycles during its design life. For the multilayered electronic device, the material interfaces at the free edge are likely to be one of the most critical locations from the point of view of component integrity and reliability. Under the linear elastic assumption, stresses near the free edge are found to be singular. In reality, materials near the edge will yield and the stresses at the edge will be very high but bounded. However, if the sizes of the local yielding (plastic zone size) near the edge are very small (small-scale yielding), the linear elastic solution is still a good approximation for material points outside the yield region and the two singular parameters $K$ and $\delta$ of the linear elastic solution can be used as design parameters.

### 4.8.1  Crack Initiation

To ensure the structural integrity of the multilayered device, the first item to check is crack initiation at the material interfaces. Hattori, Sakata, and Murakami[11] have shown that crack initiation of multilayered devices is controlled by the two singular parameters $K$ and $\delta$, where the asymptotic stress distribution near the edge is $\sigma = Kr^{\delta}$. These authors have also shown by experiment that crack-initiation can be assessed by checking against a characteristic design curve from experiment. A schematic of this characteristic design curve for crack initiation is illustrated in Fig. 4-19. With the transient thermal stress analysis methodology described in this chapter, the $\Delta K$ and $\delta$ values can easily be calculated for all transients. Then the resulting $\Delta K$ can be checked against $\Delta K_{\text{allow}}$, which is determined from $\delta$ and the design curve shown in Fig. 4-19. The crack initiation criterion is met if the transient $\Delta K$ is less than the allowable value $\Delta K_{\text{allow}}$.

The crack initiation evaluation is analogous to checking a component's strength to ensure component integrity after the first loading cycle.

### 4.8.2  Thermal Fatigue

Since the thermal loading is cyclic in most electronic applications, the next logical item to check is thermal fatigue. Although it has not been fully verified by experiment, it is quite possible that, for an interface edge between two given materials, the thermal fatigue life of the interface edge can be characterized by an equivalent $S$–$N$ curve as illustrated in Fig. 4-20. It is worth noting that although thermal fatigue curves are usually described by the Coffin–Manson equation in terms of strain, the equivalent $S$–$N$ curve shown in Fig. 4-20 is actually in terms of the generalized stress intensity factors. This is because, under the small-scale yielding assumption, plastic strains inside the very small plastic zone are functions of the generalized stress intensity factors. With the numbers of cycles ($n_i$, $i = 1, 2, 3, \ldots$) for a variety of different thermal transients (transients $1, 2, 3, \ldots$) and with the $\Delta K_i$ ($i = 1, 2, 3, \ldots$) calculated with the methodologies discussed above, the

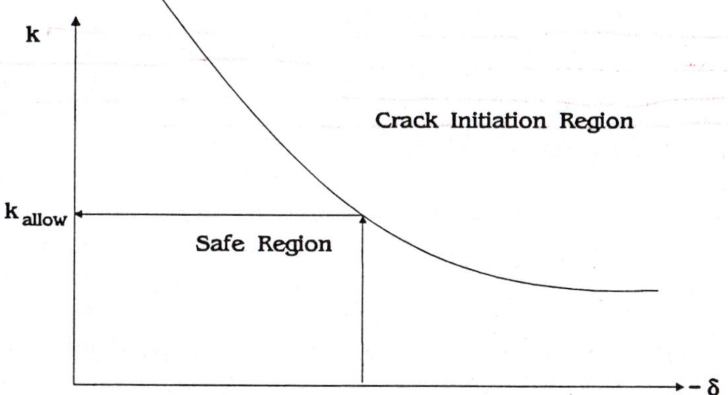



**Figure 4-19**   Two-parameter material interface crack initiation diagram.

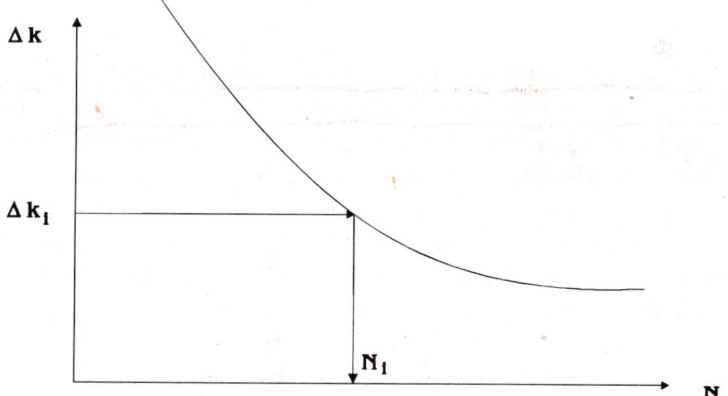



**Figure 4-20**   Resistant curve of thermal fatigue.

thermal fatigue criterion is met if

$$\left(\frac{n_1}{N_1} + \frac{n_2}{N_2} + \cdots + \frac{n_i}{N_i} + \cdots\right) \leqslant 1 \qquad (4\text{-}41)$$

where $N_i$ is the allowable number of cycles determined from the fatigue curve shown in Fig. 4-20.

## 4.9 DISCUSSION AND SUMMARY

It should be noted again that the heat treansfer, thermal stress, and fatigue/fracture evaluations discussed in this chapter are restricted to linear elastic materials only. In the electronic packaging applications, the linear elastic assumption is not always true. For instance, substantial amounts of plasticity and creep deformation have been found in transistor stacks with solder joints after a few cycles of thermal loading.[23]

To sum up, a hybrid finite element method has been successfully developed for calculating the transient thermal stresses in multilayered devices. Formulation of the special hybrid element is based on the variational principle of a modified hybrid functional. The element stresses and displacements are represented by eigenfunction solutions derived from Muskhelishvili's complex stress function theory for isotropic elasticity. The stress singularities at the free edge are included exactly in the special hybrid element along with the higher-order terms. Interelement compatibility is achieved by using independently assumed displacements along the interelement boundaries through properly selected interpolation functions. The special hybrid element can be used with the regular, isoparametric elements. The hybrid finite element method has been found to be a very accurate and efficient method for calculating transient thermal stress in multilayered devices.

The behavior of the transient thermal stresses in multilayered devices has been found to be quite different from the steady-state solutions. Under rapid thermal transients, structural evaluations based on the steady-state thermal stresses could be off by a huge factor. Depending on heat transfer coefficients and material property combinations, the transient thermal stresses could have peak stresses of different signs from the steady-state value. Applications of the transient thermal stress solutions to design against thermal crack initiation and thermal fatigue have been discussed in this chapter.

## ACKNOWLEDGMENTS

Part of this effort is supported by an in-house grant from Structural Integrity Associates, Inc., San Jose, California. The encouragement of Dr. Peter C. Riccardella of Structural Integrity Associates during the course of the study is gratefully acknowledged. The authors would also like to thank Dr. John Lau of Hewlett-Packard Company for his enthusiastic support and helpful suggestions.

## REFERENCES

1. Timoshenko, S. P., "Analysis of Bi-Metal Thermostats," *J. Optical Society of America*, **11**, 1925, pp. 233–255.
2. Chen, W. T., and C. W. Nelson, "Thermal Stress in Bonded Joints," *IBM Journal of Research and Development*, **23**, no. 2, 1979, pp. 178–188.

3. Suhir, E., "Stresses in Bi-Metal Thermostats," *J. Applied Mechanics*, **53**, 1986, pp. 657–660.

4. Eischen, J. W., C. Chung, and J. H. Kim, "Realistic Modeling of Edge Effect Stresses in Bimaterial Elements," *J. Electronic Packaging*, **112**, 1990, pp. 16–23.

5. Yin, W. L., "Thermal Stresses and Free-Edge Effects in Laminated Beams: A Variational Approach Using Stress Functions," *J. Electronic Packaging*, **113**, 1991, pp. 68–75.

6. Kuo, A. Y., "Thermal Stresses at the Edge of a Bimetallic Thermostat," *J. Applied Mechanics*, **56**, 1989, pp. 585–589.

7. Kuo, A. Y., and K. L. Chen, "Thermal Stresses at the Edge of a Multilayered Composite Strip," presented at the ASME Winter Annual Meeting, Atlanta, GA, December 1–6, 1991.

8. Kuo, A. Y., and K. L. Chen, "Transient Thermal Stress in a Multilayered Device by the Hybrid Finite Element Method," to be submitted to *Int. J. Solids and Structures*.

9. Blanchard, J. P., and N. M. Ghoniem, "An Eigenfunction Approach to Singular Thermal Stresses in Bonded Strip," *J. Thermal Stresses*, **12**, 1989, pp. 501–527.

10. Glaser, J. C., "Thermal Stresses in Compliantly Jointed Materials," *J. Electronic Packaging*, **112**, 1990, pp. 24–29.

11. Hattori, T., S. Sakata, and G. Murakami, "A Stress Singularity Parameter Approach for Evaluating the Interfacial Reliability of Plastic Encapsulated LSI Devices," *J. Electronic Packaging*, **111**, 1989, pp. 234–248.

12. Wang, Y. R., and T. W. Chou, "3-D Analysis of Transient Interlaminar Thermal Stress of Cross-Ply Composites," *ICCM-6*, 1987, **4**, 383–393.

13. Wang, Y. R., and T. W. Chou, "3-D Analysis of Transient Interlaminar Thermal Stress of Laminated Composites," Symposium on Mechanics of Composite Materials, *ASME AMD*, **92**, 1988, pp. 185–192.

14. Wang, Y. R., and T. W. Chou, "Three-Dimensional Transient Interlaminar Thermal Stresses in Angle-Ply Composites," *J. Applied Mechanics*, **56**, 1989, pp. 601–608.

15. Wang, Y. R., R. B. Pipes, and T. W. Chou, "Thermal Transient Stresses due to Rapid Cooling in Thermally and Elastically Orthotropic Medium," *Metallurgical Transactions*, **17A**, 1986, pp. 1051–1055.

16. Chen, K. L., and A. Y. Kuo, "Edge Stresses of a Multi-Layered Device under Transient Thermal Loading," presented at ASME Winter Annual Meeting, Atlanta, GA, December, 1–6, 1991.

17. Tong, P., T. H. H. Pian, and S. J. Lasry, "A Hybrid-Element Approach to Crack Problems in Plane Elasticity," *Int. J. Numerical Methods in Engineering*, **7**, 1973, pp. 297–308.

18. Wang, S. S., and F. G. Yuan, "A Hybrid Finite Element Approach to Laminate Elasticity Problems with Stress Singularities," *J. Applied Mechanics*, **50**, 1983, pp. 835–844.

19. Muskhelishvili, N. I., *Some Basic Problems of Mathematical Theory of Elasticity*, Nordhoff, Groningen, 1953.

20. Parkus, H., *Thermoelasticity*, Blaisdell, New York, 1968.

21. Muller, D. E., "A Method for Solving Algrebraic Equations using an Automatic Computer," *Mathematical Tables and Computations*, **10**, 1956, pp. 208–215.

22. Kuo, A. Y., S. S. Tang, and P. C. Riccardella, "An On-Line Fatigue Monitoring System for Power Plants: Part I—Direct Calculation of Transient Peak Stress Through Transfer Matrices and Green's Functions," *Design and Analysis Methods for Plant Life Assessment*, ed. T. V. Narayanan and S. Palusamy, ASME, 1986, pp. 25–32.
23. Pao, Y.H., K. L. Chen, and A. Y. Kuo, "A Nonlinear and Time Dependent Finite Element Analysis of Solder Joints in Surface Mounted Components under Thermal Cycling," project final report, Ford Motor Company, Dearborn, Michigan, 1991.

# 5

# Temperature Dependence of Thermal Expansion of Materials for Electronics Packages

*R. Chanchani and Peter M. Hall*

## 5.1 INTRODUCTION

One severe limitation on the reliability of modern electronic circuits is the stress caused by thermal expansion. Consequently, temperature cycling and temperature shock are standard methods of evaluating new products and processes. To understand these stresses and evaluate the tests, it is important to know the expansivity (sometimes called CTE, or coefficient of thermal expansion) of each material involved. In particular, finite element thermal stress analysis studies need the best data available. Values for the expansivities of many of the materials are available,[1] but such tables usually provide only one value for the expansivity. The temperature dependence of the expansivity is often ignored. In addition, the reported values are often averaged over a substantial temperature range, such as 25°C to 300°C. Most of the data in the literature are derived from lattice parameter measurements made by X-ray diffraction, or "$z$"-direction measurements made by dilatometers. Lattice parameter measurements are valid for single-phase materials. Dilatometer measurements depend on the vertical ("$z$"-direction) displacement of a low-expansivity piston resting on a sample of material. Dilatometer measurements may include an error if the sample "bows" when heated. In-plane measurements can be made by standing the piece on end, but then verticality must be assured and maintained, which is not easy.

This chapter presents strain gage measurements of the temperature dependence of $x$–$y$ (in-plane) expansivity for several materials used in microelectronics.[2] It is intended to help show in any given case whether the temperature dependence is truly negligible. We define the expansivity, $\alpha$, as $(1/L)\, dL/dT$, or $d \ln L/dT$. This makes it a point function of temperature, so one can speak of the expansivity at one given temperature. A table of the temperature-averaged expansivity between two temperatures is fine if the two temperatures happen to be the only ones of interest, but that is seldom the case.

## 5.2  THEORY

The general reason materials expand with increasing temperature is that when the atomic vibrations (phonons) increase in amplitude, the minimum energy occurs at greater atomic spacings. This concept is embodied in the Gruneisen relation,[3,4] which states that the expansivity of a Debye solid is proportional to the specific heat, and the constant of proportionality is practically independent of temperature. Thus (all other things being equal) a plot of expansivity as a function of absolute temperature should resemble the Debye function (Fig. 5-1). There is a rounded "knee" in the curve, such that, above the knee, the expansivity rises slowly with temperature. Below the knee it drops off markedly. The knee occurs at an absolute temperature of about $0.5\Theta$ where $\Theta$ is the Debye temperature for the material. For some

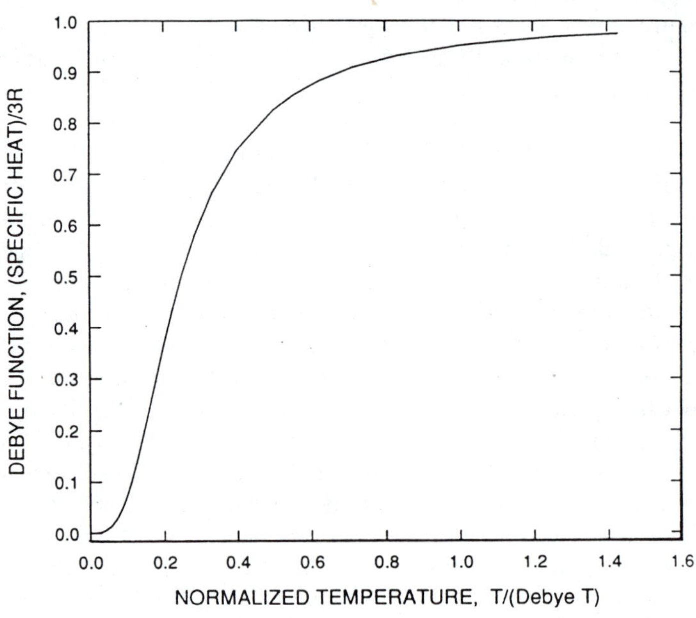


**Figure 5-1**   Debye function, expected temperature dependence of expansivity.

materials, such as copper, the knee is below the lowest temperature normally encountered in microelectronic testing. Correspondingly, the expansivity of copper changes by only 14% over the range $-50°C$ to $150°C$. For silicon[5] the knee occurs at $+50°C$. For $Al_2O_3$[6] it occurs at $+90°$. For aluminum nitride[5] it is at $+200°C$. For these materials, the temperature dependence of expansivity may become important.

The expansivity can decrease with increasing temperature if there is a phase change resulting in higher density. Examples of such materials are certain ferrous alloys, e.g., Alloy 42, Kovar™ (trademark of Carpenter Technology), Invar, and Dumet. They contain about 40% Ni, some minor additives, and the balance Fe. They are at or near a sharply defined minimum in the expansivity[7,8] when plotted as a function of Ni content. The expansivity can be made as low as 1 p.p.m./K. This comes about because of a phase change from b.c.c. to f.c.c. as the temperature is raised. In pure iron, this is the well-known ferritic–austenitic transformation,[9] and it occurs suddenly near 890°C. Over a temperature increase of less than 20 K, the length decreases by 3000 p.p.m. For most nickel-iron alloys, the same transition is still evident. In these alloys, however, the transition occurs well below 890°C, and the transition is spread over a large temperature range. If the length decrease of 3000 p.p.m. were spread over a range of 300 K, it would decrease the expansivity in that region by 10 p.p.m./K. Then the expansivity that otherwise might be about 12 p.p.m./K would be reduced to 2 p.p.m./K. This is roughly what happens. Thus the "normal" increase in length with increasing temperature can be almost completely counteracted by the "phase change" decrease in length with increasing temperature. Of course this has tremendous practical implications, and these materials have been used for 70 years for many applications. It is not always realized, however, that both above and below the transition range the expansivity of these materials is again similar to that of pure iron above and below its transition, namely 18 and 11 p.p.m./K.[7] The location of the minimum in expansivity, the value at the minimum, and the width of the minimum depend on the amount of the major and minor constituents, as well as the thermal treatment and stress state of the sample. For these materials, the temperature dependence of the expansivity is especially important. If the alloy is formulated to match the expansivity of a classical material such as alumina, it may match well for a certain temperature range but be badly mismatched at higher and lower temperatures.

Another important class of materials used in microelectronics is the ceramics family. We have already seen that alumina has a temperature-dependent expansivity. Recent developments in this field have spawned a whole series of new ceramic materials, with desirable expansivity and dielectric constant. These new materials will have different expansivity ranges. In addition, their expansivities may depend on the temperature as well as the composition, the density, their fabrication technique, and the amount of metallization contained in or on them. We have measured several of these materials, and report the results in this chapter.

Another class of materials widely used in microelectronics is commonly called printed wiring boards. These are normally made of organic resins, reinforced with fiberglass fabrics, usually containing many layers. The resins have naturally large expansivities, of the order of 60 p.p.m./K, but the glass fabrics restrict the expansivities in the plane. The result, of course, is that the $z$-axis expansivity becomes even larger, due to the constraining effect of the fabric. This puts large stresses on plated through-holes. The overall expansivity in the plane can be made as low as 10 p.p.m./K, but typically is closer to 17 p.p.m./K, to match that of the copper printed wiring that must be placed on it. Two important considerations must be mentioned concerning these materials. One is that the dimensions of the resin are a function of the moisture content. Thus, if a board is heated above 100°C for a few hours, it will contract by 30 or 40 p.p.m. as the moisture is driven out. Upon returning to room temperature, its dimensions will be smaller than they were initially, until the moisture diffuses back into the resin, a process which can take a day or two. The other important effect is the glass transition of the resin. Most common resins go through a transition in the range of 120–140°C, during which there are remarkable changes in the mechanical properties. The elastic modulus decreases markedly, and so does the expansivity. Since the resin is normally in a state of compression due to the fabric, decrease in expansivity can lead to irreversible changes in dimension and subsequent warping of the board. This is especially true if some or all of the resin layers have not been completely cured. Any additional curing can raise the glass transition temperature. In this chapter, we include some measurements of FR-4 and polyimide–quartz samples, but more work needs to be done in this area.

## 5.3 EXPERIMENTAL

These measurements were made using commercial metal-foil–polyimide strain gages (CEA-06-062UW, from Micromeasurements, Inc.). In each case, a "rosette" was adhesively attached to both sides of a sheet of the material to be measured, using AE-15 two-part epoxy adhesive from Micromeasurements. The epoxy was cured for 45 minutes at 90°C. Each rosette consisted of two gages at 90° to each other. The strain was thus recorded for both $x$ and $y$ directions on both faces of the sample. The bowing can be found from the difference between the top and bottom faces. In all cases, the four gages agreed within experimental error, so they were averaged and the average results are reported here. Use of four gages adds statistical significance to the results. Each lot of gages was calibrated by recording the strain readings of two of the rosettes mounted on a single crystal of silicon as a function of temperature in the same chamber and using the same thermocouples and methods as those used for the unknowns. Silicon is an ideal reference material because it is easily obtained with unprecedented purity and crystalline perfection. In addition, although this is secondary, its expansivity is low, only 2.5 p.p.m./K at room temperature,[5] so the calibration corrections are small. The raw data were also corrected for the temperature dependence of the gage factor and for the transverse sensitivity, both of which are small corrections.[10]

The thermal expansion for the Si reference standards, $\varepsilon_{SI}$, was taken from the literature[5] using the following formula with temperature, $T$, in degrees Celsius:

$$\varepsilon_{SI} = -59.29 + 2.368T + 0.004\,541T^2 - 1.208 \times 10^{-5}T^3 + 3.78 \times 10^{-8}T^4$$

$$(5\text{-}1)$$

where $\varepsilon_{SI}$ is in units of p.p.m. (or microstrain).

The constant term in this equation has been adjusted so that it gives $\varepsilon_{SI}$ as zero at 24°C, which is where strain gages are normally normalized. This equation agrees well with the data of Slack and Bartram[5] (Fig. 5-2), who reviewed and synthesized Si results from 25 different literature references. This equation should not be used above 140°C nor below $-140$°C, as it diverges rapidly from the data beyond that range. Silicon is unusual in that its expansivity becomes negative. It is negative for all temperatures below $-160$°C, but never more negative than $-1$ p.p.m./K.

A calibration point was determined at temperature intervals of 5 K for each lot of gages, and each raw strain reading was processed according to an interpolated calibration for the temperature recorded by the thermocouple attached to that sample at the time.

The sizes of the samples are given in Table 5-1. Each sample was cycled between $-37$°C and $+143$°C for three complete cycles (Fig. 5-3). The ramp rate was 1 K/min, so each ramp took 3 hours. The dwell time at the extremes was 1 hour. The gage readings were recorded every 5 minutes, or approximately every 5 degrees. The temperature controller was programmed to control the air temperature, and the sample temperature lagged somewhat behind. The amount of lag was normally less than 5 K, but it depended on the thermal mass and thermal conductivity of the sample, so we attached a thermocouple to both faces of each sample. The thermocouple temperature was used to process the strain data and to create the plot of Fig. 5-3.

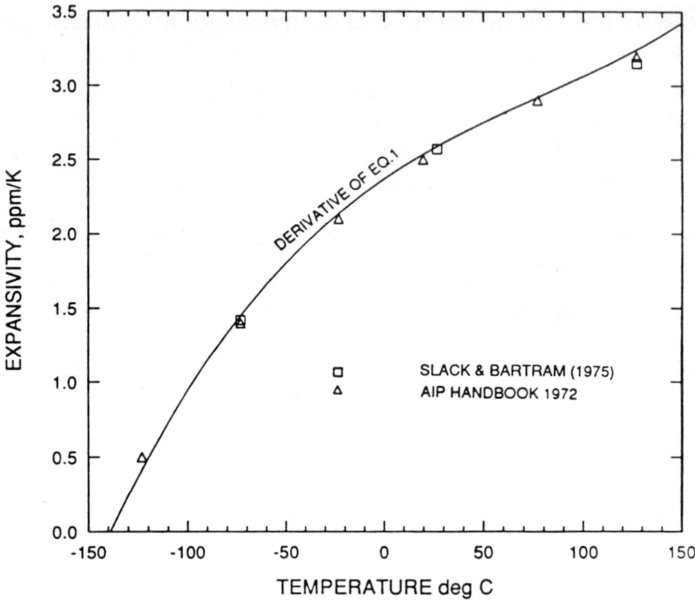



**Figure 5-2**  Expansivity of silicon.

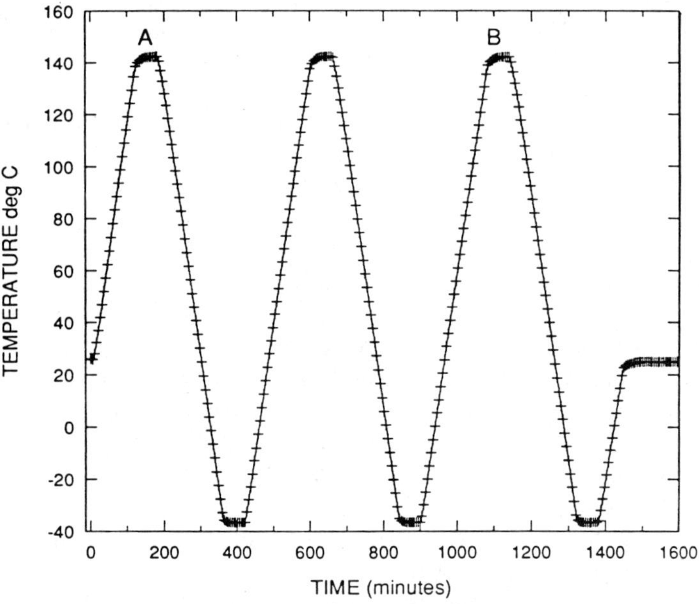



**Figure 5-3**  Temperature cycling.

**Table 5-1**    Dimensions of the Samples

| Material | Dimensions (cm) | | |
| --- | --- | --- | --- |
| | Length | Width | Thickness |
| Aluminas | | | |
|   92% tape-cast | 5.59 | 9.14 | 0.127 |
|   96% pressed | 11.43 | 9.53 | 0.069 |
|   99.5% tape-cast | 11.43 | 9.53 | 0.069 |
| Low-Temperature co-fired ceramics | | | |
|   (Green Tape™) | | | |
|   High-$\kappa$ | 3.81–4.47 | 4.47–6.10 | 0.030–0.173 |
|   Low-$\kappa$ | 4.47 | 4.47 | 0.102 |
| Aluminum nitride | 3.05 | 1.52 | 0.063 |
| Beryllium oxide | 7.62 | 3.43 | 0.07 |
| Invar | 5.08 | 5.08 | 0.061 |
| Kovar | 4.42 | 1.78 | 0.701 |
| Alloy 42 | 2.54 | 1.27 | 0.693 |
| Molybdenum | 4.57 | 3.81 | 0.058 |
| Cu–Invar–Cu | 5.08 | 5.08 | 0.102 |
| Cu–Mo–Cu | 5.08 | 3.81 | 0.076 |
| FR-4 | 11.0 | 5.0 | 0.2 |
| Polyimide–quartz | 6.0 | 2.5 | 0.25 |

The expansivity at any given point is defined as $(1/L)(dL/dT)$, where $L$ is a length and $T$ is the temperature. Our experiments provide discrete values of strain, so we cannot obtain a true derivative. We created an approximation to the derivative at any given measurement point by obtaining a linear regression of the strain vs. temperature using the reading at that time along with the readings from the three immediately preceding and the three immediately succeeding measurements. Thus, the expansivity reported here is a least-squares value of the slope over a range of 30 K, and any anomalous behavior over shorter temperature ranges has been partially smoothed out.

The expansivity values determined in this way are interpolated to obtain a value for every even 5°C. For our final values, we discarded the data of the first ramp up in temperature (i.e., up to point A in Fig. 5-3), because there is some minor additional curing of the gage adhesive, and the result is noticeably different from that of later cycles. We then averaged the data over two complete cycles (between points A and B in Fig. 5-3), and for those samples that showed no hysteresis we averaged the increasing $T$ and decreasing $T$ values for expansivity. This gave us four readings on each gage at each temperature.

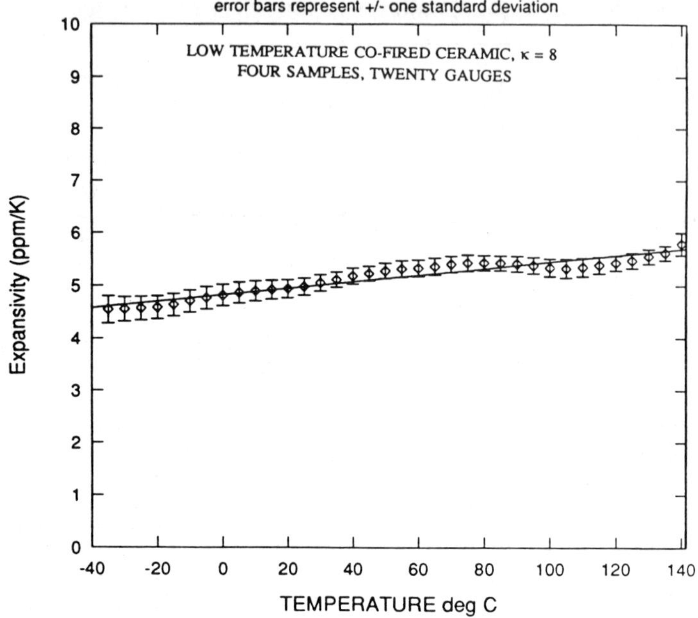


**Figure 5-4**   Reproducibility of expansivity measurements.

Figure 5-4 gives an idea of the reproducibility of these measurements. Results from four different, but equivalent, samples of low-temperature co-fired ceramic (involving 20 gages) were concatenated. The "error bars" indicate one standard deviation, and include variation due to sample-to-sample differences, gage-to-gage differences, and measurement reproducibility. All the error bar ranges shown in Fig. 5-4 are within $\pm 0.25$ p.p.m./K. The accuracy of the silicon standard is estimated by Slack and Bartram[5] as $\pm 15\%$, or slightly less than 0.4 p.p.m./K. Since all these results are based on the same calibration, however, this error does not affect their relative values. The apparent dip in the curve above 100°C appears in many samples and is not a "real" effect. Moreover, the value of the dip is within experimental error.

## 5.4 RESULTS AND DISCUSSION

### 5.4.1 Ceramics

*Alumina*   96% and 99.5% aluminas are widely used as substrate materials for thick- and thin-film hybrid integrated circuits. These substrates are made by pressing or tape-casting fine alumina powders. The defects and the

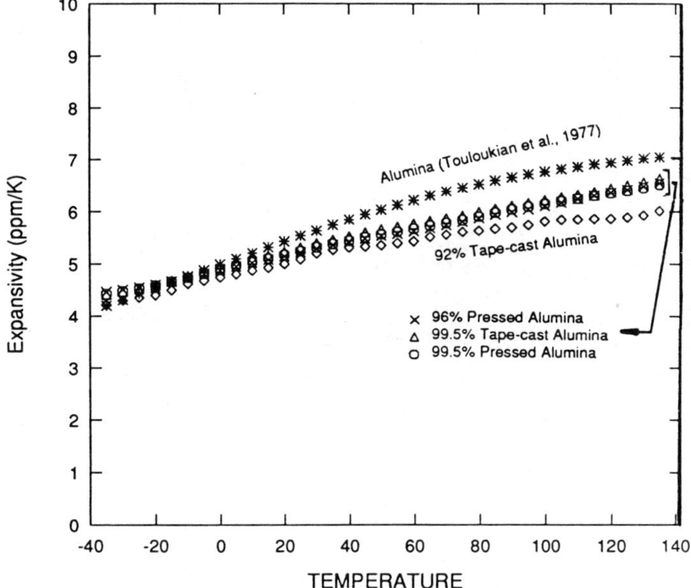



**Figure 5-5**   Expansivity of aluminas.

density of the sintered substrate can depend on the fabrication technique. Tape-cast 90–92% aluminas are used for multilayer ceramic packages for electronic and photonic devices. Multilayer aluminas are co-fired with metal conductors at high temperatures, above 1500°C, in a reducing atmosphere. Thus, these ceramic packages are also referred to as high-temperature co-fired MLCs (multilayer ceramics). Figure 5-5 shows the expansivities of commercially available aluminas used for substrates and packages: (1) 99.5% tape-cast alumina for thin-film circuits, (2) 99.5% pressed alumina for thin-film circuits, (3) 96% pressed alumina for thick-film circuits, (4) a 92% tape-cast multilayer alumina package, and (5) pure alumina as reviewed by Touloukian.[6] The expansivities increase as temperature rises. There is no significant difference in expansivity of pressed and tape-cast aluminas. There is little alumina purity dependence except for the 92% alumina sample at higher temperatures. This may be due to the presence of low thermal expansivity silica in 92% alumina samples.

***Low-Temperature Co-fired Ceramics***   In recent years, new glass ceramic-based dielectric materials have become available. These materials could be either crystallizable glasses, also called glass-ceramics, or glasses with ceramic fillers.[11] These materials are co-fired with high-conductivity metals (Au, Ag, Pd–Ag, Cu) at 850–950°C in air. The high-$\kappa$ ($\kappa = 8$) Green Tape™

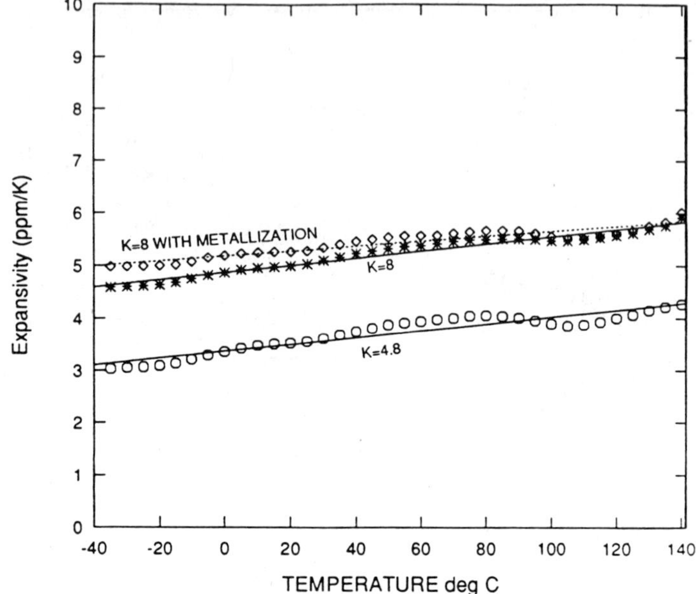


**Figure 5-6**  Expansivity of low temperature co-fired ceramics.

(Trademark of E. I. du Pont de Nemours & Co., Inc., Wilmington, DE) used in this study is made up of proprietary glassy phase and refractory filler materials. The low-$\kappa$ ($\kappa = 4.8$) Green Tape is a glass-ceramic. The constituents in the Green Tape can be tailored to obtain desirable dielectric constant and expansivity. Figure 5-6 shows the expansivity of fired Green Tape materials. The expansivity of high-$\kappa$ Green Tape matches more closely that of alumina (Fig. 5-5). In the temperature range of $-40$ to 140°C the expansivity varies from 4.5 to 6.0 p.p.m./K. The vendor's specification[11] reports the average over 25°C to 300°C to be 7.9 p.p.m./K. Thus, for room-temperature calculations, the literature value would be too high. The expansivity of low-$\kappa$ Green Tape varies from 3.1 to 4.2 p.p.m./K in the temperature range $-40$°C to 140°C.

We also studied the effect of physical design and process parameters on the expansivity of these materials. An eight-layer sample of high-$\kappa$ Green Tape was prepared with six Au metallization layers: three layers each of 36% and 64% coverage. Figure 5-6 shows that only a very small increase in expansivity of this composite was observed. There was no significant effect of $x$, $y$, and thickness dimensions on expansivity. Similarly, we saw no effect on expansivity of two post-sintering firings at 850°C for 10 minutes.

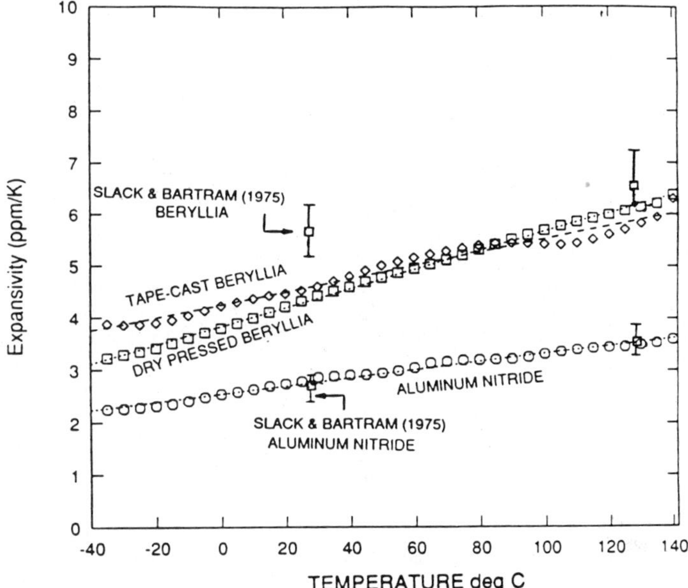


**Figure 5-7**    Expansivity of aluminum nitride and beryllium oxide.

***Aluminum Nitride and Beryllium Oxide***    Aluminum nitride and beryllia are currently being explored[12] for package, substrate, and heat sink applications because of their high thermal conductivity and expansivity match with silicon. Commercially available 99.5% tape-cast aluminum nitride and 99.5% tape-cast and dry-pressed beryllia substrates were used in this study. The expansivity (Fig. 5-7) varies from 2.3 to 3.6 p.p.m./K for aluminum nitride and 3.8 to 6.0 p.p.m./K for beryllia in the temperature range $-35°C$ to $140°C$. The expansivity of aluminum nitride matches more closely that of silicon (Fig. 5-2), whereas the expansivity of beryllia is closer to that of alumina (Fig. 5-5). As observed in the case of alumina, no significant difference between the expansivities of tape-cast and dry-pressed beryllia was observed.

The expansivity data for aluminum nitride are consistent with the data of Slack and Bartram,[5] indicated by squares on Fig. 5-7. The error bars are based on their accuracy estimate. Their data for beryllia are a little higher than our results, perhaps because of differences in purity and technique.

### 5.4.2 Metals

Figure 5-8 shows our results for four metals commonly used in electronics packaging. As expected, Invar is the lowest, while alloy 42 and Kovar are

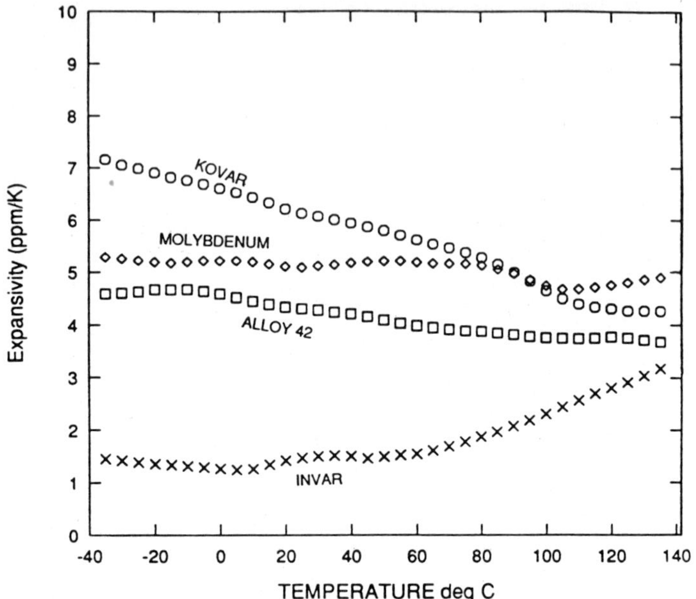


**Figure 5-8**   Expansivity of metals.

close to each other and to molybdenum. The Fe–Ni alloys are at or near the temperature at which their expansivity is a minimum, as discussed in Section 5.2, and they show a negative slope, at least for the lower temperature ranges. We observe that the Kovar and Alloy 42 match alumina well (Figs 5-5 and 5-8) at room temperature, but diverge as the temperature is lowered or raised. At some temperature well above the range of these measurements (around 400°C), the Kovar curve will again cross the alumina curve; for temperatures above 500°C, the expansivity of Kovar is well above that of alumina.[13] We show the expansivity of molybdenum as almost temperature-independent at 5.0 p.p.m./K. This is consistent with Slack and Bartram,[5] who show it as going from 4.8 p.p.m./K to 5.2 p.p.m./K over this temperature range. This is because, for molybdenum, $\Theta/2$ is $-46°C$, which is below the temperature range of these measurements. We include as Fig. 5-9 the NIST data[14] for pure copper. They are much higher than any of the other materials considered here, and they have very little temperature dependence, varying from 16.1 p.p.m./K to 17.7 p.p.m./K over this range of temperature.

### 5.4.3 Sandwiches (Cu–Invar–Cu and Cu–Mo–Cu)

Of special interest to this study are the composite structures made of two layers of copper separated by a layer of either Invar[15] or molybdenum.[1]

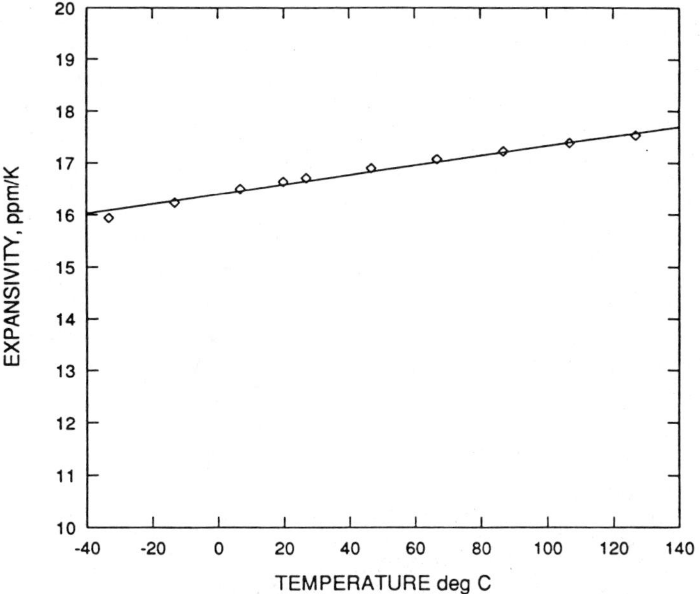



**Figure 5-9**   Expansivity of copper.

They are intended to provide electrical and thermal conductivity close to those of copper, along with an expansivity close to that of the ceramics. The center material restricts the naturally high expansivity of copper, 17 p.p.m./K at room temperature.[14] To avoid bowing of the composite, the structure is made symmetrical about its center plane, with an equal thickness of copper on the top and the bottom. If all the layers stay within their elastic limits, the overall expansivity $\alpha_{\text{ELASTIC}}$ of the composite can be expressed as[16,17]

$$\alpha_{\text{ELASTIC}} = \alpha_1 + \frac{\alpha_{\text{Cu}} - \alpha_1}{1 + A} \tag{5-2}$$

where

$$A = \frac{E_1 t_1 (1 - \mu_{\text{Cu}})}{E_{\text{Cu}} t_{\text{Cu}} t (1 - \mu_1)} \tag{5-3}$$

where the subscript 1 refers to the middle metal layer in the sandwich, and $E_1$ and $E_{\text{Cu}}$ are the elastic moduli, $\mu_1$ and $\mu_{\text{Cu}}$ are the Poisson's ratios, and $t_1$ and $t_{\text{Cu}}$ are the thicknesses of the two materials. (The copper thickness is the combined thickness of the two copper layers.) Note that if $A \gg 1$, material 1 dominates the expansivity. If $A \ll 1$, the copper dominates.

There is some question about the form of the Poisson's ratio dependence in (5-3). Some, e.g., Turner[18] use the factor $(1-2\,\mu)$ in place of $(1 - \mu)$. This is the same as replacing the elastic modulus with the bulk modulus. Others ignore it completely. In our case, as often, it makes little difference, as the three Poisson's ratios are almost identical: the Poisson's ratio of copper is 0.34,[13] that of molybdenum is 0.30,[13] and that of Invar is 0.29.[19]

Tummala and Rymaszewski[1] (eq. 13.3) present a similar equation to calculate this composite expansivity. Their equation, however, was derived for particles of one phase in a matrix of a second phase, so their equation is not applicable for layers. This is clear in the limit where the two materials have the same elastic modulus. Poisson's ratio, and volume fraction. The composite expansivity should then be the arithmetic average of $\alpha_1$ and $\alpha_2$, but their equation favors $\alpha_1$. Similarly, in the limit as $E_1$ goes to zero, the expansivity should be that of the copper. Their equation reduces to an average expansivity weighted by the volume fraction of each component.

When the temperature of such a composite is changed, the stress states of the two materials compensate for each other. That is, one undergoes tensile stress while the other is in compression. It is interesting when one of the materials is stressed beyond its elastic limit. In that case, its elastic modulus, $E$, in (5-3) must be replaced by $d\sigma/d\varepsilon$ where $\sigma$ is the stress and $\varepsilon$ is the strain. In a normal stress–strain curve this derivative starts out equal to the elastic modulus. After reaching the elastic limit, it steadily decreases, approaching zero, at least for a ductile material. When the temperature change is continued in the same direction, eventually the composite expansivity takes on the value of the expansivity of the nonductile material. But whenever the direction of the temperature change is reversed, at least initially the materials are again within their elastic limits, and (5-3) applies with the elastic moduli. When such a sandwich is cycled over a large temperature range, the expansivity curve for increasing $T$ is quite different from that for decreasing $T$.

***Cu–Invar–Cu***   Figure 5-10 shows $\Delta L/L$ for a sample of Cu–Invar–Cu, where the Invar thickness is 60% of the total thickness, over the temperature range $-35°C$ to $140°C$. We include data for two complete temperature cycles to show the reproducibility of the data. This figure shows that whenever the temperature change is reversed, either at high or at low temperature, the initial slope (which is the expansivity) is high and constant (i.e., the curve is linear). This is the elastic region. As the temperature ramp proceeds, however, the slope decreases because the copper yields. The lowest it could go would be the expansivity of the Invar, which, as we have seen, is less than 3 p.p.m./K in this range. The result is a history-dependent expansivity for the composite structure, but it is always between the Invar expansivity and the elastic expansivity of the composite.

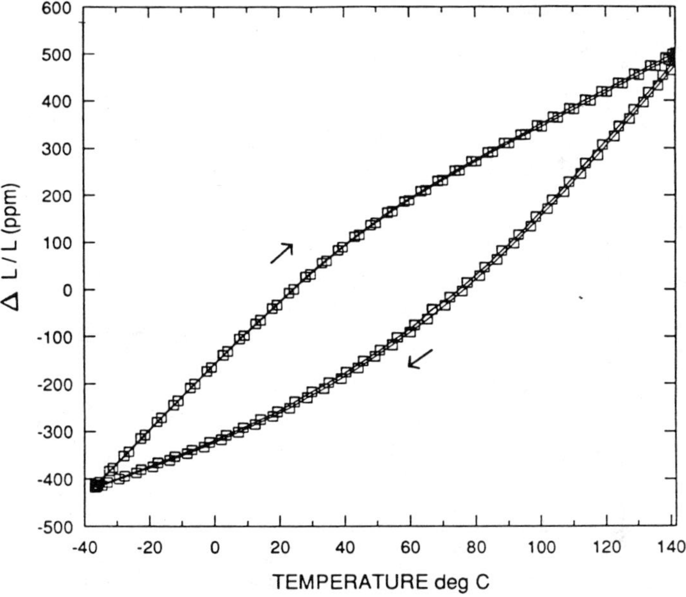



**Figure 5-10**   $\Delta L/L$ of Cu–Invar–Cu (40% Cu).

The expansivity for this structure is shown in Fig. 5-11. The bottom curve is the expansivity for Invar by itself, as given before (Fig. 5-8). The top curve is derived from (5-2) for the elastic case, using the bottom curve for the expansivity of Invar by itself. Here we have used handbook values for the elastic modulus of copper as a function of temperature.[13] We have used NIST data[14] for the expansivity of copper. For the elastic modulus of Invar, we have used data for Alloy 42 as a function of temperature.[13] The uppermost curve is nearly parallel to the lowest curve; their separation varies from 5.2 to 5.9 p.p.m./K. Thus the second term in (5-2) is relatively constant over this temperature range. The expansivity of the sandwich must fall between these two curves. As the temperature is increased from $-35°C$, we expect the expansivity of the sandwich to start out close to the elastic curve, and stay with it until the elastic limit of the copper is reached. Figure 5-11 shows that this limit is reached near room temperature. After that, the expansivity decreases, as the copper yields and $d\sigma/d\varepsilon$ decreases. This continues until the sandwich expansivity is close to that of Invar. Then it levels out because it can never go below that of pure Invar. The observation that it never reaches the Invar curve can be interpreted to mean that the slope $d\sigma/d\varepsilon$ for the copper never goes to zero. Similar considerations are relevant to the decreasing-$T$ curve. The sandwich curve for decreasing temperature starts out at 140°C close to the elastic curve, and stays with

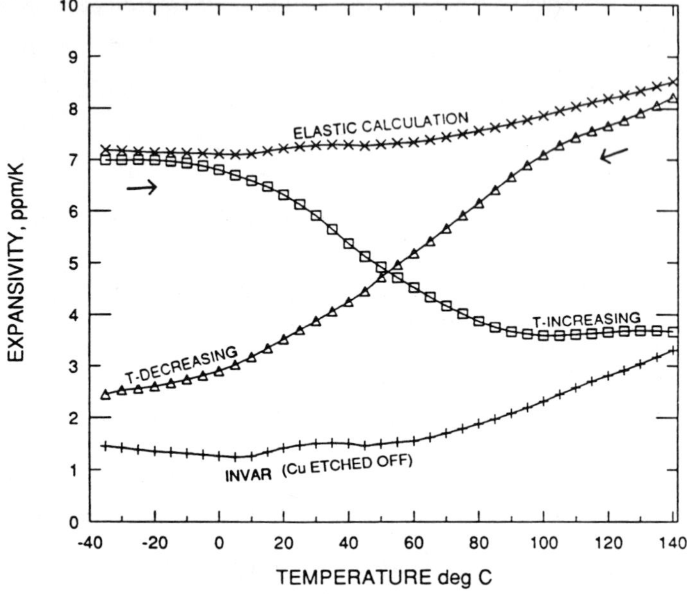


**Figure 5-11**   Expansivity of Cu–Invar–Cu (40% Cu).

it for about 40 K. But at 100°C, it begins to decrease, as the elastic limit is reached. And again it keeps on decreasing until it approaches the Invar curve, at some temperature below −40°C.

Thus the expansivity of this sandwich depends on its recent history. At room temperature, for example, if the sample's most recent temperature excursion has been above room temperature, then increasing $T$ will produce the elastic value, but decreasing $T$ will produce a substantially lower value. Conversely, if the most recent temperature excursion has been below room temperature, then decreasing $T$ will produce the elastic value, and increasing $T$ will produce a much smaller value.

***Cu–Mo–Cu***   Similar effects are present with the sandwich Cu–Mo–Cu, but on a much smaller scale. Figure 5-12 shows the expansivity curves for such a sandwich, where the copper fraction is 23%. As with the Cu–Invar–Cu, the sandwich curves approach the lowest curve after a large temperature excursion, but do not cross it. In this case, the elastic modulus of the molybdenum is 46 Mpsi (317 GPa),[13] compared to Invar, which is 20 Mpsi (138 GPa),[7] and the fraction of copper is less, 23% compared to 40%. (Since molybdenum is a better thermal conductor than Invar, not as much copper is needed in the structure as when the middle metal is Invar.) Thus the value of $A$ is now 8.6 compared to 1.6 for Cu–Invar–Cu. The second term in the

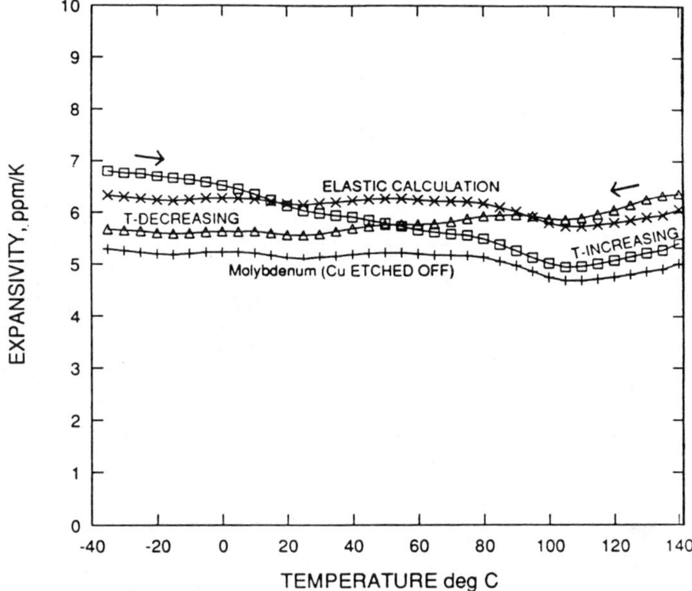


**Figure 5-12**    Expansivity of Cu–Mo–Cu (23% Cu).

elastic equation is then only 1.04 p.p.m./K compared to 5.8 p.p.m./K for Cu–Invar–Cu. This is consistent within experimental error of the hysteresis observed in Fig. 5-12. One concludes from Figs. 5-11 and 5-12 that the expansivity of Cu–Mo–Cu is much better behaved than that of Cu–Invar–Cu. That is, it does not depend on thermal history as much.

### 5.4.4 Organic Boards and Packages

Organic boards and packages are an important class of materials widely used in the industry. Compared to ceramics, organic board and package materials have up to an order of magnitude higher expansivity and generally an order of magnitude lower elastic modulus (ref. 1, pp. 36, 40 and ref. 20). This is confirmed by the expansivity data for FR-4 and polyimide–quartz boards shown in Fig. 5-13. The expansivity ranges from 17.1 to 22.7 p.p.m./K for FR-4 and from 13.3 to 16.2 p.p.m./K for polyimide–quartz. The FR-4 expansivity seems to be intentionally matched to that of copper shown in Fig. 5-9. As expected, polyimide–quartz has lower expansivity values than FR-4 because of the lower expansivity of quartz.

The expansivity curve of FR-4 seems to exhibit several features not evident in the expansivity curve of polyimide–quartz. These data represent a small

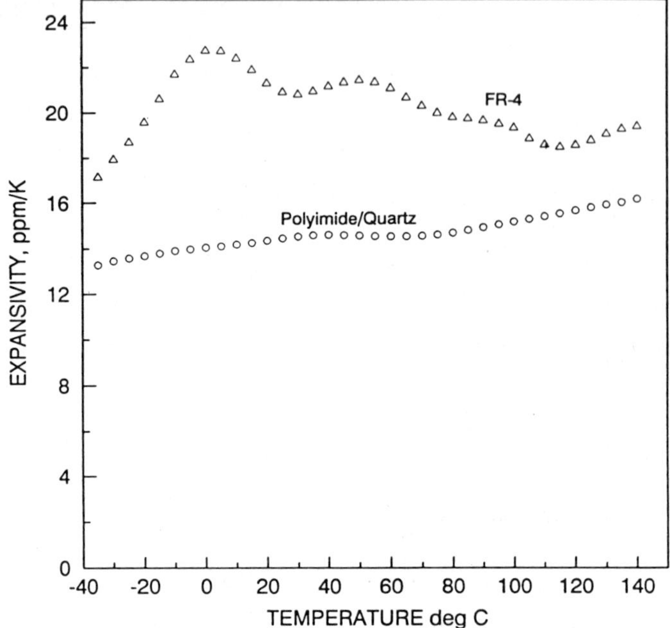


**Figure 5-13**   Expansivity of FR-4 and polyimide–quartz boards.

sampling for any generalization, especially as there are many kinds of resins, fabrics, stackups, and curing schedules used in the industry. There are, however, some known reasons for possible nonlinearities in the curve. One is the moisture coefficient of expansion, which can complicate such curves by producing hysteresis and other nonlinear effects. The other is that the temperature range is probably approaching the glass transition of the resin. This would in turn depend on the composition of the resin layers, and their degree of curing. The polyimide–quartz board has a much higher glass transition, and that would account for its well-behaved curve in this temperature range. In both cases, we have not included the first heating cycle, so moisture effects should be minimized, but the glass transition effects should be evident. This entire area needs more study with more samples of many different types of boards.

## 5.5 SUMMARY

The results of this study, summarized in Table 5-2, should be a valuable guide for materials selection to minimize thermal expansion mismatch in electronics and photonics packages. These data should be particularly useful

**Table 5-2**   Expansivities of Packaging Materials

| Material | Expansivity (p.p.m./K) | | |
|---|---|---|---|
| | $-35°C$ | $25°C$ | $140°$ |
| Alumina (pure) | 4.2* | 5.5* | 7.1* |
| 92% high-temperature co-fired alumina | 4.4 | 5.1 | 6.2 |
| Aluminum nitride | 2.3 | 2.7 | 3.6 |
| Beryllium oxide | 3.8 | 4.6 | 6.0 |
| Low-temperature co-fired ceramics | | | |
| $\kappa = 8$ | 4.6 | 5.0 | 5.8 |
| $\kappa = 4.8$ | 3.1 | 3.5 | 4.3 |
| Copper | 16.1† | 16.7† | 17.7† |
| Molybdenum | 4.8‡ | 5.0‡ | 5.2‡ |
| Silicon | 2.0‡ | 2.5‡ | 3.3‡ |
| Invar | 1.4 | 1.5 | 3.2 |
| Kovar | 7.2 | 6.1 | 4.1 |
| Cu–Invar–Cu | 2.4–7.0 | 2.9–6.7 | 3.6–8.1 |
| Cu–Mo–Cu | 5.7–6.8 | 5.7–6.5 | 5.3–6.4 |
| FR-4 | 17.1 | 20.9 | 19.4 |
| Polyimide–quartz | 13.3 | 14.4 | 16.2 |

* Touloukian et al. (1977)[6]
† Hahn (1970)[14]
‡ Slack and Bartram (1975)[5]

for the calculation of thermal stresses by finite element analysis. Table 5-2 lists our measured values of expansivity at $-35°C$, $25°C$, and $+140°C$ and a few values from the literature as well. For some materials, such as copper and molybdenum, the expansivity increases by less than 10% as the temperature is changed from $-35°C$ to $+140°C$. This indicates a low Debye temperature. Others, such as alumina, beryllia, and aluminum nitride, change by 40–70%, indicating a higher Debye temperature. This temperature dependence could be important in some calculations. The low-temperature co-fired ceramics fall between, with an increase of about 30%.

The expansivities of pressed and tape-cast aluminas were not significantly different in the temperature range studied. The expansivity of the aluminas, however, increased slightly with increasing purity, especially at the higher temperatures. In the case of low-temperature co-fired glass ceramic-based dielectrics, there was little, if any, dependence of the expansivity on the preparation method, such as the number of layers, the number of firings, the size of the sample, or the amount of metallization in the sample.

The Invar, Alloy 42 and Kovar are completely different, as their expansivity versus temperature curve goes through a minimum as a result of two competing effects: the decrease in length due to a phase change and the increase in length due to thermal vibrations as temperature is increased. Thus, our samples of Kovar and Alloy 42 showed a decrease in expansivity as the temperature is increased, because the minimum in expansivity occurs at a temperature above the range of these measurements. The minimum for the Invar, on the other hand, appears to be close to room temperature. If Kovar is selected to match the expansivity of alumina, it should be considered that Kovar's expansivity decreases with temperature, whereas that of the ceramics increases. Thus, the match may be good at room temperature, and yet diverge markedly below and above room temperature.

When a sandwich is made of Cu–Invar–Cu or Cu–Mo–Cu, the expansivity is constrained to lie between the calculated expansivity of the composite, assuming elastic properties, and the expansivity of the nonductile material. In the case of Cu–Invar–Cu, this represents a variation of over 5 p.p.m./K, so at any temperature the composite expansivity can be anywhere within this band, depending on the recent thermal history of the sample. In the case of Cu–Mo–Cu the range is much less (about 1 p.p.m./K). This is partly because the expansivity of molybdenum is closer to that of copper, but mostly because its elastic modulus is much higher than that of copper.

The organic printed wiring boards, FR-4 and polyimide–quartz, exhibited larger expansivity values than ceramics. Most of the moisture dependence of expansivity was deliberately excluded by not considering the data from the first heating cycle. Polyimide–quartz having a glass transition higher than the temperature range studied exhibited an almost constant expansivity. The FR-4 board was matched well to copper, but the curve showed some anomalous dependences not evident in the polyimide–quartz curve. This is probably related to its lower glass transition temperature.


## ACKNOWLEDGEMENTS

The authors are grateful to R. A. Deighan who wrote the data-handling routines used in this study, and to F. L. Howland, who provided insight.


## REFERENCES

1. Tummala, R. R., and E. J. Rymaszewski, *Microelectronics Packaging Handbook*, Van Nostrand Reinhold, New York, 1989, pp. 36–37.
2. Chanchani, R., and P. M. Hall, "Temperature Dependence of Thermal Expansivity of Ceramics and Metals for Electronic Packages," *IEEE Trans. Components, Hybrids, and Manufacturing Technology*, **13**, no. 4, 1990, pp. 743–750.
3. AIP, *American Institute of Physics Handbook*, 3rd edn., D. E. Gray, ed., McGraw-Hill, New York, 1971, pp. 4–113.

4. Kittel, C., *Introduction to Solid State Physics*, 3d edn., Wiley, New York, 1968, p. 183.

5. Slack, G. A., and S. F. Bartram, "Thermal Expansion of Some Diamondlike Crystals," *J. Applied Physics*, **46**, 1975, pp. 89–98.

6. Touloukian, Y. S., R. K. Kirby, R. E. Taylor, and T. Y. R. Lee, eds. *Thermophysical Properties of Matter*, Vol. 13, *Thermal Expansion*. Plenum Press, New York, 1977, pp. 176–193.

7. Guillaume, C. E., "Anomaly of the Nickel Steels," *Proceedings of the Physical Society (London)*, **32**, 1920, pp. 374–404.

8. ASM, *Metals Handbook*, Vol. 1, *Properties and Selection of Metals*, 8th edn., American Society for Metals, Metals Park, OH, 1961, pp. 816–819.

9. Smallman, R. E., *Modern Physical Metallurgy*, 4th edn., Butterworth, London, 1985.

10. Hall, P. M., and R. A. Deighan, "On Using Strain Gages in Electronic Assemblies When the Temperature is not Constant," *IEEE Trans. Components, Hybrids, and Manufacturing Technology*, **CHMT-9**, pp. 492–497.

11. Eustice, A. L., S. J. Horowitz, J. J. Stewart, A. R. Travis, and H. T. Sawhill, "Low Temperature Cofireable Ceramics: A New Approach for Electronic Packaging," *Proceedings of the Electronic Components Conference*, 1986, pp. 37–47.

12. Chanchani, R., "Processability of Thin-Film, Fine-Line Pattern on Aluminum Nitride Substrates," *IEEE Trans. Components, Hybrids, and Manufacturing Technology*, **11**, 1988, pp. 427–432.

13. King, J. A., *Materials Handbook for Hybrid Microelectronics*, Artech House, Boston, 1988.

14. Hahn, T. A., "Thermal Expansion of Copper from 20 to 800 K—Standard Reference Material 736," *J. Applied Physics*, **41**, 1970, pp. 5096–5101.

15. Dance, F. J., and J. L. Wallace, "Clad Metal Circuit Board Substrates for Direct Mounting of Ceramic Chip Carriers," *Electronic Packaging and Production*, January, 1982, pp. 226–237.

16. Howland, F. L., Personal Communication. "Thermal Stresses and Responses of an Elastic *N*-Layer Composite," 1987.

17. Hall, P. M., F. L. Howland, Y. S. Kim, and L. H. Herring, "Strains in Aluminum-Adhesive-Ceramic Tri-Layers," *ASME J. Electronic Packaging*, **112**, 1990, pp. 288–302.

18. Turner, P. S., "Thermal-Expansion Stresses in Reinforced Plastics," *J. Res. National Bureau of Standards*, **37**, 1946, pp. 239–250.

19. Nickel Development Institute, *Nickel Alloys for Electronics*. Reference Book Series, Vol. 1, 1987, p. 22.

20. Hall, P. M., "Creep and Stress Relaxation in Solder Joints of Surface-Mounted Chip Carriers," *IEEE Trans. Components, Hybrids, and Manufacturing Technology*, **CHMT-12**, 1987, pp. 556–565.

# 6

# Thermal Stress Considerations in Die-Attachment

*Goran S. Matijasevic, Chen Yu Wang, and Chin C. Lee*

## 6.1 INTRODUCTION

An important aspect of packaging the semiconductor device is how securely the die is attached to the substrate. The chip is normally bonded onto a substrate or a package using hard solder, soft solder, metal-filled epoxy, or glass.[1-3] The package together with the die-bonding layer serves the purposes of heat dissipation, mechanical support, and sometimes electrical conduction. With increasing power requirements of the chip, quality die-attach becomes a special concern. Thermal stress considerations need to be taken into account even in a perfectly executed die-attach where the constituent materials have formed a single entity with no voids. Due to thermal expansion mismatch among the die, the bonding material, and the package, stress is introduced in the cooling step of the bonding process. Stress is generated in the die that may cause cracking. Dynamic stress is also produced in the bonded devices when they are subjected to power cycling, thermal cycling, or thermal shock.

Furthermore, despite the continuous effort in the electronics industries to produce better die bonds, voids have persistently existed in the bonding layers.[4-6] These voids have been identified nondestructively using X-ray[7] and scanning acoustic microscopy,[8] or destructively by a sectioning and polishing process[9] or by etching off the entire chip.[10] Voids can occur for various reasons: air entrapment, uneven spreading of the adhesive or solder over the backside of the die, preform outgassing, contamination of the bonding layer, etc. The presence of voids reduces the reliability of the devices.

It is well known that the voids increase the chip operating temperature[11,12] and that they increase thermal and mechanical stress failures.[5]

In this chapter, we review the properties of the various materials involved in die-attach. We then consider the thermally-induced stresses in the die due to die-attachment using an analytical approach. Methods for die stress measurement are briefly described. A discussion of the quality of die-attach and its relationship to die stress then follows. Finally, we propose methods of decreasing the thermal stress effects in die-attach.

## 6.2 PROPERTIES OF DIE-ATTACH MATERIALS FOR VARIOUS APPLICATIONS

Die-attach material requirements include high adhesion, high thermal conductivity, and electrical conductivity. Furthermore, a very important consideration is that its use should not cause high stress on the IC die and the package substrate. At the same time, the material should be resistant to fatigue and creep rupture. Processing compatibility with regards to temperature, moisture, and contaminants is also a factor in material choice. Finally, cost and ease of use determine the use of a given material.

Bonding materials can be classified as hard solders, soft solders, metal-filled epoxies, and glasses. Each of these has its application advantages and we will briefly examine these as well as their disadvantages.

Epoxies and polyimides filled with precious metals have widespread use in low-cost packaging. Processing time on an automatic bonder is typically one second. They usually offer the added advantage of single-step, low-temperature (150–200°C) cure processing.[13] They do not induce high stress because they deform inelastically to accommodate the thermal displacements. However, these organic adhesives have poor thermal conductivity and thus metal fillers (usually silver) are added to increase both thermal and electrical conductivities. Minimum filler content is 38% (to allow for a thermal path in a randomly packed structure) and the content is typically 70%. This conduction mechanism can sometimes break down, causing high resistivities after burn-in.[14] The epoxies also have very poor thermal stability and are therefore not used in high-reliability packages. They are also likely to release ionic contaminants due to outgassing, which is particularly detrimental to laser diode chips.

Metal-filled glass adhesives are designed to become inorganic after firing, typically at temperatures between 400 and 450°C. This avoids outgassing and moisture problems associated with epoxy and polyimide adhesives.[15] They can also be used for bonding directly to silicon dies, where the presence of a $SiO_2$ layer aids the bonding mechanism whereas this layer is a problem with other bonding techniques, especially with eutectic bonding with Au–Si preform. However, glass has poor thermal conductivity and silver is added similarly to the epoxies. Furthermore, high stress is produced on the die backside due to high processing temperature and lack of plastic deformation. This can cause horizontal die cracking.[16] An oxidizing ambient at high temperature is needed for the glass die bonding and this may cause unwanted oxidation of other die or package metals. Use of solvents and binders for processing purposes also brings forth the concern of complete solvent removal.

The most commonly used soft solders are tin–lead (Sn–Pb) alloys. Others include various low-melting lead-, tin-, and indium-based alloys. They are generally inexpensive and have acceptable thermal conductivity, but are mechanically weak. Dies bonded using soft solders do not experience high stress because the bonding layers deform plastically to absorb the stress developed.[17] When soft solder is used to bond the die, most of the stress occurs in the bonding layer because it is much softer than the die and leadframe or substrate.[18] Their yield strengths are low, usually less than 50 MPa. The capability of plastic deformation, however, makes soft solders subject to thermal fatigue[19] and creep rupture,[20] causing long-term reliability problems. As a result, during thermal cycling the bonding layer degrades because of thermal fatigue due to the plastic deformation produced by the dynamic stress. If voids and cracks exist in the bonding layer, they will propagate during thermal cycling and thus cause the device to suffer an early failure. Another concern is the possible formation of intermetallics. The intermetallic phases are stochiometric binary compounds that are generally very brittle, and if there is a thick layer of this phase it may cause cracking of the solder bond.[21]

Since adhesive joining and soft solders are presented in greater detail in other chapters of this book, we will give more attention to hard solders. Additionally, hard solder die bonding will generate most stress in the die. Thus, die cracking is most likely to occur as a result of this type of die-attachment.

Hard solders are low-melting gold eutectics that have high strengths. Their high cost (due to gold content) restricts them to high-reliability applications. The most commonly used hard solders include gold–tin (Au–Sn), gold–germanium (Au–Ge), and gold–silicon (Au–Si) eutectic alloys. Their important properties are given in Table 6-1.[18] They all have high thermal conductivity and are free from thermal fatigue because of high yield strengths, which results in elastic rather than plastic deformation.[18] They are not subject to fatigue or creep rupture during thermal cycling as are soft solders. Since their yield strengths at room temperature are above 185 MPa, the stresses generated in the solder after bonding are not of concern. Because of the lack of plastic deformation, device dies bonded using hard solders have a certain amount of residual stress depending upon the melting temperature.

**Table 6-1**    Properties of Hard Solder Alloys

| Alloy | Melting Point (°C) | Thermal Conductivity (W/m-°C) | Coefficient of Thermal Expansion $(10^{-6}/°C)$ | Yield Strength (MPa) at | | |
|---|---|---|---|---|---|---|
| | | | | 23°C | 100°C | 150°C |
| Au–20Sn | 280 | 57.3 | 15.9 | 275 | 217 | 165 |
| Au–12Ge | 356 | 44.4 | 13.3 | 185 | 177 | 170 |
| Au–3Si | 363 | 27.2 | 12.3 | 220 | 207 | 195 |

Less stress will be generated by the eutectic with the lower melting temperature, i.e. Au–Sn.

Die cracking occurs because of the stress developed in the bonding process and the subsequent power cycling of the device. This stress development is due to the mismatch of thermal expansion coefficients among the die, the bonding layer, and the package. Table 6-2 lists the coefficients of thermal expansion (CTE) of several different bonding materials as well as substrate materials and semiconductors. Thermal conductivity is also listed for comparison.

It can be seen that with proper choice of the bonding material as well as using a suitable substrate, the coefficients of thermal expansion can be sufficiently matched as to minimize residual stress formation. This is why a lot of work is devoted to new types of adhesives and substrates.

## 6.3 ANALYTICAL CONSIDERATION OF THERMAL STRESSES IN DIE-ATTACH

Thermal expansion mismatch results in stresses in the die that can potentially exceed the ultimate strength of the semiconductor. To consider this, we will begin with a simple model which has an analytical solution for normal stress inside the die and substrate. This model will later be enhanced so that shear and peeling (transverse normal) stress can be calculated.

### 6.3.1 Timoshenko and Other Models

Timoshenko first did the analysis for a bimetallic strip structure.[22] In this model he assumed that the two layers behaved like beams capable of axial and bending deformations. An analytical solution for the maximum normal stress in the two joined metals was found. This model assumes that the stresses are uniformly distributed across the length.

**Table 6-2**   Coefficients of Thermal Expansion (CTE) and Thermal Conductivity of Important Electronic Materials

| Material | Coefficient of Thermal Expansion $(10^{-6}/°C)$ | Thermal Conductivity (W/m-°C) |
|---|---|---|
| Au–20Sn | 15.9 | 57 |
| Au–12Ge | 13.3 | 44 |
| Au–3Si | 12.3 | 27 |
| Pb–63Sn | 25.0 | 50 |
| Pb–5Sn | 29.0 | 63 |
| Epoxy–glass (in plane) | 15.0 | 0.2 |
| Epoxy–Kevlar ($x$–$y$) | 6.0 | 0.2 |
| Polyimide | 50.0 | 0.2 |
| Ge | 5.8 | 60 |
| Si | 2.6 | 150 |
| GaAs | 5.9 | 46 |
| InP | 4.5 | 68 |
| Alumina (96% $Al_2O_3$) | 6.7 | 35 |
| AlN | 3.3 | 230 |
| BeO | 8.0 | 240 |
| Silica | 0.5–0.7 | 0.5–2.0 |
| Cu | 17.0 | 380 |
| Cu–W (20%Cu) | 7.0 | 248 |
| Cu–Mo (20%Cu) | 7.2 | 197 |
| Kovar | 5.3 | 17 |
| Cu-clad Invar | 3.0 | 100 |
| Diamond | 2.3 | 2000 |

Goland and Reissner studied cemented lap joints to determine the loads at the edges of the joint and the stresses under external mechanical loads.[23] Aleck used the variational principle in two-dimensional theory of elasticity to obtain an approximate solution for the stresses induced by a uniform change in temperature of a thin rectangular plate clamped along an edge.[24] Zeyfang used Aleck's method to study an elastic plate of finite length bonded to a rigid infinite body, and determined that for long plates the stress has a uniform part valid for the central part of the plate and a nonuniform term for the edges.[25]

Taylor and Yuan studied the axial stresses caused by cooling from high-temperature bonding operations that are due to the thermal expansivity mismatch.[26] They used a one-dimensional, elastic analytical model to predict the location of the tensile stress in a shear-constrained brittle strip. Chen and Nelson expanded this work to other physical conditions and geometries, presenting several models of thermal stresses in bonded joints with and without bending.[27]

Olsen and Berg expanded the basic Timoshenko model to a three-layer structure.[18] The model used is basically one-dimensional and thus greatly simplifies the actual die-bonded specimens. However, it does provide us with an estimate of the stress developed using simple calculations. It is useful in predicting the maximum stress that the well-bonded die will incur, which can be used for comparison with ultimate die strength.[28]

### 6.3.2 Calculation of Maximum Die Stress

Figure 6-1 gives the configuration of the three-layer structure with a length $L$. The structure initially has zero stress when the bonding medium is in the molten state. Upon cooling, the bonding medium becomes solidified and stress is induced due to mismatch of thermal expansion coefficients among the die, the solder, and the substrate. If the substrate has higher thermal expansion coefficient than the die, it will contract more than the die, leading to tensile stress on the substrate which causes the structure to bend as shown in Fig. 6-1. For soft solders, which are highly ductile, large stresses do not build up, but are relieved by plastic flow. However, for hard solders, which are high-strength, much of the stress is transferred to the die.

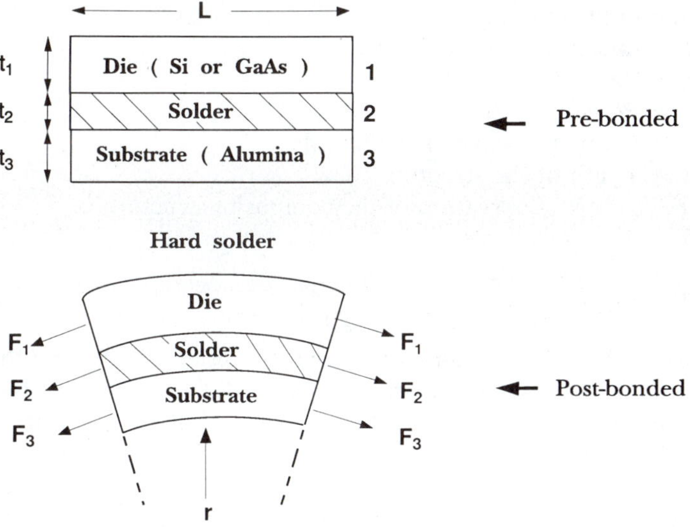


**Figure 6-1**   Schematic of bonding structure and the post-bonded configuration.

In this model, some material parameters such as elastic moduli and Poisson's ratios are assumed to be independent of the temperature. The materials involved are also assumed to incur only elastic deformation. This is a significant simplification, but the model approximately predicts the dependence of stress with varying material parameters. When we derive the equations for this model, the thermal expansion coefficients are taken to be temperature-dependent. The resulting equations are given by

$$F_1 + F_2 + F_3 = 0 \qquad (6\text{-}1a)$$

$$\frac{(t_1 + t_2)}{2r} + F_1 \frac{(1 - v_1)}{(E_1 t_1 L)} - F_2 \frac{(1 - v_2)}{(E_2 t_2 L)} = (A_1 - A_2)\, \Delta T \qquad (6\text{-}1b)$$

$$\frac{(t_2 + t_3)}{2r} + F_2 \frac{(1 - v_2)}{(E_2 t_2 L)} - F_3 \frac{(1 - v_3)}{(E_3 t_3 L)} = (A_2 - A_3)\, \Delta T \qquad (6\text{-}1c)$$

$$\frac{L\left[\dfrac{E_1 t_1^3}{(1 - v_1)} + \dfrac{E_2 t_2^3}{(1 - v_2)} + \dfrac{E_3 t_3^3}{(1 - v_3)}\right]}{12r}$$

$$+ F_1 \frac{t_1}{2} + F_2\left(t_1 + \frac{t_2}{2}\right) + F_3\left(t_1 + t_2 + \frac{t_3}{2}\right) = 0 \quad (6\text{-}1d)$$

where

$$A_i \triangleq \frac{1}{\Delta T} \int_T^{T_m} \alpha_i(T)\, dT \quad \text{is the average thermal expansion coefficient}$$

and   $F_i$ = in-plane forces acting on the various layers
   $t_i$ = thicknesses of each layer
   $E_i$ = elastic modulus
   $v_i$ = Poisson's ratio
   $\alpha_i(T)$ = thermal expansion coefficients
   $L$ = length of the structure
   $r$ = radius of curvature of the composite structure
   $\Delta T = T_m - T$
   $T_m$ = melting point
   $T$ = temperature

In Eq. (6-1) the unknown quantities are the forces $F_1, F_2, F_3$, and the radius of curvature, $r$. Timoshenko's original paper did not consider the elastic deformation in the thickness direction and thus did not have the $(1 - v_i)$ factors.[22] Olsen and Berg[18] considered elastic deformation in their analysis, but missed including the $(1 - v_i)$ factors, the length parameter $L$ in Eq. (6-1d), and assumed temperature independence of expansion coefficients.[28]

The stress on the top and the stress on the bottom surface of the first

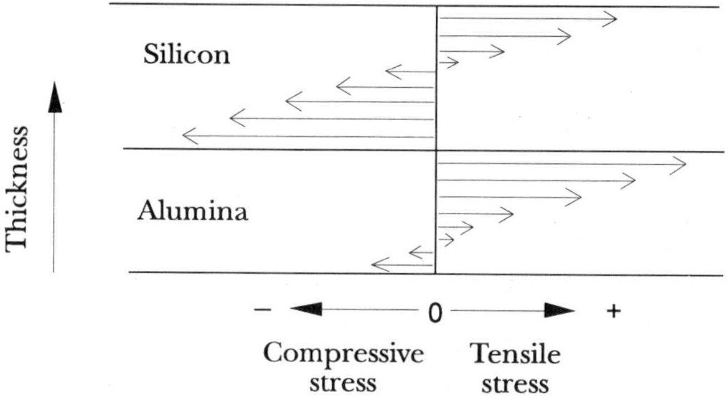



**Figure 6-2**  Stresses in die and substrate due to the bending of the structure.

layer are given by

$$\sigma_{\text{top}} = \frac{F_1}{t_1 L} - \frac{t_1 E_1}{[2r(1 - v_1)]} \tag{6-2a}$$

$$\sigma_{\text{bottom}} = \frac{F_1}{t_1 L} + \frac{t_1 E_1}{[2r(1 - v_1)]} \tag{6-2b}$$

where the second term is the stress induced by the bending of the structure. These stresses are defined so that positive value of the stress indicates tensile stress, while negative indicates compressive stress. Figure 6-2 shows the stresses due to bending of the die and the substrate.

In a sample calculation, the structure first considered consists of silicon as the first layer, Au–Si or Au–Sn eutectic as the second layer, and alumina substrate as the third layer. The thickness of each layer is 400 μm, 14 μm, and 250 μm, respectively. The length of the structure is 6 mm, the same as that of the die that was used in the actual die-attach study.[29] The materials parameters used for the calculation are given in Table 6-3. Calculation of the radius of curvature of the structures bonded with Au–Si and Au–Sn shows that the device bonded with Au–Si bends more, resulting in a smaller radius of curvature. This is due to the higher melting point of the Au–Si eutectic alloy, 363°C, compared to 280°C of Au–Sn eutectic.[28]

Figure 6-3 shows how the calculated stress on the top surface and in the center of the silicon die varies with temperature for devices bonded with Au–Si, Au–Sn, and low-temperature Au–Sn bonding processes.[30] We see that the stress on the top of the die is tensile. The stress caused by the in-plane force $F_1$ is compressive. The bending of the structure produces tensile stress on the top of the die. On the other hand, it produces a compression stress

**Table 6-3**  Material Parameters

| Material | Coefficient of Thermal Expansion at 25°C ($10^{-6}$/°C) | Young's Modulus (GPa) | Poisson's Ratio |
|---|---|---|---|
| Si | 2.6 | 187.0 | 0.25 |
| GaAs | 6.2 | 85.5 | 0.55 |
| Au–Sn | 16.2 | 69.0 | 0.40 |
| Au–Si | 12.0 | 82.7 | 0.30 |
| Alumina | 7.3 | 275.0 | 0.32 |

component on the bottom surface of the die, which makes the stress there even more compressive.

From Fig. 6-3 we see that the magnitude of the stresses is smaller for dies bonded with Au–Sn than with Au–Si eutectic. The stress on the bottom surface of the die is compressive. If the surface of the die bonded has no microcracks and voids are not present in the bonding layer, the compressive stress on the die is not adequate to cause die cracking. However, the presence

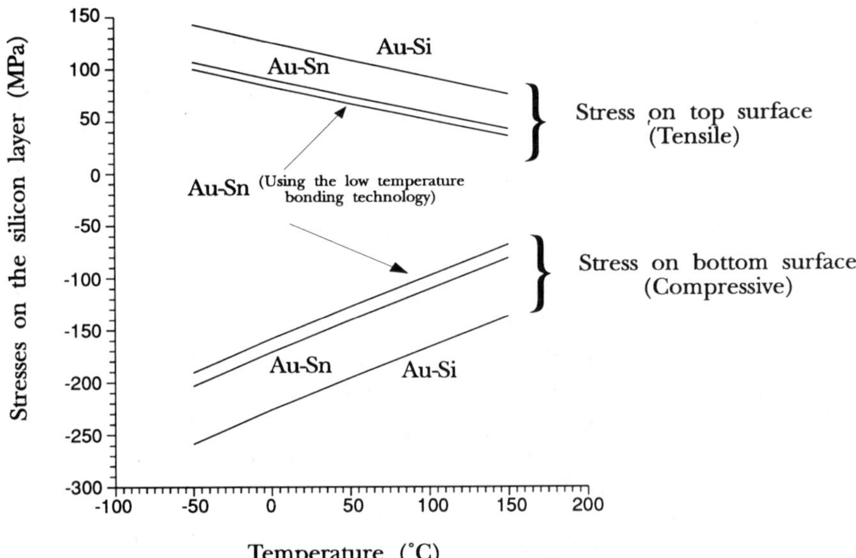


**Figure 6-3**  Calculated stresses on the top and bottom surfaces of the silicon die versus temperature. Calculations are done for Au–Si, Au–Sn, and low-temprature Au–Sn bonding. The stress on the top surface is tensile.

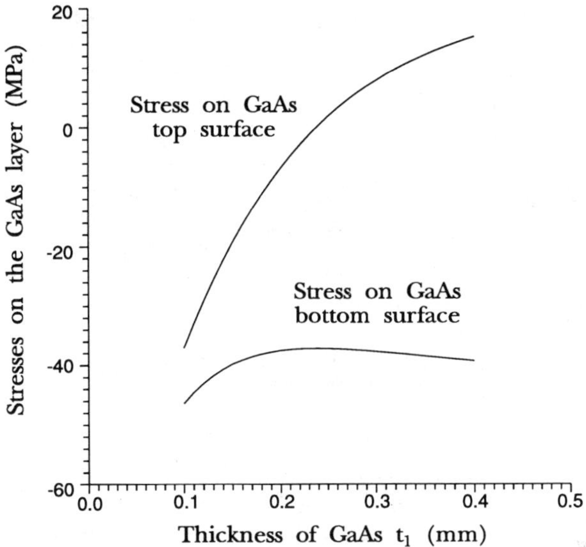


**Figure 6-4**  Maximum normal stress on the GaAs die for different die thicknesses. Thicknesses of the Au–Sn bonding layer and of the alumina substrate are 15 μm and 250 μm, respectively; assembly length is 6 mm; temperature is 25°C.

of voids would induce localized high stress, leading to possible die cracking, as will be discussed in Section 6-5.

In the above calculation we considered that the thermal expansion coefficient changes with the temperature. Since the thermal expansion coefficient will decrease when the temperature decreases, at low temperatures the mismatch between different materials depends on the slope of the thermal expansion change relative to temperature. Thus temperature dependence of the thermal expansion coefficient may be of significance in the stress calculation, depending on this slope.

The stress on the top surface of the die in the above calculation is tensile. At −50°C, this tensile stress is 107 MPa for a 400 μm die bonded with Au–Sn and 143 MPa for a die bonded with Au–Si. Silicon tensile strength depends on a number of processing factors, but with the ultimate strength reported to be as low as 130 MPa,[16] dies bonded with Au–Si eutectic are likely to have higher probability of cracking at lower temperatures.

To understand how the stress is related to different die-attach parameters, we can calculate the normal stress when the thicknesses of the die or of the bonding layer changes. Figure 6-4 shows that when the thickness of the GaAs changes from 400 μm to 100 μm, the normal stress on the top surface of the

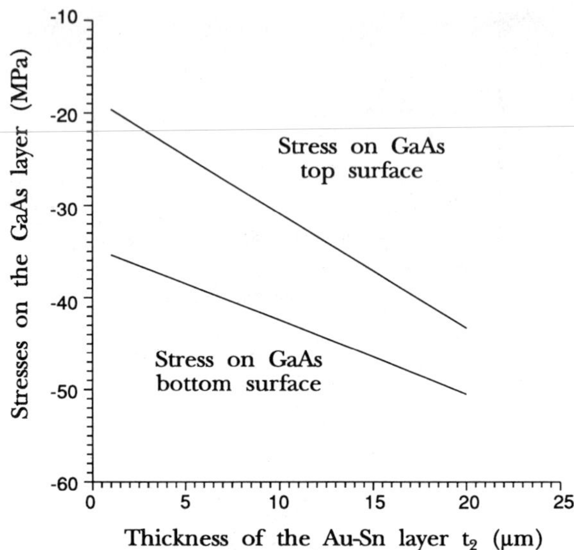



**Figure 6-5**   Maximum normal stress on a thin GaAs die for different bonding layer thickness. Thicknesses of the GaAs die and of the alumina substrate are 100 μm and 250 μm, respectively; assembly length is 6 mm; temperature is 25°C.

GaAs changes from tensile to compressive. When the device has bending, the normal stress is due to the combination of the normal force and the force due to the bending of the structure as given in Eq. (6-2). The average stress across the die thickness is compressive, but the stress on the top surface may be tensile. When the die is much thinner than the substrate, the whole device will not bend too much and the bending effect will be negligible. The stress on the top surface will then be compressive.

Figures 6-5 and 6-6 show how the maximum normal stress on the GaAs die changes with varying bonding layer thickness. It can be seen that, for a thin die, the thinner bonding layer will induce less stress on the GaAs device. For a thicker die, stress on the interface with the bonding layer decreases but tensile stress on the top surface of the die increases.

Timoshenko's model is widely used in engineering applications and can calculate the maximum normal stresses in the components. This work provides the basis for most studies on material assemblies. It provides the die strength required for a particular assembly. However, it does not give any information about the stress distribution or about the shear and transverse (peeling) stresses. These are responsible for the cohesive and adhesive strength of the die attachment.

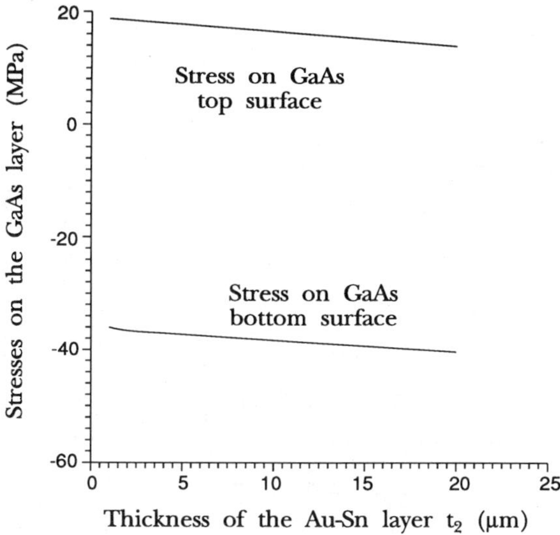


**Figure 6-6**    Maximum normal stress on a thick GaAs die for different bonding layer thickness. Thicknesses of the GaAs die and of the alumina substrate are 400 μm and 250 μm, respectively; assembly length is 6 mm; temperature is 25°C.

### 6.3.3 Suhir's Model

Suhir has developed models for the stress distributions and curvature of bimaterial[31,32] and trimaterial assemblies[33] with the objective of guiding the design of microelectronic structures, and has extended this analysis to a multilayered structure for stress analysis in elastic thin films.[34] The analytical model developed is an extension of the Timoshenko model. A trimaterial structure is shown in Fig. 6-7.

Since the three materials have different thermal expansion coefficients, as the temperature is lowered from the stress-free die-attach temperature, forces and moments shown in Fig. 6-7 will develop. In the elastic regime, equilibrium at the new temperature requires interfacial compatibility of displacements. This requires that the displacements $u_1^-(x)$ of the lowest lamella of the first layer (die) be equal to the displacements $u_2^+(x)$ of the top lamella of second layer (solder or adhesive), and that the displacements $u_2^-(x)$ of the lowest lamella of second layer be equal to the displacements $u_3^+(x)$ of the top lamella of third layer (substrate),[33] i.e.,

$$u_1^-(x) = u_2^+(x), \qquad u_2^-(x) = u_3^+(x) \tag{6-3}$$

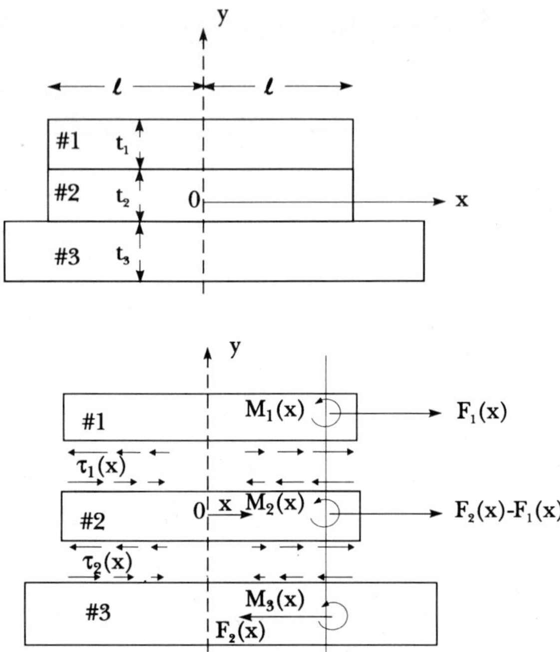



**Figure 6-7**   Stress analysis model for Suhir's analysis.[33]

The displacements are given by

$$u_1^-(x) = \alpha_1 \,\Delta T\, x - \lambda_1 \int_0^x F_1(\zeta)\, d\zeta + \kappa_1 \tau_1(x) + \frac{t_1}{2} \int_0^x \frac{d\zeta}{r(\zeta)} \qquad (6\text{-}4a)$$

$$u_2^+(x) = \alpha_2 \,\Delta T\, x + \lambda_2 \int_0^x [F_1(\zeta) - F_2(\zeta)]\, d\zeta - \kappa_2 \tau_1(x) - \frac{t_2}{2} \int_0^x \frac{d\zeta}{r(\zeta)} \qquad (6\text{-}4b)$$

$$u_2^-(x) = \alpha_2 \,\Delta T\, x + \lambda_2 \int_0^x [F_1(\zeta) - F_2(\zeta)]\, d\zeta + \kappa_2 \tau_2(x) + \frac{t_2}{2} \int_0^x \frac{d\zeta}{r(\zeta)} \qquad (6\text{-}4c)$$

$$u_3^+(x) = \alpha_3 \,\Delta T\, x + \lambda_3 \int_0^x F_2(\zeta)\, d\zeta - \kappa_3 \tau_2(x) - \frac{t_3}{2} \int_0^x \frac{d\zeta}{r(\zeta)} \qquad (6\text{-}4d)$$

where   $\lambda_i = (1 - v_i)/E_i t_i$ and $\kappa_i = t_i/3G_i$, $i = 1, 2, 3$ are axial and interfacial compliances for the three layers

$F_i$ = in-plane forces acting on the various layers, $F_i(x) = \int_{-l}^x \tau_i(\zeta)\, d\zeta$

$\tau_i$ = shearing stress

$t_i$ = thicknesses of each layer

$E_i$ = elastic modulus

$G_i$ = shear modulus, $G_i = E_i/2(1 + v_i)$
$v_i$ = Poisson's ratio
$\alpha_i$ = thermal expansion coefficients
$l$ = half-length of the structure
$r$ = radius of curvature of the composite structure
$\Delta T$ = melting point—given temperature

The first terms in Eqs. (6-4) result from the unrestricted thermal expansion of the three layers. The second terms are due to the forces $F_i(x)$ at a given cross-sectional location $x$ and are calculated under the assumption that these forces are constant over the component thickness. The third terms account for the actual nonuniform distribution of the shearing forces $F_i(x)$ in the direction normal to the interface, assuming that the additional displacements that result from this are directly proportional to the shearing stress in the given cross-section and are not affected by the stresses in the other cross sections. The final terms of the equations result from the bending of the structure.

The relationship between the forces $F_i(x)$ and the radius of curvature $r(x)$ can be found on the basis of the rotational equilibrium condition from Fig. 6-7:

$$\frac{t_1 + t_2}{2} F_1(x) + \frac{t_2 + t_3}{2} F_2(x) = M_1(x) + M_2(x) + M_3(x) \qquad (6\text{-}5)$$

where $M_i(x)$, the bending moments, are related by $M_i = -D_i/r(x)$ to flexural rigidities of the components, $D_i = E_i t_i^3/12(1 - v_i^2)$.

If the thickness of the die-attach layer is small compared to the other two layers, its thermal expansion coefficient is not involved in determining the interfacial shear stresses. Functions $\tau_1(x)$ and $\tau_2(x)$ will coincide and become

$$\tau(x) = k \frac{\Delta\alpha \, \Delta T}{\lambda \cosh kl} \sinh kx \qquad (6\text{-}6)$$

where the eigenvalue is $k = \sqrt{\lambda/\kappa}$, the axial compliance is $\lambda = \lambda_1 + \lambda_3 + \lambda_{13} = (1 - v_1)/E_1 t_1 + (1 - v_3)/E_3 t_3 + t^2/4D$, interfacial compliance is $\kappa = \kappa_1 + 2\kappa_2 + \kappa_3 = t_1/3G_1 + 2t_2/3G_2 + t_3/3G_3$, and $t$ is total thickness. Additional compliance associated with the bowing of the structure is given by $\lambda_{13}$, and $D$ is total flexural rigidity $D = D_1 + D_2 + D_3$. The difference between the thermal expansion coefficients in this case refers to the die and substrate, i.e., $\Delta\alpha = \alpha_3 - \alpha_1$. However, shear stress in the interface still depends on the compliances of the bonding layer. The size of the assembly is also important in determining the shear stress.

By using the boundary conditions $\tau(0) = 0$ and $F(l) = 0$, Eq. (6-6) for the shear stress can be solved. The first of these conditions indicates that the shear stress in the middle of the assembly is zero, since, due to the symmetry of thermal loading, there is no displacement in the middle cross-section where $x = 0$. The second condition reflects the fact that at the edge, where there are no external forces acting on the assembly, the force $F(x)$ must be zero.

The curvature of the structure then becomes

$$\frac{1}{r(x)} = \frac{t \, \Delta\alpha \, \Delta T}{2\lambda D}\left(1 - \frac{\cosh kx}{\cosh kl}\right) \tag{6-7}$$

Normal stresses are maximum at the interfaces and will be, on the bottom of the die:

$$\sigma_{1b}(x) = -\frac{\Delta\alpha \, \Delta T}{\lambda t_1}\left(1 + 3\frac{tD_1}{t_1 D}\right)\left(1 - \frac{\cosh kx}{\cosh kl}\right) \tag{6-8a}$$

and on the top of the die:

$$\sigma_{1t}(x) = -\frac{\Delta\alpha \, \Delta T}{\lambda t_1}\left(1 - 3\frac{tD_1}{t_1 D}\right)\left(1 - \frac{\cosh kx}{\cosh kl}\right) \tag{6-8b}$$

The transverse normal or peeling stresses are due to forced bending of the assembly despite differences in flexural rigidity of the components. The simplified expression for the peeling stress is given by[33]

$$p(x) = -\frac{\mu}{\kappa}\Delta\alpha \, \Delta T \frac{\cosh kx}{\cosh kl} \tag{6-9}$$

where $\mu = (t_3 D_1 - t_1 D_3)/2D$ reflects the differences in adherent thickness and flexural rigidities.

Using Eqs. (6-6), (6-8b), and (6-9), we can now calculate the distribution of the different stresses in the assembly. Figure 6-8 shows the maximum stresses in a 6 mm Si die attached to an alumina substrate using an Au–Sn eutectic solder.

The normal stresses that act on the die are maximum at the middle of the cross-section and fall to zero at the die edges. These are practically independent of the die size. Their maximum value can be calculated using the earlier given extension of the Timoshenko model. Thus to determine whether the die ultimate strength will accommodate a particular die-attachment, this is sufficient for calculations. Normal stress depends on the thickness of the die, and thinner dice will have less stress on the top of the

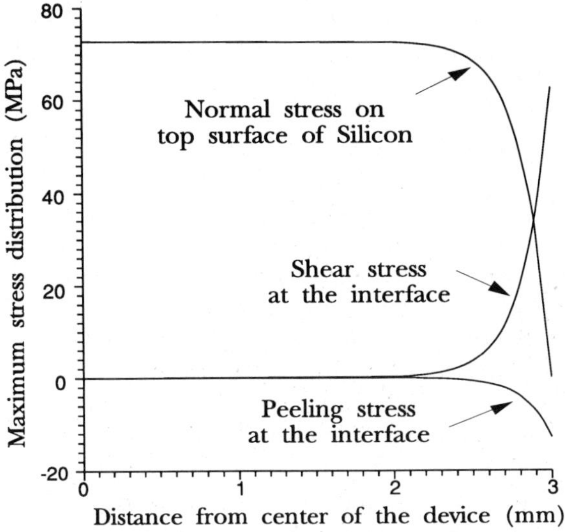


**Figure 6-8**    Maximum normal stress on top of a 400-μm thick Si die bonded to a 250-μm thick alumina substrate with 15-μm Au–Sn bonding layer. Maximum shear and peeling stress at the interface is also given across the half-length of the assembly. Temperature is 25°C.

die. This is under the assumption that the bonding is void-free. If this is not the case, voids will create localized stress and can cause die cracking, as will be discussed in Section 6-5.

The shearing and peeling stresses are responsible for the adhesive and cohesive strength of the bonding material.[33] In case of hard solder die-attach, cohesive strength of the solder is not an issue as it is with soft solders where these stresses can contribute to fatigue. The shear forces are also of concern when there is incomplete bonding or voids near the edges of the die. They can then initiate horizontal die cracking.[16]

### 6.3.4 Numerical Calculation of Die Stress

Comparisons between the solutions using the different models and the solutions using numerical methods have been carried out.[35–37] Glaser compared Suhir's model with finite element analysis (FEA) and found that Suhir's equations provide a good first-order solution for bonding stresses. Some discrepancy was found in the peeling stress calculation.[35] Eischen, Chung, and Kim also compared the model with a theory of elasticity model as well as FEA and found good agreement with peak values but some differences in the distribution of stress.[36] Pan and Pao did a calculation for a five-layer double-bonded assembly using an extended Timoshenko model and found good agreement with FEA as well as experimental measurement carried out with strain gauges.[37]

Lau pointed out that a large number of works using FEA are erroneous in calculation of thermal stresses because of averaging of the nodal values at the interface between two different materials.[38] He also compared the FEA calculation with Suhir's results and found excellent agreement. Lau has also presented the basic and the governing equations of thermoelasticity for electronics packaging.[39] Using these, along with the finite element method, calculations are possible not only for die-attachment stresses but for complicated actual chip packages.[40]

The boundary element method (BEM) has also been used to investigate the heat conduction and thermal stresses in IC packages.[41] Good agreement was found in the results obtained with this method compared to FEA.

Finally, it is worth noting that electrical modeling of the thermal stress analysis has been done by Riemer.[42] The method uses a three-terminal equivalent circuit to represent elastic tensile and shear characteristics as well as the thermal coefficient of expansion. The thermal stresses in the assembly are derived from the branch currents and the local thermal expansion coefficients are given by node voltages. Results compare well with FEA.

## 6.4 DIE STRESS MEASUREMENT

Thermally induced strains in electronic device dies need to be evaluated both under normal operating conditions and under thermal loading of the device. Analytical and numerical calculations are used to predict these thermal stresses, but actual device measurements can be of value especially in situations where the parameters become too large for full consideration in the calculations or when the loading conditions may involve a number of factors that cannot be taken into account.

Early strain measurements on a bonded semiconductor wafer were done with a photoelastic method using near-infrared light.[43] Silicon becomes birefringent when stressed and residual stress patterns can be observed using an infrared image converter and a pair of crossed prisms.

Another method used early on to study the strains in the bonded device was double-crystal X-ray diffraction microscopy.[44] Comparison with a demounted sample gave the diffraction patterns, which indicated the stress generated due to the bonding. Nonuniform stress distributions caused higher stress concentrations and were correlated with cracks in the die.

### 6.4.1 Piezoresistive Stress Sensors

The most widely used method for quantitative measurement of die stress has involved the use of piezoresistive stress sensors.[45–52] The sensors are fabricated by diffusion of a n-type or p-type dopant into the semiconductor. Sensors are laid out across the chip in locations where measurements are needed. A stress sensor with 64 measurement points on a 4.5 mm × 4.5 mm die has four resistors oriented at 0°, 45°, 90°, and 135°, allowing the calculation of the planar stress.[52] The basic equation for converting resistance to stress is[52]

$$\frac{\Delta R}{R_l} = \pi_l\,\Delta\sigma_l + \pi_t\,\Delta\sigma_t + \pi_{lt}\,\Delta\tau_{lt} + \alpha\,\Delta T \qquad (6\text{-}10)$$

where $\Delta R/R_l$ is the normalized resistance change in the longitudinal direction, $\Delta\sigma$ is the change in normal stress, $\Delta\tau$ is the change in the shear stress, and $\pi$ is the piezoresistive coefficient. Subscripts $l$ and $t$ refer to longitudinal and transverse directions. Temperature sensitivity is given with $\alpha$ as the temperature coefficient of resistance and $\Delta T$ the change in temperature. Temperature calibration can be complex, especially when temperature is not constant.[50] The resistivity is also affected by light, but that dependency can be removed by making calibration measurements in the dark.[51]

Knowing the piezoresistive coefficients of the gauge material we can thus calculate the principal stresses in a biaxial stress state. Although we are dealing with a 6 × 6 tensor,[45] we mostly need to know 2–3 coefficients. However, even with these, inaccuracy is possible since precise doping may not be known and the values are taken from previous experimental measurements. Another more accurate method of calibration is with the use of Moiré interferometry which can provide accurate strain measurement at the location of the gauges.[53] Calibration can be achieved without any need to know the piezoresistive properties of the strain gauge.

### 6.4.2 Fractional Fringe Moiré Interferometry

Fractional fringe Moiré interferometry is a full-field interferometric technique sensitive to in-plane deformations.[54,55] It has high sensitivity and excellent spatial resolution. Unlike the stress chips, which only give strain information on gauge locations, this method can provide deformation patterns covering the whole specimen as well as focusing on a local region of interest.

It is based on the formation of fringes by the interference of light wave fronts diffracted from a specimen grating of high frequency. Gratings of around 1200 lines/mm are created on the specimen surface from a master mold by depositing and imprinting an epoxy adhesive.[54,55] When the specimen is deformed, fringes are formed by the interference of the virtual gratings with light diffracted from the specimen gratings. With crossed gratings, two normal displacement components can be monitored from which strain can be obtained by differentiation. Strain contour maps can be constructed regardless of the complexity of the structure and materials used.

## 6.5 QUALITY OF DIE-ATTACH (EFFECT OF VOIDS) AND RELATIONSHIP TO DIE STRESS

### 6.5.1 Voids and Die Stress

As we have shown, the matching of thermal expansion coefficients is of great concern in large-power semiconductor devices. If we match all the materials in a package so that they have the same thermal expansion coefficients, then assuming uniform temperature distribution there would be no stresses produced if no voids existed. If temperature distribution is not uniform, stress can be generated but it will not be significant and not be localized unless there are voids located underneath the die. Likewise, for packages with thermal expansion mismatch, vertical die cracking can be prevented if the stress field generated is uniform and below the tensile strength of the die, i.e., there are no voids to cause localized stress. Accordingly, the dice with no voids in the bonding medium would not crack during the bonding process or under thermal cycling and power cycling after the bonding process, provided that nearly perfect bonding is achieved.

However, voids are fairly common in die attachment. Efforts by the industry to eliminate them have been only partially successful. Voids are either produced during die attachment or introduced later during cooling, storage, or use of the device. Built-in voids include trapped gas bubbles introduced by the reactions of the attachment process (outgassing) or by the process itself (air entrapment during the scrubbing motion). They can also be caused by various liquid or solid agents trapped during attachment or simply by uneven wetting of the surfaces (sometimes due to oxidation or organic contaminants) to be joined. Edge voids can also occur due to incomplete spreading of the die-attach material, especially in the case of large dies. Further voiding may be introduced after attachment if mechanical stresses upon cooling cause cracking or disbonding to occur. Metallurgical fatigue as well as creep and rupture may also introduce voids in the die-bonding medium. Voids, especially built-in ones, can be detected non-destructively by examining the chip-to-substrate bond with X-ray radiography or with scanning acoustic microscopy.

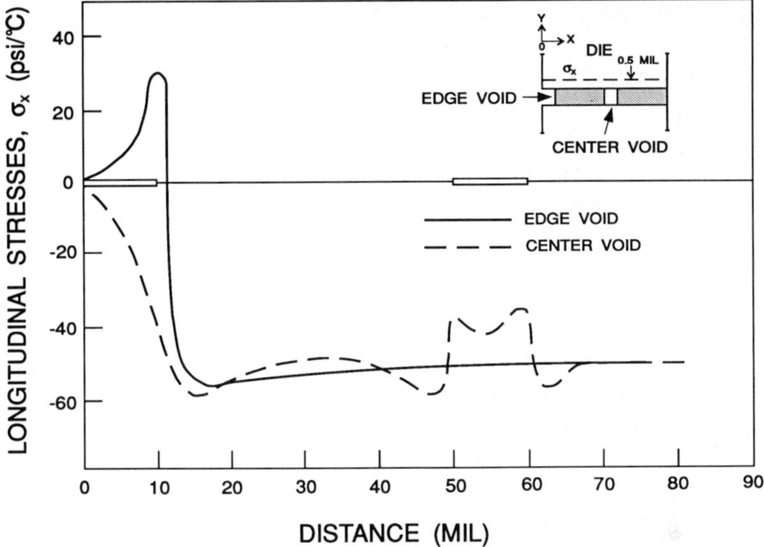


**Figure 6-9**   Variations of the longitudinal tensile stresses along the die-attach interface for the situations of a 10-mil edge and a 10-mil center void.[5]

Voids in the bonding layer are detrimental in two ways. They can cause hot spots on the surface of the die to develop as heat generated is not uniformly carried away. This is because voids cause a significant increase in the thermal resistance between the chip and the package. This has been reported, measured, and studied numerically.[10–12] Voids also cause die cracking to occur during the bonding process or thermal cycling because they generate localized stress on the backside of the chip.[4,5,56] It is this second aspect of voids that we will discuss here.

Chiang and Shukla[5] asserted that most die crack failure can be traced to imperfect die-attachment (voiding). Voids cause local stress concentration, which is very much dependent on the location of the voids. Using finite element analysis they showed that the presence of an edge void at the die-attach interface changes the local stress and creates a tensile longitudinal stress field. For a center void, the longitudinal stress at that location becomes less compressive than the average stress obtained without the void. This is shown in Fig. 6-9.[5] This leads to the conclusion that for die-attachment without voids or with some center voids there will be no die cracking, whereas specimens with voids near edges of the dice are likely to have vertical die cracking. Experimental observations of actual production samples as well as of samples with purposely formed voids confirmed this conclusion. This has been further confirmed by other studies.[4,16]

### 6.5.2 Nondestructive Determination of Die-Attach Quality

Correlation was also carried out between voids in the bonding layer and die cracking due to thermal shock test. Several specimens with a few voids ranging in size from 50 μm to 200 μm were examined.[29,57,58] GaAs and Si dies were bonded to an alumina substrate with gold–tin eutectic alloy. After examining the bonding quality with a scanning acoustic microscope (SAM), a thermal shock test was performed. For samples with voids located away from the edges of the chips, the location and size of the voids changed somewhat after the thermal shock test. However, for chips with center voids there was no disbonding or cracking of the die.

A different result was obtained for die-attach with large edge voids. One specimen that had several voids, including a 500 μm void located near the edge of the chip, is shown in Fig. 6-10.[57] This is a sample with 350 μm thick GaAs bonded to 250 μm thick alumina using an Au–Sn eutectic preform. A SAM image of this bond is shown in Fig. 6-10*a*. The chip of this particular specimen suffered vertical fracture after 40 cycles of thermal shock between −196°C and +160°C. The location of the crack is indicated by arrows in the SAM image in Fig. 6-10*b*. The SAM image of the die after cracking indicates that the remaining portion of the die remained well bonded and that the fracture occurred at the edge-void location. This observation

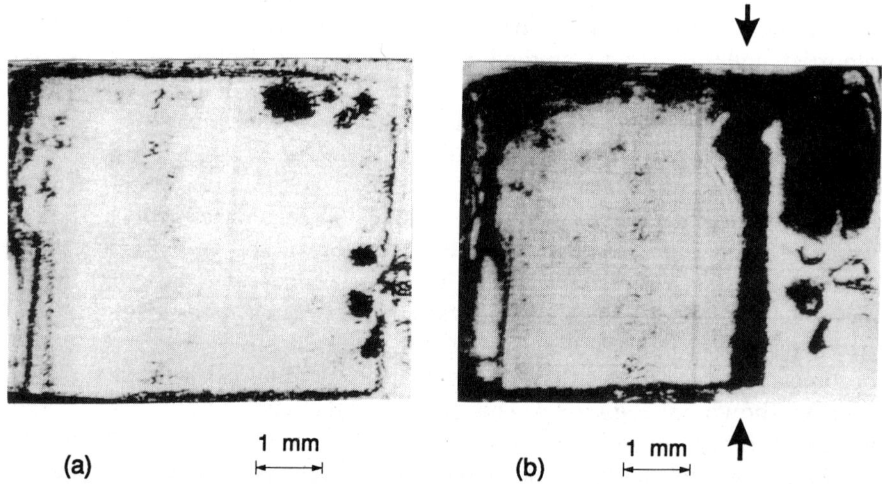


**Figure 6-10**   SAM images of GaAs die bond sample with edge voids (*a*) before and (*b*) after 40 cycles of thermal shock after which it suffered complete vertical fracture along a line indicated by the arrows.

confirms the prediction of Chiang and Shukla.[5] The acoustic microscope as a nondestructive tool was successfully used here to identify a potential failure before it occurred.

### 6.5.3 Methods of Improving Die-Attach Quality

Since a major concern in die stress formation is voids and any unevenness in the bonding layer, die-attach uniformity becomes crucial. Several methods have been employed to achieve voidless high-quality bonding. Doing the bonding in a very pure environment, without contamination of the die-attach materials, is of great importance. Application of pressure to achieve good contact and therefore adhesion is important whether or not scrubbing is used.[29,59] Voidless die-attach has also been achieved with the "vacuum-release" process.[60]

Aside from the bonding layer itself, die backside roughness can also cause cracking of the die. GaAs is found to be especially susceptible to die cracking as it is an inherently weak material.[61] Wafer back etching was found to strengthen the Si die since it removes microcracks that may cause concentrations of stress.[62,63] Backgrinding and wafer thinning of GaAs devices[64] have also shown that the strength of the die improves significantly both before and after die-attach.[59]

## 6.6 CONCLUSION

Thermal stress considerations in die-attachment are a part of the overall stress consideration of the electronics package. The major concern for stresses in the die comes in the case of hard solder bonding when the greatest stresses are generated. Timoshenko's model provides the means of ascertaining how much maximum stress will be developed in the die. This can then be used for comparison with the ultimate die strength. Suhir's model provides a much better picture of the forces acting on the die. Numerical solutions such as finite element analysis and boundary element methods also provide the distribution of stresses, which can then be used further in calculations of stresses that the whole semiconductor package exerts on the die. Discrete measurements of the stress can be done using integrated strain gauges. Full-field measurements are done using photoelasticity or fractional fringe Moiré interferometry.

The quality of the die bond is important because imperfections or voids create high localized stresses that can cause die cracking. Achieving void-free bonding as well as strengthening the die by backgrinding and thinning can prevent microcracks from forming and causing device failure.

## NOMENCLATURE

| | |
|---|---|
| $A_i$ | Average thermal expansion coefficient |
| $D_i$ | Flexural rigidity of the $i$th layer |
| $D = \sum_i D_i$ | Total flexural rigidity of the assembly |
| $E_i$ | Elastic modulus of the $i$th layer |
| $F_i$ | In-plane forces acting on the $i$th layer |
| $G_i$ | Shear modulus for the $i$th layer |
| $k$ | Stiffness parameter (eigenvalue) |
| $l$ | Half the structure length |
| $L$ | Length of the structure |
| $M_i$ | The bending moment of the $i$th layer |
| $p(x)$ | Peeling stress |
| $t_i$ | Thickness of the $i$th layer |
| $T$ | Temperature |
| $T_m$ | Melting temperature |
| $\Delta T$ | $T_m - T$ |
| $r$ | Radius of curvature of the composite structure |
| $u_i^+$ | Longitudinal displacement for the $i$th layer |
| $x$ | Abscissa of the given cross-section |
| $\alpha_i$ | Thermal expansion coefficient of the $i$th layer |
| $\Delta\alpha$ | Difference between the coefficients of thermal expansion |
| $\lambda_i$ | Axial compliance of the $i$th layer |
| $\lambda_{i,i+1}$ | Axial compliance due to bending |
| $\lambda$ | Total axial compliance |
| $\kappa$ | Elastic coefficient of compliance |
| $\mu$ | Peeling stress parameter |
| $v_i$ | Poisson's ratio for the $i$th layer |
| $\sigma_i$ | Normal stress acting over the cross-section of the $i$th layer |
| $\tau_i$ | Shear stress in the $i$th interface |
| $\zeta$ | Variable |

## REFERENCES


1. Feinstein, L. G., "Die Attachment Methods," *Electronic Materials Handbook,* Vol. 1: *Packaging,* M. L. Minges and C. A. Dostal, eds., ASM International, Materials Park, Ohio, 1989, pp. 213–223.
2. Selvaduray, G. S., "Die Bond Material and Bonding Mechanisms in Microelectronic Packaging," *Thin Solid Films,* **153**, 1987, pp. 431–455.
3. Shukla, R. K., and N. P. Mencinger, "A Critical Review of VLSI Die Attachment in High Reliability Application," *Solid State Technology,* **28**(7), 1985, pp. 67–74.
4. Van Kessel, C. G. M., S. A. Gee, and J. J. Murphy, "The Quality of Die-Attachment and Its Relationship to Stresses and Vertical Die-Cracking," *IEEE Trans. Components, Hybrids, and Manufacturing Technology,* **CHMT-6**(4), 1984, pp. 414–420.

5. Chiang, S. S., and R. K. Shukla, "Failure Mechanism of Die Cracking Due to Imperfect Die Attachment," *Proc. 34th Electronics Components Conference*, IEEE/Electronic Industries Association (EIA), 1984, pp. 195–202.

6. Pavio, J. S., "Successful Alloy Attachment of GaAs MMIC's," *IEEE Trans. Electron Devices*, **ED-34**(12), 1987, pp. 2616–2620.

7. Bascom, W. D., and J. L. Bitner, "Void Reducton in Large-area Bonding of IC Components," *Solid State Technology*, **18**(9), 1975, pp. 37–44.

8. Tsai, C. S., C. C. Lee, and J. K. Wang, "Diagnosis of Hybrid Microelectronics Using Transmission Acoustic Microscopy," *Proc. 17th International Reliability Physics Symposium*, IEEE, 1979, pp. 178–182.

9. Li, T. P. L., E. L. Zigler, and D. E. Hillyer, "AES/ESCA/SEM/EDX Studies of Die Bond Materials and Interfaces," *Proc. 22nd International Reliability Physics Symposium*, IEEE, 1984, pp. 169–174.

10. Yerman, A. J., J. F. Burgess, R. O. Carlson, and C. A. Neugebauer, "Hot Spots Caused by Voids and Cracks in the Chip Mountdown Medium in Power Semiconductor Packaging," *IEEE Trans. Components, Hybrids, and Manufacturing Technology*, **CHMT-6**(4), 1983, pp. 473–479.

11. Mahalingam, M., M. Nagarkar, L. Lofgran, J. Andrews, D. R. Olsen, and H. M. Berg, "Thermal Effects of Die Bond Voids in Metal, Ceramic, and Plastic Packages," *Proc. 34th Electronics Components Conference*, IEEE/Electronic Industries Association (EIA), 1984, pp. 469–477.

12. Palisoc, A. L., Y. J. Min, and C. C. Lee, "The Effect of Die-bond Voids on the Thermal Performance Degradation of Solid State Devices," 10th International Conference on Boundary Element Methods, in *Boundary Elements X*, Vol. 2, C. A. Brebbia, ed., Computational Mechanics Publications, Southampton, U.K., 1988, pp. 543–554.

13. Koopman, N. G., T. Reiley, and P. A. Totta, "Chip-to-package Interconnections," *Microelectronic Packaging Handbook*, R. R. Tummala and E. J. Rymaszewski, eds., Van Nostrand Reinhold, New York, 1989, pp. 361–453.

14. Opila, R. L., and J. D. Sinclair, "Electrical Reliability of Silver Filled Epoxies for Die Attach," *Proc. 23rd International Reliability Physics Symposium*, IEEE, 1985, pp. 164–172.

15. Kearney, K. M., "Trends in Die Bonding Materials," *Semiconductor International*, **11**(6), 1988, pp. 84–88.

16. Kasem, Y. M., and L. G. Feinstein, "Horizontal Die Cracking as a Yield and Reliability Problem in Integrated Circuit Devices, *Proc. 37th IEEE Electronic Components Conference*, IEEE/Electronic Industries Association (EIA), 1987, pp. 96–104.

17. Tribula, D., D. Grivas, and J. W. Morris, Jr., "Stress Relaxation in 60Sn–40Pb Solder Joints," *J. Electronic Materials*, **17**, 1988, pp. 387–390.

18. Olsen, D. R., and H. M. Berg, "Properties of Die Bond Alloys Relating to Thermal Fatigue," *IEEE Trans. Components, Hybrids, and Manufacturing Technology*, **CHMT-2**(2), 1979, pp. 257–263.

19. Frear, D., D. Grivas, and J. W. Morris, Jr., "A Microstructural Study of the Thermal Fatigue Failures of 60Sn–40Pb Solder Joints," *J. Electronic Materials*, **17**, 1988, pp. 171–180.

20. Frost, H. J., R. T. Howard, P. R. Lavery, and S. D. Lutender, "Creep and Tensile

Behavior of Lead-rich Lead–Tin Solder Alloys," *IEEE Trans. Components, Hybrids, and Manufacturing Technology*, **CHMT-11**, 1988, pp. 371–379.

21. Woychik, C. G., and R. C. Senger, "Joining Materials and Processes in Electronic Packaging," *Principles of Electronic Packaging*, D. P. Seraphim, R. C. Lasky, and C.-Y. Li, eds., McGraw-Hill, New York, 1989, pp. 577–619.

22. Timoshenko, S., "Analysis of Bi-Metal Thermostats," *J. Optical Society of America*, **11**, 1925, pp. 233–255.

23. Goland, M., and E. Reissner, "The Stresses in Cemented Joints," *J. Applied Mechanics*, **11**(1), 1944, pp. A17–A27.

24. Aleck, B. J., "Thermal Stresses in a Rectangular Plate Clamped Along an Edge," *J. Applied Mechanics*, **16**(2), 1949, pp. 118–122.

25. Zeyfang, R., "Stresses and Strains in a Plate Bonded to a Substrate: Semiconductor Devices," *Solid State Electronics*, **14**, 1971, pp. 1035–1039.

26. Taylor, T. C., and F. L. Yuan, "Thermal Stress and Fracture in Shear-Constrained Semiconductor Device Structures," *IRE Trans. Electron Devices*, **9**, 1962, pp. 303–308.

27. Chen, W. T., and C. W. Nelson, "Thermal Stress in Bonded Joints," *IBM J. Research and Development*, **23**(2), 1979, pp. 179–188.

28. Matijasevic, G. S., C. Y. Wang, and C. C. Lee, "Void-Free Bonding of Large Silicon Dice Using Gold–Tin Alloy," *IEEE Trans. Components, Hybrids, and Manufacturing Technology*, **CHMT-13**(4), 1990, pp. 1128–1134.

29. Matijasevic, G. S., C. Y. Wang, and C. C. Lee, "Extremely Reliable Bonding of Large Silicon Dice Using Gold–Tin Alloy," *Proc. 40th IEEE Electronic Components Conference*, IEEE/Electronic Industries Association (EIA), 1990, pp. 786–790.

30. Lee, C. C., and C. Y. Wang, "A Low Temperature Bonding Process Using Deposited Gold–Tin Composites," *Thin Solid Films*, **208**, 1992, pp. 202–209.

31. Suhir, E., "Stresses in Adhesively Bonded Bi-Material Assemblies Used in Electronic Packaging," *Materials Research Society Proc.*, **72**, 1986, pp. 133–138.

32. Suhir, E., "Stresses in Bi-Metal Thermostats," *J. Applied Mechanics*, **53**(3), 1986, pp. 657–660.

33. Suhir, E., "Die Attachment Design and Its Influence on Thermal Stresses in the Die and the Attachment," *Proc. 37th Electronics Components Conference*, IEEE/Electronic Industries Association (EIA), 1987, pp. 508–517.

34. Suhir, E., "An Approximate Analysis of Stresses in Multilayered Elastic Thin Films," *J. Applied Mechanics*, **55**(1), 1988, pp. 143–148.

35. Glaser, J. C., "Thermal Stresses in Compliantly Joined Materials," *Trans. ASME—J. Electronic Packaging*, **112**(1), 1990, pp. 24–29.

36. Eischen, J. W., C. Chung, and J. H. Kim, "Realistic Modeling of Edge Effect Stresses in Bimaterial Elements," *Trans. ASME—J. Electronic Packaging*, **112**(1), 1990, pp. 16–23.

37. Pan, T.-Y., and Y.-H. Pao, "Deformation in Multilayer Stacked Assemblies," *Trans. ASME—J. Electronic Packaging*, **112**(1), 1990, pp. 30–34.

38. Lau, J. H., "A Note on the Calculation of Thermal Stresses in Electronic Packaging by Finite Element Methods," *Trans. ASME—J. Electronic Packaging*, **111**(4), 1990, pp. 313–320.

39. Lau, J. H., "Thermoelasticity for Electronic Packaging," *Proc. ASM International 3rd Electronic Materials & Processing Congress,* 1990, pp. 111–119.

40. Lau, J. H., "Thermal Stress Analysis of Plastic Leaded Chip Carriers," *Proc. InterSociety Conference on Thermal Phenomena in Electronic Systems (I-THERM) II,* 1990, pp. 57–66.

41. Sato, M., R. Yuuki, and S. Yoshioka, "Boundary Element Analysis of Steady-State Heat Conduction and Thermal Stress in the LSI Package," *Japan Society of Mechanical Engineers International Journal,* **33**(3), 1990, pp. 334–341.

42. Riemer, D. E., "Thermal-Stress Analysis with Electrical Equivalents," *IEEE Trans. Components, Hybrids, and Manufacturing Technology,* **CHMT-13**(1), 1990, pp. 194–199.

43. Riney, T. D., "Residual Thermoelastic Stresses in Bonded Silicon Wafers," *J. Applied Physics,* **32**(3), 1961, pp. 454–460.

44. Zeyfang, R., "Residual Stresses in Thin Single Crystals Bonded to an Amorphous Substrate: Silicon-Integrated Circuits," *J. Applied Physics,* **42**(3), 1971, pp. 1182–1185.

45. Spencer, J. L., W. H. Schroen, G. A. Bednarz, J. A. Bryan, T. D. Metzgar, R. D. Cleveland, and D. R. Edwards, "New Quantitative Measurements of IC Stress Introduced by Plastic Packages," *Proc. 19th International Reliability Physics Symposium,* IEEE, 1981, pp. 74–80.

46. Schroen, W. H., J. A. Spencer, J. A. Bryan, R. D. Cleveland, T. D. Metzgar, and D. R. Edwards, "Reliability Tests and Stress in Plastic Integrated Circuits," *Proc. 19th International Reliability Physics Symposium,* IEEE, 1981, pp. 81–87.

47. Usell, R. J., Jr., and S. A. Smiley, "Experimental and Mathematical Determination of Mechanical Strains Within Plastic IC Packages and Their Effect on Devices During Environmental Tests," *Proc. 19th International Reliability Physics Symposium,* IEEE, 1981, pp. 65–73.

48. Van Kessel, C. G. M., S. A. Gee, and J. J. Murphy, "The Quality of Die-Attachment and Its Relationship to Stresses and Vertical Die-Cracking," *Proc. 33rd Electronics Components Conference,* IEEE/Electronic Industries Association (EIA), 1983, pp. 237–244.

49. Edwards, D. R., K. G. Heinen, J. E. Martinez, and S. Groothuis, "Shear Stress Evaluation of Plastic Packages," *Proc. 37th Electronics Components Conference,* IEEE/Electronic Industries Association (EIA), 1987, pp. 84–95.

50. Hall, P. M., and R. A. Deighan, III, "On Using Strain Gauges in Electronic Assemblies When Temperature is Not Constant," *IEEE Trans. Components, Hybrids, and Manufacturing Technology,* **CHMT-9**(4), 1986, pp. 492–497.

51. Lundström, P., and K. Gustafsson, "Mechanical Stress and Life for Plastic-Encapsulated, Large-Area Chip," *Proc. 38th Electronics Components Conference,* IEEE/Electronic Industries Association (EIA), 1988, pp. 396–405.

52. Gee, S. A., W. F. van den Bogert, V. R. Akylas, and R. T. Shelton, "Strain Gauge Mapping of Die Surface Stresses," *Proc. 39th Electronics Components Conference,* IEEE/Electronic Industries Association (EIA), 1989, pp. 343–350.

53. Bastawros, A. F., and A. S. Voloshin, "*In Situ* Calibration of Stress Chips," *IEEE Trans. Components, Hybrids, and Manufacturing Technology,* **CHMT-13**(4), 1990, pp. 888–892.

54. Bastawros, A. E., and A. S. Voloshin, "Transient Thermal Strain Measurements in Electronic Packages," *Proc. InterSociety Conference on Thermal Phenomena in Electronic Systems (I-THERM) II*, 1990, pp. 67–73.
55. Bastawros, A. F., and A. S. Voloshin, "Thermal Strain Measurements in Electronic Packages Through Fractional Fringe Moiré Interferometry," *Trans. ASME—J. Electronic Packaging*, **112**(4), 1990, pp. 303–308.
56. Johnson, J. E., "Die Bond Failure Modes," *Proc. 12th International Reliability Physics Symposium*, IEEE, 1974, pp. 150–154.
57. Matijasevic, G., and C. C. Lee, "A Reliability Study of Au–Sn Eutectic Bonding with GaAs Dice," *Proc. 27th International Reliability Physics Symposium*, IEEE, 1989, pp. 137–140.
58. Lee, C. C., and Goran Matijasevic, "Highly Reliable Die-attachment on Polished GaAs Surfaces Using Gold–Tin Eutectic Alloy," *IEEE Trans. Components, Hybrids, and Manufacturing Technology*, **CHMT-14**(3), 1989, pp. 406–409.
59. Nishiguchi, M., N. Goto and H. Nishizawa, "Highly Reliable Au–Sn Eutectic Bonding with Background GaAs LSI Chips," *IEEE Trans. Components, Hybrids, and Manufacturing Technology*, **CHMT-14**(3), 1991, pp. 523–528.
60. Mizuishi, K., M. Tokuda, and Y. Fujita, "Fluxless and Virtually Voidless Soldering for Semiconductor Chips," *IEEE Trans. Components, Hybrids, and Manufacturing Technology*, **CHMT-11**(4), 1988, pp. 447–451.
61. Vidano, R. P., D. W. Paananen, T. H. Miers, J. M. Krause-Singh, K. R. Agricola, and R. L. Hauser, "Mechanical Stress Reliability Factors for Packaging GaAs MMIC and LSIC Components," *IEEE Trans. Components, Hybrids, and Manufacturing Technology*, **CHMT-12**(4), 1987, pp. 612–617.
62. Hawkins, G., H. Berg, M. Mahalingam, G. Lewis, L. Lofgran, "Measurement of Silicon Strength as Affected by Wafer Back Processing," *Proc. 25th International Reliability Physics Symposium*, IEEE, 1987, pp. 216–223.
63. Lim, T. B., "The Impact of Wafer Back Surface Finish on Chip Strength," *Proc. 27th International Reliability Physics Symposium*, IEEE, 1989, pp. 131–136.
64. Nishiguchi, M., N. Goto, T. Sekiguchi, H. Nishizawa, H. Hayashi, and K. Ono, "Mass Production Back-grinding/Wafer-thinning Technology for GaAs Devices," *IEEE Trans. Components, Hybrids, and Manufacturing Technology*, **CHMT-13**(3), 1990, pp. 528–533.

# 7

# Die Stress Measurement Using Piezoresistive Stress Sensors

*James N. Sweet*

## 7.1 INTRODUCTION

The in-situ measurement of mechanical stress in microelectronics packages has become desirable in recent years as a result of evolving packaging technology. In the case of integrated circuit (IC) packages, stress can be produced either through the die attachment process or through encapsulation in a molding material. In the case of chip-on-board (COB) packaging, stress can be produced by the chip coating used to protect the IC against handling damage and chemical attack. As die sizes have increased, stresses from these sources have steadily grown in magnitude, resulting in such phenomena as die cracking or shear stress-induced breakage of the die passivation and metal conductors under the passivation. In order to quantitatively determine the stress magnitudes in a packaged IC it is desirable to have a test device that can simulate the actual IC chip as closely as possible.

The stress measurement is made difficult by the small size of the die and the resultant spatial resolution required. For example, typical semiconductor dies have a lateral or side dimension in the range of about 5–15 mm. It is desirable to be able to derive the stress at positions resolved to an accuracy, $\Delta x$, on the order of 1/10 the die size, or $\Delta x \leqslant 0.5$ mm $= 500$ μm. In addition, vertical dimensions in an IC plastic package are very small. Die and lead frame thicknesses are typically $\sim 0.5$ mm, and so a vertical resolution of $\Delta z \approx 0.04$ mm $= 40$ μm is desired. The small size resolutions required make the use of attached strain gages difficult for this application. Fortunately, the Si die itself furnishes an excellent capability for direct stress or strain

221

determination through the piezoresistive effect. If appropriate resistive structures are defined on the die using conventional IC processing techniques, such as ion implantation and photolithography, then stresses can be accurately determined using a test chip that is similar in size to the actual die under study and has the same mechanical properties.

This chapter is concerned with the use of Si test chips with piezoresistive stress sensors for measurement of the components of the stress tensor at the die surface. In this measurement, the change in resistance, $\Delta R$, of one or more implanted or diffused Si resistors is used to infer the stress tensor variation that produced the change. The measurement, then, has two basic requirements: (1) a reasonable and measurable relative resistance sensitivity $\Delta R/R_0$ from an initial resistance $R_0$ to a new resistance $R = \Delta R + R_0$, resulting from stress changes seen in packaging, and (2) an ability to perform a sensor calibration so that the stress tensor components can be accurately determined from the resistance-change measurements. While the achievement of the first condition has been relatively straightforward, significant improvements in calibration have only been achieved in recent years, and there is still active research in this field. At this writing, there is still no universal procedure, nor commercially available equipment for performing this calibration.

The organization of this chapter is as follows. In Section 7.2 we review the theory of piezoresistive stress sensors. The major goal is to define the coupling constants that relate stress tensor variation to resistance shift and to discuss the properties of these constants that are required for accurate stress measurement. The phenomenological theory relating resistance changes to stress changes is discussed first and is followed by a review of the theory of the coupling constants or piezoresistive coefficients. This discussion forms the basis for the remainder of the chapter. The discussion is somewhat complex. An understanding of the mathematics involved in the phenomenological theory requires some familiarity with matrix and tensor analysis. An understanding of the theory of the piezoresistive coefficients is aided by a knowledge of solid state physics, especially the theory of energy bands in solids. Some readers may want to proceed directly to Section 7.3 where experimental measurements of the coupling constants are reviewed with an emphasis on how well the theory in Section 7.2 can be used to predict their values. Stress sensor geometries are discussed in Section 7.4 and questions of chip layout and calibration are addressed in Section 7.5. A review of experimental piezoresistive stress measurements reported in the literature is given in Section 7.6 and the chapter is summarized in Section 7.7.

## 7.2 THEORY OF PIEZORESISTIVE SENSORS

### 7.2.1 Background

The piezoresistance effect in Si was first discussed extensively by Smith in 1956.[1] Using doped single-crystal Ge and Si rod-shaped samples, he applied a uniaxial stress in the direction of the sample axis and measured the resistance change in both the axis (longitudinal) direction and in the transverse direction. This is the first reported measurement of the "pi" coefficients, $\pi_{ij}$, which relate a relative change in resistance in the $i$th Cartesian coordinate direction, $\Delta R_i/R_{0i}$, to the stress tensor components as designated by the index $j$ in the six-component vector notation described below. It should be remarked here that the notation used to mathematically describe the stress response in confusing. We also note that some authors, such as Smith, use an upper case $\Pi$ for the "pi" coefficients. Smith also gave a qualitative explanation of the large piezoresistive effect in Si and Ge as being due to the variation of the anisotropic Fermi surfaces in these materials produced by mechanical stress. In 1957, Mason and Thurston presented the use of piezoresistive sensors as strain gages and described the general phenomenological theory of piezoresistance in Ge and Si.[2] This phenomenological theory relates the electrical resistivity to the stress tensor through the general theory of tensor coupling of fields to applied forces. In 1961, Pfann and Thurston gave a detailed description of piezoresistive device geometries and also discussed the effect of shear stresses on the resistive variation.[3] They also derived expressions for the $\pi$ coefficients for arbitrarily oriented axes, thus facilitating the design of complex sensors. In 1963, Tufte and Stelzer described the fabrication of piezoresistive transducers using diffused resistive structures and presented experimental $\pi$ coefficient data for a range of Si resistor structure dopings.[4] Almost all subsequent work on piezoresistance in Si devices has been based on these four pioneering papers.

Although there was sporadic activity in piezoresistance over the next decade and a half, it was not until 1981 that the use of piezoresistive stress sensors for packaging studies was reported by a group from Texas Instruments.[5] Since this initial TI work by Spencer et al.,[5] a number of groups have used piezoresistive sensors to measure stress in packaged ICs (see Section 7.6).

The theoretical description of piezoresistive sensors consists of two components: (1) the phenomenological description relating a change in material electrical resistivity or conductivity to an input stress or strain, and (2) the fundamental theoretical description of the dependence of the coupling constants ($\pi$ matrix coefficients) on Si material properties such as doping type, doping surface concentration and profile, and temperature. In the first part of this section we describe the phenomenological theory, while the second part is devoted to a brief description of the solid state physics that governs the $\pi$ matrix coefficients.

### 7.2.2 Phenomenological Theory

Electrical conduction in a linear anisotropic solid, such as single-crystal Si, can be described by a second rank conductivity or resistivity tensor, $\kappa_{ij}$ or $\rho_{ij}$, respectively. These tensors relate the electric field Cartesian components, $E_i$, to the electric current density components, $j_i$. In this discussion, we shall represent a vector, tensor, or matrix with individual elements, $A_{ijk}$, either by the element in square brackets, $[A_{ijk}]$, or by a boldface symbol, $\mathbf{A} = [A_{ijk}]$. We shall also use the convention that repeated indices indicate a summation over Cartesian components. For example, the conventional matrix product, $\mathbf{c} = \mathbf{ab}$ becomes, $c_{ij} = \sum_{j=1}^{3} a_{ij}b_{jk} \Rightarrow c_{ij} = a_{ij}b_{jk}$.

Usually, the electric current density is considered to be created through an existing or imposed electric field via the relation

$$j_i = \kappa_{ij}E_j \tag{7-1}$$

However, the field can also be considered to be established by currents from current sources via the relation

$$E_i = \rho_{ij}j_j \tag{7-2}$$

It is evident that the conductivity tensor or matrix, $\kappa$, and the resistivity tensor, $\rho$, are inverses, $\rho = \kappa^{-1}$. In general, it can be shown that both $\rho$ and $\kappa$ are symmetric tensors, regardless of the crystal symmetry.[6] For a crystal with cubic symmetry, like unstrained Si or Ge, the tensors are diagonal, with the diagonal elements equal, $\rho_{ij} = \rho\delta_{ij}$, and $\kappa_{ij} = \kappa\delta_{ij}$. The quantity $\delta_{ij}$ is defined by: $\delta_{ij} = 1$ if $i = j$, and $\delta_{ij} = 0$ if $i \neq j$. The diagonal nature of $\rho$ and $\kappa$ and the equality of the diagonal elements is a consequence of the fact that the crystal *looks the same* in each of the cubic axes or $x$-, $y$-, and $z$-directions. Following an argument of Ashcroft and Mermin,[7] if the tensors were nondiagonal, then a field in the $x$-direction could induce a current in another direction, say the $y$-direction. Using the cubic symmetry of the crystal, there would be an equal and opposite current induced in the $-y$-direction. The only consistent possibility is zero current in the $y$-direction, so $\kappa_{ij}$ must vanish for $i \neq j$.

When a generalized stress, $\sigma = [\sigma_{ij}]$, is applied to a crystal, the cubic symmetry is broken and it is reasonable to predict that off-diagonal elements will appear in $\rho$ or $\kappa$ as a result of this symmetry breaking. In addition, the magnitudes of the diagonal elements can also shift. In the following discussion of the stress-induced changes, we follow closely the discussion presented by Bittle, Suhling, Beaty, Jaeger, and Johnson.[8] The most general coupling of $\rho$ and $\sigma$ to second order in the stress tensor elements is given by

$$\rho_{ij} = \rho_{ij}^0 + \pi_{ijkl}\sigma_{kl} + \Lambda_{ijklmn}\sigma_{kl}\sigma_{mn} \tag{7-3}$$

The quantity, $\rho_{ij}^0 = \rho^0\delta_{ij}$, is the unstressed resistivity tensor, while the second two terms give the change, $\Delta\rho_{ij}$, produced by the stresses, $\sigma_{ij}$. In most cases of interest in electronics packaging, it is sufficient to consider only the term which is linear in the stress components, $\rho_{ij} = \rho_{ij}^0 + \Delta\rho_{ij}$, where

$$\Delta\rho_{ij} = \pi_{ijkl}\sigma_{kl} \tag{7-4}$$

In the general case, there are $3^4 = 81$ components of the fourth-order tensor, $\pi_{ijkl}$. Fortunately, a great reduction in the number of components required occurs as a result of symmetries. For cubic crystals, the number of independent coupling constants is reduced from 81 to 3. As discussed by Bittle et al.,[8] a reduction from 81 to 36 occurs because the resistivity and stress tensors are symmetric, $\Delta\rho_{ij} = \Delta\rho_{ji}$ and $\sigma_{kl} = \sigma_{lk}$, resulting in the dependence of $\Delta\rho_{ij}$ on $\sigma_{kl}$ being the same as its dependence on $\sigma_{lk}$. A further reduction to only three distinct $\pi$ coefficients occurs because of the cubic nature of the crystal. This is similar to the concept that only three elastic constants are required to completely specify the elastic behavior of a cubic crystal.[9]

Given the amount of compaction in the $\pi$ matrix, it is usual to write Eq. (7-4) in a more compact six-component vector notation shown in Eq. (7-5). The six-component notation makes the equations easier to write in explicit form, but the new vectors and matrices do not transform in the usual ways under a rotation of the coordinate system. In the six-component notation, the indices of the six distinct $\Delta\rho_{ij}$ or $\sigma_{ij}$ values transform as

$$
\begin{array}{cc}
\textit{Tensor} & \textit{6-Component} \\
11 = xx & \rightarrow 1 \\
22 = yy & \rightarrow 2 \\
33 = zz & \rightarrow 3 \\
23 = yz & \rightarrow 4 \\
13 = xz & \rightarrow 5 \\
12 = xy & \rightarrow 6
\end{array}
\tag{7-5}
$$

In this scheme, we shall use a Greek letter subscript, $\Delta\rho_\alpha$ or $\sigma_\alpha$, to designate the components. The transforms shown in Eq. (7-5) are used to translate between the usual second-rank tensor scheme and the six-vector scheme. In the transformation of the piezoresistance or $\pi$ tensor from a fourth-rank tensor to a second-rank tensor or matrix, it is customary to include a factor of $\rho^0$, the unstressed resistivity in the definition

$$\pi_{1111} \to \rho^0\pi_{11}; \qquad \pi_{1122} \to \rho^0\pi_{12}; \qquad 2\pi_{2323} \to \rho^0\pi_{44} \qquad (7\text{-}6)$$

In the six-component vector notation, the piezoresistance equations are equivalent to

$$\frac{\Delta\rho_\alpha}{\rho_0} = \pi_{\alpha\beta}\sigma_\beta \qquad (7\text{-}7)$$

where the $\pi_{\alpha\beta}$ elements are given by

$$\pi = \begin{pmatrix} \pi_{11} & \pi_{12} & \pi_{12} & 0 & 0 & 0 \\ \pi_{12} & \pi_{11} & \pi_{12} & 0 & 0 & 0 \\ \pi_{12} & \pi_{12} & \pi_{11} & 0 & 0 & 0 \\ 0 & 0 & 0 & \pi_{44} & 0 & 0 \\ 0 & 0 & 0 & 0 & \pi_{44} & 0 \\ 0 & 0 & 0 & 0 & 0 & \pi_{44} \end{pmatrix} \qquad (7\text{-}8)$$

A more involved and systematic discussion of notation is given by Bittle et al.[8] Smith has shown that the $\pi$ matrix form given in Eq. (7-8) is the most general that can occur for crystals with full cubic symmetry, such as Si or Ge which have a diamond crystal structure.[10] It can be noted that the elastic stiffness matrix, $[c_{\alpha\beta}]$, or the compliance matrix, $[s_{\alpha\beta}]$, which relate the six-component strain $[\varepsilon_\alpha]$ with the stress vector $[\sigma_\alpha]$ through the relations

$$\varepsilon_\alpha = s_{\alpha\beta}\sigma_\beta, \qquad \sigma_\alpha = c_{\alpha\beta}\varepsilon_\beta \qquad (7\text{-}9)$$

have the same form as the piezoresistance matrix given in Eq. (7-8). The values for the nonzero $s_{\alpha\beta}$ and $c_{\alpha\beta}$ are given by Middelhoek and Audet[11] as

$$c_{11} = 1.657 \times 10^5 \text{ MPa}$$

$$c_{12} = 0.639 \times 10^5 \text{ MPa}$$

$$c_{44} = 0.796 \times 10^5 \text{ MPa}$$

and

$$s_{11} = 0.768 \times 10^{-5} \text{ MPa}^{-1}$$

$$s_{12} = -0.214 \times 10^{-5} \text{ MPa}^{-1}$$

$$s_{44} = 1.256 \times 10^{-5} \text{ MPa}^{-1}$$

The goal of this analysis is to determine expressions for the electric field components, Eq. (7-2), as functions of the current density vector and the applied stress. Expanding Eq. (7-2) and using the six-component notation for the resistivity tensor leads to the expressions

$$E_1 = \rho_1 j_1 + \rho_6 j_2 + \rho_5 j_3$$

$$E_2 = \rho_6 j_1 + \rho_2 j_2 + \rho_4 j_3 \qquad (7\text{-}10)$$

$$E_3 = \rho_5 j_1 + \rho_4 j_2 + \rho_3 j_3$$

In all cases under consideration, measurements are made relative to the unstressed state where only $\rho_1$, $\rho_2$, and $\rho_3$ are nonzero. Using the six-vector relation, Eq. (7-7) for $\Delta\rho$ and the relation $\rho_\alpha = \rho_{\alpha 0} + \Delta\rho_\alpha$ yields the fundamental relation between the field components and the current density and stress components (in conventional two-index tensor notation)

$$E_1/\rho = [1 + \pi_{11}\sigma_{11} + \pi_{12}(\sigma_{22} + \sigma_{33})]j_1 + \pi_{44}\sigma_{12}j_2 + \pi_{44}\sigma_{13}j_3$$

$$E_2/\rho = \pi_{44}\sigma_{12}j_1 + [1 + \pi_{11}\sigma_{22} + \pi_{12}(\sigma_{11} + \sigma_{33})]j_2 + \pi_{44}\sigma_{23}j_3 \qquad (7\text{-}11)$$

$$E_3/\rho = \pi_{44}\sigma_{13}j_1 + \pi_{44}\sigma_{23}j_2 + [1 + \pi_{11}\sigma_{33} + \pi_{12}(\sigma_{11} + \sigma_{22})]j_3$$

The expressions in Eqs. (7-11) are valid in a coordinate system aligned with the principal cubic crystal axes or directions, with unit vectors

$$\hat{x} = [100], \qquad \hat{y} = [010], \qquad \hat{z} = [001] \qquad (7\text{-}12)$$

In the notation used in Eqs. (7-12), crystal directions are indicated by three indices in square brackets, $[ijk]$, while a crystal plane normal to the $[ijk]$ direction is indicated by $(ijk)$. Equivalent crystal directions, such as the ones specified in Eqs. (7-12) for a cubic crystal are indicated by $\langle ijk \rangle$. For example, the directions [100], [010], and [001] are the $\langle 100 \rangle$ directions.

In most cases of experimental interest, the coordinate systems describing both the current vector and stress tensor are rotated with respect to the system in the $\langle 100 \rangle$ directions. For these rotated systems, the $\pi_{\alpha\beta}$ coefficients can be transformed, but the transformation is not done by applying the usual rotation matrix transformation to Eq. (7-7) because the six-component

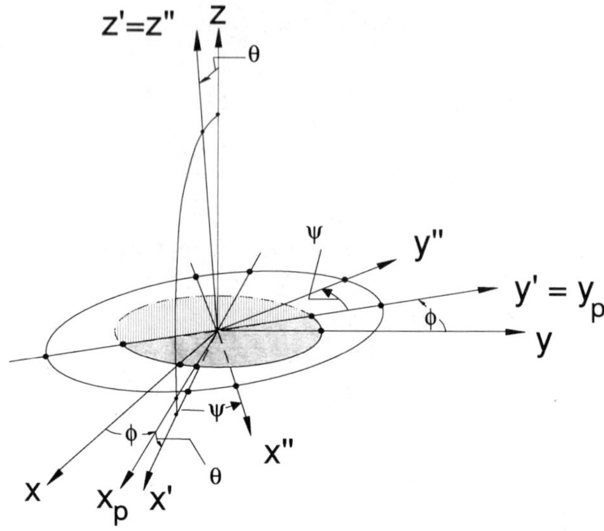


**Figure 7-1**   Euler angles, $\phi$, $\theta$, and $\psi$ used to define the orientation of a rotated coordinate system relative to the original system. The rotation is performed with the origin of the coordinate system in a constant location. The first rotation by an angle $\phi$ is performed in the $x$–$y$ plane about the $z$ axis to the $\{x_p, y_p, z\}$ coordinate system. The second rotation by an angle $\theta$ is performed in the $x_p$–$z$ plane about the $y_p$ axis to the $\{x', y', z'\}$ system. The third rotation by an angle $\psi$ is performed in the $x'$–$y'$ plane about the $z'$ axis to the $\{x'', y'', z''\}$ system.

scheme has been artificially constructed from the regular second- and fourth-rank tensor equations. The following discussion of coordinate transformations follows that of Mason and Thurston[2] and Kanda.[12]

We consider an $\{x'', y'', z''\}$ coordinate system which is rotated relative to the original $\{x, y, z\}$ system, as shown in Fig. 7-1. For simplicity, the designation $\{x_1, x_2, x_3\}$ will be used to specify the original coordinate system, with a similar notation for the primed system. There are two common methods of designating the orientation of the rotated system relative to the original system. In the first method the direction cosines are specified, $a_{ij} = \cos(\gamma_{ij})$, where $\gamma_{ij}$ is the angle between $x_i$ and $x'_j$. Although there are nine direction cosines, they are not all independent because only three numbers or rotation angles are needed to completely specify the orientation of the **x**$'$ system relative to the **x** system. This issue is discussed in detail by Goldstein.[13] The three angles $(\phi, \theta, \psi)$, called Euler angles, are used to describe sequential rotations relative to the original coordinate system. As Goldstein points out, different authors used different rotation sequences, as well as different rotation directions, making comparison of final formulas difficult. The convention we are using is the same as that used by Mason and Thurston,[2] Kanda,[12] and Thurston.[14] In their convention, Fig. 7-1, the first rotation is counterclockwise by an angle $\phi$ about the $x_3$ or $z$ axis, while the second rotation is by an angle $\theta$ about the $x_2'$ or $y'$ axis. A third rotation by an angle $\psi$ is performed in the $x_2'$–$x_3'$ or $x'$–$y'$ plane about the $x_3'$ or $z'$ axis. In the Euler angle set used by Goldstein, the $\phi$ rotation is the same as that in Fig. 7-1, but the second, or $\theta$, rotation is performed about the $x_p$ axis rather than the $y$ axis used in Fig. 7-1.

In the Mason and Thurston notation,[2] the rotation matrix, $R(\phi, \theta, \psi)$ is given by

$$R(\phi, \theta, \psi) =$$

$$\begin{pmatrix} \cos\phi\cos\theta\cos\psi - \sin\phi\sin\psi & \sin\phi\cos\theta\cos\psi + \cos\phi\sin\psi & -\sin\theta\cos\psi \\ -\cos\phi\cos\theta\sin\psi - \sin\phi\cos\psi & -\sin\phi\cos\theta\sin\psi + \cos\phi\cos\psi & \sin\theta\sin\psi \\ \cos\phi\sin\theta & \sin\phi\sin\theta & \cos\theta \end{pmatrix}$$

$$(7\text{-}13)$$

All of the physics and geometry of piezoresistive stress sensors is contained in Eqs. (7-11) and (7-13). The rotation matrix, Eq. (7-13) is used to transform the electric field and current density vectors into the direction of the resistor, as characterized by the angles $(\phi_1, \theta_1, \psi_1)$. Similarly, a transformation of the stress tensor $\sigma$ by a rotation through $(\phi_2, \theta_2, \psi_2)$ allows the stresses to be resolved along a different set of axes. Equation (7-11) relates the electric field, current density, and stress, which can be written in the symbolic form

$$\frac{\mathbf{E}}{\rho} = f(\mathbf{j}, \sigma) \tag{7-14}$$

where the function $f$ is defined by Eqs. (7-11). To transform Eq. (7-14), the field and current density components in the original system are required as functions of the components in the system oriented in the resistor direction. These transforms are given by $\mathbf{E} = \mathbf{R}^{T}(\phi_1, \theta_1, \psi_1)\mathbf{E}'$ and $\mathbf{j} = \mathbf{R}^{T}(\phi_1, \theta_1, \psi_1)\mathbf{j}'$, where $\mathbf{R}^{T}$ is the transpose of $\mathbf{R}$ and $\mathbf{E}'$ and $\mathbf{j}'$ are the electric field and current density vectors with components expressed in the rotated system. Similarly, the original stress tensor components are related to those in a rotated system, $(x_1'', x_2'', x_3'') \to (\phi_2, \theta_2, \psi_2)$, through the relation, $\sigma = \mathbf{R}^{T}(\phi_2, \theta_2, \psi_2)\sigma''\mathbf{R}(\phi_2, \theta_2, \psi_2)$. In writing these relations, we have used the fact that the inverse of the rotation matrix and transpose are equal,

$\mathbf{R}^{-1} = \mathbf{R}^{\mathrm{T}}$.[13] In component notation, the transformation equations become

$$E_i = R_{ji}(\phi_1, \theta_1, \psi_1)E_j'$$

$$j_i = R_{ji}(\phi_1, \theta_1, \psi_1)j_j' \qquad (7\text{-}15)$$

$$\sigma_{ij} = R_{li}(\phi_2, \theta_2, \psi_2)\sigma_{lm}'' R_{mj}(\phi_2, \theta_2, \psi_2)$$

When the expressions derived from Eqs. (7-15) are substituted into Eqs. (7-11), the resultant expressions can be solved for the field components in the primed system as functions of the current density components in the primed system and the stress tensor components in the double-primed system. The expressions equivalent to Eq. (7-11) but in the primed system can be written in component notation as

$$\frac{E_i'}{\rho} = (\delta_{ij} + g_{ij}(\pi_{11}, \pi_{12}, \pi_{44}, \sigma''))j_j' \qquad (7\text{-}16)$$

where the functions $g_{ij}$ are linear in both $\pi$ coefficients and the stress tensor components, as in Eqs. (7-11). Expressions similar to the ones presented here but in terms of direction cosines are presented by Bittle et al.[8]

The stress tensor measurement problem is formally contained in Eq. (7-16). This equation is inverted to find the $\boldsymbol{\sigma}''$ components, $\sigma_{ij}''$, as functions of the measured resistance changes and the $\pi$ coefficients. The sensitivity of the derived stress tensor components to the measured resistivity shifts depends in detail on the exact functional dependences given in Eq. (7-16).

Because of the complexity of the expressions that result from Eq. (7-16), it has become common to describe the piezoresistive behavior in terms of the responses to either uniaxial tensile or compressive stresses applied either in the resistor direction (longitudinal) or perpendicular to the resistor axis (transverse). The responses to these loads, when the current and electric field are both directed along the resistor axis, are characterized by a longitudinal piezoresistive constant, $\pi_l = \pi_{11}'$, and a transverse coefficient, $\pi_t = \pi_{12}'$. In the case of plane stress, the response to a shear stress, $\tau_{xy} = \sigma_{xy}$, may be characterized by a shear piezoresistive coefficient, $\pi_s = \pi_{16}'$. The general response to a state of plane stress with a longitudinal stress, $\sigma_l$, a transverse stress, $\sigma_t$, and a shear stress, $\tau_{12}'$ is given by

$$\frac{\Delta\rho}{\rho} = \pi_l\sigma_l + \pi_t\sigma_t + \pi_s\tau_{12}' \qquad (7\text{-}17)$$

In Eq. (7-17), $\tau_{12}'$ is resolved in an axes set with one axis along the resistor and the other axis perpendicular to the resistor. Through the use of Eq. (7-16), the longitudinal, transverse, and shear piezoresistive coefficients may be expressed in terms of $\pi_{11}, \pi_{12}, \pi_{44}$, and the direction cosines or Euler angles characterizing the resistor orientation. Expressions for $\pi_l$ and $\pi_t$ for a number of orientations of engineering interest are given by Pfann and Thurston.[3] These authors also give expressions for the coefficients in Eq. (7-17) in terms of $\pi_{11}, \pi_{12}, \pi_{44}$, and the direction cosines of the resistor axis and transverse stress direction with respect to a cubic axis direction. In many cases, these expressions may be used to derive the relations specified by Eq. (7-16) without using the complex rotational transforms specified by Eq. (7-15).

To connect measurement with theory, we consider a resistive strain gage of the type shown in Fig. 7-2. The design shown is typical of those used to make high-resistance resistive strain gages. In a good design, the resistance of the legs in the long direction of the gage is much greater than that of the wide connections at either end of the serpentine structure. The resistance of any leg is given by the relation

$$R = \frac{\rho L}{ab} \tag{7-18}$$

where $L$ = resistor length and the resistor cross-section is rectangular with

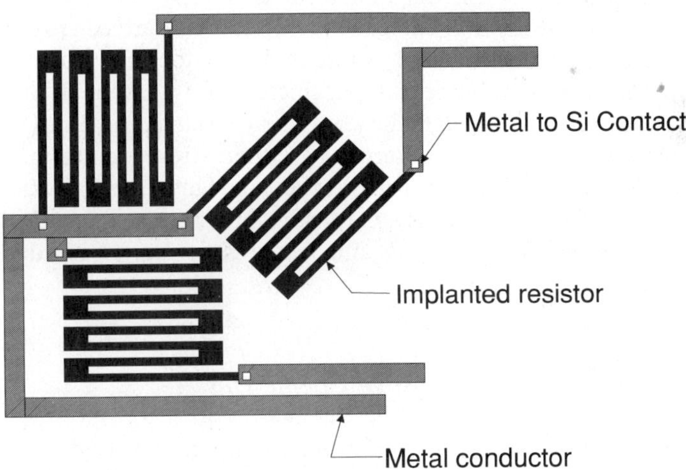


**Figure 7-2**  Layout of a group of piezoresistors used for stress measurement. The serpentine resistor layout is required to achieve a high resistance value, needed for measurement accuracy. The dark regions represent diffused or implanted resistors, while the cross-hatched regions are conductors that contact the diffused areas through contact windows, white squares.

with width $a$ and height $b$. When a stress is applied to the resistor, there is a variation $\delta\rho$ in the resistivity and also a variation in the resistor dimensions. The resultant fractional variation, $\delta R/R$ is given by

$$\delta R/R = \frac{\delta\rho}{\rho} + \frac{\delta L}{L} - \frac{\delta a}{a} - \frac{\delta b}{b} \qquad (7\text{-}19)$$

For a linear isotropic material, the cross-section strains, $\varepsilon_a = \delta a/a$ and $\varepsilon_b = \delta b/b$, are related to the longitudinal strain, $\varepsilon_L = \delta L/L$, by Poisson's ratio, $\mu = -\varepsilon_a/\varepsilon_L = -\varepsilon_b/\varepsilon_L$. Then Eq. (7-19) becomes

$$\frac{\delta R}{R} = \frac{\delta\rho}{\rho} + (1 + 2\mu)\varepsilon_L \qquad (7\text{-}20)$$

Equation (7-20) can be used to derive the gage factor that gives the fractional change in resistance per unit strain along the resistor axis, $K = (\delta R/R)/\varepsilon_L$. The result is

$$K = \frac{(\delta\rho/\rho)}{\varepsilon_L} + 1 + 2\mu \qquad (7\text{-}21)$$

For metals, only the second term in Eq. (7-21) is present, and since $\mu \approx 0.5$, $K \approx 2$. In contrast, for Si, the first term is usually much larger than the second. The explicit form of the gage factor can be derived from Eq. (7-16). It should be noted that the theory presented here does not make either stress or strain the more fundamental parameter. It is assumed that the two quantities are related through the compliance matrix via Eq. (7-9), and so Eq. (7-16) can be written in terms of either stress or strain. It should, however, also be noted that in general the gage factor for anisotropic Si will not be independent of the transverse strains or stresses, since $\delta R/R$ from Eq. (7-16) is not proportional to $\varepsilon_L$, and hence $K$ in Eq. (7-21) depends in a detailed way on the strains.

### 7.2.3 Theory of the Piezoresistive Coefficients

To complete the formal theory of piezoresistivity in semiconductors, a theory relating the $\pi$ coefficients to fundamental material properties is required. Unfortunately, this theory is complex and accurate calculations of the $\pi_{\alpha\beta}$ from first principles cannot be made. However, the theory in an approximate form does provide understanding and guidance in designing stress-sensitive and temperature-insensitive piezoresistors, and so it will be reviewed briefly.

In actual use, it is desirable to determine the $\pi_{\alpha\beta}$ experimentally, and this will be discussed below.

The development of the theory is closely connected to the pioneering measurements of the piezoresistive effect in Ge and Si by Smith.[1] The electronic energy band structure of Ge and Si was a subject of intense study during the late 1940s and early 1950s. Smith's experimental work showed that the Fermi surfaces for electrons and holes in these materials were nonspherical. Contemporary results for the energy band structures of these materials are given by Sze.[15] In the following discussion, we will concentrate on Si, as this is the main material of interest for practical measurements. The scalar conductivity, $\kappa$, of p- or n-type Si is related to the carrier concentration $n$ and the mobility $\mu$ by the relation

$$\kappa = n\mu e \tag{7-22}$$

where $e$ is the electronic charge. If there is more than one type or piece of the Fermi surface, then the conductivity is related to the carrier concentration, $n_i$, and the mobility, $\mu_i$, on the $i$th piece of the surface by

$$\kappa = \sum_i \kappa_i$$

$$= \sum_i n_i \mu_i e \tag{7-23}$$

When stress is applied to an Si crystal, Eq. (7-23) shows that the conductivity can vary from either a change in the carrier concentrations, $n_i$ or the mobilities, $\mu_i$. A calculated band structure, $E(\mathbf{k})$, for Si is shown in Fig. 7-3.[16] This figure shows the variation of the electron energy $E$ with wavevector $\mathbf{k}$ along two lines, one from the zone center $\Gamma$ along the [100] direction and one from $\Gamma$ along the [111] direction. The inverse effective mass tensor, $1/m^*_{ij}$, is related to the energy band structure through the relation

$$\left(\frac{1}{m^*}\right)_{ij} = \frac{1}{\hbar^2} \frac{\partial^2 E(\mathbf{k})}{\partial k_i \, \partial k_j} \tag{7-24}$$

In Eq. (7-24), $\hbar$ is Planck's constant divided by $2\pi$ and the $k_i$ are the Cartesian components of the wavevector $\mathbf{k}$. Equation (7-24) shows that energy bands that have a high curvature or second derivative of $E(\mathbf{k})$ have a small effective mass. The effective mass and the conductivity are closely related. In a simple relaxation time model in which the electronic scattering is characterized by a single relaxation time, $\tau$, and there is a scalar effective mass, $m^*$, the

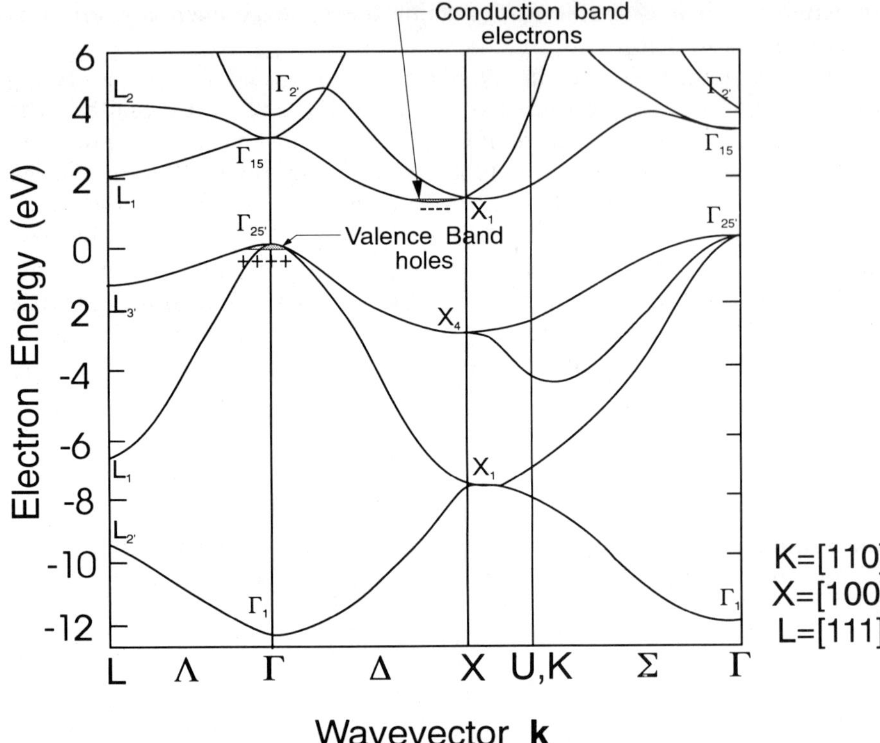


**Figure 7-3**    Energy band structure, $E(\mathbf{k})$, for Si along several paths in $\mathbf{k}$ space. The center of the zone in $\mathbf{k}$ space is designated as $\Gamma$, $\Lambda$ represents a path along the [111] direction in $\mathbf{k}$ space, while $\Delta$ represents a path along the [100] direction. $\Sigma$ is a path along the [110] direction. Valence band holes are located at the zone center in the $\Gamma_{25'}$ band while conduction band electrons are in the valleys of the $\Gamma_{15}$ band along the $\Delta$ direction.

conductivity is related to $m^*$, $\tau$, and $n$ by

$$\kappa = \frac{ne^2\tau}{m^*} \tag{7-25}$$

or, equivalently,

$$\mu = \frac{e\tau}{m^*} \tag{7-26}$$

This equation shows that high mobility is associated with small effective masses.

Figure 7-3 shows that the electrons in the conduction band occupy a small pocket or valley in $\mathbf{k}$ space along the $\langle 100 \rangle$ axis. There are six equivalent

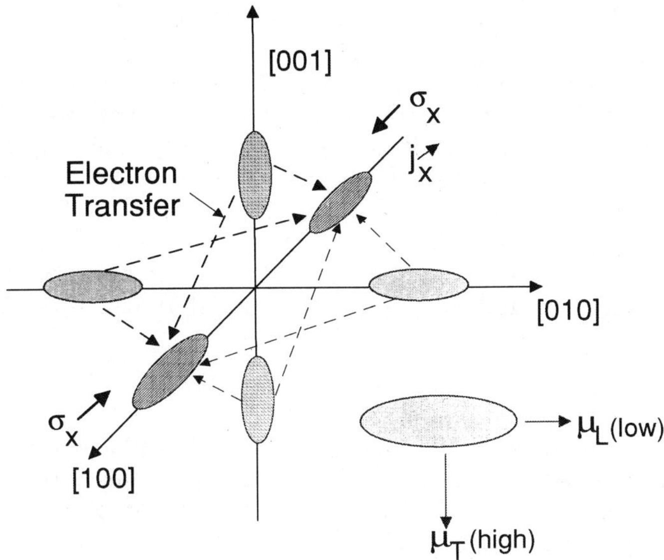


**Figure 7-4**    The **k** space conduction electron distribution in n-type Si. The conduction electrons occupy "valleys" or ellipsoids of revolution located on the prinxipal axes, $\langle 100 \rangle$ directions. Also shown is the transfer of electrons produced by a uniaxial stress, $\sigma_x$, along the [100] direction. The mobility, $\mu_L$, associated with motion along the major axis is low relative to that associated with motion along a minor axis, $\mu_T$.

valleys, two on each direction of the primary axes. A diagram of the conduction electron distribution in **k** space is shown in Fig. 7-4. The six equivalent valleys are ellipsoids of revolution, a requirement of the cubic symmetry. For these ellipsoids, the effective mass along a $\langle 100 \rangle$ direction, called the longitudinal effective mass, is $m_L \approx 1.0 m_e$, while the effective mass for motion transverse to the ellipsoid axis is the transverse effective mass, $m_T \approx 0.2 m_e$, where $m_e$ is the free-electron mass. This band structure is primarily responsible for the piezoresistance effect in n-type Si, as first explained by Herring[17] and by Herring and Vogt.[18]

In the case of p-type Si, Fig. 7-3 shows that there are two hole energy bands which are degenerate or have the same energy at the zone center, $\mathbf{k} = 0$. The lower or steepest band is the light hole band, with an effective hole mass $\approx 0.16 m_e$, while the flatter band is the heavy hole band with an effective mass $\approx 0.49 m_e$. This large difference in effective masses for these bands is thought to be the cause of the large piezoresistive effect in p-type Si. However, the transport theory is more complex for hole motion in the valence bands than for the conduction bands of n-type Si, and so, historically, the piezoresistive effect in n-type Si was studied first. We shall discuss the n-type effect in some detail and then return to a brief analysis of p-type material.

### n-Type Silicon

According to the deformation potential theory discussed by Herring and Vogt[18] and by Keyes,[19] the major effect of a uniaxial compression is to raise or lower all of the energy levels in a conduction band valley by a constant amount. For example, a uniaxial compression in the [100] direction lowers the energy levels in the two longitudinal ellipsoids on the [100] axis and raises the energy levels of the four transverse ellipsoids on the [010] and [001] axes (Fig. 7-4). As a result, electrons transfer from the four transverse ellipsoids to the longitudinal ellipsoids on the [100] axis. From Eq. (7-23)

$$\kappa = 2n_L\mu_L e + 4n_T\mu_T e \tag{7-27}$$

where $L$ = longitudinal (in the stress and field direction) and $T$ = transverse (perpendicular to the stress and field direction). In the unstrained state, $n_L = n_T = n/6$, where $n$ = total number of conduction electrons per unit volume. Substituting in Eq. (7-27) yields

$$\kappa = \tfrac{1}{3}n\mu_L e + \tfrac{2}{3}n\mu_T e \tag{7-28}$$

Assuming that there is no stress-induced change in the mobilities, $\mu_L$ and $\mu_T$, then the change in $\kappa$ is given by

$$\delta\kappa = 2\delta n_L\mu_L e + 4\delta n_T\mu_T e \tag{7-29}$$

Since all of the electrons stay in the conduction band the variations $\delta n_L$ and $\delta n_T$ are related by the constraint, $\delta n = 0$, or

$$2\delta n_L + 4\delta n_T = 0 \tag{7-30}$$

With the use of Eqs. (7-28), (7-29), and (7-30), and the definition $K = m_L/m_T$, the following expression can be derived for the quantity, $\delta\kappa/\kappa$:

$$\frac{\delta\kappa}{\kappa_0} = -\frac{2\delta n_L}{n}\frac{(K-1)}{(1+2K)} \tag{7-31}$$

Equation (7-31) contains the basics of the theory of piezoresistance in n-type Si. When a uniaxial compression is applied in the [100] direction, the energy levels in the [100] valleys are lowered and the energies of the [010] and [001] valleys are increased. As a result the conduction electrons redistribute, with electrons flowing from the high-energy [010] and [001] valleys into the

[100] valleys. As a result, $\delta n_L$ is positive, and Eq. (7-31) shows that $\delta\kappa/\kappa < 0$. The relation $\delta\kappa/\kappa = -\delta\rho/\rho$ shows that the resistivity increases. Since the strain $\varepsilon_x$ is negative in a compression, we see that n-type Si has a negative gage factor, as experimentally observed.

In calculations based on the theory outlined above, Herring[17] and Herring and Vogt[18] derived expressions for the n-type Si piezoresistive coefficients $\pi_{11}$, $\pi_{12}$, and $\pi_{44}$. Although these calculations assume low or nondegenerate doping, they can be formally extended to the high or degenerate doping case of interest by using Fermi distribution functions in place of the Boltzmann distribution used by Herring. The resultant equations have been expressed by Tufte and Stelzer[20] as

$$\pi_{11}(T) = \pi_{11}(0)\,\frac{\mathscr{F}_{j-1/2}(\eta)}{\mathscr{F}_{j+1/2}(\eta)}$$

$$\pi_{11}(0) = \frac{2}{3}\,\frac{\Xi_u}{(c_{11}-c_{12})k_B T}\,\frac{(K-1)}{(2K+1)} \tag{7-32}$$

$$\pi_{12} = -\pi_{11}/2$$

$$\pi_{44} = 0$$

In the second of Eqs. (7-32), $T$ is the absolute temperature, $k_B$ is Boltzmann's constant, $c_{11}$ and $c_{12}$ are elastic constants, and $\Xi_u$ is a constant with units of energy known as the deformation potential. The quantity $\mathscr{F}_j(\eta)$ in the first of Eqs. (7-32) is a Fermi integral, defined by

$$\mathscr{F}_j(\eta) = \frac{1}{\Gamma(j+1)}\int_0^\infty \frac{\varepsilon^j\,d\varepsilon}{1+e^{\varepsilon-\eta}} \tag{7-33}$$

In Eq. (7-33), $\Gamma(x)$ is the gamma function with the properties: $\Gamma(x+1) = x!$, $\Gamma(1/2) = \pi^{1/2}$, and $\eta$ is the Fermi energy measured relative to the conduction band energy in units of $k_B T$, $\eta = (E_F - E_C)/k_B T$. The parameter $j$ is called the scattering parameter and characterizes the dependence of the scattering or relaxation time, $\tau$, from Eq. (7-25) on energy. The relation is assumed to be of the form, $\tau = \tau_0(E/E_0)^j$, where $\tau_0$ and $E_0$ are constants. Although Eqs. (7-32) have been derived assuming that the only effect of strain is to change the electron densities in the six conduction band valleys, they do contain much of the physics of the problem and also provide a theoretical basis for defining a reduced set of calibration measurements for n-type Si. The quantity $\pi_{11}(0)$ is the zero-temperature $\pi_{11}$ coefficient for lightly doped Si, in which the Fermi level is below the conduction band and hence $\eta < 0$. In this limit, $\mathscr{F}_k(\eta) \to \exp(\eta)$, independent of $k$, and so the ratio of Fermi integrals in the first of Eqs. (7-32) goes to 1. The second of Eqs. (7-32) predicts that $\pi_{11}(0)$ will be independent of doping density and that it will vary inversely as the absolute temperature, assuming that the deformation potential and mobility are temperature-independent. At high doping densities, the Fermi level enters the conduction band and $\eta$ becomes $> 0$. In this case, the first of Eqs. (7-32) can be used to show that the magnitude of $\pi_{11}$ decreases and that the variation with temperature is reduced significantly.

Using experimental data on the piezoresistance of P-doped and As-doped Si over a doping density range from $N_D \approx 1 \times 10^{15}$ to $1 \times 10^{20}$ impurity atoms/cm$^3$, Tufte and Stelzer[20] evaluated how well the simple theory in Eqs. (7-32) could explain their results. In the high doping density limit, with an assumed parabolic conduction band, they show that the carrier concentration, $n \approx N_D$, is related to the Fermi level by $\eta = Bn^{2/3}$, where $B$ is a constant, given by[21]

$$B = \frac{\hbar^2}{2m_d} \frac{(3\pi^2)^{2/3}}{k_B T} \tag{7-34}$$

where $m_d$ is a quantity called the density of states effective mass and can be derived from the band structure. The resultant dependence of $\pi_{11}$ on $n$, $T$, and the scattering parameter $j$ is given by

$$\pi_{11} = \frac{\pi_{11}(0)(j + \tfrac{3}{2})}{Bn^{2/3}} \tag{7-35}$$

Equation (7-35) explicitly predicts the dependence of $\pi_{11}$ on doping density or carrier concentration and temperature. By fitting their $T = 77$ K data for $\pi_{11}$ vs. $N_D$, Tufte and Stelzer verified the $n^{-2/3}$ dependence of $\pi_{11}$ on doping at $n > 10^{19}$ cm$^3$ and also determined that $j = -0.5$ provided the best fit between theory and experiment. However, at higher temperatures the fit became worse, indicating that the theory is an oversimplification. As Herring had previously pointed out, scattering between conduction band valleys, called intervalley scattering, becomes important at higher temperatures and must be properly accounted for in a complete theory.[17]

The other major predictions of the simple theory are that $\pi_{12}$ has half the magnitude of $\pi_{11}$ with an opposite sign, and that $\pi_{44} = 0$. The experimental ratio $|\pi_{44}/\pi_{12}|$ is frequently taken as a measure of how well the data is explained by the simple theory. Another common metric is the hydrostatic pressure coefficient, defined by $\pi_p = (1/p)\, dp/d\rho$. This coefficient defines the response to a hydrostatic loading corresponding to a pressure $p$; $\sigma_{11} = \sigma_{22} = \sigma_{33} = -p$, and $\sigma_{ij} = 0$, for $i \neq j$. Inspection of Eq. (7-11) shows that $\pi_p = \pi_{11} + 2\pi_{12}$, and thus the simple theory predicts a zero pressure coefficient, since $\pi_{12} = -\pi_{11}/2$ from the third of Eqs. (7-32). This result is in agreement with the argument that the major effect of pressure is to reduce the unit cell size uniformly, and thus the population of the conduction valleys will remain unchanged. Deviations of $\pi_p$ from zero are produced by other effects such as a shift in the energy gap with pressure. After a consideration of piezoresistance in p-type Si, we shall examine to what extent the experimental data supports the simple theory.

## p-Type Silicon

Conduction in p-type Si results from the motion of holes at the top of the valence band or, more precisely, bands. A good description of this band structure is given by Boer.[22] The valence band in Si develops from the $p_{3/2}$ and $p_{1/2}$ orbital levels of the free Si atom. The $p_{3/2}$ levels are four-fold degenerate (hold four electrons), while the $p_{1/2}$ levels are doubly degenerate. The $p_{1/2}$ levels are lowered in energy by the spin–orbit interaction by about 0.045 eV at the zone center, $\mathbf{k} = 0$. The behavior of these bands is shown in Fig. 7-5, for three directions. Most of the holes in the valence band will be in the upper, or $j = 3/2$ derived bands. However, the split-off $j = 1/2$ band is close enough in energy to have a significant effect on transport (at 300 K, the thermal energy $k_B T = 0.025$ eV). From Eq. (7-24), the band with the greatest curvature will be associated with the smallest effective mass, or the light hole band. This is the lower of the two $j = 3/2$ derived bands in Fig. 7-5. The upper $j = 3/2$ band is called the heavy hole band. Since the band

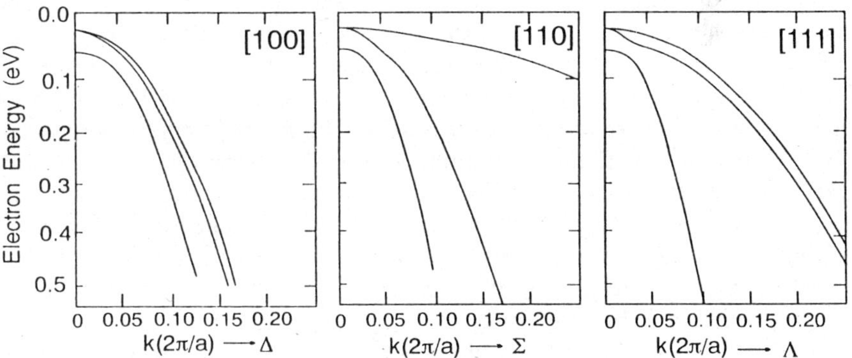


**Figure 7-5**  Magnification of the $E(\mathbf{k})$ valence band structure in Si near $\Gamma_{25'}$ for three directions, [100], [110], and [111], showing the light and heavy hole bands (upper two bands) and the split-off, $j = 1/2$, band (lower band). The magnitude of the wavevector, $k = |\mathbf{k}|$, is plotted in units of $2\pi/a$, where $a$ is the lattice constant. $a = 5.43$ Å for Si. For the upper two bands, the light hole band has the greatest curvature and the difference between light and heavy hole effective masses is greatest for motion in the [110] direction.

curvature varies with direction, it is not possible to uniquely define effective masses for the two $j = 3/2$ derived bands. Boer quotes a light hole average mass $m_{lh} = 0.15m_e$, and a heavy hole mass $m_{hh} = 0.54m_e$ in the [100] directions.[22]

Piezoresistance results from the shifting or movement in **k** space of these bands with applied stress. From Fig. 7-5 it is evident that the largest curvature or effective mass difference is in the [110] direction, while there is a small difference in the cubic axis or [100] direction. Thus we would predict a large piezoresistive coefficient for p-type resistors oriented in a [110] direction. Experimentally, the gage factor for p-type Si is found to be positive, indicating that a positive strain produced by a tensile stress causes the resistance to increase. This indicates that the high-mobility (low-mass) band moves down in energy with respect to the low-mobility (high-mass) band. As a result, holes flow from the low-mass band to the high-mass band and the conductivity decreases. Although the complexity of the band structure does not enable the making of simple theoretical predictions, as in the case of n-type Si, experimentally it is found that $\pi_{44}$ is much larger in magnitude than the other two coefficients. We now turn to a review of experimental determinations of the $\pi$ coefficients and examine to what extent the experimental data verifies the theoretical predictions.

## 7.3 EXPERIMENTAL MEASUREMENTS OF PIEZORESISTIVE COEFFICIENTS

The question of interest to users of diffused or implanted Si strain gages is how well the piezoresistive coefficients can be predicted from a knowledge of the resistor orientation and the doping parameters. In this section, we review reported measurements of the $\pi$ coefficients and address the question of prediction. A summary of reported measurements is given in Table 7-1. The first measurements are those reported by Smith.[1] His measurements were made on lightly and uniformly doped bulk single-crystal specimens. These data confirmed the prediction of the simple theory, $\pi_{11}^n \approx -2\pi_{12}^n$ and also confirmed by direct measurement that the pressure coefficient, $\pi_{11} + 2\pi_{12}$, was small relative to the major $\pi$ coefficient for both n and p doping. Tufte and Stelzer[4] measured the piezoresistive response of Si with heavily doped n- and p-type diffused resistive layers. These measurements showed that both the magnitude and the temperature dependence of the major coefficients, $\pi_{11}^n$ and $\pi_{44}^p$, decreased as the doping density increased. These data also showed a small pressure effect, indicating that the relation $\pi_{11}^n \approx -2\pi_{12}^n$ remained true even at high doping. The Tufte and Stelzer data for bulk doped n-type Si are reproduced in Fig. 7-6. The interesting message in these data is that moderately large and relatively temperature-independent $\pi$ coefficients can be obtained at surface concentrations $n_s \approx 1-5 \times 10^{20}$ cm$^{-3}$,

**Table 7-1**  Experimental Values of Piezoresistive Coefficients

| Si Type | Doping Level (cm$^{-3}$) | $\pi_{11}$ (10$^{-5}$ MPa$^{-1}$) | $\pi_{12}$ (10$^{-5}$ MPa$^{-1}$) | $\pi_{44}$ (10$^{-5}$ MPa$^{-1}$) | $\pi_{11} + 2\pi_{12}$ (10$^{-5}$ MPa$^{-1}$) | Ref. |
|---|---|---|---|---|---|---|
| n | $3.7 \times 10^{14}$ | $-102.2$ | 53.4 | $-13.6$ | 4.6 | 1 |
| p | $1.8 \times 10^{15}$ | 6.6 | $-1.1$ | 138.1 | 4.4 | 1 |
| n | $1.0 \times 10^{17}$ | $-80$ | | $-12$ | 2 | 4 |
| p | $1.0 \times 10^{17}$ | | | 114 | | 4 |
| n | $1.0 \times 10^{20}$ | $-43$ | | | | 4 |
| p | $1.0 \times 10^{20}$ | | | 71 | | 4 |
| n | $1.0 \times 10^{16}$ | $-105$ | | $-12.5$ | 2 | 20 |
| n | $1.0 \times 10^{20}$ | $-38$ | | $-19$ | 2.5 | 20 |
| n | | $-48.4$ | 25.3 | $-11.3$ | | 24 |
| p | | 5.7 | $-2.3$ | | | 24 |
| n | | $-29.3$ | 16.7 | $-12.7$ | | 25 |
| n | | $-28.6$ | | $-14.7$ | | 32 |
| p | | | | 83 | | 26 |
| p | Low | | | 99 | | 27 |
| p | High | | | 77.5 | | 27 |

typical of integrated circuit source and drain ion implantations. In Fig. 7-7, $\pi_{11}$ for n-type and $\pi_{44}$ for p-type are shown as a function of doping surface concentration (diffused samples) or bulk doping density (uniformly doped sample) at a constant temperature, $T = 27°C$. The data on diffused samples is from ref. 4 and on bulk doped n-type material from ref. 20. It can be seen that there is good agreement between the values for $\pi_{11}$ obtained for bulk and diffusion doped n-type samples.

There is, however, a question of how much the $\pi$ coefficient magnitudes depend on the details of the doping profile. In a typical integrated circuit process, an impurity ion, P or As for n-type or B for p-type, is implanted into a region of the opposite doping density. These impurity ions may then be driven or diffused into the Si by a high-temperature annealing step. A model profile of diffusion doping, described below, and an actual measured profile on an ion-implanted Sandia Laboratories IC are shown in Fig. 7-8. The junction depth, $x_J$, is defined as the depth at which the doping from the implantation, $n$, is equal in magnitude to the base region doping of the opposite sign, $n_B$. For the implanted layer in Fig. 7-8, $x_J \approx 0.4$ μm. For an idealized case, the doping profile is characterized by diffusion for a time $t$

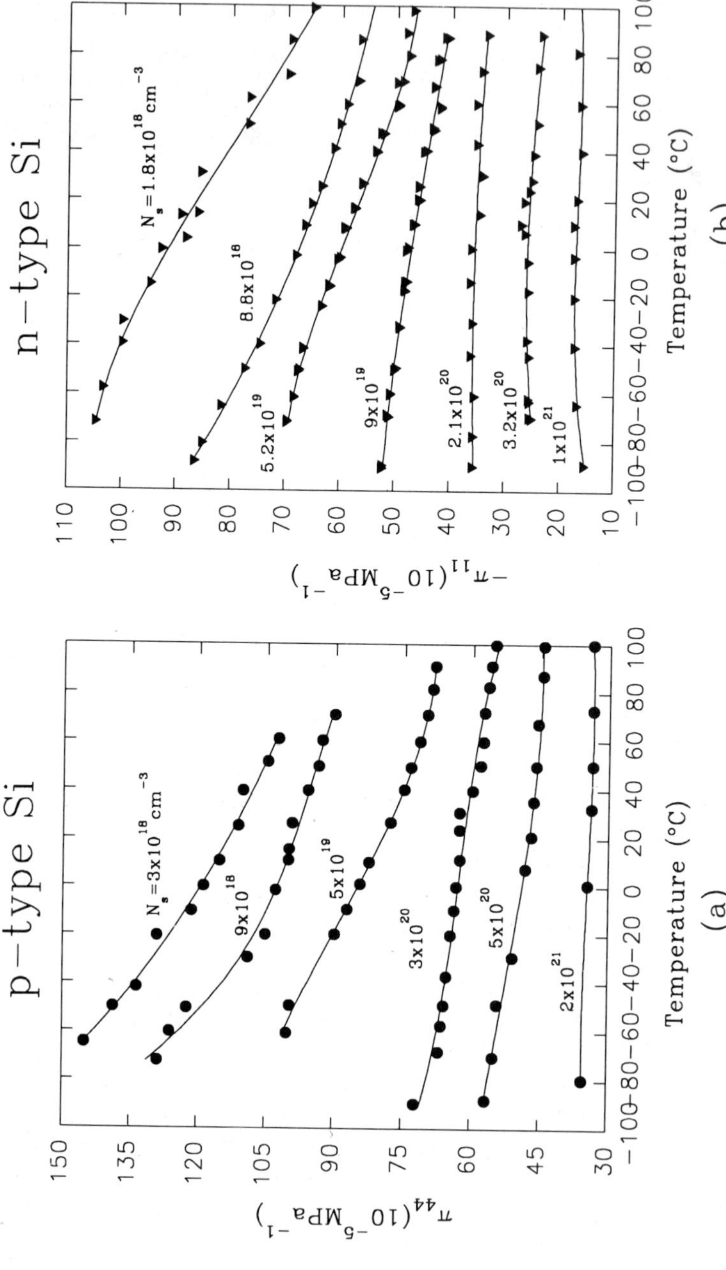


**Figure 7-6** Piezoresistance coefficient data (a) $\pi_{44}$ for p-type Si and (b) $\pi_{11}$ for n-type Si vs. temperature obtained by Tufte and Stelzer[4] for various doping densities, as indicated by the surface concentrations, $N_s$.

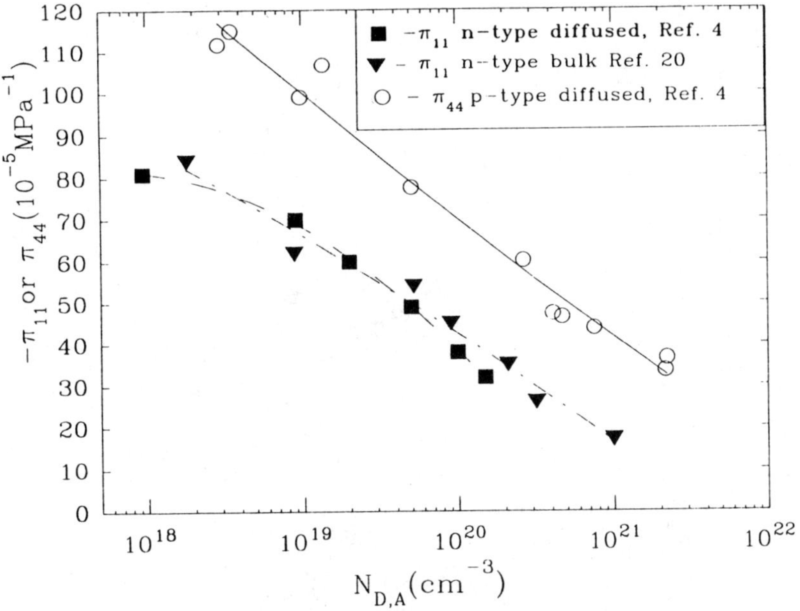



**Figure 7-7**    Room-temperature $\pi_{11}$ for n-type and $\pi_{44}$ for p-type data obtained by Tufte and Stelzer for n- and p-type diffused and n-type bulk-doped samples. For bulk-doped samples, $N_{D,A}$ is the donor doping density for n-type or the acceptor doping density for p-type bulk-doped samples. For diffused samples, $N_{D,A}$ is the surface dopant concentration.

with a constant diffusion constant $D$. The doping density at depth $x$ and time $t$ can be described by a function $n(x, t)$ of the form

$$n(x, t) = n_s f(x/2\sqrt{Dt}) \tag{7-36}$$

The function $f$ is found from a solution of the diffusion equation. For example, for diffusion from a source with a constant surface density, $n_s$, $f$ is the complementary error function,[23]

$$f(u) = \text{erfc}(u) = 1 - \left(\frac{2}{\sqrt{\pi}}\right) \int_0^u e^{-\beta^2} \, d\beta \tag{7-37}$$

In this case, the junction depth is given by

$$x_J = 2\sqrt{Dt} \, f^{-1}(n_B/n_s) \tag{7-38}$$

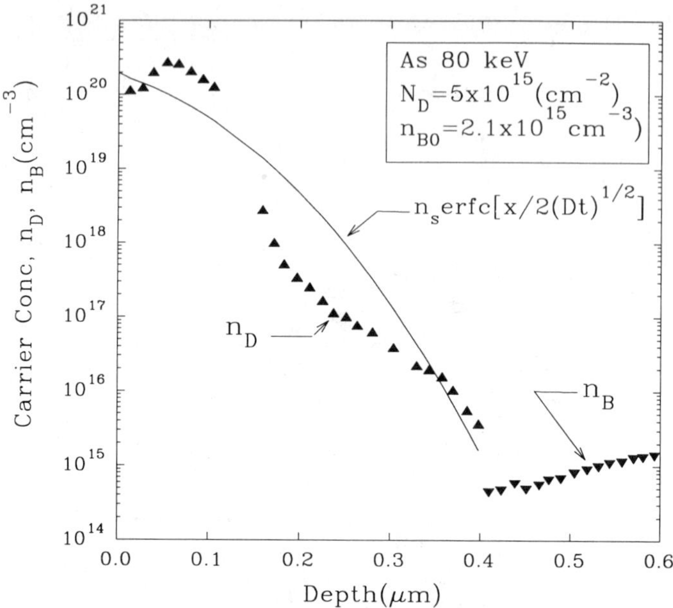


**Figure 7-8**   Carrier concentration or donor atom density of an ion-implanted layer used to make n-type piezoresistors. This profile resulted from an As implant at 80 keV. $n_D$ is the implanted atom density and $n_B$ is the background p-type acceptor or hole density. A model diffusion profile is also shown (solid line) characteristic of diffusion from a source with constant density $n_s = 2 \times 10^{20}$ cm$^{-3}$ for a time $t$ with diffusion constant $D$. $N_D$ is the total implant dose and $n_{BO}$ is the background doping of the p-type region into which the n-type impurities are implanted.

where $f^{-1}$ is the inverse of $f$. In typical semiconductor diffusions or implants, such as the one shown in Fig. 7-8, $n_B \ll n_s$. The effective conductivity of a nonhomogeneous conducting layer is found by adding or integrating the conductivity for each differential slice in the layer, since the slices act like resistors in parallel. If we assume that the mobility varies relatively slowly with doping density, then the effective or average conductivity, $\langle \kappa \rangle$ is directly proportional to the average carrier concentration, $\langle n \rangle$, through Eq. (7-22). $\langle n(t) \rangle$ is then calculated by averaging $n(x, t)$ over the junction depth:

$$\langle n(t) \rangle = \left( \frac{1}{x_J} \right) \int_0^{x_J} n(x, t)\, dx$$

$$= \frac{n_s}{f^{-1}(n_B/n_s)} \int_0^{u_J} f(u)\, du \qquad (7\text{-}39)$$

In Eq. (7-39), $u_J = x_J/[2(Dt)^{1/2}]$. The condition, $n_B \ll n_s$, implies that $u_J$ will be relatively large. For example, if $f(u) = \mathrm{erfc}(u)$, then for $n_B/n_s \approx 10^{-5}$, $u_J = f^{-1}(n_B/n_s) \approx 3.2$. The net result is that the integral in the second line of Eq. (7-39) is practically independent of $u_J$ and since $f^{-1}$ is a slowly varying function of $n_B/n_s$, $\langle n \rangle$ is essentially directly proportional to and dependent only on $n_s$, independent of the actual profile. This is frequently an excellent approximation, even for cases such as that shown in Fig. 7-8, in which the profile is not one created by concentration-independent diffusion.

Tufte and Stelzer argue that the average value of the piezoresistive coefficient should be calculated by averaging it with the depth-dependent conductivity:

$$\langle \pi \rangle = \frac{\int_0^{x_J} n(x) e\mu(x)\pi(x)\, dx}{\int_0^{x_J} n(x) e\mu(x)\, dx}$$

$$\approx \frac{(1/x_J) \int_0^{x_J} n(x)\pi(x)\, dx}{\langle n \rangle}$$

$$= \left( \frac{1}{x_J \langle n \rangle} \right) \int_{n_s}^{n_B} \left( \frac{n\pi(n)}{|dn/dx|} \right) dn \qquad (7\text{-}40)$$

In the second line of Eq. (7-40) the mobility has been assumed to be constant, $\mu(x) = \mu$, while in the third, the variable of integration has been changed from $x$ to $n$. The third line shows why only the near-surface region is important in determining $\langle \pi \rangle$.

Considering the doping profile shown in Fig. 7-8, it can be seen that $n$ decreases rapidly with increasing depth and that $|dn/dx|$ also becomes large. As a result, almost all of the contribution to the integral comes from the near-surface region where $n \geqslant 10^{20}/\mathrm{cm}^3$.

Tufte and Stelzer repeated their high doping density measurements on bulk n-type samples and made a detailed comparison between simple theory and experiment,[20] as discussed above. They found relatively good agreement between $\pi_{11}^n(n)$ for bulk doped samples and $\pi_{11}^n(n_s)$ for diffused samples, confirming the arguments about the diffused region $\pi$ coefficient depending only on the surface concentration.

The remaining data in Table 7-1 are from calibration of actual stress sensor test chips. Miura and co-workers reported on the calibration of a chip with both p- and n-type resistors in different orientations, using a four-point bending rig for application of uniaxial stress and a pressure system for hydrostatic loading.[24] Their geometry will be discussed in more detail below. Here we only note that the orientation of the p-type resistors in the [100] directions made them insensitive to $\pi_{44}^p$ and hence, for the p-type material, only the small $\pi_{11}^p$ and $\pi_{12}^p$ components are reported.

The value $\pi_{11}^n = -48.4 \times 10^{-5}/\text{MPa}$ is consistent with a surface concentration of $n_s \approx 10^{20}/\text{cm}^3$.

Natarajan and Bhattacharyya reported the data in the next row of Table 7-1 for n-type material that they used to make a stress sensor test chip.[25] They do not present their method of measurement but the results are consistent with those of the previous workers. Gee, Akylas, and van den Bogert[32] reported the determination of longitudinal, transverse, and shear piezoresistive coefficients for n-type material. They used a four-point bending rig to apply uniaxial stress to resistors oriented in four directions on (111) Si, two parallel to the chip edges and two at 45° with respect to the edges. Using their reported results, the values of $\pi_{11}^n$ and $\pi_{44}^n$ were calculated, with the assumption that $\pi_{12}^n = -\pi_{11}^n/2$.

Lundstrom and Gustafsson[26] reported the data on the next line for $\pi_{44}^p$. They used a p-type (111) Si substrate, but they did not report their method of performing the calibration. They do remark that "... a large uncertainty concerning the values of the piezoresistive constants remains, and that no big effort has been carried out in order to make an exact determination of $\pi_{44}^p$."

The data on the last two lines of Table 7-1 was reported by Beaty and co-workers from Auburn University.[27] They measured the value of $\pi_{44}^p$ using a p-type (100) Si wafer and a four-point bending rig for application of uniaxial stresses. Unfortunately, they did not report the surface doping concentration for their resistors, but instead plotted the value of $\pi_{44}$ vs. the resistor unstressed resistance. The values listed *low* and *high* are those found for the lowest- and highest-value resistors investigated. The value marked *high* agrees reasonably well with those reported by other investigators for highly doped p-type Si.

We can summarize this review by drawing some conclusions about the probable values of the 300 K piezoresistive coefficients and the *best* way of determining them if calibration equipment is not available. For n-type material with $n_s \approx 10^{20}/\text{cm}^3$, $\pi_{11}^n$ is obtained from a surface concentration measurement and Fig. 7-7. $\pi_{12}^n$ is derived from the relation, $\pi_{12}^n = -\pi_{11}^n/2$. For $\pi_{44}^n$, a value in the range $-11.5$ to $-19 \times 10^{-5}/\text{MPa}$ is assumed, with the larger $|\pi_{44}^n|$ being associated with the higher $n_s$.[4] For p-type material, $\pi_{44}^p$ is determined knowing $n_s$ by Fig. 7-7. There is no obvious best way for finding the small components, $\pi_{11}^p$ and $\pi_{12}^p$. However, the two measurements of these components reported in Table 7-1 are in relatively good agreement, even though the doping densities are quite different. If no other data are available, an average of these measurements, yielding $\pi_{11}^p = 6.2 \times 10^{-5}/\text{MPa}$ and $\pi_{12}^p = -1.7 \times 10^{-5}/\text{MPa}$ could be used. The required accuracy of stress sensor measurements, together with a sensitivity analysis, will determine how important a calibration is. We will discuss this again when we consider some specific geometries.

## 7.4 STRESS SENSOR GEOMETRIES

In this section, we discuss the various resistor geometric orientations that can be used for stress sensing and comment on their various assets and liabilities. Much of this discussion is based on the detailed analysis of Bittle et al.[8] There are two primary wafer geometries used in Si device fabrication, as shown in Fig. 7-9. Metal–oxide–semiconductor (MOS) chips are fabricated on a (100) crystal plane, normal to the [001] cubic axis.[28] The primary wafer flat is perpendicular to a $\langle 110 \rangle$ direction, as shown, and the chip edges can be considered to lie in the [110] and [$\bar{1}$10] directions. For bipolar ICs, (111) Si is sometimes used.[25] In this case, the primary flat is perpendicular to [$\bar{1}$10] and the other axis is in the [$\bar{1}\bar{1}$2] direction.

Considering first the (100) Si case, the [110] and [$\bar{1}$10] directions parallel to the chip edges are the directions in which we wish to resolve the stress tensor, $[\sigma_{ij}]$. Conventional resistor orientations used include the directions parallel to the chip edges and the [100] and [010] directions located along the 45° diagonals, as shown in Fig. 7-10. Equations (7-15) and (7-16) can be used to derive the required expressions for the resistances, using $\theta_1 = \theta_2 = 0$, while $\phi_1 = $ resistor orientation relative to [100] and $\phi_2 = \pi/4$, the stress tensor $\sigma_{11} = \sigma_x$ orientation relative to [100]. The resultant expressions for the four relative resistance variations, $\Delta R_i / R_{i0}$ are

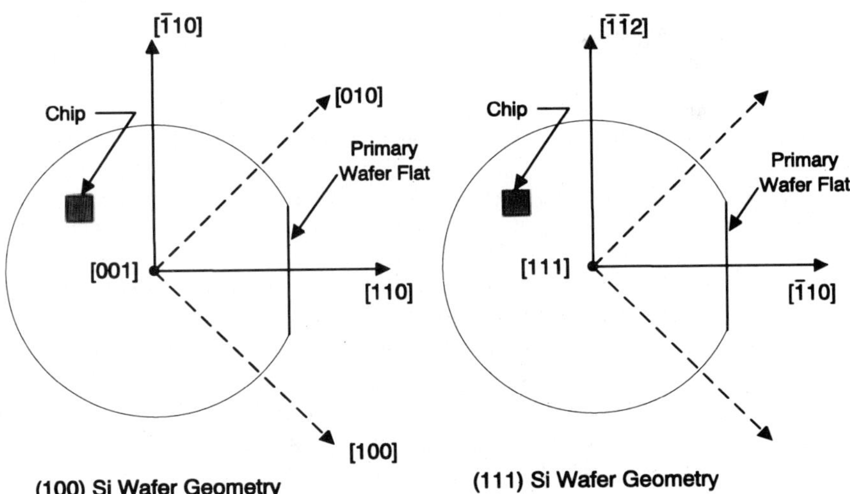


**Figure 7-9**  Two primary wafer geometries used in Si water processing. The (100) geometry is used for MOS ICs, while the (111) geometry is sometimes used for bipolar ICs. The crystallographic directions associated with the chip edges and vertical are shown.

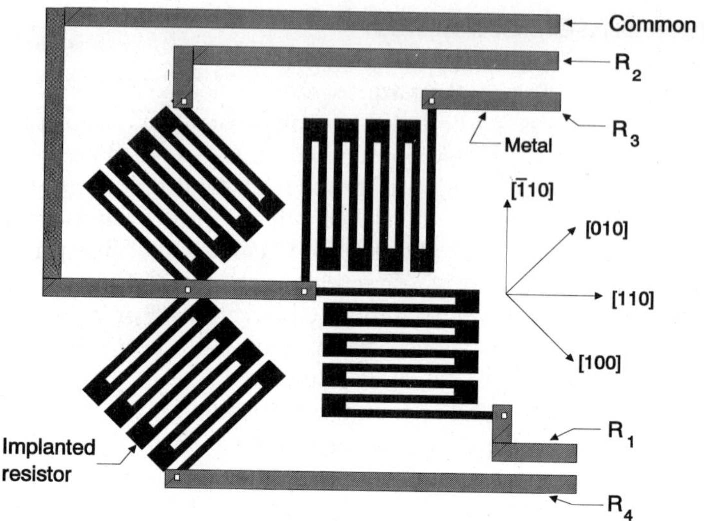


**Figure 7-10**    Four resistors laid out at 45° angles on (100) Si, as described by Eqs. (7-41). The resistances are measured between The $R_i$ connections and Common.

$$\frac{\Delta R_1}{R_{10}} = \left(\frac{\pi_{11} + \pi_{12} + \pi_{44}}{2}\right)\sigma_x + \left(\frac{\pi_{11} + \pi_{12} - \pi_{44}}{2}\right)\sigma_y + \pi_{12}\sigma_z \qquad [110]$$

$$\frac{\Delta R_2}{R_{20}} = \left(\frac{\pi_{11} + \pi_{12}}{2}\right)\sigma_x + \left(\frac{\pi_{11} + \pi_{12}}{2}\right)\sigma_y + \pi_{12}\sigma_z + (\pi_{11} - \pi_{12})\tau_{xy} \qquad [100]$$

$$\frac{\Delta R_{30}}{R_{30}} = \left(\frac{\pi_{11} + \pi_{12} - \pi_{44}}{2}\right)\sigma_x + \left(\frac{\pi_{11} + \pi_{12} + \pi_{44}}{2}\right)\sigma_y + \pi_{12}\sigma_z \qquad [\bar{1}\bar{1}0]$$

$$\frac{\Delta R_4}{R_{40}} = \left(\frac{\pi_{11} + \pi_{12}}{2}\right)\sigma_x + \left(\frac{\pi_{11} + \pi_{12}}{2}\right)\sigma_y + \pi_{12}\sigma_z - (\pi_{11} - \pi_{12})\tau_{xy} \qquad [010]$$

$$(7\text{-}41)$$

In Eqs. (7-41), we have used a natural or *chip* coordinate system, with the $x$ axis along [110], the $y$ axis along [$\bar{1}$10], and the $z$ axis along [001]. We have also designated the in-plane shearing stress, $\sigma_{xy}$, by the symbol $\tau_{xy}$ for clarity and to conform to usual practice. Several important results can be obtained from a study of these equations. First, $\Delta R$ has no dependence on shearing stresses with a vertical component, $\tau_{xz}$ or $\tau_{yz}$. Thus, only four of the six independent stress tensor components determine the resistance changes. Second, if there is no $xy$ shearing stress present, only two independent coefficients or constants are needed, the sum $\pi_{11} + \pi_{22}$ and $\pi_{44}$. Third, in the case of plane stress, $\sigma_z = 0$, any three of the resistors can be used to resolve the three in-plane stress components. The plane stress approximation expressions for these stresses have been given by Bittle et al.,[8]

$$\sigma_x = \frac{\pi_{44}\left(\dfrac{\Delta R_1}{R_{10}} + \dfrac{\Delta R_3}{R_{30}}\right) + (\pi_{11} + \pi_{12})\left(\dfrac{\Delta R_1}{R_{10}} - \dfrac{\Delta R_3}{R_{30}}\right)}{2\pi_{44}(\pi_{11} + \pi_{12})}$$

$$\sigma_y = \frac{\pi_{44}\left(\dfrac{\Delta R_1}{R_{10}} + \dfrac{\Delta R_3}{R_{30}}\right) - (\pi_{11} + \pi_{12})\left(\dfrac{\Delta R_1}{R_{10}} - \dfrac{\Delta R_3}{R_{30}}\right)}{2\pi_{44}(\pi_{11} + \pi_{12})} \tag{7-42}$$

$$\tau_{xy} = \left(\frac{1}{\pi_{11} - \pi_{12}}\right)\left[\frac{\Delta R_2}{R_{20}} - \frac{1}{2}\left(\frac{\Delta R_1}{R_{10}} + \frac{\Delta R_3}{R_{30}}\right)\right]$$

Although these equations are relatively simple in appearance, it is not obvious whether an n-type or a p-type resistor set would provide the highest accuracy. As discussed above, the $\pi$ coefficients that are both the smallest in absolute value and known with the least precision are $\pi_{11}$ and $\pi_{12}$ for p-type Si, so it appears that n-type Si is the best choice, especially if an accurate value of $\tau_{xy}$ is desired.

The expressions in Eqs. (7-42) have been used by several investigators to analyze data obtained with piezoresistive test chips with resistors of one dopant type. Spencer and co-workers used a chip with two resistors, $R_1$ and $R_3$ in Eqs. (7-41) and (7-42), aligned along the chip primary directions, [110] and [$\bar{1}\bar{1}0$].[5] Equations (7-42) show that this resistor arrangement enables the determination of only $\sigma_x$ and $\sigma_y$, and then only if a plane stress situation or an estimate of $\sigma_z$ is assumed. These investigators used an estimated value of $\sigma_z$ to reduce their data.

The first two of Eqs. (7-41) show that only the quantities, $\pi_{11} + \pi_{12}$ and $\pi_{44}$ are required for the determination of $\sigma_x$ and $\sigma_y$ for the case where $\sigma_z$ and $\tau_{xy}$ are zero. These quantities can be determined experimentally from a uniaxial stress measurement. Inspection of Eqs. (7-41) shows that if only $\sigma_x$ is unequal to zero, a condition of uniaxial stress, then a measurement of $\Delta R_1/R_{10}$ and $\Delta R_3/R_{30}$ yields the quantities $\pi_{11} + \pi_{12} + \pi_{44}$ and $\pi_{11} + \pi_{12} - \pi_{44}$ directly.

Stress sensor measurements with four resistors oriented in the directions specified in Eqs. (7-41) have been reported by Natarajan and Bhattacharyya.[25] Although there are four resistors, Eqs. (7-41) cannot be manipulated to yield the four stress tensor components represented in the equations because all depend on $\sigma_z$ in the same way. If a plane stress situation (or a fixed value of $\sigma_z$) is assumed, then the experimental measurements can be used to resolve the unknown stresses through Eqs. (7-42). However, from inspection of

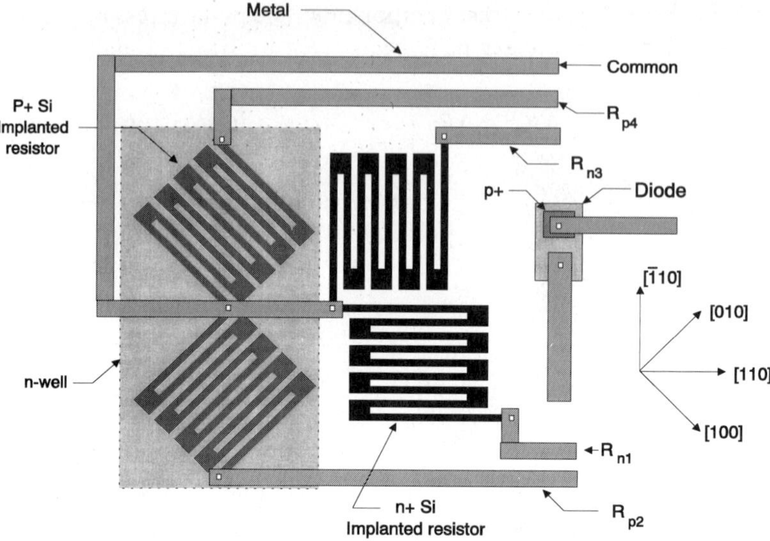


**Figure 7-11**  CMOS stress sensor cell of the type reported by Miura et al.[24] The n-type resistors are aligned with the chip edges, $\langle 110 \rangle$ directions, and the p-type resistors are aligned along $\langle 100 \rangle$ directions. The substrate is n-type. The diode is used for temperature measurement.

Eqs. (7-42) it can be seen that individual values of $\pi_{11}$ and $\pi_{12}$ are now needed, not just the sum of these quantities. If n-type Si is used, then the relation, $\pi_{11}^n \approx -2\pi_{12}^n$, together with a uniaxial stress calibration can yield $\pi_{11}^n$ and $\pi_{12}^n$.

If both n-type and p-type resistors are used on the same chip, then more information can be derived about the stress state because the differences in the $\pi$ coefficients for the two types of resistors will yield a different stress response, even for resistors in the same orientation. Miura and co-workers have described measurements with a CMOS stress sensor cell with the geometry shown in Fig. 7-11.[24] Bittle's analysis shows how the resistance measurements can be manipulated to yield four stress tensor components: $\sigma_x$, $\sigma_y$, $\sigma_z$, and $\tau_{xy}$. They also show that a uniaxial calibration can be used to derive the quantities: $\pi_{11}^n + \pi_{12}^n$, $\pi_{11}^p + \pi_{12}^p$, and $\pi_{44}^n$. Unfortunately, the quantities $\pi_{11}^p$ and $\pi_{12}^p$ are required individually, and there is no theoretical relation relating them. Thus, an additional calibration step is required for complete experimental specification of the piezoresistive coefficients. Miura et al. performed this calibration step using hydrostatic pressure.[24] This is a difficult measurement to make experimentally because substantial pressure, $\sim 20$ MPa ($\sim 3000$ psi) is needed to achieve the required accuracy.

The question of how much accuracy may be achieved in measuring stress components using piezoresistive sensors is an interesting and complex one.

One way of studying the measurement problem is to determine the sensitivity of the derived stresses to variation in either the $\pi$ coefficients or the measured resistance variations, $\Delta R_i/R_{i0}$. The accuracy with which the resistance variations can be measured is dependent on the chip design and the initial resistance value, $R_0$, as well as the variation in $R$ due to temperature. Generally, the temperature coefficient for heavily doped Si is $\leqslant 0.2\%/°C$.[29] This shows that it is important to maintain the measurement temperature as constant as possible to obtain an accurate measurement.

The measurement sensitivity to uncertainties or variations in either parameters or measured variables can be evaluated through study of the sensitivity coefficients, which give the relative variation in a derived stress component produced by a relative variation in one of the factors on the right-hand side of Eqs. (7-42).[30] In the case of the $\pi$ coefficients, we define a sensitivity coefficient as $S_{ijk} = (\pi_{ij}/\sigma_k) \, d\sigma_k/d\pi_{ij} = d(\ln \sigma_k)/d(\ln \pi_{ij})$. $S_{ijk}$ is the fractional change in the derived $\sigma_k$ value per unit fractional change in $\pi_{ij}$.

As an example of the use of sensitivity analysis, we consider the use of the $R_1 \rightarrow R_4$ resistors, defined in Eqs. (7-41), for measuring a uniaxial stress, $\sigma_x$, in the [110] direction, with all other stress components being zero. In this case, from Eqs. (7-41), the resistance variations, defined as $\delta R_i = \Delta R_i/R_{i0}$, satisfy the relations

$$\frac{\delta R_1}{\delta R_3} = \left(\frac{\pi_{11} + \pi_{12} + \pi_{44}}{\pi_{11} + \pi_{12} - \pi_{44}}\right), \qquad \delta R_2 = \delta R_4 \qquad (7\text{-}43)$$

Using these relations in Eqs. (7-42), the stress gage factors and sensitivity coefficients can be calculated. The results for the sensitivity coefficients are given by the expressions in Eqs. (7-44) and (7-45).

For resistors in the [110] directions, the sensitivity factors are

$$S_{111} = -\left(\frac{\pi_{11}}{2(\pi_{11} + \pi_{12})}\right)$$

$$S_{121} = -\left(\frac{\pi_{12}}{2(\pi_{11} + \pi_{12})}\right) \qquad (7\text{-}44)$$

$$S_{441} = -\tfrac{1}{2}$$

For resistors in the [100] directions, the sensitivity factors are given by

**Table 7-2**   Estimated Values of Piezoresistive Coefficients for a Surface Doping $n_s = 10^{20}/cm^3$

| Si Type | $\pi_{11}$ $(10^{-5}\,\mathrm{MPa}^{-1})$ | $\pi_{12}$ $(10^{-5}\,\mathrm{MPa}^{-1})$ | $\pi_{44}$ $(10^{-5}\,\mathrm{MPa}^{-1})$ |
|---|---|---|---|
| n | $-38$ | 19 | 19 |
| p | 5 | $-2$ | 70 |

$$S_{111} = \frac{-\pi_{11}}{(\pi_{11} + \pi_{12})}$$

$$S_{121} = \frac{-\pi_{12}}{(\pi_{11} + \pi_{12})} \tag{7-45}$$

$$S_{441} = 0$$

In order to understand the magnitudes of these sensitivity coefficients, we consider a model calculation for diffused resistors with a surface doping density of $n_s = 10^{20}/cm^3$. The assumed values of the $\pi_{\alpha\beta}$ are given in Table 7-2. Using Eqs. (7-42) and the stress gage factors in Table 7-3, the sensitivity factors in Table 7-4 are found. The gage factor, GF, is defined as the % change in resistance per unit stress, or GF (%/MPa) = $[\Delta R/R(\%)]/\sigma_x(\mathrm{MPa})$. To appreciate the magnitude of the numbers involved, we note that the reported breaking stress of (111) Si is $\sim 350$ MPa and that (100) Si is weaker than (111) Si.[5] Stresses that are a significant fraction of this maximum are not uncommon in die-attach or encapsulation situations. Table 7-3 shows that a 100 MPa stress will typically cause resistance shifts of a few percent or less. From Table 7-3, a 10 MPa stress in the [110] direction imposed on a [1$\bar{1}$0] n-type resistor will produce a $\Delta R/R \approx 0.2\%$, about the same shift as produced by a temperature variation of 1°C.

**Table 7-3**   Stress Gage Factors for the Four Fundamental Resistors for (100) Si, Uniaxial Stress in the [110] Direction

| Resistor No. | Direction | $GF_n$ (%/MPa) | $GF_p$ (%/MPa) |
|---|---|---|---|
| 1 | [110] | 0.000 | 0.037 |
| 3 | [$\bar{1}$10] | $-0.019$ | $-0.068$ |
| 2 | [100] | $-0.010$ | 0.002 |
| 4 | [010] | $-0.010$ | 0.002 |

**Table 7-4**   Sensitivity Factors for the (100) Si Stress Sensors

| Resistor No. | Doping Type | $S_{111}$ | $S_{121}$ | $S_{441}$ |
|---|---|---|---|---|
| 1, 3 | n | $-1$ | 0.5 | $-0.5$ |
| 1, 3 | p | $-0.83$ | $-0.5$ | 0.33 |
| 2, 4 | n | $-2$ | 1 | 0 |
| 2, 4 | p | $-1.67$ | 0.67 | 0 |

An examination of Table 7-3 indicates that a p-type resistor in the [110] direction has the highest gage factor for this stress, but that the p-type resistors are not as selective as n-type resistors laid out in the same orientations. The p-type resistors respond to both longitudinal and transverse stresses, while the n-type resistors have a GF $\approx 0$ for longitudinal stresses.

The sensitivity factors in Table 7-4 give the fractional change in derived stress for a given fractional change in a $\pi$ coefficient. For example, a 1% increase in $\pi_{11}$ for a p-type resistor laid out in a [110] direction, $R_1$ or $R_3$, will produce a $-0.83\%$ decrease in the derived stress. The $\pi$ coefficients known with the least precision are the small components, $\pi_{11}^p$ and $\pi_{12}^p$. Table 7-3 shows that p-type resistors in the [110] directions are somewhat less sensitive to the values of $\pi_{11}^p$ and $\pi_{12}^p$ then the same resistors in [100] directions. However, the high sensitivity of p-type resistors to these coefficients limits their utility unless accurate calibration information is available.

Analysis of stress sensors for a (111)-oriented wafer proceeds in much the same fashion as that for the (100) orientation. In this case, the first Euler angle rotation is through an angle, $\phi = \pi/4$ about the $z$ or [001] axis. The second rotation about the $y'$ axis through an angle $\theta = \cos^{-1}(1/\sqrt{3})$ tips the [001] axis into the [111] direction. A third rotation about the $z''$ or [111] axis through an angle $\psi = \pi/2$ aligns the new $x$ and $y$ axes with [$\bar{1}10$] and [$\bar{1}\bar{1}2$], respectively. This wafer orientation is chosen by some investigators because the longitudinal and transverse piezoresistive coefficients, Eq. (7-17), are isotropic in the (111) plane. Thurston discusses this in detail and also shows that the shear piezoresistance coefficient is zero for all resistor orientations in the plane.[31] Gee, Akylas, and van den Bogert discuss a four-arm strain gage rosette fabricated on n-type (111) Si.[32] They also point out that (111) wafers are usually cut in an orientation offset from the (111) plane by 4° and that this can cause the shear piezoresistance coefficient to deviate from its theoretical zero value.

Bittle et al. have derived the resistance response for resistors oriented as shown in Fig. 7-12. Their results are given in Eqs. (7-46).

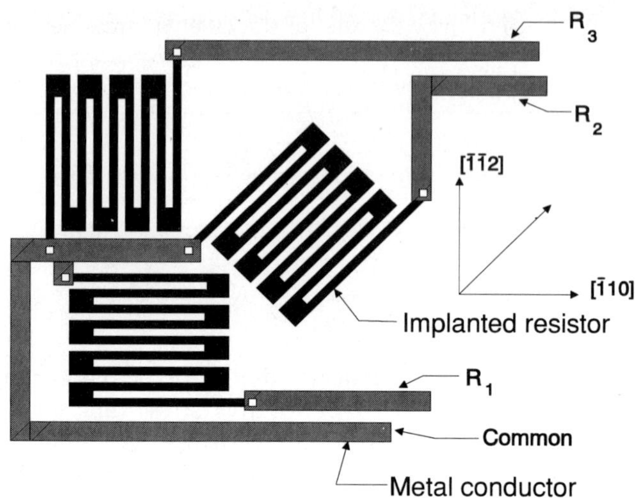



**Figure 7-12**    Three-element stress sensor cell for (111) Si, with resistors along $[\bar{1}\bar{1}0]$, $[\bar{1}\bar{1}2]$, and at 45° with respect to the $[\bar{1}\bar{1}0]$ resistor.

$$\frac{\Delta R_1}{R_{10}} = B_1\sigma_x + B_2\sigma_y + B_3\sigma_z + 2\sqrt{2}(B_2 - B_3)\tau_{yz}$$

$$\frac{\Delta R_2}{R_{20}} = \left(\frac{B_1 + B_2}{2}\right)(\sigma_x + \sigma_y) + B_3\sigma_z + 2\sqrt{2}(B_2 - B_3)\tau_{xz} + (B_1 - B_2)\tau_{xy} \quad (7\text{-}46)$$

$$\frac{\Delta R_3}{R_{30}} = B_2\sigma_x + B_1\sigma_y + B_3\sigma_z - 2\sqrt{2}(B_2 - B_3)\tau_{yz}$$

The $B$ coefficients are defined in terms of the $\pi$ coefficients by the expressions in Eqs. (7-47).

$$B_1 = \frac{\pi_{11} + \pi_{12} + \pi_{44}}{2}$$

$$B_2 = \frac{\pi_{11} + 5\pi_{12} - \pi_{44}}{6} \quad (7\text{-}47)$$

$$B_3 = \frac{\pi_{11} + 2\pi_{12} - \pi_{44}}{3}$$

In the case where all shearing stresses, $\tau_{ij}$, are zero, the expression for $\Delta R_1/R_{10}$ shows that $B_1$ is a longitudinal piezoresistive coefficient, while $B_2$ and $B_3$ are transverse coefficients. A comparison with the expressions given by Pfann and Thurston for the longitudinal and transverse coefficients shows that they are the same as those given by Eqs. (7-47).[3]

Bittle also shows that for a plane stress situation ($\tau_{xz} = \tau_{yz} = \sigma_z = 0$), Eqs. (7-46) can be inverted to give the planar stresses, $\sigma_x$, $\sigma_y$, and $\tau_{xy}$ as functions of the measured $\Delta R$ values, along with $B_1$ and $B_2$. He also shows that $B_1$ and $B_2$ can be determined experimentally from a uniaxial stress measurement. For example, if only $\sigma_x$ is unequal to zero, then Eqs. (7-46) show that

$$\frac{\Delta R_1}{R_{10}} = B_1 \sigma_x$$

$$\frac{\Delta R_2}{R_{20}} = \left(\frac{B_1 + B_2}{2}\right)\sigma_x \qquad (7\text{-}48)$$

$$\frac{\Delta R_3}{R_{30}} = B_2 \sigma_x$$

If a calibration of n-type resistors is performed using Eqs. (7-48), then the simple theoretical relation, $\pi_{11}^n \approx -2\pi_{12}^n$, can be used to derive $B_3$. This is, perhaps, the biggest advantage of using n-type diffused resistors in (111) Si. A complete calibration can be performed using uniaxial stress measurements alone, without the need to perform an additional hydrostatic calibration.

Bittle et al. also show that use of p-type and n-type three-resistor rosettes in a CMOS cell on (111) Si makes determination of all six stress tensor components possible, at least in principle. They present explicit expressions for the $\sigma_{ij}$ as functions of the six resistance shifts, $\Delta R_i^{n,p}$, and the six parameters, $B_i^{n,p}$. The expressions are complex, and a sensitivity analysis would be necessary to determine the measurement accuracy that could be achieved with their scheme for the various stress tensor components. At the time this chapter was written, this design had not been implemented in a test chip design.


## 7.5 EXPERIMENTAL DESIGNS AND CALIBRATION

### 7.5.1 Chip Layout

The major requirements for the piezoresistors are: (1) the resistance should fall in a convenient range for measurement, and (2) the resistance should be much higher than any parasitic series resistance. Generally, this parasitic resistance will be in the range of about 100–500 $\Omega$, depending on the exact circuit design. This parasitic resistance consists of three components; the resistance of metal-to-Si contacts, the resistance of metal lines used to connect

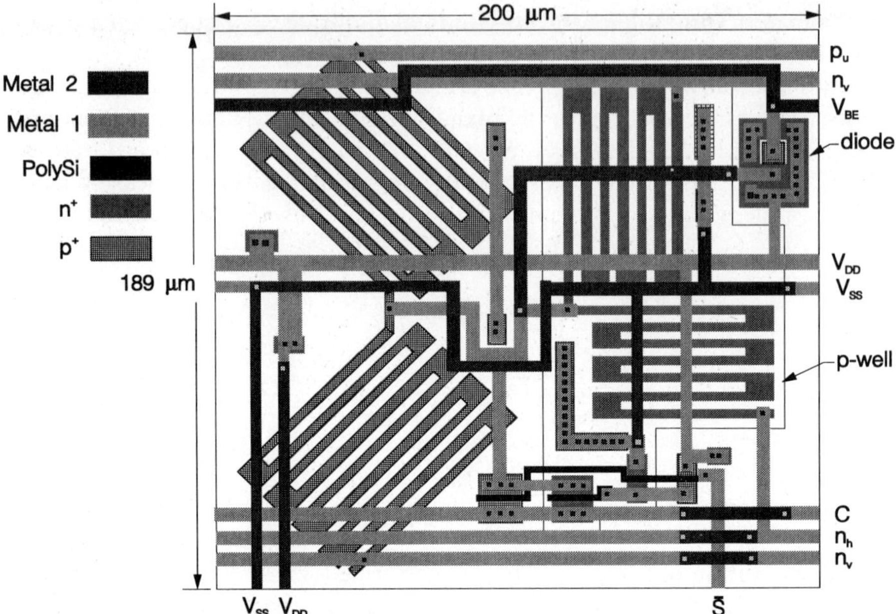



**Figure 7-13**    An actual CMOS layout of a stress sensor cell showing diffused resistors, metal lines, and contacts.[35] Two levels of metal are used for the interconnect. $V_{SS}$ and $V_{DD}$ are the power supply ground and high voltage, respectively. The $\bar{S}$ or complementary select signal bit is used to connect the cell common line to the common or C bus for measurement.

the resistor to bondpads, and the resistance of any semiconductor switches, such as CMOS transmission gates, used to connect the resistor to an output measurement bus. An actual CMOS layout of a piezoresistance cell is shown in Fig. 7-13. Metal resistance will usually be in the range of about 10–30 $\Omega$ for long runs. Contacts to Si will be $\sim 10\,\Omega$/contact for Al to highly doped Si contacts which are $\sim 1.0\,\mu$m on a side.

The resistance will usually be dominated by the transmission gate resistance, if present, The resistance of a MOS transistor in the linear region is given by[33]

$$G = (W/L)\mu_s C_0(V_G - V_T) \tag{7-49}$$

where $W$ and $L$ are the width and length of the conducting channel, respectively, $\mu_s$ = the surface mobility of the conducting channel, $C_0$ = capacitance per unit area of the gate oxide, $C_0 = \varepsilon_{ox}/t_{ox}$, where $\varepsilon_{ox}$ = gate oxide dielectric constant and $t_{ox}$ = gate oxide thickness. $V_G$ and $V_T$ are the gate and threshold voltages, respectively. For a width-to-length ratio $W/L = 0.1$, the resistance $R = 1/G \approx 100\ \Omega$. With this in mind, it is desirable to have the design resistance in the range of about 10–40 k$\Omega$. However, this may be difficult to attain if a small cell is required. For 50 $\Omega/\square$ material, a 10 $k\Omega$ resistor requires 200 $\square$. For a technology with a 1 μm minimum feature size, a 200 $\square$ resistor will be 200 μm long.

Usually, the resistor is layed out in a serpentine geometry, as shown in Fig. 7-2. It is important to minimize the resistance associated with the turnarounds because this resistance will have a different stress sensitivity than the resistance of the legs. This is usually done by making the width of the turnarounds about three times the leg width. In addition, the sharp corners at the turnarounds add extra resistance and sometimes rounded corner designs are used. This is difficult if minimum width geometry resistors are used, so most reported designs use square corners.

The effect of parasitic resistance can be analyzed approximately in the following way. We assume that the total resistance, $R_T$, is given by the sum of $R_i$, the resistance of the implanted resistor, and $R_p$, the parasitic resistance,

$$R_T = R_i + R_p \tag{7-50}$$

The piezoresistive shift is assumed to occur only in the implanted resistor, $R_i$, and the resistance shift is related to the stress tensor components, $\sigma_\beta$, by a relation of the form

$$\frac{\Delta R_i}{R_i} = \sum_\beta c_\beta([\pi_{\alpha\beta}])\sigma_\beta \tag{7-51}$$

The constants, $c_\beta$, are linear functions of the $\pi_{\alpha\beta}$, such as those given in Eq. (7-1). The quantity measured is $\Delta R_T/R_{T0}$ and this related to the above quantity by

$$\frac{\Delta R_T}{R_{T0}} = \frac{\Delta R_i}{R_{i0}(1 + R_p/R_{i0})} \approx \frac{\Delta R_i}{R_{i0}}\left(1 - \frac{R_p}{R_{i0}}\right) = \sum_\beta c_\beta([\pi_{\alpha\beta}])\sigma_\beta\left(1 - \frac{R_p}{R_{i0}}\right) \tag{7-52}$$

If the resistor is calibrated, the factor $(1 - R_p/R_{i0})$ is absorbed in the calibration and should not affect the measurement accuracy, although it will affect the derived or reported values of the $\pi_{\alpha\beta}$. If estimated values of the $\pi_{\alpha\beta}$ are used, then this factor represents a constant correction that would have to be applied to derive the true stresses. However, in this case the probable errors in the $\pi_{\alpha\beta}$ would likely be larger than the error produced by not making the parasitic resistance correction. This analysis shows that the effect of parasitic resistance should be small if the quantity $R_p/R_{i0}$ is kept to a few percent.

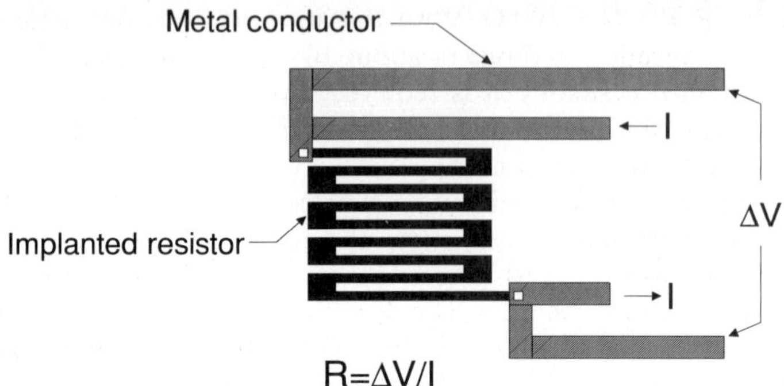


**Figure 7-14**    Four-terminal (Kelvin) arrangement for making resistance measurements. The current *I* is supplied by a current source and the voltage drop $\Delta V$ is measured with a high-input-impedance voltmeter.

Although the above analysis suggests that parasitic series resistance at the few percent level is not significant in affecting overall measurement accuracy, the error can be reduced by using four-terminal measurement techniques. A schematic diagram of a four-terminal resistance measurement is shown in Fig. 7-14. The key to making this measurement is to connect the voltage measurement leads as close to the resistor as possible. Since the voltmeter or $\Delta V$ circuit has an extremely high input impedance ($\geqslant 10$ MΩ) only a negligible current flows in this circuit, and so the ohmic voltage drop across parasitic resistance (except for the tw metal–Si contacts) is negligible. Because the voltmeter circuit measures only the voltage drop across the resistor, any parasitic resistance in the current circuit does not contribute to the overall measured resistance. If a small cell or resistor size is desired, then it may not be possible to make resistors with a small $R_p/R_{i0}$. In this case, four-terminal measurement techniques could be useful. The chip described by Gee, Akylas, and van den Bogert utilized four-terminal measurement circuitry.[32]

We have already discussed the effect of temperature variations on measurement accuracy. In the general case, there is both an intrinsic temperature coefficient of resistivity and also a dependence through the $\pi$ coefficients. The form assumed by Miura et al.[24] and by Gee et al.[32] for the dependence of resistance on stress and temperature was of the form

$$\frac{\Delta R([\sigma_\beta], T)}{R_0(T_0)} = [\alpha_1(T - T_0) + \alpha_2(T - T_0)^2]$$

$$+ \sum_\gamma c_\gamma [\pi_{ij}(T_0)(1 + \beta_{ij}(T - T_0))]\sigma_\gamma(T) \qquad (7\text{-}53)$$

In Eq. (7-52), the stresses, $\sigma_y$, are those at the measurement temperature,

$T$, and $R_0$ is the no-stress resistance at a reference temperature $T_0$. The $c_\gamma$ functions are linear in the $\pi_{ij}$, as in Eq. (7-41), and thus there is a linear dependence on the three $\beta_{ij}$ coefficients. The repeated $ij$ index does not represent a summation in this case. It indicates a functional dependence. A detailed analysis of the piezoresistance equations by Suhling, Carey, Johnson, and Jaeger has confirmed that this form is theoretically valid.[34] Gee et al. assumed that the $\beta_{ij}$ were zero for their heavily doped samples,[29] while Miura et al. measured $\beta$ for the linear combinations, $\pi_{11}^n + \pi_{12}^n + \pi_{44}^n$ and $\pi_{11}^n + \pi_{12}^n - \pi_{44}^n$.[24]

Experimentally, it is desirable to utilize the same measurement temperature, $T_0$, for all stress measurements in order to minimize temperature effects. For high-accuracy measurements, the temperature should be measured with a semiconductor diode located near the associated piezoresistor. This is done in the cell described by Miura et al.[24] and is also done in a similar CMOS cell used on a multifunction test chip, ATC03, described by Sweet, Tuck, Peterson, and Palmer.[35] This cell is shown in Fig. 7-13. The diode is in the upper right-hand corner of Fig. 7-13.

### 7.5.2 Calibration

As we have discussed above, calibration involves applying known stresses and measuring the resistance shifts produced by these stresses. Then the governing equations, such as Eqs. (7-41) for (100) Si, are used to express the $\pi_{\alpha\beta}$ as functions of the stresses and resistance shifts. The easiest way to apply a known and uniform stress is with uniaxial stressing. Another, but far less easy way experimentally, is to apply hydrostatic pressure. The first description of calibration by application of uniaxial stress was given by Gee et al.[32] These authors developed a four-point bending rig and associated apparatus that could be used to place a strip of die in a state of uniform tension. Miura et al. also report using a four-point bending system but give no details on the experimental technique or equipment construction.[24] More recently, Beaty et al.[27] have reported calibration measurements made with a four-point bending system. These authors present a detailed description of the experimental technique and also give a detailed error analysis. One potential source of error they found was the stress induced by electrical probes used to connect instruments to the IC on test. They minimized this source of error by using loading weights much larger than the force associated with the probes. Their estimate of the overall error associated with determination of the applied stress was $\pm 5\%$.

Subsequent to the work described in ref. 27, the same group developed a uniaxial pull fixture to facilitate experimental calibration measurements at higher temperatures.[34] In this fixture, one end of a strip of die is clamped between asbestos pads and Al blocks in a fixed support. The other end is clamped in a similar fixture that is free to slide and then weights are hung from the movable end to apply tension to the Si strip. This rig is much more compact than the four-point bending rig and can easily be placed in an oven. It is also easier to contact the sample with electrical probes or wires. As of this writing, there is no commercial equipment available to make this type of measurement.

Another calibration technique has recently been described by Bastawros and Voloshin.[36] In this method, an optical measurement technique, called Moiré interferometry, is used to measure one component of the strain tensor directly. The strain is produced by a somewhat arbitrary load induced in a sectioned packaged part. This is accomplished by forming a grating on the sectioned surface in an adhesive film applied to the section with a mold. The grating is then metallized to make it reflective. The grating surface is illuminated by two coherent laser beams, with angles of incidence equal in magnitude but opposite in sign with respect to the normal. A virtual fringe pattern is formed when the sample and its associated grating are strained. From this pattern, the strain perpendicular to the grating ruling direction and in the plane of the grating may be determined. If the sample resistance shift is measured along with the strain, a gage factor, of the form given by Eq. (7-21), may be determined.

This calibration method has the advantage that the applied stress need not be uniform, although it should be uniaxial. However, there is the experimental difficulty of sectioning a sample precisely along a known direction relative to the resistor axis and also the requirement to have an electrical connection to the resistor being measured after the part is sectioned. Since a load must be applied, the die must be encapsulated or molded in some fashion.

Bastawros and Voloshin give no details on the resistor or die orientations used in their experiments, so a detailed comparison of their results with those of other investigators reported in Table 7-1 is impossible. They report an n-type gage factor, $K_n \approx 24$ and a p-type gage factor, $K_p \approx 69$. If we assume that their resistors were on (100) Si and that their axes were in the sectioning direction, then a rough comparison can be made. If it is assumed that the load cell only applies an axial stress, $\sigma_1 = \sigma_x$, then the first of Eqs. (7-9) can be used to find the strain. In this case, $\varepsilon_x = s_{11}\sigma_x = 0.768\sigma_x \, (10^5 \text{ MPa})$. Using the first of Eqs. (7-41), the gage factor is given by $K = (\pi_{11} + \pi_{12} + \pi_{44})/(2s_{11})$. Using the estimated $\pi$ coefficient values in Table 7-2, we find $K_p = 47$ and $K_n \approx 0$. The p-type factor agrees somewhat with that of Bastawros and Voloshin but the n-type factor is not in agreement. If the resistors were perpendicular to the plane of the section (in the $y$ direction), then we would estimate, $K_p = -43$ and $K_n = -23$. The data on resistance change vs. strain presented in ref. 35 shows that the measured gage factor is positive, and so this does not appear to explain the data. It would appear that this technique is promising but that more careful experimentation is required to derive gage factors that can then be used to derive $\pi$ coefficients that in turn can be compared to those measured in uniaxial uniform stress experiments.

## 7.6 EXPERIMENTAL STRESS MEASUREMENTS

In this section we review some of the experimental stress measurements that have been reported in the literature. In each case, a piezoresistive stress sensor was used to derive stress data from experimental resistance shifts. This review is not meant to be exhaustive but rather to illustrate the type of information that can be obtained with piezoresistive stress sensors. Before discussing the experimental measurements, it is worthwhile to consider the sources of mechanical stress in modern electronics packages.

The first major source of stress is the thermally induced stress caused by the die-attachment process. This is discussed in some detail in a recent review by Suhir.[37] The attachment process is illustrated in Fig. 7-15. The die is

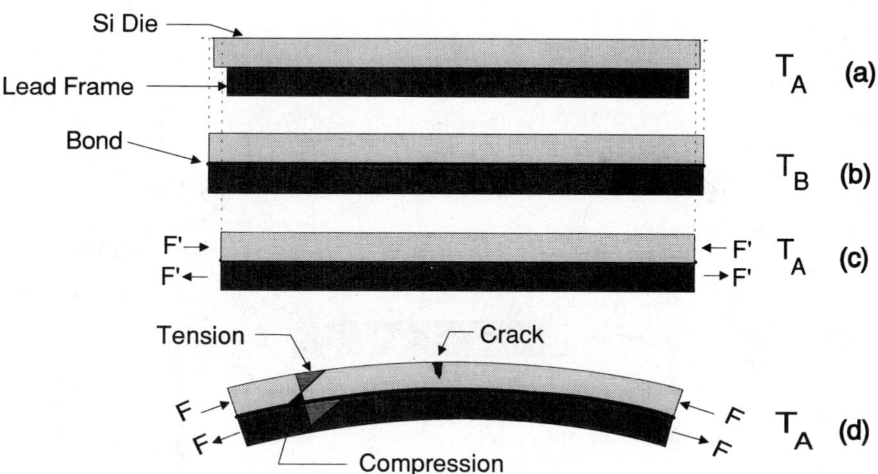


**Figure 7-15**  Schematic diagram of the die-attachment process showing the development of tensile stresses at the die surface. (a) Unattached die and lead frame at room ambient temperature, $T_A$. (b) Die and lead frame with bonding material at the bonding temperature, $T_B$. (c) Upon cooling to $T_A$ without bending, the Si die is in compression and the lead frame in tension. A force couple with force magnitude $F'$ is required to maintain the lead frame–die assembly in a nonbending state. (d) With a smaller force couple of magnitude $F < F'$ applied by other parts of the package, the assembly bends, as shown. The upper surface of the die goes into tension, possibly producing a crack.

attached to a metallic lead frame at an elevated temperature, $T_B$, either by epoxy or by an Au–Si eutectic. When the die and lead frame cool down to ambient temperature, $T_A$, the die is put in a state of compression as a result of its lower thermal expansion coefficient relative to the lead frame, Fig. 7-15c. Suhir presents calculated results for the stress distribution that show that it is highest in the die center and goes to zero at the edges. Relaxation of the stress-induced forces allow bending of the die and lead frame, as shown in Fig. 7-15d. This results in a nonuniform stress distribution throughout the die thickness, with the bottom of the die in compression and the top in tension, as discussed by van Kassel, Gee, and Murphy.[38] In extreme cases, a crack in the Si can form, as shown in Fig. 7-15d. A piezoresistive sensor on the die surface will thus measure a tensile stress after die-attach and it would be expected that the highest stress magnitude would be measured at the die center.

The second stress source is produced by the differential thermal expansion in molding or encapsulation. Again, if it is assumed that the die and the plastic encapsulant are at thermal equilibrium during the encapsulation process, then on return to ambient the die will be put in a state of compression, as shown in Fig. 7-16. Typically, plastics have thermal expansion coefficients in the range, $\alpha_{\text{plastic}} \approx 16\text{–}25$ ppm/°C, while $\alpha_{\text{Si}} = 2.3$ ppm/°C. The compressive stress is accompanied by a shear stress, $\tau_{xy}$, at the die

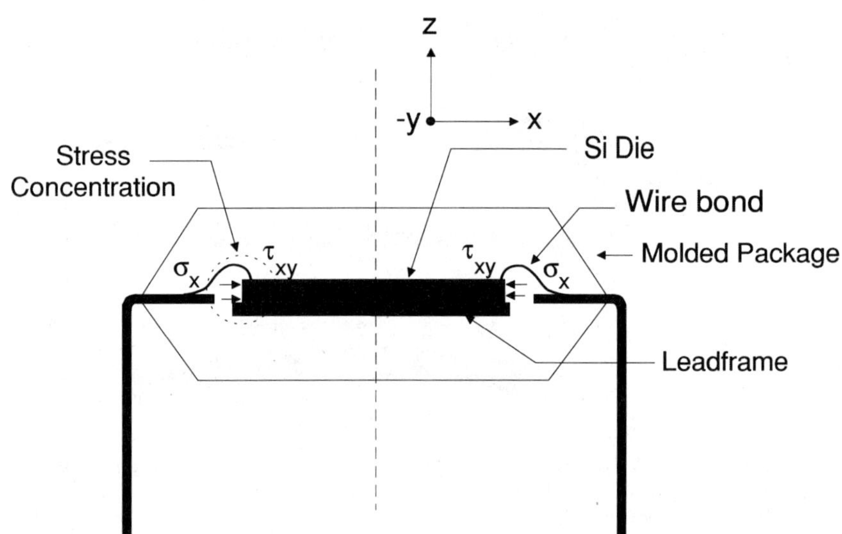


**Figure 7-16**   Cross-section of a molded plastic package, showing stresses that develop at the die surface resulting from mold compound shrinkage. If the die is considered to be a thin plate, then it will be in a state of plane stress, with only $\sigma_x$, $\sigma_y$, and $\tau_{xy}$ being significant.

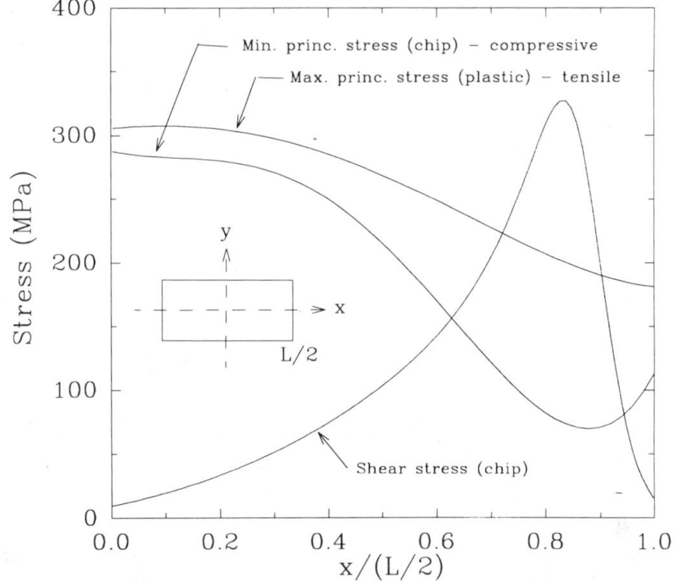


**Figure 7-17**    Calculated principal stresses and shear stress for a molded plastic package, as reported by Groothius et al.[39] The calculation was done by the finite element method. The coordinate $x$ is measured along the centerline of the die of length $L$, from the center of the die to the edge located at $L/2$.

surface, and this shear stress is reponsible for damage to chip passivation and metal conductor lines. Some results of finite element calculations of molding-induced stresses in plastic packages have been given by Groothuis, Schroen, an Murtuza.[39] Their calculations show that $\tau_{xy}$ is a minimum at the die center and a maximum at a corner. The principal compressive stress, on the other hand, is a maximum at the die center and a minimum at the edge. Their results for the stress distributions along the die centerline, from the die center to its edge, are shown in Fig. 7-17.

The correlation between the principal compressive stress and $\tau_{xy}$ has frequently been used in piezoresistive stress measurements. For example, Eq. (7-41) shows that resistors aligned in the direction of the chip edges on (100) Si are insensitive to the value of $\tau_{xy}$ but they do respond to the three compressive stresses. Thus, their resistance shifts can be used to infer the expected shear-induced damage from various molding compounds.

The first detailed description of piezoresistive stress measurements was given by Spencer et al.[5] These investigators used (100) Si with resistors aligned along the chip edges, [110] and [$\bar{1}\bar{1}0$]. With only two resistor orientations, Eq. (7-43) shows that they could not uniquely determine all the compressive stresses from two measurements. The data are presented as $\Delta R$

as a function of position on the die and mold compound used. They find the highest $\Delta R$ values at the chip center, with the sign being consistent with compressive stress. A later comparison with finite element calculations showed good agreement between theory and experiment.[39] This experiment was typical of many in that stresses produced by a number of different molding compounds were compared. Four mold compounds were evaluated. The highest stress was produced by a compound with a crystalline silica filler, as compared to those with fused silica fillers. It was also found that about 75% of the maximum stress occurred after molding, but before post-mold cure. The remaining 25% occurred after curing and the consequent densification of the mold material.

In a study of die attachment, van Kassel, Gee, and Murphy used a (111) Si chip with p-type resistors, 0.15 cm on a side, oriented along the chip edges.[38] To simulate a larger die, they used arrays of the fundamental chip. They also included bipolar transistors for temperature measurement. A strain gage factor $K = 55$ was used for the longitudinal gage factor. In most cases, tensile stress was observed, as predicted from the model of lead frame and die bowing after die-attach and stress relaxation. In a few cases, compressive stresses were observed and ascribed to warping of the die surface as a result of nonuniform die-attach adhesion. For this study, three types of die-attach (Au–Si, epoxy, polyimide) and two types of leadframes (Alloy 42 Au plated and Cu 194 Ag plated) were used. The lowest stresses were found for the Alloy 42 lead frame with an Au–Si eutectic die-attach. The higher stresses for the epoxy and polyimide attached parts was felt to be caused by nonuniformities in the thickness and/or adhesion of the layers for these materials.

A detailed study of die stresses in a molded plastic package was reported by Natarajan and Bhattacharyya.[25] They used a (100) Si with four n-type resistors, oriented as described by Eq. (7-1). For data analysis, they assumed plane stress, $\sigma_z = 0$. Two types of die-attach materials were used, epoxy and polyimide. In both cases, tensile stress was observed after die-attach, with the epoxy producing lower stress by about a factor of 2. This was explained by the lower elastic modulus, coefficient of thermal expansion (CTE or $\alpha$), and glass transition temperature ($T_g$) for the epoxy as compared to the polyimide. Three different mold compounds were investigated, with varying Young's modulus and coefficient of expansion. In all cases, compressive stresses were observed after molding, with the lowest value being associated with the compound that had the lowest CTE and Young's modulus. Stress measurements were also made at various temperatures and it was found that the stress shifted from compressive to tensile at higher temperatures of about 100–125°C. Presumably, this occurs because of stress relief caused by mold compound expansion. This measurement was done by measuring the resistance as a function of temperature for both a free and a molded part and assuming that the difference in temperature coefficients was all due to the variation of stress with temperature. This method of data reduction ignores the variation of the $\pi$ coefficients with temperature, Eq. (7-52).

Another detailed investigation of shear stress effects in plastic packages was done by Edwards, Heinen, Groothuis, and Martinez.[40] They used a chip of the same type but somewhat larger than that used in their previous work.[5] In this study, the stress was measured as a function of mold compound (three types), die attach (paste or film, Ag-filled polyimide), lead frame (8-mil Alloy 42, 6-mil Alloy 42, and 8-mil Cu), and die premold coating (none, polyimide, silicone gel). All samples were molded and in almost all the measurements compressive stresses were observed. In the case of a Cu lead frame, tensile stress was observed at the die corners, even after molding. The gel-coated samples demonstrated significantly lower stress levels but increased levels of wire breakage after temperature cycling. They found that the stress measurements allowed them to select between low-stress molding compounds that had very similar physical properties. They also found that parts with higher compressive stress gradients across the chip after temperature cycling experienced the greatest shear stress damage of metal lines in the die corners. Another result was that die-attach layers with more uniform bonding produced lower and more uniform stresses and more stability during temperature cycling.

Measurements on a plastic encapsulated large-area chip have been reported by Lundström and Gustafsson.[26] They used a (111) Si chip with p-type resistors and temperature measurement diodes. They report an experimental value of $\pi_{44}$, as given in Table 7-1 but do not give any details as to how the calibration was performed. They do, however, state that there is an uncertainty in $\pi_{44}$, but that the major purpose of the work was to measure relative rather than absolute stress levels. They measured tensile stresses after die-attach and compressive stress after transfer molding. Four different mold compounds and two different lead frame materials (Alloy 42 and Cu) were evaluated. They found that after die-attach the Alloy 42 lead frames had the lowest stress and this condition persisted after molding. However, after 1000 temperature cycles from $-55°C$ to $125°C$, the stress levels were nearly the same. Stress is reported as a function of temperature but no description of the data reduction technique is given. This paper contains extensive and detailed data on the packages and materials investigated. When a protective precoat of silicone rubber was used before molding, very low stress levels were attained.

A detailed mapping of stress over a die surface has been reported by Gee, van den Bogert, and Akylas.[41] They used a (111) Si chip with an array on n-type resistors.[32] In their data analysis they assumed plane stress. This assumption was based on results of finite element calculations that showed that $\tau_{xz}$ and $\tau_{yz}$ were small compared to $\tau_{xy}$ and on the basis that $\sigma_{zz}$ would be small because of the small package dimensions. A 28-pin plastic DIP was used with several molding compounds and thermal shock testing. In a principal stress decomposition, the stresses were fairly uniform in magnitude across a die profile, decreasing somewhat at the die edge. The three different molding materials used could be differentiated by the magnitude of the principal stresses they produced. In a maximum shear stress decomposition, the shearing stress was found to be highest at the die corners, in agreement with other investigators. Although the shearing stress magnitude was much smaller than the principal stress magnitudes, the authors show that the shearing stress measurement is not sensitive to temperature variations and thus it was felt to be measurable to a higher precision. After thermal shock testing, $-65°C$ to $150°C$, the corner shearing stresses were found to decrease significantly. This decrease was ascribed to plastic delamination from the sides of the die and from the lead frame.

In a recent study, Miura, Nishimura, Kawai, and Murakami reported on stress sensor measurements made on a molded chip-on-lead (COL) dual in-line package.[42] They used a previously described chip with both p-type and n-type resistors, together with a temperature-sensing diode.[24] The authors claim a 0.1 MPa "sensitivity" for this chip. This claim cannot be taken too seriously because a temperature variation of only $\sim 10^{-4}°C$ would produce the same resistance variation as a 0.1 MPa variation. They examined two types of die-attachment materials (epoxy paste, rubber paste), three types of lead frames (Alloy 42, Alloy 50, Cu), and three types of molding materials. After die-attachment they observed tensile stress at the die center. Some of their data is shown in Fig. 7-18. The tensile stress is plotted vs. the thermal expansion coefficient of the lead frame material for each type of die attach. It is evident that the rubber paste produced significantly lower stress levels. After molding, compressive stresses were observed at the die center. Figure 7-19 shows a plot of the measured stress for each type of lead frame with a rubber paste die-attach. The stresses are plotted against a "resin parameter", $R_p$ defined by, $R_p = (\alpha_{resin} - \alpha_{Si})E_{resin}$, where $\alpha_{resin}$ and $\alpha_{Si}$ are the thermal expansion coefficients for the molding resin and Si, respectively and $E_{resin}$ is the Young's modulus for the resin. The authors feel that $R_p$ is a good parameter for predicting molding stresses and their data tend to support this hypothesis.

The general trend that can be observed in the literature on the application of piezoresistive stress sensors is that the measurements have become more accurate and detailed. In addition, the claimed accuracy has increased. In most cases, the information on experimental techniques and the number of samples used (statistics) are insufficient to allow a judgment on the claimed accuracy. It is, however, clear that detailed information about packaging stresses has been obtained by a number of different investigators using piezoresistive test chips fabricated by their respective companies.

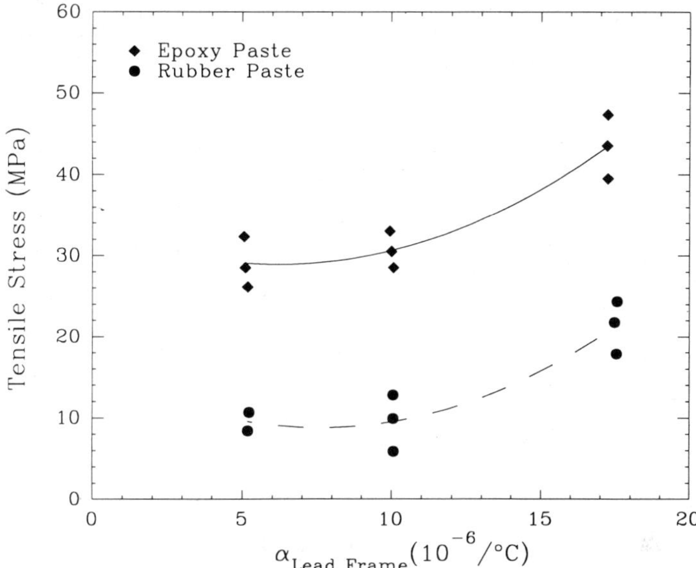



**Figure 7-18**    Data of Miura et al.[42] showing the piezoresistive measured tensile stress after die-attach for three types of lead frames and two types of die attach. The lead frames are characterized by their thermal expansion coefficient values.

## 7.7 SUMMARY

In this review, we have discussed the properties of Si piezoresistive stress sensors and given some examples of their use from the literature. The underlying principles are well understood and the discussion presented in this chapter and the references should be sufficient to allow design of test chips for measurement.

These sensors are clearly a valuable tool for measuring packaging-induced mechanical stress but, typically, the experimental techniques are difficult and the measurement accuracy is hard to determine. Thus, we should consider piezoresistive stress sensing as, at best, a semiquantitative technique at the present time. Although sensor calibration measurements have been reported, they are not *routine* and no commercial equipment is made for this purpose. Also, there is no commercial source of test chips available at this writing. An examination of the literature shows that only large, vertically integrated semiconductor manufacturers have reported making and using piezoresistive test chips. As new packaging schemes, such as multichip modules and chip-on-board, come on the scene, there will probably be an increased demand from other types of manufacturers and researchers for piezoresistive

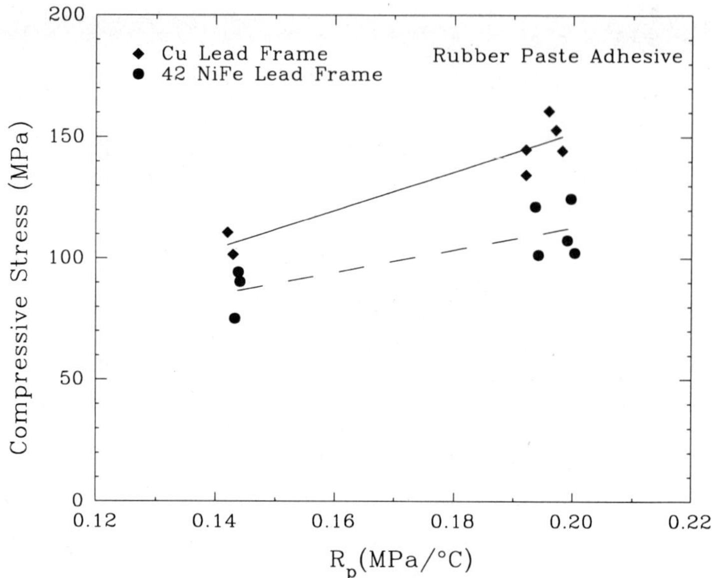


**Figure 7-19**   Data of Miura et al.[42] showing measured compressive stress for two types of lead frames and three types of mold compounds as a function of the mold parameter, $R_p$, as defined in the text.

sensors. Hopefully, sources of die, calibration equipment, and measurement techniques will appear to satisfy this demand.

## ACKNOWLEDGMENTS

The author thanks Robert Nasby, David Peterson, and David Palmer of Sandia National Laboratories for critical reviews of this manuscript. This work was performed at Sandia National Laboratories supported by the U.S. Department of Energy under Contract DE-AC04-76DP00789.

## REFERENCES

1. Smith, C. S., "Piezoresistance Effect in Germanium and Silicon," *Phys. Rev.*, **94**, 1956, p. 42.
2. Mason, W. P., and R. N. Thurston, "Use of Piezoresistive Materials in the Measurement of Displacement, Force, and Torque," *J. Acoustical Soc. Am.*, **29**, 1957, p. 1096.

3. Pfann, W. G., and R. N. Thurston, "Semiconducting Stress Transducers Utilizing the Transverse and Shear Piezoresistance Effects," *J. Appl. Phys.*, **32**, 1961, p. 2008.

4. Tufte, O. N., and E. L. Stelzer, "Piezoresistive Properties of Silicon Diffused Layers," *J. Appl. Phys.*, **34**, 1963, p. 313.

5. Spencer, J. L., W. H. Schroen, G. A. Bednarz, J. A. Bryan, T. D. Metzgar, R. D. Cleveland, and D. R. Edwards, "New Quantitative Measurements of IC Stress Introduced by Plastic Packages," *Proc. 19th Annual Reliability Physics Symposium*, 1981, pp. 74–80.

6. Landau, L. D., and E. M. Lifshitz, *Electrodynamics of Continuous Media*, Pergamon Press, Oxford, 1984, pp. 86–88.

7. Ashcroft, N. W., and N. D. Mermin, *Solid State Physics*, Holt, Rinehart and Winston, New York, 1976, pp. 250–251.

8. Bittle, D. A., J. C. Suhling, R. E. Beaty, R. C. Jaeger, and R. W. Johnson, "Piezoresistive Stress Sensors for Structural Analysis of Electronic Packages," *J. Electronic Packaging*, **113**, 1991, p. 203.

9. Landau, L. D., and E. M. Lifshitz, *Theory of Elasticity*, Addison Wesley, Reading, MA, 1959, pp. 36–41.

10. Smith, C. S., "Macroscopic Symmetry and Properties of Crystals," *Solid State Physics*, Vol. 6, F. Seitz and D. Turnbull, eds. Academic Press, New York, 1958, pp. 175–249.

11. Middelhoek, S., and S. A. Audet, *Silicon Sensors*, Academic Press, New York, 1989, pp. 105–151.

12. Kanda, Y., "A Graphical Representation of the Piezoresistance Coefficients in Silicon," *IEEE Trans. Electronic Devices*, **ED-29**, 1982, p. 64.

13. Goldstein, H., *Classical Mechanics*, 2d edn., Addison Wesley, Reading, MA, 1981, pp. 128–148.

14. Thurston, R. N., "Use of Semiconductor Transducers in Measuring Strains, Accelerations, and Displacements," *Physical Acoustics*, Vol. I, Part B, W. P. Mason, ed., Academic Press, New York, 1964, pp. 215–235.

15. Sze, S. M., *Physics of Semiconductor Devices*, 2d edn., Wiley, New York, 1981, pp. 12–16.

16. Chelikowsky, J. R., and M. L. Cohen, "Nonlocal Pseudopotential Calculations for the Electronic Structure of Eleven Diamond and Zinc-Blende Semiconductors," *Phys. Rev.*, **B14**, 1976, p. 556.

17. Herring, C., "Transport Properties of a Many-Valley Semiconductor," *Bell System Technical J.*, **34**, 1955, p. 237.

18. Herring, C., and E. Vogt, "Transport and Deformation-Potential Theory for Many-Valley Semiconductors with Anisotropic Scattering," *Phys. Rev.*, **101**, 1956, p. 944.

19. Keyes, R. W., "The Effects of Elastic Deformation on the Electrical Conductivity of Semiconductors," *Solid State Physics*, Vol. 11, F. Seitz and D. Turnbull, eds., Academic Press, New York, 1960, pp. 149–221.

20. Tufte, O. N., and E. L. Stelzer, "Piezoresistive Properties of Heavily Doped n-Type Silicon," *Phys. Rev.*, **133A**, 1964, p. 1705.

21. Ref. 7, pp. 36–37.

22. Boer, K. W., *Survey of Semiconductor Physics*, Van Nostrand Reinhold, New York, 1990, pp. 210–222.

23. Jaeger, R. J., *Introduction to Microelectronic Fabrication,* Vol. V in Modular Series on Solid State Devices, G. W. Neudeck and R. F. Pierret, eds., Addison Wesley, Reading MA, 1988, pp. 49–62.

24. Miura, H., A. Nishimura, S. Kawai, and K. Nishi, "Development and Application of the Stress Sensing Test Chip for IC Plastic Packages," *Proc. 64th Annual Meeting of the Japanese Society of Mechanical Engineers,* pp. 1826–1832.

25. Natarajan, B., and B. Bhattacharyya, "Die Surface Stresses in a Molded Plastic Package," *Proc. 36th Electronic Components Conference, IEEE,* 1986, pp. 544–551.

26. Lundström, P., and K. Gustafsson, "Mechanical Stress and Life for Plastic-Encapsulated Large Area Chip," *Proc. 38th Electronic Component and Technology Conference, IEEE,* 1988, pp. 396–405.

27. Beaty, R. E., J. C. Suhling, C. A. Moody, D. A. Bittle, R. W. Johnson, R. D. Butler, and R. C. Jaeger, "Calibration Consideratons for Piezoresistive-based Stress Sensors," *Proc. 40th Electronic Component and Technology Conference, IEEE,* 1990, pp. 797–806 and "Piezoresistive Coefficient Variation in Silicon Stress Sensors Using a Four-Point Bend Text Fixture," *IEEE Trans. Components, Hybrids, Manuf. Technol.,* **CHMT-15,** 1992, p. 904.

28. Sze, S. M., *Semiconductor Technology,* 2d edn., McGraw-Hill, New York, 1988, Chap. 11.

29. Beadle, W. E., J. C. C. Tsai, and R. D. Plummer, *Quick Reference Manual for Silicon Integrated Circuit Technology,* Wiley, New York, 1985, pp. 4–10.

30. Beck, J. V., and K. J. Arnold, *Parameter Estimation in Engineering and Science,* Wiley, New York, 1977.

31. Thurston, R. N., "Use of Semiconductor Transducers," *Physical Acoustics,* Vol. 1 Part B, W. P. Mason, ed., Academic Press, New York, 1964, pp. 215–235.

32. Gee, S. A., V. R. Akylas, and W. F. van den Bogert, "The Design and Calibration of a Semiconductor Strain Gauge Array," *1988 IEEE Proc. Microelectronic Test Structures., IEEE,* 1988, pp. 185–191.

33. Grove, A. S., *Physics and Technology of Semiconductor Devices,* Wiley, New York, 1967, pp. 321–324.

34. Suhling, J. C., M. T. Carey, R. W. Johnson, and R. C. Jaeger, "Stress Measurement in Microelectronic Packages Subjected to High Temperatures," ASME Winter Annual Meeting, Dec. 1991, ASME publication AMD-Vol. 131/EEP-Vol. 1, *Manufacturing Processes and Materials Challenges in Microelectronic Packaging,* ASME, 1991, pp. 143–152.

35. Sweet, J. N., M. R. Tuck, D. W. Peterson, and D. W. Palmer, "Short and Long Loop Manufacturing Feedback Using a Multisensor Assembly Test Chip," *IEEE Trans. Components, Hybrids, Manuf. Technol.,* **CHMT-14,** 1991, p. 529.

36. Bastawros, A. F., and A. S. Vososhin, "In-Situ Calibration of Stress Chips," *Proc. 40th Electronic Component and Technology Conference, IEEE,* 1990, pp. 791–795.

37. Suhir, E., "Thermal Stress Failures in Microelectronic Components—Review and Extension," *Advances in Thermal Modeling of Electronic Components,* Vol. 1, A. Bar-Cohen and A. D. Krans, eds., Hemisphere Publishing, New York, 1988, pp. 337–412.

38. van Kassel, C. G. M., S. A. Gee, and J. J. Murphy, "The Quality of Die-

Attachment and Its Relationship to Stresses and Vertical Die-Cracking," *IEEE Trans. Components, Hybrids,* **CHMT-6**, 1983, p. 414.

39. Groothuis, S., W. Schroen, and M. Murtuza, "Computer Aided Stress Modeling for Optimizing Plastic Package Reliability," *Proc. 23d Annual Reliability Physics Symposium,* 1985, pp. 184–191.

40. Edwards, D. R., K. G. Heinen, S. K. Groothuis, and J. E. Martinez, "Shear Stress Evaluation of Plastic Packages," *IEEE Trans. Components, Hybrids, Manuf. Technol.,* **CHMT-12**, 1987, p. 618.

41. Gee, S. A., W. F. van den Bogert, and V. R. Akylas, "Strain-Gauge Mapping of Die Surface Stresses," *IEEE Trans. Components, Hybrids, Manuf. Technol.,* **CHMT-12**, 1989, p. 587.

42. Miura, H., A. Nishimura, S. Kawai, and G. Murakami, "Structural Effect of IC Plastic Package on Residual Stress in Silicon Chips," *Proc. 40th Electronic Component and Technology Conference,* IEEE, 1990, pp. 316–321.

# 8

# Analysis of the Thermal Loading on Electronics Packages by Enhanced Moiré Interferometry

*Arkady Voloshin*

## 8.1 INTRODUCTION

In the area of thermal stress analysis, there has been a persisting need for reliable tools to better understand and characterize material behavior, reveal structural response to thermal loads, and determine distributions of stresses and strains in different geometries under various boundary conditions. Different approaches—analytical, numerical, or experimental—have been applied to handle these problems.

Reliable analytical models have been developed for a relatively large class of problems; however, with the increasing complexity of geometry and boundary conditions numerical algorithms soon became necessary to handle the solution. Numerical methodologies such as the finite element method or the boundary element method can handle a wider class of problems that analytical models cannot solve. However, for such techniques extreme care should be given to the choice of element size and shape, mesh design, and boundary conditions, which cannot always be reliably prescribed. Validation of both analytical and numerical models is always required to provide confidence in their findings. This is usually achieved via experimental measurements of the same quantities that were determined analytically or numerically. Experimental evaluations of stresses and strain usually provide realistic solutions since they are not affected by assumptions made to facilitate analytical and numerical procedures. However, direct measurements of the stresses are not feasible; they are usually calculated on the basis of material properties and the knowledge of the strain field. The experimental evaluation of the strain tensor components is customarily based on direct measurement of the associated displacement field.

Thus, it is the objective of this chapter to introduce an accurate and sensitive experimental methodology for measurement of displacement fields resulting from thermal loading of microelectronics packages.

### 8.1.1 Displacement Measurements

Displacement measurement tools can be divided into techniques for local and for full-field displacement measurements. Local displacement measurements techniques are relatively simple to use and provide an accurate way to evaluate displacement at the given point. This type of measurement, while attractive for some simple, quality-control-oriented applications, is not suitable for the full-field analysis required for evaluation of thermally induced deformations in microelectronics packages.

Full-field displacement measurement methods include, among others, the grid method,[1,2] speckle,[3] holography,[4] and moiré.[5,6] All these methods may be used to monitor both in-plane orthogonal displacement components. Holography and moiré may also be used to monitor out-of-plane displacements.

Determining displacements with grids is one of the oldest methods of experimental mechanics. Various techniques have been developed to apply the grid and to record and analyze its deformations. Distances between discrete points on the grid are measured before and after loading to yield displacements at grid points. The grids are measured and analyzed using microscopes, digitizing tablets, or digital image processing. The sensitivity of the grid method is dependent on the ability of the measuring instrument to detect the new location of a grid point relative to its original location and on the grid density.

Speckle is an extension of the grid method. It is basically a grid method in which the grid is random and identified by the characteristics and details of the object's surface. Two speckle images are usually produced before and after deformation of the body. Displacements are extracted by comparison of the two images. Due to the randomness of speckle images, the only practical way to process them is by digital image processing. Speckle is convenient for automated measurements but is still limited in its resolution, which is slightly better than that of the grid method, because the data points are not limited to specific grid lines but rather represent any general point in the camera view. Speckle is suitable for applications where deformations are relatively large.

Holography is a very sensitive and accurate technique for displacement measurement. Displacements as small as half the wavelength of light used may be detected. However, it should be noted that relatively long exposure times required to record the holograms limit the use of holography in many applications. Holography is mostly used for quasi-static and out-of-plane displacement measurements where there is adequate time for hologram recording and where the required illumination angles may be easily accomplished. Such angles are rather difficult to achieve experimentally when the in-plane displacements are considered. The need for longer exposure time was overcome by double-exposure dynamic holography; however, the problems are generally more complex due to the requirements of a coherent laser with short pulse duration. The main advantage of holography is that specimens do not require special surface preparation.

Another attractive approach to measurement of the in-plane as well as the out-of-plane displacements is the use of moiré methods. There are two types of moiré—geometric moiré and interferometric moiré. Since the interferometric moiré is the technique of choice for in-situ displacement measurement in microelectronic packages subjected to thermal loading, it is presented in detail in the next section.

## 8.2 ESSENTIALS OF MOIRÉ INTERFEROMETRY

Moiré interferometry combines the concepts and techniques of geometrical moiré and optical interferometry. Geometric moiré as generated by low frequency bar-and-space gratings may be explained on the basis of obstruction or mechanical interference; it has been shown[7] that all moiré phenomena can be treated as optical interference. Here, high-sensitive moiré interferometry is utilized as a method for the whole-field, in-plane displacement measurements. This technique offers a unique combination of high sensitivity, optical contrast and range, as well as good spatial resolution. This technique was introduced as a high-sensitivity displacement measuring technique, along with procedures for producing high-frequency specimen gratings in 1980 by Post.[8] Since its introduction several improvements have been made to the technique. Those include simpler optical systems,[9] setups for measurements of both displacement components ($U$ and $V$),[10,11] optical separation of $U$ and $V$ fields,[12] better methodologies for specimen grating production,[13] and measurements of out-of-plane displacements as well as in-plane measurements.[14]

### 8.2.1 Specimen Grating

Specimen grating is produced from the photographic mold[13] by a simple replication technique. This is schematically illustrated in Fig. 8-1. A pool

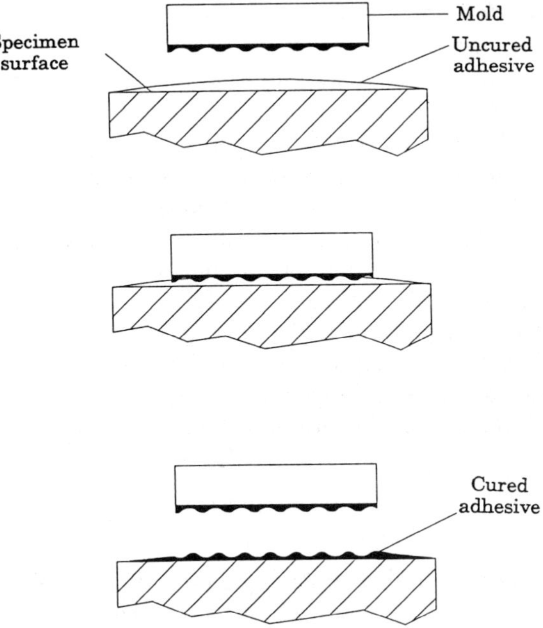


**Figure 8-1**   Replication of specimen grating.

of liquid adhesive is poured on the specimen and squeezed into a thin film by pressing against the mold. After curing, the photographic mold is pried off—only a small prying force is required—leaving a reflective diffraction grating bonded to the surface of the specimen. The weakest interface in the system occurs between the gelatin of the photographic plate and the evaporated aluminum or gold, which accounts for the transfer of the reflective film to the specimen. The result is a thin reflective high-frequency phase-type diffraction grating formed on the specimen.

In such a replication scheme the mold is lost in the process. An alternative procedure that preserves the mold (or master) and makes it reusable is to add an intermediate replication step that does not harm the mold. In this three-step process, a submaster is produced by replicating the phase grating surface in silicon rubber, in much the same way as in Fig. 8-1. The submaster is then used to replicate the phase grating in an adhesive on the specimen surface.[15] Silicon rubber of the liquid type has the distinct virtue for the intermediate step that it is a nonadhesive replicating material. It permits easy separation from the master and also from the final specimen grating; it bonds well, however, to specially primed surfaces. Details of this procedure have been described elsewhere.[16]

### 8.2.2 Moiré Interferometry

A schematic description of moiré interferometry is shown in Fig. 8-2. A high-reflection, symmetrical, phase-type diffraction grating is reproduced on the specimen surface. When loads are applied to the specimen the grating moves and deforms together with the specimen surface.

Two beams of coherent light with wavelength $\lambda$ illuminate the specimen grating obliquely from angles $+\beta$ and $-\beta$ (Fig. 8-2). The two-beam interference creates walls of constructive and destructive interference called a virtual grating in the zone of their intersection; the virtual grating is cut by the plane of the specimen surface, where an array of parallel and closely spaced fringes are formed. These fringes are arrays of the bright and dark bars that act as the reference grating. Its frequency, $F$, is given by the governing equation for two beam interference as

$$F = \frac{2}{\lambda} \sin \beta \tag{8-1}$$

The specimen grating and virtual reference grating interact to form a moiré pattern, which is viewed and can be recorded by the camera. Its interaction with the specimen grating produces the $U$-displacement field. In practice, another pair of coherent incident beams can be used to form a reference grating perpendicular to the $y$-axis, to interact with the corresponding array of lines of the crossed-line specimen grating and produce the $V$-displacement field, as shown in Fig. 8-3.

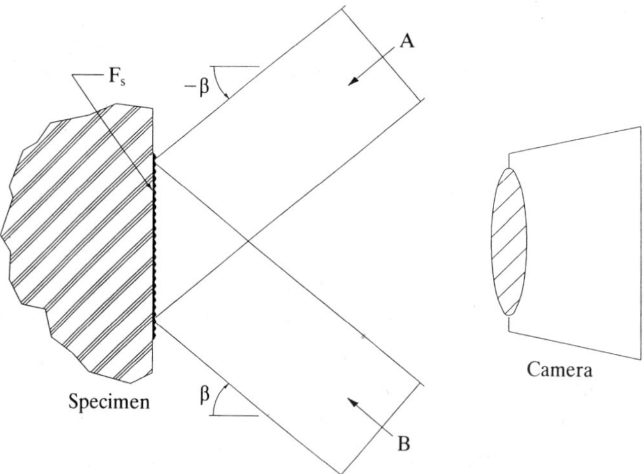


**Figure 8-2**  Schematic diagram of moiré interferometry.

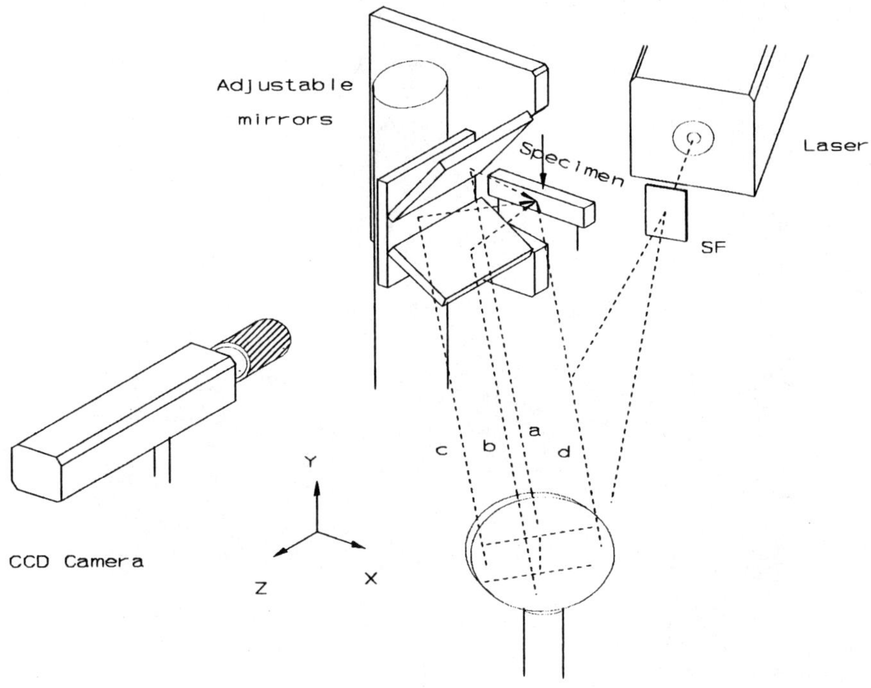


**Figure 8-3**  Four-beam interferometer.

Representative patterns of moiré interferometry fringes are shown in Fig. 8-4. They represent the in-plane displacements of every point on the specimen surface as contour maps of equal displacement fringes. Quantitatively, for each point in the fringe pattern

$$U = \frac{N_x}{F}, \qquad V = \frac{N_y}{F} \tag{8-2}$$

where $U$ and $V$ are components of displacement in $x$- and $y$-directions, respectively; $N_x$ and $N_y$ are fringe orders when lines of the reference grating are perpendicular to the $x$- and $y$-directions, respectively; and $F$ is the frequency of the reference grating. In the example (Fig. 8-4), the reference grating frequency $F$ was 2400 lines/mm, which corresponds to a sensitivity $(1/F)$ of 0.417 µm per fringe order. $N_x$ and $N_y$ are determined at the fringe centers either manually or with the aid of digitizing tablets. Data collection is restricted to fringe centers, whether dark or bright.

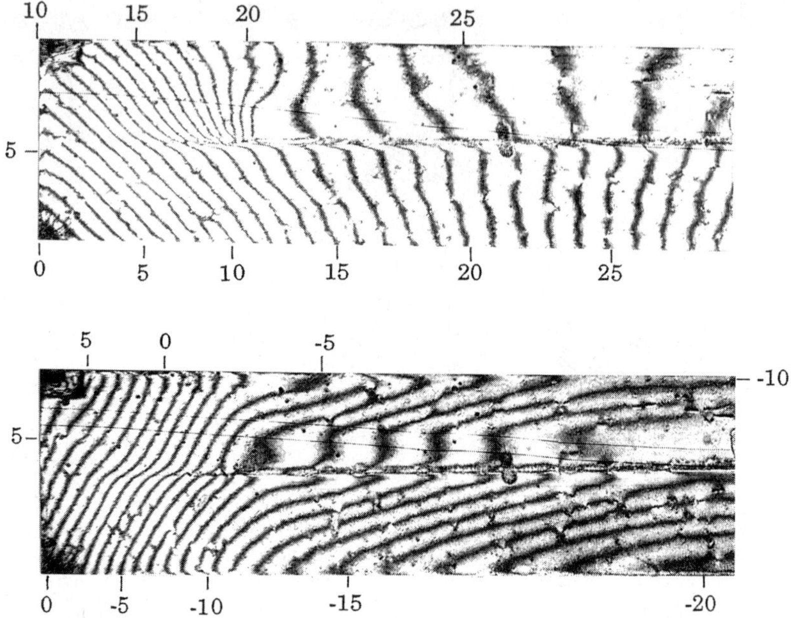

**Figure 8-4** $U$ and $V$ fringe patterns for the cross-section of a thermally loaded electronics package.

To assign fringe orders it is necessary to define a point of zero displacement as a reference point and assign $N = 0$ to the fringe passing through that point and then number the adjacent fringes relative to that fringe. The reference point may be any point whose displacement is known. For strain analysis, relative displacements rather than the displacements themselves are of interest. Thus, knowledge of the location of a reference point is immaterial, any arbitrary (and convenient) point can be considered a reference and the fringe passing through it may be assigned zero order. All other fringes may then be numbered relative to that fringe.

Once the displacement map has been constructed, strains may be determined from strain–displacement relations as follows:

$$\varepsilon_x = \frac{\partial U}{\partial x}, \qquad \varepsilon_y = \frac{\partial V}{\partial y}$$

$$\gamma_{xy} = \frac{\partial U}{\partial y} + \frac{\partial V}{\partial x} \tag{8-3}$$

The derivatives in the right-hand side of Eq. (8-3) may be determined from measured displacements and the equation may be rewritten as

$$\varepsilon_x = \frac{\Delta U}{\Delta x}, \qquad \varepsilon_y = \frac{\Delta V}{\Delta y}$$

$$\gamma_{xy} = \frac{\Delta U}{\Delta y} + \frac{\Delta V}{\Delta x}$$

$$(8\text{-}4)$$

Use of Eq. (8-4) requires that $\Delta x$ and $\Delta y$ are sufficiently small. To satisfy this requirement it is necessary to collect displacement data from the fringe pattern at points that are relatively closely spaced. This is possible only if the fringe density of the pattern is high. Such fringe densities in moiré patterns are usually generated when the displacements are high, such as in the cases of large loads and compliant materials. However, cases of low fringe densities are often encountered, usually when stiff structures made of rigid materials are considered and when applied loads are limited to small values that are inadequate for inducing high displacement levels. A more complicated situation arises when both high and low fringe densities exist at different regions of the same field. An example of such a situation is shown in Fig. 8-4, where the lower left section of the field has high fringe density and would therefore yield reliable strain predictions while the upper right section of the field contains very few fringes and they cannot be reliably used for strain analysis. Low fringe density does not yield sufficient displacement data for reliable strain analysis when conventional fringe counting is used to acquire the displacement data.

Another issue that has to be considered when using the moiré method for displacement and strain analysis is the existence of a null field. In general, the specimen grating does not have lines that are perfectly straight and uniformly spaced. Also, the optical elements used to form the reference grating are not so accurate that a perfect reference grating is formed. The result of this is that a few fringes usually appear in the field of view before the specimen is loaded. These must be subtracted from the pattern obtained after loading in order to determine the load-induced displacements. This can be done manually by subtracting the fringe orders at corresponding points in the load and no-load patterns. Doing this significantly reduces the amount of displacement data available, since for each point considered in the load pattern there should exist a corresponding point that lies on a fringe center in the no-load pattern. This condition is hard to satisfy and can only be circumvented if the null field is made to contain an initial high-density fringe pattern (carrier fringes).[17] The introduction of high-density initial fringe patterns may make the fringe density in the load pattern too high, beyond the resolution of the recording or photographing device. The most commonly adopted approach in coping with the null field is to ignore it, assuming that the initial pattern is sparse, while the full-load pattern exhibits a significantly higher number of fringes. However, this will not work for cases where the full-load pattern has a low fringe density.

The practical implementation of the technique of moiré interferometry just described has a sensitivity corresponding to a moiré grating with 2400 lines/mm, i.e., 0.417 μm per fringe order or 0.208 μm, if half-fringes are counted instead of full fringes. The fringes produced by the technique described above have good contrast. This makes it simple to detect, record and analyze them. Another feature of the moiré method is that the location of the fringe pattern is coincident with the specimen. It is a real-time method that allows study of both the steady-state and transient situations. Two orthogonal displacement components can be monitored simultaneously by the same interferometer with minor modifications if crossed-line specimen gratings are used.

In the majority of applications of moiré interferometry, high fringe densities were always sought, and the null field was usually ignored. The choice of specimen shape, size, and material, and the load level applied to it dictates the density of the resulting fringe pattern. Sometimes it is not possible to set these parameters such that high fringe densities are produced. This is the case when stiff structures are examined and when the applied loads are small. Here, the small amount of displacement information gathered by simple fringe counting would be insufficient for accurate strain analysis. The situation is further complicated if the initial pattern contains a considerable number of fringes. Ignoring the initial pattern in such cases would introduce erroneous displacement data.

To extend the application of moiré interferometry to patterns with low fringe density and to develop a capability to account for the initial patterns properly, the approach of fractional fringe analysis[18,19] was introduced to compute the displacements at any point in the field. Digital image processing is used to accurately collect light intensity information needed for the computations. The approach eliminates ambiguities associated with the determination of the locations of fringe centers for fringe counting, thus reducing random errors in the collected data. Moreover, it enables fast and automated analysis of moiré fringe patterns while significantly increasing the sensitivity of the method.

## 8.3 DIGITAL IMAGE ANALYSIS ENHANCED MOIRÉ INTERFEROMETRY

It has been already mentioned that all moiré phenomena are cases of optical interference.[7] The mechanism of fringe formation in moiré methods accounts for diffraction of light and focuses on how the different diffraction orders emerging from a specimen grating recombine and interfere to create moiré fringes. The underlying value of analyzing the actual mechanism of moiré fringe formation is that it can be combined with the wave theory of light to arrive at the exact interpretation of moiré fields in terms of the displacements that produced them. Displacements may be represented as continuous functions of the light intensity distribution in the whole field, not just at the fringe centers as suggested by the casual interpretation of fringe formation.

### 8.3.1 Mechanism of Fringe Formation

When a beam of light goes through a slit, it diffracts into a number of diffraction orders depending on how wide the slit is. The same effect occurs whenever light goes through or reflects from a grating (any grating, coarse or fine). The number and orientation of the different diffraction orders are given by

$$\sin \phi_n = \sin \beta + n\lambda F \tag{8-5}$$

where $\phi_n$ = diffraction angle for the $n$th order
$\beta$ = angle of the incident beam
$F$ = frequency of the grating
$\lambda$ = wavelength of the light used

Coarse gratings give a very large number of closely spaced diffraction orders, while fine gratings give few widely separated diffraction orders.

When the specimen is deformed, the specimen grating will deform with it. The frequency of the grating will no longer be uniform; it will change locally from point to point according to the local deformation. This will introduce rotational separations between beams that were originally parallel, and the conditions of two-beam interference are created. A fringe pattern of destructive and constructive interference will be formed by each pair of intersecting beams. This fringe pattern will not be uniform since, in general, the deformation is not uniform. It can easily be seen that many fringe patterns are formed simultaneously by the different pairs of diffraction orders. Each of these fringe patterns carries essentially the same displacement information, despite the fact that patterns created by higher diffraction orders contain higher numbers of fringes. An observer receives the summation of all these patterns superposed. The average pattern received is basically composed of a strong clear fringe pattern plus some ghost patterns. The clear pattern is caused by the strongest diffraction beams emerging from the specimen, while the other patterns are created by the diminishing higher-order diffractions. Normally, those higher-order patterns are not recognizable when the light source used is not very powerful. They can easily be eliminated completely by simple optical arrangements.[20]

It is clear that the amount of light arriving at a certain point in the camera plane depends on the diffraction beams impinging at that point and on their strength. But these are dictated by deformation of the surface of the specimen and, therefore, the manner in which light intensity is distributed in a moiré field is related to the displacements at the grating surface. Such a relation was developed by Sciammarella in 1965.[21,22] The wave theory of light has been utilized to follow light wavetrains emerging from the specimen grating in directions determined by the diffraction equation. These directions indicate the local changes that occurred in the specimen grating frequency due to displacement. The analysis, basically the most general case of two-beam interference, reduces to the following simple relation:

$$I(x) = I_0 + I_1 \cos 2\pi F U(x)$$
$$+ I_2 \cos 4\pi F U(x)$$
$$+ I_3 \cos 6\pi F U(x)$$
$$+ \cdots$$
$$+ I_n \cos 2\pi n F U(x) \tag{8-6}$$

where
$$F = \text{reference grating frequency}$$
$$U(x) = \text{displacement at point } x \text{ in a direction perpendicular to the grating lines}$$
$$I_0 = \text{a background intensity}$$
$$I_1, I_2, I_3, \ldots, I_n = \text{harmonic components corresponding to the different diffraction orders contributing to the moiré pattern, and}$$
$$n = \text{maximum number of contributing diffraction orders}$$

The value $n$ is determined by the grating equation (8-5). It is evident from Eq. (8-5), that higher grating frequencies are desirable since they will give a limited number of diffraction orders and the overall quality of moiré patterns will be better.

Since the grating frequencies used for moiré interferometry are very high, only few diffraction orders are present and moiré fringes are produced by one diffraction order (in most practical cases). Figure 8-5 shows the optical path of the different orders emerging from the specimen grating in moiré interferometry. Here, the two beams A and B obliquely illuminate the specimen at angles $+\beta$ and $-\beta$. Each of the incident beams is diffracted at the specimen surface into diffraction directions described by Eq. (8-5). As shown in Fig. 8-5, light from beam A, diffracted in the $+1$ order of the specimen grating, emerges perpendicular to the specimen. Also, light from beam B that is diffracted in the $-1$ order emerges perpendicular to the

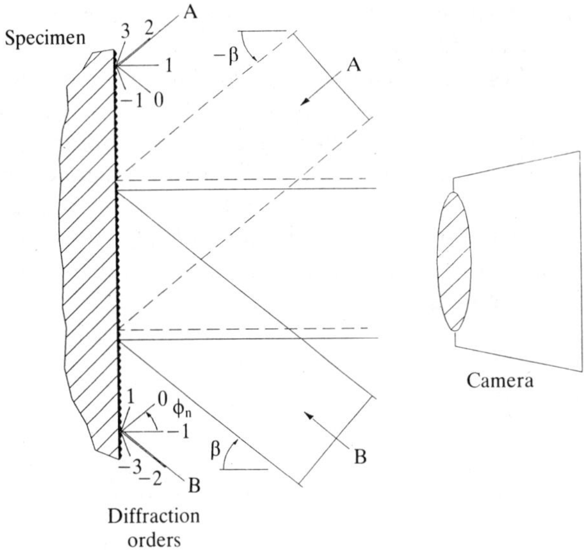


**Figure 8-5**    Diffraction scheme in moiré interferometry.

specimen. Those two diffraction beams coexist in space with no angular separation as long as the specimen is not loaded and the grating frequency is uniform. The camera will receive a uniform light intensity throughout the field. When the specimen is loaded, the local frequency changes and the diffracted beams are no longer perpendicular to the specimen; they will have an angular separation representative of the deformation. Due to their interference a moiré fringe pattern is formed at the camera plane. The fringe pattern formed is due to interference of one diffraction order only, as shown in Fig. 8.5. Other diffraction orders do not contribute to the pattern. This is a very important characteristic of moiré interferometry that explains why its patterns are of high quality and contrast. Moreover, this further simplifies Eq. (8-6) to include a term representing only one diffraction pattern. The optical law for moiré interferometry becomes

$$I(x) = I_0 + I_1 \cos 2\pi F U(x) \tag{8-7}$$

The definition of all parameters is the same as for Eq. (8-6).

In Fig. 8-5, the angle $\beta$ is such that the incident beams A and B coincide with the $+2$ and $-2$ diffraction orders, respectively. This is not a chance coincidence. Angle $\beta$ could theoretically be any value. However, practical limitations exclude almost all other possibilities but the one shown in the figure. The arrangement shown prescribes the virtual grating frequency, $F$, to be twice the specimen grating frequency (simple manipulations using Eq. (8-5) reveal this condition).

This choice makes the specimen a well-defined reference that can easily be used to align all other optical components relative to it by a simple optical observation of the paths of the different diffraction orders. Thus, the only parameters needed to completely define the arrangement shown in Fig. 8-5 are the specimen grating frequency, $F$, and the wavelength of the light, $\lambda$. For example, when $F$ is 1200 lines/mm and $\lambda$ is 632.8 nm (for helium–neon lasers), the angle $\beta$ is 49.4°. These conditions are used in all examples presented in the current work.

### 8.3.2 Fractional Fringe Analysis

The optical law for moiré interferometry may then be expressed in terms of the continuous fringe order $\psi(x)$, rather than the displacement as

$$I(x) = I + I \cos 2\pi\psi(x) \tag{8-8}$$

It can be seen that each time $\psi(x) = n$, where $n$ is an integer, the intensity will be a maximum (center of a bright fringe). When $\psi(x) = \frac{1}{2}(2n + 1)$, the intensity will be a minimum (center of a dark fringe). This is graphically represented in Fig. 8-6. The bright fringes are the loci of points where the displacements, in the $x$-direction, of the specimen grating with respect to the reference grating are equal to an integer number multiplied by the pitch of the master grating. The dark fringes have similar interpretation, but in terms of the half-pitches.

When $\psi(x)$ is an integer or an odd multiple of $\frac{1}{2}$, the displacement information is basically the same as would be obtained by simple fringe counting. This is where fringe counting stops. No displacement information is available about other points in the field, for example, at point $a$ in Fig. 8-6, between the bright fringe at $\psi = 1$ and the dark fringe at $\psi = \frac{3}{2}$. This is where Eq. (8-8) picks up. Knowledge of the light intensity, $I(a)$, at that point together with the amplitudes $I_0$ and $I_1$ renders $\psi(a)$ known. The value of $\psi(a)$ will be a fraction between 0 and $\frac{1}{2}$ and will yield the displacement at point $a$ when multiplied by the pitch of the reference grating. The same argument is true for any intermediate point between any two fringes. This is the reason for the name *fractional fringe analysis*. The fringe orders need not be integers (or odd multiples of $\frac{1}{2}$). They may be any fractional number!

Equation (8-8) may be rewritten to yield the displacements directly:

$$U(x) = \frac{1}{2\pi F} \arccos\left(\frac{I(x) - I_0}{I_1}\right) \tag{8-9}$$

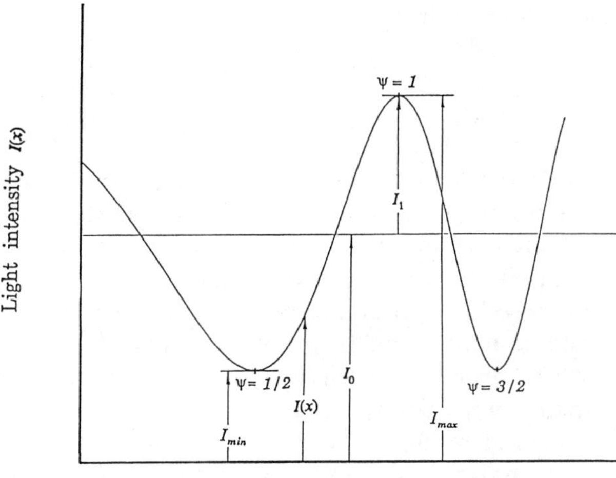



**Figure 8-6**   Light-intensity distribution in moiré interferometry.

For simplicity the above equation may be applied over each half-fringe separately, thus determining the displacements relative to the starting point. The actual cumulative displacements are found by adding the displacement at the startint points to all values determined. Mathematically, this is expressed as follows:

$$U(x) = U_0 + \frac{1}{2\pi F} \arccos\left(\frac{I(x) - I_0}{I_1}\right) \tag{8-10}$$

where $U_0$ = displacement at the starting point
$x$ = zero at that point

$U_0$ is the reference grating pitch multiplied by either an integer or an odd multiple of $\frac{1}{2}$, depending whether the starting point is a center of a bright or a dark fringe. The amplitudes $I_0$ and $I_1$ are easily determined from light-intensity distributions, such as that in Fig. 8-6. $I_0$ is equal to $\frac{1}{2}(I_{max} + I_{min})$, and $I_1$ is equal to $\frac{1}{2}(I_{max} - I_{min})$.

The task of determining the displacement field has become that of determining the light-intensity field. Recent advances in the field of digital image analysis have led to significant progress in automated measurements of light intensity in photoelasticity over fairly large fields and with high accuracy[18] and also in geometric moiré for displacement measurements.[19] The following section will extend such an approach to full-field moiré interferometry.

### 8.3.3 Digital Image Processing

The two main components of a digital image processing system are the image acquisition device and the image processor. A CCD (charge coupled device) video camera is used as the image acquisition device. CCD cameras are superior to many other light-sensing systems for a variety of reasons. The radiometric response of CCDs is extremely linear, and the devices are quite sensitive to small changes in light intensity. The spectral response of these devices is broader than that of the vidicon or film systems. The fixed position of each pixel provides high geometric precision in the sampled image. The devices are small and lightweight. Typical CCD systems provide array sizes of up to 500 by 500 pixels. CCD reliability is also quite high.

The second main component of the system is the image processor. This is a computer-based system that receives the image information from the acquisition device, converts the information to a digital format, and stores the information in an image buffer. Then, in conjunction with the host computer, it enables access to the digitized image information, and supports performance of mathematical operations on the image.

In the present work the image processor used is a PC-based system consisting of a 480-by-512 frame grabber and a high-speed frame processor, with a maximum light-intensity resolution of 256 gray levels. Together with software developed in-house, the digital image analysis enhanced moiré interferometry approach described above allows one to circumvent the high fringe density requirement for accurate strain analysis.

The sensitivity of the method is theoretically increased by more than two orders of magnitude. This is explained by the ability to find intermediate displacements between fringe centers when fractional fringe analysis is used. Typically, a PC-based image processor has a maximum light intensity resolution of 256 gray levels, which means that the image processor is capable (ideally) of identifying 256 light-intensity values in any one image. If the image is magnified such that one-half fringe occupies the whole screen, and assuming that the image has maximum quality and perfect contrast, which means that all gray levels between 0 and 255 will exist in the image; then fractional fringe analysis (which is applicable to one half-fringe) will give 256 different displacement values over the half fringe, which is equivalent to 512 times fringe multiplication. However, in practice an image of a moiré pattern does not contain all gray levels. Due to loss of contrast and quality of laser source used it is common to see only 100 to 200 gray levels in any one image. This still keeps the sensitivity two orders of magnitude higher than that of regular moiré interferometry. The optical magnification of the camera's lens system may always be used to achieve best sensitivity by focusing the analysis on any desired region of a smaller number of fringes. In such a case, the effective fringe multiplication factor will depend on the number of full fringes in the image and the total number of gray levels in the field. Now the method can be applied to stiff structures or materials that would normally show small deformations in response to a load.

The digital image processing enhanced technique introduced also relaxes the limitation of getting the displacements only at the fringe centers; now displacements at any point in the field (whether at a fringe center or not) can be determined. This leads to another important advantage that the null field need no longer be ignored. No matter how few are the fringes in the null field, they can be properly taken care of. The null field displacements are determined by fractional fringe analysis at each and every point in the field and are subtracted from the final displacement values at the same points. Taking the null field into account improves the accuracy of the final data, and does not necessitate a perfect null field to start with—any unavoidable initial pattern can easily be handled by enhanced moiré interferometry.

The validity of the procedure discussed above was proved in a series of experiments where both traditional and digital image processing-based approaches were applied to the same structure.[23] The results obtained confirm the claim of high resolution for digital image analysis enhanced moiré interferometry. This technique was also applied to the analysis of the displacement field in the near proximity of the crack tip and good agreement with analytical and numerical predictions was observed.[24]

In summary, digital image processing combined with fractional fringe analysis increased the sensitivity of the method, broadened its applicability, and significantly improved data management. Practical applications of digital image analysis enhanced moiré interferometry to the analysis of the thermal loading on the microelectronics packages are presented in the following sections.


## 8.4 FULL-FIELD ANALYSIS OF THERMALLY INDUCED DEFORMATIONS

Fractional fringe moiré interferometry has been applied for strain analysis of various microelectronic components subjected to thermal load. With recent advances in microelectronics and the increasing trends in use of new materials such as ceramics, polymers, and alloys in highly populated electronic components running at high power levels, the mechanical design considerations in electronic devices began to acquire a lot of attention. Unfortunately, the size, the complexity, and the coexistence of materials of diverse mechanical properties in a small electronic device represent a complicated mechanical design problem for which little or no knowledge is available about the behavior of the device under load, in particular thermal load.

The existing numerical approaches to predicting device behavior are heavily dependent on accurate knowledge of the thermomechanical properties of all components of the microelectronic package, and this information is not readily available. Thus, the full field in-situ methodology of digital image analysis enhanced moiré interferometry became a tool of choice for measurement of the thermal indices of deformations in microelectronics packages.

### 8.4.1 Thermal Strain Measurements in IC-Packages

Detection of thermally induced strains in integrated circuit electronics packages (IC-packages), during either normal operating conditions or testing, has become of great concern lately. Due to mismatch in the coefficients of thermal expansion of the different materials coexisting in a package, severe strain concentrations occur, leading to ultimate mechanical/ electrical failure of the package. A number of analytical and numerical assessments of packages under such loading conditions were carried out in an effort to predict thermally induced strains.[25–28] However, these have been limited by the complexity of the package details and the variety of simplifying assumptions that had to be made to enable the analysis. This has limited the reliability of such models in representing actual packages. The relatively small size of the device and the small order of deformations rendered classical experimental procedures incapable of handling the strain analysis of electronics packages. The best tool currently available for actual strain measurements is the use of stress chips.[29–31] These, however, suffer from inaccurate calibration methodologies and, consequently, poor interpretation of the collected data. Moreover, the information provided is limited to strain gage locations on the surface of the silicon chip.

The in-situ full-field approach was applied to analysis of thermally loaded packages.[32] The specimens were prepared from AT&T 1 MB DRAM and 64 K SRAM packages by slicing the device along selected planes to expose the chip and the lead frame (Fig. 8-7). Crossed gratings of a frequency of 1200 lines/mm in the horizontal and vertical directions were replicated on the specimens. They were placed at room temperature in a specially designed oven, capable of heating the specimen uniformly.

At room temperature the specimen was inserted into the oven and the moiré interferometry apparatus was adjusted for proper alignment of all optical components and specimen orientation. Some deviations from the ideal setting nearly always exist, and they yield an initial fringe pattern of a relatively small number of fringes (null or zero field). If such a field exists, its effect has to be subtracted from the final displacement field to yield the net thermal load-induced displacements. The region of interest was then determined on the specimen and the magnification of the camera was adjusted such that this region occupied most of the viewing screen to

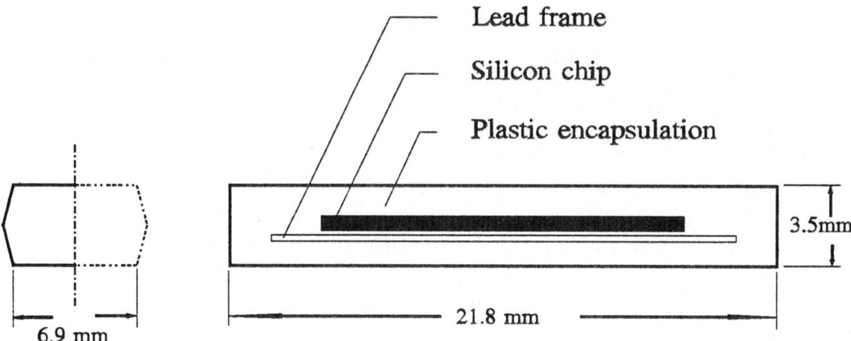



**Figure 8-7**    Schematic of AT&T 1 MB DRAM device.

achieve maximum possible spatial resolution. For each specimen the null field, for each of the displacement components, was recorded; then the heaters were turned on and the specimen was allowed to heat up to the final desired temperature. Fringe patterns for both displacement components ($U$ and $V$) were recorded continuously throughout the heating history for analysis. The selective frames representing the deformation history, e.g., at every 10°C increase in temperature, were analyzed later using digital image processing combined with fractional fringe analysis as discussed earlier. Once displacement data were available, the total strains were computed by numerical differentation of the distributions with respect to the horizontal ($x$) and vertical ($y$) directions. Equations (8-4) were used for this purpose. They provide the total normal strain components, which include the free thermal expansion of the material, to get the net mechanical strains, i.e., strains due to the mismatch of coefficients of thermal expansion of the different materials in the package. The free expansion strain for each material ($\alpha\,\Delta T$) has to be subtracted from the total strain, giving the net normal mechanical strains as

$$
\begin{aligned}
\varepsilon_{x,m} &= \varepsilon_x - \alpha\,\Delta T \\
\varepsilon_{y,m} &= \varepsilon_y - \alpha\,\Delta T
\end{aligned}
\tag{8-11}
$$

A typical example of the fringe patterns at 90°C for the left half of a 1 MB DRAM specimen is shown in Fig. 8-8. For this specimen, room temperature was 24.5°C and the null field had no fringes for the horizontal direction and thus the pattern for the horizontal component shown in Fig. 8-8$a$ is the final contour map of axial displacements due to heating. The null field for the vertical direction, Fig. 8-8$b$, had some fringes that had to be taken into consideration when processing the final fringe pattern for the verrtical component at 90°C, Fig. 8-8$c$. Each fringe (dark line) is a line of constant displacement, i.e., all points on that line have undergone the same amount

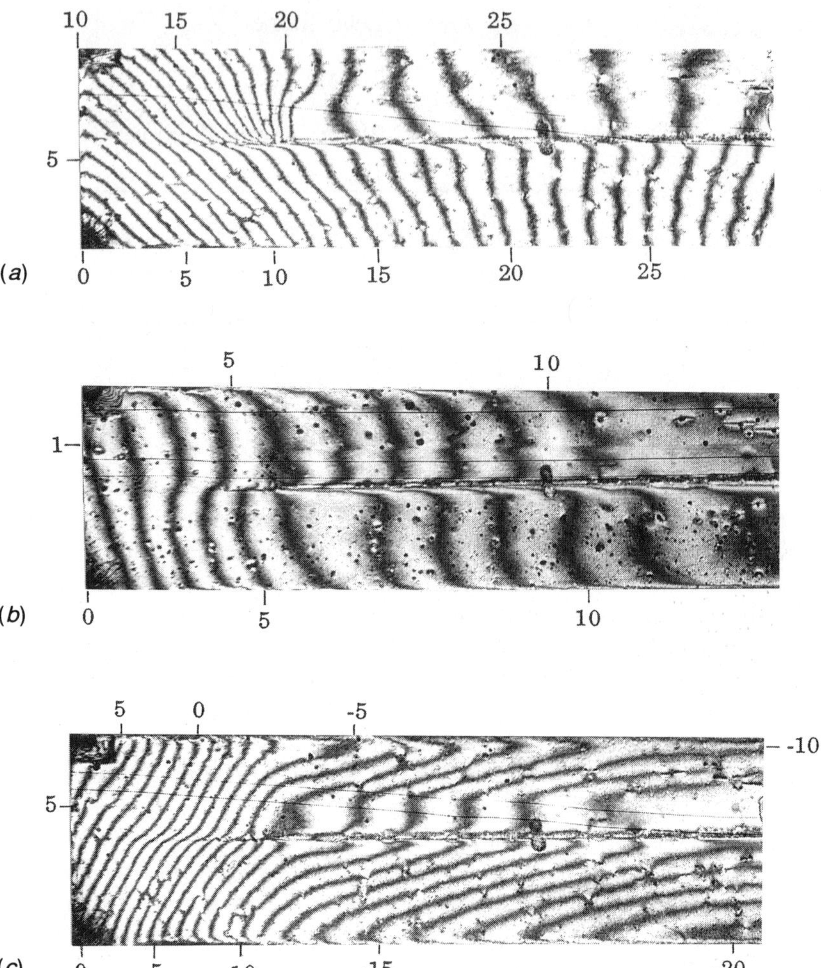

**Figure 8-8**   Typical fringe patterns in a 1 MB DRAM specimen. (*a*) *U*-fringe pattern at 90°C. (*b*) *V*-fringe pattern at 24.5°C (null field). (*c*) *V*-fringe pattern at 90°C.

of displacement in the direction under consideration. To assign fringe orders, a point of zero at the lower left corner of the package was considered the reference point in this example. Fringes passing through it were assigned the order zero and all other fringe orders are counted relative to that point, with orders increasing in the positive directions of the reference axes. The assigned fringe orders are also shown in Figs. 8-8*a*, *b*, and *c*.

Total horizontal strains $\varepsilon_x$ were derived from deformation fields according to Eq. (8-4). Upon subtraction of the free thermal expansion term, the net mechanical strain $\varepsilon_{x,m}$ distribution is as shown in Fig. 8-9. One can easily

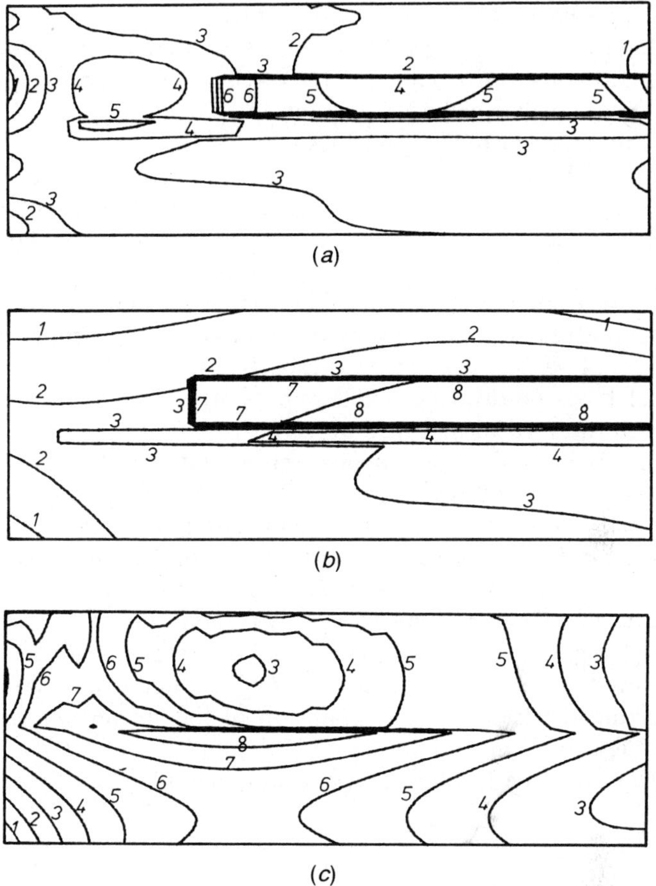

(a)

(b)

(c)

**Figure 8-9**  Net strain map at 90°C for a 1 MB DRAM device. (a) Horizontal strain $\varepsilon_x$. Contour values: (1) $-1.055 \times 10^{-3}$; (2) $-7.344 \times 10^{-4}$; (3) $-4.134 \times 10^{-4}$; (4) $-9.236 \times 10^{-4}$; (5) $2.286 \times 10^{-4}$; (6) $5.497 \times 10^{-4}$. (b) Vertical strain $\varepsilon_y$. Contour values: (1) $-4.974 \times 10^{-4}$; (2) $-2.807 \times 10^{-4}$; (3) $-6.412 \times 10^{-5}$; (4) $1.525 \times 10^{-4}$; (5) $3.692 \times 10^{-4}$; (6) $5.858 \times 10^{-4}$; (7) $8.025 \times 10^{-4}$; (8) $1.019 \times 10^{-3}$. (c) Shear strain $\gamma_{xy}$. Contour values: (1) $-6.270 \times 10^{-4}$; (2) $-3.571 \times 10^{-4}$; (3) $-8.725 \times 10^{-5}$; (4) $1.826 \times 10^{-4}$; (5) $4.525 \times 10^{-4}$; (6) $7.224 \times 10^{-4}$; (7) $9.923 \times 10^{-4}$; (8) $1.262 \times 10^{-3}$.

see that the silicon chip and the lead frame are in tension while the plastic encapsulant is in compression. Similarly, the strains for the vertical field based on the net vertical deformation field (after subtracting the null field deformations) are shown in Fig. 8-9b. The corresponding shear strains are shown in Fig. 8-9c.

Figures 8-9a, b, and c reflect the thermal strain picture in the examined plane of the package at the steady temperature of 90°C and show areas and levels of high and low strains. Stresses may be derived from these strains using material constitutive relations (Hooke's law) thus providing realistic and accurate data that are necessary for any design, redesign, testing, and evaluation of microelectonics packages.

For the AT&T 64 K SRAM, shown in Fig. 8-10, the deformation history as the specimen heats up from room temperature to 80°C is shown in Figs. 8-11a and b. Since the package is symmetrical, the patterns shown are for the left half only. The fringe patterns in Fig. 8-11a are the observed horizontal fields (U-displacement component). The top picture gives the null field at room temperature (20°C), while subsequent pictures show fringe patterns for temperatures of 40, 50, 70, and 80°C, respectively. Similarly, Fig. 8-11b gives the vertical fields (V-displacement component) at the same temperatures starting with the vertical null field at 25°C.

Numerical values for displacements were found from the patterns shown using fractional fringe analysis, and were stored in displacement data files for the stated temperatures. Displacement patterns may be used to study the deformation history or to determine strains in the whole package or in a local region of interest. As an example, strains were determined at all listed temperatures in a small region (0.5 mm by 0.5 mm) at the left corner of the silicon chip, Fig. 8-10. Local displacement contours in that region at 80°C were measured and net mechanical strains were found from the total strains by subtracting the free thermal expansion of the silicon ($\alpha \Delta T$); they are shown in Figs. 8-12a, b, and c.

Strain data may be further retrieved for a specific line of interest.[33] An

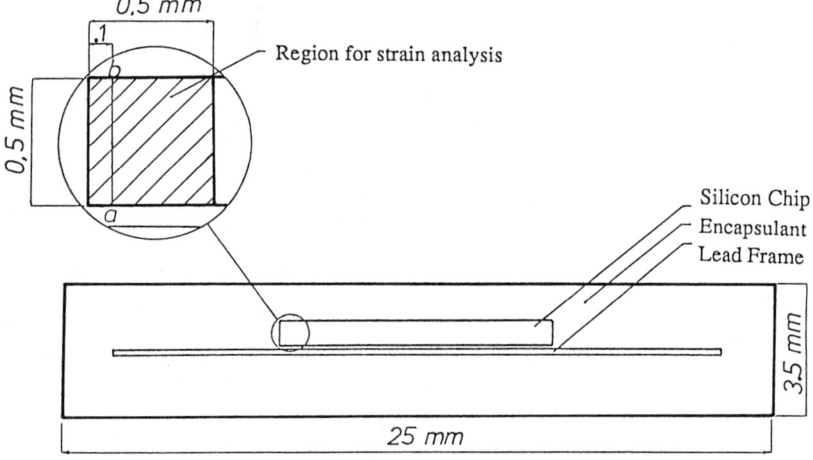


**Figure 8-10**   Schematic of AT&T 64 K SRAM device.

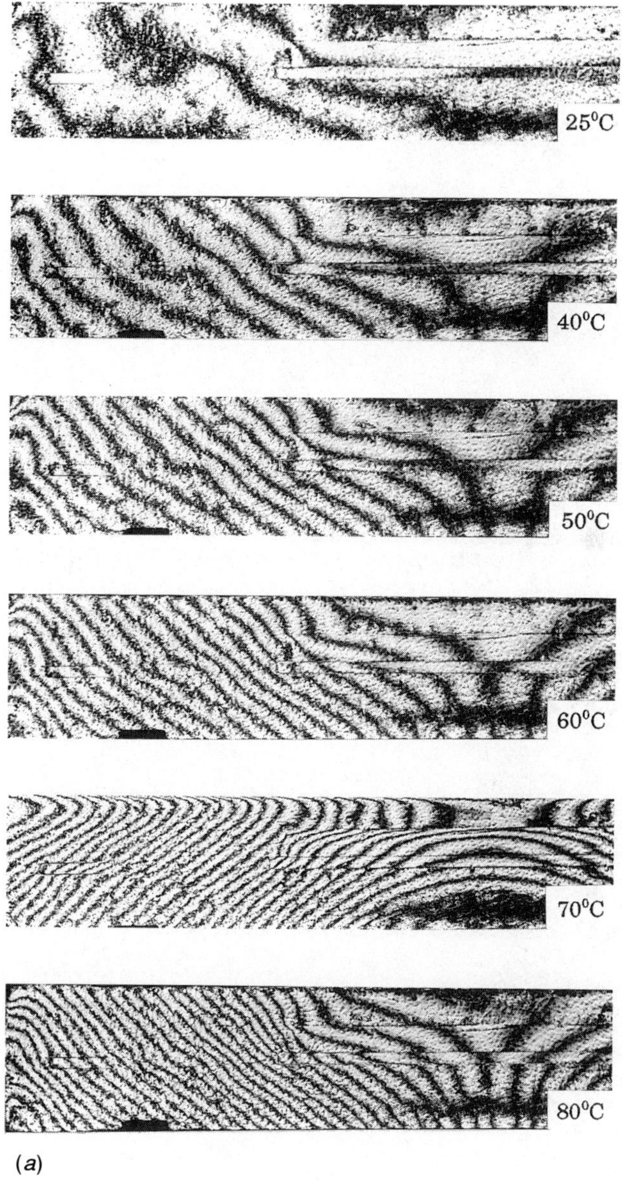


**Figure 8-11** (a) Fringe patterns for the horizontal field (U) as the temperature changes from +25°C to 80°C. (*Continued*)

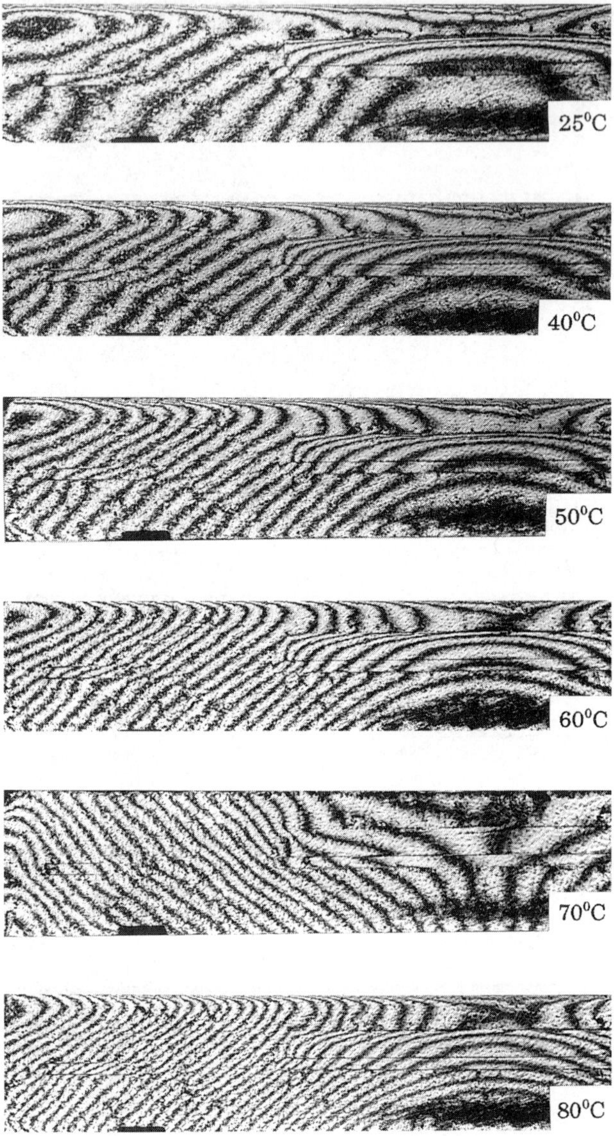

(b)

**Figure 8-11** (*continued*)  (*b*) Fringe patterns for the vertical field (*V*) as the temperature changes from $+25°C$ to $+80°C$.

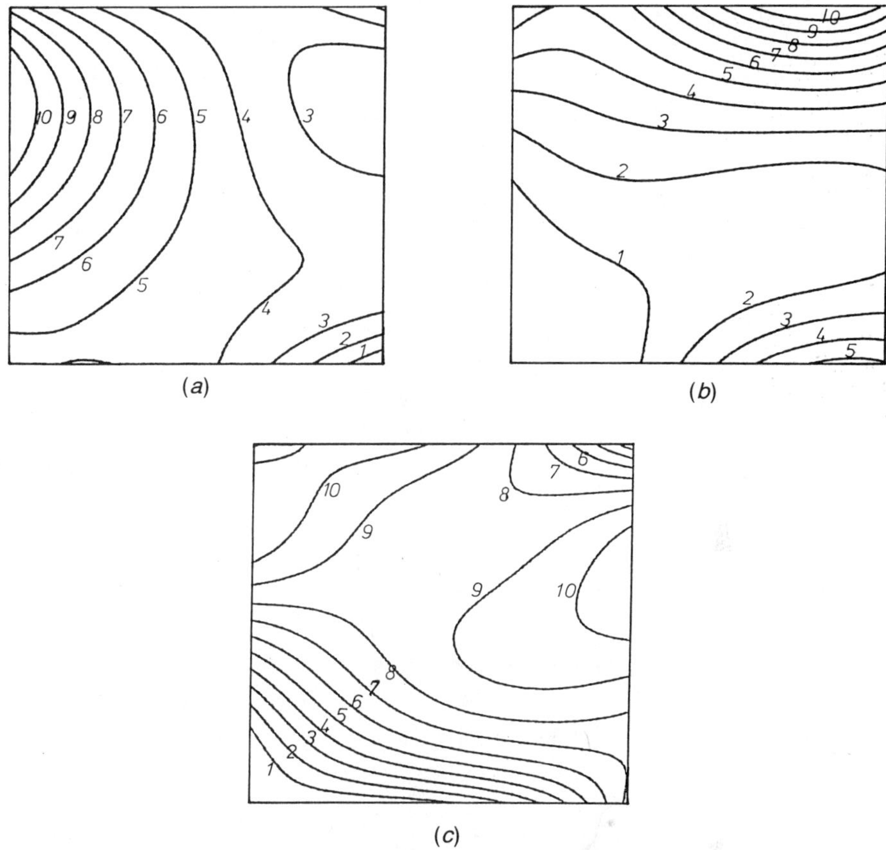

**Figure 8-12**  Net strain contours at the chip corner at 80°C. (a) $\varepsilon_x$ contour values: (1) 1.290 × 10$^{-4}$; (2) 2.455 × 10$^{-4}$; (3) 3.619 × 10$^{-4}$; (4) 4.783 × 10$^{-4}$; (5) 5.948 × 10$^{-4}$; (6) 7.112 × 10$^{-4}$; (7) 8.276 × 10$^{-4}$; (8) 9.441 × 10$^{-4}$; (9) 1.060 × 10$^{-4}$; (10) 1.177 × 10$^{-3}$. (b) $\varepsilon_y$ contour values: (1) 1.246 × 10$^{-3}$; (2) 1.752 × 10$^{-3}$; (3) 2.256 × 10$^{-3}$; (4) 2.762 × 10$^{-3}$; (5) 3.267 × 10$^{-3}$; (6) 3.773 × 10$^{-3}$; (7) 4.278 × 10$^{-3}$; (8) 4.783 × 10$^{-3}$; (9) 5.288 × 10$^{-3}$; (10) 5.794 × 10$^{-3}$. (c) $\gamma_{xy}$ contour values: (1) −0.216 × 10$^{-3}$; (2) −1.898 × 10$^{-3}$; (3) −1.675 × 10$^{-3}$; (4) −1.451 × 10$^{-3}$; (5) −1.228 × 10$^{-3}$; (6) −1.005 × 10$^{-3}$; (7) −7.820 × 10$^{-4}$; (8) −5.588 × 10$^{-4}$; (9) −3.355 × 10$^{-4}$; (10) −1.123 × 10$^{-4}$.

example for the axial strain distributions along a vertical line *a–b*, shown in Fig. 8-10, is given in Fig. 8-13. This normal strain is tensile, which is expected from the knowledge that the coefficient of thermal expansion for silicon is much less than that of the molding compound. The growth and distribution patterns of such strains are revealed by the presented technique for the first time.

Figure 8-13 shows an interesting phenomenon—the significant departure from the trend of change of strain with temperature. The strain increases

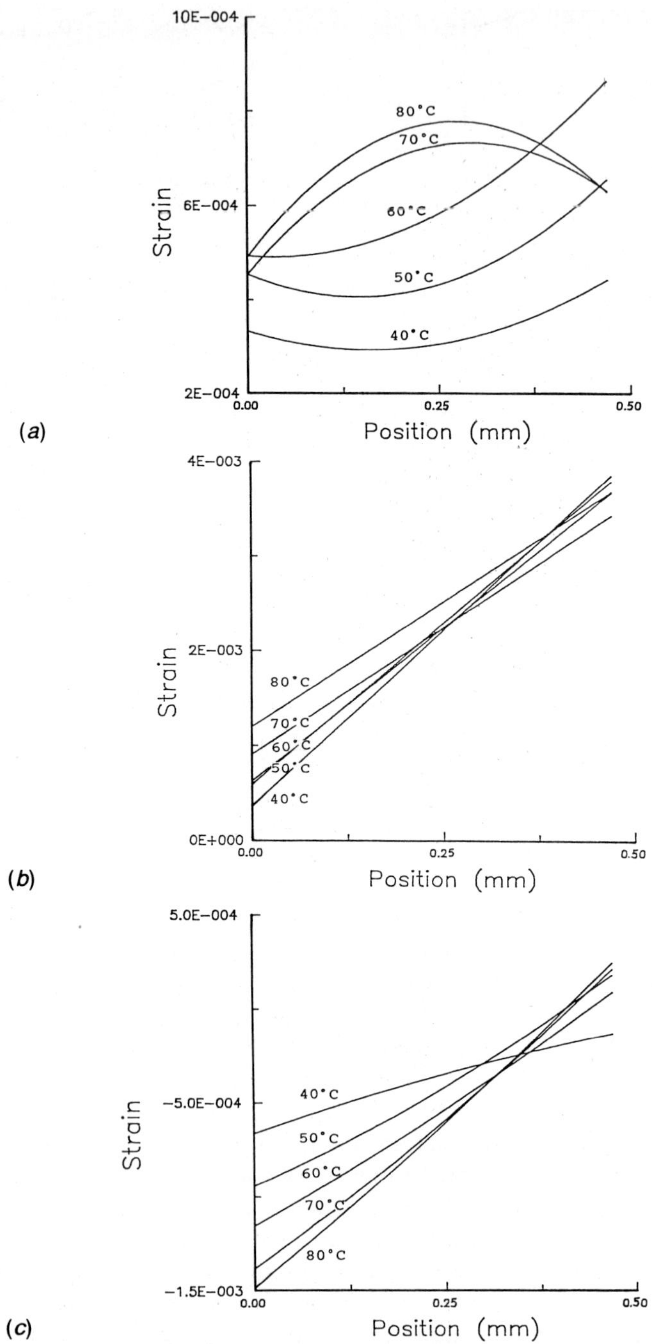


**Figure 8-13**  Normal strain along line $a$–$b$: (a) $\varepsilon_x$; (b) $\varepsilon_y$; (c) $\varepsilon_{xy}$.

nearly linearly with temperature, as one might expect, but only up to 60°C. Beyond this temperature there appears the effect of nonlinear material behavior of at least one of the package components. An additional contribution to the nonlinear strain behavior is due to possible axial slip that may have taken place by the softening of the adhesive layers between the chip and the volume of the package beneath it. The observed displacement patterns for the whole package, Fig. 8-11, endorse this possibility, as the displacements beneath the chip are seen to be much higher than those above it particularly at high temperatures. However, further thorough investigation is necessary to account properly for the change in strain development trends.

A special cooling apparatus able to provide temperatures as low as $-70°C$ was also built that can accommodate a sliced package with a moiré grating. The package was cooled uniformly from room temperature to $-50°C$. A schematic of the cross-section of the package under investigation is shown in Fig. 8-14. The moiré patterns were recorded at room temperature and at $-50°C$; they were processed as described earlier and resulting displacement fields in the small area close to the end of the chip are shown in Figs. 8-15a and b. Analysis of the principal strain distribution along the line a–b (Fig. 8-14) shows definite nonlinear behavior of the strain with decrease of the package temperature (Fig. 8-16). Such information may be of great interest to the designer of microelectronics packages.

### 8.4.2 Effect of Conformal Coating on Strain Relief in Packages

Due to mismatch in the coefficients of thermal expansion of materials used in the package, severe strain concentrations can be developed, as was shown

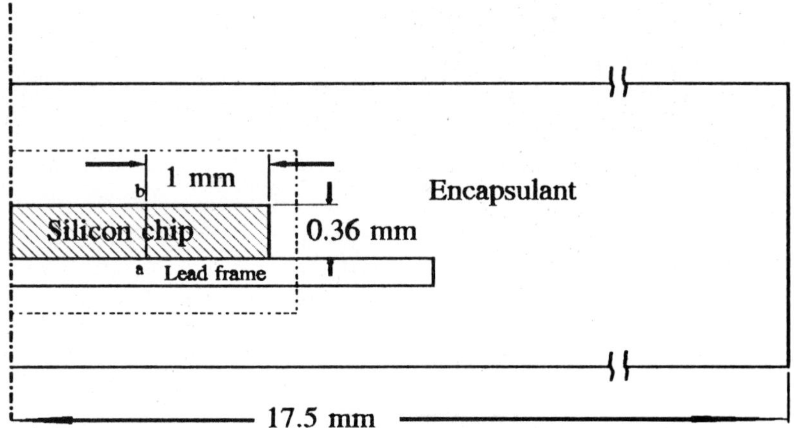


**Figure 8-14**   Schematic of the right half of the package.

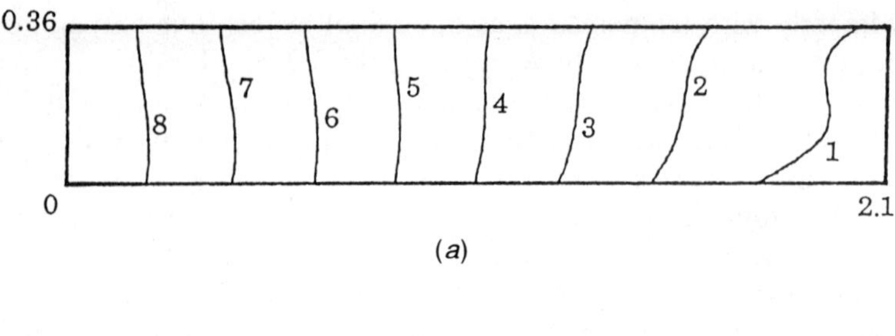

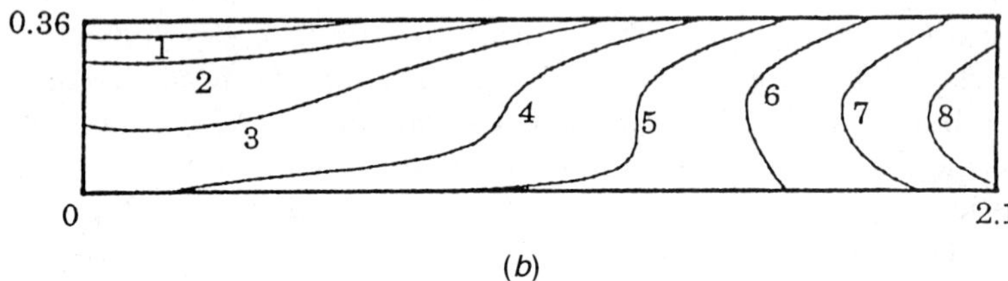

**Figure 8-15** Displacement distribution in the right half of the silicon chip cooled to $-50°C$. (a) $U$ component contour values (mm): (1) $-1.052 \times 10^{-3}$; (2) $-0.919 \times 10^{-3}$; (3) $-0.785 \times 10^{-3}$; (4) $-0.651 \times 10^{-3}$; (5) $-0.518 \times 10^{-3}$; (6) $-0.384 \times 10^{-3}$; (7) $-0.251 \times 10^{-3}$; (8) $-0.117 \times 10^{-3}$. (b) $V$ component contour values: (1) $1.182 \times 10^{-4}$; (2) $0.672 \times 10^{-4}$; (3) $0.162 \times 10^{-4}$; (4) $-0.348 \times 10^{-4}$; (5) $-0.858 \times 10^{-4}$; (6) $-1.368 \times 10^{-4}$; (7) $-1.878 \times 10^{-4}$; (8) $-2.389 \times 10^{-4}$.

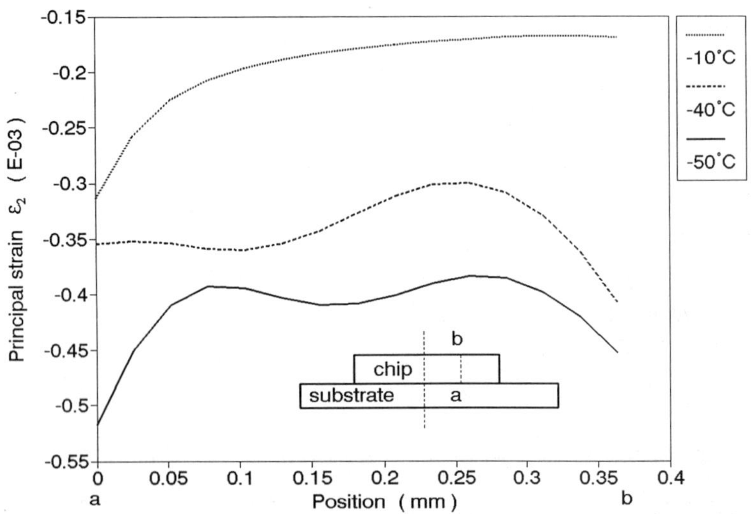


**Figure 8-16** Principal strain distribution along the line $a$–$b$.

in the preceding section. One of the ways to relieve this load on the package, and especially on the chip, is introduction of conformal coatings between the chip and encapsulant material. Industrial experience shows that such an approach does actually reduce failure rate of the packages. However, it is of great importance to be able to evaluate the effect of various material combinations used for conformal coatings on the reduction of strain levels in the package. One of the ways to accomplish this goal is to evaluate the behavior of the particular package subjected to temperature change with and without conformal coating.[34]

Two types of packages were prepared, the difference being presence or absence of conformal coating. The packages were sliced along the longitudinal axis and a moiré grating was applied to the exposed surface. Both types of packages were subjected to similar temperature changes. The moiré patterns were recorded and analyzed as discussed in Section 8.3. The cross-section of the microelectronics package used for study of the effect of conformal coating at elevated temperature is shown in Fig. 8-17. Two sections were evaluated, *a–b* and *c–d* (Fig. 8-17). The principal strain distributions calculated on the basis of moiré patterns are shown in Fig. 8-18 for section *a–b*, and in Fig. 8-19 for section *c–d*. The results clearly show that conformal coating reduces the strain on the chip on average by 25% along section *a–b* and by 10% along section *c–d*.

Analysis of the package shown in Fig. 8-14 subjected to cooling from $+20$ to $-50°C$, performed for both cases with and without conformal coating, shows reduction of strain levels along section *a–b* between 0 and 60%, as is obvious from the results presented in Fig. 8-20.

Results of the experimental evaluation of the effects of conformal coating on strain reduction in microelectronics packages show that it can act as a strain-relief agent. Moreover, the presented technique of digital image analysis enhanced moiré interferometry is capable of quantifying the effect of the conformal coating used in this study.

## 8.5 CONCLUSIONS AND FUTURE TRENDS

An experiment technique of digital image analysis enhanced moiré inter-ferometry has been introduced for in-situ analysis of microelectronics packages subjected to thermal load. The increased sensitivity of the method was achieved through the enhancement of moiré interferometry by digital image processing. The enhancement eliminated many of the existing limitations of using moiré interferometry for strain analysis, and has, therefore, broadened its applicability to include situations that were often avoided due to inherent difficulties in displacement data collection.

While moiré interferometry itself is a reliable whole-field displacement

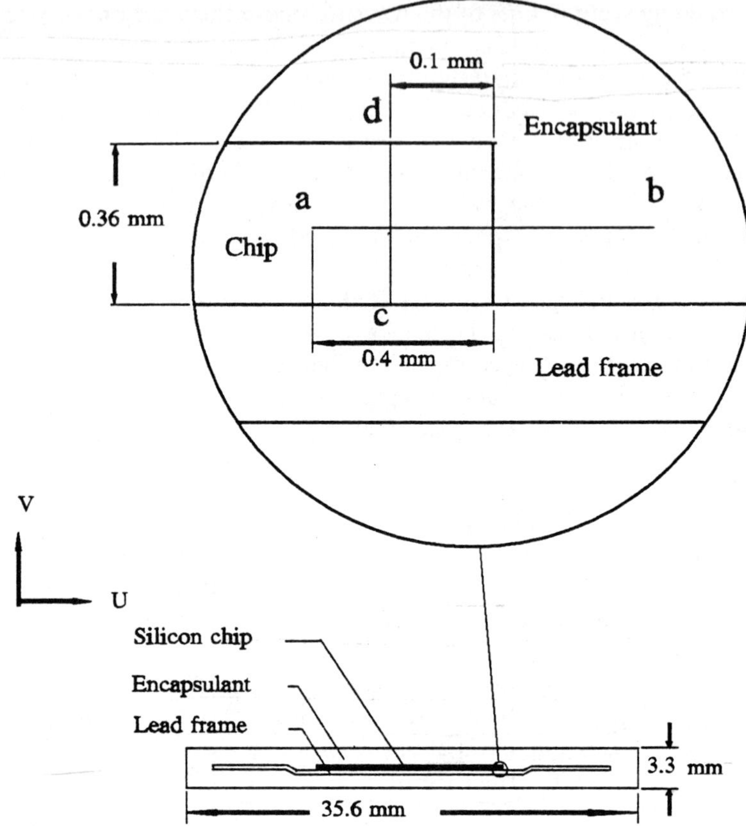


**Figure 8-17**   Schematic of the package cross-section.

monitoring technique that produces fringe patterns indicative of the in-plane displacements, the technique has been somewhat handicapped by the methodologies of quantifying the patterns in terms of actual displacements. Simple fringe counting methods were the only means used to extract displacements from fringe patterns. These methods required high fringe densities in moiré fields in order to collect displacement data sufficient for strain analysis. The displacement was given only at fringe centers and therefore the number of displacement data points along a certain line was limited to the number of fringes intersecting that line. Another limitation of the fringe counting methods is that the null field had to be ignored due to inability to extract null-field data at the same points used from the final load-induced pattern.

The technique introduced in this work relaxes the above limitations. The technique, which is based on fractional fringe analysis of moiré patterns,

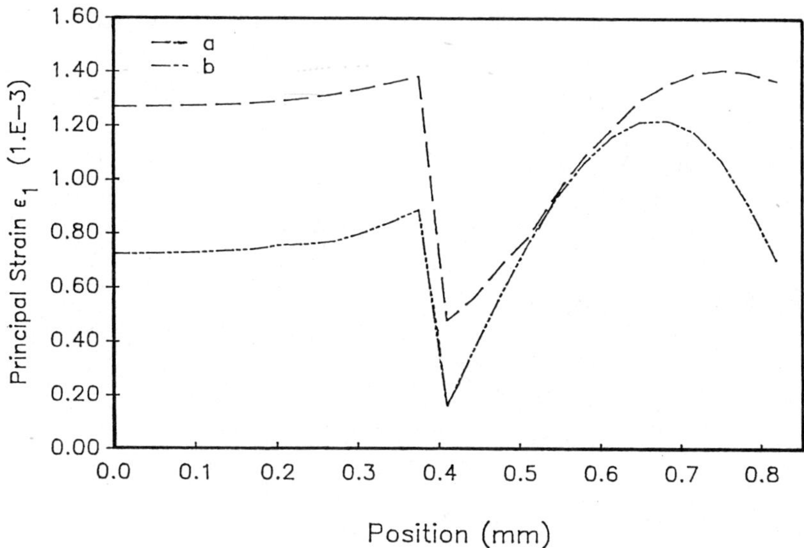

**Figure 8-18**    Principal strain distribution along line *a–b*: dashed line (*a*) uncoated chip specimen; chain line (*b*) coated specimen.

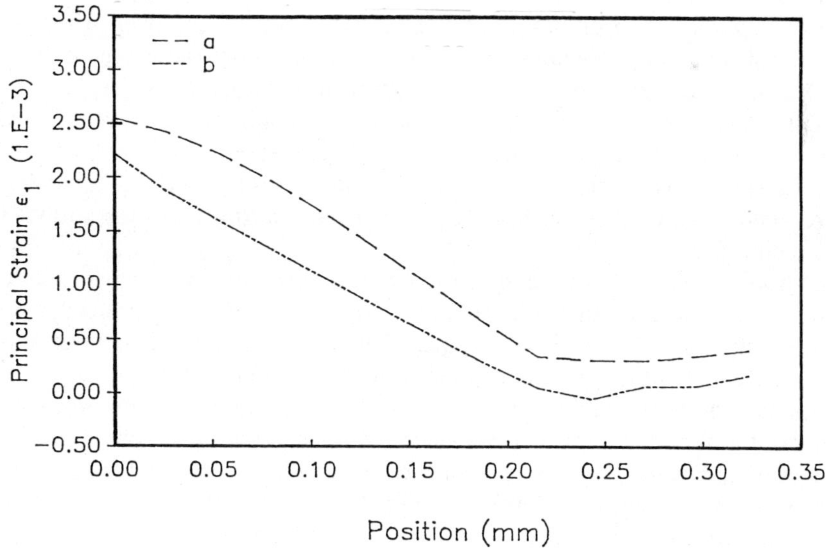

**Figure 8-19**    Principal strain distribution along line *c–d*: dashed line (*a*), uncoated chip specimen; chain line (*b*), coated specimen.

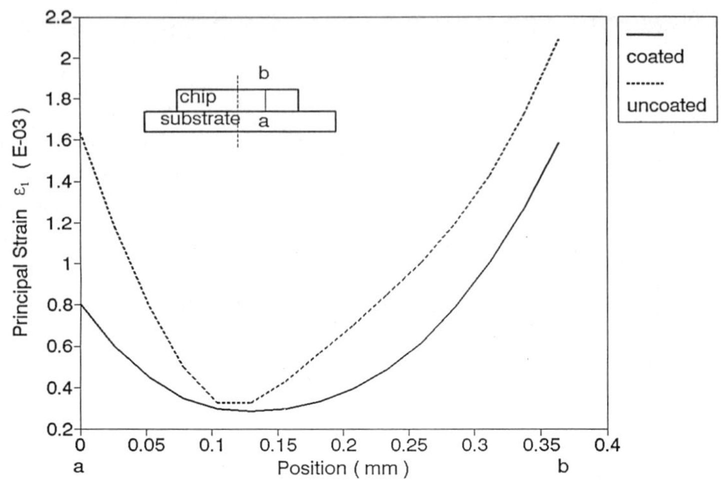


**Figure 8-20**  Effect of the temperature on principal strain distribution.

is capable of obtaining displacement information at any point of the field on the basis of the light intensity at that point. A computerized image-processing measuring system has been developed to acquire and digitize moiré patterns and to quantify the light-intensity values needed for displacement determination.

The capabilities of enhanced moiré interferometry have been utilized in the important areas of analysis of thermally induced deformations in microelectronics packages. The effects of temperature change in the wide region $-50°C$ to $+90°C$ were analyzed for several areas of interest. Full-field strain maps were developed for electronics packages at elevated temperatures. Details and levels of the net strains in the different components of the devices were visible. The strain relief of the conformal coating was also shown by this technique.

Digital image analysis enhanced moiré interferometry possesses excellent potential in the area of experimental analysis of thermally induced deformations in microelectronics packages. Being an accurate, convenient, computerized full-field displacement measurement technique of high sensitivity, it can be applied to almost any structure undergoing small deformations. The potential of the approach is evident in microelectronics. The increasing concerns regarding complexity and reliability in microelectronic devices necessitate the need for reliable techniques such as enhanced moiré interferometry for proper assessment of device mechanical performance under thermal load. Some of the areas where it can be immediately applied include delamination problems in plastic IC-packages, monitoring of power dissipation in high-power modules, strain analysis of circuit boards, and strain analysis in solder bumps.

## REFERENCES


1. Parks, V. J., "The Grid Method," *Experimental Mechanics*, **9**(7), 1969, pp. 27N–33N.
2. Andrews, H. C., and C. L. Patterson, "Correspondence," *IEEE Trans. Computers*, **C-25**(2), 1976, pp. 196–202.
3. Peters, W. H., and W. F. Ranson, "Digital Image Techniques in Experimental Stress Analysis," *Optical Engineering*, **21**(3), 1982, pp. 427–431.
4. Kobayashi, A. S., ed., *Handbook on Experimental Mechanics*, Prentice-Hall, Englewood Cliffs, NJ, 1987, chap. 8.
5. Tollenaar, D., *Moiré Interference by Screen Printing*, Amsterdam Institute for Graphics Technology, 1945.
6. Morse, S. A., A. J. Durelli, and C. A. Sciammarella, "Geometry of Moiré Fringes in Strain Analysis," *J. Engineer Mechanics Div.*, *ASCE*, **86**(EM4), 1960, pp. 105–126.
7. Guild, J., *The Interference System of Crossed Diffraction Gratings*, Clarendon Press, Oxford, 1956.
8. Post, D., "Optical Interference for Deformation Measurements—Classical, Holographic and Moiré Interferometry," *Mechanics of Nondestructive Testing*, W. W. Stinchcomb, ed., Plenum Press, New York, 1980, pp. 1–53.
9. Post, D., and W. A. Baracat, "High-Sensitivity Moiré Interferometry—A Simplified Approach," *Experimental Mechanics*, **21**(3), 1981, pp. 100–104.
10. Weissman, E. M., and D. Post, "Full-Field Displacement Rosette by Moiré Interferometry," *Experimental Mechanics*, **22**(9), 1982, pp. 324–328.
11. Post, D., "Moiré Interferometry at VPI and SU," *Experimental Mechanics*, **23**(2), 1983, pp. 203–210.
12. Dadkhah, M. S., F. X. Wang, and A. S. Kobayashi, "Simultaneous On-Line Measurement of Orthogonal Displacement Fields by Moiré Interferometry," *Experimental Techniques*, **12**(7), 1988, pp. 28–29.
13. Basehore, M. L., and D. Post, "High-Frequency, High-Reflectance Transferable Moiré Gratings," *Experimental Techniques*, **8**(5), 1984, pp. 29–31.
14. Basehore, M. L., and D. Post, "Displacement Fields (U, W) Obtained Simultaneously by Moiré Interferometry," *Applied Optics*, **21**(4), 1982, pp. 2558–2562.
15. McDonach, A., J. McKelvie, P. MacKenzie, and C. A. Walker, "Improved Moiré Interferometry and Applications in Fracture Mechanics, Residual Stress and Damaged Composites," *Experimental Techniques*, **23**(2), 1983, pp. 20–24.
16. Post, D., "Moiré Interferometry," *Handbook of Experimental Mechanics*, A. S. Kobayashi, ed., Prentice-Hall, Englewood Cliffs, NJ, 1987, pp. 314–487.
17. Guo, Y., D. Post, and R. Czarnek, "The Magic of Carrier Fringes in Moiré Interferometry," *Experimental Techniques*, **29**(2), 1989, pp. 169–173.
18. Burger, C. P., and A. S. Voloshin, "A New Instrument for Whole Field Stress Analysis," *ISA Trans.*, **22**(2), 1983, pp. 85–95.
19. Voloshin, A. S., C. P. Burger, R. E. Rowlands, and T. S. Richard, "Fractional

Moiré Strain Analysis using Digital Imaging Techniques," *Experimental Mechanics*, **26**(1), 1986, pp. 254–258.

20. Post, D., "New Optical Methods of Moiré Fringe Multiplication," *Experimental Mechanics*, **8**(2), 1968, pp. 63–68.

21. Sciammarella, C. A., "Basic Optical Law in the Interpretation of Moiré Patterns Applied to the Analysis of Strains—Part I," *Experimental Mechanics*, **5**(5), 1965, pp. 154–160.

22. Sciammarella, C. A., B. E. Ross, and D. L. Sturgeon, "Basic Optical Law in the Interpretation of Moiré Patterns—Part II," *Experimental Mechanics*, **5**(6), 1965, pp. 161–166.

23. Bastawros, A. F., A. S. Voloshin, and P. Rodogoveski, "Experimental Validation of Fractional Fringe Moiré Interferometry," *Proc. 1989 Society of Experimental Mechanics Spring Conference*, Cambridge, MA, May 28–June 1, 1989, pp. 401–406.

24. Bastawros, A. F., and A. S. Voloshin, "Mixed Mode Stress Intensity Factors by Fractional Fringe Moiré Interferometry," *Proc. 1990 Society of Experimental Mechanics Spring Conference*, Albuquerque, NM, June 3–6, 1990, pp. 69–75.

25. Lau, J. H., "Thermal Stress Analysis of SMT PQFP Package and Interconnections," ASME paper 88-WA/EEP-9, November 1988.

26. Royce, B. S. H., "Differential Thermal Expansion in Microelectronic Systems," *IEEE Trans. Components, Hybrids, and Manufacturing Technology*, **CHMT-11**(4), 1988, pp. 454–463.

27. Sullivan, T., J. Rosenberg, and S. Matsuoka, "Photoelastic and Numerical Investigation of Thermally Induced Restrained Shrinkage Stresses in Plastics," *IEEE Trans. Components, Hybrids, and Manufacturing Technology*, **CHMT-11**(4), 1988, pp. 473–480.

28. Kokkas, A. G., "Thermal Analysis of Multiple-Layer Structures," *IEEE Trans. Electron Devices*, **ED-21**, Nov. 1974, pp. 674–681.

29. Gee, S. A., W. F. van den Bogert, V. R. Akylas, and R. T. Shelton, "Strain Gauge Mapping of Die Surface Stresses," in *Proc. IEEE 39th Electronic Components Conference*, 1989, pp. 343–350.

30. Spencer, J., "Calculating Stress and Mobility in Silicon Chips Using Strain Gauge Measurements," *Semiconductor Engineering Journal*, **1**, 1981, pp. 34–37.

31. Usell, R. J., and S. A. Smiley, "Experimental and Mathematical Determination of Mechanical Strains within Plastic IC Packages and their Effect on Devices During Environmental Tests," *Proc. IEEE 19th IRPS*, 1981, pp. 64–73.

32. Bastawros, A. F., and A. S. Voloshin, "Thermal Strain Measurements in Electronic Packages through Fractional Fringe Moiré Interferometry," *J. Electronic Packaging*, **112**(4), 1990, pp. 303–308.

33. Bastawros, A. F., and A. S. Voloshin, "Transient Thermal Strain Measurements in Electronic Packages," *IEEE Trans. Components, Hybrids, and Manufacturing Technology*, **13**(4), 1990, pp. 961–966.

34. Voloshin, A. S., and P. H. Tsao, "Effect of the Conformal Coating on Electronic Package Deformation," *Proc. 1991 Japan International Electronic Manufacturing Technology Symposium*, June 26–28, Tokyo, Japan, 1991, pp. 268–271.

# 9

# Correlation of Analytical and Experimental Approaches to Determination of Thermally Induced Printed Wiring Board (PWB) Warpage

*C.-P. Yeh, C. Ume, R. E. Fulton, K. W. Wyatt, and J. W. Stafford*

## 9.1 INTRODUCTION

Thermomechanical design effects in the printed wiring board (PWB) design process are becoming increasingly important due to ever more stringent electronic product requirements. In the past few years, the finite element method (FEM) has become a vital and effective tool to support many facets of the PWB design process. Despite its increasing popularity in PWB design, the FEM has seldom been validated for its appropriateness and accuracy in modeling PWB thermomechanical behavior. We have conducted a research project in developing advanced FEM-oriented capabilities to simulate thermally induced PWB warpage. The FE analysis results are validated by correlating them with measurements obtained from a separate experimental approach using the shadow moiré method.

Printed wiring board (PWB) technology has become one of the critical technologies in the design and manufacture of electronic systems. Most PWBs are composite structures of epoxy-based fiberglass reinforced dielectric layers that are interleaved with copper foils used for conducting electric current (power and ground planes). These layers are stacked and cured under certain prescribed temperatures and pressures during the lamination process. Residual stresses caused by the coefficient of thermal expansion (CTE) mismatch between different board materials, combined with hydrothermal effects, can introduce warpage.[1] This effect can be further aggravated by more thermal cycles during fabrication and component assembly processes such as solder masking, soldering, etc. The excessive warpage can not only adversely create various manufacturing difficulties (for example, misregistration) but can also cause serious reliability problems during normal operation (for example, solder joint fatigue due to thermal cycling).

Advanced packaging technologies such as surface mount technology (SMT), ULSI (ultra-large-scale integrated circuit), VHSIC (very high-speed integrated circuit), etc., have dramatically increased PWB power and packaging densities. However, the thermomechanical design aspects, especially warpage design, are traditionally overlooked in the PWB design process. The electronics industry still follows a so-called "design–build–test" scheme, i.e., measuring the bow and twist on the prototype boards. This scenario has several drawbacks. First, building prototypes is a time-consuming and expensive process. Second, if redesigns or repairs are necessary to reduce excessive warpage, it will further prolong product development cycle time. As such, electronics companies may incur loss of market share and even financial collapse. Third, tests do not offer sufficient insight and understanding about the behavior of PWB structures and materials. Test alone tends to lead to "trial-and-error" and "ad hoc" experimental procedures rather than a logical, rational, scientific basis underlying the physical nature of the phenomenon.[2] Furthermore, these test procedures are often limited to certain specific configurations and domains; thus, they are not suitable for new designs. Designers need to be able to utilize computer-based software tools to analytically simulate the physical behavior of PWBs in the early design phases of the design process.

In collaboration with Motorola, Inc., Georgia Tech has carried out research to develop advanced analytical capabilities for investigating thermally-induced PWB warpage during the manufacturing process. In a complementary effort a comprehensive, rigorous experimental approach has also been carried out to determine the validity, accuracy, robustness, and limits of the analytical approach. This chapter documents the procedures for both approaches. The correlation philosophy is also illustrated and the initial results from both approaches are compared and discussed.

## 9.2 FINITE ELEMENT ANALYSIS FOR PWB WARPAGE

In recent years, the FEM has become a vital, effective tool for PWB thermomechanical design. First, the FEM is capable of mathematically simulating irregular, complex geometry[3] where a closed-form solution is almost impossible to obtain.[4] The development of state-of-the-art finite element techniques and advances in high-performance digital computers allow static/dynamic, linear/nonlinear, and steady-state/transient analyses to be done in a rather economical and rapid manner, which is crucial to the PWB design process. Furthermore, the FEM is especially advantageous in conducting sensitivity analyses since parametric changes in the model (dimensions, material properties, and loading characteristics, etc.) can be easily incorporated.

## 9.2.1 PWB Geometric Configurations

Typical PWB configurations consist of interleaving copper foils and dielectric composites (FR-4) which are made of woven fabrics impregnated with epoxy-based resin. The woven fabrics are essentially a two-dimensional construction with two sets of yarns or fiber bundles interlaced in mutually orthogonal directions. Due to these biaxially woven fabrics, the FR-4 can be considered orthotropic. To determine the FR-4 geometric parameters that affect PWB warpage and to measure the warpage itself during thermal cycles, several simplified configurations of PWBs were carefully designed and modeled. Seven different configurations were made through combinations of three different FR-4 layers including prepreg in various stacking sequences. Figures 9-1 and 9.2 show two such configurations, denoted configurations A and B. The warp (or grain) direction of each FR-4 lamina is indicated by an arrow. Two outermost copper foils on configuration A are etched to form two sets

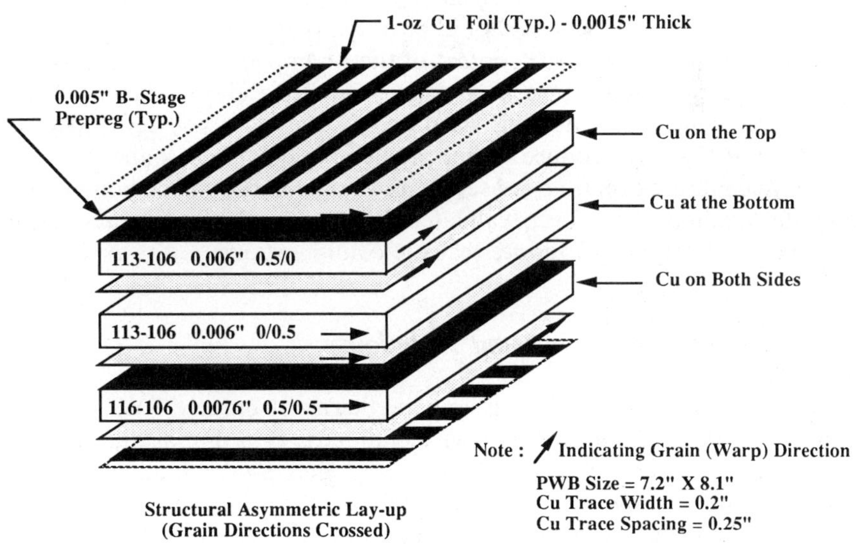


**Figure 9-1**  Geometric configuration A.

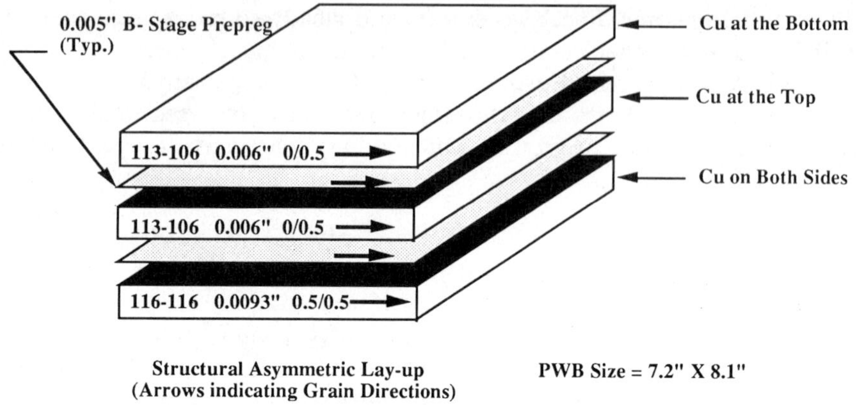


**Figure 9-2**   Geometric configuration B.

of equispaced parallel traces that are perpendicular to each other. This kind of configuration is usually referred to as the Manhattan style configuration. The purpose of the Manhattan configuration is to produce a predictable saddle-like deformed shape that consists of both positive and negative curvatures when a temperature loading is applied. Configuration B is a simple layered structure with solid copper foil on one side. When heated, a bowl-like deformed shape occurs due to its asymmetric lay-up. This configuration produces relatively large warpage for a small temperature change and is later used to perform a sensitivity study for the material properties of the PWBs (Section 9.2.3). It should be noted that these configurations are far from a "real" PWB in the context of their complexity and functionality. However these configurations were appropriate for the study since (1) they are easy to build for measuring warpage, (2) they are easy to model and analyze, and (3) they produce rather predictable deformed shapes that can be used to calibrate experimental setups and expedite the correlation process. After the methods have been validated for the simple configurations, the concepts can be applied to more realistic configurations.

### 9.2.2 Modeling Assumptions and Techniques

A major goal of this research is to simulate thermally induced PWB warpage using finite element analysis (FEA) techniques. It is important that the FEA system have advanced capabilities in modeling (a) a laminate layered structure, (b) orthogonality of the composites, and (c) temperature-dependent nonlinearity of material properties. The ANSYS finite element code[5] contains these capabilities and was selected and used throughout the project. To simplify the modeling complexity, the following assumptions were made.

1.  A composite laminate is considered to be a stack of orthotropic laminae or layers. Microscopically the woven fabrics make the FR-4 layer material nonhomogeneous. For the purposes of analyzing PWB global warpage effect, these layers are assumed to be quasi-homogeneous since fabrics are positioned in an equispaced and organized manner. In addition, the crimp effects or interactions between the interlacing fabrics in the fill and warp directions are ignored.
2.  All material properties in a PWB are linear, elastic, and temperature-independent over the temperature range 25–110°C). Only geometric nonlinearity, i.e., large deflections, is considered.
3.  The Kirchhoff–Love hypothesis for plates is valid. The bonds between laminae are presumed infinitesimally thin and the shear deformation is negligible compared to the bending deformation. This assumption requires that displacements be continuous across layer boundaries and that no slip exists between layers.
4.  The thickness of specimens is constant and the specimen is reasonably flat before temperatures are applied.
5.  Due to its symmetry about the two horizontal ($X$ and $Y$) axes, only a quarter of the specimen needs to be modeled for economic reasons. The behavior for the other three-quarters of the specimen can be obtained by taking mirror images of the model.
6.  The same property values were used throughout the analysis for all dielectric layers (FR-4 glass fabric styles 116, 113 and prepreg).
7.  The effect of Z-direction (out-of-plane) properties was neglected because they were found to have no effect on PWB warpage.

An ANSYS layered shell element type was selected for building the FE models. Uniform temperatures at various increments were applied across the FE model to simulate the isothermal heating situations. Appropriate boundary conditions were also imposed on the model to account for the "cut boundary" of the quarter of the specimens modeled. The following mechanical properties are used to build the FE models: Young's moduli ($E_x$, $E_y$), shear modulus ($G_{xy}$), Poisson's ratio ($v_{xy}$), coefficients of thermal expansion (CTEs; $\alpha_x$, $\alpha_y$). A comprehensive literature survey was conducted to characterize material mechanical properties. It is found that considerable discrepancies exist among various material property values (especially for FR-4) documented in different sources. In addition, few publications provide information regarding temperature-dependent properties. These published material properties and their corresponding references along with a set of representative properties[1,6–10] used in the FE models are shown in Table 9-1.

**Table 9-1** Reference Material Mechanical Properties

| Material | Property | Published | Used | Sources | Remark |
|---|---|---|---|---|---|
| **FR-4** | $E_x$ (psi) | $1.6–3.7 \times 10^6$ | $2.2 \times 10^6$ | Greene, Daniel | |
| | $E_y$ (psi) | $1.6–3.7 \times 10^6$ | $1.8 \times 10^6$ | Greene, Daniel | |
| | $G_{xy}$ (psi) | $0.3–1.0 \times 10^6$ | $0.4 \times 10^6$ | Daniel, DeBra | |
| | $v_{xy}$ | 0.02–0.4 | 0.16 | Daniel, DeBra | For 113/106 layer only |
| | $\alpha_x$ (in./in.-°C) | $11–28 \times 10^{-6}$ | $20.45 \times 10^{-6}$ | Broutman (measured) | For 113/106 layer only) |
| | $\alpha_y$ (in./in.-°C) | $11–28 \times 10^{-6}$ | $25.11 \times 10^{-6}$ | Broutman (measured) | |
| **Copper foil** | $E_x$ (psi) | $12–17.3 \times 10^6$ | $12 \times 10^6$ | Oak/Mitsui | |
| | $E_y$ (psi) | $12–17.3 \times 10^6$ | $12 \times 10^6$ | Oak/Mitsui | |
| | $v_{xy}$ | 0.3–0.35 | 0.33 | Daniel, DeBra | |
| | $\alpha_x$ (in./in.-°C) | $15–20 \times 10^{-6}$ | $18.92 \times 10^{-6}$ | Broutman (measured) | For $\frac{1}{2}$ oz. copper foil only |
| | $\alpha_y$ (in./in.-°C) | $15–20 \times 10^{-6}$ | $16.33 \times 10^{-6}$ | Broutman (measured) | For $\frac{1}{2}$ oz. copper foil only |

Discrepancies in properties may arise from the following causes:

1. *Different specifications.* The glass fiber of FR-4 has many different kinds of glass compositions, filament types, strand counts, and weave patterns, which can produce very different material properties. Most references have only employed a generic product (for example, epoxy/glass, epoxy/E-glass, woven/glass/epoxy, etc.) without specifying specific glass styles.

2. *Imprecise composition.* An FR-4 composite lamina consists of many ingredients (resins, fabrics, accelerators, curing agents, contaminants, etc.) and there is no consensus regarding the composition percentages of these ingredients. Different vendors/suppliers might have their own specifications. To make the matter even worse, some vendors carry out their fabrication process in an imprecise, ad hoc manner. It is not uncommon to find that ingredient percentages may vary for the same kind of materials in different lots supplied by the same vendor. Any variations in the ingredient contents may result in different material properties.

3. *Complex fabrication process.* The PWB fabrication is a very complex process including numerous baking, pressing, stacking, and curing processes at certain prescribed sequences under a set of well-controlled parameters (temperatures, pressure, time durations, etc.). Any dissimilarity in the process and control parameters may generate changes in material properties. The most common problems in the PWB lamination process are the resin flow and the thickness yields, which are extremely difficult to predict and control. Even with the most stringent control, the thicknesses of a laminate remain inconsistent. As a result, the test materials based on the same lot might be different in thickness and resin content.

4. *Misconceptions.* Metal foils used to create the circuit patterns are usually made of copper. There are two kinds of copper foils used in the PWB industry, namely, electrodeposited (ED) copper and cold-rolled copper. The grain structures and, hence, the material properties for both the ED copper and the cold-rolled copper, especially for thin copper foils, are fundamentally different from those for bulk copper ingots whose material properties are customarily used for PWB design and analysis. Moreover, material properties vary with other factors such as the thickness of the foils, the class of copper foils,[11] etc. These factors are seldom taken into consideration during the design process.

Little work has been done to characterize the material properties used for multilayer PWBs by either suppliers or manufacturers.[7] The following discusses the sensitivity of the PWB behavior relative to various mechanical properties.

### 9.2.3 Sensitivity Analysis for Mechanical Properties

To effectively model the PWB warpage, the following orthotropic, temperature-dependent, in-plane mechanical material properties over the desirable span are required for the FE models: Young's moduli, shear modulus, Poisson's ratio, and CTEs. Out-of-plane properties were found to be of little significance through a sensitivity analysis,[12] and therefore are not tested. A test plan for material property characterization has been designed and scheduled. Since the material property characterization can be very expensive and time-consuming,[7] the test plan should be executed in a careful, systematic, and trackable manner. To (a) better understand the significance of each material property used in the FE models and their relative influence on the board warpage, and (b) help to prioritize the test schedule, a sensitivity study was first carried out over a certain range of property values. Sensitivity studies determine the rate of change of a given quantity (PWB warpage) with respect to a set of control or design parameters (material properties). Classical lamination theory (CLT)[1,13–15] instead of FEA was selected for this sensitivity study since CLT provides a direct representation between warpage and the material properties. Thus, a closed-form solution can be obtained and implemented by a personal computer in a simple and economic manner. The FEA approach, in contrast, is only capable of obtaining pointwise solutions for each run although the geometric model is virtually unchanged. This limitation, coupled with relatively low computational cost, makes the closed-form solution desirable for the sensitivity study.

Based upon CLT, the warpage (out-of-plane displacement) can be solved from the following three basic equations:

**(a) Stress–Strain Relationship**   The stress–strain relationship for a single lamina is represented in the following equation:

$$\begin{bmatrix} \alpha_1 \\ \alpha_2 \\ \tau_{12} \end{bmatrix} = \begin{bmatrix} Q_{11} & Q_{12} & 0 \\ Q_{12} & Q_{22} & 0 \\ 0 & 0 & Q_{66} \end{bmatrix} \begin{bmatrix} \varepsilon_1 \\ \varepsilon_2 \\ \gamma_{12} \end{bmatrix} \tag{9-1}$$

where $\quad 1, 2$ = principal directions

$\quad E_1, E_2$ = Young's modulus in directions 1 and 2

$\quad\quad G_{12}$ = in-plane shear modulus

$\quad v_{12}, v_{21}$ = major and minor Poisson's ratios

$\quad\quad \alpha_1, \alpha_2$ = CTE in directions 1 and 2

$\quad\quad\quad Q_1 = E_1/(1 - v_{12}v_{21})$

$\quad\quad\quad Q_{22} = E_2/(1 - v_{12}v_{21})$

$\quad\quad\quad Q_{12} = v_{21}E_1/(1 - v_{12}v_{21}) = v_{12}E_2/(1 - v_{12}v_{21})$

$\quad\quad\quad Q_{66} = G_{12}$

**(b) Equilibrium Equation**    The equilibrium equation is obtained by integrating the stresses in Eq. (9-1) for all layers through the board thickness minus the thermal resultants:

$$\begin{bmatrix} N \\ M \end{bmatrix} = \begin{bmatrix} A & B \\ B & D \end{bmatrix}\begin{bmatrix} e \\ k \end{bmatrix} - \begin{bmatrix} N^T \\ M^T \end{bmatrix} = 0 \tag{9-2}$$

where  $N, M$ = net edge force and moment resultant matrices
$N^T, M^T$ = thermal force and moment resultant matrices
$A, B, C$ = extensional, coupling, and bending stiffness matrices
$e, k$ = strain and curvature matrices

**(c) Displacement–Curvature Relationships**    The bending for a uniformly heated laminate plate produces constant curvatures,[16] therefore the curvature–displacement relationships can be written as

$$k_x = -\frac{\partial^2 w}{\partial x^2}$$

$$k_y = -\frac{\partial^2 w}{\partial y^2} \tag{9-3}$$

$$k_{xy} = -2\frac{\partial^2 w}{\partial x\, \partial y} = 0$$

where $w$ = out-of-plane displacement (warpage).

By substituting the curvatures obtained from Eqs. (9-3) and integrating the differential equations in (9-4) with appropriate boundary conditions, the warpage at any point $(x, y)$ on a rectangular plate can be found.

To better understand the relative impact of individual material property parameters used in the FE models on the specimen warpage, a sensitivity study was carried out for configuration B in Fig. 9-2. Four materials were considered: copper foil, FR-4 116/116 layer, FR-4 106/113 layer, and cured prepreg layer. Seven key design parameters (layer thickness, $E_x$, $E_y$, $G_{xy}$, $v_{xy}$, $\alpha_x$, and $\alpha_y$) for each material were investigated. A set of baseline parameter values were identified from various sources and are listed in the third column of Table 9-1. Each parameter was varied independently by scaling factors of 0.25 to 1.75 in 0.05 increments, with the warpage being calculated for each case. These scaling factors were used so that up to $\pm 75\%$ variations in each material property parameter were considered in the study. This CLT algorithm was implemented using the MATLAB mathematical software,[17] a high-performance interactive software package for scientific and engineering numeric computations. After warpages were calculated, the sensitivity of the parameter, $S$, was then calculated as follows:

$$S = \frac{\Delta w/w}{\Delta b/b} \times 100 \qquad (9\text{-}4)$$

where $w$ = baseline warpage
$\Delta w$ = change in warpage
$b$ = baseline parameter value
$\Delta b$ = change in parameter value

A large sensitivity value indicates that the warpage is more sensitive to the parameter for that parameter value. A negative sensitivity value means that an increase (decrease) in parameter value results in a decrease (increase) in warpage.

Both calculated warpage and sensitivity versus the corresponding parameter values were plotted using a graphics routine provided by MATLAB. Two typical results are shown in Figs. 9-3 and 9-4. The overall results show that the CTEs are the most sensitive parameters while the Young's moduli and layer thickness are moderately sensitive parameters. The Poisson's ratio and the shear modulus are the least sensitive parameters with respect to specimen warpage. Some selected results were also compared to results obtained from a separate FEA approach. The maximum discrepancy between these two approaches was less than 0.5%, indicating that excellent agreement was achieved. The sensitivity analysis for material properties can not only provide qualitative insight and understanding about the PWB warpage but also prioritizes the ongoing test plan for basic specimen material properties.

### 9.2.4 Mechanical Property Measurements

To correlate the above sensitivity analysis results, thermal expansions per unit length for two materials, $\frac{1}{2}$ oz. copper foil and FR-4 glass style 113/106, was first tested[18] using strain gauges over a temperature range from 23 to 162.8°C. A single-element high-elongation strain gauge capable of measuring strains to 30% was adhered to the specimen surface with a suitable adhesive of similar strain capabilities. The warp and fill directions of the FR-4 layer were properly discerned and carefully aligned with strain gauges. The surface of each material was prepared with degreaser and cleaner prior to strain gauge installation. To compensate for the thermal elongation of the gauge itself, a dummy strain gauge from the identical lot was also adhered to a

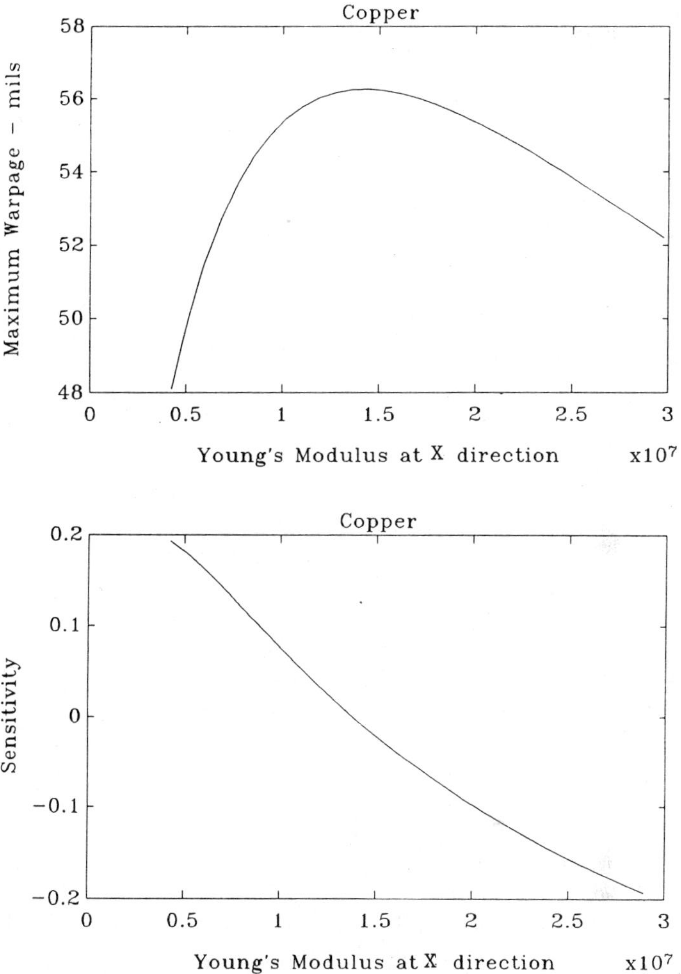

**Figure 9-3**  Sensitivity analysis results for the Young's modulus of copper in the X-direction (warp direction).

bar of titanium silicate material with very low CTE. Lead wires were then attached to each gauge and a protective coating was applied to the assemblage. The temperature was raised in increments. At each increment, the temperature was held constant until no further change in strain was noted so as to ensure that the test specimen was at equilibrium temperature. The strain was then recorded. The results are shown in Figs. 9-5 and 9-6. The CTEs used in the FEA were obtained from a linear curve-fitting process for the tested thermal expansion values below 110°C. The material properties for 1 oz. copper foil, FR-4 116/116 layer, and prepreg layer were assumed

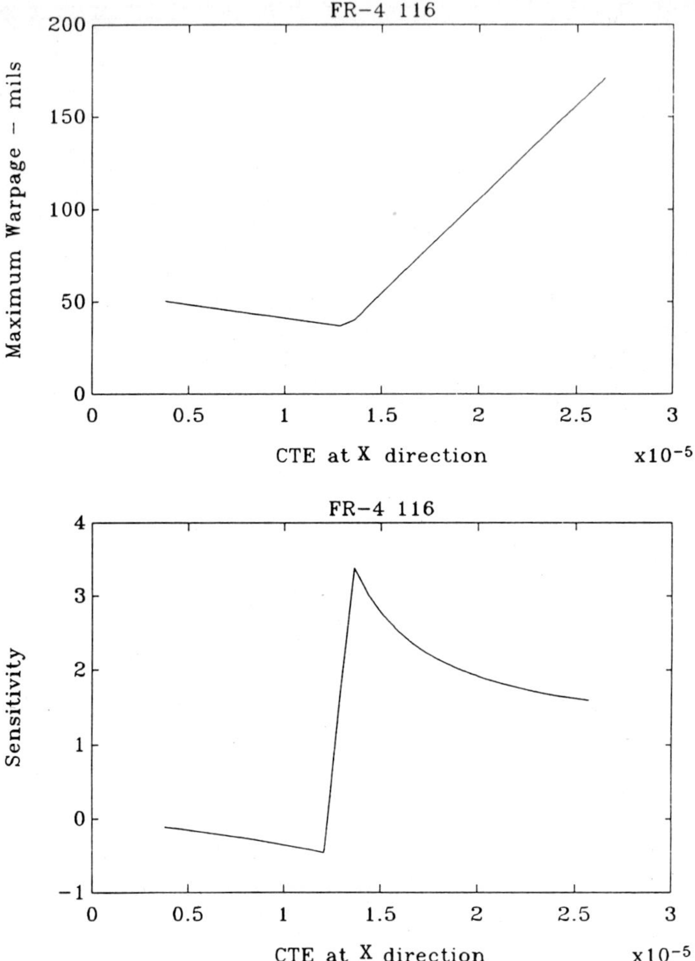



**Figure 9-4**   Sensitivity analysis results for the CTE of FR-4 (116/116) in the *X*-direction (warp direction).

to be the same as those tested. The corresponding property values along with their references are listed in Table 9-1.

### 9.2.5 Discussion of Analytical Results

ANSYS FEA was carried out for all specimen models on a VAX 6440 computer. The warpage (out-of-plane displacement) and stress results for nodes of interest were computed and printed. The deformed shape as well

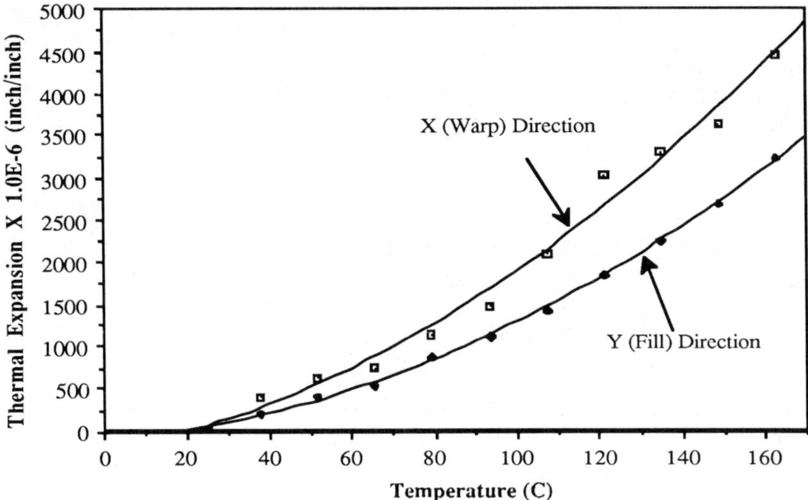


**Figure 9-5**   Thermal expansion measurements per unit length for $\frac{1}{2}$ oz. copper foil.

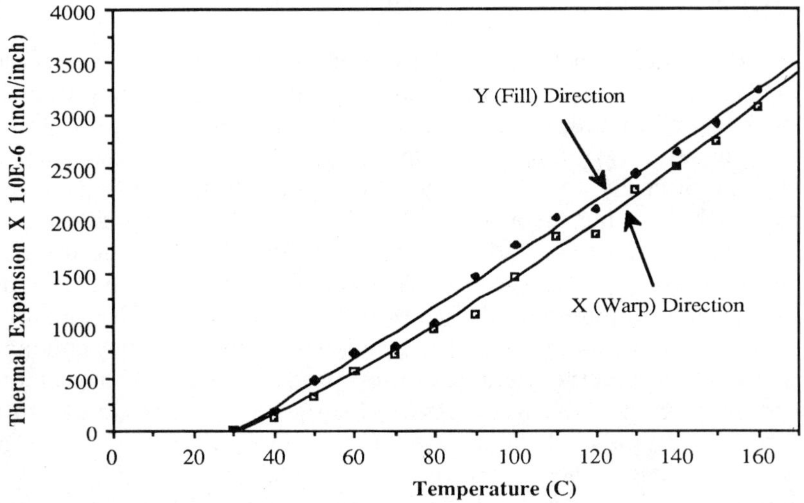


**Figure 9-6**   Thermal expansion per unit length measurements for FR-4 113/106 layer.

as stress distribution results can be plotted using the ANSYS post-processor facility. Figure 9-7 shows the FE models for specimen configuration A, before and after a uniformly distributed temperature load is applied. These FEA results were compared with corresponding experimental results as discussed below.

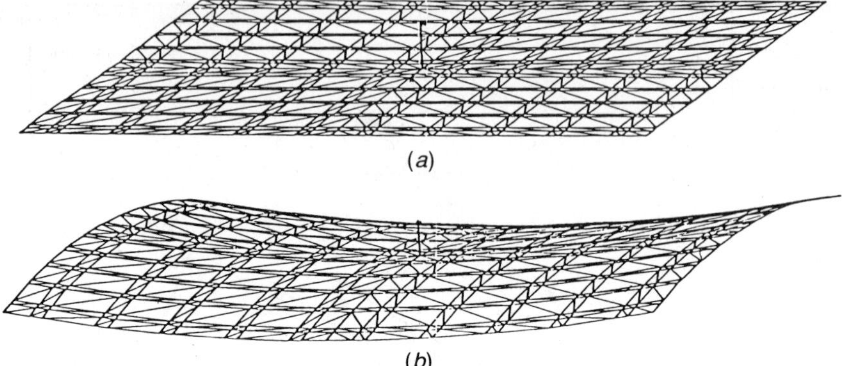

**Figure 9-7**  Finite element model for configuration A: (a) before a temperature load is applied; (b) after a temperature load is applied.


## 9.3 EXPERIMENTAL VERIFICATION OF PWB WARPAGE

### 9.3.1 Overview of Experimental Technique—Shadow Moiré

The shadow moiré method[19,20] was used in this research to measure thermally induced PWB warpage because the experimental setup is simple, inexpensive, and capable of making full-field measurements.[21] Moiré fringes appear whenever two sets of parallel, alternate light and dark lines are overlapped optically together. Physically, moiré fringe patterns can be interpreted by relating them to the displacement field. The shadow moiré method uses a reference grating (or Ronchi rulings) that is placed directly in front of a test object or specimen. The reference grating is made of glass and is typically composed of parallel equispaced black lines. The shadow of the reference grating on the specimen, generated by transmitting a collimated beam of light through the reference grating, produces a virtual image, termed the specimen grating. When the observed surface is flat and parallel to the reference grating, the shadows are parallel to the grating lines and the spacing of the shadow lines is identical to the grating spacing. As a result, no fringes are present and this is called a "zero" fringe order. When the examined surface is inclined or curved, shadow moiré fringes are formed by the reference grating and the specimen grating. This moiré fringe pattern can then be analyzed in order to determine the out-of-plane displacement of the specimen. A typical experimental setup consists of a camera, a light source, a specimen, a reference grating, an oven, and a mounting fixture (Fig. 9-8). The reference grating is placed directly in front of the specimen surface. The light is directed at an angle $\alpha$ through the grating onto the specimen. When viewed by eye or a camera at an incident angle, the governing equation for

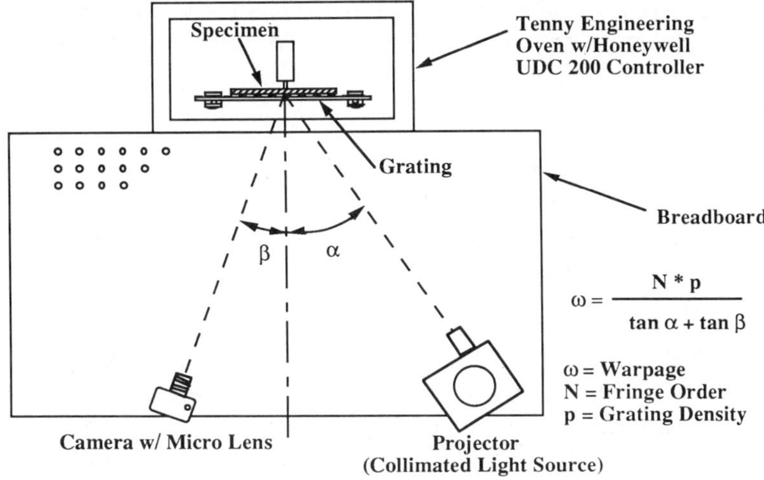


**Figure 9-8**    Schematic view of shadow moiré experimental method.

the out-of-plane displacement is

$$w = \frac{N_p}{\tan \alpha + \tan \beta} \tag{9-5}$$

where $N$ = fringe order or fringe count at the corresponding point
  $p$ = grating pitch
  $w$ = out-of-plane displacement at the $N$th fringe order

Each black and white fringe has a constant elevation to the reference grating. The black fringes are customarily used for counting fringe orders. Fringe pattern analysis and interpretation can be an error-prone, crucial process, and may require considerable skill.

### 9.3.2 Sample Preparation

To validate the FEA techniques and results above, all seven configurations were laminated by Motorola, Inc. to serve as test vehicles. As indicated earlier, these configurations have been carefully designed and analyzed at the proper elevated temperatures using FE software to ensure that the specimens produce (1) sufficient warpages that can be measured with ease, and (2) predictable deformed shapes that can be used to facilitate the correlation process. The lamination process was carefully monitored to ensure the accurate sequence of stack-up and orientation of layers. To improve the contrast of the moiré fringe pattern, a nonglossy, temperature-resistant white paint is uniformly and lightly applied to the specimen source.[22,23]

### 9.3.3 Experimental Setup

The experimental setup for measuring PWB warpage as shown in Fig. 9-8 is briefly discussed in the following.

***Grating***    Three 10 in. × 10 in. gratings were considered, with grating densities of 200, 300, and 500 lines/in., respectively, and low-CTE tinted glass substrates. The resolution and maximum measuring range for each grating were carefully determined in a separate calibration test[20] and the 300 lines/in. grating was selected in the experiment for its suitability in the warpage ranges produced by these specimens. The CTEs of the gratings are relatively small (0.9 p.p.m./°C) compared with those of the specimens (24 p.p.m./°C) and the thermal expansion of the grating due to the temperature changes is therefore ignored.

***Mounting Fixture***    The mounting fixture was designed not only to support both the specimen and the grating but also to maintain spacing stability. To assign fringe orders, a reference point on the specimen should remain stationary relative to the reference grating throughout the measurement. The specimen is allowed to expand freely in all directions. Due to the symmetry, the center of the PWB is used as the reference point without loss in generality. A flat-headed screw with a thin washer with a specially calibrated thickness is used to allow enough space to avoid interference between the specimen and the grating. An aluminum rod is fastened to a spring-loaded uniaxial translational positioner and a pre-tension is applied to the positioner to ensure that the flat-headed screw is in contact with the grating. The entire mounting fixture is placed and aligned properly in the oven heating chamber.

***Oven***    To produce an isothermal, steady-state thermal condition, a Tenny Engineering laboratory oven with a Honeywell UDC 200 Minipro Digital controller was used. For viewing purposes, two panes of transparent, high-temperature glass were mounted to the oven door. Two panes were used in order to minimize the heat loss.

***Light Source***    A standard slide projector used as a collimated light source was found acceptable for producing a reasonably uniform illumination field.

***Camera***   A high-quality Nikon 35-mm camera equipped with a specialized f2.5 micro lens has been used as the viewing system. The aperture of the camera lens was set small to increase the depth of focus. For stability and alignment purposes, the camera is supported by a tripod equipped with a multi-axis bubble level.

### 9.3.4  Experimental Procedures

The specimen was heated from 25°C (room temperature) to various temperatures in a stepwise manner. The upper temperature limit of 100°C was chosen to avoid $T_g$ (approximately 125°C). This allows us to assume that the material properties stay constant within this temperature range. It should be noted that temperatures up to 250°C will be applied later as the research progresses, to simulate various fabrication and assembly processes such as solder reflow. For each temperature step, a 15-minute stabilizing time was allowed in order to ensure isothermal conditions. The fringe pattern was continuously monitored and photographs were taken at different temperature steps.

Figure 9-9 shows changes in the fringe patterns at different elevated temperatures for configurations A and B (Figs. 9-9, 9-10). It was found that there is a considerable initial warpage in each specimen at room temperature. Several heating–cooling cycles were applied to each tested specimen before the measurements were taken in order to release "built-in" residual stresses as much as possible. As the temperature rises, the fringe pattern becomes denser and the board continues to warp. The initial warpage serves as a datum surface when calculating the net thermally induced warpage. The net warpage due to a temperature rise is obtained by subtracting the measured warpage from the initial warpage. Since the specimens are assumed to be structurally symmetric about horizontal and vertical axes, only a quarter of the specimen was considered. Three points on the specimen are particularly of interest: two midpoints at neighboring edges and a corner point. These points are easily discerned in the photographs. The resulting warpages and their corresponding temperatures for configurations A and B are plotted in Figs. 9-11 and 9-12, respectively. A quasi-linear relationship exists, which validates the assumption of constant material properties over the temperature range tested. It was also observed from both configurations that one midpoint seems to have significantly less warpage than the corner and the other midpoint. This is probably because FR-4 is an orthotropic composite material, and the CTE mismatch in one axis between FR-4 and copper is much greater than that in the other axis. It was also noticed that in both the heating and cooling processes all warpage curves follow the same paths; therefore, one can conclude that the warpage falls within the elastic range of the material.

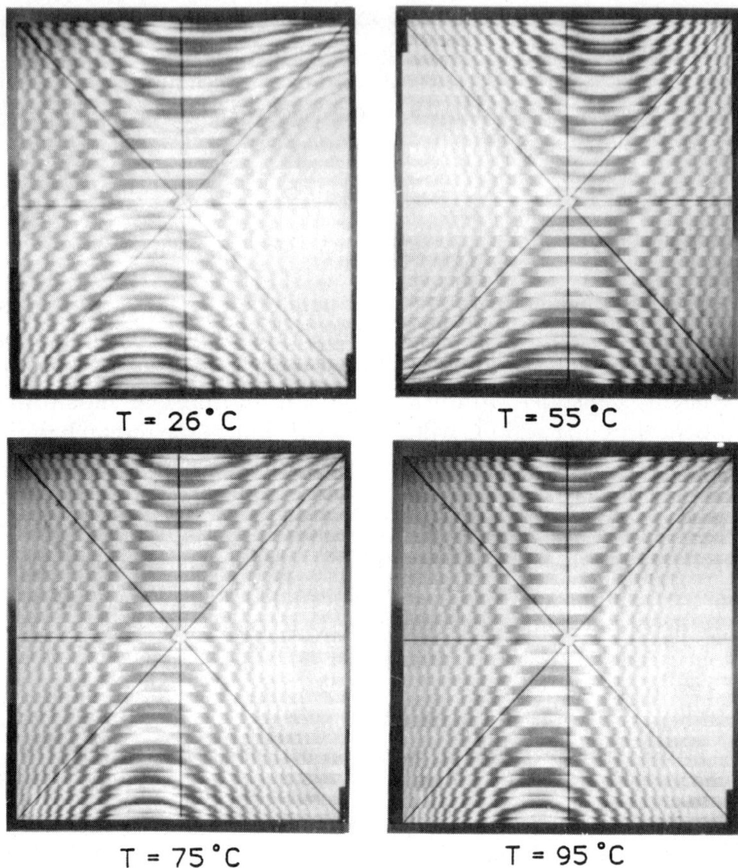



**Figure 9-9**  Moiré fringe patterns for configuration A at various temperatures.

### 9.3.5 Comparison of Experimental and Analytical Results

The FEA results are compared with experimental results for specimen configurations A and B. As shown in Fig. 9-13, a fairly good correlation has been obtained for specimen configuration B. Although there existed a considerable discrepancy between FEA and experimental results for configuration A, the overall trends in specimen warpage were correctly predicted (see Fig. 9-14). This discrepancy may originate from the following causes:

1. The CTEs for 1 oz. copper foil, FR-4 116/116 layer, and the prepreg layer are different from those tested, especially for 1 oz. copper, which exists only in configuration A.

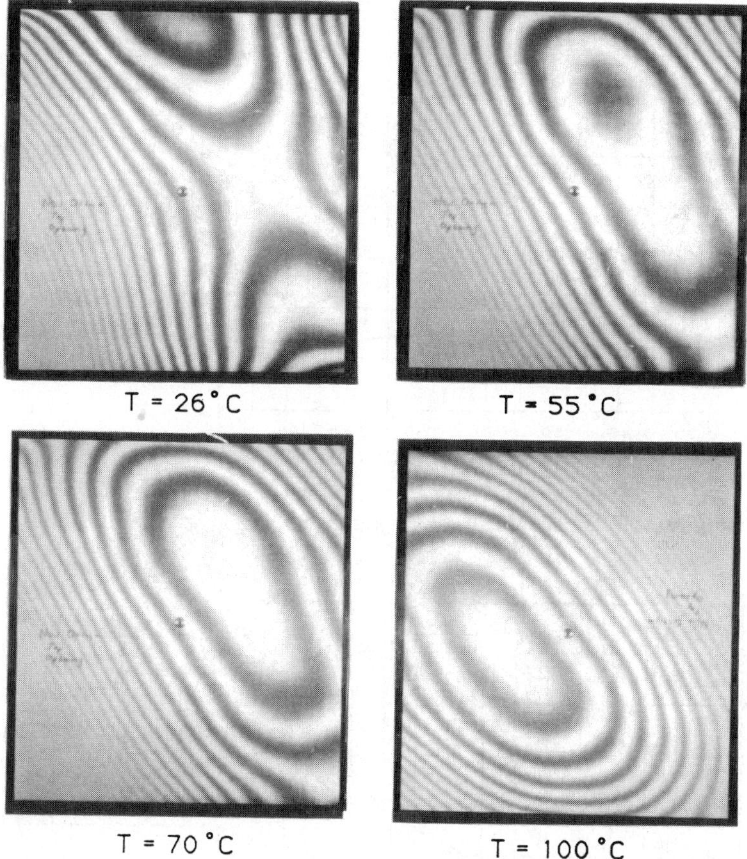



**Figure 9-10**   Moiré fringe patterns for configuration B at various temperatures.

2. The Young's moduli, shear modulus, and Poisson's ratio used in the models were obtained from various references, and therefore may not be the true mechanical properties.
3. The specimen thickness across each board is not uniform. This effect has not been modeled in the FE models due to impracticability.

It should be stressed that comparisons of experimental and analytical results are inconclusive because true material properties were not available. Published generic material properties had to be substituted for true properties. Material characterization tests are needed and are underway in the studies. It is believed that better correlation can be achieved when more accurate mechanical properties become available.

Both the experimental and analytical results obtained from this research

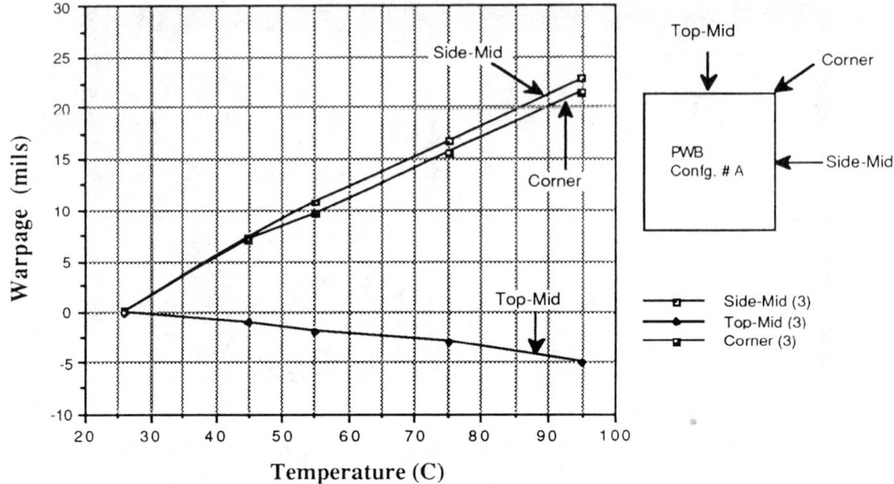


**Figure 9-11**   Warpage measurement results for specimen configuration A.

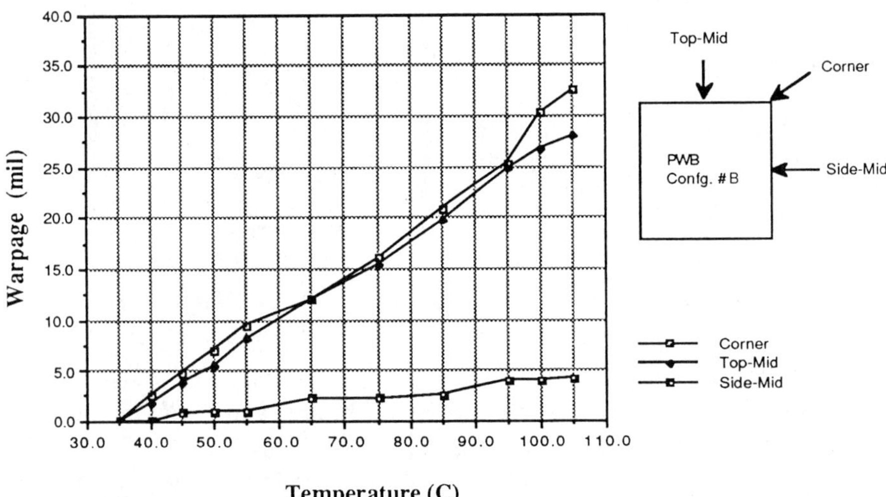


**Figure 9-12**   Warpage measurement results for specimen configuration B.

have shown that the maximum thermally induced warpages for the configurations A and B are approximately 25 mils and 35 mils, respectively. It should be noted that these are relative warpages, measured from the initial warpage due to lamination. This shows that even a relatively small range of temperature excursion (20–100°C) produces substantial PWB warpage. These thermally induced warpages are further compared with the bow and

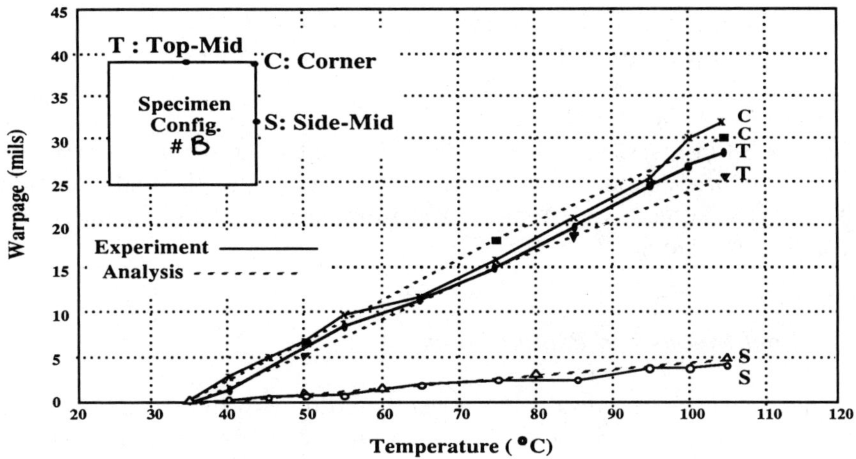


**Figure 9-13**    Correlation between FEA analysis and measurement results for specimen configuration B.

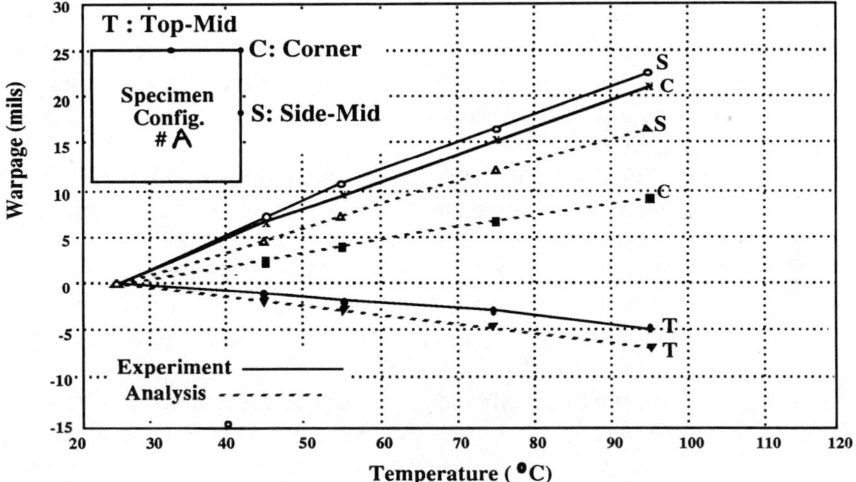


**Figure 9-14**    Correlation between FEA analysis and measurement results for specimen configuration A.

twist tolerances given by IPC specification.[11] The bow and twist tests are widely used in the PWB industry to determine the board flatness. For the Class C, glass-based multilayer boards, the warpage should satisfy

$$w/L = 100 \leqslant 1.0 \qquad (9\text{-}6)$$

where $w$ = out-of-plane displacement or warpage
  $L$ = edge length of board

From Eq. (9-6), the allowable maximum warpage for the test vehicle (7.2 in. × 8.1 in.) is 72 mils. It is conceivable that if the warpage due to the lamination process had been taken into account, and thermal excursion taken beyond a mere 80°C range, PWB warpage in excess of 72 mils would have been measured.


### 9.3.6 Implications and Ramifications

The above correlation results have been based on two simple test specimen configurations. Although these warpage results may not be directly applicable to real PWB specimens, the underlying knowledge and implications resulting from this research work can be very useful to PWB warpage design. From the sensitivity study described in Section 9.3.3, for example, design rules relevant to the thermally-induced PWB warpage can be deduced and generalized. Some such design rules are shown in the following.

1. *Geometric symmetry.* Asymmetric lay-ups, in general, create larger PWB warpage. The PWB designer should always select and use symmetric configurations about the board midplanes whenever possible to avoid excessive warpage.
2. *CTE match.* The CTE mismatch between different board materials can generate serious board warpage. When selecting board materials, the PWB designer should consider those materials that have similar CTEs.
3. *Thickness tolerance.* It has been found that thermally induced PWB warpage is very sensitive to the layer thickness tolerance. To minimize warpage, the layer thickness tolerance needs to be carefully controlled.

In order to obtain the thermally induced warpage for realistic PWBs, much more sophisticated and detailed finite element analyses and experiments are required. The above generalized design rules, however, can still be applied as rules of thumb to minimize many warpage-oriented design/manufacturing problems in the late stages of the PWB design process.

In addition to these rules, certain important implications and findings can also be obtained from this research project. First, an FEA system such as ANSYS, together with proper modeling techniques, provides a powerful tool for simulating and analyzing PWB mechanical behavior under thermal loadings. Through such capabilities, the designer can electronically explore and evaluate design concepts without physically building time-consuming and expensive prototypes. Finally, the PWB warpage consideration is comparable in importance to other issues such as cross-talk analysis and impedance control in PWB electrical design. More research is needed on the effect on warpage of many design and manufacturing tolerances.

## 9.4 CONCLUSIONS

This chapter reports on an analytical and experimental study of the effects of warpage in PWB design. Correlation of results indicates that a satisfactory agreement was achieved between the analytical and experimental approaches. It is believed that the correlation can be further improved when more accurate mechanical properties become available. As added confidence is developed in the modeling and experimental techniques, warpage measurements for higher temperatures (up to 250°C) will also be made in order to study PWB warpage during the soldering process.

## ACKNOWLEDGMENT

The authors acknowledge the support of the Georgia Tech Manufacturing Research Center (MaRC) and Motorola, Inc., for providing the necessary funding and equipment to carry out this research.

## REFERENCES


1. Daniel, I. M., T. M. Wang, and J. T. Gotro, "Thermo-mechanical Behavior of Multilayer Structures in Microelectronics," *Journal of Electronic Packaging*, **112**, March 1990, pp. 11–15.
2. Suhir, E., "Analytical Modeling in Electronic Packaging Structures: Its Merits Shortcomings and Iteration with Experimental and Numerical Techniques," *Journal of Electronic Packaging*, **111**, June 1989, pp. 157–161.
3. Yao, S. C., "PCB Thermal Analysis," *Printed Circuit Design*, February 1989, pp. 36, 38, 42, 45.
4. Bocci, W. J., "Finite Element Engineering Analyses Applied to Microelectronics" *Proc. 1986 National Aerospace and Electronics Conferences (NAECON '86)*, Vol. 4, May 1986, pp. 1150–1153.
5. *ANSYS User Manual*, Vols. 1 and 2, Version 4.4A, Swanson Analysis Systems, Inc., Houseton, PA, 1990.
6. DeBra, L., "Finite Element Modeling of Plated-Through-Hole in Surface-Mount-Technology Printed Wiring Boards," *Boeing Electronics Report for USAF Contract No. F33615-82-C-5072*, July 1988.
7. Gasparaitis, D. D., and M. W. Lauroesch, "Determining Coefficients of Thermal Expansion of Multilayer Printed Circuit Boards," *Proc. Sixth Annual International Electronics Packaging Conference*, 1986, pp. 127–139.

8. Greene, D. et al., "Printed Wiring Board Laminates for Multi-plane Applications," *Proc. First International SAMPE Electronics Conference*, May 1989, pp. 623–625.

9. Moran, J., Letter from Oak/Mitsui, Inc. to Chao-pin Yeh of Georgia Tech, dated September 5, 1989.

10. Newton, T. D., Letter from Norplex/Oak Inc. to Mr. F. Juskey of Motorola, Inc., dated March 21, 1989.

11. Coombs, Jr., C. F., *Printed Circuits Handbook*, 3d edn., McGraw-Hill, New York, 1988.

12. Fulton, R. E. et al., *Multidisciplinary Approach to Printed Wiring Board Design*, Manufacturing Research Center (MaRC) Report, Georgia Tech, September 1990.

13. Ashbee, K., *Fundamental Principles of Fiber Reinforced Composites*, Technomic Publishing Company, Lancaster, PA, 1989.

14. Carlsson, L. A., *Tailoring Thermal Expansion Characteristics of Composite Laminates*, Department of Mechanical Engineering, Florida Atlantic University, 1987.

15. Jones, R. M., *Mechanics of Composite Materials*, McGraw-Hill, New York, 1975.

16. Timoshenko, S., and S. Woinowsky-Krieger, *Theory of Plates and Shells*, 2d edn., Wiley, New York, 1959.

17. *MATLAB User Manual*, The Math Works, Inc., South Natick, MA, 1989.

18. Skaper, G. N., and M. Montero, "Thermal Expansion Determination of Copper/ Fiberglass Material in a Multilayer Circuit Board Application," L. J. Broutman & Associates, Ltd. Report File Number 52-573, dated June 18, 1991.

19. Chiang, F. P., "Moire Methods of Strain Analysis," *The Third Edition of the SESA's Manual on Experimental Stress Analysis*, August 1979, pp. 290–308.

20. Martin, T. et al., "Experimental Measurement of PWB Warpage," *Proc. 1991 ASME Winter Annual Meeting (ASME/WAM)*, Atlanta, GA, December 1991.

21. Yeh, C. P. et al., "Experimental and Analytical Investigation of Thermally Induced Warpage for PWBs," *Proc. IEEE 41st Electronic Component Technology Conference (ECTC)*, Oakland, CA, 1991.

22. Mousley, R. F., "A Shadow Moire Technique for the Measurement of Damage in Composites," *Journal of Composite Structures*, **4**, 1985, pp. 231–244.

23. Redner, A. S., "Shadow-Moire Surface Inspection," *Materials Evaluation*, July 1990, pp. 873–878.

# 10

# Thermal Stress-Induced Open-Circuit Failure in Microelectronics Thin-Film Metallizations

*Q. Guo, L. M. Keer, and Y.-W. Chung*

## 10.1 INTRODUCTION

Very large-scale integration (VLSI) circuits have been achieved through the use of narrow metallizations in thin-film form. These on-chip interconnects play the key role in the overall chip size. With the ever-decreasing size of the microchip, narrow thin-film metallizations become the most important interconnects in large-scale integrated circuits. Among the large number of metals that have been investigated, aluminum is the most widely used interconnect material for VLSI devices because of its low resistivity, ability to form low-resistance contacts to p-type and n-type Si, and ease of deposition by either sputtering or evaporation. The trend to smaller circuitry is currently an active area of development.[1,2]

In the manufacturing process of thin-film interconnects in VLSI, the aluminum alloy metallizations are fabricated on a silicon substrate and covered with passivation films at high temperatures. The aluminum interconnects are commonly heat treated at 400°C, and for soft materials such as aluminum and its alloys intrinsic stresses are relaxed through diffusion during the heat treatment and only thermal stresses are important. When the multilayer structure is cooled to a lower temperature, thermal stresses are generated because the thermal expansion coefficient of aluminum alloy is much larger than that of its surroundings. Thermal stress modeling for this process by Jones,[3] Jones and Basehore,[4] Yosh et al.[5] using the finite element method, and Niwa et al.[6] and Korhonen et al.[7] based on the Eshelby theory of inclusions shows that the thermal stresses in the thin-film interconnects at room temperature could be well above 0.5 GPa. Experimental measurements of Shute,[8] Flinn and Chiang,[9] and Korhonen et al.,[10] among others, also show that residual stresses in thin-film interconnects can be as high as 300 MPa. An overview on the fundamentals of residual stress measurements, mechanical properties, and time-dependent deformation processes of thin-films used in integrated circuits and magnetic disks was given by Nix.[11]

Several failure mechanisms associated with high residual stresses are becoming increasingly important. Open circuits from voiding (Fig. 10-1a) and short circuits from hillocks due to poor adhesion between films and substrates (Fig. 10-1b) are phenomena related to residual stresses in aluminum thin-film interconnects in integrated circuits. It is important to understand the mechanisms that control the failure processes so that integrated circuit structures can be designed and manufactured for mechanical reliability as well as for electronic device performance. Only the open-circuit failure from diffusion (shown in Fig. 10-1a) will be treated in detail in this chapter. Failures of interconnects from hillocks and packaging encapsulation resulting from debonding between films (Robock and Nguyen[12]) due to interfacial cracks will not be addressed here. Theoretical analyses of failures from interface cracks have been given by the authors elsewhere.[13-15]

Voids in interconnects can lead to open-circuit failure when a void completely severs the line. Historically, void formation and propagation due to electromigration was and still is a major problem that limited the

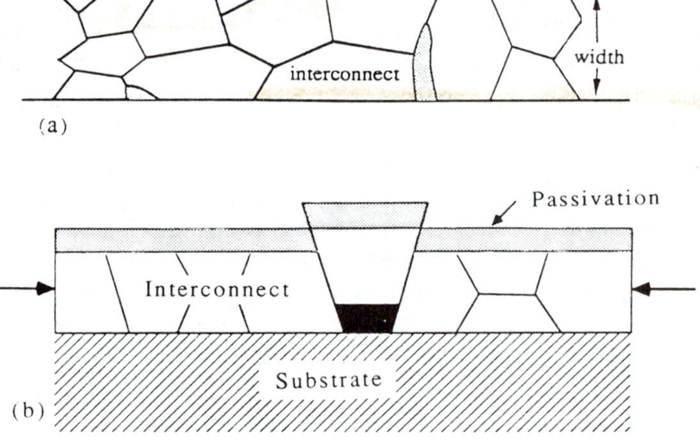


**Figure 10-1** Mechanisms of deformation in thin-film interconnects leading to open-circuits from cracking (shaded area in (a)) and short circuits from hillocks (b). The black area is a void due to a hillock.

reliability of electronic devices.[16] In order to increase the electromigration resistance of aluminum as the cross-sectional area of aluminum interconnects decreases, a small percentage of Si or Cu is added to the Al.[17] However, new failure mechanisms of diffusive cavitation in narrow aluminum metallizations were reported in 1984 by Curry et al.[18] This open-circuit failure occurs through void formation and propagation along grain boundaries in the absence of applied voltage or current, and hence is not related to electro-migration. Since then, voiding of the interconnects has been reported by Mayumi et al.,[17] Hinode et al.,[19] and McPherson and Dunn,[20] among many others in isothermal aging tests where constant temperatures were maintained during tests, and by Whitman and Chung[21] and Guo et al.[22] in thermal fatigue tests. The conductor lines appear shiny under a light microscope, suggesting that these voids are not the result of corrosion.[23]

These voids have two distinct morphologies—rounded wedgelike and narrow cracklike as shown in Fig. 10-2—which were known previously in a different context,[24] as equilibrium- and nonequilibrium-shaped voids, respectively. Theoretical analyses of Chuang et al.[24] and Martinez and Nix[25] have shown that rounded wedge-shaped voids occur when surface diffusion is sufficiently rapid, while narrow cracklike voids occur when grain boundary diffusion dominates. Although both wedgelike and cracklike voids have been observed experimentally in aluminum alloy narrow conductor lines, most open-circuit failures of the lines have been reported to be cracklike voids.[17,19]

The effects of many factors, such as temperature, line width, passivation, etc., on the failure rates of interconnect lines have been studied experi-mentally. Some of the key observations reported in the literature have been summarized by Guo et al.[22] It should be noted that divergent results have

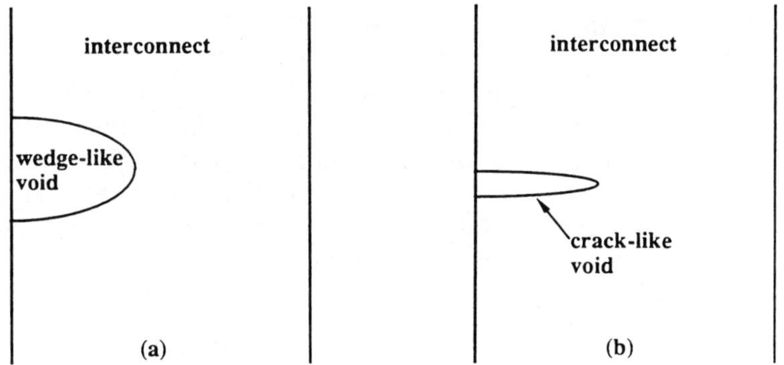


**Figure 10-2**   Schematic illustration of void shapes: (a) a wedgelike void; (b) a cracklike void.

been reported from different experimental investigations. In one instance the time to failure was found to be proportional to $w^{1.3}$, where $w$ is the line width, ranging from 1 μm to 1.8 μm,[17] while in another instance it was found to be proportional to $w^{2.7}$ with $w$ in the range from 1 μm to 2.5 μm.[19] A maximum failure rate was observed by McPherson and Dunn[20] in one study to occur at about 165°C, after which it decreased. In another study, Hinode et al.[19] found that the failure rate increased continuously from 200°C to 295°C. The causes and explanations for such divergent trends will be discussed later according to the model constructed by Guo et al.[22] Stress-induced voiding and cracking models due to thermally activated diffusion of vacancies have been proposed by many researchers.[17–22,26–31]

The problem discussed above is not a mathematically well-formulated one. The methodology used here for the modeling of the new open-circuit failure process is in terms of the general thermodynamics of a stressed solid (Section 10.2), making the necessary assumptions from experimental observations; therefore, the major factor responsible for the failure can be modeled theoretically. The mathematical model will in turn be tested by experiment. A theoretical model based on grain boundary and surface diffusion and fracture mechanics is constructed for thin-film interconnections under both isothermal aging and thermal fatigue in Section 10.3. The effects of aging temperature, line width, passivation, stress and other factors in isothermal aging are discussed in Section 10.4. Divergent experimental observations can be explained well according to the model. Thermal fatigue tests performed by Whitman and Chung[21] on typical Al–Au interconnects in real devices will be used to assess the theoretical model. The effects of film thickness, frequency, thermal fatigue temperatures, and hold time on the number of cycles to failure are studied in Section 10.5. Readers whose primary concern is how to use the present model can skip Section 10.2 and part of Section 10.3.


## 10.2 THERMODYNAMICS OF STRESSED SOLIDS

Submicrometer structures must be analyzed from the combined points of view of solid mechanics and materials science. In a submicrometer structure the conditions for thermomechanical equilibrium in each solid phase and at an interface can be determined by considering the change in total energy associated with virtual variations of the actual two-phase solid system. The total energy $E$ of the system may be expressed[32–34] as

$$E = \int_{V_{\mathrm{I}}} e(\mathbf{F}^{\mathrm{I}}, s, \rho_h)\, dV + \int_{V_{\mathrm{II}}} e(\mathbf{F}^{\mathrm{II}}, s, \rho_h)\, dV + \int_{\Sigma} \chi(\mathbf{F}'^{\mathrm{I}}, \mathbf{A}, \eta, \Lambda_h)\, dS \quad (10\text{-}1)$$

where $e$ is the energy density, $s$ is the entropy density, the $\rho_h$ are the densities of the independent constituent components, measured per unit volume in an appropriate reference state, and $\mathbf{F}$ is the deformation gradient from the reference state to the actual state. Here, $\chi$ is the surface energy $\mathbf{F}'^1$ is the surface deformation gradient, referred to the $\alpha$ phase, $\mathbf{A}$ is the difference between the surface deformation gradients referred to the I and II phases, $\eta$ is the excess surface entropy and the $\Lambda_h$ are the densities of the constituent components at the surface. For nondiffusing and elastic solids one can easily choose a reference state to define deformation, say strain. But for diffusing solids, deformation of an element may not be allowed because its boundaries become "diffuse." Following the approach developed by Larche and Cahn,[35] an appropriate reference for solid solutions in which atoms diffuse by a vacancy mechanism is the lattice structure, regardless of whether atoms or vacancies occupy given atomic sites.

Through the calculation of the first variation for the energy $E$, under conditions of conservation of total number of atoms $N_h$ $(h = 1, \ldots, n)$ of each component species and the consideration of total entropy $S$, the following equations are obtained:[32–35]

$$\operatorname{div} \boldsymbol{\sigma} = 0 \qquad \text{(mechanical equilibrium)} \qquad (10\text{-}2)$$

$$\frac{\partial e}{\partial s} = \theta \qquad \text{(thermal equilibrium)} \qquad (10\text{-}3)$$

$$\frac{\partial e}{\partial \rho_h} = \lambda_h = \mu_h \qquad (h = 1, \ldots, m, \text{ for interstitial phases}) \qquad (10\text{-}4a)$$

$$\frac{\partial e}{\partial \rho_h} - \frac{\partial e}{\partial \rho_n} = \lambda_h - \lambda_n = \mu_h \qquad (h = m + 1, \ldots, n, \text{ for substitutional phases})$$
$$(10\text{-}4b)$$

where for both phases $\boldsymbol{\sigma}$ is the stress tensor and $\theta$, $\lambda_h$ are Lagrange multipliers. Equations (10-4) are the conditions for chemical equilibrium, which must be constant in order to retain equilibrium. When not in the equilibrium state, the $\mu_h$ are not constant throughout the solid and diffusion can occur in the directions determined by the gradient of $\mu_h$ if the stress is tensile and large enough to surmount the diffusion barrier. For this reason $\mu_h$ is called the diffusion potential.

The energy density $e$ is seen from Eqs. (10-2)–(10-4) to have the form

$$de = \sigma_{ij} \, d\varepsilon_{ij} + \theta \, ds + \rho_0 \mu_h \, dc_h \qquad (10\text{-}5)$$

where $\rho_0$ is the total number of lattice sites or atoms per unit volume in the reference state, $c_h = \rho_h/\rho_0$ or the relative volume fraction of each component species, and $\varepsilon_{ij}$ is the strain tensor. We adopt the summation convention, which states that whenever the same letter subscript occurs twice in a term, the repeated subscript indicates the summation over all its possible values. Define the Helmholtz free energy as

$$\phi = e - \theta s \tag{10-6}$$

Therefore,

$$d\phi = \sigma_{ij}\,d\varepsilon_{ij} - s\,d\theta + \rho_0 \Sigma \mu_h\,dc_h \tag{10-7}$$

By application of a Legendre transformation,

$$\varphi = \phi - \sigma_{ij}\,d\varepsilon_{ij} \tag{10-8a}$$

or

$$d\varphi = -\varepsilon_{ij}\,d\sigma_{ij} - s\,d\theta + \rho_0 \mu_h\,dc_h \tag{10-8b}$$

From Eq. (10-8b) the Maxwell relation

$$-\rho_0\frac{\partial \mu_i}{\partial \sigma_{ij}} = \frac{\partial \varepsilon_{ij}}{\partial c_i} \qquad \text{(no summation on } i) \tag{10-9}$$

is developed, and assuming elastic stresses, from Hooke's law,

$$\varepsilon_{ij} = \varepsilon_{ij}^c + S_{ijkl}\sigma_{kl} \tag{10-10}$$

where $\varepsilon_{ij}^c$ is the strain produced by the change of composition from $c_0$ at the stress-free state to $c$ and $S_{ijkl}$ is the compliance tensor. The general expression for the diffusion potential under multiaxial stresses can be obtained from the Maxwell equation (10-9) and Hooke's law as

$$\mu(\sigma_{ij}, c) = \mu(0, c) - \frac{1}{\rho_0}\left[\left(\frac{\partial \varepsilon_{ij}^c}{\partial c}\right)\sigma_{ij} + \frac{1}{2}\left(\frac{\partial S_{ijkl}}{\partial c}\right)\sigma_{ij}\sigma_{kl}\right] \tag{10-11}$$

where $\mu(\sigma_{ij}, c)$ is the chemical potential of the stressed state and $\mu(0, c)$ is the chemical potential of the stress-free state. Here, subscripts $i$ for $\mu$ and $c$ in Eq. (10-11) are omitted for simplicity. Equation (10-11) can be rewritten as

$$\mu(\sigma_{ij}, c) = \mu(0, c) - \Omega\eta_{ij}\sigma_{ij} - \tfrac{1}{2}\Omega D_{ijkl}\sigma_{ij}\sigma_{kl} \tag{10-12}$$

where $\Omega$ is the molar volume of the lattice or the volume of the atom at the reference state and

$$\eta_{ij} = \frac{\partial \varepsilon^c_{ij}}{\partial c}, \qquad D_{ijkl} = \frac{\partial S_{ijkl}}{\partial c} \qquad (10\text{-}13a, b)$$

Equation (10-12) has the same form as that given by Li et al.[36] Since $\varepsilon^c_{ij}$ is a symmetric tensor, so also is $\eta_{ij}$. In terms of the principal directions, the principal values, $\eta_i$, $i = 1, 2, 3$, can be measured through the lattice parameters.

In general the experimental data on $D_{ijkl}$ are sparse. For dilute Fe–Al, the values listed in the Landolt–Bornstein tables (see also Larche and Cahn[37]) are of the same order as $S_{ijkl}$; therefore, the contribution of the deviatoric stress tensor to diffusion is small. However, for metals the plastic deformations are determined by the deviatoric stress tensor, while the hydrostatic stress provides no contribution to plastic deformation. Both finite element calculations[3–5] and experimental measurements[9,10] show that, although the stresses in a passivated interconnect are very high, they are in a nearly hydrostatic state. Therefore, plastic deformation has little to do with the open-circuit failure from cracking in interconnects. On the other hand, the large hydrostatic stress is sufficient to surmount the mass-moving barrier and the hydrostatic stress gradient provides the driving force to move mass in interconnects to create voids. This explains why stress-assisted diffusion, rather than plastic deformation, is the major mechanism for the open-circuit failure of aluminum interconnects.

If the second-order terms in Eq. (10-12) can be neglected, then one has

$$\mu(\sigma_{ij}, c) = \mu(0, c) - \Omega \eta_{ij} \sigma_{ij} \qquad (10\text{-}14)$$

The flux is therefore proportional to the gradient of the diffusion potential:

$$J_i = -B_{ij} \frac{\partial \mu}{\partial x_j} \qquad (10\text{-}15)$$

where $J_i$ is the mass flux in the $i$th direction and $B_{ij}$ is the diffusion coefficient.


## 10.3 A STRESS-INDUCED DIFFUSION FAILURE MODEL

Consider a cracklike grain boundary void located in the center or at an edge of the interconnect (Fig. 10-3). The coordinates are shown in Fig. 10-3 with the $y$-axis along the lengthwise direction of the strip conductor line and the $z$-axis normal to the surface. The $x$-axis is directed along the width and away from the void tip. The origin of the coordinates is chosen such that the position of the void tip can be defined by $x = a$ for a void at the edge

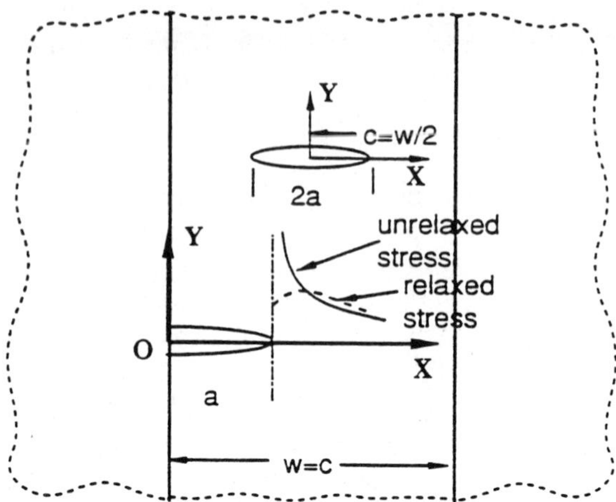


**Figure 10-3**   An edge or a central cracklike void in a strip interconnect line of width *w*. Unrelaxed stress refers to the stress obtained when the diffusion process is neglected; relaxed stress refers to that obtained when the interaction between elastic stress and diffusion is considered.

and the center (Fig. 10-3), where *a* is the void size. Here, only cracklike voids are considered because both experimental observations of Muyami et al.[17] and Hinode et al.[19] and theoretical studies of Chuang et al.[24] and Martinez and Nix[25] show that cracklike voids occur more frequently in interconnects. In this case, matter is transported from the crack tip to the grain boundary, resulting in stress relaxation of the crack tip. Stresses in interconnects with cavitations can be analyzed by classical elasticity theory. In order to clarify the locations of voids, edge voids or cracks refer to voids at the edges and central voids or cracks refer to voids at the centers of interconnects.

Consider now the addition of a matter layer of thickness $dt$ to a flat grain boundary of area $\delta A$ subject to tensile stress $\sigma$ perpendicular to the boundary. The boundary must be separated by the distance $dt$ to accommodate the new matter, resulting in a strain $d\varepsilon_{yy}^{c} = dt/\delta_b$ due to the material layer, where $\delta_b$ is the effective width of a grain boundary. The relative volume fraction of the added layer $dt$ is $dc = (\delta A)\, dt/(\delta A\, \delta_b) = dt/\delta_b$. Therefore, $\eta_{yy} = d\varepsilon_{yy}^{c}/dc = 1$. From Eq. (10-14) one obtains that the chemical potential $\mu$ of the stressed state is related to the chemical potential $\mu_0$ of the stress-free state as

$$\mu = \mu_0 - \Omega\sigma \tag{10-16}$$

where $\sigma$ is the stress that acts normal to the grain boundary and $\Omega$ is the atomic volume. This result agrees with that given by Herring,[38] Coble,[39] and Li et al.[36] among others. The last term of Eq. (10-12) is of the order of $\Omega\sigma^2/E$.[5] Since the Young's modulus $E$ for aluminum is about 70 GPa, this term is negligible. This linearized equation has been used by many authors, e.g., Chuang et al.,[24] Weertman,[40] Raj and Ashby,[41] and Evans et al.,[42] to study grain boundary diffusive cavitation.

The finite element analysis of Yost et al.[5] shows that the stress $\sigma$ (Fig. 10-4) in an interconnect is not uniform across the interconnect. This results in a chemical potential gradient. The mass diffusion along a grain boundary according to Eq. (10-15) can be specified further as

$$J = -\frac{D_b}{\Omega k T}\frac{\partial \mu}{\partial x}\tag{10-17}$$

where the flux $J$ is the number of atoms diffusing along the grain boundary per unit time, $D_b$ is the grain boundary diffusion coefficient, $k$ is Boltzmann's

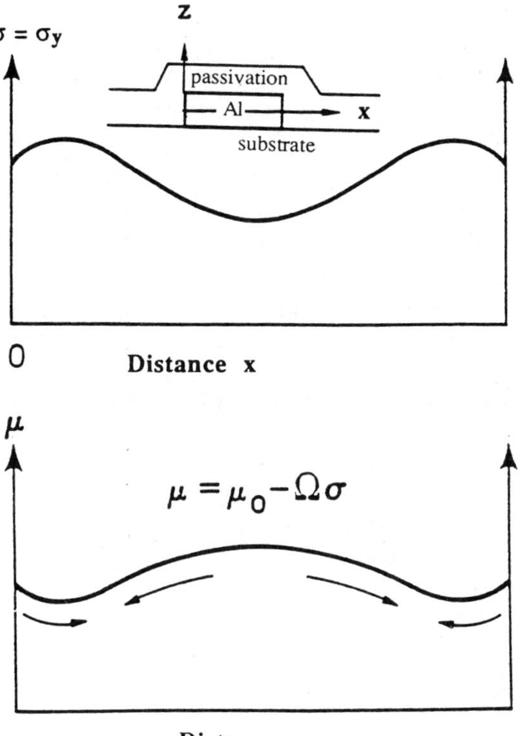


**Figure 10-4**   Stress distribution in an interconnect and chemical potential distribution.

constant, and $T$ is the temperature. Mass conservation for grain boundary diffusion gives

$$\frac{\Omega \delta_b}{\partial x} \frac{\partial J}{} + v_y = 0 \tag{10-18}$$

where, as before, $\delta_b$ is the effective width of a grain boundary and $v_y$ is the rate of material deposition on the grains.

It follows from Eqs. (10-16)–(10-18) that

$$-\frac{D_b \delta_b \Omega}{kT} \frac{\partial^2 \sigma}{\partial x^2} = v_y \tag{10-19}$$

For periodic voids at a grain boundary, it has been assumed previously by Chuang et al.,[24] Raj and Ashby,[41] and Rice[43] that the material deposition rate $v_y$ is uniform along the boundary. Finite element calculations of Needleman and Rice[44] show that $v_y$ decreases with the distance from the void tip. For the present case of an encapsulated Al alloy interconnect, the diffusion in the passivation layer is not active. The material deposition rate $v_y$ can be regarded as zero at the interface between the interconnect and the passivation layer from the continuity of displacements ($x = c$, see Fig. 10-3). Therefore, a linear distribution of $v_y$ with maximum value at the void tip and zero at $x = c$ may be a reasonable assumption, i.e.,

$$v_y = v_0(c - x) \tag{10-20}$$

where $v_0 c$ is the material accumulation rate at the void tip, $c = w/2$ for a central boundary crack, $c = w$ for a boundary crack at the edge, and $w$ is the width of the Al interconnect line.

Solving for the stress $\sigma$ along the grain boundary from Eqs. (10-19) and (10-20) yields

$$\sigma_y(x) = \sigma_c + \frac{\sigma_c - \sigma_a}{2} \frac{c-x}{c-a} \left( \frac{(c-x)^2}{(c-a)^2} - 3 \right) - \frac{\sigma_a'}{2} \left( \frac{c-x}{c-a} - 1 \right)(c-x) \tag{10-21}$$

where $\sigma_c$ is the stress at $x = c$, and $\sigma_a$ and $\sigma_a'$ are the stress and stress derivative, respectively, with respect to $x$ at the void tip $x = a$. From Eq. (10-21) the average stress $\bar{\sigma}$ over the uncavitated part of the interconnect can be obtained as

$$(c - a)\sigma_a' + 5\sigma_a = 8\bar{\sigma} - 3\sigma_c \tag{10-22}$$

where

$$\bar{\sigma} = \frac{1}{c-a} \int_a^c \sigma_y(x)\, dx \tag{10-23}$$

Due to diffusion, stresses at void tips are relaxed and elastic stress analysis is not appropriate near the void tip. However, from the analysis of Chuang and Rice[45] it is known that $\sigma_a$ and $\sigma_a'$ are related to the void growth rate $da/dt$. If one can obtain $\sigma_c$ and the average stress $\bar{\sigma}$ over the uncavitated part of the interconnect, then Eq. (10-22) can be used to solve for the void growth rate $da/dt$.

For a cracklike void in the *steady state* (i.e., the void growth profile retains a constant shape near the tip), the stress $\sigma_a$ and stress gradient $\sigma_a'$ at the void tip are given by Chuang and Rice[45] from the continuity of chemical potential and flux at the void tip as

$$\sigma_a = 2\gamma_s \sin(\psi/2)\left(\frac{da/dt}{B}\right)^{1/3} \tag{10-24}$$

and

$$\sigma_a' = 4\gamma_s \sin(\psi/2)\, d\left(\frac{da/dt}{B}\right)^{2/3} \tag{10-25}$$

where $da/dt$ is the void growth rate and $\gamma_s$ is the specific surface energy. The tip angle $\psi$ is determined from the grain boundary energy $\gamma_b$ as

$$\cos\psi = \frac{\gamma_b}{2\gamma_s} \tag{10-26}$$

$$d = \frac{D_s \delta_s}{D_b \delta_b} = \frac{D_{0s}\delta_s e^{-Q_s/kT}}{D_{0b}\delta_b e^{-Q_b/kT}} \tag{10-27}$$

and

$$B = \frac{D_s \delta_s \Omega \gamma_s}{kT} \tag{10-28}$$

where $D_s$ and $\delta_s$ are the surface diffusion coefficient and the effective thickness of the surface diffusion layer, respectively. $Q_s$ and $Q_b$ are the surface diffusion and grain boundary diffusion activation energies, respectively. By substituting Eqs. (10-24) and (10-25) into Eq. (10-22) and solving for $da/dt$, one obtains for a void in steady state

$$\left(\frac{da/dt}{B}\right)^{1/3} = \frac{-5 + \sqrt{25 + 4(c - a)d(8\bar{\sigma} - 3\sigma_c)/[\gamma_s \sin(\psi/2)]}}{4(c - a)d} \tag{10-29}$$

The analyses of Evans et al.[42] and Vitek[46] show that, although elastic stress analysis is not appropriate at the void tip, it is still valid at points away from the tip and, therefore, $\bar{\sigma}$ and $\sigma_c$ in Eqs. (10-22), (10-29) can be estimated from fracture mechanics. From elastic stress analysis for a cracklike void in a passivated interconnect, one has (see Appendix 10A)

$$8\bar{\sigma} - 3\sigma_c = 5\sigma_0 f(a/c) \tag{10-30}$$

where

$$f(a/c) = 1 + \frac{13}{5\sqrt{2}} \frac{\kappa[1 - k(a/c)^2]}{\sqrt{1 - a/c}} \sqrt{a/c} \tag{10-31}$$

and $\kappa = 0.93$ if the void is located at the edge of an interconnect and $\kappa = 1$ for a void located at the center. It should be noted that the stress $\sigma_0$ is the interconnect stress, which can be obtained either experimentally[9,10] or numerically[3-5] as if there were no voids in the interconnect.

Substituting Eq. (10-30) into Eq. (10-31) gives

$$\left(\frac{da/dt}{B}\right)^{1/3} = \frac{-5 + \sqrt{25 + 20(c - a)d\sigma_0 f(a/c)/[\gamma_s \sin(\psi/2)]}}{4(c - a)d} \tag{10-32}$$

If the grain boundary diffusion is much faster than the surface diffusion (i.e., $d \ll 1$), then

$$\frac{da}{dt} = \frac{B\sigma_0^3}{[2\gamma_s \sin(\psi/2)]^3} [f(a/c)]^3 \tag{10-33}$$

If the opposite is true, then

$$\frac{da}{dt} = B\left(\frac{5\sigma_0 f(a/c)}{4(c - a)d\gamma_s \sin(\psi/2)}\right)^{3/2} \tag{10-34}$$

According to the theoretical analyses of Chuang et al.[24] and Martinez and Nix,[25] Eq. (10-33) is appropriate for cracklike voids, while Eq. (10-34) is for wedgelike voids. If one assumes that failure rate is proportional to $\sigma_0^N$, Eqs. (10-33) and (10-34) show that $1.5 \leqslant N \leqslant 3$. As an example, a best fit to the experimental results of McPherson and Dunn[20] gives $N = 2.33$.

Experimental observations of Mayumi et al.[17] and Hinode et al.[19] have shown that open-circuit failures in interconnects are mainly caused by cracklike voids. Based on these experimental observations and the finite difference analyses of Martinez and Nix[25] for a void under coupled grain boundary and surface diffusion, it is believed that the ratio $d$ of the surface diffusivity to grain boundary diffusivity is small. Considering these experimental and theoretical results, Eq. (10-33), which is obtained under the condition that grain boundary diffusion is much faster than surface diffusion (i.e., $d \ll 1$), should be more appropriate than Eq. (10-34). By comparing their experimental results with predictions from both Eq. (10-33) and Eq. (10-34), Whitman and Chung[21] showed that prediction based on Eq. (10-33) is indeed better. Accordingly, the following discussion is mainly based on Eq. (10-33).

Before we compare the theory with experimental results, we need to know the surface and grain boundary diffusion activation energies $Q_s$ and $Q_b$. Although the measured grain boundary diffusion activation energy for pure Al films varies from 0.3 to 1.2 eV, in most cases values were found between 0.5 and 0.7 eV.[47] The grain boundary activation energy was reported by Bagnoli et al.[48] and Berry[49] to be about 0.55 eV for pure Al thin films and 0.95 eV for Al–0.17%Cu and Al–1%Si films. The surface diffusion activation energy $Q_s$ for pure Al is given by Gjostein[50] to be about 0.48 eV for the temperature range 50–295°C. It is not known how $Q_s$ will be changed upon the addition of a few percent Cu or Si. If it is assumed that $Q_s/Q_b$ for pure Al is the same as that for Al–Si and Al–Cu, then $Q_s$ for Al–Si and Al–Cu should be about 0.82 eV. We will use these values for $Q_b$ and $Q_s$ in subsequent analysis. It is also not known how the preexponential factors $D_{0s}$ and $D_{0b}$ for surface and grain boundary diffusion coefficients will be changed upon the addition of a few percent Cu or Si. Only relative lifetimes or failure rates are compared, so that $D_{0s}$, $D_{0b}$ and many other uncertain factors can be removed.

## 10.4 DISCUSSION OF EXPERIMENTAL RESULTS OF ISOTHERMAL AGING

The effects of temperature, line width, line thickness, passivation, interlayer dielectric (ILD) stress, etc., on failure rates have been experimentally investigated. Experimental results obtained under isothermal aging will be discussed according to the above model.

### 10.4.1 Temperature Effect

The effect of aging temperature on failure can be studied according to Eq. (10-33). Figure 10-5 shows a typical stress–temperature plot from

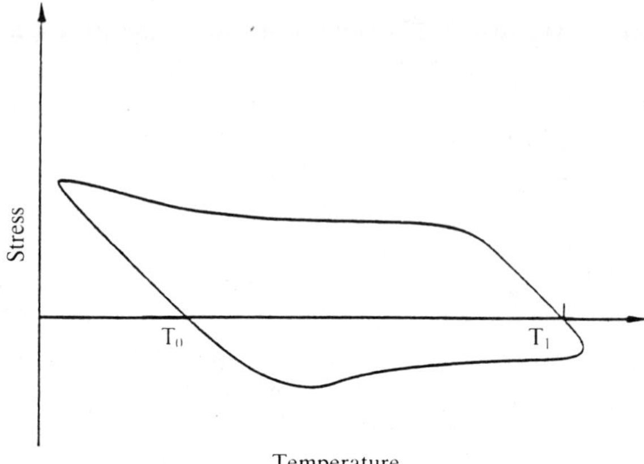


**Figure 10-5**   A typical temperature dependence of interconnect stress.

experimental measurements of Shin[51] and Gardner and Flinn.[52] When aging temperature $T$ is within the linear stress–temperature portion as shown in Fig. 10-5, the stress $\sigma_0$ in an interconnect may be written as

$$\sigma_0(T) = M(T' - T) \tag{10-35}$$

where the stress-free temperature $T' = T_0$ or $T_1$ (Fig. 10-5), depending on the testing condition, and $M$ is a function of the Young's modulus and Poisson ratio of the Al line and the difference of the thermal expansion coefficients of the Al interconnect and the substrate and passivation below and above the line.[11] In the linear stress–temperature portion, the temperature dependence of the failure rate is given from Eqs. (10-33) and (10-35) as

$$\frac{da}{dt} \propto \frac{\exp(-Q_s/kT)(T' - T)^3}{T} \tag{10-36}$$

McPherson and Dunn[20] performed aging tests from 100°C to about 200°C and found that a maximum failure rate occurs at about 165°C, with $T' = 232°C$. Substituting $T' = 232°C$ and $Q_s = 0.82\ \text{eV}$ for Al–1%Si into Eq. (10-36) yields a maximum failure rate temperature of about 169°C, which agrees well with the experimental result of McPherson and Dunn.[20] In another investigation by Hinode et al.[19] for an aging temperature range from 200°C to 295°C, the samples were heat treated at 450°C for one hour after fabrication before being subjected to an aging test. If $T' = T_0 \approx 225°C$ is used, the stress in the interconnect for the aging tests from 200°C to 295°C would be near zero or compressive (see Fig. 10-5). Hinode et al.[19] pointed out that the Al line is sufficiently relaxed around 400°C because of heat treatment before isothermal aging tests and that $T' = T_1 = 400°C$ should be used. Substituting $T' = T_1 = 400°C$ into Eq. (10-36), one finds that the failure rate varies monotonically over the temperature range 200°C to 295°C, as observed experimentally by Hinode et al.[19]


## 10.4.2 Line Width Effect

The effects of line width on failure rates have been experimentally studied by Mayumi et al.[17] and Hinode et al.[19] for different line width ranges. If it is assumed that the failure has a power dependence on the line width $w$, the study of Mayumi et al.[17] for line widths from 1 µm to 1.8 µm showed that the relation between the mean time to failure (MTF) and the line width was MTF $\propto w^{1.3}$, while the study of Hinode et al.[19] for line widths from 1 µm to 2.5 µm found that the failure time was proportional to $w^{2.7}$. Since no experimental data on the interconnect stress dependence of the line width are available, finite element results of Yost et al.[5] and Jones and Basehore[4] for the stresses of different line widths are used in this discussion. Finite element stress analyses for a passivated interconnect line show that stresses at 25°C after cooling from the stress-free temperature of 400°C are 536, 500, 422, 285, and 222 MPa for line widths 1, 1.5, 2, 3, and 4 µm, respectively. Figure 10-6 shows the line width dependence of the failure time from Eq. (10-33) normalized by the failure time of 1-µm wide interconnect. Experimental results of Mayumi et al.[17] and Hinode et al.[19] are also shown in the figure. It can be seen that theoretical and experimental results agree reasonably well.


## 10.4.3 Line Thickness Effect

The film thickness dependence of the stress has been studied experimentally by Shute et al.[53] for thicknesses ranging from 0.5 µm to 2 µm for both passivated and unpassivated Al–Cu films. The stresses were found, respectively, for passivated 0.5-µm thick and 1-µm thick films to be 2.8 and 1.6 times the stress in a 2-µm thick passivated film. In a more recent study of Shute,[8] she found that the thickness dependence of stress in Al–Cu films is mainly contributed by the thickness dependence of yield stress of the films. The result that thinner films have higher yield strengths for Al–Cu films agrees the results of Kuan and Murakami[54] for Pb thin films and Doerner et al.[55,56] for Al films. By definition, yield strength is the stress required to move dislocations. For a planar thin film under passivation, much of the yield strength can be attributed to the fact that the substrate and passivation constrain dislocation motion in the film. Based on the solution of Freund[57]

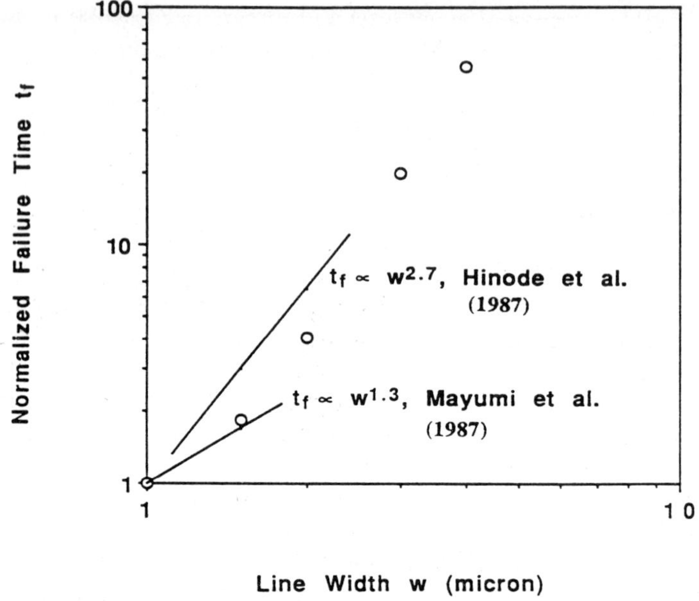


**Figure 10-6**    Interconnect line width dependence of the normalized failure time from Eq. (10-33). The open circles represent the calculated results from Eq. (10-33) and the finite element stress analyses of Jones[3–4] and Yost et al.[5] After ref. 22.

for a dislocation in an epitaxial film, Nix predicted that the yield strength of a thin film is approximately proportional to the inverse of the film thickness.[11] This explains why thinner films have higher yield strengths. The film thickness dependence of the normalized failure time is plotted in Fig. 10-7 according to Eq. (10-33) together with the experimental results obtained by Mayumi et al.[17] In Fig. 10-7 two theoretical prediction lines are given; one is based on Nix's theoretical prediction of the film-thickness dependence of yield stress and the other is based on Shute's experimental result of the thickness dependence of the stress.[8,53] It can be seen from Fig. 10-7 that the theoretical predictions agree well with the experimental results. It is important to note that, for unpassivated films, a stress gradient may exist across the film thickness and may be significant for films greater than or equal to a few micrometers. The current theory does not address this situation.

### 10.4.4 Passivation Effect

Experimental results on the effects of passivation on failure rates have been reported by Hinode et al.[19] for 0.9-μm wide conductors with and without

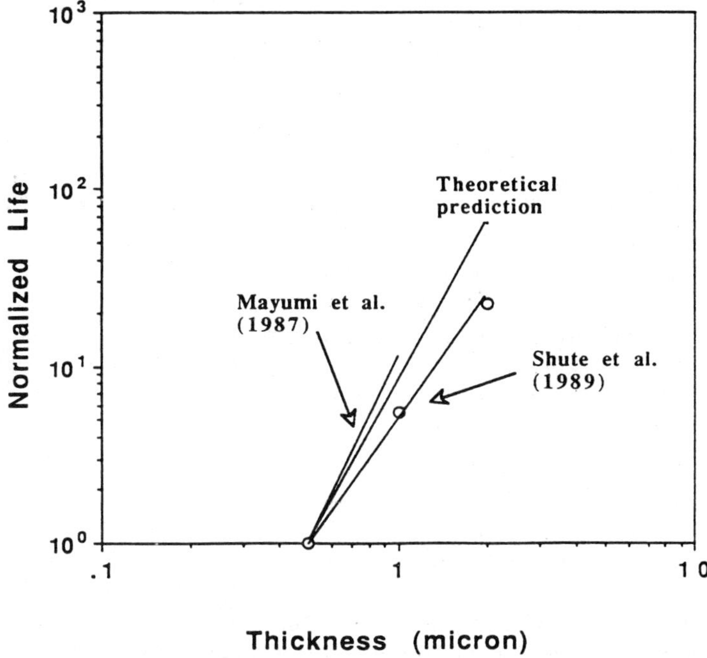


**Figure 10-7**   Interconnect line thickness dependence of the normalized failure time from Eq. (10-33) and the experimental results of Mayumi et al.[17] Theoretical results are the results obtained from theoretical stress analysis of Nix[11] based on Eq. (10-33). The circles are the results predicted from Eq. (10-33) based on the stress measurements of Shute et al.[8,53] After ref. 22.

passivation layers. Failure rates of conductors with passivation are shown to be at least two orders of magnitude higher than those without passivation. Finite element calculations of Jones[3] and Jones and Basehore[4] show that the average stress $\sigma_0$ in a 1-µm wide line is 5.2 times the stress without passivation. The stress ratio with and without passivation from the rigid box approximation, in which the Al interconnect line is assumed to be in a rigid box with finite thermal expansion coefficient, could be higher.[3–5] Experimental measurements for passivated and unpassivated interconnects of Flinn and Chiang[9] and Shute[8] show that stresses in passivated interconnects are about 2 to 6 times those without a passivation layer, depending on fabrication procedures and materials of the dielectric interlayers. Therefore, according to Eq. (10-33) the failure rates for interconnects with passivation are much higher than for those without passivation.

From the above discussion, one can see that our theoretical failure model has captured the main experimental results, and divergent reported experimental observations can be explained well according to the model. Other effects on the failure rates can also be studied by using Eq. (10-33). For example, the effect of interlayer dielectric (ILD) on the failure of interconnects has been studied experimentally by Shin[51] with ILD stresses varying from high compressive to low compressive to tensile from different manufacturing processes. Stress measurements suggest that the stresses at room temperature in the encapsulated metal interconnect are highest with highly compressive ILD, lower with tensile ILD, and lowest with low compressive ILD. Therefore, according to Eq. (10-33), the damage to interconnects ranges from least with low compressive ILD to greatest with highly compressive ILD. This is exactly the results reported in table III of Shin.[51] Therefore, the present model also provides the basis for the manufacturing process to minimize the failure rate of interconnects by reducing the stress.

## 10.5 FAILURE OF INTERCONNECTS UNDER THERMAL FATIGUE

Failure of thin-film interconnects in microelectronics is primarily attributed to thermal fatigue caused by changes in operating temperature due to either power on/off or external temperature fluctuations. Although the failure process of diffusive voiding modeled above for interconnects has been proved in many experimental investigations under isothermal aging as discussed above, few experimental studies have been performed on interconnects under thermal fatigue. Experimental investigations of interconnects under thermal cycling were performed by Whitman and Chung[21] to determine the actual failure modes. In their study, Al–Cu interconnects of thickness 0.25 μm were thermally cycled to investigate the failure mechanism under thermal fatigue. The complete details of their experimental procedures have been reported by Whitman et al.[21,58] so only a brief summary of these details will be given here.

### 10.5.1 Sample Preparation

The procedure for sample preparation is shown in Fig. 10-8. A silicon wafer was thermally oxidized to give an $SiO_2$ layer of the desired thickness. Using photolithography, 3-μm wide steps spaced 27 μm apart were etched into the $SiO_2$ using a buffered HF solution. An Al–Cu alloy of 0.25 μm thickness was then evaporated onto the wafer. Turning the mask 90° allowed Al lines to be etched perpendicular to the steps with the same size and spacing. The etchant used was a 13% $HNO_3$, 30% $H_3PO_4$, 57% $CH_3COOH$ solution. The result was a cross-hatch as shown in Fig. 10-9. The film thickness was about twice that of the oxide step. All samples were annealed at 400°C for 1 hour

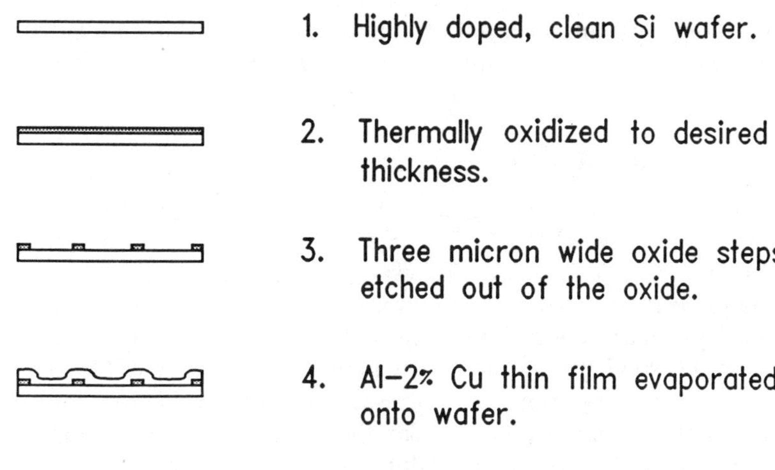

1. Highly doped, clean Si wafer.

2. Thermally oxidized to desired thickness.

3. Three micron wide oxide steps etched out of the oxide.

4. Al–2% Cu thin film evaporated onto wafer.

5. Three micron wide lines etched out of film. Lines and steps are separated by 27 microns.

**Figure 10-8**   Photolithographic process used by Whitman and Chung[21] to make Al–Cu samples.

**Figure 10-9**   Top view of a sample. The Al–Cu lines are horizontal, running perpendicularly to $SiO_2$ steps.

in $10^{-6}$ Torr vacuum. Thermal cycling experiments were conducted in air to study the failure mode, temperature range, and hold time effects.

### 10.5.2 Testing and Results

***Test I***   The first test was performed on a 0.25-μm Al–6%Cu sample. The temperature range was 25–125°C, with a triangular waveform, and the period was 40 minutes. After 3000 cycles, cracklike voids appeared in the lines as shown in Fig. 10-10*a*. No oxidation was evident, and the lines were still shiny under a light microscope, suggesting that the cracks were not the result of corrosion. Thermal cycling had resulted in the precipitation of Cu-rich phase. However, the precipitation was not associated with voided regions. Often, no precipitations were observed near the crack, as is seen in Fig. 10-10*a*. Also, no "debris" was present along a crack. Energy-dispersive X-ray analysis was performed in the region near the crack tip for several cracks, and far away from the crack. The Cu concentration near the crack tip was no different than for the rest of the line.[21]

***Test II***   The previous test was repeated with a 0.25-μm Al–3%Cu sample, cycled between 50°C and 150°C using a triangular waveform with a period of 20 minutes. The sample was inspected approximately every 400 cycles. Again after 3000 cycles, voiding was observed (see Fig. 10-10*b*). Little precipitation was observed. This was probably due to a lower Cu concentration. The voids initiated at the edge and propagated toward the center of the line.

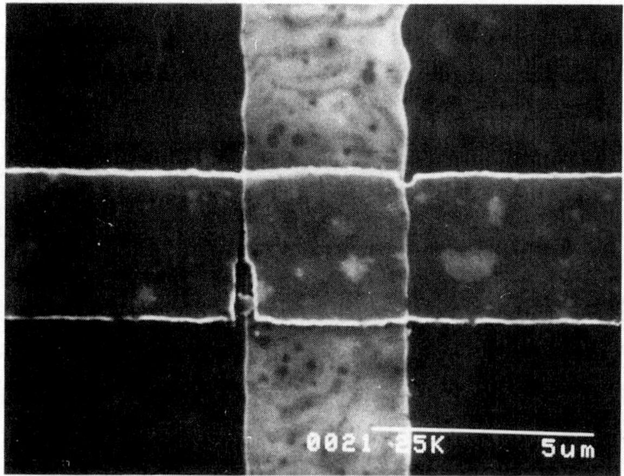


(*a*)

**Figure 10-10**   (*a*) Photograph showing voids in a 0.25-μm Al–6%Cu line after 3000 cycles from 25°C to 125°C with a triangular wave form for a period of 40 minutes. (*Continued*)

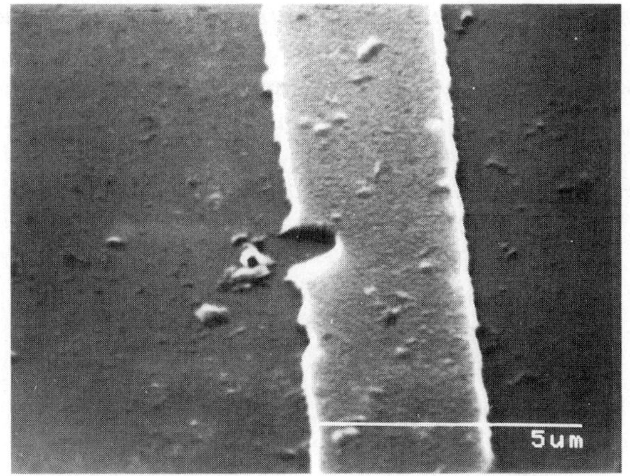

(*b*)

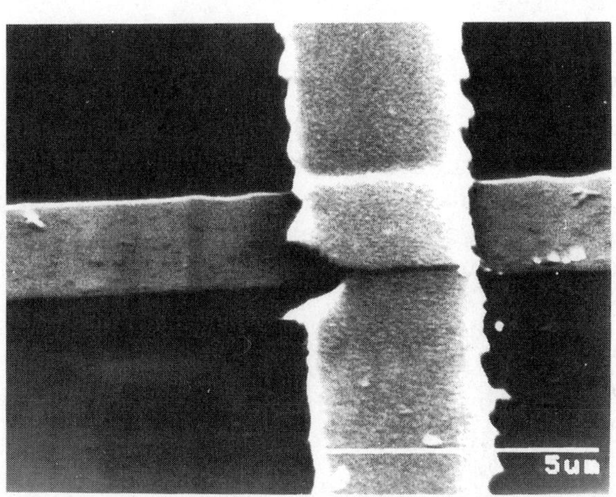

(*c*)

**Figure 10-10** (*continued*)   (*b*) Photograph showing voids in a 0.25-μm Al–3%Cu line after 3000 cycles from 50°C to 150°C with a triangular wave form for a period of 20 minutes. (*c*) Photograph showing voids in a 0.25-μm Al–3%Cu after 1400 cycles from 65°C to 150°C; the period is 30 minutes with 10-minute hold time at 150°C.

***Test III***   Another 0.25-μm Al–3%Cu sample was thermally cycled from 65°C to 150°C. The period for this test was 30 minutes, but a 10-minute hold time was applied at 150°C to study hold time effects. After 1400 cycles, voids appeared (Fig. 10-10*c*) at only half the number of cycles to failure as compared to the sample cycled with no hold time.

### 10.5.3  Comparison of Experiment and Theoretical Prediction

The above experimental study of Whitman and Chung[21] under thermal fatigue shows that the failure mechanisms for interconnects are the same under isothermal aging and thermal fatigue. Under a cyclic temperature loading, the void growth rate per cycle $da/dN$ over one cycle can be obtained from Eq. (10-33) as

$$\frac{da}{dt} = \frac{f(a/c)^3}{[2\gamma_s \sin(\psi/2)]^3} \frac{\Omega\delta_s\gamma_s}{k} \int_0^{\Delta t} \frac{\sigma_0(T)^3}{T(t)} D_{0s} e^{-Q_s/kT(t)} \, dt \qquad (10\text{-}37)$$

for the case where the grain boundary diffusion is much faster than surface diffusion; and from Eq. (10-34) as

$$\frac{da}{dt} = \left(\frac{5D_{0b}\delta_b f(a/c)}{4(c-a)D_{0s}\delta_s\gamma_s \sin(\psi/2)}\right)^{3/2} \frac{\Omega\delta_s\gamma_s}{k}$$

$$\times \int_0^{\Delta t} \frac{\sigma_0(T)^{3/2}}{T(t)} D_{0s} e^{-(3Q_b - Q_s)/2kT(t)} \, dt \qquad (10\text{-}38)$$

for the case where the surface diffusion is much faster than the grain boundary diffusion. In Eqs. (10-37) and (10-38), the temperature $T$ is a function of time $t$ during each thermal cycle. If a triangular waveform of temperature variation from $T_i$ to $T_f = T_i + \Delta T$ is used, then Eq. (10-37) and Eq. (10-38) become

$$\frac{da}{dN} = \frac{f(a/c)^3}{[2\gamma_s \sin(\psi/2)]^3} \frac{\Omega\delta_s\gamma_s}{k} \frac{D_{0s}\Delta t M^3}{\Delta T} \int_{T_i}^{T_f} \frac{(T_0 - T)^3}{T} e^{-Q_s/kT} \, dT \qquad (10\text{-}39)$$

and

$$\frac{da}{dN} = \left(\frac{5MD_{0b}\delta_b f(a/c)}{4(c-a)D_{0s}\delta_s\gamma_s \sin(\psi/2)}\right)^{3/2} \frac{\Omega D_{0s}\delta_s\gamma_s}{k} \frac{\Delta t}{\Delta T}$$

$$\times \int_{T_i}^{T_f} \frac{(T_0 - T)^{3/2}}{T} e^{-(3Q_b - Q_s)/2kT} \, dT \qquad (10\text{-}40)$$

respectively. In deriving Eqs. (10-39) and (10-40) the linear stress–temperature relation (10-35) has been used. Although there are some uncertain factors in Eq. (10-39) or Eq. (10-40), such as $D_{0s}$ and $\gamma_s$, only $T_0$, $Q_s$, and $Q_b$ will be needed if relative lifetime $N_f$ is compared. $T_0 \approx 225°C$ has been used by many authors and, as before, $Q_b = 0.95$ eV and $Q_s$ is estimated to be about 0.82 eV for Al–Cu.

Let us now compare test I with test II. All material and geometric parameters before the integrals in Eqs. (10-39) or (10-40) are identical except for the period $\Delta t$, and can be eliminated when ratios are taken. Using the values of $T_0 = 225°C$, $Q_s = 0.82$ eV, and $Q_b = 0.95$ eV for thin Al–Cu or Al–Si films, one can obtain that $N_f(\text{II})/N_f(\text{I}) = 0.9$ from Eq. (10-39) and $N_f(\text{II})/N_f(\text{I}) = 0.5$ from Eq. (10-40), where $N_f(\text{I})$ and $N_f(\text{II})$ are the number of cycles to failure for test I and test II, respectively. Since only relative lives are compared, failure of interconnects is defined when the voids have comparable size. The predicted result of $N_f(\text{II})/N_f(\text{I}) = 0.9$ for cracklike voids from Eq. (10-39) agrees very well with experimental values ranging from 0.7 to 1.

The third test was performed with a ramp time of 10 minutes and a hold time of 10 minutes. From Eq. (10-37) one obtains the ratio of the number of cycles to failure with hold $N_{f,h}$ to that with no hold $N_{f,nh}$ as

$$\frac{N_{f,h}}{N_{f,nh}} = \frac{2 \int_{T_i}^{T_f} (T_0 - T)^3 T^{-1} e^{-Q_s/kT} \, dT}{2 \int_{T_i}^{T_f} (T_0 - T)^3 T^{-1} e^{-Q_s/kT} \, dT + (T_f - T_i)(T_0 - T_f)^3 T_f^{-1} e^{-Q_s/kT_f}}$$

$$(10\text{-}41)$$

Substituting $T_0 = 225°C$, $Q_s = 0.82$ eV, and the temperatures $T_i$ and $T_f$ for test II without hold and for test III with hold into Eq. (10-41) yields $N_f(\text{III})/N_f(\text{II}) = 0.42$ for cracklike voids. This also agrees well with the experimental results of 0.33 to 0.5. If surface diffusion is dominant, then wedgelike voids are favorable. An equation similar to Eq. (10-41) can be obtained from Eq. (10-38) for this case. The calculated result for this case gives $N(\text{III})/N(\text{II}) = 0.31$ (see also Whitman and Chung[21]).

It should be noted that there are no adjustable parameters in the above calculations, other than the activation energies $Q_s$, $Q_b$, and the stress-free temperature $T_0$. Those numbers can be estimated reasonably well from experiment. Therefore, the present model can be considered to be predictive. Comparison with experimental results of Whitman and Chung[21] under thermal fatigue shows that Eq. (10-33) for the case of cracklike voids appears to be more appropriate than Eq. (10-34). This also agrees with the experimental observations of Mayumi et al.[17] and Hinode et al.[19] and the theoretical analyses of Chuang et al.[24] and Martinez and Nix.[25]

## 10.6 SUMMARY

It has been well known that high tensile residual stresses in Al interconnects can be developed due to the thermal mismatch between the interconnects and their surroundings. It is now recognized that the open-circuit failure of interconnects in microelectronic devices is a stress-assisted diffusion process. A failure model based on diffusion theory and fracture mechanics is constructed in the present study for a narrow interconnect line. Based on the present model, divergent reported experimental observations on failure rate dependence of aging temperature and line width can be explained well. Predicted effects of the temperature, stress, material properties, passivation, interlayer dielectric stress, interconnect thickness, and width on the failure rate based on this model agree well with observed experimental results.

The mechanism of voiding of patterned Al–Cu thin-film interconnects deposited on a silica substrate under thermal fatigue was studied experimentally by Whitman and Chung.[21] It was found that the failure mechanisms and modes of interconnects under isothermal aging and thermal fatigue are the same. A failure model for an interconnect under a cyclic temperature loading is given. Comparison with experimental results under thermal fatigue also shows excellent agreement. Based on the thermal fatigue results of Whitman and Chung[21] and Guo et al.,[22] it is shown that the model for cracklike voids is a good one for both isothermal aging and thermal fatigue. Since there is no adjustable parameter in the model, it can be considered to be predictive.

The model provides the basis for new interconnect alloy design and manufacturing processes. Based on the model, failure rates of interconnects can be suppressed by reducing the surface and grain boundary diffusion coefficients and the stress in the interconnects. The stress in interconnects can be changed by varying the stress state of the interlayer dielectric through different manufacturing processes as shown by Shin.[51]

# Appendix 10A
# Elastic Stress Analysis of a Crack-Like Void in a Passivated Interconnect

Suppose that the average stress $\sigma_y$ (Fig. 10-3) in an uncavitated passivated interconnect is $\sigma_0$; the stress distribution along the grain boundary in front of the void can be obtained by using the superposition scheme as shown in Fig. 10-11. Both central and edge cracks are shown in the same figure. The problem, as shown in Fig. 10-11c, is a sandwich strip containing an edge or a central crack normal to the interface under uniform internal normal stress $\sigma_0$. For a crack in an elastic solid as shown in Fig. 10-10, from elastic analysis the stress at $y = 0$ in front of the void tip can be written approximately as

$$\sigma(x) = \sigma_0 + \frac{K}{\sqrt{2\pi(x - a)}} \tag{10-A1}$$

where the stress intensity factor $K$ is a function of geometry and materials. Equation

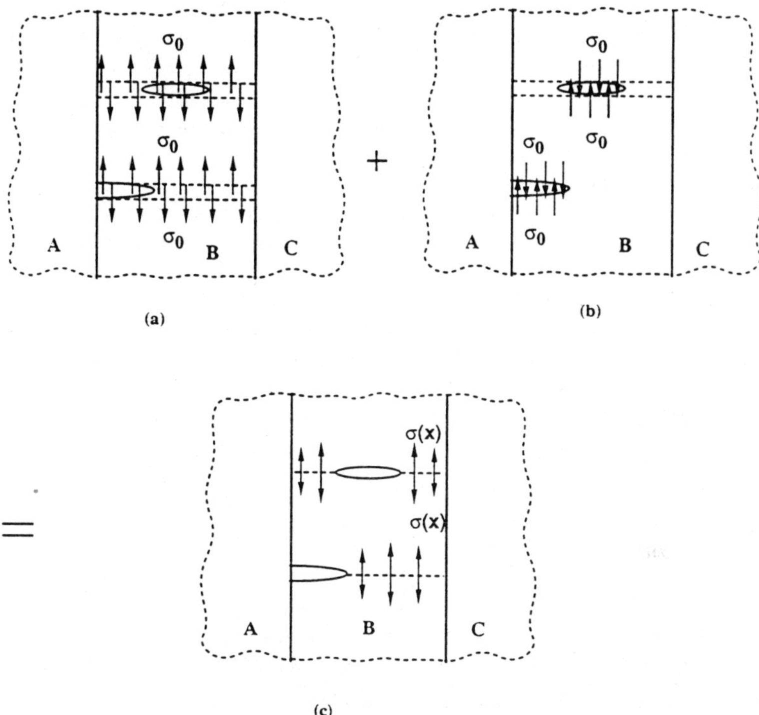



**Figure 10-11**   Stress analysis of a central crack or an edge crack in an interconnect line through superposition.

(10-A1) is the result of classical elastic analysis, and it is not valid at the void tip due to diffusion.[42,46]

A sandwich layer containing a central crack normal to the interface has been studied by Hilton and Sih.[59] From the reported numerical results of Murakami[60] and Hilton and Sih,[59] the stress intensity factor $K$ for $a/c$ between 0 and 0.9 can be approximated as

$$K = [1 - k(a/c)^2]\sigma_0\sqrt{\pi a} \tag{10-A2}$$

where $k$ depends on the ratios of the Young's moduli $E$ and Poisson ratios $v$ of the materials or the more general Dundurs' constants.[61] For identical materials, $k = 0$. For the present case, the Young's moduli of the passivation layer and the Si substrate are about twice the Young's modulus of the Al interconnect, while the Poisson ratios are nearly the same; hence, from the numerical results of Hilton and Sih[59] for a central crack in a sandwiched layer, one has $k \approx 0.25$.

From Figures 4, 7, and 9 of Lu and Erdogan[62] for two bounded elastic layers containing an edge crack, the stress intensity factor for an edge crack may be approximated as

$$K = 0.93[1 - k(a/c)^2]\sigma_0\sqrt{\pi a} \tag{10-A3}$$

where $k \approx 0.25$ is the same as that for a central crack, but $a$ and $c$ have different meanings for central and edge cracks.

Although the stress at the void tip is less than that given by Eq. (10-A1) because of stress relaxation near the tip, the analyses of Evans et al.[42] and Vitek[46] show that the classical elasticity solution remains valid away from the tip. Therefore, at $x = c$, Eq. (10-A1) gives

$$\sigma_c = \sigma_0 + \frac{K}{\sqrt{2\pi(c - a)}} \tag{10-A4}$$

From Eqs. (10-23) and (10-A1)–(10-A4) one can obtain Eqs. (10-30) and (10-31) in the text.


## NOMENCLATURE

| | |
|---|---|
| $a$ | Dimensions of cracks or voids |
| $B$ | A surface diffusion-related quantity defined by Eq. (10-28) |
| $c$ | Half of the interconnect width for a central void and the width of interconnect for a void at the edge |
| $d$ | Ratio of surface diffusivity to the grain boundary diffusivity as defined by Eq. (10-27) |
| $D_b$ | Grain boundary diffusion coefficient |
| $D_s$ | Surface diffusion coefficient |
| $D_{0b}$ | Grain boundary diffusion coefficient preexponential factor |
| $D_{0s}$ | Surface diffusion coefficient preexponential factor |
| $e$ | Energy density as defined through Eqs. (10-1) and (10-5) |
| $E$ | Total energy of a system, defined in Eq. (10-1) |
| $f$ | Nondimensional function of $a/c$ as defined by Eq. (10-31) |
| $J$ | Mass flux |
| $k$ | Boltzmann constant |
| $M$ | A material constant as used in Eq. (10-35) |
| $N$ | Number of cycles during cycling |
| $N_f$ | Number of cycles to failure |
| $Q_b$ | Grain diffusion activation energy |
| $Q_s$ | Surface diffusion activation energy |
| $t$ | Time |
| $T$ | Temperature |
| $T'$ | Stress-free temperature $T_0$ or $T_1$ as shown in Fig. 10-5 |
| $T_i, T_f$ | Temperature variation in thermal fatigue |
| $v_y$ | Rate of material deposition on the grains |
| $w$ | Width of interconnects |
| $x, y, z$ | Coordinates as defined in Fig. 10-3 |

| | |
|---|---|
| $\chi$ | Surface energy density |
| $\delta_b$ | Effective width of a grain boundary |
| $\delta_s$ | Effective width of a surface diffusion layer |
| $\Delta t$ | Ramp time for a triangular waveform |
| $\Delta T$ | Temperature difference between $T_f$ and $T_i$ |
| $\phi$ | Helmholtz free energy |
| $\gamma_s$ | Surface energy |
| $\gamma_b$ | Grain boundary energy |
| $\eta$ | Excess surface entropy |
| $\varphi$ | A Legendre transformation in Eq. (10-8) |
| $\lambda_i$ | Lagrange multipliers in Eq. (10-4) |
| $\mu_i, \mu$ | Chemical potentials |
| $\mu_0$ | Chemical potential of the stress-free state |
| $v$ | Poisson ratio |
| $\theta$ | A Lagrange multiplier |
| $\rho_0$ | Total number of lattice sites per unit volume |
| $\rho_i$ | Densities of independent constituent components |
| $\sigma$ | Stress normal to the grain boundary |
| $\sigma_0$ | Average stress in an interconnect as if there were no voids |
| $\sigma_a$ | Stress at the void tip |
| $\sigma_a'$ | Stress derivative with respect to $x$ at the void tip |
| $\Omega$ | Volume of an atom |
| $\psi$ | Void tip angle defined by Eq. (10-26) |
| $B_{ij}$ | Diffusion coefficients |
| $D_{ijkl}$ | Defined by Eq. (10-13) |
| $\varepsilon_{ij}$ | Strain tensor |
| $\varepsilon_{ij}^c$ | Strain produced by composition change |
| $S_{ijkl}$ | Compliance tensor |
| $\sigma_{ij}$ | Stress tensor |
| $\eta_{ij}$ | Defined by Eq. (10-13) |

## REFERENCES


1. Sinha, A. K., "Interconnects and Contacts for VLSI Applications," *Materials Research Society Symposium Proceedings*, **74**, 1986, pp. 735–391.
2. Diesburg, D. E., and E. D. Castel, "Refractory Metals in Submicron IC Architecture," *Journal of Metals*, **41**, 1989, pp. 23–26.
3. Jones, Jr., R. E., "Line Width Dependence of Stresses in Aluminum Interconnect," *Proc. 1987 IEEE International Reliability Physics Symposium*, 1987, pp. 9–14.
4. Jones, Jr., R. E., and M. L. Basehore, "Stress Analysis of Encapsulated Fine-Line Aluminum Interconnect," *Applied Physics Letters*, **50**, 1987, pp. 725–727.
5. Yost, F. G., A. D. Romig, Jr., and R. J. Bourcier, "Stress Driven Diffusive Voiding of Aluminum Conductor Lines: A Model for Time Dependent Failure," Sandia National Laboratories Report SAND88-0946, Albuquerque, NM, 1988.
6. Niwa, H., H. Yagi, H. Tsuchikawa, and M. Kato, "Stress Distribution in an

Aluminum Interconnect of Very Large Scale Integration," *Journal of Applied Physics*, **68**, 1990, pp. 328–333.

7. Korhonen, M. A., R. D. Black, and C.-Y. Li, "Stress Relaxation of Passivated Aluminum Line Metallizations on Silicon Substrates," *Journal of Applied Physics*, **69**, 1991, pp. 1748–1755.

8. Shute, C. J., "Study of Aluminum-Copper Thin Film Properties Via X-Ray Diffraction," Ph.D. Thesis, Northwestern University, Evanston, IL, 1991.

9. Flinn, P. A., and C. Chiang, "X-Ray Diffraction Determination of the Effect of Various Passivations on Stress in Metal Films and Patterned Lines," *Journal of Applied Physics*, **67**, 1990, pp. 2927–2937.

10. Korhonen, M. A., C. A. Paszkiet, R. D. Black, and C.-Y. Li, "Stress Relaxation of Continuous Film and Narrow Line Metallizations of Aluminum on Silicon Substrates," *Scripta Metallurgica et Materialia*, **24**, 1990, pp. 2297–2302.

11. Nix, W. D., "Mechanical Properties of Thin Films," *Metallurgica Transactions*, **20A**, 1989, pp. 2217–2245.

12. Robock, P. V., and L. T. Nguyen, "Plastic Packaging," *Microelectronics Packaging Handbook*, R. R. Tummala and E. J. Rymaszewski, eds., Van Nostrand Reinhold, New York, 1989, pp. 523–672.

13. Guo, Q., and Keer, L. M., "A Crack at the Interface of Two Power-law Hardening Materials—Antiplane Case," *Journal of the Mechanics and Physics of Solids*, **38**, 1990, pp. 183–194.

14. Guo, Q., and Keer, L. M., "A Crack at the Interface Between an Elastic-Perfectly Plastic Solid and a Rigid Substrate," *Journal of the Mechanics and Physics of Solids*, **38**, 1990, pp. 1649–1654.

15. Keer, L. M., and Q. Guo, "Stress Analysis for Symmetrically Loaded Bonded Layers," *International Journal of Fracture*, **43**, 1990, pp. 69–81.

16. Arzt, E., and W. D. Nix, "A Model for the Effect of Line Width and Mechanical Strength on Electromigration Failure of Interconnects with 'Near-Bamboo' Grain Structures," *Journal of Materials Research*, **6**, 1991, pp. 731–736.

17. Mayumi, S., T. Umemoto, M. Shishino, H. Nanatsue, S. Ueda, and M. Inoue, "The Effect of Cu Addition to Al–Si Interconnects on Stress Induced Open-Circuit Failure," *Proc. 1987 IEEE International Reliability Physics Symposium*, 1987, pp. 15–21.

18. Curry, J., G. Fitzgibbon, Y. Guan, R. Muollo, G. Nelson, and A. Thomas, "New Failure Mechanisms in Sputtered Aluminum–Silicon Films," *Proc. 1984 IEEE International Reliability Physics Symposium*, 1984, pp. 6–8.

19. Hinode, K., N. Owada, T. Nishida, and K. Mukai, "Stress-Induced Grain Boundary Fractures in Al–Si Interconnects," *Journal of Vacuum Science and Technology*, **B5**, 1987, pp. 518–522.

20. McPherson, J. W., and C. F. Dunn, "A Model for Stress-Induced Metal Notching and Voiding in Very Large-Scale-Integrated Al–Si(1%) Metallization," *Journal of Vacuum Science and Technology*, **B5**, 1987, pp. 1321–1325.

21. Whitman, C. S., and Y.-W. Chung, "Thermomechanically Induced Voiding of Al–Cu Thin Films," to be published in *Journal of Vacuum Science and Technology*, **A9**, 1991, pp. 2516–2522.

22. Guo, Q., C. S. Whitman, L. M. Keer, and Y.-W. Chung, "A Stress Induced

Diffusion Model for Failure of Interconnects in Microelectronic Devices," *Journal of Applied Physics*, **69**, 1991, pp. 7572–7580.

23. Turner, T., and K. Wendel, "The Influence of Stress on Aluminum Conductor Life," *Proc. 1985 IEEE International Reliability Physics Symposium*, 1985, pp. 142–174.

24. Chuang, T. J., K. I. Kagawa, J. R. Rice, and L. B. Sills, "Nonequilibrium Models for Diffusive Cavitation of Grain Interfaces," *Acta Metallurgica*, **27**, 1979, pp. 265–284.

25. Martinez, L., and W. D. Nix, "A Numerical Study of Cavity Controlled by Coupled Surface and Grain Boundary Diffusion," *Metallurgica Transactions*, **13A**, 1982, pp. 427–437.

26. Groothuis, S. K., and W. H. Schroen, "Stress Analysis of Encapsulated Fine-Line Aluminum Interconnect," *Proc. 1987 IEEE International Reliability Physics Symposium*, 1987, pp. 1–8.

27. Koyama, H., Y. Mashiko, and T. Nishioka, "Suppression of Stress Induced Aluminum Void Formation," *Proc. 1986 IEEE International Reliability Physics Symposium*, 1986, pp. 24–29.

28. Li, C. Y., R. D. Black, and W. R. LaFontaine, "Analysis of Thermal Stress-Induced Grain Boundary Cavitation and Notching in Narrow Al–Si Metallizations," *Applied Physics Letters*, **53**, 1988, pp. 31–33.

29. Sullivan, T. D., "Thermal Dependence of Voiding in Narrow Aluminum Microelectronic Interconnects," *Applied Physics Letters*, **55**, 1989, pp. 2399–2401.

30. Yost, F. G., "Voiding Due to Thermal Stress in Narrow Conductor Lines," *Scripta Metallurgica*, **23**, 1989, pp. 1323–1328.

31. Kato, M., H. Niwa, H. Yagi, and H. Tsuchikawa, "Diffusion Relaxation and Void Growth in an Aluminum Interconnect of Very Large Scale Integration," *Journal of Applied Physics*, **68**, 1990, pp. 334–338.

32. Johnson, W. C., and J. I. D. Alexander, "Interfacial Conditions for Thermomechanical Equilibrium in Two-Phase Crystals," *Journal of Applied Physics*, **59**, 1986, pp. 2735–2746.

33. Larche, F. C., and J. W. Cahn, "A Linear Theory of Thermochemical Equilibrium of Solids Under Stress," *Acta Metallurgica*, **21**, 1973, pp. 1051–1063.

34. Voorhees, P. W., and W. C. Johnson, "The Thermodynamics of a Coherent Interface," *Journal of Chemical Physics*, **90**, 1989, pp. 2793–2801.

35. Larche, F. C., and J. W. Cahn, "The Interactions of Composition and Stress in Crystalline Solids," *Acta Metallurgica*, **33**, 1985, pp. 331–357.

36. Li, J. C. M., R. A. Oriani, and L. S. Darken, "The Thermodynamics of Stressed Solids," *Zeitschrift für Physikalische Chemie Neue Folge*, **49**, 1966, pp. 271–290.

37. Larche, F. C., and J. W. Cahn, "Thermomechanical Equilibrium of Multiphase Solids Under Stress," *Acta Metallurgica*, **26**, 1978, pp. 1579–1589.

38. Herring, C., "Diffusional Viscosity of a Polycrystalline Solid," *Journal of Applied Physics*, **21**, 1950, pp. 437–445.

39. Coble, R. L., "A Model for Boundary Diffusion Controlled Creep in Polycrystalline Materials," *Journal of Applied Physics*, **34**, 1963, pp. 1679–1682.

40. Weertman, J., "Theory of High Temperature Intercrystalline Fracture Under Static or Fatigue Loads, With or Without Damage," *Metallurgica Transactions*, **5**, 1974, pp. 1743–1751.

41. Raj, R., and M. F. Ashby, "Intergranular Facture at Elevated Temperature," *Acta Metallurgica*, **23**, 1975, pp. 653–666.

42. Evans, A. G., J. R. Rice, and J. P. Hirth, "Suppression of Cavity Formation in Ceramics: Prospects for Superplasticity," *Journal of the American Ceramic Society*, **63**, 1980, pp. 368–375.

43. Rice, J. R., "Constraints on the Diffusive Cavitation of Isolated Grain Boundary Facets in Creeping Polycrystals," *Acta Metallurgica*, **29**, 1980, pp. 675–681.

44. Needleman, A., and J. R. Rice, "Plastic Creep Flow Effects in the Diffusive Cavitation of Grain Boundaries," *Acta Metallurgica*, **28**, 1980, pp. 1315–1332.

45. Chuang, T. J., and J. R. Rice, "The Shape of Intergranular Creep Cracks Growing by Surface Diffusion," *Acta Metallurgica*, **21**, 1973, pp. 1625–1628.

46. Vitek, V., "A Theory of Diffusion Controlled Intergranular Creep Crack Growth," *Acta Metallurgica*, **26**, 1978, pp. 1345–1356.

47. Schreiber, H.-U., and B. Grabe, "Electromigration Measuring Techniques for Grain Boundary Diffusion Activation Energy in Aluminum," *Solid-State Electronics*, **24**, 1981, pp. 1135–1146.

48. Bagnoli, P. E., A. Diligenti, B. Neri, and S. Ciucci, "Noise Measurements in Thin-Film Interconnects: A Nondestructive Technique to Characterize Electromigration," *Journal of Applied Physics*, **63**, 1988, pp. 1448–1451.

49. Berry, B. S., "Anelastic Relaxation and Diffusion in Thin-Layer Materials," *Diffusion Phenomena in Thin Films and Microelectronic Materials*, D. Gupta and P. S. Ho, eds., Noyes Publications, NJ, 1988, pp. 73–145.

50. Gjostein, N. A., *Diffusion*, American Society for Metals, Metals Park, OH, 1973, pp. 241–274.

51. Shin, H., *Effect of ILD Stress and Additional Annealing Treatments on the Stress Notching and Voiding*, Technical Report #130, Motorola Inc., 1990.

52. Gardner, D. S., and P. A. Flinn, "Mechanical Stress as a Function of Temperature in Aluminum Films," *IEEE Transactions on Electron Devices*, **35**, 1988, pp. 2160–2169.

53. Shute, C. J., J. B. Cohen, and D. A. Jeannotte, "Residual Stress Analysis of Al Alloy Thin Films by X-Ray Diffraction as a Function of Film Thickness," *MRS Society Proceedings*, **130**, 1989, pp. 29–34.

54. Kuan, T. S., and M. Murakami, "Low Temperature Strain Behavior of Pb Thin Films on a Substrate," *Metallurgica Transactions*, **13A**, 1982, pp. 383–391.

55. Doerner, M. F., Ph.D. Dissertation, Stanford University, Stanford, CA, 1987.

56. Doerner, M. F., D. S. Gardner, and W. D. Nix, "Plastic Properties of Thin Films on Substrates as Measured by Submicron Indentation Hardness and Substrate Curvature Techniques," *Journal of Materials Research*, **1**, 1986, pp. 845–851.

57. Freund, L. B., "The Stability of a Dislocation Threading a Strained Layer on a Substrate," *Journal of Applied Mechanics*, **54**, pp. 553–557.

58. Whitman, C. S., Y.-W. Chung, and D. A. Jeannotte, "Thermomechanical Induced Voiding of Al–Cu Thin Films," *Morris E. Fine Symposium*, P. K. Liaw, J. R. Weertman, H. L. Marcus, and J. S. Santer, eds., TMS, 1991, pp. 451–458.

59. Hilton, P. D., and G. C. Sih, "A Laminate Composite With a Crack Normal to the Interfaces," *International Journal of Solids and Structures*, **7**, 1971, pp. 913–930.

60. Murakami, Y., "Stress Intensity Factor Handbook," Vol. 1, Pergamon Press, Oxford, 1987, pp. 499–595.

61. Dundurs, J., "Edge-Bonded Dissimilar Orthogonal Elastic Wedges Under Normal and Shear Loading," *Journal of Applied Mechanics*, **36**, 1969, pp. 650–652.

62. Lu, M. C., and F. Erdogan, "Stress Intensity Factors in Two Bonded Elastic Layers Containing Cracks Perpendicular to and on the Interface—II. Solutions and Results," *Engineering Fracture Mechanics*, **28**, 1983, pp. 507–528.

# 11

# Thermal Stress and Stress-Induced Voiding in Passivated Narrow Line Metallizations on Ceramic Substrates

*M. A. Korhonen, P. Børgesen, and Che-Yu Li*

## 11.1 INTRODUCTION

Thermal stresses in metallizations are of current concern in the microelectronics industry. Stress-induced void and hillock formation are the main causes of interconnect failure before service.[1-4] Recently there has been a growing concern that stress-induced voids in the interconnect lines, formed during fabrication, may enhance subsequent electromigration damage during the use of the microchips.[5,6] For definitiveness we address pure aluminum metallizations on oxidized silicon substrates. The general principles are applicable to other metal hard substrate systems as well, most notably to aluminum and copper alloys on various ceramic substrates. The emphasis of the treatment is on the formation of tensile thermal stresses in the metallization during cooldown from elevated temperatures, and on the relaxation of them by void growth. However, much of the treatment is also applicable to hillock formation under compressive thermal stresses.

There are two general methods for the measurement of stresses in thin films:

1. Wafer curvature techniques, based on bending of the substrate due to the stresses in the metallizations.
2. X-ray diffraction techniques, based on directly determining the interatomic distances.

Both techniques of stress measurement are outlined in Section 11.2.

Thermal stresses in metallizations on ceramic substrates are caused by the different thermal expansion coefficients of the metal and the substrate. In the following we suppose that the substrate is much thicker than the metallization, hence there will be only negligible stresses on the substrate side. The temperature change $\Delta T$ induces a thermal strain of

$$\varepsilon_{11}^T = \varepsilon_{22}^T = \varepsilon_{33}^T = \Delta\alpha\,\Delta T \tag{11-1}$$

in three mutually perpendicular directions, see Fig. 11-1. Often there is an order of magnitude difference in thermal expansion $\Delta\alpha$ between the metallization and the substrate: for aluminum we have $\alpha_{Al} = 23 \times 10^{-6}/K$ while for silicon $\alpha_{Si} = 3 \times 10^{-6}/K$. The thermal stress in a continuous thin film is found by multiplying the thermal strain by the biaxial modulus $E/(1 - v)$,

$$\sigma = -\frac{E\,\Delta\alpha\,\Delta T}{1 - v} \tag{11-2}$$

The estimation of the triaxial thermal stresses that arise in passivated line metallizations turns out to be a tedious task in the general case. Triaxial thermal stresses can be calculated numerically by finite element methods (FEM)[7,8] or by analytic means.[9-11] From the point of view of physical understanding of the effects of different parameters on stress formation and relaxation, analytic means appear to offer an advantage. In the particular case we are going to study—long lines of metallization (Fig. 11-1)—the strains or stresses do not depend on the axial coordinate, which makes

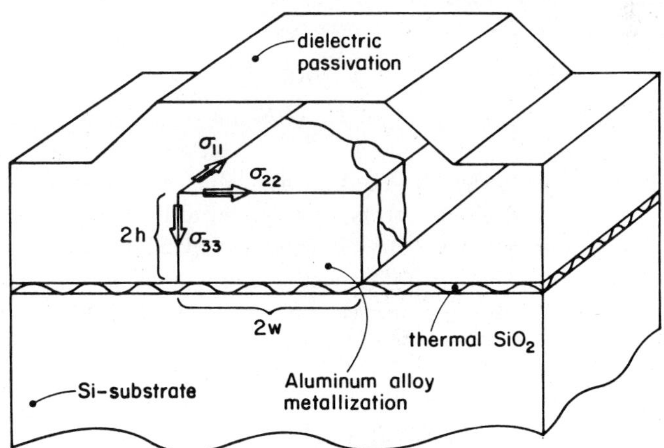


**Figure 11-1**   Schematic of a passivated interconnect line on silicon substrate, displaying the principal stresses.

analytic calculations more tractable. In Section 11.3 we estimate the thermal stresses in confined line metallizations based on the Eshelby theory of inclusions.[12]

The Eshelby theory of inclusions provides a powerful technique for dealing with all free strains, including the strains due to plastic or diffusional flow.[12] Hence the Eshelby theory of inclusions can serve as a general framework for analyzing stress relaxation also. During cooldown from elevated temperatures, high tensile stresses develop in passivated line metallizations. Stress redistribution by plastic flow, constrained by the ceramic substrate and passivation, creates stress concentrations that favor void nucleation in the metallization. However, in confined line metallizations plastic flow cannot relax the volume strain: hence stress relaxation will eventually proceed concomitantly with void growth, as dealt with in Section 11.4.

## 11.2 MEASUREMENT OF STRESSES IN METALLIZATIONS

There are two widely applied methods for measuring stresses in thin films residing on (relatively) thick substrates: (1) wafer curvature and (2) X-ray diffraction techniques.

### 11.2.1 Wafer Curvature Methods

The biaxial stress $\sigma$ in a continuous metallization bends the substrate, resulting in a curvature $R$. In the approximation that the substrate is much thicker than the metallization, $t_s \gg t_m$, we can write for the stress in the metallization

$$\sigma = \frac{E t_s^2}{6(1 - v)R t_m} \tag{11-3}$$

where $E$ and $v$ are, respectively, Young's modulus and Poisson's ratio of the substrate.[13,14] For a uniaxial stress, $(1 - v)$ should be replaced by 1. This formula is generally presumed to be valid independently for the different layers deposited on the substrate. However, some investigators have recently suggested that this requirement may not always be fulfilled;[15,16] however, with $t_m$ set to the total thickness of the film layers, Eq. (11-3) still yields the right average stress.

There are several ways to measure the curvature $R$. A simple (if not very accurate) way is to prepare a beam of length $L$ and to measure the deflection in one end, $\delta = L^2/2R$, by an optical microscope when the other end is held fixed. Higher accuracy can be achieved by using an optical lever: the location of a beam reflected from sample is recorded by a position-sensitive detector. By using a charge coupled device (CCD), for example, stress relaxation can be monitored essentially in real time. This technique has been recently reviewed in connection with microelectronic metallizations.[13,14]

Besides light, X-rays can be used to probe the curvature of single crystalline substrates. The method is based on the Bragg reflection condition, Eq. (11-4) below. The Bragg condition is very precise for single crystals of high perfection; the reflection condition can be met, or destroyed, by a rotation of a few thousandths of a degree of the crystal. Placing the wafer in a well-collimated beam and moving it sideways by $\Delta x$, and measuring the rotation $\Delta\theta$ to keep the crystal in reflection, gives the radius of curvature as $R = \Delta x/\Delta\theta$. This is usually done in a double crystal spectrometer (e.g., ref. 17) or by X-ray topographic techniques (e.g., ref. 18). Although the topographic methods are inferior to double crystal methods in terms of precision,[18] they offer potential for fast inspection of stresses in large areas of integrated circuits, especially when using synchrotron radiation.[19]

### 11.2.2 X-Ray Diffraction Stress Measurement

In the X-ray stress measurement the reflection angle $\theta$ yields the interatomic distance $d$ from Bragg's law:

$$2d \sin \theta = \lambda \tag{11-4}$$

where $\lambda$ is the wavelength of the radiation used. By differentiating Eq. (11-4) we find that the strain $\varepsilon = \Delta d/d$ can most conveniently be determined as

$$\varepsilon = \frac{\sin \theta^*}{\sin \theta} - 1 \tag{11-5}$$

where $\theta^*$ is the Bragg angle when the stress is zero.

From the stress–strain relation for cubic single crystals due to Moeller and Martin,[20] it follows that the strain in the direction specified by the polar angles $(\phi, \psi)$ (Fig. 11-2) can be given as

$$\varepsilon_{\phi\psi} = s_{12}(\sigma_{11} + \sigma_{22} + \sigma_{33}) + \frac{s_{44}}{2} [(\sigma_\phi - \sigma_{33}) \sin^2 \psi + \sigma_{33}]$$

$$+ s_0(A_1\sigma_{11} + A_2\sigma_{22} + A_3\sigma_{33}) \tag{11-6}$$

where $s_{11}$, $s_{12}$, and $s_{44}$ are the single-crystal elastic compliances (e.g., ref. 21),

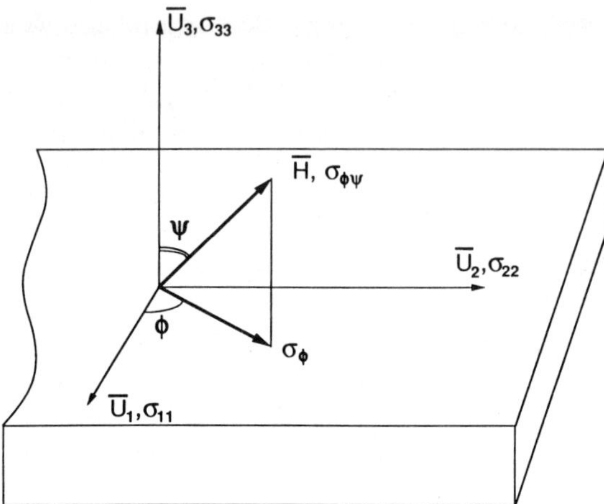


**Figure 11-2**   The coordinates used in the X-ray measurement of stresses: $U_1$, $U_2$, and $U_3$ denote the principal stress directions.

and the anisotropy factor is defined as $s_0 = s_{11} - s_{12} - s_{44}/2$. The anisotropy coefficients $A_1$, $A_2$, and $A_3$ will be dealt with in a moment. Within the approximation of isotropic elasticity the formula is applicable with $s_{12} = -v/E$, $s_{44} = 2(1 + v)/E$, and $s_0 = 0$. Based on a minimum of three readings, $\varepsilon_{1\psi}$, $\varepsilon_{2\psi}$, and $\varepsilon_{33}$, we can solve for the stresses $\sigma_{11}$, $\sigma_{22}$, and $\sigma_{33}$. In the case of the biaxial stress state, $\sigma_{33} = 0$ and $\sigma_{11} = \sigma_{22}$, one reading, $\varepsilon_{33}$, is enough to yield the stress state when the stress-free Bragg angle $\theta^*$, Eq. (11-5), is known. In this case, the stress can be recorded essentially in real time by using such fast position-sensitive detectors as CCDs or photodiode arrays.[22]

In the case of strong texture, the assumption $s_0 = 0$ may lead to grossly inaccurate stress estimates. However, the main problem coming from the texture may be due not to the anisotropy of the elastic properties but to the fact that the textured metallizations yield X-ray reflections only in a few discrete directions. It appears advisable to measure the strains by complying with the conditions dictated by the texture, i.e., to determine the strains along the directions **H** corresponding to the strong reflections.[23] Generally, the texture in the metallizations on rigid substrates may be described as the fiber texture where one prominent lattice direction, $U_3$ in the crystallographic indices, is oriented perpendicular to the substrate, so that there is a rotational symmetry around the $U_3$ axis, Fig. 11-2. When we now select the reflection vector to be used, **H**, the geometry is completely defined: the principal stress directions $U_1$ and $U_2$, Fig. 11-2, are given in the crystallographic indices as

$$\mathbf{U}_1 \cos\phi + \mathbf{U}_2 \sin\phi = \frac{\mathbf{H} - \mathbf{U}_3 \cos\psi}{\sin\psi}, \qquad \mathbf{U}_2 = \mathbf{U}_3 \times \mathbf{U}_1 \qquad (11\text{-}7)$$

Based on the single-crystal formula by Moeller and Martin,[20] it can readily be shown that the anisotropy factors can be represented as the square sum

$$A_i = U_{ij}^2 H_j^2 \tag{11-8}$$

where $U_{ij}$ denotes the $j$th component of the vector $\mathbf{U}_i$. When the type of the fiber texture, $\mathbf{U}_3$, is known, the anisotropy factors can be evaluated from Eq. (11-8) for the given reflection $\mathbf{H}$. It is remarkable that for [111]-type reflections all cubic crystals appear isotropic-like: the third term on the right of Eq. (11-6) reduces to $s_0(\sigma_{11} + \sigma_{22} + \sigma_{33})/3$, which is independent of the direction of the measurement. Korhonen et al. have given a formula corresponding to Eq. (11-6) for [111]-fiber texture in the case of a general reflection $\mathbf{H}$.[11] Further examples for [111]-textured films are provided by Flinn and coworkers.[14,24]

## 11.3 ESTIMATION OF THERMAL STRESSES IN PASSIVATED LINE METALLIZATIONS

We consider long passivated line metallizations lying on a thick silicon substrate, Fig. 11-1. The elastic properties of the passivation material are supposed to be close enough to the values of silicon. In practice, it turns out that the most important property, as far as the thermal stresses are concerned, of the passivation layer is its coefficient of thermal expansion. Because for the common ceramic passivation materials used, $SiN_x$, $SiO_2$, and phosphosilicate glass, the coefficient of thermal expansion is much smaller than for aluminum or copper, we can assume to a fair approximation that the lines in question lie in a silicon matrix, as shown schematically in Fig. 11-3.

### 11.3.1 Eshelby Theory of Inclusions

In this section we use the tensor notation for the stresses and strains in order to comply with the notations introduced by Eshelby.[12] We consider a metal line ("inclusion") lying in an infinite elastic medium ("matrix"); the inclusion suffers uniform free strains $\varepsilon_{ij}^T$, due to temperature change, for example. By a free strain $\varepsilon_{ij}^T$ we mean that in the absence of constraining surroundings, the sample would suffer the shape change $\varepsilon_{ij}^T$, Fig. 11-4. In case of no constraint, no elastic strains (or stresses) arise in the inclusion, $\varepsilon_{ij} = 0$. However, often the free shape or size change of material bodies is constrained by a surrounding medium which itself becomes

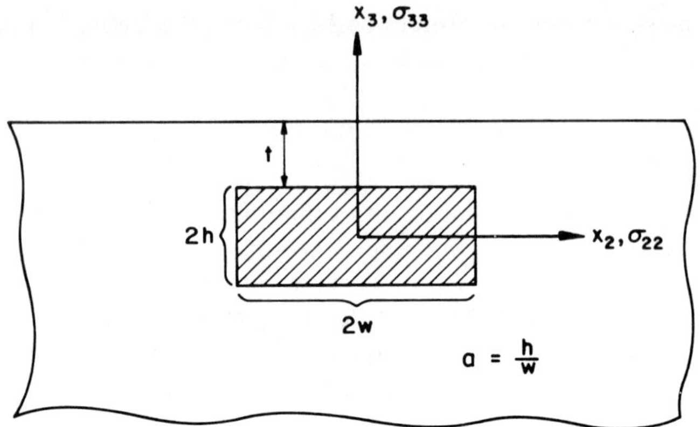



**Figure 11-3**   Modeling of the interconnect line: a cross-section of the aluminum metallization embedded in the silicon matrix.

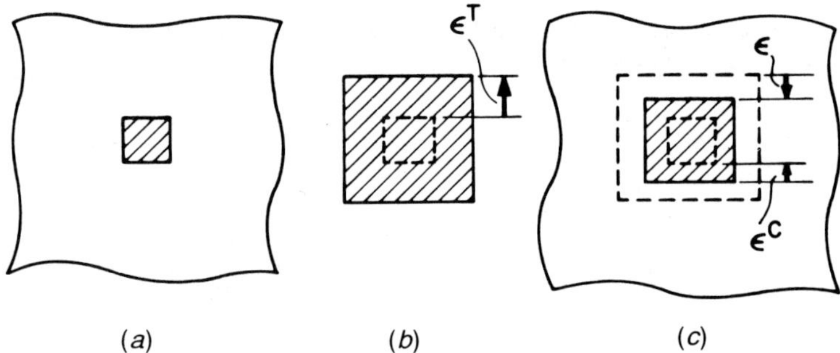



**Figure 11-4**   The inclusion is visualized as being removed from the matrix, suffering the free strain, and then being brought back: (a) the original size of the inclusion and the hole in the matrix; (b) the free strain $\varepsilon^T$ in the inclusion; (c) the resulting elastic stress in the inclusion, $\varepsilon$, and in the matrix, $\varepsilon^C$.

strained:

$$\varepsilon_{ij}^C = \varepsilon_{ij}^T + \varepsilon_{ij} \tag{11-9}$$

see Fig. 11-4. Note that $\varepsilon_{ij}^T$ and $\varepsilon_{ij}^C$ are counted from the zero-stress state of the constraining medium, while $\varepsilon_{ij}$ is counted from the zero-stress state of the inclusion. The elastic stress in the inclusion is then given as

$$\sigma_{ij} = c_{ijkl}\varepsilon_{kl} = c_{ijkl}(\varepsilon_{kl}^C - \varepsilon_{kl}^T) \tag{11-10}$$

where $c$ is the elastic stiffness tensor.

Eshelby showed that the constrained strains $\varepsilon_{ij}^C$ are related to the free strains $\varepsilon_{ij}^T$ as

$$\varepsilon_{ij}^C = S_{ijkl}\varepsilon_{kl}^T \qquad (11\text{-}11)$$

where $S$ is the Eshelby tensor.[12,25] Equivalently, this result can be written directly between the stresses and free strains as

$$\varepsilon_{ij}^T = T_{ijkl}\sigma_{kl} \qquad (11\text{-}12)$$

where $\mathbf{T}^{-1} = \mathbf{s}(\mathbf{S} - \mathbf{I})^{-1}$, and $\mathbf{I}$ denotes the unitary matrix and $\mathbf{s}$ the compliance matrix, $\mathbf{s} = \mathbf{c}^{-1}$.[11] If the free strains are known the stress can directly be found from the inverse of Eq. (11-12). The reason we prefer Eq. (11-12) to its inverse lies in its suitability to the forthcoming treatment of plastic flow, void growth, and the heterogeneity of elastic properties.

The coefficients $S_{ijkl}$ can be given in terms of a harmonic and a biharmonic potential function, $\Phi$ and $\Psi$.[12,25] It follows from the definitions of Eshelby[12] that

$$S_{ijkl} = \frac{1}{8\pi(1-v)}[\Psi_{ijkl} - 2v\Phi_{ij}\delta_{kl} - (1-v)(\delta_{ik}\Phi_{jl} + \delta_{il}\Phi_{jk} + \delta_{jk}\Phi_{il} + \delta_{ji}\Phi_{ik})]$$

$$(11\text{-}13)$$

where the indices in the potentials $\Phi$ and $\Psi$ denote the derivatives with respect to the coordinates of the principal axes (Fig. 11-1). We emphasize that Eq. (11-13) is quite general, applicable to an inclusion of arbitrary shape. In evaluating the derivatives of the potential functions it is helpful to apply the connecting formulas

$$\nabla^2\Phi = -4\pi, \qquad \nabla^2\Psi = 2\Phi \qquad (11\text{-}14)$$

which follow from the basic properties of harmonic functions.[12,25]


### 11.3.2 Reduction to Two-Dimensional Problem*

In two dimensions (2D) the problem of finding the Eshelby tensor $S$ is greatly simplified, and analytic solutions are possible for elliptic and rectangular cross sections. In long lines of metallization embedded in an elastic matrix, the stress and strain do not depend on the axial coordinate: hence the problem is reducible to 2D. We immediately see that all stresses and strains with indices $(i, j) = (1, 2)$ or $(1, 3)$ must be zero. For the dilational strain

---

* Reduction to 2D was suggested by W. T. Chen, 1991.

in the axial direction we can write

$$\varepsilon_{11} = \frac{\sigma_{11}}{E} - \frac{v}{E}(\sigma_{22} + \sigma_{33}) \tag{11-15}$$

Because the free strain in the axial direction, $\varepsilon_{11}^T$, does not cause any shift of the metallization with respect of the underlying substrate, $\varepsilon_{11}^C = 0$, the axial strain in the metallization is simply $\varepsilon_{11} = -\varepsilon_{11}^T$. We see that if $\sigma_{22}$ and $\sigma_{33}$ are known based on a plane problem, $\sigma_{11}$ can immediately be found from Eq. (11-15). However, the problem is not really two-dimensional, because $\varepsilon_{11}^T$ affects stresses $\sigma_{22}$ and $\sigma_{33}$. Consider the case when $\varepsilon_{11}^T = -\varepsilon_{11}$ is the only free strain different from zero. Obviously, $\varepsilon_{11}$ gives rise to Poisson contractions, $\varepsilon_{22} = \varepsilon_{33} = v\varepsilon_{11}^T$. Hence we can reduce the problem to 2D by using the apparent free strains,

$$\begin{aligned}
\varepsilon_{22}^T &= \varepsilon_{22}^T(\text{real}) + v\varepsilon_{11}^T(\text{real}) \\
\varepsilon_{33}^T &= \varepsilon_{33}^T(\text{real}) + v\varepsilon_{11}^T(\text{real})
\end{aligned} \tag{11-16}$$

in all our 2D-calculations.


**Elliptic Cross-section**

The potential functions for the elliptic cross-section can be deduced from the original works of Eshelby;[12] the results for the interior of the inclusion are

$$\Phi = \frac{-2\pi}{1+a}(ax_2^2 + x_3^2), \qquad \Psi_{12} = \frac{-4\pi ax_2x_3}{(1+a)^2} \tag{11-17}$$

where $a = h/w$ is the aspect ratio. It is readily seen by differentiation of Eqs. (11-17) that all coefficients $S_{ijkl}$ (Eq. (11-13)), become constant: consequently the stresses and strains in an elliptic inclusion are uniform. Generally, this applies for any ellipsoidal inclusion in an infinite, elastic matrix, as pointed out by Eshelby.[12] Based on Eqs. (11-12), (11-13), and (11-17) we can solve the stress state in the elliptic inclusion in terms of the free strains:

$$\sigma_{22} = \frac{-E}{2(1-v^2)(1+a)^2}[(2+a)\varepsilon_{22}^T + a\varepsilon_{33}^T]$$

$$\sigma_{33} = \frac{-E}{2(1-v^2)(1+a)^2}[a\varepsilon_{22}^T + a(2a+1)\varepsilon_{33}^T] \tag{11-18}$$

$$\sigma_{23} = \frac{-E}{(1-v^2)(1+a)^2}[a\varepsilon_{12}^T]$$

In the case of thermal free strains, $\varepsilon_{12}^T = 0$ and $\varepsilon_{22}^T = \varepsilon_{33}^T = (1 + v)\,\Delta\alpha\,\Delta T$, Eqs. (11-16).

## Rectangular Cross-section

The potential functions for the rectangular cross-section are

$$\Phi = -\int_{-w}^{w}\int_{-h}^{h} \ln(r^2)\,dx_2'\,dx_3', \qquad \Psi_{12} = -\left|_{-w}^{w}\right.\left|_{-h}^{h}\right. r^2 \ln r \quad (11\text{-}19)$$

where $r$ is the distance between the point of observation $(x_2, x_3)$ and the point of integration $(x_2', x_3')$. The functions have to be evaluated over the rectangle in Fig. 11-3, i.e., between the limits $x_2' = -w$ to $w$ and $x_3' = -h$ to $h$. The present 2D-potential functions can be considered as a particular case of the expressions for a cuboidal inclusion given by Lee and Johnson.[26] For the case of pure dilatation, $\varepsilon_{22}^T = \varepsilon_{33}^T$, we find the particularly simple expressions

$$\sigma_{22} = -\frac{E\varepsilon_{22}^T}{2\pi(1 - v^2)}\left|_{-w}^{w}\right.\left|_{-h}^{h}\right. \arctan\!\left(\frac{x_2 - x_2'}{x_3 - x_3'}\right)$$

$$\sigma_{33} = -\frac{E\varepsilon_{22}^T}{2\pi(1 - v^2)}\left|_{-w}^{w}\right.\left|_{-h}^{h}\right. \arctan\!\left(\frac{x_3 - x_3'}{x_2 - x_2'}\right) \qquad (11\text{-}20)$$

$$\sigma_{23} = \frac{E\varepsilon_{22}^T}{2\pi(1 - v^2)}\left|_{-w}^{w}\right.\left|_{-h}^{h}\right. \ln[(x_2 - x_2')^2 + (x_3 - x_3')^2]$$

from Eqs. (11-12), (11-13), and (11-19). Following the same procedure we could find the corresponding simple but lengthy expressions for general free strains. Equations (11-20) are very similar in form to the expressions given previously by Hu[9] except that our $\sigma_{22}$ replaces his $-\sigma_{33}$ and vice versa. The difference comes from the fact that Hu's formulas apply for the outside of the inclusion, while Eqs. (11-20) give both the inside and outside stresses.

In comparison with Eqs. (11-18), the most notable features of Eqs. (11-20) are that the stresses depend on the position $(x_2, x_3)$, and that the purely dilatational strains give rise to the shear stress. Figure 11-5 displays the stresses for an aluminum line of square cross-section, along the line $x_2 = 0$ and along the diagonal $x_3 = x_2$ for the case where the line is embedded in a ceramic matrix whose elastic constants are similar to aluminum. It can be noted that the stress in the width direction, $\sigma_{22}$, decreases towards the line edge along the line $x_2 = 0$, while the thickness stress $\sigma_{33}$ increases. Most importantly, the shear stress is zero along the axes $x_2 = 0$ and $x_3 = 0$, and

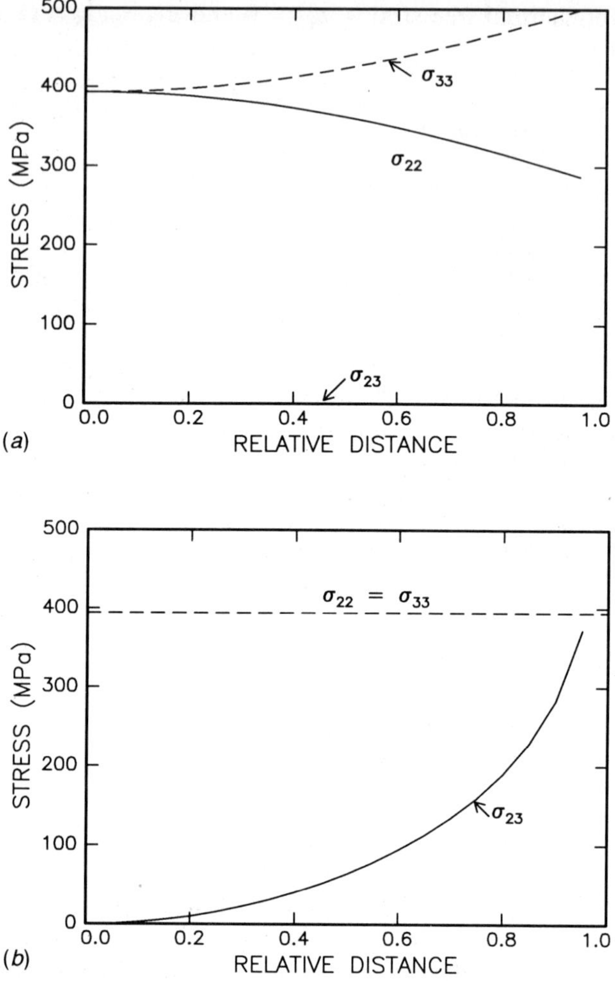


**Figure 11-5**  Stresses in a line of square cross-section, from the center of the line (a) in the $x_2$-direction, (b) in the diagonal direction, $x_3 = x_2$. The average stresses are the same as in a line of circular cross-section.

reaches a local maximum along the diagonal $x_3 = x_2$. At the line edge the shear stress approaches infinity. The average stresses over the whole cross-section are the same as the stress values at the origin. The average stresses for the square cross-section are also the same as the uniform stresses for the circular cross-section, Eqs. (11-18) with $a = h/w = 1$. It is remarkable that, quite generally, the average stresses of a rectangular cross-section agree closely with stresses for the corresponding elliptic cross-section, as can readily be shown from Eqs. (11-18) and (11-20). In what follows we will take advantage of this fact. From Eqs. (11-16), (11-18), and (11-20) it can easily be shown that in the case of thermal stresses the axial stress is simply $\sigma_{11} = \sigma_{22} + \sigma_{33}$ when the metal line and the surrounding ceramic matrix have the same elastic constants; the result is true for any cross-section.

### 11.3.3 The Problem of the Heterogeneous Inclusion

The case when the inclusion and the matrix have different elastic properties is known as the problem of the heterogeneous inclusion. To solve this problem Eshelby devised the method of the "equivalent" inclusion.[12,25] Consider the stress state of an inclusion that has suffered free strains of $\varepsilon_{ij}^T$, Eq. (11-12). Assume now that the elastic coefficients $c_{ijkl}$, and the corresponding compliances, $s_{ijkl}$, of the inclusion are transformed to $c_{ijkl}^*$ and $s_{ijkl}^*$. In order for this transformation to leave the displacements and tractions felt by the matrix unchanged, the stresses $\sigma_{ij}$ and the strains $\varepsilon_{ij}^C$ must remain the same:

$$\sigma_{ij} = c_{ijkl}(\varepsilon_{kl}^C - \varepsilon_{kl}^T) = c_{ijkl}^*(\varepsilon_{kl}^C - \varepsilon_{kl}^{T*}) \tag{11-21}$$

It turns out that the free strains $\varepsilon_{ij}^{T*}$ of the transformed inclusion can always be selected so that Eq. (11-21) is fulfilled. Combining Eqs. (11-12) and (11-22) we obtain the connection between the stress state and free strains of the heterogeneous inclusion:[11]

$$\varepsilon_{ij}^{T*} = (T_{ijkl} + s_{ijkl} - s_{ijkl}^*)\sigma_{kl} \tag{11-22}$$

This relation shows that the free strains change proportionally to the change in the elastic compliances. Given the thermal strains $\varepsilon_{ij}^{T*}$ of a heterogeneous inclusion, the stress is readily found from the inverse of Eq. (11-22).

The application of Eq. (11-22) for lines of elliptic cross-section poses no difficulty because the transformation $T$ for the corresponding homogeneous problem is known. However, for lines of rectangular cross-section a difficulty arises. Even if the free strains $\varepsilon_{ij}^{T*}$ were uniform over the cross-section, it follows from Eq. (11-22) that $\varepsilon_{ij}^T$ for the corresponding homogeneous problem would depend on the position $(x_2, x_3)$, while the Eshelby formulation, Eq. (11-11), is valid only for uniform free strains. We take recourse to the observation that the average stresses over a line of rectangular cross-section are about the same as the uniform stresses in an elliptic interconnect line on the same aspect ratio. Accordingly, when we are interested only in the average stresses in a line, we can use Eqs. (11-18) from the outset to a fair approximation. When, in addition, we wish to know the stress distribution, we can select as a reasonable estimate the stress distribution for the homogeneous rectangular line that produces the already calculated average stresses.

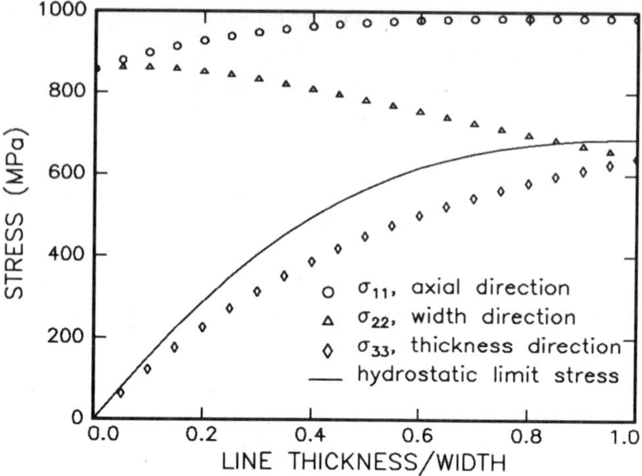


**Figure 11-6**   The average thermal stresses in an aluminum line embedded in a silicon matrix, after heat treatment at 400°C. The solid line gives the hydrostatic stress after complete relaxation of shear stresses.

Figure 11-6 gives the thermal stresses at room temperature for aluminum lines embedded in silicon, after a heat treatment at 400°C, as based on Eqs. (11-12), (11-18), and (11-22). The most important consequence of the silicon confinement is the overall high stress state. Comparison with the square cross-section of Fig. 11-5 shows that $\sigma_{33}$ is about 60%, and $\sigma_{11}$ about 20% larger because of silicon confinement (instead of a ceramic, like $SiO_2$, with compliances similar to aluminum). This gives a large contribution to the hydrostatic component of the stress state, which turns out to be of paramount importance in stress-induced voiding, to be dealt with in Section 11.4. Figure 11-6 also displays the hydrostatic stress state (solid line) that results when plastic shear redistributes the stresses.[10]

### 11.3.4 The Effects of Finite Passivation

In actual interconnects the passivation thickness is finite, of course. The mathematical model for this case would be that of an inclusion in a semispace, with a finite passivation thickness $t$, Fig. 11-3. Hu has given the stresses outside a rectangular line embedded in a semispace, arising due to thermal expansion of the line.[9] The total stress inside an inclusion in the semispace can be considered to be formed of two parts:[27] (1) the inside stress as based on the assumption of an infinite space (Eqs. (11-20)), and (2) the outside, correction stress field due to an imaginary "mirror" inclusion. We combine Hu's "mirror" field[9] with the inside stress field (Eqs. (11-20)), to

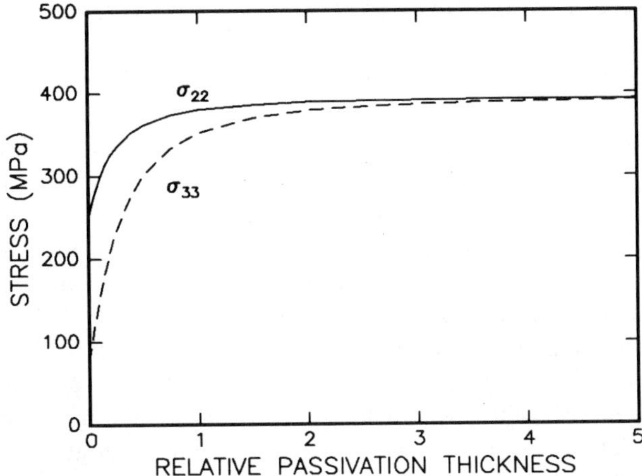


**Figure 11-7**    The effect of finite passivation layer thickness on the stresses in the line center, for the case of a square cross-section. The passivation thickness $t$ is given in the units of line height $2h$ (Fig. 11-3).

produce the total stress state in the metal line. Figure 11-7 gives an example of the average stresses for the square cross-section dealt with in Fig. 11-5, at different depths from the surface. As can be seen, the stresses correspond to the infinite case once the passivation thickness $t$, Fig. 11-3, is of the order of the line height $h$. Hence the solution for the infinite matrix gives a reasonable upper bound estimate for the thermal stresses in passivated, narrow line metallizations on hard substrates. The stress estimates based on Eqs. (11-18) and (11-20) should be particularly accurate for the interconnects in the lower layers of a multilayered structure.

## 11.4 STRESS RELAXATION AND VOID FORMATION

In confined line metallizations, plastic deformation can only redistribute the stresses. Figure 11-6 displays the thermal stresses in the case where no stress redistribution has taken place during cooldown from the heat treatment. Because plastic deformation conserves the volume, the sum of the free strains

$$\varepsilon^T = \varepsilon_{11}^T + \varepsilon_{22}^T + \varepsilon_{33}^T \tag{11-23}$$

or the volume free strain remains constant at $3\Delta\alpha \, \Delta T$. The hydrostatic stress state, shown in Fig. 11-6 (solid line), is the result of complete stress redistribution by plastic or diffusional flow,[10] in the hypothetical case when shear stresses are reduced to zero.

### 11.4.1 Stresses After Redistribution

The stresses in confined metallizations may never reach the elastic stress levels, $\sigma_{11}$, $\sigma_{22}$, and $\sigma_{33}$, indicated in Fig. 11-6. However, the actual stress level after cooldown must lie between the estimated elastic stresses and the hydrostatic stress (solid line) in the absence of void formation. Experimentally, for a well-bonded 300-nm continuous aluminum film, where the grain size is of the same order as the film thickness, we measure a biaxial stress of about 350 MPa at room temperature immediately after cooldown.[28] According to the Tresca yield criterion we expect, and also measure,[28,29] about the same axial stress in 300-nm aluminum lines. For a 300-nm continuous aluminum film, bonded to a silicon substrate and passivated by a 300-nm layer of $SiN_x$, we observe a biaxial stress of about 500 MPa immediately after the cooldown;[30] evidently the strength increase from 350 MPa can be attributed to the additional constraint due to the passivation. On the other hand, for free-standing thin films of aluminum of similar grain size, Steinwall and Johnson[31] report yield strengths of the order of 120–180 MPa. Although these are large values as compared to the strengths encountered in bulk metal samples, they still remain much smaller than in the constrained case.

In addition to the above constraint effects, passivated narrow metal lines can stand even higher stresses without plastic yielding in the presence of a hydrostatic stress component. From Fig. 11-6 we see that the stress in the thickness direction increases with the aspect ratio, from 0 MPa at $h/w = 0$ up to about 650 MPa at $h/w = 1$. Suppose now that yielding in the triaxial case is limited by the same effective shear strength $\tau_y = 250$ MPa as observed in the case of passivated continuous films. When we allow for Eq. (11-23) and the Tresca law,

$$\tau_y = \frac{\sigma_{max} - \sigma_{min}}{2} \tag{11-24}$$

we can estimate (with Eqs. (11-18) and (11-22)) the highest stresses possible in the passivated, 300-nm thick lines after cooldown from 400°C.

For an aspect ratio $h/w = 0.25$ we estimate that, after yielding, the stress in the thickness direction is about 300 MPa, while the width and axial stresses are about 800 MPa. One day after cooldown from 400°C heat treatment, we measure lateral stresses of the order of 600 MPa in aluminum lines with $h/w = 0.25$.[30] It thus appears quite feasible that the maximum stresses immediately after cooldown may have been of the order of 800 MPa.

In the case of lines of square cross-section, $h/w = 1$, the Eshelby theory predicts elastic stresses of 640 MPa in the thickness and width directions of the lines while the maximum stress of 950 MPa is reached in the axial

**Table 11-1**    Maximum Stresses and Strain Energy Densities in 300-nm Thick Aluminum Films on Silicon, With and Without SiN$_x$ Passivation. Estimates for Bulk Aluminum and Free 300-nm Aluminum Film Are For Comparison

| Metal Type | Max. Stress (MPa) | Energy Density (MJ/m$^3$) | Note |
| --- | --- | --- | --- |
| Annealed bulk metal | 20 | 0.003 | Ref. 31 |
| Unpassivated free line | ~150 | 0.16 | Ref. 31 |
| Unpassivated bonded line | 350 | >0.9 | Ref. 28 |
| Unpassivated bonded film | 350 | 1.8 | Ref. 28 |
| Passivated film, $h/w = 0$ | 500 | 3.6 | Ref. 30 |
| Passivated line, $h/w = 0.25$ | 800 | 9.8 | Predicted |
| Passivated line, $h/w = 1$ | 1000 | 13.2 | Predicted |

direction, Fig. 11-6. Substitution of these values in Eq. (11-24) results in an effective shear stress of 155 MPa, which is much smaller than the observed shear strength of about 250 MPa in passivated continuous films of the same thickness.[30] Hence, in the passivated thin ($<300$ nm) aluminum lines of square cross-section we expect no large-scale yielding during cooldown from a heat treatment at 400°C. We suggest that the relatively small stresses recently reported for square cross-sections[32,33] are likely to result from voiding. We summarize our stress data in Table 11-1, where we also indicate the elastic energy densities expected, $(\sigma_1^2 + \sigma_2^2 + \sigma_3^2)/2E$.

### 11.4.2 Void Nucleation

Stress-induced grain boundary voids are generally observed in passivated aluminum-based metallizations after annealing treatments at 300–450°C. Voids are occasionally seen also in continuous films, both passivated and unpassivated, but practically never in unpassivated lines.[3,34] This suggests that the stresses and elastic energies in unpassivated lines lie below the critical level needed for void nucleation, Table 11-1, while in the continuous films they lie close enough to the critical level that adverse processing conditions may lead to void formation. We presume that high strain energy density, not easily relieved by plastic flow, is the decisive factor in void nucleation during cooldown. Contributing to this is the fact that plastic flow becomes increasingly difficult with stress redistribution: besides giving rise to local back stresses, plastic flow makes the overall stress state more hydrostatic.

Experimentally we observe that voids always form on grain boundaries,[35–37] which suggests grain boundary sliding (GBS) as the nucleation mechanism.

Grain boundary sliding is likely to accompany dislocation glide during cooldown in the case of constrained deformation, and provides the initial flaws from which the voids then grow to the thermodynamically stable size. It has recently been suggested that atomic-size flaws next to a highly constraining interface might grow plastically, in an unstable fashion, to a size of tens of times the original size.[38] The elastic stresses of Table 11-1 compare well with the high stress values implied in this study.

It appears that all stress-induced voids nucleate during cooldown.[36] Strong evidence on the significance of constraint can be seen in that the voids are likely to nucleate at line edges;[2,36,37] here also the shear stresses are largest, see Fig. 11-5. Further, a thicker passivation means stronger constraint, and results in increasingly triaxial stress state. Thicker passivations are, indeed shown to increase the number voids per unit volume up to the limit corresponding to an "infinite" passivation thickness,[36] see Fig. 11-7.

### 11.4.3 Stress Relaxation by Void Growth

Stress redistribution by plastic deformation or diffusional flow can lower the stresses only close to the hydrostatic level, Fig. 11-6, which can be quite high in the cases of practical interest, $h/w > 0.25$. Evidently, then, void growth remains the sole possibility for continued stress relaxation. Experimental observations suggest that void growth becomes the rate-determining factor within some hours after cooldown,[11,35] well before the stresses reach the hydrostatic stress level. In the following we are going to deal with three identified, distinct regimes of void growth:

1. Rapid initial growth during cooldown
2. Slower, exponentially decaying growth governed by grain boundary diffusion
3. Final logarithmic growth governed by dislocation creep in the bulk

First we must distinguish between the contributions of stress redistribution by plastic flow and stress relaxation due to void growth. A clear-cut differentiation can be based on the volume free strain $\varepsilon^T$, Eq. (11-23). Because stress redistribution by plastic flow conserves $\varepsilon^T$, initially $3\Delta\alpha\,\Delta T$, we have a direct measure for the relative void volume:

$$\varepsilon_v = -3\Delta\alpha\,\Delta T + \varepsilon^T \tag{11-25}$$

where the present volume strain, $\varepsilon^T$, is a linear function of the stresses through Eqs. (11-12) and (11-22). In the case of complete relaxation of stresses by void growth it is possible to recover the whole thermal strain as the void volume. By measuring the stresses, by X-ray diffraction, for example,[24,29,39] it is possible to calculate the void volume from Eq. (11-25) when the transformation tensor $T$, Eq. (11-12), is known.

## Rapid Initial Void Growth

We suppose with Herring that the flux of atoms from the void tip to the grain boundary is given by

$$J = \frac{D}{kT} \frac{\partial \sigma}{\partial x} \tag{11-26}$$

where $D$ is the effective grain boundary diffusion coefficient (normalized over the cross-section), $kT$ is the thermal energy, and $x$ measures the distance along the boundary.[40] Hull and Rimmer[41] were the first to show that in the steady state the derivative $\partial \sigma / \partial x$ is proportional to $(\sigma - \sigma_B)$, or to the average stress $\sigma$ normal to the grain boundary, minus the back stress $\sigma_B$ due to the capillarity effect (e.g., ref. 42). It is evident that this conclusion is valid also for the case when $\sigma$ is changing slowly with time. We write for the average volumetric strain rate

$$\dot{\varepsilon}_v = \frac{J\Omega}{l} = \frac{D\Omega}{kT} \frac{(\sigma - \sigma_B)}{lw} \tag{11-27}$$

where $w$ is the line width and $l$ is the distance between voids along the line.

During cooldown the stress $\sigma$ and the diffusivity $D$ are high, and the capillarity stress, $\sigma_B$, becomes small once the voids have grown large enough; hence void growth is initially very rapid. However, eventually the continuing atomic flux to grain boundaries creates the kind of back stress proposed by Dyson,[43] which contributes to $\sigma_B$ of Eq. (11-27). Jackson and Li extended Dyson's concept to thin films constrained by rigid substrates, and showed that grain boundary diffusion alone cannot relieve the stresses in the bulk.[44] Hence, constrained grain boundary thickening creates back stresses that eventually, when $\sigma = \sigma_B$, will inhibit further atomic flux from voids to grain boundaries.

By using the void growth law, Eq. (11-27), with the constitutive equations for plastic flow (e.g., ref. 45), it is possible to simulate stress relaxation and void growth during cooldown once the material parameters and cooling rates are known. We do not pursue this line here, however.

## Exponentially Decaying Void Growth

Within some hours after the cooldown the stress relaxation rate is likely to be governed by grain boundary void growth and plastic deformation exists no more as an independent mechanism. We limit ourselves to two cases of practical interest:

1. Columnar grain structure ($h/w < 0.3$) where the stresses in the axial and width direction, $\sigma_{11}$ and $\sigma_{22}$, are about equal,[11,24,39] and significantly larger than the thickness stress $\sigma_{33}$.[11,39] Because the mass transport occurs to grain boundaries lying perpendicular to the thin film, the stresses $\sigma_{11}$ and $\sigma_{22}$ relax at a much faster rate than the stress $\sigma_{33}$.[11]
2. Bamboo grain structure ($h/w \approx 1$) where the stresses $\sigma_{22}$ and $\sigma_{33}$ are about equal, and smaller than the axial stress $\sigma_{11}$. Now most grain boundaries lie perpendicular to the axial direction, hence the axial stress $\sigma_{11}$ relaxes at a much faster rate than $\sigma_{22}$ and $\sigma_{33}$, as verified recently by Moske et al.[46]

Stress relaxation of these particular lines can be described by one stress component, $\sigma$, that is normal to grain boundaries; hence Eq. (11-25) reduces to a particularly simple form:

$$\varepsilon_v = -3\Delta \alpha \Delta T - s_{\text{eff}} \sigma \tag{11-28}$$

where $s_{\text{eff}}$ is the effective compliance, of the order of $E$.

We assume that the back stress $\sigma_B$ builds up proportionally to the flux of atoms to grain boundaries, $\dot{\sigma}_B \propto j$. Combination of Eqs. (11-27) and (11-28), subject to this condition, now yields the time dependence of the relative void volume as

$$\varepsilon_v = \varepsilon_{v0} + s_{\text{eff}}(\sigma_0 - \sigma_B)[1 - \exp(-t/\tau)] \tag{11-29}$$

where $\tau \approx s_{\text{eff}} kTwl/D\Omega$ is a time constant, typically of the order of 100–1000 minutes for pure aluminum at room temperature. The initial values $\sigma_0$ and $\varepsilon_{v0}$ are the (hypothetical) back extrapolated stress and the void volume at zero time (immediately after taking the sample from the heat treatment), based on the assumption that all stress relaxation is governed by void growth. Actual void growth data on passivated, pure aluminum lines, described elsewhere in more detail,[47] provides a test for the validity of the proposed void growth model. Figure 11-8 shows that these data fit reasonably well to the predictions of Eq. (11-29). Figure 8-11 also displays the corresponding stress data.[37] It is remarkable that both stress and void volume data can be fitted with the same time constant $\tau$, which shows that the void volume is, indeed, linearly related to the stress as required by Eq. (11-28).

We find values of $\sigma_0$ of the order of 400–500 MPa, while the actual lateral

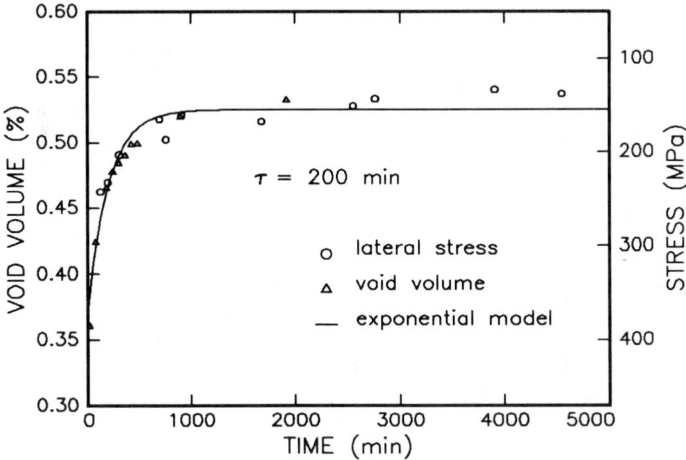


**Figure 11-8**   Experimentally observed void volume[47] and residual stress[37] as a function of time.

stresses during cooldown are likely to have been substantially larger, Table 11-1. The extrapolated void volume at zero time, about 0.35% in the present example, is indicative of fast initial void growth. The fact that the stresses during cooldown are actually higher than $\sigma_0$ and the diffusivity $D$ is larger at elevated temperatures qualitatively explains the observed high initial void volumes.

## Logarithmic Void Growth

Equation (11-29) predicts that relaxation stops once the stress reaches the back stress level $\sigma_B$. This prediction is valid only as far as grain boundary processes are concerned. Bulk processes can further disperse the atoms crowding the grain boundary regime. According to our experimental results, the back stress is, indeed, reduced with time.[11] Hence void growth continues, although at a much slower rate, toward the limiting value of $3\Delta\alpha\,\Delta T$.

Dislocation creep is likely to be the mechanism to reduce the back stress, and determine the void growth rate. Figure 11-9 shows the stress relaxation behavior for 300-nm thick metallizations of pure aluminum at room temperature from some hours up to a year. The stress of an unpassivated, continuous film (and line as well[28]) is found to decay logarithmically with time; Flinn and coworkers have published data covering shorter times.[13,14] The logarithmic relaxation behavior is believed to be governed by dislocation creep.[28,35] The medium-term (up to one month) relaxation behavior of the passivated line is seen to be relatively fast, and corresponds to the exponential

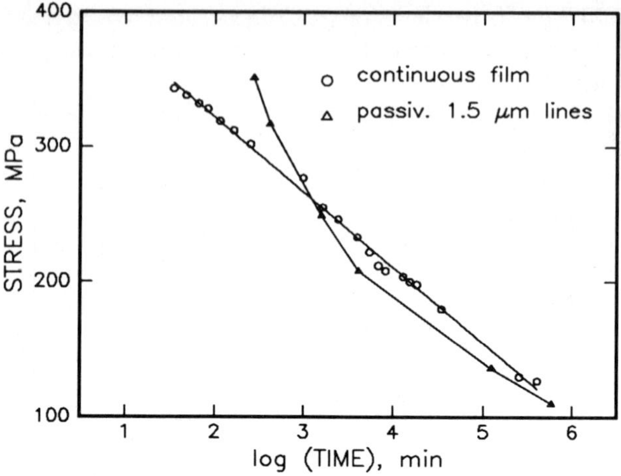


**Figure 11-9**  Experimentally observed relaxation of the lateral stress behavior for 300 nm thick aluminum metallizations.[35]

decay stage of Fig. 11-8. The long-term stress relaxation behaviors of passivated and unpassivated metallizations are found to eventually converge, Fig. 11-9. Hence, it appears that the long-term stress relaxation in the passivated lines is also governed by dislocation creep. If this is so, void growth also should proceed logarithmically with time. Indeed, the experimental results (although covering a very limited time span) give some indication of this behavior, Fig. 11-10.

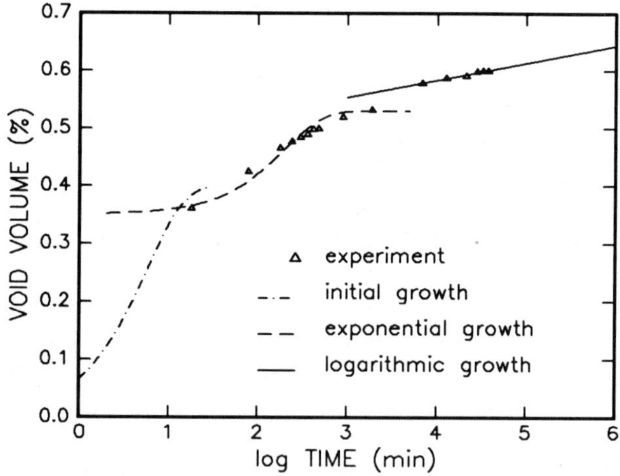


**Figure 11-10**  Schematic of the three stages of void growth and stress relaxation.

The fact that $\sigma_B$ can be quite large suggests that voids still have a large growth potential during matrix creep-controlled relaxation processes. However, the logarithmic stress decay guarantees that for any time span of practical interest (say, 10 years) the void volume at room temperature increases no more than about 20% of the value reached within some weeks after the heat treatment. Hence practically all line failures in pure aluminum due to stress voiding at room temperature should take place within a month or two after the fabrication. Subsequent line failures are likely to be connected with the use of the microchips. Electromigration can lead to void coalescence[5] as well as to a further increase in the void volume.[6] It appears that in most practical cases stress-induced voids act as a precursor to the eventual electromigration damage.[48]

In conclusion, in passivated aluminum lines stresses relax and voids grow rapidly initially, as shown schematically in Fig. 11-10. Later, all stress relaxation is connected to continuing void growth. From about one hour after the heat treatment to some weeks, relaxation is governed by the flux of atoms from voids to grain boundaries, and the void volume increases toward an apparent saturation value, Fig. 11-10. Finally, in the long run, dislocation processes take over to decrease the back stresses due to grain boundary thickening; now the stresses decay, and voids grow, logarithmically with time. Hence, the main body of interconnect failures due to stress-induced voiding at room temperature should occur within a month or two after fabrication.

## 11.5  SUMMARY

High tensile stresses arise in metallizations bonded to rigid substrates after excursions to elevated temperatures because of the differences in the thermal expansion coefficients. The most important methods of measurement of residual stresses in thin metallizations are briefly reviewed. The thermal stresses in passivated line metallizations can be estimated based on the Eshelby theory of inclusions. Analytic models are derived for lines of both elliptic and rectangular cross-section: the average stresses turn out to be about the same for the same aspect ratio. However, in rectangular lines the stresses vary over the cross-section; in particular, large shear stress concentrations arise at line edges. The high thermally induced stresses, coupled with severely constrained plastic flow, favor void nucleation at line edges. In passivated, narrow line metallizations on rigid substrates, plastic flow or diffusional creep can only redistribute the stresses because the volume change is constrained. For the relaxation of volume strain, void growth is required: hence stress relaxation and void growth are directly connected. In passivated pure aluminum metallizations there appear to be three distinct stages of void growth and stress relaxation. During cooldown and immediately thereafter at room temperature, voids grow and stresses relax rapidly due to stress-directed grain boundary diffusion of atoms from voids to grain boundaries. However, diffusion of atoms to grain boundaries soon becomes constrained by the back stress created by grain boundary thickening; void growth rate begins to decay exponentially. In the final stage, the logarithmically decaying void growth rate is determined by diffusion of atoms from grain boundaries to the bulk.

## ACKNOWLEDGMENTS

This work was supported in part by IBM Corporation, the Materials Science Center at Cornell University, the National Nanofabrication Facility at Cornell, and the Academy of Finland.

## REFERENCES


1. Turner, T., and K. Wendel, *Proc. Int. Reliability Physics Symposium 25*, IEEE, New York, 1985, p. 142.
2. Yue, J. T., W. P. Funsten, and R. V. Taylor, *Proc. Int. Reliability Physics Symposium 25*, IEEE, New York, 1985, p. 126.
3. Hinode, K., N. Owada, T. Nishida, and K. Mukai, *J. Vac. Sci. Technol.*, **B5**, 1987, p. 518.
4. Li, Che-Yu, R. D. Black, and W. R. LaFontaine, *Appl. Phys. Lett.*, **53**, 1988, p. 31.
5. Li, Che-Yu, P. Børgesen, and T. Sullivan, *Appl. Phys. Lett.*, **59**, 1991, p. 1464.
6. Børgesen, P., M. A. Korhonen, D. D. Brown, and Che-Yu Li, in *Stress-Induced Phenomena in Metallizations*, C.-Y. Li, P. Totta, and P. Ho, eds., AIP, Conference Proceedings 263, New York, 1992, p. 219.
7. Jones, R. E., *Proc. Int. Reliability Physics Symposium 25*, IEEE, New York, 1987, p. 1.
8. Sauter, A. I., and W. D. Nix, in *Thin Films: Stresses and Mechanical Properties I*, MRS Symp. Proc. 188, Pittsburgh, PA, 1990, p. 15.
9. Hu, S. M., *J. Appl. Phys.*, **66**, 1989, p. 2741.
10. Niwa, H., H. Yagi, H. Tsuchikawa, and M. Kato, *J. Appl. Phys.*, **68**, 1990, p. 328.
11. Korhonen, M. A., R. D. Black, and Che-Yu Li, *J. Appl. Phys.*, **69**, 1991, 1748.
12. Eshelby, J. D., *Proc. Roy. Soc. A*, **241**, 1957, p. 376; also *Proc. Roy. Soc. A*, **251**, 1959, p. 561.
13. Flinn, P. A., D. S. Gardner, and W. D. Nix, *IEEE Trans.*, **ED-34**, 1987, 689.
14. Flinn, P. A., in *Stress-Induced Phenomena in Metallizations*, C.-Y. Li, P. Totta, and P. Ho, eds., AIP, Conference Proceedings 263, New York, 1992, p. 73.
15. Chen, S. T., C. H. Yang, F. Faupel, and P. S. Ho, *J. Appl. Phys.*, **64**, 1988, p. 6690.
16. Moske, M. A., P. S. Ho, D. J. Mikaelsen, J. J. Cuomo, and R. Rosenberg, in *Electronic Packaging Materials Science V*, MRS Symp. Proc. 203, Pittsburgh, PA, 1991, p. 77.

17. Adamczewska, J., and T. Budzynski, *Thin Solid Films*, **113**, 1984, p. 271.
18. Hearn, E. W., *Adv. X-ray Anal.*, **20**, 1977, p. 273.
19. Simomaa, K., T. Tuomi, and J. Partanen, *Proc. European Workshop for Synchrotron Radiation Sources*, Aussois, France, 1991.
20. Möller, H., and G. Martin, *Mitt. Kaiser Wilhelm Inst. Eisenforsch. Düsseldorf*, **21**, 1939, p. 261.
21. Segmüller, A., and M. Murakami, *Treat. Mater. Sci. Technol.*, **27**, 1988, p. 1434.
22. Korhonen, M. A., V. K. Lindroos, and L. S. Suominen, *Adv. X-ray Anal.*, **32**, 1989, p. 407.
23. Korhonen, M. A., and C. A. Paszkiet, *Scripta Metall.*, **23**, 1989, p. 1449.
24. Flinn, P. A., and C. Chiang, *J. Appl. Phys.*, **67**, 1990, p. 2927.
25. Mura, T., *Micromechanics of Defects in Solids*, Martinus Nijhoff, The Hague, 1982.
26. Lee, J. K., and W. C. Johnson, *Scripta Metall.*, **11**, 1977, p. 477.
27. Aderogba, K., *Math. Proc. Camb. Phil. Soc.*, **90**, 1976, p. 555.
28. Korhonen, M. A., C. A. Paszkiet, R. D. Black, and Che-Yu Li, *Scripta Metall.*, **24**, 1990, p. 2297.
29. Korhonen, M. A., L. S. Suominen, and Che-Yu Li, in *Nondestructive Characterization of Materials IV*, C. O. Ruud, J. F. Bussiere, and R. E. Green, eds., Plenum Press, New York, 1991, p. 15.
30. Paszkiet, C. A., M. A. Korhonen, and Che-Yu Li, in *Mechanical Behavior of Materials and Structures in Microelectronics*, MRS Symp. Proc. 226, Pittsburgh, PA, 1991, p. 419.
31. Steinwall, J. E., and H. H. Johnson, in *Thin Films: Stresses and Mechanical Properties II*, MRS Symp. Proc. 188, Pittsburgh, PA, 1990, p. 177.
32. Hosoda, T., H. Niwa, H. Yagi, and H. Tsuchikawa, *Proc. Int. Reliability Physics Symposium 29*, IEEE, New York, 1991, p. 77.
33. Yagi, H., H. Niwa, T. Hosoda, M. Inoue, and T. Tsuchikawa, *Proc. Int. Workshop on Stress-Induced Phenomena in Metallizations*, Ithaca, New York, 1991.
34. Paszkiet, C. A., M. A. Korhonen, and Che-Yu Li, in *Thin Films: Stresses and Mechanical Properties I*, MRS Symp. Proc. 188, Pittsburgh, PA, 1990, p. 153.
35. Korhonen, M. A., C. A. Paszkiet, and Che-Yu Li, *J. Appl. Phys.*, **69**, 1991, p. 8083.
36. Korhonen, M. A., W. R. LaFontaine, P. Børgesen, and Che-Yu Li, *J. Appl. Phys.*, **70**, 1991, p. 6774.
37. Paszkiet, C. A., M. A. Korhonen, and Che-Yu Li, in *Electronic Packaging in Materials Science V*, MRS Symp. Proc. 203, Pittsburgh, PA, 1991, p. 381.
38. Tvergaard, V., *Acta Metall.*, **39**, 1991, p. 419.
39. Tezaki, A., T. Mineta, H. Egawa, and T. Noguchi, *Proc. Int. Reliability Physics Symposium 28*, IEEE, New York, 1990, p. 221.
40. Herring, C., in *Physics of Powder Metallurgy*, W. E. Kingston, ed., McGraw-Hill, New York, 1951, p. 143.
41. Hull, D., and D. E. Rimmer, *Phil. Mag.*, **4**, 1959, p. 673.
42. Riedel, H., *Fracture at High Temperatures*, Springer, Berlin, 1987.
43. Dyson, B. F., *Can. Met. Quart.*, **18**, 1979, p. 31.
44. Jackson, M. S., and Che-Yu Li, *Acta Metall.*, **30**, 1982, p. 1993.
45. Korhonen, M. A., S.-P. Hannula, and Che-Yu Li, in *Unified Constitutive*

*Equations for Creep and Plasticity*, A. K. Miller, ed., Elsevier Applied Science, Amsterdam, 1987, p. 89.

46. Moske, M. A., P. S. Ho, C. K. Hu, and S. M. Small, in *Stress-Induced Phenomena in Metallizations*, C.-Y. Li, P. Totta, and P. Ho, eds., AIP, Conference Proceedings 263, New York, 1992, p. 195.
47. Børgesen, P., J. K. Lee, M. A. Korhonen, and Che-Yu Li, in *Mechanical Behavior of Materials and Structures in Microelectronics*, MRS Symp. Proc. 226, Pittsburgh, PA, 1991, p. 407.
48. Korhonen, M. A., P. Børgesen, and Che-Yu Li, in *Thin Films: Stresses and Mechanical Properties III*, MRS Symp. Proc. 239, Pittsburgh, PA, 1992.

# 12

# Predicted Bow of Plastic Packages of Integrated Circuit (IC) Devices

*E. Suhir*

## 12.1 INTRODUCTION

Plastic packages play a major role in the further growth of the micro-electronic industry.[1] This is due to the low cost and the reliability of plastic packages, as well as to the excellent compatibility of plastic package designs with mass production techniques.[2] There are, however, several serious concerns in the rapid evolution of plastic packaging technology, and one of the most critical of them is elevated residual bow of plastic packages.

In this study we develop analytical stress models for the evaluation of thermally induced bows in the following two types of plastic packages: thin elongated packages with large chips, known as thin small-outline packages (TSOPs), and high-lead-count large square packages with relatively small chips, known as plastic quad flat packages (PQFPs). In the case of TSOP packages (Fig. 12-1) we consider the package warpage caused by the thermal contraction mismatch of the constituent materials: silicon chip, metal lead frame, and molding compound. The temperature change is assumed to be the same throughout the package. In the case of PQFP packages (Fig. 12-2) we evaluate the bow due to the temperature gradient (nonuniform distribution of temperature) in the through-thickness direction of the molded body. The models developed enable one to evaluate the effect of the package geometry and materials properties on its bow, and, more importantly, to design a plastic package with a sufficiently low residual warpage.

**Figure 12-1**   Thin small outline package (TSOP).

**Figure 12-2**   Plastic quad flat package (PQFP).

## 12.2 THIN PLASTIC PACKAGE

### 12.2.1 Basic Equations

From the standpoint of structural analysis, a thin plastic package can be treated as an elongated trimaterial composite plate manufactured at an elevated temperature and subsequently cooled down to the room or testing temperature. The cross-section of such a plate is shown schematically in Fig. 12-3.

Let this package be subjected to a change $\Delta t$ in temperature. The temperature-induced forces $T_i$, $i = 1, 2, 3, 4$, in the components and the

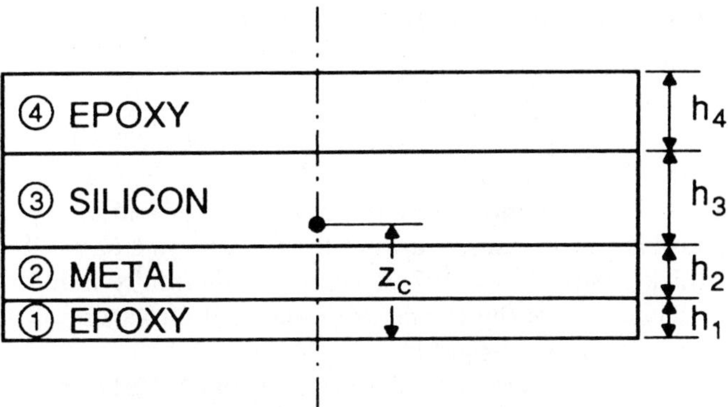


**Figure 12-3**   Transverse cross-section of a thin plastic package.

curvature $\kappa$ of the composite structure can be evaluated, assuming perfect adhesion, from the compatibility conditions for the interfacial strains,

$$-\alpha_1 \Delta t + \lambda_1 T_1 + \frac{h_1}{2}\kappa = -\alpha_2 \Delta t + \lambda_2 T_2 - \frac{h_2}{2}\kappa$$

$$-\alpha_2 \Delta t + \lambda_2 T_2 + \frac{h_2}{2}\kappa = -\alpha_3 \Delta t + \lambda_3 T_3 - \frac{h_3}{2}\kappa \qquad (12\text{-}1)$$

$$-\alpha_3 \Delta t + \lambda_3 T_3 + \frac{h_3}{2}\kappa = -\alpha_1 \Delta t + \lambda_4 T_4 - \frac{h_4}{2}\kappa$$

and the equations of equilibrium for the forces,

$$T_1 + T_2 + T_3 + T_4 = 0 \qquad (12\text{-}2)$$

and bending moments,

$$\left(\frac{h_1}{2} + h_2 + h_3 + \frac{h_4}{2}\right)T_1 + \left(\frac{h_2}{2} + h_3 + \frac{h_4}{2}\right)T_2 + \left(\frac{h_3}{2} + \frac{h_4}{2}\right)T_3 = EI\kappa \quad (12\text{-}3)$$

In these equations $\alpha_1$, $\alpha_2$, and $\alpha_3$ are the coefficients of thermal expansion for the molding compound, metal lead frame, and silicon chip, respectively;

$$\lambda_1 = \frac{1}{E_1^0 h_1}, \qquad \lambda_2 = \frac{1}{E_2^0 h_2}, \qquad \lambda_3 = \frac{1}{E_3^0 h_3}, \qquad \lambda_4 = \frac{1}{E_1^0 h_4} \qquad (12\text{-}4)$$

are the in-plane compliances of the material layers;

$$E_i^0 = \frac{E_i}{1 - v_i^2}, \qquad i = 1, 2, 3 \qquad (12\text{-}5)$$

are generalized Young's moduli of the materials; $E_i$ and $v_i$, $i = 1, 2, 3$, are Young's modulus and Poisson's ratio of the $i$th material; $h_i$, $i = 1, 2, 3, 4$, are the layers' thicknesses; $T_i$, $i = 1, 2, 3, 4$, are the thermally induced forces in the material's layers; $\kappa$ is the temperature-induced curvature of the molded body; and $EI$ is its flexural rigidity.

The first terms in either part of the conditions (12-1) are unrestricted (stress-free) thermal contractions. The second terms are the strains due to the thermally induced forces. The third terms are due to bending. Clearly, the bending strains have opposite signs on the convex and the concave sides of the given layer. It should be pointed out that the equations (12-1) reflect an assumption that the curing temperature of the chip-to-lead frame attachment material is close to the molding temperature of the entire package, so that the temperature change $\Delta t$ can be assumed the same throughout the molded body. Equation (12-2) simply states that, since no external forces act on the package, the thermally induced forces arising in the constituent materials must be self-equilibrated. As to the equilibrium condition for the moments, it can be formed with respect to any horizontal axis in the plane of the cross-section. In our analysis this equation has been formed with respect to the midplane of the fourth ($i = 4$) layer. This results in the Eq. (12-3).

Equations (12-1), (12-2), and (12-3) can be rewritten as

$$\lambda_1 T_1 - \lambda_2 T_2 + \beta_{12}\kappa = \alpha_{12}\,\Delta t$$

$$\lambda_2 T_2 - \lambda_3 T_3 + \beta_{23}\kappa = \alpha_{23}\,\Delta t$$

$$\lambda_3 T_3 - \lambda_4 T_4 + \beta_{34}\kappa = -\alpha_{13}\,\Delta t \qquad (12\text{-}6)$$

$$T_1 + T_2 + T_3 + T_4 = 0$$

$$\beta_{14} T_1 + \beta_{24} T_2 + \beta_{34} T_3 - EI\kappa = 0$$

where the following notation is used:

$$\alpha_{12} = \alpha_1 - \alpha_2, \qquad \alpha_{23} = \alpha_2 - \alpha_3, \qquad \alpha_{13} = \alpha_1 - \alpha_3$$

$$\beta_{12} = \frac{h_1 + h_2}{2}, \qquad \beta_{23} = \frac{h_2 + h_3}{2}, \qquad \beta_{34} = \frac{h_3 + h_4}{2} \qquad (12\text{-}7)$$

$$\beta_{14} = \beta_{12} + \beta_{23} + \beta_{34}, \qquad \beta_{24} = \beta_{23} + \beta_{34}$$

The five equations (12-6) enable one to determine the four thermally induced forces $T_i$, $i = 1, 2, 3, 4$, and the curvature $\kappa$ of the package.

### 12.2.2 Curvature

Solving the system (12-6) for the curvature, we obtain

$$\kappa = -\frac{\Delta t}{D}\left[\alpha_{12}\lambda_3(\beta_{12}\lambda_4 - \beta_{24}\lambda_1) + \alpha_{13}\lambda_2(\beta_{13}\lambda_4 - \beta_{34}\lambda_1) + \alpha_{23}\beta_{23}\lambda_1\lambda_4\right]$$

$$(12\text{-}8)$$

where

$$D = EI\lambda + \beta_{12}^2\lambda_3\lambda_4 + \beta_{13}^2\lambda_2\lambda_4 + \beta_{14}^2\lambda_2\lambda_3 + \beta_{23}^2\lambda_1\lambda_4 + \beta_{24}^2\lambda_1\lambda_3 + \beta_{34}^2\lambda_1\lambda_2$$

$$(12\text{-}9)$$

and the parameter $\lambda$ is expressed as

$$\lambda = \lambda_1\lambda_2\lambda_3 + \lambda_2\lambda_3\lambda_4 + \lambda_3\lambda_4\lambda_1 + \lambda_4\lambda_1\lambda_2 \qquad (12\text{-}10)$$

### 12.2.3 Maximum Bow

Within the length $2l_s$ of the chip–lead frame assembly (Fig. 12-4), the elastic curve of the molded body is a parabola. Beyond this length it is simply a straight line. Assuming small deflections, we proceed from the following equation for the deflection function $w(x)$ (see, for instance, ref. 3):

$$w''(x) = \kappa, \qquad 0 \leqslant x \leqslant l_s \qquad (2\text{-}11)$$

The first integration yields

$$w'(x) = \kappa x \qquad (12\text{-}12)$$

where the constant of integration is put equal to zero, since the curve $w(x)$ must be symmetric with respect to the origin. The next integration results in the equation

$$w(x) = \tfrac{1}{2}\kappa x^2 \qquad (12\text{-}13)$$

where the constant of integration is put equal to zero again, so that the maximum ordinate of the curve $w(x)$ occurs at the ends of the package.

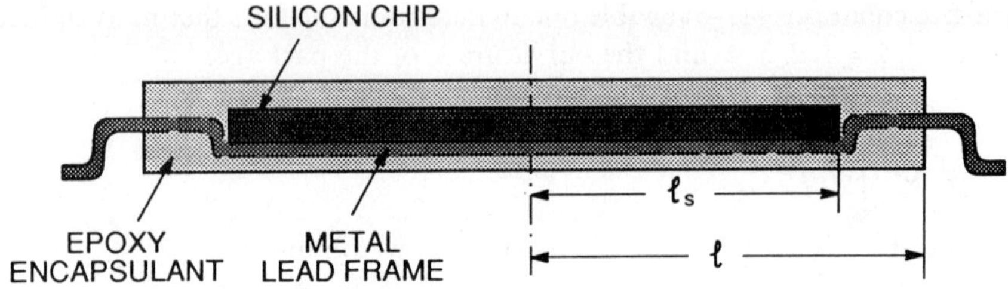


**Figure 12-4**   Longitudinal cross-section of a thin plastic package.

The deflection at $x = l_s$ is

$$w(l_s) = \tfrac{1}{2}\kappa l_s^2$$

Then the maximum deflection at the end $(x = l)$ is

$$w_0 = w(l) = \tfrac{1}{2}\kappa l_s^2 + w'(l_s)(l - l_s) = \kappa l l_s\left(1 - \frac{l_s}{2l}\right) \tag{12-14}$$

### 12.2.4 Zero Bow Condition

As is evident from (12-8), the induced curvature is zero if the following condition is fulfilled:

$$\alpha_{12}\lambda_3(\beta_{12}\lambda_4 - \beta_{24}\lambda_1) + \alpha_{13}\lambda_2(\beta_{13}\lambda_4 - \beta_{34}\lambda_1) + \alpha_{23}\beta_{23}\lambda_1\lambda_4 = 0 \tag{12-15}$$

This condition can be used, particularly, to determine whether the chip–lead frame assembly can be positioned within the molding compound in such a way that no residual bow occurs. Of course, such an optimal position of the assembly may turn out to be unrealistic if it results in an unacceptably thin epoxy layer on either side of the package.

Solving Eq. (12-15) for the thickness $h_1$, we obtain

$$h_1 = \frac{E_1^0\xi[\alpha_{12}\lambda_3(h + h_3) + \alpha_{13}\lambda_2(\xi + h_3)] - \alpha_{23}(h - \xi)}{2hE_1^0(\alpha_{12}\lambda_3 + \alpha_{13}\lambda_2)} \tag{12-16}$$

where

$$h = h_1 + h_2 + h_3 + h_4$$

is the total thickness of the package, and

$$\xi = h_1 + h_4 = h - h_2 - h_3 \qquad (12\text{-}17)$$

is the total thickness of the molding compound (on both sides of the chip).

Let us show, as an illustration, that in a bimaterial body, when the material $i = 2$ (metal lead frame) is removed, the only way to achieve zero curvature of the package is to position the component $i = 3$ (silicon chip) in the midplane of the molding compound. Indeed, when the thickness $h_2$ is negligibly small, the terms containing the compliance $\lambda_2$ become significantly larger than the other compliance terms, and the formula (12-16) yields

$$h_1 = \frac{\xi(\xi + h_3)}{2h} = \frac{(h_1 + h_4)h}{2h} = \frac{h_1 + h_4}{2}$$

so that $h_1 = h_4$.

Another means to eliminate the residual bow is to employ a molding compound whose coefficient $\alpha_1$ of thermal expansion is related to its generalized Young's modulus $E_1^0$ by the equation

$$\alpha_1 = \frac{\alpha_2\lambda_3(h_1\beta_{12} - h_4\beta_{24}) + \alpha_3\lambda_2(h_1\beta_{13} - h_4\beta_{34}) - \alpha_{23}\beta_{23}/E_1^0}{\lambda_3(h_1\beta_{12} - h_4\beta_{24}) + \lambda_2(h_1\beta_{13} - h_4\beta_{34})} \qquad (12\text{-}18)$$

This formula can be obtained from (12-15) by solving that equation for $\alpha_1$. It should be pointed out that application of high-expansion molding compounds for lower warpage inevitably results in higher thermally induced stresses in the package. This, however, is thought to be acceptable, as long as these stresses are still sufficiently low.

Obviously, both measures (i.e., increasing the ratio of the epoxy layer thicknesses above and below the chip–lead frame assembly, and employing a molding compound with an elevated thermal expansion) can be applied concurrently, to bring the curvature of the package down. In addition, application of very thin and/or low-expansion lead frames can also contribute essentially to a low residual curvature, since the bow of the silicon–lead frame assembly decreases with the decrease in the thickness and the thermal expansion coefficient of the lead frame. Finally, it is always advisable to use silicon chips that are thinner than conventional 20-mil thick chips. This will not only result in a lower warpage of the silicon–lead frame assembly, and hence in a smaller bow of the entire package, but will reduce the total thickness of the package as well.

### 12.2.5 Special Case: Bimaterial Assembly

As a special case we examine the problem of the bow of an "isolated" chip–lead frame assembly. When the thicknesses $h_1$ and $h_4$ of the molding compound layers are small, the compliances $\lambda_1$ and $\lambda_4$ of these layers become large, compared to the compliances $\lambda_2$ and $\lambda_3$. Then we obtain

$$\lambda \cong (\lambda_2 + \lambda_3)\lambda_1\lambda_4$$

$$D \cong EI\lambda + \beta_{23}^2\lambda_1\lambda_4 \cong [EI(\lambda_2 + \lambda_3) + \beta_{23}^2]\lambda_1\lambda_4$$

and the curvature of the assembly can be evaluated by the formula

$$\kappa = \frac{\alpha_{23}\beta_{23}\,\Delta t}{EI(\lambda_2 + \lambda_3) + \beta_{23}^2} \tag{12-19}$$

Here the flexural rigidity $EI$ should be calculated for components 2 and 3 only.

### 12.2.6 Numerical Examples and Discussion

The numerical examples are carried out for a 14-mm long, 1-mm thick elongated surface mounted plastic package (Figs. 12-1 and 12-4). The calculations of the thermally induced bow, as well as the zero bow thickness $h_1$, are given in Table 12-1. The residual curvature, determined for the temperature change $\Delta t = 150°C$, is $\kappa = 0.723 \times 10^{-3}/\text{mm}$. With the half package length of $l = 7$ mm and half chip length of $l_s = 4.6$ mm, the formula (12-14) yields $w_0 = 0.005\,57$ mm $\cong 5.6$ μm. Calculation using a finite element computer program (ANSYS) resulted for this package in maximum bow of about 6.0 μm, which is in good agreement with the prediction based on an analytical stress model.

   The bow of the package in question becomes zero if the thickness of layer 1 of the molding compound, determined in accordance with (12-16), is about 10 μm. Clearly, this value is too small and cannot be permitted in an actual package design. This result indicates, however, that the thickness $h_1$ of the epoxy layer on the lead frame side of the package should be made, for given materials and given total thickness of the package, as small as possible, to minimize the package bow. The calculated bows versus thickness $h_1$ are plotted in Fig. 12-5.

   For an "isolated" chip–lead frame assembly, using the data of Table 12-1, we obtain that the elevation of the centroid above the lower surface is $z_c = 0.273$ mm, and therefore the flexural rigidity of the package is

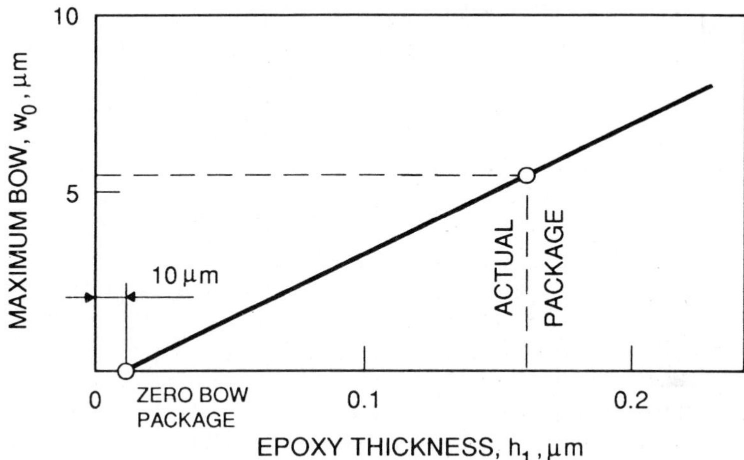



**Figure 12-5**    Calculated bow vs. thickness of the epoxy layer under the lead frame.

$EI = 146.8$ kg–mm. Then Eq. (12-19) yields $\kappa = 2.125 \times 10^{-3}$/mm. If layers 1 and 4 of the molding compound were of the same thickness, i.e., 0.23 mm, then the calculated curvature would be $\kappa = 1.076 \times 10^{-3}$/mm. Thus, the flexural rigidity provided by the molding compound results in a smaller curvature of the package by a factor of 1.97. The curvature $\kappa = 0.723 \times 10^{-3}$/mm of the "actual" package, where layer 1 is essentially thinner than layer 4, is smaller by a factor of 2.93 than the curvature of the silicon–lead frame assembly. Clearly, this is due to both the additional flexural rigidity of the epoxy encapsulant and the favorable effect of the thermal contraction mismatch between the molding material and the silicon–lead frame assembly. Thus, such a mismatch can be effectively utilized for "straightening-out" the composite structure if the thickness of the molding compound on the chip side of the silicon–lead frame assembly is made essentialy larger than the thickness of the molding material layer on the lead frame side. This effect can be enhanced considerably by employing a molding compound with an elevated coefficient of thermal expansion. Based on the data of Table 12-1 for all the package characteristics except the coefficient $\alpha_1$ of thermal expansion of the molding compound, and using Eq. (12-18), we find that the $\alpha_1$ value, resulting in zero bow of the molded body, is $\alpha_1 = 37.4 \times 10^{-6}$/°C. Note that this value is obtained under the assumption that Young's modulus of the epoxy material remains unchanged. In reality, polymeric materials with higher expansion have lower Young's moduli, and therefore the actual coefficient of thermal expansion of the molding compound, resulting in zero curvature, is somewhat larger than the above value.

**Table 12-1**  Bow of a Thin Surface Mounted Plastic Package (Calculation Sheet)

| | COMPONENT'S NUMBER | 1 | 2 | 3 | 4 | $\Delta t$, °C | 150 |
|---|---|---|---|---|---|---|---|
| | MATERIAL'S NAME | EPOXY | METAL | SILICON | EPOXY | h, mm | 1.00 |
| 1 | THICKNESS, h, mm | 0.160 | 0.200 | 0.340 | 0.300 | $\ell$, mm | 7.00 |
| 2 | Young's Modulus, E, kg/mm$^2$ | 1680 | 11000 | 12300 | 1680 | $\ell_2$, mm | 4.6 |
| 3 | (1) (2) | 268.80 | 2200.00 | 4182.00 | 504.00 | $\Sigma_1$, kg/mm | 7154.8 |
| 4 | COMP. ORDINATE, z, mm | 0.080 | 0.260 | 0.530 | 0.850 | | |
| 5 | (3) (4) | 21.50 | 572.00 | 2216.46 | 428.40 | $\Sigma_2$, kg | 3238.36 |
| 6 | 1/12 (3) (1)$^2$, kg/mm | 0.5734 | 7.3333 | 40.2866 | 3.7800 | $z_c = \dfrac{\Sigma_2}{E_1}$, mm | 0.4526 |
| 7 | (3) $(z-z_c)^2$, kg/mm | 37.3177 | 81.6085 | 25.0534 | 79.5951 | | |
| 8 | (6) + (7), kg/mm | 37.8911 | 88.9418 | 65.3400 | 83.3751 | EI, kg x mm | 275.5480 |
| 9 | Poisson's Ratio, $\upsilon$ | 0.40 | 0.30 | 0.24 | 0.40 | | |
| 10 | Compliance $\lambda = (1-\upsilon^2)/(3)$, mm/kg | $3.125 \times 10^{-3}$ | $0.4136 \times 10^{-3}$ | $0.2253 \times 10^{-3}$ | $1.6667 \times 10^{-3}$ | $\alpha_{12}$, 1/°C<br>$\alpha_{23}$, 1/°C | $4.0 \times 10^{-6}$<br>$8.8 \times 10^{-6}$ |
| 11 | CTE, $\alpha$, 1/°C | $16.0 \times 10^{-6}$ | $12.0 \times 10^{-6}$ | $3.20 \times 10^{-6}$ | $16.0 \times 10^{-6}$ | $\alpha_{13}$, 1/°C | $12.8 \times 10^{-6}$ |

| $\lambda_1\lambda_2\lambda_3$ | $\lambda_2\lambda_3\lambda_4$ | $\lambda_3\lambda_4\lambda_1$ | $\lambda_4\lambda_1\lambda_2$ | $\lambda$, mm$^3$/kg$^3$ | | $A = \alpha_{12}(\beta_{12}\lambda_3\lambda_4 - \beta_{24}\lambda_1\lambda_3) =$ |
|---|---|---|---|---|---|---|
| $0.2912 \times 10^{-9} + 0.1553 \times 10^{-9} + 1.1735 \times 10^{-9} + 2.1542 \times 10^{-9} = 3.7742 \times 10^{-9}$ | | | | | | $=-1.3912 \times 10^{-12}$ mm$^4$/kg$^2$ °C |
| $\beta_{12} = \dfrac{h_1 + h_2}{2}$ | $\beta_{23} = \dfrac{h_2 + h_3}{2}$ | $\beta_{34} = \dfrac{h_3 + h_4}{2}$ | $\beta_3 = \beta_{12} + \beta_{23}$ | $\beta_{24} = \beta_{23} + \beta_{34}$ | $\beta_{14} = \beta_{13} + \beta_{34}$ | $B = \alpha_{13}(\beta_{13}\lambda_2\lambda_4 - \beta_{34}\lambda_1\lambda_2) =$ <br> $=-1.3235 \times 10^{-12}$ mm$^4$/kg$^2$ °C |
| 12   0.18 | 0.27 | 0.32 | 0.45 | 0.59 | 0.77 | $C = \alpha_{23}\beta_{23}\lambda_1\lambda_4 + A + B =$ |
| $\lambda_3\lambda_4$ | $\lambda_1\lambda_4$ | $\lambda_1\lambda_2$ | $\lambda_2\lambda_4$ | $\lambda_1\lambda_3$ | $\lambda_2\lambda_3$ | $=9.6607 \times 10^{-12}$ mm$^4$/kg$^2$ °C |

| 13 |  | $0.3755 \times 10^{-6}$ | $5.2084 \times 10^{-6}$ | $1.2925 \times 10^{-6}$ | $0.6893 \times 10^{-6}$ | $0.7041 \times 10^{-6}$ | $0.0932 \times 10^{-6}$ | $D = EI\lambda + \Sigma_3 =$ | |
|---|---|---|---|---|---|---|---|---|---|
| 14 | (12)(13) | $0.0676 \times 10^{-6}$ | $1.4063 \times 10^{-6}$ | $0.4136 \times 10^{-6}$ | $0.3102 \times 10^{-6}$ | $0.4154 \times 10^{-6}$ | $0.0718 \times 10^{-6}$ | $= 2.0043 \times 10^{-6}$ mm$^4$/kg$^2$ | |
| 15 | (12)(14) | $0.0122 \times 10^{-6}$ | $0.3797 \times 10^{-6}$ | $0.1324 \times 10^{-6}$ | $0.1396 \times 10^{-6}$ | $0.2451 \times 10^{-6}$ | $0.0553 \times 10^{-6}$ | $\Sigma_3$, mm$^4$/kg$^2$ | $0.9643 \times 10^{-6}$ |

$$\text{CURVATURE, } \kappa = \Delta t \, \frac{C}{D} = 0.7230 \times 10^{-3} \text{ 1/mm}$$

$$\text{BOW } w_0 = \kappa \, \ell\ell \left( 1 - \frac{\ell}{2\ell_s} \right) = 5.567 \times 10^{-3} \text{ mm}$$

| THICKNESS $h_1$ FOR ZERO PKGE BOW | | |
|---|---|---|
| 1 | $\alpha_{12}\lambda_3$, mm/kg x °C | $0.9012 \times 10^{-9}$ |
| 2 | $\alpha_{13}\lambda_3$, mm/kg x °C | $5.2941 \times 10^{-9}$ |
| 3 | $2h\, E_1^0 \left[ (1) + (2) \right]$, 1/°C | $2.4781 \times 10^{-5}$ |
| 4 | $(1)\,(h+h_3)$, mm$^2$/kg °C | $1.2076 \times 10^{-9}$ |
| 5 | $(2)\,(\xi+h_3)$, mm$^2$/kg °C | $4.2353 \times 10^{-9}$ |
| 6 | $E_1^0\,\xi\left[ (4) + (5) \right]$, mm/°C | $0.5007 \times 10^{-5}$ |
| 7 | $(6) - \alpha_{23}\,(h-\xi)$, mm/°C | $0.0255 \times 10^{-5}$ |
| 8 | $h_1 = (7) + (3)$, mm | 0.0103 |
| $E_1^0 = \dfrac{E_1}{1-\upsilon_1^2}$, kg/mm$^2$ | 2000 | $\xi = h - h_2 - h_3$, mm |
|  |  | 0.46 |

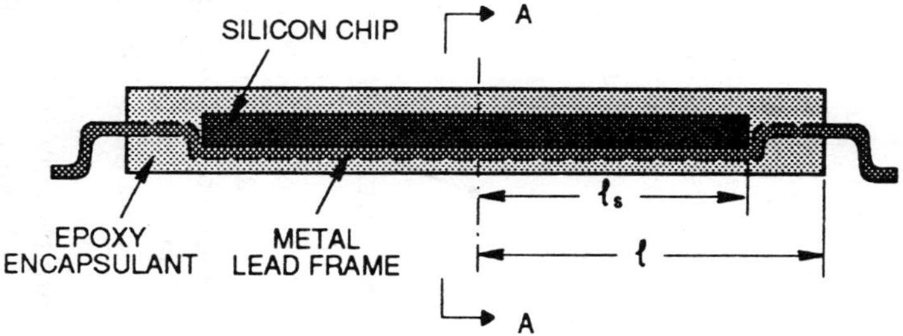

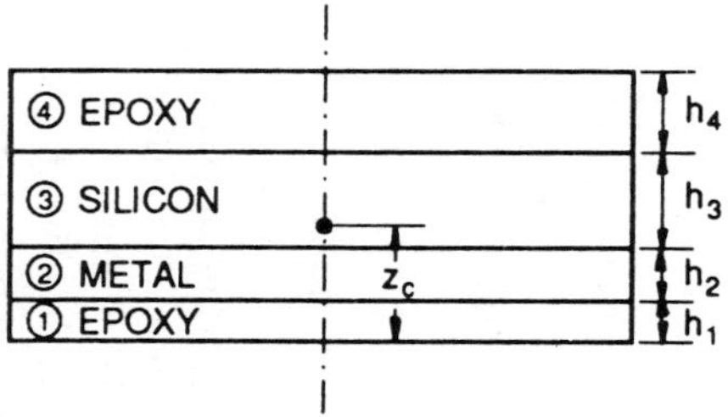

## 12.3 LARGE PLASTIC PACKAGE

The problem considered in this section concerns the situation when, during a typical molding operation, the packages, after having been ejected from the molding tool, are dropped onto a metal table. This results in a temperature gradient within the package, and its bow. Such bow can be significant if the package is large and the glass transition temperature of the plastic material falls within the temperatures of the "cold" (ambient) temperature of the package surface at the plate side and the "hot" (molding) temperature of the free surface of the package.

### 12.3.1 Basic Equations

We treat a large plastic package as a thin rectangular plate and consider the effect of a nonuniform distribution of temperature in the through-thickness ($z$) direction. We assume that the distribution of temperature in the $x$, $y$ plane is uniform, and that the chip is so small (compared to the size of the package), and that the lead frame is so thin (compared to the package thickness), that the bow of the molded body can be evaluated without considering the flexural rigidity of the chip and the lead frame. The effects of thermal expansion mismatch between dissimilar materials in the package are also not taken into account. These effects are thought to be small and have been treated previously in great detail (see, for instance, ref. 4). Finally, we assume that all the major hypotheses of the technical theory of thin plates are valid (see, for instance, refs. 3, 5).

The displacement of an arbitrary point of a plate in the direction of the coordinate axes $x$ and $y$ can be presented, in accordance with the hypotheses of straight normals,[3,5] in the form

$$u = u_0 - z \frac{\partial w}{\partial x}, \qquad v = v_0 - z \frac{\partial w}{\partial y}, \qquad w = w_0 \qquad (12\text{-}20)$$

where $u = u_0(x, y, z)$, $v = v_0(x, y, z)$, and $w = w_0(x, y, z)$ are the displacements of a point located in the main plane ($z = 0$) in the directions $x$, $y$, $z$, respectively, and $z$ is the through-thickness coordinate of the given point with respect to this plane. In our analysis we place the main plane at the distance

$$z_c = \frac{\int_0^h E z_1 \, dz_1}{\int_0^h E \, dz_1} \qquad (12\text{-}21)$$

from the lower surface of the plate. In this formula, $h$ is the plate's thickness,

$E$ is Young's modulus of the (molding) material, and

$$z_1 = z + z_c \tag{12-22}$$

is the vertical coordinate of the given point with respect to the lower surface of the plate. Clearly, Eq. (12-22) is equivalent to the condition

$$\int_{-z_c}^{h-z_c} Ez\, dz = 0 \tag{12-23}$$

From Eqs. (12-20) we obtain the following formulas for the strains (see, for instance, ref. 3):

$$\varepsilon_x = \frac{\partial u}{\partial x} = \varepsilon_x^0 - z\frac{\partial^2 w}{\partial x^2}$$

$$\varepsilon_y = \frac{\partial v}{\partial y} = \varepsilon_y^0 - z\frac{\partial^2 w}{\partial y^2} \tag{12-24}$$

$$\gamma_{xy} = \frac{\partial v}{\partial x} + \frac{\partial u}{\partial y} = \gamma_{xy}^0 - 2z\frac{\partial^2 w}{\partial x\, \partial y}$$

where

$$\varepsilon_x^0 = \frac{\partial u_0}{\partial x}, \qquad \varepsilon_y^0 = \frac{\partial v_0}{\partial y}, \qquad \gamma_{xy}^0 = \frac{\partial v_0}{\partial x} + \frac{\partial u_0}{\partial y} \tag{12-25}$$

are the strains of the points located in the main plane. The strains can be expressed through the normal stresses $\sigma_x$ and $\sigma_y$ and the shear stress $\tau_{xy}$ on the basis of Hooke's law equations:

$$\varepsilon_x = \frac{1}{E}(\sigma_x - v\sigma_y) + \alpha\, \Delta t$$

$$\varepsilon_y = \frac{1}{E}(\sigma_y - v\sigma_x) + \alpha\, \Delta t \tag{12-26}$$

$$\gamma_{xy} = \frac{1}{G}\tau_{xy} = \frac{2(1+v)}{E}\tau_{xy}$$

Here $v$ is Poisson's ratio of the (molding) material, $G$ is its shear modulus, $\alpha$ is the coefficient of thermal expansion, and $\Delta t$ is the temperature change. Equations (12-26) correspond to the two-dimensional state of stress.

After excluding the strains $\varepsilon_x$, $\varepsilon_y$, $\gamma_{xy}$ from (12-24) and (12-26) and solving the obtained relationships for the stresses, we obtain

$$\sigma_x = \frac{E}{1 - v^2}\left[\varepsilon_x^0 + v\varepsilon_y^0 - \left(\frac{\partial^2 w}{\partial x^2} + v\frac{\partial^2 w}{\partial y^2}\right)z\right] - \frac{E}{1 - v}\alpha\,\Delta t$$

$$\sigma_y = \frac{E}{1 - v^2}\left[\varepsilon_y^0 + v\varepsilon_x^0 - \left(\frac{\partial^2 w}{\partial y^2} + v\frac{\partial^2 w}{\partial x^2}\right)z\right] - \frac{E}{1 - v}\alpha\,\Delta t \qquad (12\text{-}27)$$

$$\tau_{xy} = G\gamma_{xy}^0 - \frac{E}{1 + v}z\frac{\partial^2 w}{\partial x\,\partial y}$$

These stresses result in the following forces and moments acting in the plate's cross-sections:

$$N_x = \int_{-z_c}^{h-z_c}\sigma_x\,dz = a_1\varepsilon_x^0 + a_2\varepsilon_y^0 + a_3\frac{\partial^2 w}{\partial x^2} + a_4\frac{\partial^2 w}{\partial y^2} - N_T$$

$$N_y = \int_{-z_c}^{h-z_c}\sigma_y\,dz = a_1\varepsilon_y^0 + a_2\varepsilon_x^0 + a_3\frac{\partial^2 w}{\partial y^2} + a_4\frac{\partial^2 w}{\partial x^2} - N_T \qquad (12\text{-}28)$$

$$N_{xy} = \int_{-z_c}^{h-z_c}\tau_{xy}\,dz = b_1\gamma_{xy}^0 - b_2\frac{\partial^2 w}{\partial x\,\partial y}$$

$$M_x = \int_{-z_c}^{h-z_c}\sigma_x z\,dz = a_3\varepsilon_x^0 + a_4\varepsilon_y^0 + a_5\frac{\partial^2 w}{\partial x^2} + a_6\frac{\partial^2 w}{\partial y^2} + M_T$$

$$M_y = \int_{-z_c}^{h-z_c}\sigma_y z\,dz = -a_3\varepsilon_y^0 + a_4\varepsilon_x^0 + a_5\frac{\partial^2 w}{\partial y^2} + a_6\frac{\partial^2 w}{\partial x^2} + M_T \qquad (12\text{-}29)$$

$$M_{xy} = -\int_{-z_c}^{h-z_c}\tau_{xy}\,dz = b_2\gamma_{xy}^0 - b_3\frac{\partial^2 w}{\partial x\,\partial y}$$

where the following notation is used:

$$a_1 = \int_{-z_c}^{h-z_c}\frac{E}{1 - v^2}\,dz, \qquad a_2 = \int_{-z_c}^{h-z_c}\frac{Ev}{1 - v^2}\,dz, \qquad a_3 = -\int_{-z_c}^{h-z_c}\frac{E}{1 - v^2}z\,dz$$

$$a_4 = -\int_{-z_c}^{h-z_c}\frac{Ev}{1 - v^2}z\,dz, \qquad a_5 = \int_{-z_c}^{h-z_c}\frac{E}{1 - v^2}z^2\,dz, \qquad a_6 = \int_{-z_c}^{h-z_c}\frac{Ev}{1 - v^2}z^2\,dz$$

$$b_1 = \frac{1}{2}\int_{-z_c}^{h-z_c}\frac{E}{1 + v}\,dz, \qquad b_2 = \int_{-z_c}^{h-z_c}\frac{E}{1 + v}z\,dz, \qquad b_3 = \frac{1}{2}\int_{-z_c}^{h-z_c}\frac{E}{1 + v}z^2\,dz$$

$$(12\text{-}30)$$

and

$$N_T = \int_{-z_c}^{h-z_c} \frac{E\alpha\,\Delta t}{1-v}\,dz, \qquad M_T = \int_{-z_c}^{h-z_c} \frac{E\alpha\,\Delta t}{1-v}\,z\,dz \qquad (12\text{-}31)$$

Since Poisson's ratio $v$ changes in quite a narrow range, and its dependence on temperature is seldom known with sufficient accuracy, one may assume, for the sake of simplicity, that this ratio is constant throughout the plate. Then, considering (12-23), we obtain $a_3 = a_4 = b_2 = 0$, and Eqs. (12-28) and (12-29) yield

$$N_x = \frac{S}{1-v^2}(\varepsilon_x^0 + v\varepsilon_y^0) - N_T$$

$$N_y = \frac{S}{1-v^2}(\varepsilon_y^0 + v\varepsilon_x^0) - N_T \qquad (12\text{-}32)$$

$$N_{xy} = \frac{S}{2(1+v)}\,\gamma_{xy}^0$$

$$M_x = D\left(\frac{\partial^2 w}{\partial x^2} + v\frac{\partial^2 w}{\partial y^2}\right) + M_T$$

$$M_y = D\left(\frac{\partial^2 w}{\partial y^2} + v\frac{\partial^2 w}{\partial x^2}\right) + M_T \qquad (12\text{-}33)$$

$$M_{xy} = D(1-v)\frac{\partial^2 w}{\partial x\,\partial y}$$

where

$$S = \int_{-z_c}^{h-z_c} E\,dz \qquad (12\text{-}34)$$

is the in-plane stiffness of the plate, and

$$D = \frac{1}{1-v^2}\int_{-z_c}^{h-z_c} Ez^2\,dz \qquad (12\text{-}35)$$

is its flexural rigidity.

### 12.3.2 Deflection Surface

A molded body is, in effect, an unsupported plate. This means that no external forces and moments act on it, and therefore

$$N_x = N_y = N_{xy} = 0$$

and

$$M_x = M_y = M_{xy} = 0$$

Then the relationships (12-33) yield

$$\frac{\partial^2 w}{\partial x^2} + v\frac{\partial^2 w}{\partial y^2} = -\frac{M_T}{D}$$
$$v\frac{\partial^2 w}{\partial x^2} + \frac{\partial^2 w}{\partial y^2} = -\frac{M_T}{D}$$

(12-36)

and

$$\frac{\partial^2 w}{\partial x\,\partial y} = 0$$

(12-37)

From (12-36), we find

$$\frac{\partial^2 w}{\partial x^2} = \frac{\partial^2 w}{\partial y^2} = -\frac{M_T}{D(1 + v)}$$

(12-38)

Hence, the thermally induced curvatures are the same in the $x$ and $y$ directions. After integrating Eqs. (12-38), we find

$$\frac{\partial w}{\partial x} = -\frac{M_T x}{D(1 + v)} + f_1(y)$$
$$\frac{\partial w}{\partial y} = -\frac{M_T y}{D(1 + v)} + f_2(x)$$

(12-39)

where $f_1(y)$ and $f_2(x)$ are functions of the variables $x$ and $y$. Substitution of Eqs. (12-39) into (12-37) yields

$$f_1'(y) = 0, \qquad f_2'(x) = 0$$

and therefore

$$f_1(y) = C_1, \qquad f_2(x) = C_2$$

Since, however, bending must be symmetric with respect to the origin $x = y = 0$, then one should put $C_1 = C_2 = 0$. The next integration of Eqs. (12-39) yields

$$w = -\frac{M_T x^2}{2D(1 + v)} + g_1(y) = -\frac{M_T y^2}{2D(1 + v)} + g_2(x) \qquad (12\text{-}40)$$

The deflection function $w$ must have the same constant value $w = w_0$ at the origin $x = y = 0$, whichever of the two formulas in (12-40) is used, so that

$$g_1(y) = g_2(x) = w_0$$

Then Eqs. (12-40) can be combined as follows

$$w = w_0 - \frac{M_T}{4D(1 + v)}(x^2 + y^2) \qquad (12\text{-}41)$$

In a square plate whose sides are equal to $2a$, the maximum bow $w_0$ can be determined from the condition

$$w(a, a) = 0$$

so that

$$w_0 = \frac{M_T a^2}{2D(1 + v)} \qquad (12\text{-}42)$$

Then formula (12-41) results in the following equation for the deflection surface:

$$w = w_0\left(1 - \frac{x^2 + y^2}{2a^2}\right) \qquad (12\text{-}43)$$

As one can see from this equation, only the corners of the molded body touch the horizontal surface, while the sides of the package have gaps with this surface. The maximum gaps occur at the midpoints of the package sides:

$$w_g = w(a, 0) = \frac{M_T a^2}{4D(1 + v)} \qquad (12\text{-}44)$$

and are twice as small as the maximum bow at the origin.

With the formula (12-35) and the second formula in (12-31), we present the formula (12-42) for the maximum bow as

$$w_0 = \frac{a^2}{2} \frac{J_T}{J} \tag{12-45}$$

where the integrals $J_T$ and $J$ are expressed as follows:

$$J_T = \int_{-z_c}^{h-z_c} E\alpha\, \Delta t\, z\, dz = \int_0^h E\alpha\, \Delta t\, z_1\, dz_1 - z_c \int_0^h E\alpha\, \Delta t\, dz_1$$

$$J = \int_{-z_c}^{h-z_c} Ez^2\, dz = \int_0^h Ez_1^2\, dz_1 - z_c^2 \int_0^h E\, dz_1 \tag{12-46}$$

### 12.3.3 Special Cases

**1.**   Examine first a special case where Young's modulus and the coefficient of thermal expansion are temperature-independent, and the temperature is distributed linearly over the thickness of the body:

$$t = t_0 + \frac{t_1 - t_0}{h}\, z_1 \tag{12-47}$$

Here $t_0$ is the temperature of the lower surface of the plate $(z = 0)$, and $t_1$ is the temperature of the upper surface $(z = h)$. In this case the integrals (12-46) are expressed as

$$J_T = \frac{E\alpha h^2(t_1 - t_0)}{12}, \qquad J = \frac{Eh^3}{12} \tag{12-48}$$

and formula (12-45) yields

$$w_0 = \frac{a^2}{2}\, \kappa \tag{12-49}$$

where

$$\kappa = \frac{\alpha(t_1 - t_0)}{h} \tag{12-50}$$

is the thermally induced curvature of the plate. After comparing formula (12-49) with (12-45), we conclude that instead of examining the bow of an actual plate whose Young's modulus and Poisson's ratio are temperature dependent while the temperature is distributed in an arbitrary manner across the plate's thickness, we can evaluate the bow of an equivalent plate with a linearly distributed temperature and constant (equivalent) coefficient of thermal expansion determined as

$$\alpha_e = \frac{h}{t_1 - t_0} \frac{J_T}{J} \tag{12-51}$$

**2.**   As a second special case, we consider a situation where the temperature of the entire plate is $t_1$ while the temperature of a thin layer near its lower surface is $t_0$ $(t_1 > t_0)$. Such a situation occurs, for instance, during the first moments of cooling, when a "hot" package is placed on a "cold" surface (at room temperature). In such a case, the condition (12-4) yields

$$E_0 \int_{-z_c}^{\varepsilon} z\, dz + E_1 \int_{\varepsilon}^{h-z_c} z\, dz = 0 \tag{12-52}$$

where $\varepsilon$ is the thin thickness of the cold lower layer of the plate, $E_0$ is Young's modulus of the molding material at the temperature $t_0$ of the cold layer, and $E_1$ is Young's modulus of the main body of the package. From (12-52) we find

$$E_0(\varepsilon^2 - z_c^2) + E_1[(h - z_c)^2 - \varepsilon^2] = 0$$

Since the thickness $\varepsilon$ is small, its square can be neglected in comparison with the expected ordinate $z_c$ squared. Then we have

$$E_1(h - z_c^2) = E_0 z_c^2$$

so that

$$z_c = \frac{h}{1 + \sqrt{E_0/E_1}} \tag{12-53}$$

The integrals (12-46) result in the following formulas:

$$J_T = E_0 \alpha_0 t_0 \int_{-z_c}^{\varepsilon} z\, dz + E_1 \alpha_1 t_1 \int_{\varepsilon}^{h-z_c} z\, dz \cong \frac{h^2 E_0 (\alpha_1 t_1 - \alpha_0 t_0)}{2(1 + \sqrt{E_0 E_1})^2}$$

$$J = E_0 \int_{-z_c}^{\varepsilon} z^2\, dz + E_1 \int_{\varepsilon}^{h-z_c} z^2\, dz \cong \frac{h^3 E_0}{3(1 + \sqrt{E_0/E_1})^2} \tag{12-54}$$

where $\alpha_0$ and $\alpha_1$ are the coefficients of thermal expansion for the upper "cold" and the bottom "hot" parts of the package, respectively. Then the formula (12-45) results in the following expression for the maximum bow:

$$w_0 = \frac{3}{4} \frac{a^2}{h} (\alpha_1 t_1 - \alpha_0 t_0) \tag{12-55}$$

Note that this bow is independent of the Young's modulus of the material.

### 12.3.4 Numerical Examples

**1.** Let a large flat square plastic package experience linear temperature gradient and have the following geometric characteristics: $a = 1.7$ cm, $h = 0.4$ cm. Young's modulus and coefficient of thermal expansion of the molding compound are shown in Table 12-2,[6] in which the computations of the thermally induced bow are performed. The integrals are calculated on the basis of the trapezoid rule. The calculated maximum bow is about $w_0 = 7.8$ mils. This is quite close to what can be found in actual packages. The calculated equivalent coefficient of thermal expansion is $\alpha_e = 4.370 \times 10^{-5}/°C$, which is within the limits of the change of the actual coefficient of thermal expansion.

According to the experimental data of Bair et al.,[6] the glass transition temperature for the molding compound in question is $T_g = 150°C$. As evident from the data of Table 12-2, the corresponding coefficient of thermal expansion is very close to the calculated equivalent coefficient of thermal expansion. The computed ordinate of the point whose temperature $t$ is equal to the glass transition temperature $T_g$ is $z_g = 0.320$ cm. Note that this point is located substantially higher on the $z_1$ axis than the neutral plane.

**2.** Examine now a case where the bottom surface of the package is quenched. Assuming that the temperature of the bulk of the molded body is $t_1 = 175°C$, and the temperature of the quenched layer is $t_0 = 50°C$, and using the formula (12-55), with $\alpha_0 = 2 \times 10^{-5}/°C$ and $\alpha_1 = 6 \times 10^{-5}/°C$, we obtain that the maximum bow is $w_0 = 26.7$ mils. This value is larger by a factor of 2.6 than the bow calculated for the case of linear distribution of temperature. The elevation $z_c$ of the main plane above the low surface, calculated according to (12-53), is $z_c = 0.1345$ cm. Hence, this plane is located somewhat lower than in the case of a linear temperature gradient. Unlike the maximum bow, the $z_c$ value depends on Young's moduli of the "cold" and the "hot" parts of the molded body, but is independent of the coefficients of thermal expansion.

**Table 12-2**   Thermally Induced Bow (Calculation Sheet)

| No. | $T$ (°C) | $E$ (dyne/cm$^2$) | (1) (3) | (1) (4) | $\alpha$ (1/°C) | $T - T_0$ (°C) | (3) (6) (7) | (1) (8) |
|---|---|---|---|---|---|---|---|---|
| (1) | (2) | (3) | (4) | (5) | (6) | (7) | (8) | (9) |
| 0 | 50 | $19.5 \times 10^{10}$ | 0 | 0 | $2 \times 10^{-5}$ | 0 | 0 | 0 |
| 1 | 75 | $18.5 \times 10^{10}$ | $18.5 \times 10^{10}$ | $18.5 \times 10^{10}$ | $2 \times 10^{-5}$ | 25 | $0.0925 \times 10^9$ | $0.0925 \times 10^9$ |
| 2 | 100 | $17 \times 10^{10}$ | $34 \times 10^{10}$ | $68 \times 10^{10}$ | $2.2 \times 10^{-5}$ | 50 | $1.870 \times 10^9$ | $3.74 \times 10^9$ |
| 3 | 125 | $16 \times 10^{10}$ | $48 \times 10^{10}$ | $144 \times 10^{10}$ | $3.1 \times 10^{-5}$ | 75 | $0.372 \times 10^9$ | $1.116 \times 10^9$ |
| 4 | 150 | $12 \times 10^{10}$ | $48 \times 10^{10}$ | $192 \times 10^{10}$ | $4.6 \times 10^{-5}$ | 100 | $0.552 \times 10^9$ | $2.208 \times 10^9$ |
| 5 | 175 | $5 \times 10^{10}$ | $25 \times 10^{10}$ | $125 \times 10^{10}$ | $6.0 \times 10^{-5}$ | 125 | $0.375 \times 10^9$ | $1.875 \times 10^9$ |
| Sum | | $88 \times 10^{10}$ | $173.5 \times 10^{10}$ | $547.5 \times 10^{10}$ | | | $3.2615 \times 10^9$ | $9.0315 \times 10^9$ |
| Correction | | $12.25 \times 10^{10}$ | $12.5 \times 10^{10}$ | $62.5 \times 10^{10}$ | | | $0.1815 \times 10^9$ | $0.937 \times 10^9$ |
| Corrected Sum | | $\Sigma_3$ $75.75 \times 10^{10}$ | $\Sigma_4$ $161 \times 10^{10}$ | $\Sigma_5$ $485 \times 10^{10}$ | | | $\Sigma_8$ $3.074 \times 10^9$ | $\Sigma_9$ $8.094 \times 10^9$ |

$h = 0.4$ cm, $a = 1.7$ cm, $\Delta h = 0.08$ cm, $\Delta T = 125°$C, $T_g = 150°$C

$$z_c = \Delta h \frac{\Sigma_4}{\Sigma_3} = 0.170 \text{ cm}$$

$$\alpha_e = \frac{h}{\Delta T z_c} \frac{(\Delta h/z_c)\Sigma_9 - \Sigma_8}{(\Delta h/z_c)^2 \Sigma_5 - \Sigma_3} = 4.3705 \times 10^{-5}/°\text{C}$$

$$w_0 = \frac{a^2}{2} \frac{\alpha_e \Delta T}{h} = 0.019\,73 \text{ cm} \cong 7.77 \text{ mils}$$

$$z_g = h \frac{T_g - T_0}{\Delta T} = 0.320 \text{ cm}$$

### 12.3.5 Approximate Formula for Maximum Bow

Let us assume that the molding compound is characterized by the values $E_1$ and $\alpha_1$ of Young's modulus and coefficient of thermal expansion below glass transition temperature, and by the values $E_2$ and $\alpha_2$ above glass transition. These values are usually presented in the specification data for molding materials. The $z_c$ value, which determines the location of the neutral surface, is, in accordance with (12-2),

$$z_c = \frac{E_1 \int_0^{z_g} z_1 \, dz_1 + E_2 \int_{z_g}^{h} z_1 \, dz_1}{E_1 z_g + E_2(h - z_g)} = \frac{1}{2} \frac{E_1 z_g^2 + E_2(h^2 + z_g^2)}{E_1 z_g + E_2(h - z_g)} \tag{12-56}$$

where

$$z_g = h \frac{T_g - t_0}{t_1 - t_0} \tag{12-57}$$

is the ordinate of the plane corresponding to $t = T_g$.

The integrals (12-46) yield

$$J_T = E_1 \alpha_1 z_g \left[ \frac{t_1 - t_0}{3h} z_g^2 + \frac{1}{2}\left(t_0 - \frac{z_c}{h}(t_1 - t_0)\right)z_g - z_c t_0 \right]$$

$$+ E_2 \alpha_2 (h - z_g)\left[ \frac{t_1 - t_0}{3h} z_g^2 + \frac{1}{2}\left(\frac{t_0 + 2t_1}{3} - \frac{z_c}{h}(t_1 - t_0)\right)(h + z_g) \right] \tag{12-58}$$

$$J = E_1 z_g (\tfrac{1}{3} z_g^2 - z_c^2) + E_2(h - z_g)[\tfrac{1}{3}(z_g^2 + h z_g + h^2) - z_c^2]$$

These formulas can be used for a tentative evaluation of the integrals entering the formulas (12-45) for the maximum bow. Using data from Table 12-2, we have $E_1 = 16.6 \times 10^{10}$ dyne/cm$^2$, $\alpha_1 = 2.8 \times 10^{-5}/°C$, $E_2 = 8.2 \times 10^{10}$ dyne/cm$^2$, $\alpha_2 = 5.3 \times 10^{-5}/°C$. With $z_g = 0.320$ cm, using (12-56), we obtain $z_c = 0.1827$ cm. Then the computed values of the integrals (12-58) are $J_T = 143.2 \times 10^5$ dyne, and $J = 69.8 \times 10^7$ dyne-cm. With $a = 1.7$ cm, and the calculated integrals $J_T$ and $J$, formula (12-45) yields $w_0 = 11.7$ mils. This result is about 50% greater than a more accurate value obtained using numerical integration.

Examine the extreme cases of very high and very low glass transition temperatures, so that the thermomechanical characteristics of the molding compound do not change within the package. In the case of very high $T_g$, the region above the glass transition temperature has no effect on the mechanical behavior of the package, so that one may put $E_2$ equal to zero, and assume $z_g$ significantly larger than $z_c$. Then we obtain

$$w_0 = \frac{a^2(t_1 - t_0)}{2h} \alpha_1$$

Similarly, in the case of a very low glass transition temperature, we find

$$w_0 = \frac{a^2(t_1 - t_0)}{2h} \alpha_2$$

Since the coefficient $\alpha_2$ of thermal expansion in the region above the glass transition temperature is considerably larger than the coefficient $\alpha_1$ below this temperature, for a smaller thermally induced bow one should try to employ molding compounds with high glass transition temperatures. Clearly, in all the cases, a small temperature change from the curing (cross-linking) temperature to the room (testing) temperature, and greater thickness of the package, lead to lower bows. The size of the package, however, has the largest effect on the thermally induced bow.

## 12.4  SUMMARY

1. Simple and easy-to-apply calculation techniques have been developed to evaluate the bow of thin plastic packages subjected to a uniform temperature change, as well as of large flat packages experiencing through-thickness temperature gradients.
2. The technique developed for the prediction of the thermally induced bow in a thin plastic package structure enables one to carry out an optimal physical design of such a package, so that the residual bow remains sufficiently small. Minimum bow can be achieved by a rational combination of the following measures:

   Proper positioning of the chip-lead frame assembly with respect to the midplane of the package
   Application of thinner chips, and thin and low-expansion lead frames
   Use of moderate- or even high-expansion molding compounds

3. The technique developed for the prediction of the thermal bow in a large flat package enables one to determine the effect of the package geometry and material properties on its warpage. It has been found that Young's modulus of the epoxy material has a small effect on the package bow, while the coefficient of thermal expansion, in contrast,

strongly affects the bow. The size of the package has the greatest effect on the bow, which is proportional to the package size squared.


## ACKNOWLEDGMENT

The author acknowledges, with thanks, useful discussions with, and valuable comments made by, L. T. Manzione and H. E. Bair.


## NOMENCLATURE

| | |
|---|---|
| $a$ | Flat package dimension |
| $D$ | PQFP flexural rigidity; determinant of a system of algebraic equations |
| $E$ | Young's modulus |
| $EI$ | TSOP flexural rigidity |
| $G$ | Shear modulus |
| $h$ | Thickness |
| $l$ | Half the TSOP package length |
| $l_s$ | Half the TSOP chip length |
| $M_x, M_y$ | Bending moments due to the normal stress $\sigma_x, \sigma_y$ |
| $M_{xy}$ | Bending moment due to the shearing stress $\tau_{xy}$ |
| $M_T$ | Bending moment due to the thermal expansion (contraction) |
| $N_x, N_y$ | Forces due to the normal stresses $\sigma_x, \sigma_y$ |
| $N_{xy}$ | Force due to the shearing stress $\tau_{xy}$ |
| $N_T$ | Force due to the thermal expansion (contraction) |
| $S$ | In-plane stiffness of the plate (package) |
| $t$ | Temperature |
| $T_g$ | Glass transition temperature |
| $T$ | Thermally induced force |
| $u, v, w$ | Displacements of the given point in $x, y, z$ directions, respectively |
| $u_0, v_0, w_0$ | Displacements of a point located in the main plane ($z = 0$) in $x, y, z$ directions, respectively |
| $w$ | Lateral deflections |
| $w_0$ | Maximum lateral deflection |
| $x, y, z$ | Rectangular coordinates |
| $z_c$ | Distance of the main plane ($z = 0$) from the lower surface of the plate |
| $z_1$ | Vertical coordinate of the given point with respect to the lower surface of the plate |
| $z_g$ | Vertical coordinate of the plane whose temperature $t$ is equal to the glass transition temperature $T_g$ |
| $\alpha$ | Coefficient of thermal expansion |
| $\gamma_{xy}$ | Shearing strain |
| $\gamma_{xy}^0$ | Shearing strain in the main plane |
| $\varepsilon_x, \varepsilon_y$ | Normal strains in the directions $x, y$, respectively |
| $\varepsilon_x^0, \varepsilon_y^0$ | Normal strains in the main plane |

| | |
|---|---|
| $\varepsilon$ | Thickness of the "cold" layer in a PQFP package |
| $\xi$ | Total thickness of the molding compound (on both sides of the chip) |
| $\kappa$ | Curvature |
| $\lambda$ | In-plane compliance |
| $\nu$ | Poisson's ratio |
| $\sigma_x, \sigma_y$ | Normal stresses in the directions $x$, $y$, respectively |
| $\tau_{xy}$ | Shearing stress |

## REFERENCES


1. Manzione, L. T., *Plastic Packaging of Microelectronic Devices*, Van Nostrand Reinhold, New York, 1990.
2. Goosey, M. T., *Plastics for Electronics*, Elsevier, New York, 1985.
3. Suhir, E., *Structural Analysis in Microelectronic and Fiber-Optic Systems*, Vol. 1, Van Nostrand Reinhold, New York, 1991.
4. Suhir, E., "Thermal Stress Failures in Electronic Components—Review and Extension," in *Advances in Thermal Modeling of Electronic Equipment*, A. Bar-Cohen and A. Kraus, eds., Hemisphere, New York, 1989.
5. Timoshenko, S. P., and S. Woinowski-Krieger, *Theory of Plates and Shells*, McGraw-Hill, New York, 1959.
6. Bair, H. E. et al., "Thermomechanical Properties of IC Molding Compounds," *Polymer Engineering and Science*, **30**(10), 1990, pp. 609–617.

# 13

# Thermal and Moisture Stresses in Plastic Packages

*Michel Mermet-Guyennet*

## 13.1 INTRODUCTION

Increased package density and reliability expectations require a better understanding of the mechanical behavior of plastic packages under process and use conditions. This chapter reviews potential failure modes and a combined analytical/experimental approach developed under ESPRIT project 5033—PLASIC for stress analysis in plastic packages. This includes (1) a failure criterion for moisture-induced stresses; (2) a finite-element approach for thermal stress analysis; and (3) a piezoresistive technique for stress measurements and model calibration. The latter is applied to the particular problem of metal line displacement at the surface of a silicon die.

Design rules can be derived for package optimization, based on the package models, the failure mode analysis and database.

One of the main functions of an electronic package is to protect integrated circuit (IC) chips against environmental aggression. As ICs are not stand-alone devices, packages must also allow for communication with the next level of interconnections: signal input–output, power in, and power dissipation. Several package types are currently used: ceramic, glass metal, TAB-based packages, etc., but to date the cheapest type is the plastic package. In plastic packages, silicon chips are completely molded in a polymer along with power and signal leads.

Most component manufacturers using plastic packages have driven costs down by improving materials and fabrication processes. Reliability and performance were adequate for common use. In the 1980s, technological evolution has led to stress effects being an issue due to the package structure and fabrication processes, e.g.,

Evolution of chip size towards larger chips
Trend to using fine metal lines that are much more sensitive to stress effects
Wider use of surface mount assembly, imposing severe thermal shocks onto the assembled package.

These three factors have increased failure types at chip level (passivation cracks, metal line shift, corrosion) and package level (resin cracks). At the same time, the market has become more demanding in terms of reliability and package size. For instance, military applications require more and more plastic packages for cost reduction and the general trend is towards miniaturization of electronic devices (hand portable phones, thin small outline packages for memory, and so on).

The demand for new packages that meet market requirements drives the development of new packaging materials and package structure optimization. Stresses generated during package fabrication and assembly onto a substrate must be accurately evaluated and measured and failure mechanisms related to stress effect have to be clearly understood.

This chapter gives an overview of these problems (Sections 13.2 and 13.3). Models for stress effect evaluation are presented (Section 13.4), including stress measurement techniques.

## 13.2 PLASTIC PACKAGE STRUCTURE AND FABRICATION

### 13.2.1 Structure

The general structure of a plastic package is shown in Fig. 13-1. Body sizes as large as 40 × 40 mm are available; plastic quad flat packages (PQFP) with

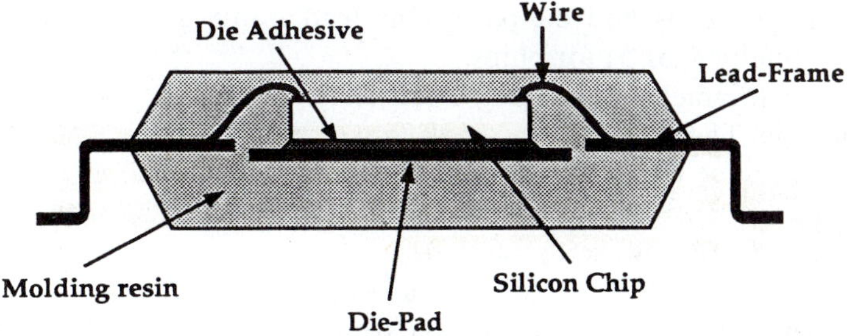


**Figure 13-1**   Cross-section of a standard plastic package.

a body size of 28 × 28 mm are widely used with lead pitches (1, 0.8, 0.65, or 0.5 mm) leading to I/O counts of 100, 120, 144–160, 208. A typical value of thickness is 2.8 mm. Other types of plastic packages are available—dual in line, single in line, plastic leaded chip carrier—but the PQFP with gull-wing leads is one widely dedicated to large chips with a high I/O count; silicon chips of 2 cm$^2$ can be packaged. Therefore, stress evaluation and optimization become a critical issue for this type of package. The lead frame is 6 mil (150 µm) thick and the material is generally Alloy 42 (Fe–Ni alloy). Copper is also used for its thermal and electrical properties but requires a thicker lead frame. The molding compound is epoxy resin filled with fused silica; the mechanical characteristics of this material are the most significant to evaluate stresses induced in silicon.

Thin small outline packages (TSOP) sometimes called very small outline packages (VSOP), are very sensitive to stress inside the resin encapsulation of chips. These packages are dedicated to memory chips (commonly 7 mm × 12 mm and low I/O count) and need to have the minimum size. The most common size is 14 mm × 20 mm with a thickness of 1 mm. Therefore, package strength is a critical issue, especially during assembly.


### 13.2.2 Fabrication

Detailed process steps are beyond the scope of this chapter. Only the main steps are reviewed, with emphasis on those generating stresses in the end product. There are four main steps in the fabrication of a plastic package:

Die-attach
Wire bonding
Molding
Lead forming and finishing

The first three operations involve the package lead frame. The lead frame is a flat metal strip that is punched out or etched to provide the base structure of the package (leads and die-pad). One lead frame generally has several positions (usually 4 or 5) for chips.

Gold wires (diameter 25–33 µm) electrically connect the die and the leads of the package. Thermosonic wire bonding is widely used; it impacts locally on silicon stress, at the pad level. The failure mode that can be encountered is cracks under the pad, due to thermal shock combined with ultrasonic vibration as melted gold wets the pads.

Lead forming is the final mechanical operation to excise and form the external leads and separate the package from the lead frame. Lead finishing is done, for example, by tin–lead electroplating. These operations do not generate significant stresses. However, they could damage the resin–lead interface locally and so provide a path for moisture penetration into the package.

Die-attach and molding are the two operations generating the highest stresses inside the package.

***Die attach***   A blob of epoxy adhesive is deposited onto the die-pad. The volume is calibrated and a special tool (fish-tail type) is used to deposit a pattern of adhesive. The silicon die is then pushed down at constant pressure. Glue thickness is uniform (20–50 μm) and voids are minimized when the die attach process has been tuned in and is under control. The second stage is curing; lead frames with mounted dies are placed inside an oven at about 180°C for 1 to 2 hours.

***Molding***   Lead frames are placed in a mold and preheated, the mold is then closed with several tons of pressure. Volume-calibrated pellets of resin are preheated and then placed in the mold where an injection system pushes the liquefied molding compound to fill the cavities up. Mold temperature is maintained at 180°C during this operation. The lead frame and resin are left for a few seconds for a first stage of resin curing. Lead frames are removed and placed in an oven for post-curing at 180°C for several hours.

This molding and curing operation generates most of the stress inside the silicon. Moreover, as mold cavities fill up, the resin solidifies. This transient behavior is not well understood. Adjustment of process parameters and mold design (size and position of injection holes) is a matter of knowhow. Significant work remains to be done to better understand resin flow and the impact of solidification on the end product.

## 13.3 STRESS-RELATED FAILURES

Some stress-related failures appear during product life due to slow temperature variations that fatigue materials. Others are more sudden (e.g., brittle fracture) and are revealed by abrupt changes in temperature. Acceleration tests are used to study these failures. Thermal cycles lead to fatigue failures and thermal shocks test the other failure mode. To test the effect of moisture content, packages are thermally shocked by dipping in an oil bath at 200°C after long exposure in a water-saturated atmosphere.

The following failure modes are encountered:

Die/adhesive cracks
Passivation cracks
Displacement of metal lines

Resin cracks
Corrosion

The last is a long-term effect and is generally preceded by a stress effect (passivation cracks or corrosion accelerated by stress). Corrosion is not addressed here, only direct stress effects are considered.

### 13.3.1 Die/Adhesive Cracks

The possible failure mechanisms are

Die cleaving in a plane parallel to the surface of the chip
Die cracking
Adhesive cracking
Adhesive pad delamination

All these failure modes are related to die-attach.[1] Actually, the mismatch in coefficient of thermal expansion (CTE) between silicon (3 p.p.m./°C) and die pad metal (7 p.p.m./°C for Alloy 42) creates a bimetallic strip effect. Silicon is brittle and fractures are generally due to notches giving rise to stress concentration effects.

Van Kessel[2] has studied fracture surfaces and initiation in silicon die. He also discusses this failure mechanism when generated by flaws on the chip backside and shows that cracks are due to tensile stresses caused by local defects in the die attach.

Failures related to adhesive are due to a lack of adhesion at silicon–adhesive or die-pad–adhesive interfaces. The metallization and contamination types sharply influence the adhesion strength.

The issue of die-attach stresses is well known and can be addressed by reducing the CTE mismatch between silicon and metal, and using soft adhesives that yield in the plastic domain at the level of stress reached in use (glass transition temperature as low as possible). Therefore, shear stresses cause plastic deformation in the adhesive and relax stresses in silicon and die-pad. Figure 13-2 shows this behavior for a set of different die adhesives. In this case, the stress relaxation parameter measured is the overall deformation of a 30-mm long silicon–copper bimetallic strip.

### 13.3.2 Metal Displacement and Passivation Cracks

Passivation cracks are initiated after the molding operation and increase in size with thermal cycles. Metal displacements appear after a few thermal

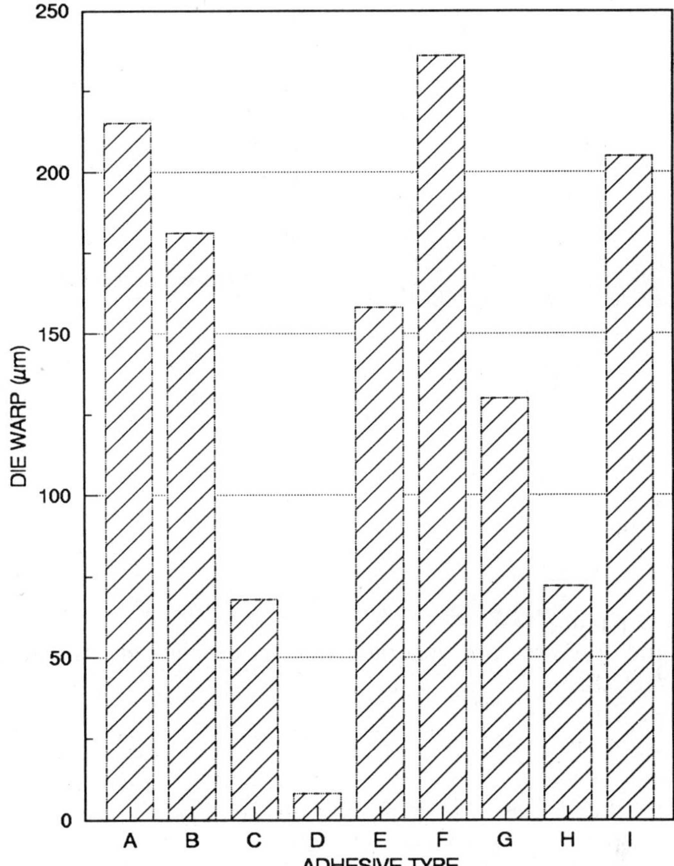



**Figure 13-2**    Maximum warp for various adhesive materials—silicon chip on 30-mm copper clad.

cycles. These phenomena are illustrated in Figure 13.3, where the growth of cracks and displacements is shown after the molding operation and thermal cycling.

Obviously, these failures relate to stresses induced by molding and particularly stresses at the resin–passivation interface. Passivation cracks are measured by the number of cracks per unit area. They lead to moisture penetration and corrosion of metal lines. One method for avoiding passivation cracks is to use thicker and harder passivation or soft passivation that absorbs stresses by plastic deformation (polyimide). Metal displacement is then less spread; thick metal lines (2–3 μm) are used to reveal it.

R. E. Thomas[3] has shown that metal deformation is linked to stresses on

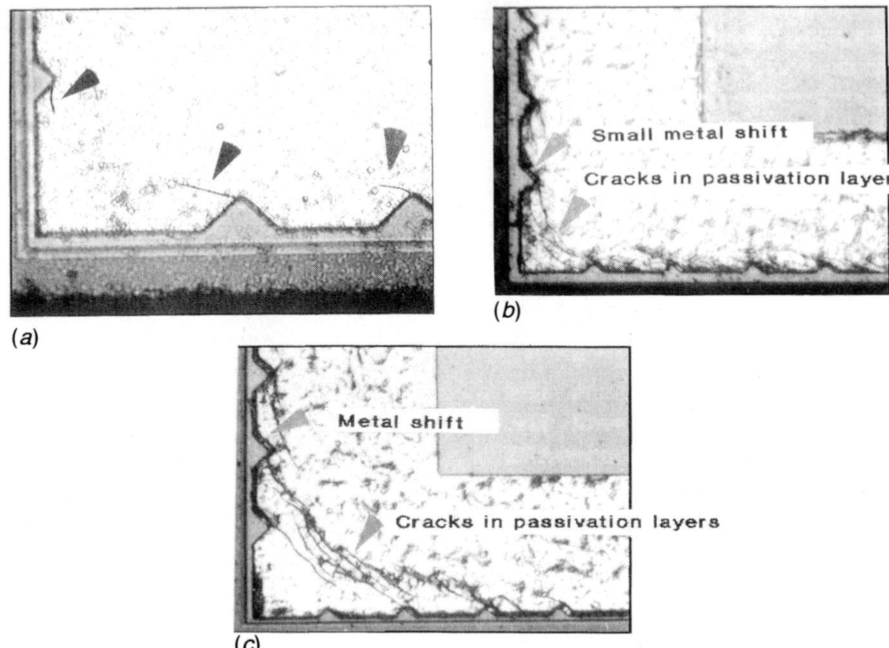



(a)

(b)

(c)

**Figure 13-3**  Evolution of passivation cracks and metal displacement at die corner after molding (a), after thermal cycles —55°C/150°C (b) and after 200 thermal cycles —55°C/150°C (c).

the chip. Non-passivated chips mounted in plastic packages were used to estimate the amount of stress that these packages were subjected to during thermal cycling. Thomas showed that, just after curing, local delamination occurs at the chip–resin interface. Delamination is initiated at the chip edge and moves towards the center of the chip with increasing number of thermal cycles. He also proposed the use of a soft film passivation at the silicon surface. The effect is positive for metal line damage but induces failures at wire bond level. The conclusion is that stresses generated by fabrication processes lead only to slight delamination, not metal deformation; thermal cycling induces metal deformation.

Two mechanisms for increasing metal displacement are summarized as follows. The first mechanism, progressive delamination, in accordance with the conclusions of Thomas,[3] is shown in Fig. 13-4. Three stages are involved:

*Stage 1:* After molding, resin and passivation are in contact, with good adhesion. During the first cycles, resin + passivation + silicon move together.

*Stage 2:* After a few cycles, slight delamination occurs. Therefore, high

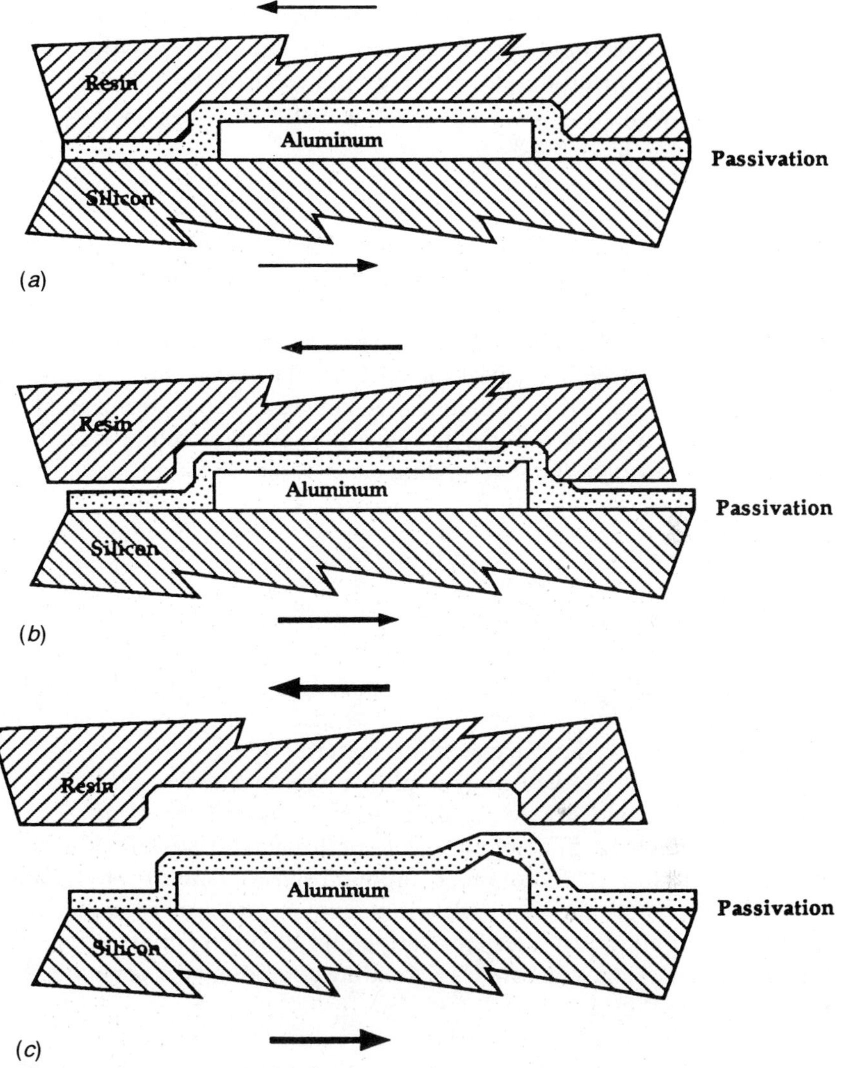



**Figure 13-4**  Mechanism of progressive delamination during thermal cycling. (*a*) After molding. (*b*) Partial delamination. (*c*) Total delamination.

forces push the metal as thermal cycling goes on. Delamination increases gradually.

*Stage 3:* In the last stage, delamination is extensive and there is no longer a bond between resin and passivation.

This progressive delamination is well illustrated in refs. 4 and 5.

A second mechanism is when metal lines bow. This cannot be explained by progressive delamination. In this case, an alternative hypothesis is to consider that the resulting global force applied to metal is larger than the adhesion of metal to silicon. Thus, when metal shift occurs there is local plastic deformation or cracking of resin.

### 13.3.3 Resin Cracks

Resin cracks generally initiate at the lower corner or side of the die-pad and propagate toward the bottom of the package. This area is where resin stress is concentrated due to the CTE mismatch between die pad, resin, and a sharp interface. Two types of failures are generally encountered: a slow fracture propagation through thermal cycling or a rapid crack propagation after thermal shocks.

**Resin Cracks Due to Fatigue**

During thermal cycles, cracks inside packages are generated at the lowest corner of the die-pad and propagate gradually. This slow phenomenon is related to fatigue of the encapsulating material.

Nishimura[6] has studied several plastic molding resins and their crack propagation behaviors. Two molding compounds with the same mechanical properties but with a different coefficient of thermal expansion are compared. Packages with low-CTE molding compound show a better behavior and can stand 10 times more thermal cycles. Initial cracks are present with high-CTE molding compound just after fabrication and before thermal cycling (length 0.1–0.3 mm). Cracks with low-CTE molding compound are generated during the first thermal cycles.

These experimental results were correlated with 2D plane stress finite element analysis. Nishimura[6] concludes that the most significant factor in propagation rate is the lowest temperature of thermal cycling and is directly linked to a high stress intensity factor in the molding compound at low temperature.

In another paper,[7] Nishimura carried out experimental and analytical studies to assess the effect of package structure on cracking caused by molding compound fatigue. He observes that the crack path is perpendicular to maximum principal stress. Lead frame material influence is studied and he concludes that crack generation is always preceded by delamination. Therefore, perfect adhesion avoids crack propagation.

## Resin Cracks Due to Thermal Shock

This failure is known in its most spectacular form when packages crack open during the solder reflow process ("popcorn phenomenon").[8,9] It has been observed that moisture content in the resin, die-pad size, resin thickness under the die pad, and mechanical properties of the resin are predominant factors. Tests to evaluate sensitivity to thermal shocks have been developed.[10] No clear experimental correlation has been found and the best way to avoid it is to use all means to keep popcorn-sensitive packages moisture-free before the thermal shock caused by surface mounting.

Some facts are clearly established:

1. Resin is very sensitive to moisture. Figure 13-5 shows a measure of moisture penetration as a function of time in a vapor-saturated atmosphere. It has been shown that this phenomenon follows a "pseudo-Fick" behavior.[11]
2. There is always local misadhesion between die-pad and resin. As packages are thermally shocked, moisture is suddenly released from the resin and the package is internally subjected to a high internal pressure. The problem in modeling this phenomenon is the lack of information on adhesion and the kinetics of water released from resin. A mechanism is described in Fig. 13-6.[9] An example of modeling following this

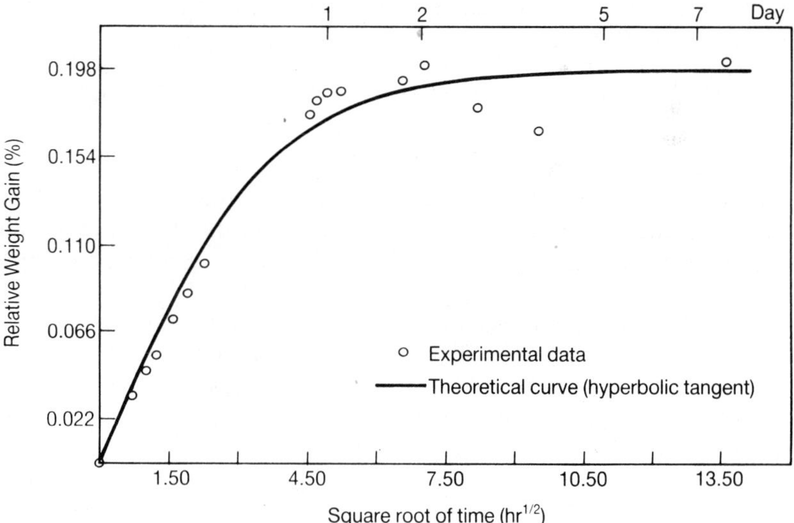


**Figure 13-5**   Curve of the relative gain in mass epoxy resin vs. square root of time. Samples are exposed to a vapor-saturated atmosphere (1 atm at ambient temperature of $-20°C$).

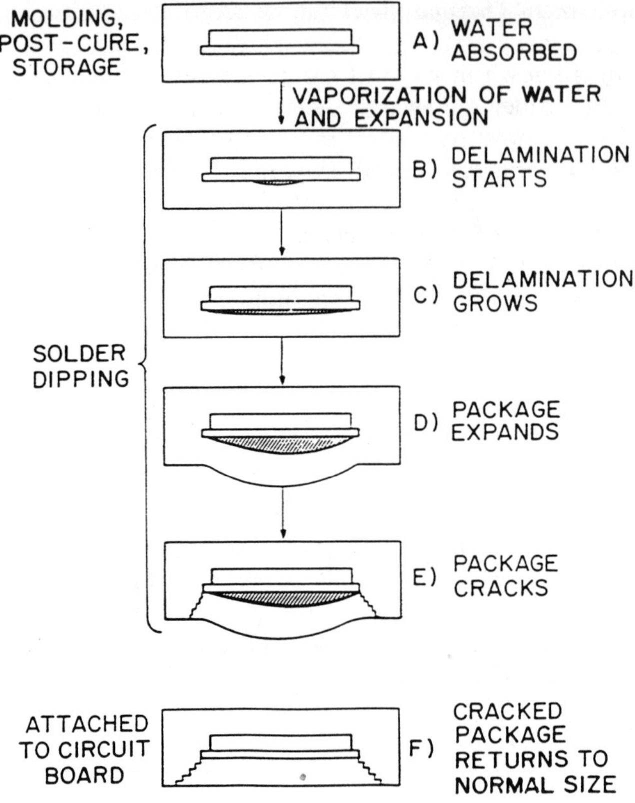


**Figure 13-6**  Mechanism of resin cracks during thermal shock. From Ohizumo.[9]

mechanism is described in Section 13.3.1, dealing with the extreme case when delamination is complete and saturated pressure is reached. Some experimental models including transient factors (moisture solubility coefficient, moisture diffusion coefficient, etc.) have also been developed.[12]

This issue has been addressed and now the spectacular effect of "popcorn" rarely occurs in assembly lines. Resin manufacturers have carried out studies to produce less sensitive resins.[13] Component vendors supply their sensitive packages in dry bags and assemblers usually bake them before mounting when they have been exposed to ambient for a long line.

Resin delamination and moisture release can lead to other subtle failures. For instance, devices have been observed to fail electrically after several days, and then work again after several hours of baking. This is a side-effect of the popcorn phenomenon for devices sensitive to charge effects: delamination and water create local charge effects that are eliminated during baking.

This example shows that, even if the popcorn effect is not evident for nonsensitive packages, long-term effects of local delamination due to surface mounting can be detrimental to system reliability.


## 13.4 THERMAL STRESS MEASUREMENT AND ANALYSIS

One of the first papers on plastic package modeling was published in 1977 by Dale.[14] He studied the impact of the fabrication process on internal stresses. His conclusion is that die bonding is the operation generating most stresses. An analytical model is developed using laminated plate theory as described in Timoshenko.[15]

In the early 1980s, reliability concerns pushed several manufacturers to develop analytical and experimental models.[16,17] One of the common issues is material characteristics, which are very process-dependent and thus require experimental validation of numerical and analytical models.

Finite element modeling has been used.[18,20] Groothuis[18] computed stress distribution at the surface of silicon. He emphasizes that all failures (passivation cracks, metal line shifts, and ball bond shear) are directly correlated to in-plane shear stress level. Edwards[19] also discusses the importance of shear stress and investigates the use of die coating. His conclusion is that, even if passivation cracks and metal line shifts are suppressed, the stress problem is shifted to the resin–coating interface interface and causes wirebond failures. It is clear that shear stresses at the resin–passivation interfaces are the fundamental components to be studied in order to understand the damage caused to the chip; but as described in Section 13.3.5, the most significant component of shear stress to be used must be clearly understood.

Natarajan[20] performed finite element analysis using the superposition principle in order to separate the die-attach effect and the encapsulation process. Interesting work has been carried out by Liechti[21] to compare the applicability of photoelastic, analytical (laminated plate theory), and finite element techniques to stress modeling of plastic packages. Special interface elements simulate the poor adhesion at interfaces. A good level of agreement was found between the plane strain model and photoelastic analysis, whereas laminated plate theory did not give a good correlation.

Several other studies of stress modeling can be found in refs. 22–27. These papers mainly deal with stress and stress-induced effects on silicon.

The distribution of die surface stresses in integrated circuits has been determined experimentally using specially designed semiconductor strain gauges.[28,29] Gee[28] concludes that the principal stresses are useful for molding compound ranking, while shear stress level mainly influences mechanical behavior during aging. For the purpose of resin ranking, Tiziani[29] also uses principal stresses, but stress effects on metal line are studied by using specific one-metal-layer test chips. As discussed later, individual stress components can be measured with strain gauges. Bittle[30] has designed a new stress sensor rosette on (111) silicon that can measure the complete three-dimensional stress state on the surface of a die.

The following sections present an example of analytical stress modeling in the case of resin cracks due to moisture pressure. Then, two tools for stress and stress effect measurement are described, and finally a methodology of modeling based on these tools is presented.

### 13.4.1 Criterion for Crack Risk Under Thermal Shock

**Assumptions and Validity of the Model**

During the soldering operation, the package is dipped in melted solder for a few minutes. The package reaches a temperature of about 200–240°C. This temperature is very close to the stress-free temperature of the structure. If the package cracks at this temperature, it means that another mechanism is involved since the temperature gradient stresses are negligible. Moreover, the glass transition temperature of the molding compound is exceeded and the molding compound is much more compliant. These two facts confirm that thermal stress is negligible.

The mechanism shown in Fig. 13-8 is used to model this risk criterion. In this case, it is supposed that delamination starts and expands under the die pad (phases B and C in Fig. 13-6). This is possible because the glass transition temperature of the resin is exceeded and contact between resin and die pad is weakened. At the same time, water contained in resin is suddenly released. The model presented here is a static model when the delamination is complete and the saturated pressure in the gap between resin and die-pad is reached.

The criterion of risk is determined using the theory of flat thick plates. In this model, only the stresses generated by pressure are considered, the thermal gradient stresses are neglected as explained before. The following assumptions are made:

1. The resin under the die-pad is considered as a plate with all edges fixed (Fig. 13-7).
2. The pressure of water is uniform at the top of the plate and the value is the pressure of saturated vapor at the temperature of thermal shock. This assumption will give the worst case. In reality, the gap between die-pad and resin does not cover the entire surface. Moreover, it supposes that the water content in the molding compound is at

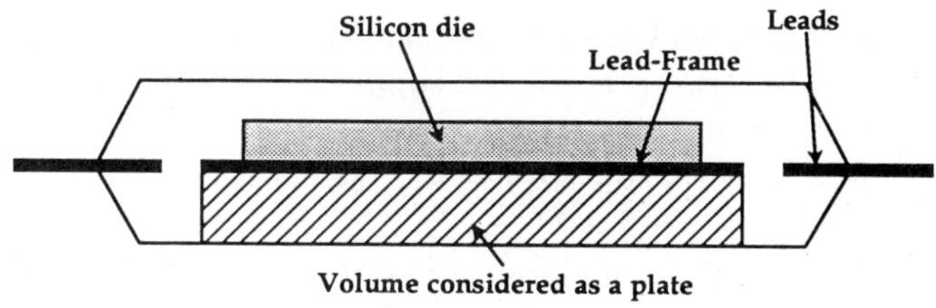


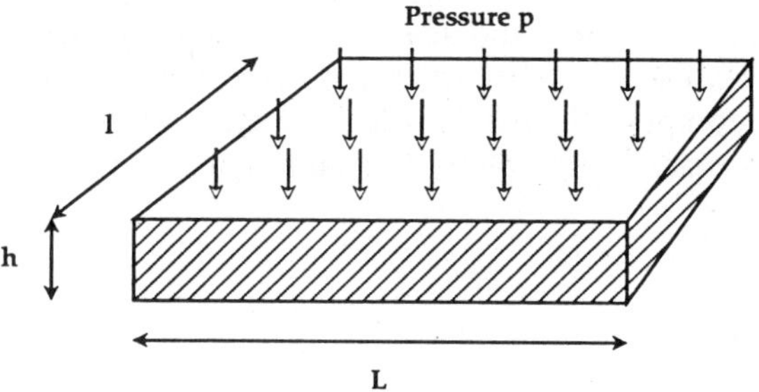


**Figure 13-7**   Cross-section of a plastic package with the plate submitted to high pressure.

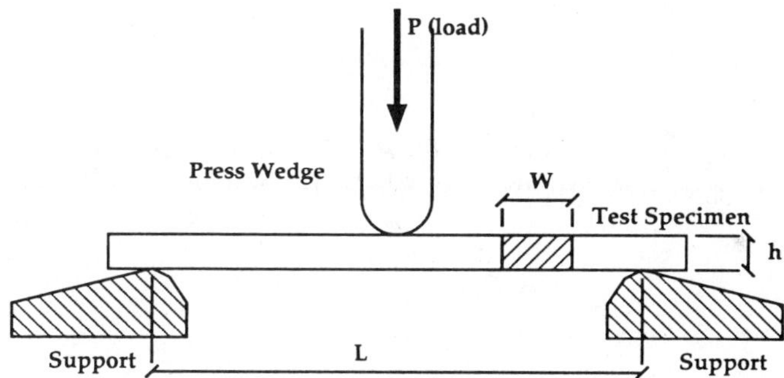


**Figure 13-8**   Flexural modulus and flexural strength measurement apparatus.

saturation. It is reasonable to suppose that just before cracking these conditions are met.

3. A crack will initiate where the energy of deformation reaches a critical value. This critical value is determined by considering the test method for the measurement of flexural strength and flexural modulus (Fig. 13-8). The flexural strength is given by

$$\sigma_f = \frac{3PL}{2Wh^2} \tag{13-1}$$

where   $P$ = load on the specimen at break point
$L$ = distance between the supports
$W$ = width of the specimen
$h$ = height of the specimen

From pure bending of a bar, it can easily be shown that the flexural strength is the maximum tensile stress reached at the bottom of the bar under test. This assumption is true under two conditions: (1) that fracture mode is of type I (due to tensile stress), and (2) that the deformation energy is calculated in the linear case, but before fracture the material yields. As a consequence, we can compare the limit in the two cases if the measurement of flexural strength is made with exactly the same conditions, especially of temperature.

**Criterion Determination**

In the case of a uniformly loaded rectangular plate, only the vertical displacement $w$ is considered and is found by solution of[31]

$$D \times \Delta(\Delta w) = p \tag{13-2}$$

where $D$ = is the flexural rigidity and $p$ is the applied pressure.
As the plate is fixed at the four edges, the elasticity solution of (13-2) is

$$w = C(1 - \alpha^2)^2(1 - \beta^2)^2 \tag{13-3}$$

where

$$\alpha = \frac{2x}{L}, \qquad \beta = \frac{2y}{l}$$

$L$ and $l$ are the sides of the rectangular plate in the $x$ and $y$ directions, respectively, and the origin of the $xy$ Cartesian coordinates is at the center of the plate. $C$ is given in ref. 31, and is the maximum deflection in the center of the plate:

$$C = (1 - v^2)\frac{pl^4}{Eh^3}\frac{0.032}{(1 + (l/L)^4)} \tag{13-4}$$

where $v$ is Poisson's ratio, and $E$ is Young's modulus.

The strain components are

$$\varepsilon_{xx} = -z\,\frac{\partial^2 w}{\partial x^2}, \qquad \varepsilon_{xy} = -z\,\frac{\partial^2 w}{\partial x\,\partial y}$$

$$\varepsilon_{yy} = -z\,\frac{\partial^2 w}{\partial y^2}, \qquad \varepsilon_{xz} = 0 \tag{13-5}$$

$$\varepsilon_{zz} = 0, \qquad \varepsilon_{zy} = 0$$

thus

$$\varepsilon_{xx} = -z\,\frac{16C}{L^2}\,(3\alpha^2 - 1)(1 - \beta^2)^2$$

$$\varepsilon_{yy} = -z\,\frac{16C}{l^2}\,(3\beta^2 - 1)(1 - \alpha^2)^2 \tag{13-6}$$

$$\varepsilon_{xy} = -z\,\frac{64C}{lL}\,\alpha\beta(1 - \alpha^2)(1 - \beta^2)$$

and

$$\sigma_{xx} = A\{(3\alpha^2 - 1)(1 - \beta^2)^2 + va^2[(3\beta^2 - 1)(1 - \alpha^2)^2]\}$$

$$\sigma_{yy} = A\{v(3\alpha^2 - 1)(1 - \beta^2)^2 + a^2[(3\beta^2 - 1)(1 - \alpha^2)^2]\} \tag{13-7}$$

$$\sigma_{xy} = 4A(1 + v)a\alpha\beta(1 - \alpha^2)(1 - \beta^2)$$

where

$$A = -\frac{16EC}{(1 - v^2)}\,z, \qquad a = \frac{L}{l}$$

Von Mises stress is defined from the second invariant of the stress tensor as

$$2\sigma_{VM}^2 = (\sigma_{xx} - \sigma_{yy})^2 + (\sigma_{xx} - \sigma_{zz})^2 + (\sigma_{yy} - \sigma_{zz})^2 + 6(\sigma_{xy}^2 + \sigma_{xz}^2 + \sigma_{yz}^2) \tag{13-8}$$

In this case, von Mises stress varies proportionally as the square of $z$. The maximum is reached at $z = h/2$ and $z = -h/2$.

The variation of the square normalized von Mises is shown over a quarter-plate in Fig. 13-9 for $z = h/2$ where $x$ and $y$ are normalized as $\alpha$ and $\beta$ vary from 0 to 1. The maximum is reached at the center of the longest side. The value of this maximum is equal for two points at $z = +h/2$ and $z = -h/2$. At $z = +h/2$ stress is tensile, whereas at $z = z/2$ stress is

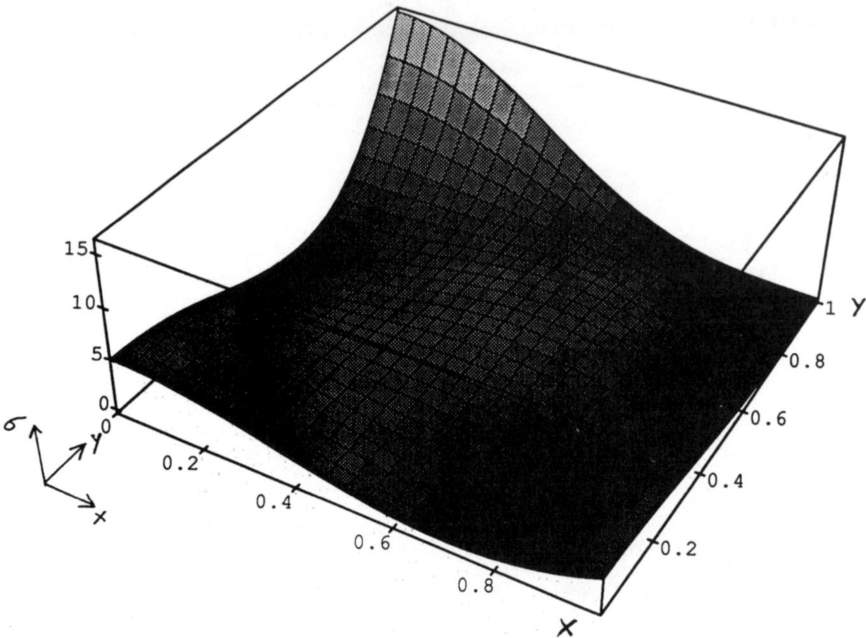

**Figure 13-9**  Distribution of $\sigma/A^2$ at $z = h/2$ for a quarter of the die-pad. The reference origin is at the center of the die-pad. The $x$-axis is parallel to the longest side.

compressive. From the analogy with the flexural strength measurement, the crack failure will occur at $z = h/2$.

Two conclusions can be drawn from this model:

1. Crack initiation in the case of thermal shock is located along the edge of the die-pad–resin interface at the center of the longest die-pad size.
2. The risk criterion can be determined from the maximum value of von Mises stress. As discussed previously in the case of flexural strength measurement, the stress is tensile at the bottom of the bar and is equal to the flexural strength just before breaking. As it is the unique stress component, it is also von Mises stress. Therefore, applying $\alpha = 0$, $\beta = 1$, and $z = h/2$ leads to

$$\sigma_f = 0.512\sqrt{1 - v + v^2}\ \frac{pl^2}{h^2}\ \frac{1}{(1 + (l/L)^4)} \tag{13-9}$$

where  $\sigma_f$ = the flexural strength of resin at temperature of thermal shock
$v$ = the Poisson coefficient at the same temperature

$p$ = the saturation pressure of water at the same temperature
$L$ and $l$ = length and width of the die pad
$h$ = the thickness of resin under the die pad

## Examples

The saturated vapor pressure at this temperature is about 25 kg/cm$^2$ at 215°C. With a die-pad of 1 cm × 1 cm, the criterion is

$$\sigma_f = \frac{5.68}{h^2}$$

where $h$ is in cm and $\sigma_f$ is in kg/cm$^2$.

A resin thickness of 1.4 mm under the pad gives a maximum stress of 290 kg/cm$^2$. The flexural strength at 215°C of regular resin is in this range; therefore, the package will not crack whatever is the degree of delamination. Low-stress resins have flexural strengths in the range 130–150 kg/cm$^2$ at 215°C, and in this case it is clear that the limit is exceeded. It does not mean that the crack will occur systematically, but a low level of moisture content inside the package must be kept.

Molding compounds with high strength at high temperature prevent cracking due to thermal shocks. The $1/h^2$ dependence of maximum stress indicates that increasing the package thickness under the die may be an effective way of retarding crack initiation.

Another significant parameter for package cracking is the temperature of the thermal shock; reducing this temperature to 200°C leads to a saturated vapor pressure of about 18 kg/cm$^2$. In an actual case, this pressure is less than saturated water pressure as the package is not saturated. A more accurate evaluation of this pressure refines the risk estimation of cracking.

### 13.4.2 Stress Measurement

In this section, two ways of measuring stress and stress effects at the surface of the die are presented; the first uses a piezoresistive gauge, the second uses a metal displacement test die.

### Piezoresistive Gauge

The piezoresistive effect is a stress-induced variation in the components of the resistivity tensor. Its formulation is complex and involves a four-dimensional tensor.[30,32] In the case of silicon, this tensor is simplified and can be expressed in reduced index notation:

$$\frac{\Delta\rho_\alpha}{\rho_{0\alpha}} = \sum_{\alpha=1}^{6} \Pi_{\alpha\beta}\sigma_\beta \qquad (13\text{-}10)$$

where $\alpha$ and $\beta$ vary from 1 to 6 and are the reduced indexes ($1 = 11, 2 = 22, 3 = 33, 4 = 12, 5 = 13, 6 = 23$)

$\rho_\alpha$ are the components of the conductivity tensor with reduced index notation

$\rho_{0\alpha}$ are the components of the reference conductivity tensor (no stress applied)

$\Pi_{\alpha\beta}$ is the piezoresistivity tensor with reduced index notation

If the reference system follows the principal axis of the silicon crystal, then

$$\Pi_{\alpha\beta} = \begin{bmatrix} \Pi_{11} & \Pi_{12} & \Pi_{12} & 0 & 0 & 0 \\ \Pi_{12} & \Pi_{11} & \Pi_{12} & 0 & 0 & 0 \\ \Pi_{12} & \Pi_{12} & \Pi_{11} & 0 & 0 & 0 \\ 0 & 0 & 0 & \Pi_{44} & 0 & 0 \\ 0 & 0 & 0 & 0 & \Pi_{44} & 0 \\ 0 & 0 & 0 & 0 & 0 & \Pi_{44} \end{bmatrix} \qquad (13\text{-}11)$$

The values of resistance change in other directions are computed using the standard change of coordinate system of a tensor.[30,32]

Figure 13-10 shows a rosette type that has been used for stress measurement. Resistors are low-doped n-type and are electrically isolated from the substrate. Typical values of resistors are in the range 1000–1200 ohms. Four resistors are combined to form a unit that allows the measurement of the

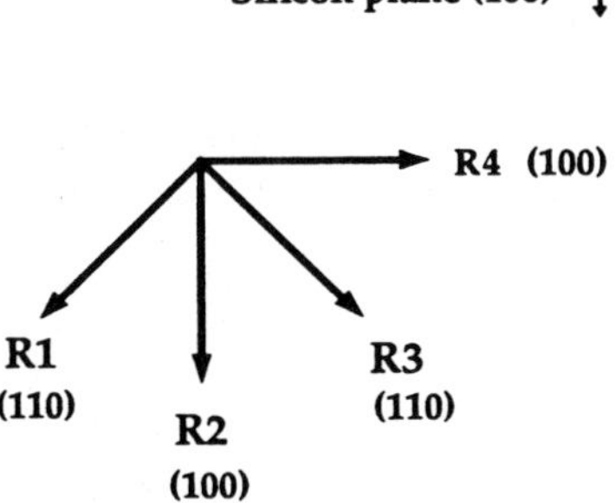


**Figure 13-10**   Rosette for stress measurement. Crystal directions are indicated considering the symmetry.

three in-plane stress components. The plane of silicon is (100), thus two resistors are referred as (100) and the two others as (110). This leads to the equation system

$$\Delta R_1/R_1 = \tfrac{1}{2}\sigma_x(\Pi_{11} + \Pi_{12}) + \tfrac{1}{2}\sigma_y(\Pi_{11} + \Pi_{12}) + \sigma_z\Pi_{12} - \tau_{xy}(\Pi_{11} - \Pi_{12})$$

$$\Delta R_2/R_2 = \tfrac{1}{2}\sigma_x(\Pi_{11} + \Pi_{12} - \Pi_{44}) + \tfrac{1}{2}\sigma_y(\Pi_{11} + \Pi_{12} + \Pi_{44}) + \sigma_z\Pi_{12}$$

$$\Delta R_3/R_3 = \tfrac{1}{2}\sigma_x(\Pi_{11} + \Pi_{12}) + \tfrac{1}{2}\sigma_y(\Pi_{11} + \Pi_{12}) + \sigma_z\Pi_{12} + \tau_{xy}(\Pi_{11} - \Pi_{12})$$

$$\Delta R_4/R_4 = \tfrac{1}{2}\sigma_x(\Pi_{11} + \Pi_{12} + \Pi_{44}) + \tfrac{1}{2}\sigma_y(\Pi_{11} + \Pi_{12} - \Pi_{44}) + \sigma_z\Pi_{12}$$

$$(13\text{-}12)$$

From this set of equations, stresses are determined for each rosette. It is important to note that this system of equations is not independent; thus only three components are accessible. In most cases, the stress component perpendicular to the silicon plane is negligible; thus, normally, the three in-plane components are computed.

From the experimental point of view, this measurement is not easy for several reasons: (1) The piezoresistive coefficients depend on temperature and the measurement must be done at constant temperature before die attach, after die attach and after molding. (2) The variation of resistance for the stress variation we consider is about 30 ohms and the discrepancy between samples is large. Therefore, only a statistical measurement is possible. As the shear stress is often low, its measurement is not very reliable.

Figure 13-11 shows a view of a basic cell of the piezoresistive chip developed

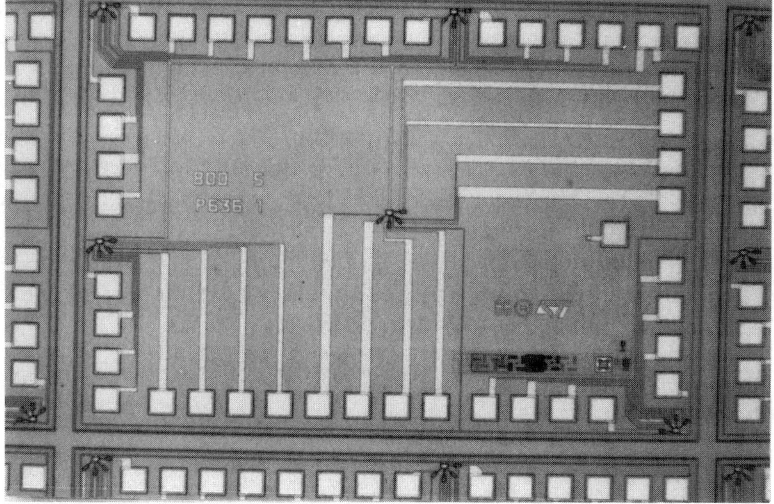

**Figure 13-11**    View of a basic cell of the piezoresistive chip (P636).

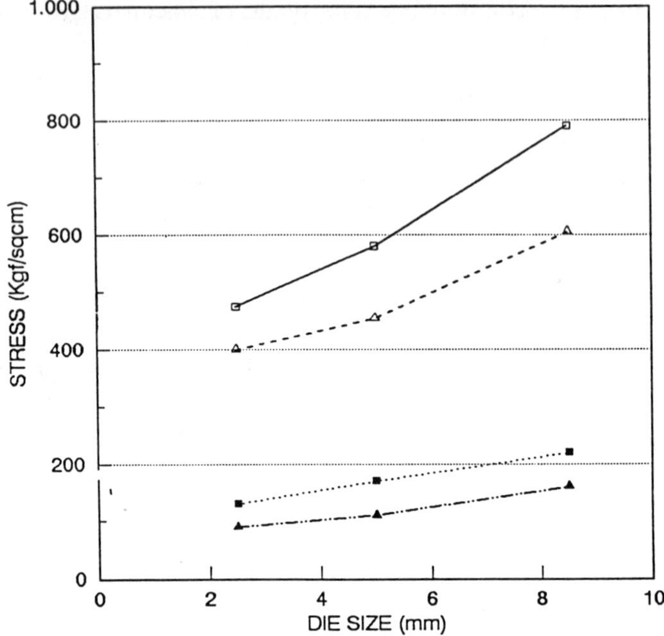



**Figure 13-12**   Compressive stresses at die center ($\square$, $\triangle$) and in-plane shear stresses at die corner ($\blacksquare$, $\blacktriangle$) vs. die size for two resin types: $\square$, $\blacksquare$, resin A; $\triangle$, $\blacktriangle$, resin B.

at SGS-Thomson for stress measurement. The size of this cell is 114 mils × 82 mils. Using different arrays of these basic cells, the evolution of stress as a function of die size is possible. This has been done for a PQFP package with 28 mm × 28 mm body size and two different low-stress resins.

Figure 13-12 shows the stress curves as a function of die size measured after molding. Resin A induced more stresses than resin B. The level of compressive stress at die center follows proportionally the level of in-plane shear stress at the die corner whatever the die size. Another interesting aspect of these curves is the variation of stresses vs. die size. The variation of stresses is quasi-linear in the range of 2 mm$^2$ to 9 mm$^2$ of die size. The increase of stress is about 50%. Utilization of low-stress (B-type) resin reduces the stress level in the same proportion.

## Metal Displacement Chip

In order to correlate data from the piezoresistive gauge, a specific chip (SD1) was developed with only one metal and one passivation layer. Figure 13-13 shows a basic cell (1.3 mm × 1.8 mm) of this chip; modules are made

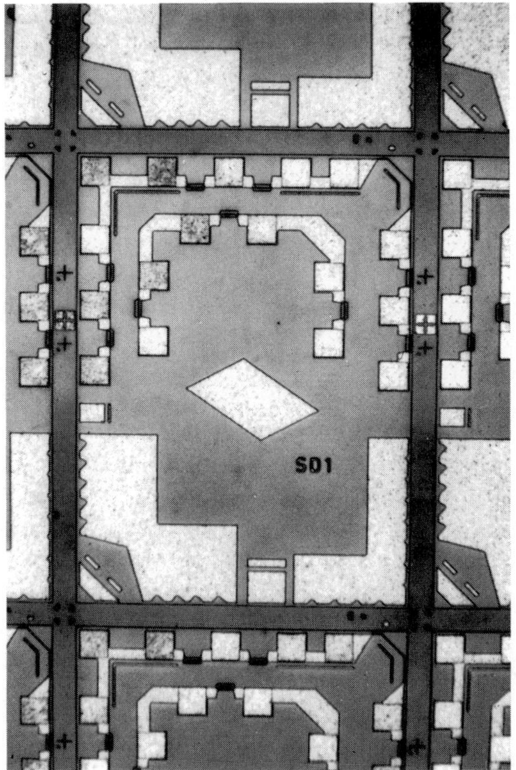


**Figure 13-13**   View of a basic cell of the metal displacement chip (SD1).

with arrays of these basic cells and allow the study of passivation cracks and metal displacements as a function of die size. The geometry on the right side includes pads and aluminum lines without passivation. These metal lines act as fuses when they are exposed to moisture. The other geometries are large metal patterns used for visual inspection of passivation cracks and measurement of metal displacements. Some notches along the metal patterns increase the stress sensitivity. Metal patterns have various shapes to simulate several types of geometry used in integrated circuits. Metal thickness is 3 μm and pattern width varies from 100 μm to 300 μm. Metal lines are 5 and 10 μm wide. Fuses are 5 μm wide.

These chips are mounted in plastic packages following the standard process and are opened with acid after post-molding curing and after thermal cycling. Visual inspections are done for passivation cracks and metal displacements are measured. This chip has been used to study the behavior of different resins.

Failure mode analysis and visual inspection indicate that:

1. Just after molding and post-curing, cracks in passivation appear only in module corners and at metal pattern notches. Crack size and number depend on the resin. The crack density is correlated with the stress level measured by a P636 piezoresistive chip. Figure 13-14 shows the difference between two resins; one is so-called low-stress while the other is a standard resin. The results obtained with a large die (4 × 5 units) and also with a small die (1 unit) demonstrate that internal stresses generated by the low-stress compound are large enough to cause cracks on the passivation layer.
2. After a few thermal cycles ($-55°C/150°C$), cracks appear all around the die: the number of cracks and their size increase with thermal cycles. Metal displacements appear all around the die after hundreds of thermal cycles. Figure 13-15 shows the metal shift distribution after 1000 thermal cycles on a module containing 30 units (5 × 6 units). The maximum shift in the metal line is 10 μm and passivation damage is limited to a band about 500 μm wide around the chip.

Figure 13-16 illustrates the width of damaged area vs. the die size. The die size is a very significant parameter in determining the damaged area.


### 13.4.3 Stress Analysis

Next, package modeling and stress analysis are conducted for correlation with experimental results on metal shift displacements. The aim of modeling

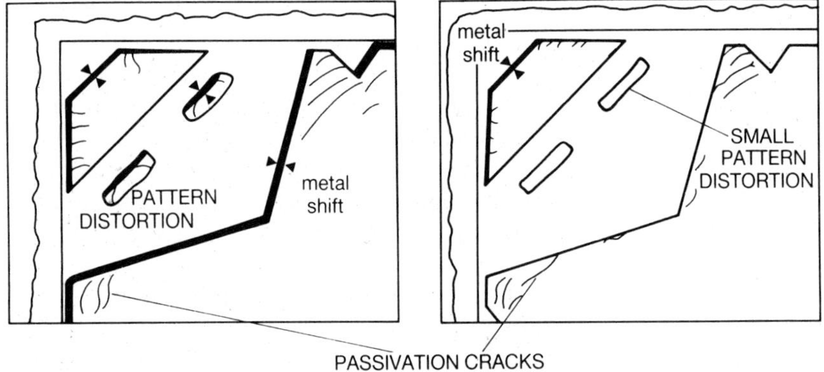


**Figure 13-14** Passivation cracks and metal displacement for two types of resins under the same conditions.

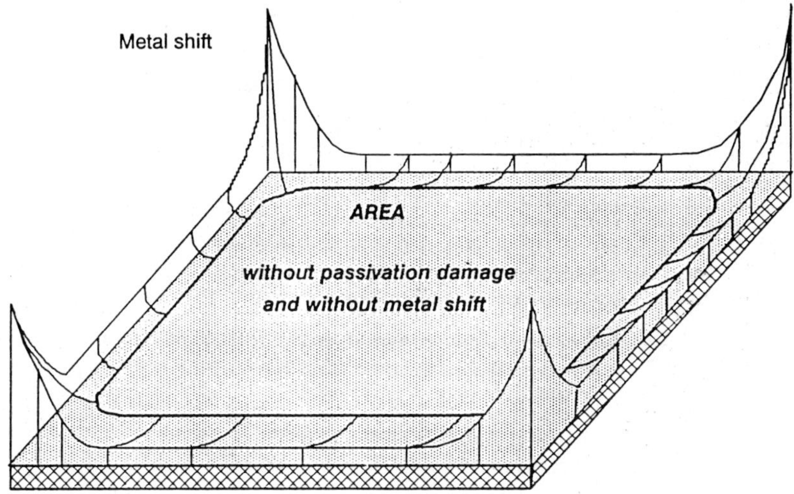



**Figure 13-15**    Map of metal shift distribution at the surface of the silicon.

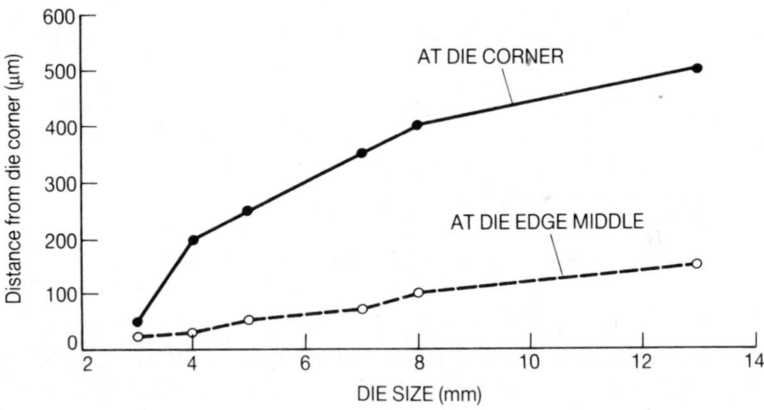



**Figure 13-16**    Width of damaged area vs. die size. Solid line, at die corner; dashed line, at die edge middle.

is threefold:

To find the stress distribution inside the package (after fabrication and through its life)

To correlate the stress distribution with experimental data and known failures

From the above, to design more reliable packages

There is a common belief that sophisticated modeling leads to a better prediction of mechanical behavior. This may be true if all variables can be accurately managed, as in research. In engineering, it is a matter of fast decision and reliable evaluation. A problem always has several variables and, prior to conducting finite element analysis, useless or nonsignificant variables for the parameters under investigation must be eliminated.

The experimental design is an example of this principle. First, target functions and operating variables are defined within their working range. Secondly, their importance is determined in their local domains with a first-order approximation (linear). Thirdly, verification of the prediction accuracy is done. These steps are iterated with higher polynomial degrees until the prediction accuracy is acceptable. The basic purpose of this method is to give not a universal model of a physical phenomenon but a local and accurate model for optimizing a physical process.

This methodology is valid as long as the analysis stays in the same domain with the same materials and equipment. Recipes are available for experimental design; however, recipes are not available for mechanical modeling as a more complex understanding of physical phenomena is required.

A gradual approach is necessary for a static thermomechanical analysis:

Which stresses are to be examined for the failure mechanism under study?
Is there an analytical solution to calculate stresses?
Is 2D finite element modeling sufficient? In which plane? Is linear analysis sufficient? Nonlinear?
Is 3D finite element linear modeling required?
Is 3D finite element nonlinear modeling required?
Are specific elements needed (e.g., friction)?
Are exotic models needed (e.g., specific rheological model)?

At each step, the availability of material data is a major problem and often experimental studies are needed to determine basic material data (Young's modulus, coefficient of thermal expansion as a function of temperature, local rheological models). If the failure model is generated by a temperature gradient (steady or transient behavior), the same approach can be followed.

It is beyond the scope of this paper to detail all these points; only the example of metal displacement in plastic packages is treated in the following sections.


### 13.4.4 Basic Assumptions for Plastic Package Modeling

The diagram in Fig. 13-17 shows the temperature variations through the assembly process of plastic packages. Stresses are generated by two assembly

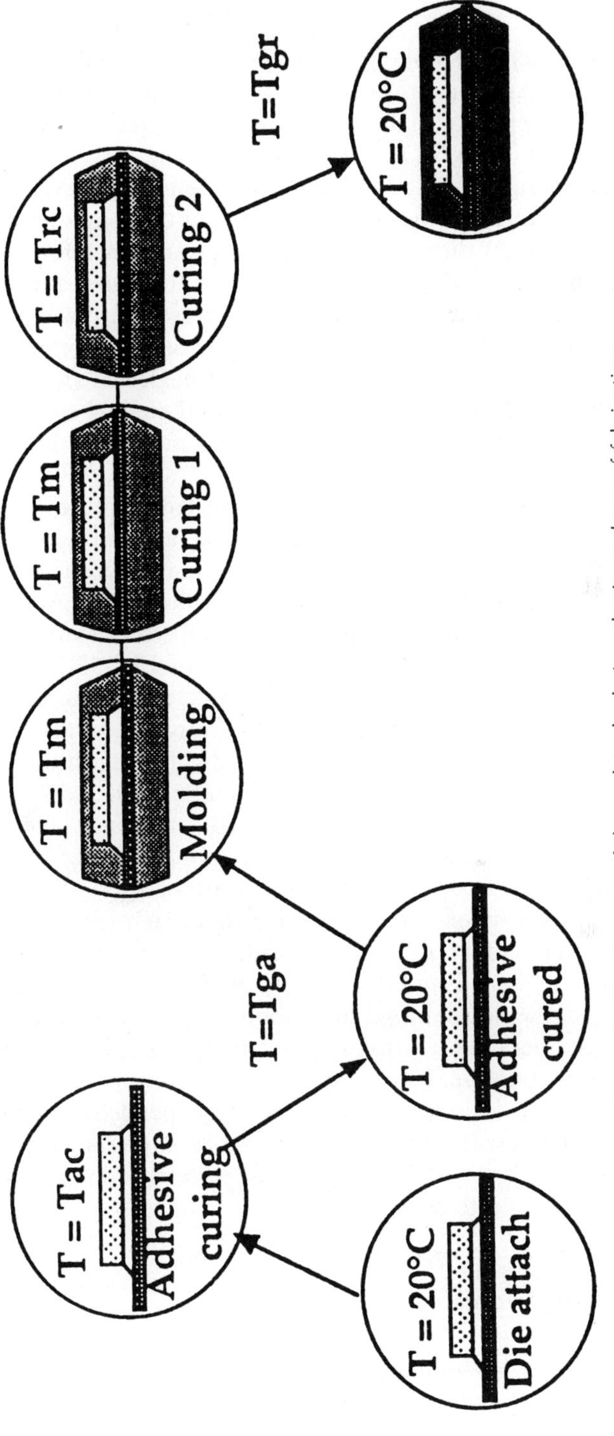


**Figure 13-17**  Diagram of thermal cycle during plastic package of fabrication.

operations: die-attach and molding. Several materials with different coefficients of temperature expansion are attached at high temperature; thus, at this stage, they are supposed to be stress-free. After the structure is cooled, differential contractions give rise to stresses.

The structure is supposed to be at a constant temperature (20°C). Actually, under working conditions, the surface of a circuit dissipates heat and creates a temperature gradient throughout the package and, therefore, a stress field is superposed on the existing steady stress field. It is important to note that if the temperature of the silicon surface increases, stresses at the silicon surface relax and thus the constant-temperature assumption is a worst case.

In this first approach local stress effects are neglected (wire bonding, IC multiplayer structure). Local analysis is a second step when global analysis is not sufficient and when global behavior does not match local effects.

Curing temperature of the die adhesive is $T_{ad}$ (e.g., 160°C) and curing temperature of resin is $T_{re}$ (e.g., 180°C). These temperatures could be used as stress-free reference temperatures, but resin and adhesive are organic materials and their internal structure and mechanical behavior change at a temperature called the glass transition temperature ($T_g$). Above this temperature they are much more compliant and temperature variation does not generate high stresses. This fact is illustrated in Fig. 13-2 where the die adhesives generating very low bending are low-$T_g$ adhesives. The actual case is that the stress-free temperatures are slightly above $T_g$ to take into account the difference between curing temperature and glass transition temperature. Glass transition temperatures of die adhesive and resin are $T_{gad}$ and $T_{gre}$.

As there are two steps of curing, the set chip + adhesive + lead frame has a stress-free reference temperature $T_{gad}$ and a second state is superposed when molding with a stress-free reference temperature of $T_{gre}$.

If a high curing temperature is used for die-attach (polyimide adhesive) the two stress-free temperatures are considered. But often, epoxy adhesive and epoxy resin are utilized for plastic packages with glass transition temperatures in the same range and about 10°C below the resin curing temperature. In this case, taking $T_{re}$ as a unique stress-free reference is a reasonable approximation.

It is important to account for temperature dependence of the material. The cooling from stress-free temperature covers a large range of temperature and if a unique value is kept it must be properly integrated over the temperature range of interest.


### 13.4.5 Metal Displacement

This example is an attempt at modeling the metal displacement shown in the previous sections and trying to find a limit value of the stress

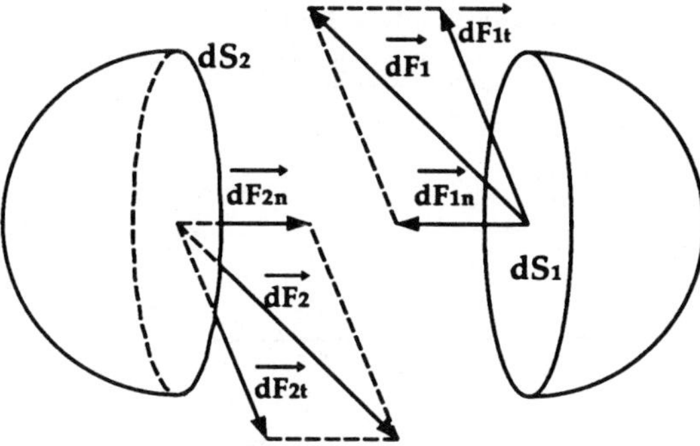


**Figure 13-18**    Small volume of matter cut by a plane with forces acting on each face.

components that can be evaluated by finite element modeling of the structure.

Metal shifts and passivation cracks occur during thermal cycling but the assumption to be checked is that this metal displacement is correlated with the shear stress level reached just after molding. First, the correct component of shear stress must be determined.

The basic definition of normal stress and shear stress is explained from a small volume of matter cut by a plane as shown in Fig. 13-18. Two opposite forces $d\mathbf{F}_1$ and $d\mathbf{F}_2$ apply to the opposite faces. The absolute values of these forces are equal as the matter is locally in equilibrium. If $\mathbf{n}$ is the vector normal to the surface $dS$ (direction shown in Fig. 13-18) and $d\mathbf{F}_1$ is broken up into two components, one normal and the other parallel to $dS$, $d\mathbf{F}_{1n}$, and $d\mathbf{F}_{1t}$, the normal stress is

$$\frac{d\mathbf{F}_{1n}\cdot\mathbf{n}}{dS}$$

and the shear stress is

$$\frac{d\mathbf{F}_{1t}}{dS}$$

This defines which stresses apply in a given direction and helps the determination of the continuity of stress components at interfaces as described in ref. 33.

Figure 13-19 shows a quarter of a silicon die with the direction of metal displacement at the top surface. Metal displacement occurs towards die center. The plane to be considered is the Al/Si interface (plane $xy$). The shear stress is defined by the force in the plane with the direction shown in Fig. 13-19. (Note that this stress component is continuous across the interface.)

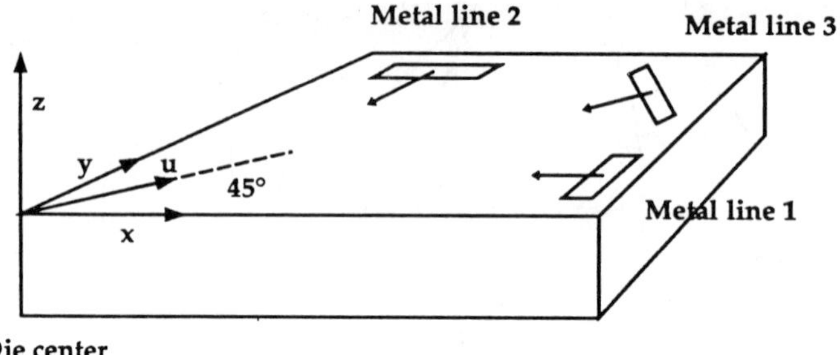


**Figure 13-19**   Quarter of a silicon chip showing the direction of metal displacement.

Therefore

For metal line 1, the $xz$ shear stress ($\tau_{xz}$) causes metal shift.
For metal line 2, the $yz$ shear stress ($\tau_{yz}$) causes metal shift.
For metal line 3, the $uz$ shear stress ($\tau_{uz}$) causes metal shift.

These considerations lead to two conclusions:

1. Often, only the $xy$ shear stress is considered. Applying the definition of shear stress, $\tau_{xy}$ is related to a plane perpendicular to the surface of the die. It can only create a fracture in silicon related to crystal plane shift and it is not a failure induced by plastic packaging.
2. Piezoresistive gauges measure $\tau_{xy}$ and can be only used for experimental correlation of the model.

Figures 13-20 to 13-22 are the results of a finite element model (linear case) for a 3 mm × 3 mm silicon chip molded in the PQFP 160 which has been used for the measurement described before. The molding compound is the type B described in Fig. 13-12. The parameters for modeling are extracted from manufacturer's data sheet (e.g., $T_{ref} = T_g = 170°C$ for molding compound). The unit for stress values is N/mm$^2$ and must be multiplied by 10 to be converted to approximately kg/cm$^2$.

Figure 13-20 shows the $\tau_{xz}$ distribution at the silicon surface. The $\tau_{yz}$ distribution is the same along the $y$ axis. The superposition of the two distributions matches the metal shift map in Fig. 13-15. This is not true for the $\tau_{xy}$ distribution shown in Fig. 13-21. Figure 13-22 shows the compressive stress $\sigma_{xx}$.

From these results, the correlation can be done with the experimental results. From Fig. 13-12, in-plane shear stress at the die corner is 100 kg/cm$^2$ and corresponds to the value in Fig. 13-21. (Notice that the maximum

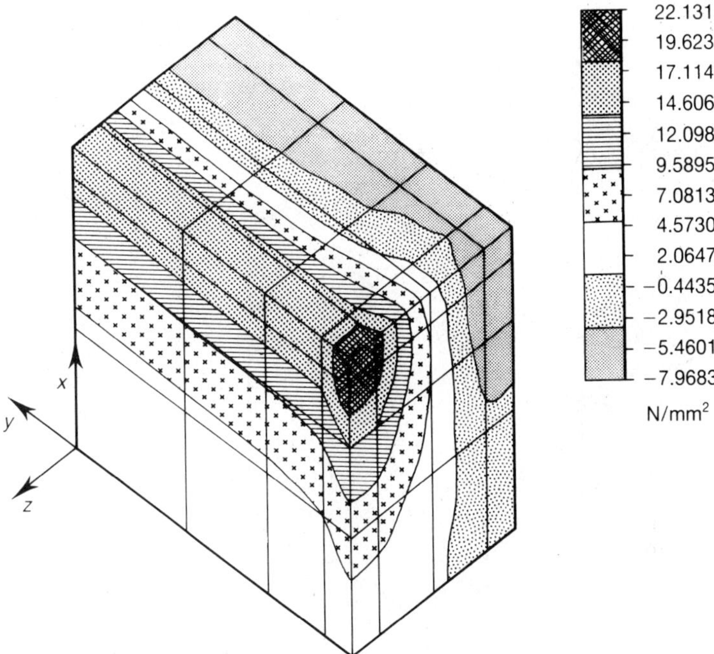



**Figure 13-20**  $\tau_{xz}$ distribution at the silicon surface. (One-quarter of silicon chip is shown; reference origin is located at the center of the die; the plane *xy* is the active side of the IC where metal lines are located.)

126 kg/cm² is exactly at the corner, the rosette on piezoresistive gauge is located at about 50 μm from the corner.) The value of compressive stress at the center of the die is 400 kg/cm² and the modeled value is 350 kg/cm². This difference comes from the meshing, which is very large in this area. From Fig. 13-16, the damaged area is located in a band 50 μm wide around the chip. From Fig. 13-20, this band corresponds to a shear stress of 150 kg/cm². For this technology, which is not representative of actual technology (thick metal line), metal displacement occurs after thermal cycling if the shear stress exceeds 150 kg/cm². The shear stress is defined by the direction of metal displacement and direction perpendicular to the metal line plane. This value can be considered as a worst-case limit. The determination of this value can be refined for a specific technology.

The procedure for ensuring reliable packaging for a specific IC technology and determining the defect-free area for this technology is the following:

1. Make a prototype lot with chips in the technology under evaluation and determine the stress-sensitive area.

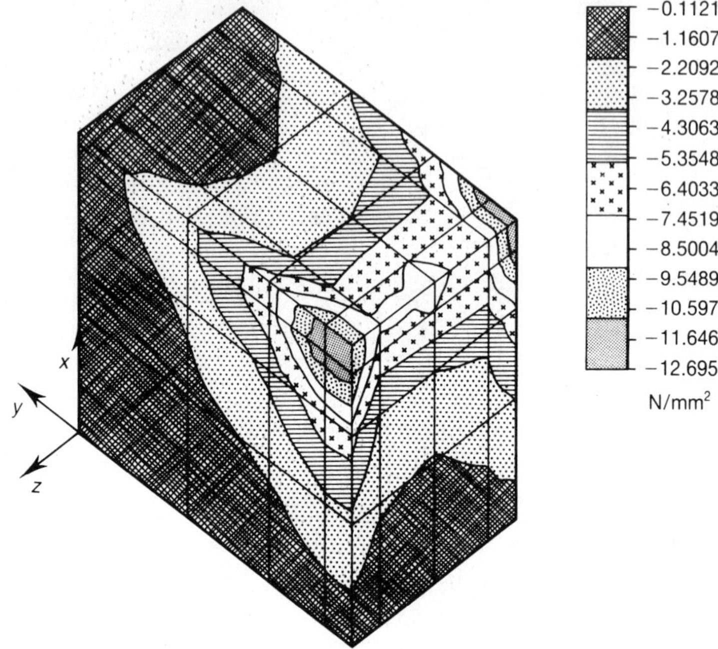



**Figure 13-21** $\tau_{xy}$ distribution at the silicon surface. (One-quarter of silicon chip is shown; reference origin is located at the center of the die; the plane $xy$ is the active side of the IC where metal lines are located.)

2. Mount a prototype lot with a piezoresistive chip and measure the three stress components ($\sigma_{xx}$, $\sigma_{yy}$, $\tau_{xy}$).
3. Model the structure and correlate with piezoresistive gauge measurements.

At this step the shear stress limit can be determined. This limit allows the determination of defect-free area at the surface of the silicon for each new die size. In the same way, the introduction of new materials can be assessed and structural optimization is possible.


## 13.5 SUMMARY

After a rapid review of plastic package structure and fabrication, several stress-related failures have been reviewed: die/adhesive cracks, passivation cracks, displacement of metal lines, and resin cracks. In order to correlate these failures with stress level and to optimize package structures, some tools for stress measurement and stress effects have been described.

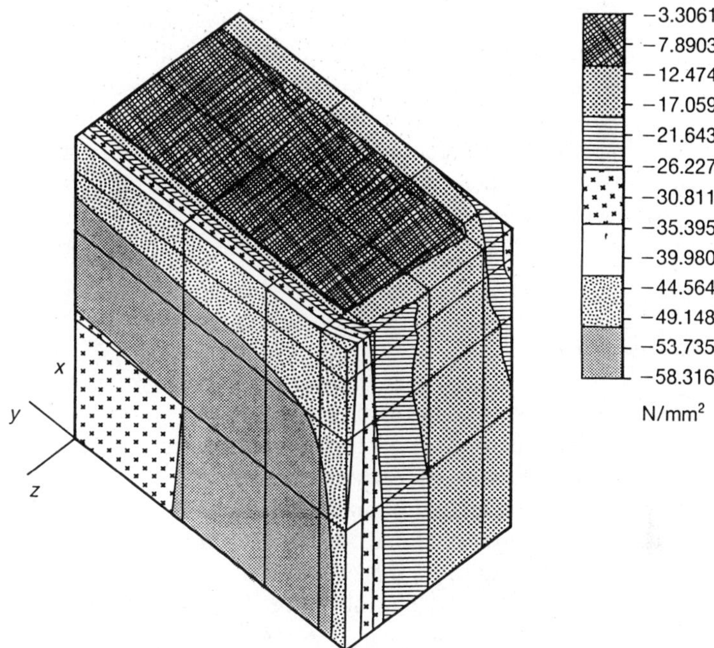



**Figure 13-22**  $\sigma_{xx}$ distribution at the silicon surface. (One-quarter of silicon chip is shown; reference origin is located at the center of the die; the plane $xy$ is the active side of the IC where metal lines are located.)

The piezoresistive chip gives only three stress components at the die surface. The study of these three stress components for various die sizes shows that the in-plane shear stress at the corner does not vary much compared to the variation between regular and low-stress resin. The choice of low-stress resin is the best way to minimize residual stresses.

The metal displacement chip allows the mapping of passivation cracks and metal displacement. Passivation cracks occur after molding and increase with thermal cycling. Crack generation is not very sensitive to the choice of the resin. This chip was used to map metal displacement after thermal cycling. The metal displacements occur after thermal cycling and are located in a band 500 µm wide around the chip for a 12-mm square chip.

Two examples of modeling were presented, the first to evaluate crack sensitivity due to moisture content and thermal shocks; the second is a methodology using stress measurement and finite element modeling to determine defect-free area at the surface of the silicon.

The model developed for crack sensitivity due to moisture absorption shows a $1/h^2$ ($h$ is the resin thickness under the die pad) dependence of maximum stress, which indicates that increasing the package thickness under the die is one effective way of retarding crack initiation.

The other model is a way of correlating the metal displacement as measured by the metal displacement chip with the level of shear stress component. The most significant shear stress is defined by the direction of metal displacement and direction perpendicular to the metal line plane. Correlation with metal displacement chip gives a limit of 150 kg/cm$^2$ for this shear stress. This value is a worst case as the metal displacement chip is very sensitive.

## ACKNOWLEDGMENTS

This work was carried out while the author was at SGS-Thomson, Corporate Package Development Team. The author gratefully thanks SGS-Thomson and in particular the following individuals: Carlo Cognetti and Jean-Pierre Moscicki for authorizing the use of internal SGS-Thomson results; Juan Exposito, Francois Lamourelle, and Roberto Tiziani for very useful discussions on die-attach, molding, and thermomechanical characterization. The author also thanks Jean-Paul Clech and Colin Paterson for correcting the manuscript.

## REFERENCES

1. Van Kessel, C. G. M., S. A. Gee, and J. Murphy, "The Quality of Die Attachment and Its Relationship to Stresses and Vertical Die Cracking," *IEEE Trans. Components, Hybrids, and Manufacturing Technology*, **CHMT-6**(4), 1983, pp. 414–420.
2. Van Kessel, C. G. M., and S. A. Gee, "The Use of Fractography in the Failure Analysis of Die Cracking," *10th Int. Symp. Testing and Failure Analysis*, Los Angeles, October 1984, pp. 258–264.
3. Thomas, R. E., "Stress-induced Deformation of Aluminum Metallisation in Plastic Molded Semiconductor Devices," *IEEE Trans. Components, Hybrids, and Manufacturing Technology*, **CHMT-8**(4), 1985, pp. 427–434.
4. Oizumo, S., N. Imanura, H. Tabata, and H. Suzuki, "Stress Analysis of Si-chip and Plastic Encapsulant Interface," *Nitto Technical Reports*, **51**, September 1987.
5. Manzione, L. T., *Plastic Packaging of Microelectronic Devices*, Van Nostrand Reinhold, New York, 1990, pp. 308–309.
6. Nishimura, A., A. Tatemichi, H. Miura, and T. Sakamota, "Life Estimation for IC Plastic Packages Under Temperature Cycling Based on Fracture Mechanics," *IEEE Trans. Components, Hybrids, and Manufacturing Technology*, **CHMT-12**(4), 1987, pp. 637–642.
7. Nishimura, A., and S. Kawai, "Effect of Leadframe Material on Plastic Encapsulated IC Package Cracking Under Temperature Cycling," *Proc. 39th Electronics Component Conference*, Houston, TX, 1989, pp. 524–530.

8. Fukuzawa, I., S. Ishiguro, and S. Nanbu, "Moisture Resistance Degradation of Plastic LSIs by Reflow Soldering," *Proc. IEEE Int. Reliability Physics Symposium*, CH2113-9/85/0000-0192, 1985.

9. Ohizumo, S., S. Ito, and H. Suzuki, "Analysis of Reflow Soldering by Finite Element Method," *Nitto Technical Reports*, **40**, 1987.

10. IPC-SM-786, *Impact of Moisture on Plastic IC Package Cracking*, Institute for Interconnects and Packaging Electronic Circuits, Lincoln Wood, IL, 1989.

11. Lamourelle, F., "Plastic Encapsulation of Microelectronic Devices, Study of Moisture Penetration," in French, PhD thesis, Université de Bordeaux I—IBM Compec, 1987.

12. Ohizumo, S., S. Ito, M. Nagasawa, K. Igarashi, and M. Kohmoto, "Analytical and Experimental Study for Designing Molding Compounds for Surface Mounting Devices," *Proc. 40th Electronic Components and Technology Conference*, Las Vegas, 1990, pp. 625–631.

13. Nishioka, T., S. Ito, M. Nagasawa, K. Igarashi, and M. Kohmoto, "Special Properties of Molding Compound for Surface Mounting Devices," *Proc. 40th Electronic Components and Technology Conference*, Las Vegas, 1990, pp. 632–640.

14. Dale, J. R., and R. C. Oldfield, "Mechanical Stresses Likely to be Encountered in the Manufacture and Use of Plastically Encapsulated Devices," *Microelectronics and Reliability*, **16**, 1977, pp. 255–258.

15. Timoshenko, S. P., and J. N. Goodier, *Theory of Elasticity*, McGraw-Hill, New York, 1983, pp. 284–288.

16. Usell, R. J., and S. A. Smiley, "Experimental and Mathematical Determination of Mechanical Strains Within Plastic IC Packages and Their Effects on Devices During Environmental Tests," *Proc. 19th Annual Int. Reliability Symposium*, IEEE, 1981, pp. 65–73.

17. Howel, J., "Reliability Study of Encapsulated Copper Leadframe/Epoxy Die Attach Packaging System," *Proc. 19th Annual Int. Reliability Symposium*, IEEE, 1981.

18. Groothuis, S., W. Schroen, and M. Murtuza, "Computer Aided Stress Modeling for Optimizing Package Reliability," *Proc. 23rd Annual Int. Reliability Symposium*, IEEE, 1985, pp. 184–191.

19. Edwards, D., K. G. Heinen, S. K. Groothuis, and J. E. Martines, "Shear Stress Evaluation of Plastic Packages," *IEEE Trans. Components, Hybrids, and Manufacturing Technology*, **CHMT-12**(4), 1987.

20. Natarajan, B., and B. Bhattacharayya, "Die Surface Stresses in a Molded Plastic Package," *Proc. 36th Electronics Component Conference*, IEEE, 1986, pp. 544–551.

21. Liechti, K. M., "Residual Stress in Plastically Encapsulated Microelectronic Devices," *Experimental Mechanics*, September 1985, pp. 226–231.

22. Schroen, W. H., P. S. Planton, and D. R. Edwards, "Finite Elements Analysis Application to Semiconductor Devices," IEEE, TH0238-6/88/0000-35, 1988.

23. Shoraka, F., C. A. Gealer, and E. Bettes, "Research Reveals Differences in Coating Effects on Die Stress," *Semiconductor International*, October 1988, pp. 110–113.

24. Lundstrom, P., and K. Gustafsson, "Mechanical Stress and Life for Plastic Encapsulated Large Area Chip," *Proc. 38th Electronics Component Conference*, Los Angeles, 1988, pp. 396–405.

25. Glasser, J. C., and M. P. Juaire, "Thermal and Structural Analysis of a PLCC

Device for Surface Mount Processes," *J. Electronic Packaging, Trans. ASME*, **111**(3), September 1989, pp. 172–178.
26. Simon, B. R., Y. Yuan, J. R. Umaretiya, J. L. Prince, and Z. J. Staszak, "Parametric Study of a VLSI Plastic Package Using Locally Refined Element Models," *5th IEEE Semi-Therm Symposium*, 1989, pp. 52–58.
27. Simon, B. R., Y. Yuan, J. R. Umaretiya, R. Bavirisetty, and J. L. Prince, "Thermal and Mechanical Finite Element Analysis of a VLSI Package Including Spatially Varying Thermal Contact Resistance," *6th IEEE Semi-Therm Symposium*, 1990, pp. 74–81.
28. Gee, S. A., W. F. Van Der Bogert, V. R. Akylas, and R. T. Shelton, "Strain Gage Mapping of Die Surface Stresses," *IEEE Trans. Components, Hybrids, and Manufacturing Technology*, **12**(4), 1989, pp. 587–593.
29. Tiziani, R., M. Mermet-Guyennet, and V. Motta, "Plastic Package Reliability Study by Means of Integrated Test Structures," *Proc. 8th Int. Electronic Manufacturing Technology Symposium*, IEEE CH2833-2/90/89-82530, 1990.
30. Bittle, D. A., J. C. Suhling, R. E. Beaty, R. C. Jaeger, and R. W. Johnson, "Piezoresistive Stress Sensors for Structural Analysis of Electronic Packages," *J. Electronic Packaging, Trans. ASME*, **113**(3), September 1991, pp. 203–215.
31. Boresi, A. P., and O. M. Sidebottom, *Advanced Mechanics of Materials*, 4th edn., Wiley, New York, 1985.
32. Nye, J. F. *Physical Properties of Crystals*, Oxford University Press, Oxford, 1957.
33. Lau, J. H., "A Note on the Calculation of Thermal Stresses in Electronic Packaging by Finite Element Methods," *J. Electronic Packaging, Trans. ASME*, **111**(4), December 1989, pp. 313–320.

# 14

# Solutions to Moisture Resistance Degradation During Solder Reflow of Plastic Surface Mount Components

*Suresh V. Golwalkar*

## 14.1 INTRODUCTION

Plastic packages are exposed to extreme temperatures in a surface mount process environment of infrared reflow, vapor phase reflow, hot air reflow, or when the packages are immersed in molten solder during wave soldering. When the internal temperature of the package exceeds 210°C, the risk of encountering different structural weaknesses such as package cracking, delamination, or bond cratering is increased.[1] The electrical functionality of the package is not threatened immediately on the onset of these physical weaknesses. However, the potential impact of such physical weaknesses on the long-term reliability is a matter of concern.

The category of package types affected are all plastic surface mount components—J bend and gull wing leaded packages such as plastic leaded chip carriers (PLCCs), small outline integrated circuits (SOICs), plastic quad flat packs (PQFPs), large molded tantalum capacitors, etc. Applications that do not subject packages to high internal temperatures are not at risk, e.g., socketing, through-hole attachment or reflow processes that concentrate heat on package leads only (e.g., hot bar reflow, laser soldering, or hand soldering).

The structural integrity of the plastic package is challenged by the combination of high moisture content (typically exceeding 0.13%[2,3]) and these high thermal excursions because (a) the local stresses at the metal–plastic, die–plastic interfaces exceed the adhesion strengths between the two adjoining materials and the net result is delamination; (b) the local stress

445

intensity factors near the die paddle edges exceed the fracture toughness of the plastic, resulting in package cracking; or (c) the local tensile and shear stress concentrations exceed the bond strength, culminating in bond cratering.

The factors that affect these phenomena are[2,4]

1. Geometrical configuration (die paddle size, minimum plastic thickness, etc.)
2. Amount of moisture absorbed
3. Mechanical properties of the materials (coefficient of thermal expansion, fracture strength, etc.)
4. Molding compound adhesion
5. Reflow process parameters (temperature, ramp, etc.)

Current understanding of the mechanisms and the roles of the different variables is described in the sections below.

Since moisture has been identified as the key variable, current practice in the industry is to control the moisture by shipping the baked surface mount components in a dessicant bag and finishing the board attachment operation by maintaining the packages "dry." The component suppliers also evaluate the compatibility of new packages with the boards by means of "pre-conditioning,"[1,4] i.e., simulating surface mount processes including rework (Fig. 14-1) and assessing the hazards. The further search for better solutions is progressing through improvements targeted in the design, materials, and process areas.

## 14.2 MOISTURE-INDUCED PHENOMENA

***Package Cracking***   Plastic package cracking occurs in one of two categories (Fig. 14-2):

*Type I:*   "Popcorn cracking," i.e., bottomside cracks from die paddle, typically observed in PLCC or PQFP packages

*Type II:*  Lateral cracks between the die paddle and the lead fingers, typically observed in SOIC/SOJ packages

Figure 14-3 describes the phenomenon of "popcorn cracking" (Type I). Cracking occurs when trapped moisture inside the plastic package expands rapidly. The package expands due to the pressure build-up from the water vapor envelope and then collapses closer to its original dimension leaving cracks in the plastic shell. Threshold moisture levels of 0.13–0.2% have been reported in the literature,[2,3] beyond which package cracking is encountered. Note that die pad delamination is often a precursor of such

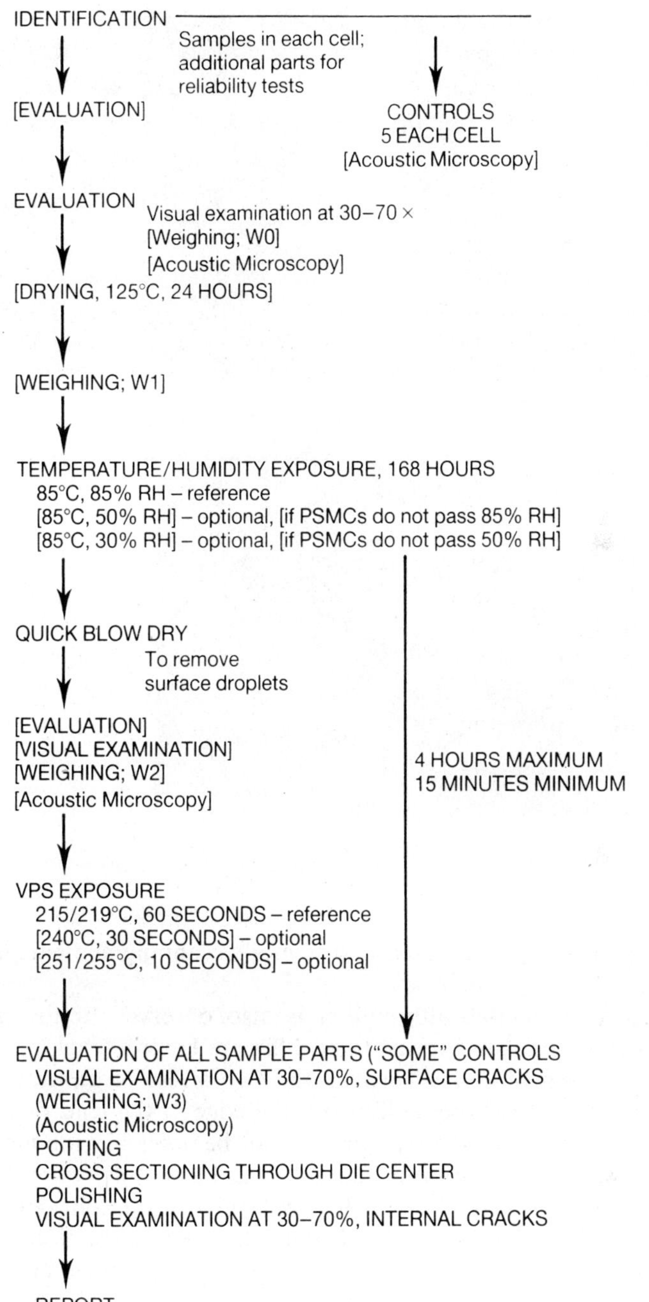


**Figure 14-1**  Preconditioning flow to evaluate moisture sensitivity of surface mount plastic packages.[1,4]

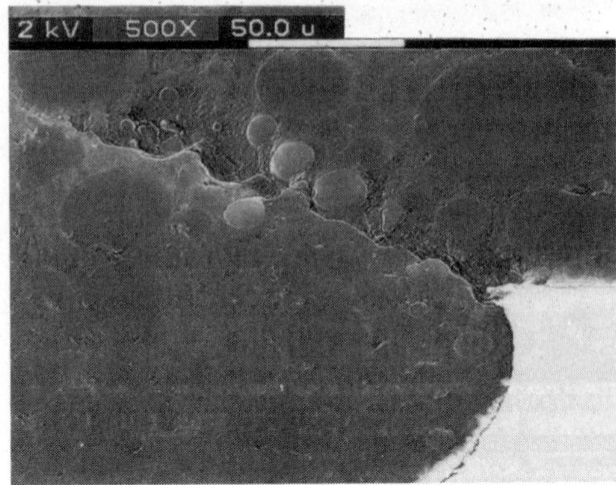


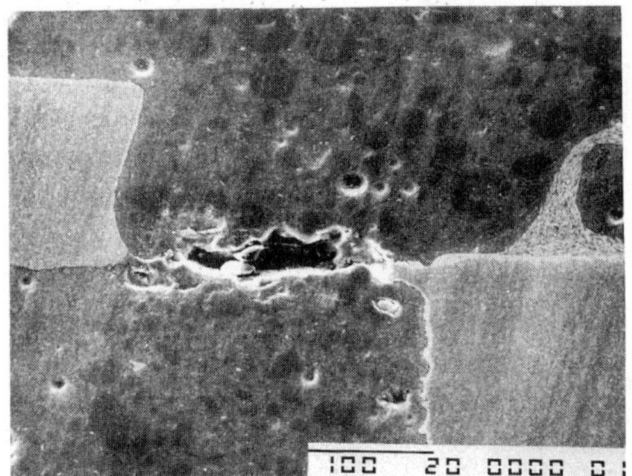


**Figure 14-2**   Examples of Type I (above) and Type II (below) cracking.

cracking and has been observed in a C-mode scanning acoustic microscope (C-SAM).[5]

For Type II cracking, although it is also observed in the presence of moisture, the mechanism appears to be different. Lateral cracking is probably driven by swelling of epoxy between the two stress-concentrating regions, namely, the edge of the die paddle and the edge of the lead finger. The delamination at the die paddle may or may not be necessary for such cracking.

Even if the crack formation is internal, board assembly rework procedures used to remove defective components may subject adjacent components to additional temperature excursions beyond the glass transition temperature of the package. This can further propagate previously formed internal cracks, if the cracks have not reached the surface until then. The ingress of flux contaminants along the crack reaching the IC surface can potentially cause bond pad corrosion leading to long-term reliability failure.

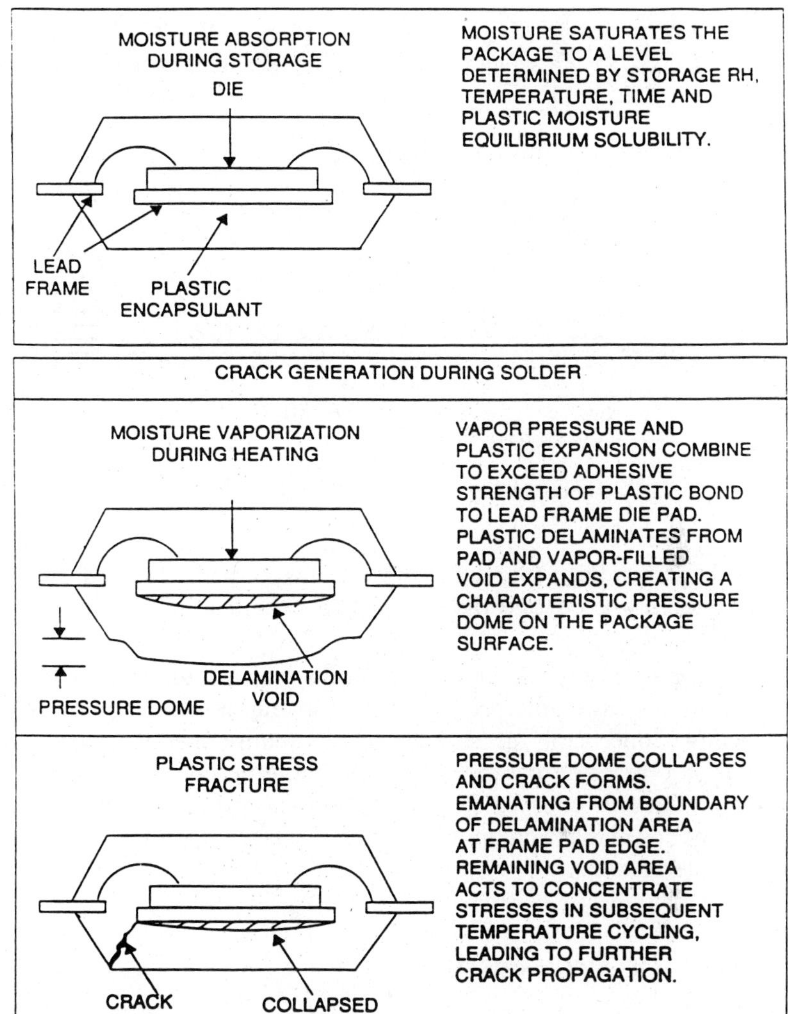



**Figure 14-3**   Phenomenon of "popcorn cracking" showing progression of delamination, bulging, and cracking.

***Bond Cratering***[6,7]   Figure 14-4 shows typical mechanism of bond cratering. The combination of severe bulging stresses, as described above in the presence of moisture, with shear stresses can cause bond lift or bond cratering. In cratering, ball bond lifts, taking with it a portion of bond pad metallization and the underlying oxide or silicon. Even though no specific moisture thresholds have been reported for bond cratering, control of moisture using

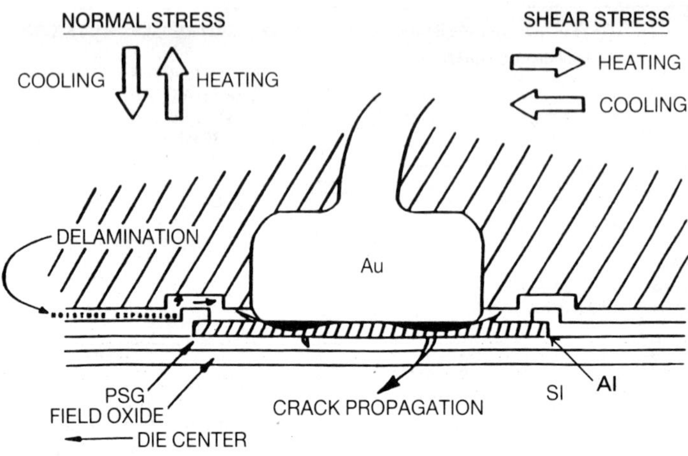



**Figure 14-4**  Schematic of bond cratering.

dessicant bags has been found to alleviate this hazard. Detailed discussion on bond cratering is beyond the scope of this chapter.

***Delamination*[5]**  The role of delamination is least understood to date. Advances in acoustic microscopy[8] are drawing more attention to this phenomenon. Conceivably, gross delamination (entire interface delamination) at the die pad–molding compound or die–molding compound interface is detrimental since it leads to further stress concentrations near the edges or at the bonds. However, more work is required in this area to reach industry-wide consensus.

## 14.3  REVIEW OF MECHANISMS

### 14.3.1  Thermal Stresses

The solder reflow temperature is always above the glass transition temperature $T_g$ (140–170°C), the temperature at which the coefficient of thermal expansion (CTE) of the plastic increases sharply. This maximizes the mismatch between the CTEs of adjoining materials—silicon, lead frame and the molding compound. This is a primary source of the thermal stress at the various interfaces. The second source of very high stresses is moisture. If plastic has absorbed excessive moisture, the resulting stresses caused by moisture flashing into steam and the swelling of the plastic will largely amplify the original plain thermal stresses.

In order to assess magnitude of the thermal stresses, the induced temperature distribution during a soldering reflow process in the plastic package is of primary importance. Temperature gradients across plastic packages during surface mounting can be simulated using 2D, plane stress FEA models.[9,10] Most authors recognize the temperature-dependent nature of the materials and the visco-elastic and visco-plastic behavior of the polymeric materials—adhesives and plastics. However, for the purpose of FEA analyses all assume isotropic linear-elastic models with temperature-independent properties and a stress-free package at the glass transition temperature of the molding compound. This is a common practice, since more complex analyses are precluded by lack of material data.

Glaser et al.[9] have measured the temperature distributions in a 68LD PLCC package during two solder reflow processes as shown in Table 14-1. These temperatures, temperature gradients, and material properties were used to predict the stress distributions via finite element analysis. The maximum stress at the peak temperature of 228°C was estimated to be close to 0.254 GPa (36 837 psi). Note that the value of this peak stress will vary with the die size. At all times during the reflow process the corners of the die and the die paddle were susceptible to higher stress intensities. The authors have emphasized an important factor that if the stress-free reference temperature is assumed to be room temperature instead of $T_g$, i.e., post mold cure and storage cause stress relaxation, then the magnitude of the above stress level is even higher.

**Table 14-1**  Typical Temperature Variation Within a 68LD PLCC Package as a Function of Time in the Solder Reflow Process[9] (Die Size = 7.2 mm)

| Process | Time (sec) | Max. Temp. (°C) | Min. Temp. (°C) | Temp. Diff. (°C) |
|---|---|---|---|---|
| IR preheat and vapor | 40 | 86.5 | 72.75 | 13.75 |
| phase reflow | 140 | 228.0 | 222.81 | 5.19 |
| | 155 | 193.06 | 176.40 | 16.66 |
| | 210 | 105.2 | 104.79 | 0.21 |
| | 275 | 184.03 | 146.39 | 37.64 |
| | 280 | 207.36 | 180.63 | 26.73 |
| | 305 | 220.45 | 220.13 | 0.31 |
| | 485 | 111.56 | 55.53 | 56.03 |
| IR reflow | 90 | 166.0 | 113.0 | 53.0 |
| | 180 | 181.0 | 149.0 | 32.0 |
| | 290 | 206.0 | 193.0 | 13.0 |
| | 325 | 187.0 | 113.0 | 74.0 |
| | 355 | 172.0 | 91.0 | 81.0 |

Table 14-1 also indicates that package internal temperatures could often be lower than the peak reflow environment temperatures. Most of the calculations in the literature, however, are based on worst-case assumptions, i.e., isothermal (peak temperature) conditions at all interfaces within the package. These assumptions indirectly accommodate the manufacturing environment, where a certain amount of variation is to be expected in the process parameters such as time, temperature, and thermal mass.

### 14.3.2 Kinetics of Moisture Absorption and Desorption

The intention of developing models for depicting kinetics and diffusion of moisture is to predict the amount of absorbed moisture content of a package at a given time and under known environmental conditions. This enables one to judge the impact of storage in a humid environment and to decide upon a suitable time window for surface mount operations after opening the dessicant bag. Such models also provide an insight into moisture control procedures, e.g., the efficiency of the high-temperature bake in removing the moisture prior to packing in dessicant bags or of a low-temperature dryout process in the presence of a dessicant such as silica gel.

The estimates of moisture absorption are based on Fickian diffusion theories and experimental data of moisture weight gain vs. time under different temperature and relative humidity conditions. Weight gain is a measure of total moisture intake into the package; it does not provide information about the distribution of moisture in the package.

The rate of moisture absorption is a strong function of temperature (Fig. 14-5), i.e., the absorption at 85°C is faster than at 40°C. The amount of moisture weight gain at equilibrium, however, is a strong function of humidity and shows very little dependency on temperature.

The process of moisture diffusion in plastic packages is typically diffusion-controlled. Moisture is first adsorbed on the surface of the package, followed by the solution process, and then transported through the plastic by bulk diffusion processes. The first two steps are considerably faster than the third process. A one-dimensional solution to Fick's diffusion equation is often used to model the moisture absorption since the surface mount components are significantly thinner than their length and width. Lin et al.[2] used actual plastic thickness instead of total plastic thickness since the die and the lead frame will not absorb moisture. The diffusivity at 85°C was found to be about $1.0 \times 10^{-4}$ cm²/h and the activation energy for diffusion to be about 0.45 eV. The variation in diffusivity for different plastics was small, typically less than 15%.

Bhattacharya et al.[11] have derived the diffusion coefficient and the total mass of water absorbed by the plastic package from first principles and

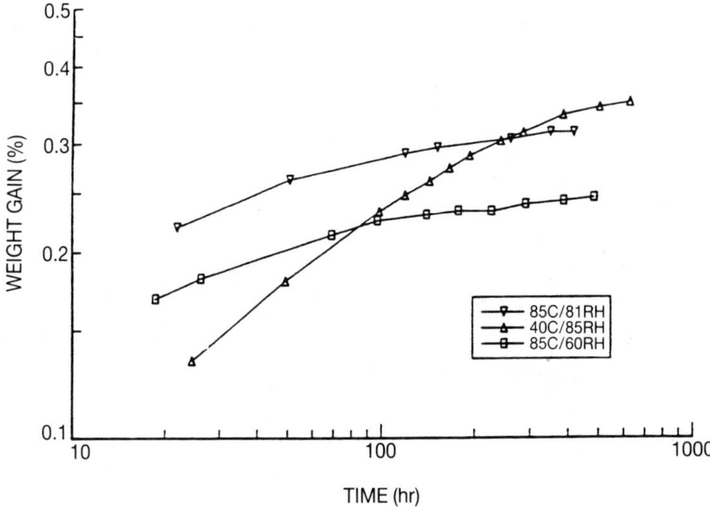



**Figure 14-5**    Moisture absorption as a function of temperature and relative humidity.[2]

through nonlinear regression analysis of mass gain vs. time data. Their models can successfully predict some of the key parameters necessary to assess the package cracking susceptibility. These parameters are total volume of water vapor molecules inside the plastic and the critical pressure at solder reflow temperatures; for example, in the case of a 68LD PLCC package, the estimated saturation density of water molecules inside the plastic is $0.3 \pm 0.09$ $g/cm^3$ at 30°C/60%RH; the estimated volume of water molecules at 85°C/85%RH is $2.08 \times 10^{-2}/cm^3$ for a saturation density of 0.963 $g/cm^3$ and the saturation pressure of vapor at reflow temperature is close to 300 psi. They have attributed the variation in the diffusivity and activation energy values to the different processing of the plastic packages. These processing variations can arise from structural differences such as degree of cross-linking, molding compound filler materials, and filler distributions.

Figure 14-6 depicts typical absorption and desorption curves for a 68LD PLCC package and corresponding danger zones where package cracking is encountered.

### 14.3.3 Interfacial Adhesion and Delamination

Kim[12] has reported that adhesive strength of the molding compound (Novolac epoxy) to copper monotonically decreases with increasing moisture content (Fig. 14-7). This has been supported by thermodynamic calculations of work of adhesion in presence and absence of moisture. The work

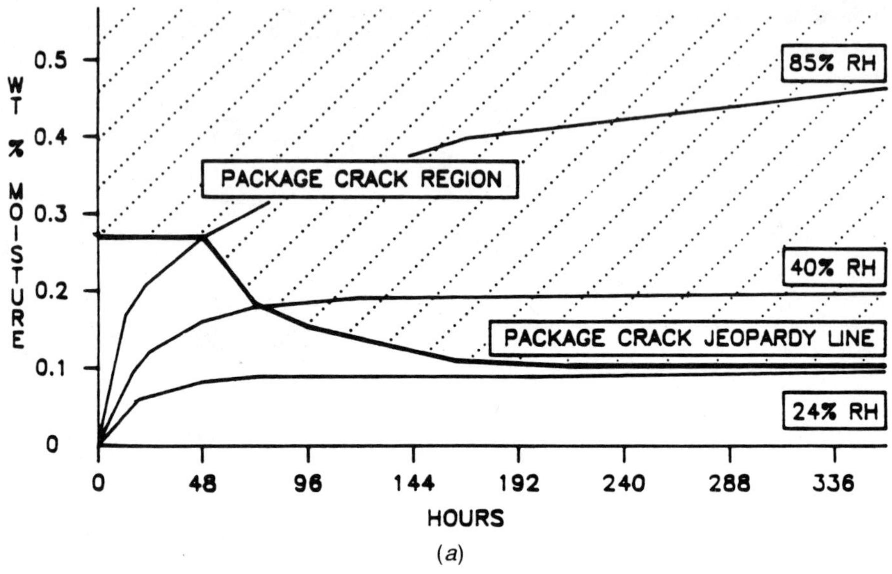


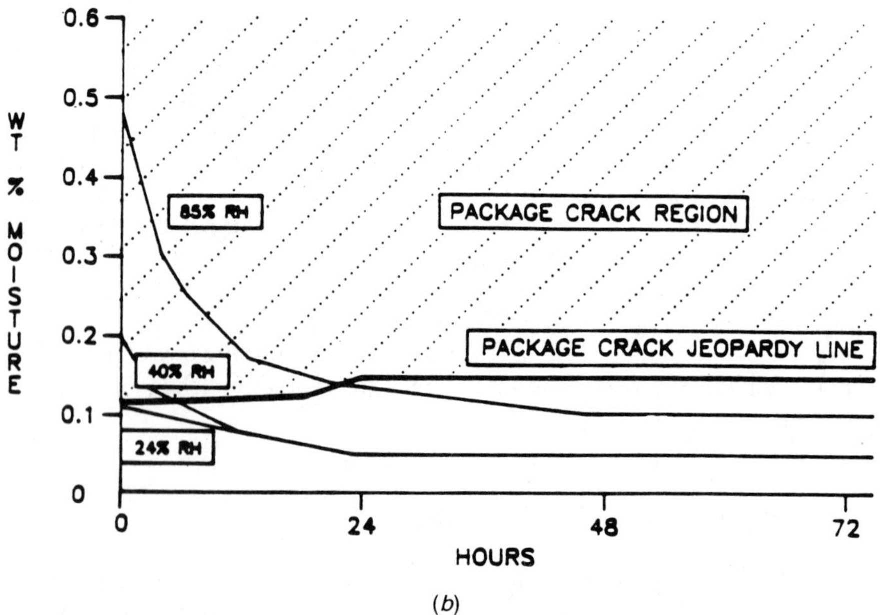


**Figure 14-6**  (a) Moisture absorption and (b) moisture desorption and critical domains for surface mount plastic package cracking.

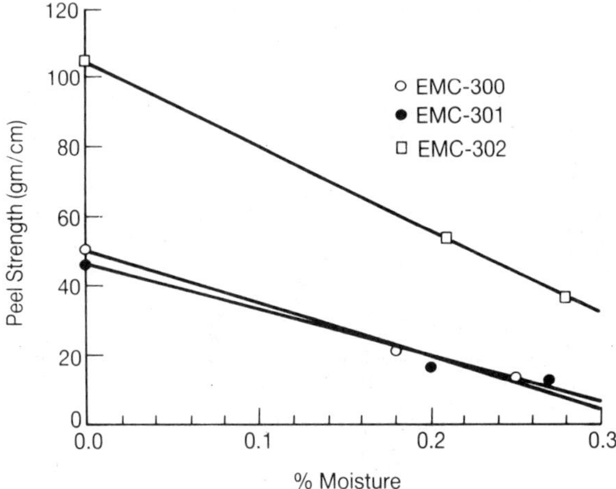


**Figure 14-7**   Monotonic decrease in peel strength with an increase in moisture content for three molding compounds.[12]

of adhesion substantially decreases in the presence of moisture (Table 14-2).

Adhesion strength degradation or percentage delamination (as a result of exposure to 85°C/85%RH followed by three vapor phase solder (VPS) reflows), as measured by scanning acoustic microscopy (SAM), has been found to increase with decreasing peel strength due to absorbed moisture (Fig. 14-8).[12] The results demonstrate that interface adhesion plays a vital role in minimizing delamination during surface mount operation.

Kitagawa et al.[13] have reported that when a vapor pressure vent on the lead frame was coated with polyimide, several packages cracked due to vent

**Table 14-2**   Calculated Work of Adhesion Between Copper and Molding Compounds in an Inert Atmosphere and in Presence of Water[12]

| Sample | $\gamma^d$ | $\gamma^p$ | $W_a$ | $W_{aw}$ |
|---|---|---|---|---|
| Cu-110 (acid cleaned) | 37.9 | 2.0 | | |
| EMC-300 | 31.3 | 2.2 | 73.0 | 67.7 |
| EMC-301 | 31.5 | 7.3 | 76.7 | 53.7 |
| EMC-302 | 31.8 | 7.3 | 77.1 | 53.7 |

$W_{aw}$ is the work of adhesion in presence of water.
$W_a$ is the work of adhesion in an inert atmosphere.
$\gamma^d$, $\gamma^p$ are dispersion and polar components of surface energies of the adherend and the substrate.

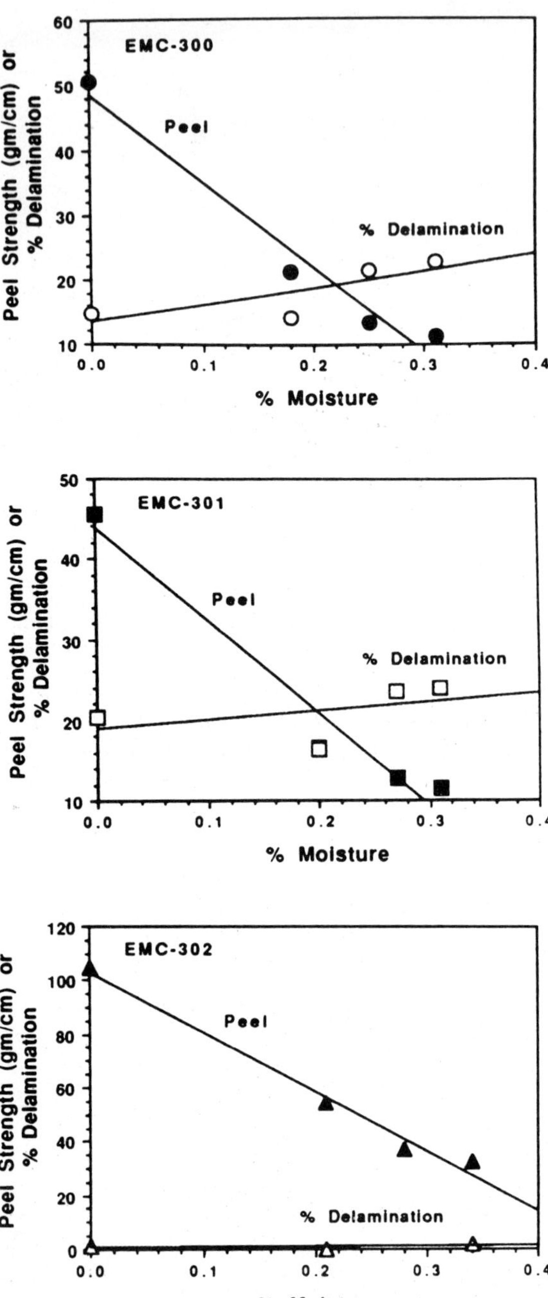


**Figure 14-8**  Relationship between adhesion strength or percentage delamination upon VPS reflow as a function of moisture content for three different molding compounds.[12]

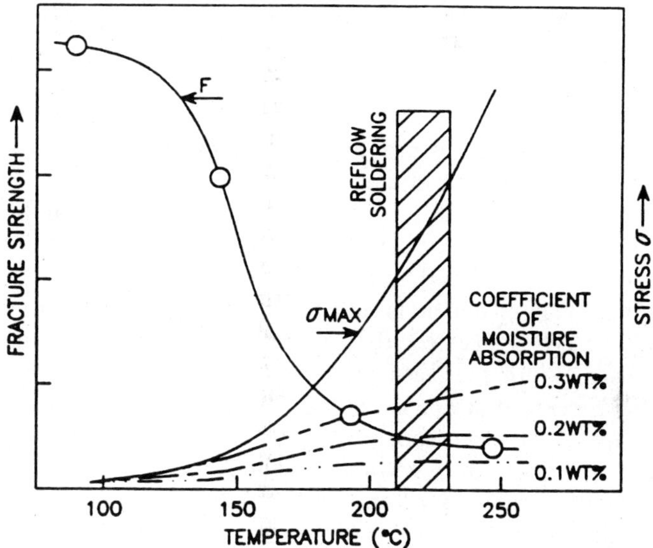



**Figure 14-9**   Strength degradation as a function of temperature; stress generated during solder reflow exceeds plastic resin strength and leads to cracking.[14]

blockage. This also explains why solder dipping operations at 260°C induce catastrophic cracking in spite of good peel strength and good interfacial adhesion.[12] The stress generated by steam pressure at 260°C exceeds the sum total of flexural strength and adhesive strength of the molding compound (Fig. 14-9).

Bhattacharya et al.[11] have estimated the gap between the molding compound and the lead frame as close to 0.32 mils on the basis of CTE mismatches alone. The doming effect seen at the bottom of the package due to saturation pressures associated with the reflow temperatures is estimated to be as high as 22 mils. Since only doming up to a few mils has been reported, it is believed that a certain portion of the flashed stream is probably dissipated across the package. Thus the doming effect is not exaggerated under the saturation pressure.

### 14.3.4 Fracture and Mechanical Behavior of Plastic

Kitano et al.[14] have measured the flexural strength of plastic as a function of temperature (Fig. 14-9). They have also observed hysteresis in terms of package cracking as a function of moisture absorption or desorption. The bending strengths reduce by an order of magnitude at temperatures above

200°C. Thus if the stress from moisture inside the package exceeds the bending strength, package cracking occurs.

Kornblum and Glaser[15] have proposed that a new mechanism, which does not refer to the pressurized steam related dome under the pad. Their contention is that the moisture gradients convert the initially isotropic material into an anisotropic material, leading to uneven displacements in the body. With the application of heat and the consequent temperature gradients, these displacements amplify, thus causing large local strain that leads to high stresses and consequently to failure.

Fracture behavior of plastic is influenced by the polymerization process. Suhl et al.[16] have suggested that increasing viscosity of the molding compound due to polymerization is the cause of increased strain in the package. The increased strain is believed to manifest as package cracking. This was supported by two separate cases: one in which the side opposite to the plastic injection port within each package was found to be more susceptible to package cracking than other sides, and another in which the components at the end of the lead frame strip away from the primary injection port showed increased susceptibility to package cracking. Attempts to alter the molding parameters (pressure, time, etc.) to compensate for such effects were not successful.

## 14.4 MEASUREMENT TECHNIQUES

Moisture absorption and desorption kinetics are primarily determined by gravimetry. Packages are exposed to different temperature/relative humidity conditions and are weighed to obtain the moisture intake.

Similarly, packages saturated with moisture are maintained at bake/room temperatures to determine the weight loss due to the loss of moisture.

Preconditioned packages (i.e., high temperature/humidity exposure followed by infrared reflow (IR)/VPS reflow) are inspected using a scanning acoustic microscope to identify delamination of various interfaces and also package cracking. Ultrasonic waves reflect from the gaps within the package.

Packages are cross-sectioned and inspected by scanning electron microscopy to look for the internal package cracks.

Peel strengths are measured by peeling off a thin lead frame foil at 180° angle from the transfer-molded rigid encapsulant using a tensile testing machine at a known rate (typically 5 inches/minute). Flexural strengths are measured using a standard three-point bend test.

## 14.5 SOLUTIONS

Several methods have been proposed in the literature to prevent package cracking. They fall into the following categories:

1. Moisture control
2. Package design (e.g., die-pad to plastic ratio)
3. Adhesion enhancement
4. Process modifications
5. Improvement of the high-temperature strength of plastics

## Moisture Control

Moisture control is currently an accepted practice industry wide. Surface mount components are baked (typically 125°C, 24 hours) and packed in dessicant bags. Manufacturing discipline is required to ensure that the surface mount operation is finished within a limited time after opening the bags, i.e., that components stay dry up to the reflow operation.

Another approach to moisture control has been in the area of next-generation molding compounds. Some of the newer molding compounds absorb moisture at lower rates than their predecessors.[17,18] Even though this does not eliminate package cracking, it certainly opens up the time window (beyond 48 hours) for use of SMCs for the surface mount operation after removal from the dessicant bags.

## Design Solutions

An empirical relationship has been reported between package cracking and the $L$ and $H$ package dimensions, where $L$ represents the short side of die paddle and $H$ represents the minimum plastic thickness below the die paddle.[1,2] Figure 14-10 shows safe and unsafe packages.

For relatively small die size, die paddle length to plastic thickness ratio (Figs. 14-10 and 14-11) can be maintained below 6 to prevent package cracking. For large die sizes, it has been proposed that the die paddle be slotted in such a way that the delamination envelope is reduced significantly.[19] The individual die paddle segments after slotting can then fulfill the condition $L/H < 6$.

## Adhesion Enhancement

Adhesion enhancement can be pursued by two approaches, mechanical and chemical. Mechanical adhesion can be improved by dimpling and/or slotting the die paddle. These additional features in the die paddle provide a mechanical anchor for the molding compound and restrict the delamination to localized regions. The dimple spacing limits the delamination envelope and the dimple depth provides the interlocking, see Fig. 14-12.[2,4,20] The

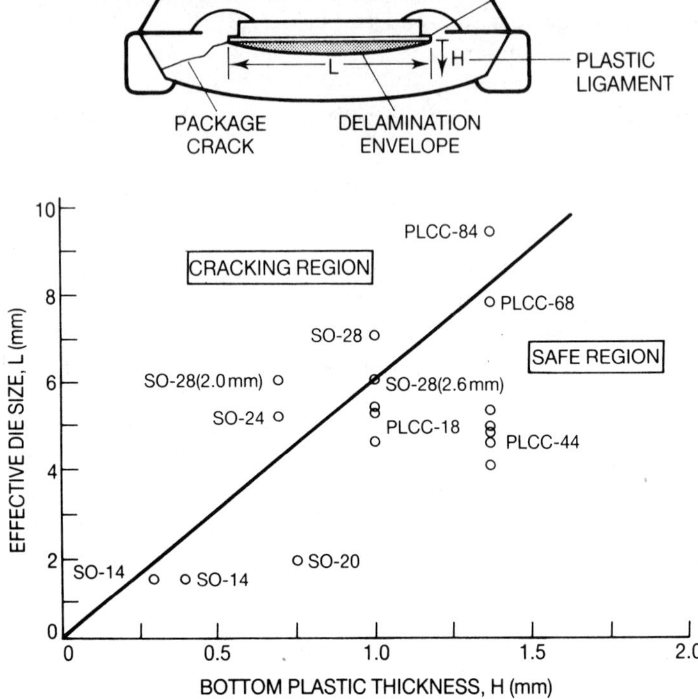


**Figure 14-10**  "Popcorn cracking" susceptibility of plastic packages preconditioned at 85°C/85% prior to VPS.[1,2]

dimple spacing also fulfills the condition $L/H < 6$. This argument helps explain the excellent performance of thin small outline packages (TSOPs), see Fig. 14-11.[4] As mentioned earlier, the steam pressure at 260°C (for a solder dip case) exceeds flexural strength and the adhesive strength. This results in shearing off of the molding compound in the dimples, see Fig. 14-13.

Chemical adhesion is enhanced by adding modifiers to the molding compound. Unfortunately, these modifiers also stick to the metallic mold during the molding operation. As a result this approach requires a balancing act between the modifiers and mold-releasing waxes.


**Process Solutions**

Suhl et al.[16] have suggested that plastic packages that did not go through the chemical deflashing operation performed better than those that underwent deflash processes. They also observed that the parts made with prolonged

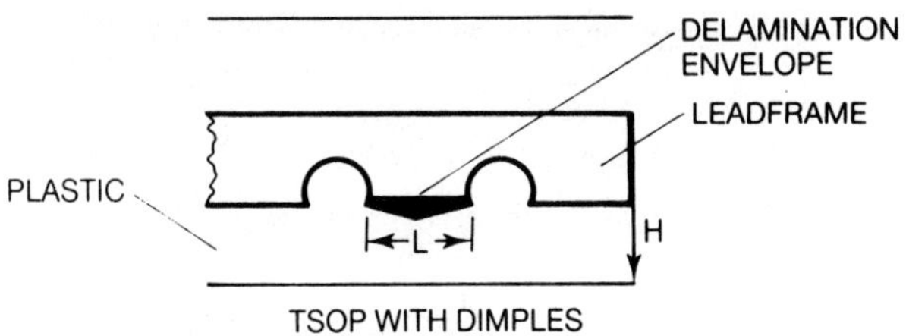



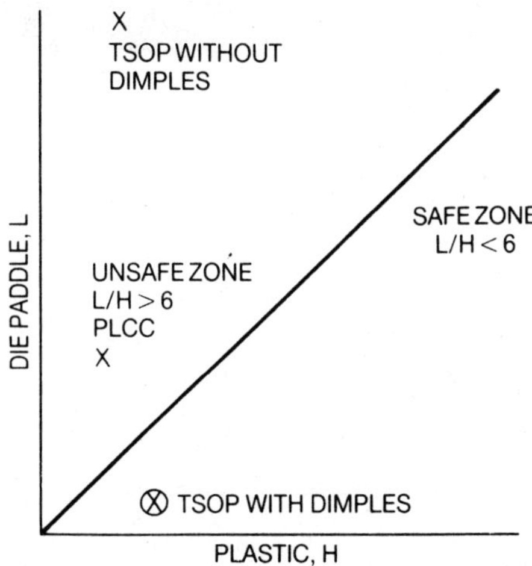



**Figure 14-11**    Schematic of safe and unsafe regions as a function of $L/H$ ratio and effect of die paddle features such as dimples or slits that reduce $L$.[2,4]

in-mold cure times showed better results than those with normal cure processes. Further work is needed in these areas to elucidate the role of such process variables.

Kornblum and Glaser[15] suggest that the thermal gradient in the package can be eliminated by preheating the devices before solder reflow. The smaller thermal gradients may inhibit the failure rates even if the moisture is not baked out. It is also believed that the smaller gradient will help if the failure is due to a brittle fracture from the thermal shock.

### Surface Treatments

Another approach to eliminating the moisture at the epoxy to die or lead frame interfaces would be to apply a chemical adhesion promoter to the

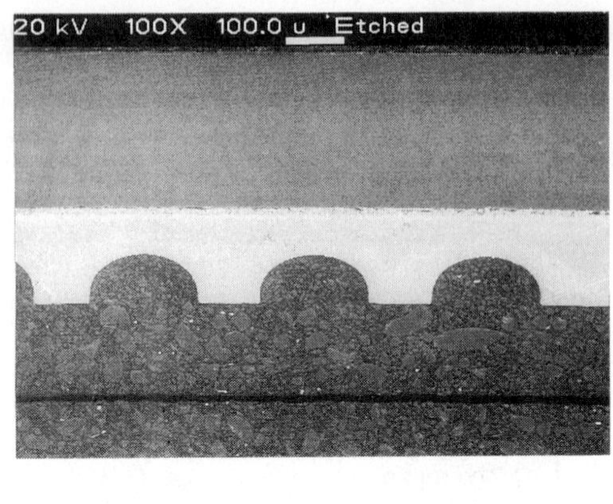


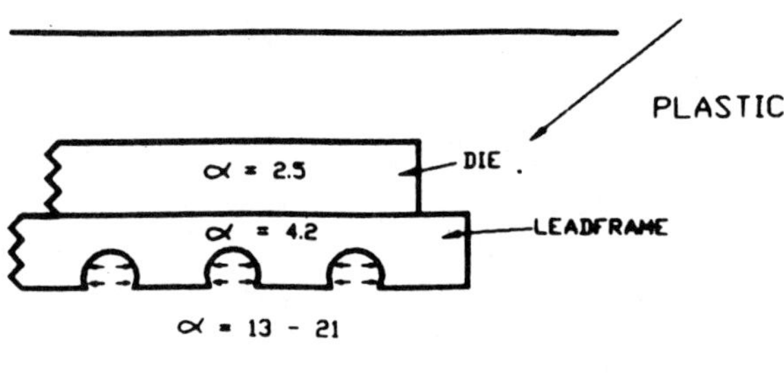


**Figure 14-12**   Mechanical adhesion enhancement with dimples. Molding compound "locks" into the dimples.[4]

lead frame and the die just prior to the molding operation.[16] One commonly used surface treatment is hexamethyldisilazene (HMDS). After wire bonding, the lead frame strip was placed in a vacuum bake oven and exposed to HMDS vapors. Then the molding operation was performed. The cracking rate was reduced from 100% to 5%. HMDS was optimized to work on Novolac and silicon dioxide and not for copper to epoxy. It is believed that a hunt for such an agent will eliminate package cracking.

It has been reported that coating of the die paddle with polyimide (thickness $> 30\,\mu m$ after hardening) eliminates package cracking.[21] Polyimide adheres well to both materials—die paddle and molding compound— and has a high glass transition temperature. As a result, it is believed that molding compound no longer peels off from the die paddle and the moisture absorbed no longer condenses. The technique has been found useful for SOP, TQFP, and TSOP packages.

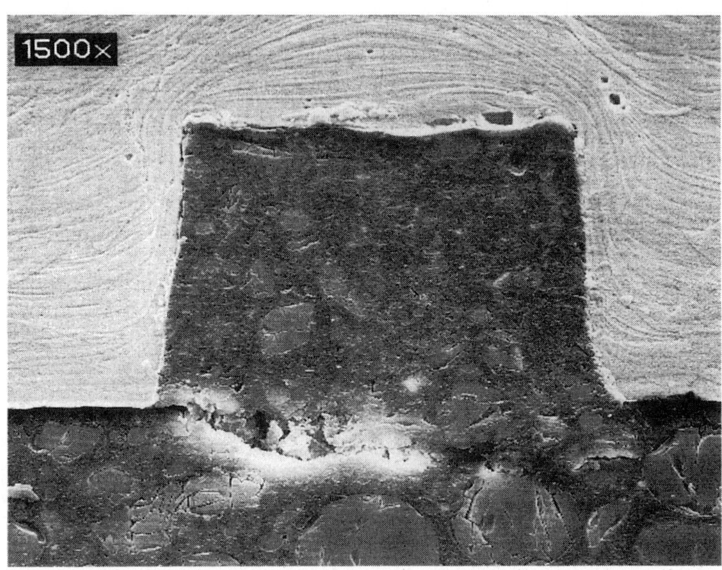

**Figure 14-13**   The stress from steam pressure at 260°C solder dipping far exceeds the strength of plastic and the adhesion strength. This results in shearing off of the molding compound at the dimples.[20]

## Material Solutions

Increased cross-link density buys package reliability in several ways. The first is the decrease of moisture diffusivity and lower solubility limits resulting from the lower porosity in the network structure.[22–24] The second is the potential for increasing the dry value of $T_g$,[24,25] decreasing CTE,[26] and increasing fracture toughness.[27] This in turn limits package cracking and controls mechanical degradation as a result of limited lowering of the elastic modulus and strength of the resin with increasing water content.[28,29] However, pursuit of higher cross-link density can result in higher stresses in the package body. Such an increased stress then demands use of polymer buffer coatings on die surfaces to protect against other hazards such as thin-film cracking.

Increasing the fracture toughness, $K_{IC}$, the cohesive strength of the compound at elevated temperatures will be a step forward in reducing the need for special handling, i.e., dessicant pack discipline, of SMCs. The fracture toughness of the molding compound can be increased by increasing the synthetic rubber modifiers (Fig. 14-14) such as silicones and related compounds.[30] This second phase in the composite matrix acts as pinning points for crack propagation, cohesive tearing, and phase separation between silica fillers. These modifiers and fillers both, however, increase the water solubility of the composite.[23,27] Increased silicone modifiers often increase

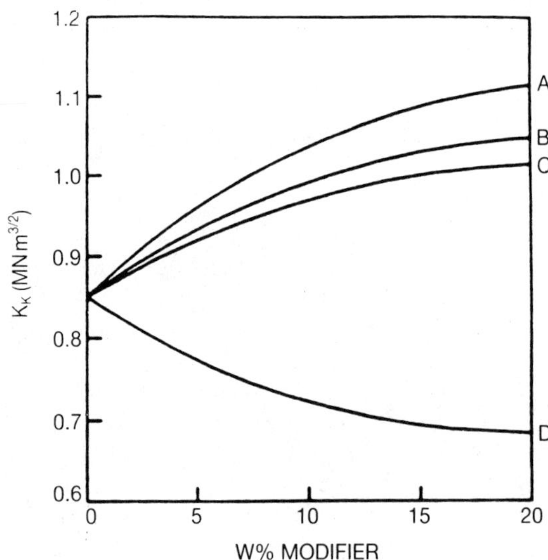


**Figure 14-14**   Fracture toughness as a function of modifier content of the molding compound.[30]

the viscosity of the resin and can adversely affect the adhesion to lead frame materials because of their poor wetting characteristics.

Molding compound modification will thus continue to be a challenge since it involves balancing the trade-offs in performance requirements. Better performance will require new formulations of molding compounds that are tougher, stronger, have higher operating temperature stability, and have lower moisture solubility and diffusivity.

## REFERENCES


1. *Impact of Moisture on Plastic I/C Package Cracking,* IPC-SM-786, 1990, pp. 1–29.
2. Lin, R., E. Blackshear, and P. Serisky, *Proc. IEEE International Reliability Physics Symposium,* 1988, pp. 83–89.
3. Steiner, T., and D. Duhl, *IEEE Trans. Components, Hybrids, and Manufacturing Technology,* **10,** 1987, pp. 209–216.
4. Golwalkar, S., P. Boysan, and R. Foehringer, *Proc. IEEE Electronic Components and Technology Conference,* 1991, pp. 745–749.
5. Moore, T., R. McKenna, and S. J. Kelsall, *Proc. IEEE International Reliability Physics Symposium,* 1991, pp. 160–166.
6. Koch, T., W. Richling, J. Whitlock, and D. Hall, *Proc. IEEE International Reliability Physics Symposium,* 1986, pp. 55–60.
7. Koyama, H., H. Shiozaki, I. Okumura, S. Mizugashira, H. Higuchi, and T. Ajiki, *Proc. IEEE International Reliability Physics Symposium,* 1988, pp. 59–63.
8. Moore, T., *Proc. 16th International Symposium for Testing and Failure Analysis,* 1990, pp. 61–67.
9. Glaser, J. C., and M. P. Juaire, American Society of Mechanical Engineers, 88-WA/EEP-4, November 1988.
10. Miyake, K., H. Suzuki, and S. Yamamoto, *IEEE Trans. Reliability,* **R-34**(5), December 1985, pp. 402–409.
11. Bhattacharya, B. K., W. A. Huffman, W. E. Jahsman, and B. Natarajan, *Proc. IEEE Electronic Components and Technology Conference,* 1988, pp. 49–58.
12. Kim, S. *Proc. IEEE Electronic Components and Technology Conference,* 1991, pp. 750–758.
13. Kitagawa, H. et al., *Proc. IEEE Electronic Components and Technology Conference,* 1989, pp. 445–449.
14. Kitano, M., A. Nishimura, and S. Kawai, *Proc. IEEE International Reliability Physics Symposium,* 1988, pp. 90–95.
15. Kornblum, Y., and J. C. Glaser, *J. Electronic Packaging, Trans. ASME,* **111,** December 1989, pp. 249–254.
16. Suhl, D., M. Kirloskar, and T. O. Steiner, *IEEE Trans. Components, Hybrids, and Manufacturing Technology,* **11,** 1988, pp. 129–132.
17. Nishioka, T., S. Oizumi, and S. Ito, *Nitto Technical Reports,* January 1991, pp. 82–93.
18. Suzuki, H., *Sumitomo Reports,* 1990, pp. 287–312.

19. *Nikkei Microdevices*, **9**, 1988, pp. 115–120.
20. Altimari, S., S. Golwalkar, P., Boysan, and R. Foehringer, *Proc. IEEE Electronic Components and Technology Conference*, 1992, pp. 945–950.
21. "The Reliability of TSOP and TQFP," *Semiconductor World*, June 1991, pp. 108–112.
22. Prough, S., and D. E. Pope, *Materials Research Society Symposium Proc.*, **167**, 1990, pp. 5–21.
23. McMaster, M. G., and D. S. Soane, *IEEE Trans. Components, Hybrids, and Manufacturing Technology*, **12**, 1989, pp. 373–386.
24. Kong, E. S. W., *Epoxy Resins and Composites*, *II*, K. Dusek, ed., Springer, New York, 1986, pp. 125–172.
25. Kamon, T., and H. Furukawa, *Epoxy Resins and Composites*, *II*, K. Dusek, ed., Springer, New York, 1986, p. 179.
26. Nakamura, Y. et al., *Nitto Technical Reports*, September 1987, pp. 23–31.
27. Drzal, L. T., *Epoxy Resins and Composites*, *II*, K. Dusek, ed., Springer, New York, 1986, pp. 1–32.
28. Porto, A. et al., *Materials Research Society Symposium Proc.*, **108**, 1988, pp. 169–174.
29. Apicella, A., and L. Nicolais, *Epoxy Resins and Composites*, *II*, K. Dusek, ed., Springer, New York, 1985, pp. 69–78.
30. Yorkgitis, E. M. et al., *Epoxy Resins and Composites*, *II*, K. Dusek, ed., Springer, New York, 1982, p. 97.

# 15

# Thermomechanical Fatigue of 63Sn–37Pb Solder Joints

*Peter L. Hacke, Arnold F. Sprecher,*
*and Hans Conrad*

Thermal fluctuation experienced by an electronics package during normal operation causes strains, and in turn stresses, in the solder joints. These result from differences in coefficient of thermal expansion between the various components of the package; see, for example, Figure 15-1. The thermal fluctuation can be produced by heat dissipated in the electronic components or by environmental temperature changes. The repetition of such fluctuations produces cyclic strains and stresses, which ultimately lead to the failure of the solder joints in fatigue.

A number of models have been developed to describe the stress–strain behavior of solder joints in fatigue and the resulting crack nucleation and growth that leads to failure. Notable among these is the "damage integral" approach of Li and coworkers[1,2] and the "integrated matrix creep" approach of Fox and coworkers.[3,4] Other approaches are given in refs. 5–8. This chapter reviews the work we have performed to date on this subject.

For our studies we constructed a loading frame and temperature cycling chamber to determine the fatigue behavior of near-eutectic 63Sn–37Pb solder joints in shear during thermal cycling in the temperature range of $-30°C$ to $130°C$. Stress–strain hysteresis loops were measured and the resulting fatigue damage (cracking) was determined. The measured stress–strain hysteresis loops during the initial stages of thermomechanical fatigue were computer-simulated by employing a constitutive equation based on the deformation kinetics in monotonic loading of bulk solder alloy specimens. The growth of fatigue cracks was measured and found to correlate with the maximum shear load in the hysteresis loop.

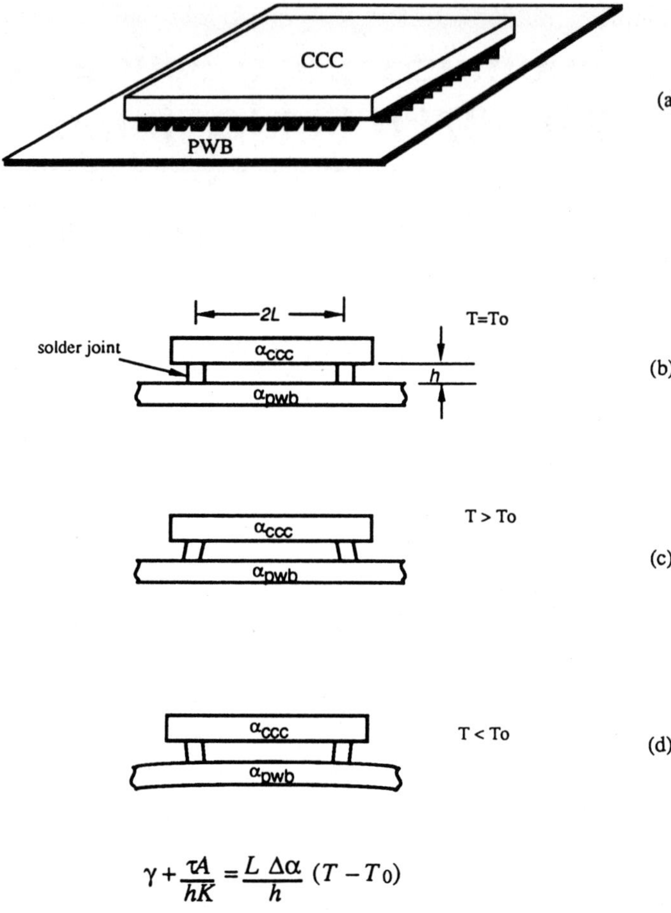



$$\gamma + \frac{\tau A}{hK} = \frac{L \, \Delta\alpha}{h} (T - T_0)$$

**Figure 15-1**    Schematics of (*a*) ceramic chip carrier and printed wiring board assembly and (*b, c, d*) strains experienced in the assembly because of temperature changes and differences in coefficient of thermal expansion of the components.

## 15.1 CONSTITUTIVE EQUATION

### *15.1.1 General*

Most information on the inelastic deformation of Pb–Sn alloys pertains to plastic deformation, much less has been obtained for anelastic deformation. Since plastic deformation is the major component of the inelastic deformation which occurs in Pb–Sn alloys at the relatively large cyclic strain amplitudes and slow loading rates, we will only deal with plastic deformation. However, at small strain amplitudes anelastic deformation may be important.[9]

In the literature plastic deformation is sometimes separated into a creep component and a time-independent plastic flow component. This is an artificial separation that can be misleading, and the distinction is not made in our analysis. All plastic deformation is time-dependent, only the rate-controlling mechanism varies, which depends on stress, strain rate, and temperature. In polycrystalline materials these mechanisms include those governing dislocation motion within the grains, grain boundary sliding, and deformation by mass transport, i.e., diffusional flow. All of these mechanisms include time as a parameter. At very high stresses or strain rates, the dislocation velocity is controlled by electron or phonon drag. At lower stresses and rates, the dislocation velocity may be controlled by thermally-activated overcoming of obstacles on the glide plane or by cross-slip of screw dislocations past them. At temperatures above about $0.5T_M$ ($T_M$ is the melting temperature in kelvins) the rates of diffusion become significant and dislocation motion may be diffusion-controlled either through the climb of edge dislocations or through the drag of solute atoms by the dislocation. Also, at high temperatures (and especially at low stresses), grain boundary sliding and diffusion flow can contribute to the plastic deformation, both of which are diffusion-controlled. Hence, it is clear that all plastic deformation is time-dependent; it is only the rate-controlling mechanism that varies.

It is well established that the plastic shear strain rate $\dot{\gamma}_p$ of metals at a given shear stress $\tau$ and temperature $T$ is given by

$$\dot{\gamma}_p = f(\text{str.}, \tau, T) \tag{15-1}$$

where str. refers to the structure and includes all of the crystal defects (point, line, surface, and volume) that are present. The structure in turn is dependent on the strain history and is given by

$$\text{str.} = f(\text{str.}_0, \gamma_p, \dot{\gamma}_p, \tau, T) \tag{15-2}$$

where $\text{str.}_0$ is the initial structure of the material and $\gamma_p$ is the plastic shear strain. According to Eq. (15-1), for a constant structure the deformation behavior in a constant strain rate tensile test should be the same as that in a creep test (constant stress) or in a stress relaxation test (constant strain). This is generally observed for tensile and creep tests on Pb–Sn alloys at high homologous temperatures.[9] The behavior in stress relaxation may differ from the other two if the inelastic deformation during the stress relaxation contains a large anelastic component.[9]

Temperature extremes of $-50°C$ and $+150°C$ would be considered harsh for electronic equipment; nevertheless, Fig. 15-2 shows that for the eutectic Pb–Sn alloy this temperature range is in the high homologous temperature regime. Plastic deformation in the temperature range $T/T_M > 0.5$, where $T_M$

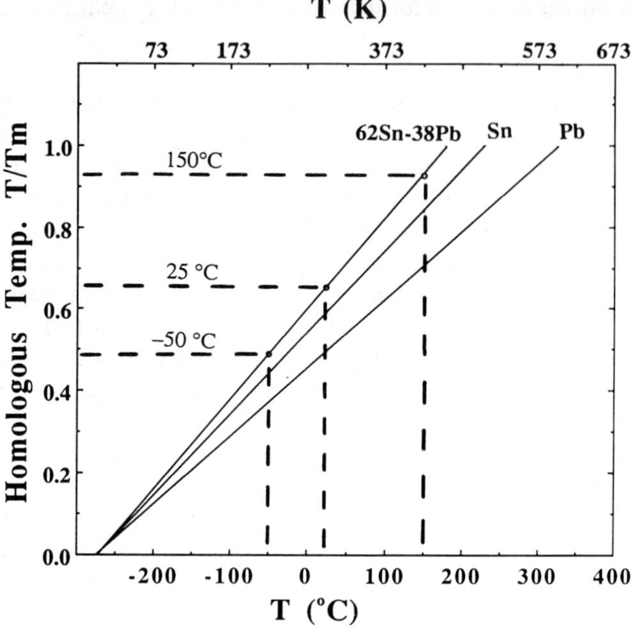



**Figure 15-2**   Homologous temperature vs. test temperature in °C for the Pb–Sn alloy system. The temperature range −50°C to 150°C is shown.

is the melting temperature in degrees Kelvin, is often called "high-temperature creep." The deformation in this regime is generally diffusion-controlled.[10] The steady-state plastic deformation rate for many metals and alloys in the high homologous temperature regime is generally given by the Dorn equation[10]

$$\frac{\dot{\gamma}_p kT}{D\mu b} = A\left(\frac{\tau}{\mu}\right)^n \left(\frac{b}{d}\right)^p \tag{15-3}$$

where $\dot{\gamma}_p$ is the plastic shear rate, $D$ is the appropriate diffusion coefficient, $\mu$ is shear modulus, $b$ is the Burgers vector, $\tau$ is the applied shear stress, $d$ is the grain size (or interphase spacing), and $kT$ has the usual meaning. The diffusion coefficient is given by

$$D = D_0 \exp(-\Delta H/kT) \tag{15-4}$$

where the preexponential $D_0$ and the activation enthalpy (energy) $\Delta H$ refer to the specific diffusion mechanism, e.g., lattice (volume) diffusion, grain boundary diffusion, or dislocation core diffusion. The magnitudes of the constants $A$, $n$, and $p$, and that of $D$ reflect the particular deformation mechanism that is rate-controlling.

### 15.1.2 Mechanisms

In the case of the eutectic Pb–Sn alloy, four regions have been identified in log–log plots of $\dot{\gamma}_p$ vs. $\tau$ in the range normally considered in the fatigue of solder joints in electronics packages.[9] Three of these regions are illustrated in Fig. 15-3, the fourth occurring at higher shear strain rates. Values for the constants in Eqs. (15-3) and (15-4), which have been obtained in monotonic loading (uniaxial and shear), stress relaxation tests, and constant strain rate tests, are given in Table 15-1. Also given are the proposed rate-controlling plastic deformation mechanism for each region and the activation energies for lattice diffusion in pure Pb and Sn. The activation energy for Regions I, III, and IV is approximately equal to, or slightly less, than that for lattice diffusion; that for Region II is approximately that for grain boundary diffusion, which is approximately one-half that for lattice diffusion. A detailed discussion of the test data pertaining to each of the regions and its analysis and significance is given by Arrowood et al.[9]

An important factor in the rate-controlling mechanism is the microstructure.[9,11] Faster cooling rates of near-eutectic Pb–Sn alloy inhibit the formation of the regular lamella and colony structure. This degenerate eutectic structure favors deformation by grain boundary sliding at certain strain rates, leading to longer fatigue life under isothermal conditions.[12] Quenching results in a morphology consisting of a finely dispersed lead-rich phase in a tin-rich matrix.[11] Mei et al.[13] showed that both liquid

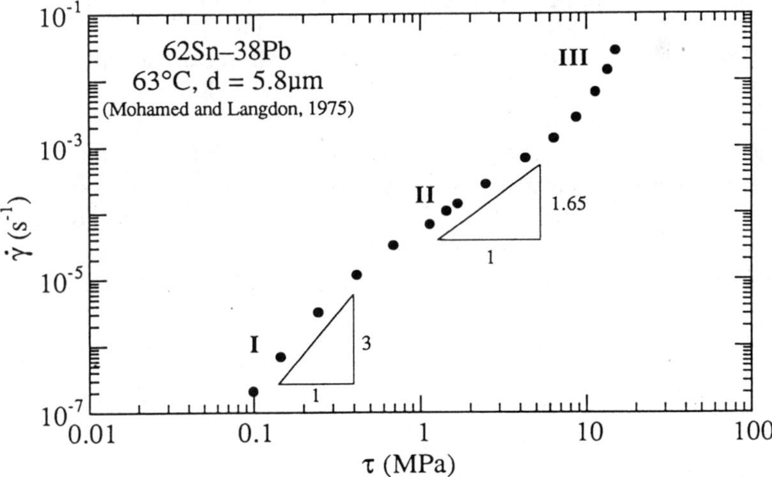


**Figure 15-3**   Log–log plot of shear rate vs. applied shear stress for eutectic Pb–Sn alloy: with a grain size of 5.8 μm at 63°C. Data from F. A. Mohamed and T. G. Langdon, *Phil. Mag.*, **32**, 1975, pp. 697–709.

**Table 15-1**  Plastic Deformation Kinetics for Eutectic Pb–Sn Alloys at $T/T_M \gtrsim 0.5$

| Region | Relative Stress (or Temperature) Level | $n$ | $p$ | $\Delta H$ (kJ/mole)* | Proposed Deformation Mechanism |
|---|---|---|---|---|---|
| I | Low (High) | 1.7–3 | 0 | 84 | (Viscous glide?) |
| II | Intermediate (Intermediate) | 1.6–2.4 | 1.6–2.3 | 44–47 | Superplasticity: grain boundary sliding |
| III | High (Low) | 3–10 | 0 | 81–100 | Power-law creep (glide and climb) |
| IV | Very high (Very low) | >10 | 0 | ~100 | Power-law breakdown (cross slip; glide of jogged screw dislocations) |

*Source:* J. Askim, *Tracer Diffusion Data*, Plenum Press, New York, 1970.
* $\Delta H_1$ (Pb) = 101 kJ/mole; $\Delta H_1$ (Sn) = 94 kJ/mole.

nitrogen-quenched and air-cooled solder joints exhibited superplastic (Region II) behavior within a certain strain rate and temperature range. In contrast, a more slowly cooled solder joint results in a lamellar eutectic morphology. Such structure was shown to deform by conventional plasticity (Region III).[11] Superplasticity is not favored in a lamellar microstructure because this structure inhibits the grain boundary sliding process. It has been found that the Pn–Sn eutectic with large lamellar phases deforms by conventional plasticity with the stress exponent $n = 6$–7 and $\Delta H = \sim 84$ kJ/mole.[11] Coarsening of an initially fine microstructure will occur during heating and plastic deformation. A coarser microstructure favors Region III deformation at the expense of Region II.[3]

In Region II, the contribution of grain boundary sliding to the total strain is $\varepsilon_{gb}/\varepsilon_t \approx 50$–60% in uniaxial tension. This ratio is much less in Regions I and III, being of the order of 20%. It is believed that grain boundary sliding is an important mechanism in the superplastic deformation of the Pb–Sn eutectic. Measurements of the grain boundary sliding indicate that the largest sliding offsets occur at the Sn–Sn boundaries within the Sn phase, less at the Pb–Sn interfaces, and very little at the Pb–Pb crystal boundaries in the Pb phase.[14]

### 15.1.3 Constitutive Equations

Grivas et al.[15] investigated the steady-state stress in shear at 0°C to 160°C of bulk eutectic Pb–Sn alloy with mean phase sizes of 5.5 to 9.9 μm at strain rates in Regions II and III. They concluded that the alloy exhibited both conventional and superplastic deformation, with both occurring simultaneously via independent mechanisms. The superplastic deformation was deduced to be rate-controlling at the lower stresses, and conventional plastic deformation controlling at the higher stresses. They assumed that the contributions of the two mechanisms to the total shear rate were additive, i.e.,

$$\dot{\gamma}_p = \dot{\gamma}_{p,\text{II}} + \dot{\gamma}_{p,\text{III}} \tag{15-5}$$

Further, their results indicated that $\dot{\gamma}_{p,\text{II}}$ and $\dot{\gamma}_{p,\text{III}}$ were each given by an equation of form of Eq. (15-3). By combining the contributions of the two mechanisms, they obtained

$$\dot{\gamma}_p = \frac{D_0 \mu b}{kT}\left[ A_{\text{II}}\left(\frac{\tau}{\mu}\right)^{1.96}\left(\frac{b}{d}\right)^{1.8}\exp\left(-\frac{48\,157}{RT}\right) + A_{\text{III}}\left(\frac{\tau}{\mu}\right)^{7.1}\exp\left(-\frac{81\,239}{RT}\right)\right] \tag{15-6}$$

where $D_0 = 0.08\ \text{cm}^2/\text{sec}$ (that for Sn self-diffusion), $A_{\text{II}} = 900$, $A_{\text{III}} = 1.3 \times 10^{15}$, $d$ is the interphase spacing, and $R$ is the gas constant ($8.314\ \text{J mol}^{-1}\ \text{K}^{-1}$).

Equation (15-6) is based on results obtained in monotonic loading tests on bulk solder specimens. The results of Fox and coworkers[3,4] indicate that an equation of this form also applies to solder joints undergoing fatigue cycling. Their results, using steady-state creep tests performed on 63Sn–37Pb solder joints with 16 I/O surface mounted packages, shows the two well-defined regions of creep (Fig. 15-4). Based on this behavior, they determined that the fatigue damage that occurred during isothermal cycling correlated well with the Region III plasticity (matrix deformation) rather than Region II deformation (grain boundary sliding). To further check the equivalence of behavior in cyclic and monotonic load tests, Guo et al.[16] determined the deformation kinetics (focusing on the stress exponent) of eutectic Pb–Sn solder joints at room temperature in isothermal mechanical fatigue tests and compared the results with those determined in monotonic loading. Reasonable agreement between the two occurred for the test conditions employed, namely 300 K and a frequency of 0.01–0.1 Hz; see Fig. 15.5.

## 15.2 EXPERIMENTAL

The loading frame and temperature cycling chamber shown in Figs. 15-6 to 15-8 were constructed to perform the desired thermomechanical fatigue tests. The bimetallic loading frame in Figs. 15-6 to 15-8 operates on the principle of a mismatch between the thermal coefficients of expansion, $\alpha$, of the

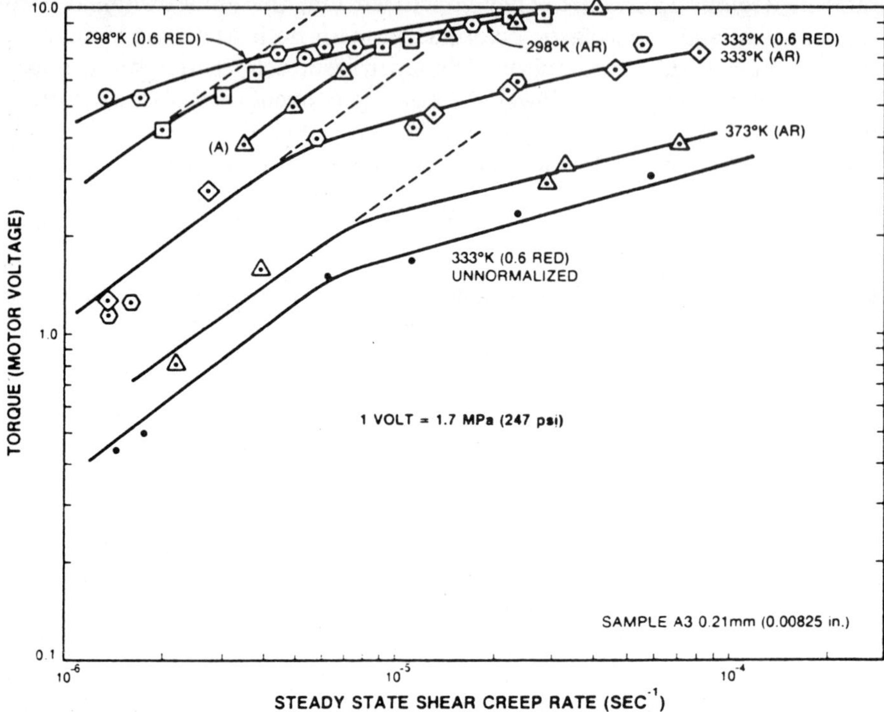


**Figure 15-4**  Torque vs. steady-state creep rate at several temperatures for 0.21-mm thick solder joints: AR = as-received, RED = after fatigue. From ref. 3.

components. A displacement is imparted to the solder specimen by virtue of the mismatch between the 4140 steel ($\alpha = 11.7 \times 10^{-6}\ \mathrm{K}^{-1}$), which is one leg of the frame, and the 360 brass ($\alpha = 11.7 \times 10^{-6}\ \mathrm{K}^{-1}$) that constitutes the remainder of the frame. The frame was built such that the specimen is loaded in shear. However, a normal component of stress will also occur due to bending moment in the solder joint and the lateral thermal mismatch between the copper, steel, and brass. The normal strain imparted to the specimen resulting from the lateral deflection was estimated to be of the order of 0.5% of the applied shear strain.

In the temperature chamber (Figs. 15-7 and 15-8), the loading frame is sandwiched by two banks of Chromel A heating elements supplying heat to the chamber. Power is pulsed to the heating elements with a temperature controller. For cooling, liquid nitrogen enters into the chamber by means of a diffuser panel located above the loading frame. The flow of liquid nitrogen is metered into the chamber with a solenoid valve operated by the temperature controller. The panel size is 24 cm × 24 cm, drilled with 64 holes of 0.5-mm

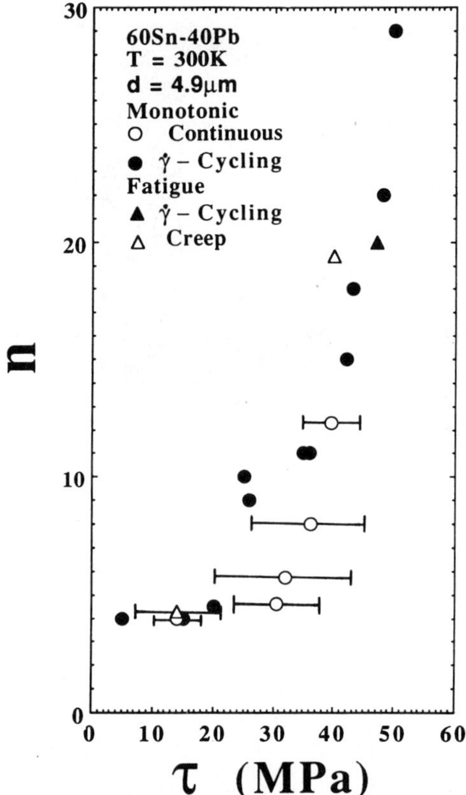



**Figure 15-5**   A comparison of the stress exponent $n$ obtained in fatigue tests with that in monotonic loading for 60Sn–40Pb with interphase spacing $d = 4.9$ μm tested at 300 K. From ref. 16.

diameter for the coolant to flow over the loading frame, which is mounted horizontally to minimize temperature gradients. Temperature in the frame is monitored with thermocouples at three places. One is encapsulated in solder and placed near the solder joint. The remaining thermocouples are placed in drilled holes; one is embedded in the brass, the other is in the steel member of the cycling frame.

The system was instrumented such that accurate measures of load and displacement in the system could be obtained. A resistance strain gauge bridge is attached to the top member of the frame. With calibration, the load and displacement to which the specimen is subjected can be determined. The dimensions of the beam were chosen so that the output signal of the strain gauge bridge was maximized. However, it was decided to limit the maximum deflection in the frame to about fifty percent of the displacement over the solder joint to limit the compliance of the frame.

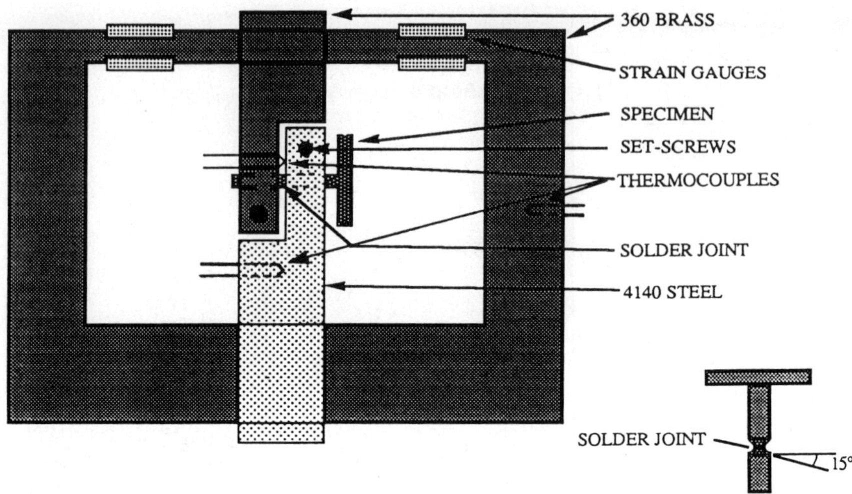


**Figure 15-6**  Schematic of the bimetallic loading frame with specimen inserted. The inset is a schematic of the test specimen consisting of the copper cylinders and the reflowed solder joint.

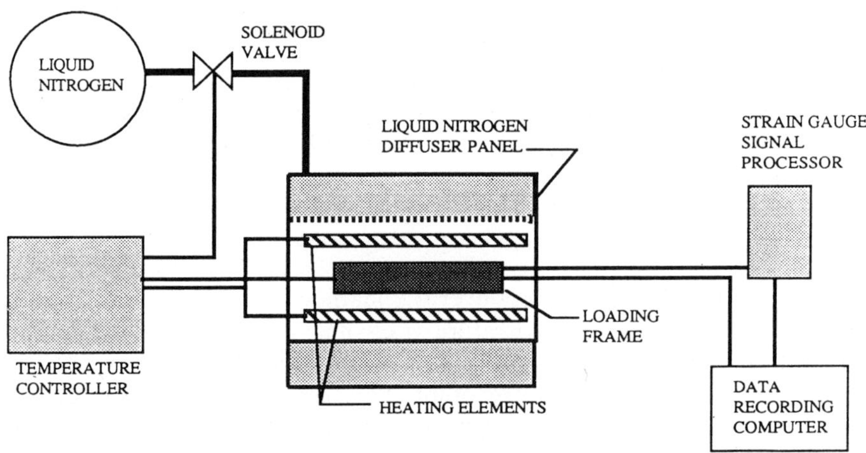


**Figure 15-7**  Schematic of the thermal cycling chamber and apparatus.

The loading frame was calibrated by simultaneously measuring displacement over the two specimen sleeves (without a solder joint) with an external clip gage and the output of the strain gage bridge as a function of an externally applied load at room temperature. By this means, the stiffness of the frame was determined to be $K = 1.08 \times 10^6$ kg/m. The relationship between strain gage bridge output and applied load was found to be 0.0498 V/kg.

**Figure 15-8**    Photograph of the cycling chamber and loading frame with specimen inserted. A reflective cover fits over the chamber opening and an insulating cover is placed over the chamber during tests (covers not shown).

The displacement $\delta_T$ resulting from the mismatch between the brass and steel for a temperature change must be accommodated by displacement over the specimen grips ($\delta_{\text{spec}}$) and elasticity in the frame ($\delta_F$), which deforms under load. This can be written as

$$\delta_T = \delta_{\text{spec}} + \delta_F \tag{15-7}$$

The temperature-induced deflection was determined by measuring the strain gage output of the bridge with a hardened steel pin clamped into the specimen sleeve of the frame. It was assumed that the steel pin did not deform and that all of the thermal mismatch was accommodated by elasticity in the frame; i.e., $\delta_{\text{spec}} = 0$ under this condition. By this means, the relationship between load and temperature was found to be 0.144 kg/°C. From this relationship and the frame stiffness, the displacement of the two sleeves with no specimen mounted (i.e., $\delta_F = 0$) as a function of temperature was calculated to be $1.33 \times 10^{-7}$ m/°C. In a temperature excursion of 160°C, as was the case in our tests, the temperature-induced displacement is $2.13 \times 10^{-5}$ m.

During a fatigue test, load is measured with the strain gage bridge, their relationship having been determined by calibration. Displacement over the specimen is calculated via Eq. (15-7), where $\delta_T$ is determined from the relationship between displacement and temperature (m/°C), knowing the temperature. $\delta_F$ is determined from the stiffness $K$ (kg/m) and the strain gage output as a function of load (V/kg), knowing the strain gage output. In this way, the stress and strain that the specimen undergoes can be determined throughout the fatigue test.

The test specimens used in these tests, shown in Fig. 15-6 (inset), were based on the Iosipescu pure-shear specimen used by Solomon[17] and adapted by Bae et al.[18] The test specimens consist of two cylindrical copper grips, one with an enlarged base. Solder blanks were formed by rolling bulk 63Sn–37Pb solder to a thickness of 0.2 mm and punching out cylinders 2.0 mm diameter. The solder was fluxed (Alpha 100, type R nonactivated) and placed between the cylindrical copper grips using a jig to maintain alignment and to establish reproducible solder joint thicknesses of 0.203 $\pm$ 0.013 mm. The specimen–jig assembly was then placed on a hot plate and heated to 35°C above the liquidus of the solder alloy. After the specimen reached the desired temperature the assembly was placed on a steel plate to cool to room temperature. The time to cool from just above the liquidus (184°C) to half-way to room temperature was 57 seconds. The resultant microstructure is shown in Fig. 15-9 at two levels of magnification. The mean phase intercept spacing is $d = 0.9$ μm (standard deviation 0.9 μm). The standard deviation is large because the distribution of sizes is greatly skewed. Occasionally large lead-rich dendrites are seen in the micrographs (Fig. 15-9$a$); however, the microstructure largely consists of smaller dispersed lead-rich regions within the tin matrix (Fig. 15-9$b$). The eutectic structure appears to be strongly degenerate, typical of quickly cooled, near-eutectic solder joints. Relatively little of the fine alternating layers of tin- and lead-rich phases characteristic of lamellar eutectic structures is seen. A colony structure is not clearly distinguishable in the reflowed solder joint.

For the thermomechanical fatigue tests the specimen was clamped into the fixture as shown in Figs. 15-6 and 15-8. The fatigue tests were performed by ramping the temperature in triangular waveform between the extremes of $-30$°C and $+130$°C. This temperature range is similar to, albeit slightly greater than, that employed by other investigators. For example, Hall[19] employed the range $-25$°C to 125°C, whereas Liljestrand and Andersson[20] used $-10$°C to 100°C. A triangular waveform is a simplification of the actual temperature–time profile experienced in a real solder joint. It was chosen as a starting point for experimental and modeling simplicity. Dwells at the low-temperature extreme would result in a slow transfer of elastic strain from the loading frame to plastic deformation in the solder joint, giving an increase in cyclic plastic strain range and a lower number of cycles to failure. A dwell at the high-temperature extreme would coarsen the microstructure; but since the joint is essentially stress-relaxed at the end of the ramp up to 130°C,

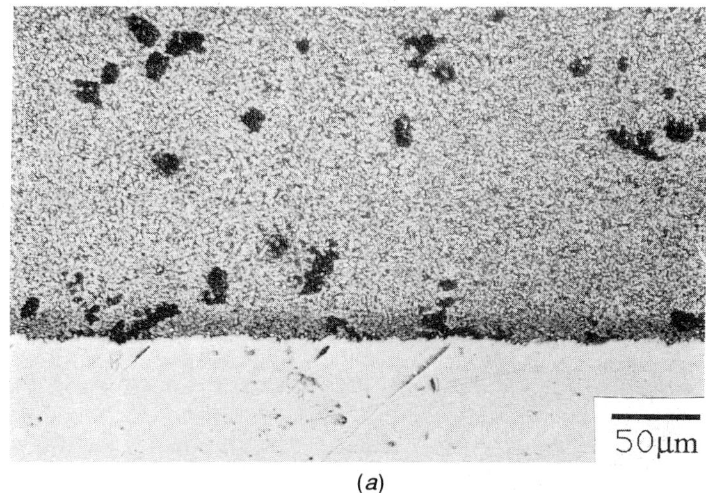

(a)

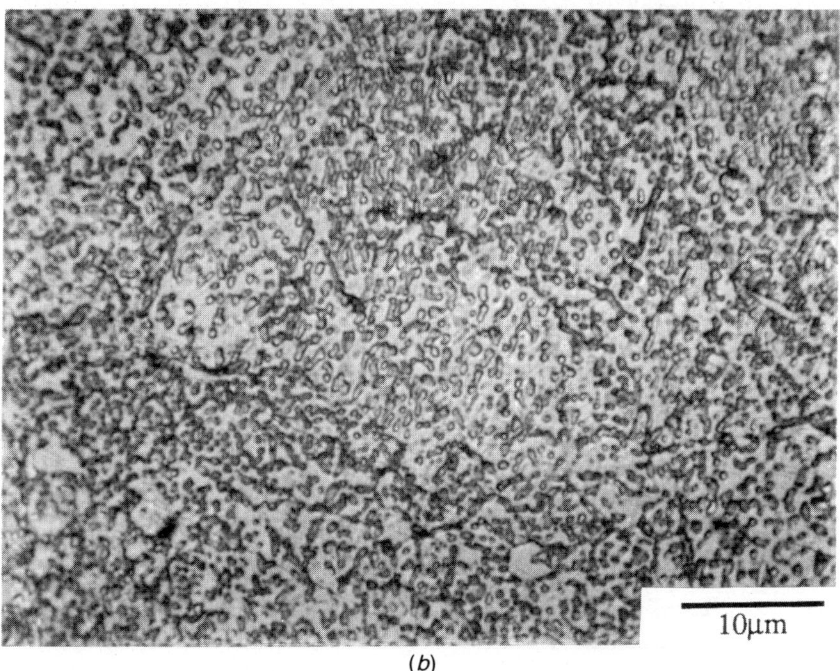

(b)

**Figure 15-9**    As-solidified microstructure (optical micrograph) of the 63Sn–37Pb alloy used in the present tests: (a) low magnification and (b) high magnification. Dark phase is the Pb-rich phase, light phase the Sn-rich phase.

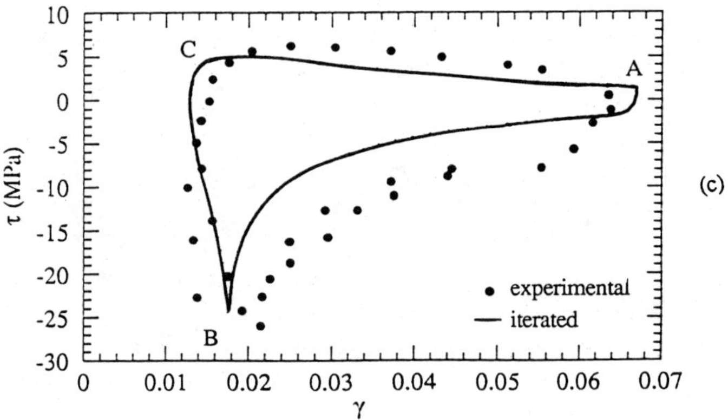



**Figure 15-10**  Typical hysteresis loops generated during thermal cycling with 12-min period and simulation: (*a*) nominal shear stress vs. temperature; (*b*) shear strain vs. temperature; (*c*) nominal shear stress vs. shear strain.

little additional plastic deformation would result. Temperature ramp cycling periods of 12, 24, 48, and 300 minutes were used in our tests. A new specimen (copper cylindrical grips and solder joint) was used in each test to examine the initial stress–strain hysteresis behavior.

Typical hysteresis loops obtained for the various temperature cycle frequencies are presented in Figs. 15-10 to 15-13. They represent the third complete cycle undone by the specimen. The manner in which the simulated (iterated) curves were determined is given below.

The deformation behavior along the $\tau$ vs. $\gamma$ hysteresis loops of Figs. 15-10 to 15-13 is interpreted as follows. Referring to Fig. 15-10$a$, at point A, the high-temperature extreme of the cycle (130°C, $0.88T_M$), the stress in the joint is near zero because the flow stress for plasticity is low at this temperature. As the temperature is ramped down to $-30$°C (point B), accommodation of the displacement due to thermal mismatch in the assembly gradually changes from plastic deformation in the solder to elastic deformation in the assembly as the solder flow stress increases. Upon heating to 130°C between points B and C, the applied stress is below the flow stress of the solder and the slope largely reflects the elastic deformation of the frame. During the heating, the plastic deformation of the solder is small until point C is reached, where the temperature is now high enough for significant plastic deformation to occur. Upon further increase in temperature from C to A, the flow stress of the solder decreases with the increase in temperature and plastic flow now occurs in the solder joint, gradually reducing the applied stress level.

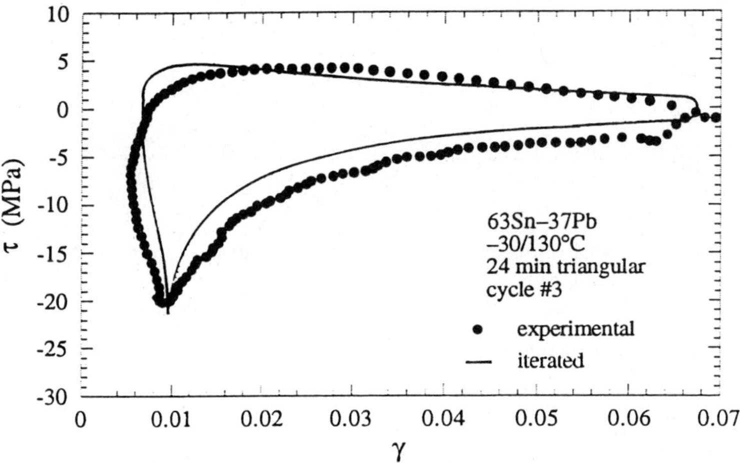


**Figure 15-11**  Typical nominal shear stress vs. shear strain hysteresis loop generated during thermal cycling with a 24-min cycle period and simulation.

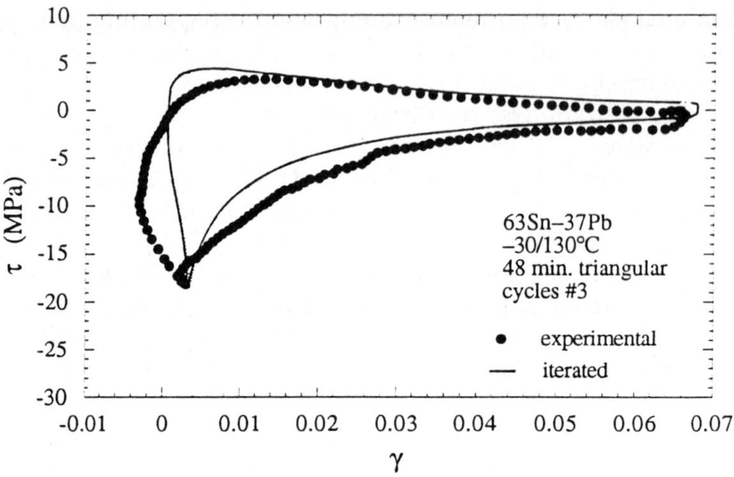

**Figure 15-12**   Typical nominal shear stress vs. shear strain hysteresis loop generated during thermal cycling with a 48-min cycle period and simulation.

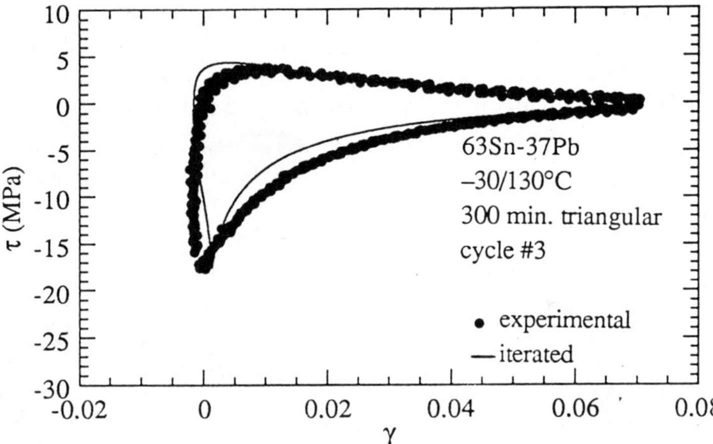

**Figure 15-13**   Typical nominal shear stress vs. shear strain hysteresis loop generated during thermal cycling with a 300-min cycle period and simulation.

## 15.3 COMPUTER SIMULATION

Following refs. 19 and 21, the shear strain in the solder joint test assembly is taken to be

$$\gamma + \frac{\tau A}{K} = \frac{L}{H}\Delta\alpha(T - T_0) \qquad (15\text{-}8)$$

where $\gamma = \gamma_e + \gamma_p$ is the total shear strain (elastic + plastic) in the solder, $\tau$ is the shear stress, $A$ is the solder specimen cross-sectional area, $K$ is the stiffness of the loading assembly (force/displacement), $L$ is the distance over which thermal expansion of the frame takes place, $h$ is the solder thickness, $\Delta\alpha$ is the difference in thermal coefficient of expansion of the frame materials, and $T - T_0$ is the excursion temperature from room temperature. $\gamma_e = \tau/\mu(T)$, where $\mu(T)$ is the shear modulus as a function of temperature. If the loading frame is instantaneously subjected to a small increment in temperature $\Delta T$ above $T_0$, the assembly will change dimensions according to the coefficients of thermal expansion of brass and steel, thus creating a stress and strain in the system. Plastic deformation in the solder joint, which is time-dependent, will not occur instantaneously. Hence, at the instant of temperature change the strain is absorbed elastically in the solder $\gamma_e$ and in the surrounding assembly given by the term $\tau/K$. The resultant instantaneous stress change is then

$$\Delta\tau = \frac{L\,\Delta\alpha\,\Delta T}{A}\left[\frac{1}{\left(\dfrac{1}{K} + \dfrac{h}{A\mu(T)}\right)}\right] \qquad (15\text{-}9)$$

Assuming that all of the resulting instantaneous deformation is absorbed elastically in the solder and assembly, the solder will subsequently deform plastically over an increment of time $\Delta t$ following the increment in temperature $\Delta T$. The rate of plastic deformation will be a function of the applied stress, the temperature, the solder structure, and the controlling deformation mechanism. The magnitude of the plastic strain can be determined via the product of $\Delta t$ and the result of a constitutive equation such as Eq. (15-6) above. The direction of the plastic shear strain in the solder joint will be such that the magnitude of stress in the system is reduced. The change in stress due to plastic shear strain over time $\Delta t$ is thus given by

$$\Delta\tau_p = -\text{sgn}(\tau)|\Delta\gamma_p|\left[\frac{1}{\left(\dfrac{A}{hK} + \dfrac{1}{\mu(T)}\right)}\right] \qquad (15\text{-}10)$$

where $\text{sgn}(\tau)$ is a function that returns the sign (direction) of the shear stress. The mechanical state of the system after an increment in temperature $\Delta T$ and time $\Delta t$ can thus be characterized. The above equations can be iterated with the rate of temperature change and the material parameters of the test assembly to simulate the stress–strain behavior in a thermomechanical cycle. For the present analysis, the calculation of the elastic strain is at the start of each iteration. The state of strain in the solder joint at the end of each iteration is taken as the total strain, without regard to relative amounts of elastic or plastic strain. A result of this procedure is the absence of any visible elastic unloading at the low-temperature extreme of the cycle as the temperature begins to ramp up. The effects of this are negligible when the rate of temperature change is slow, because the plastic strain greatly exceeds the elastic strain component.

The constitutive equation employed in our modeling of the $\tau$ vs. $\gamma$ hysteresis loops is based on Eq. (15-6) but with the preexponential constants chosen to best fit our experimental results. This is acceptable, since these constants depend on the fine details of the theory and cannot generally be predicted with a high degree of accuracy. Our results gave

$$\dot{\gamma} = \frac{C_{II}\, \tau^{n_{II}}}{T\; d^{p_{II}}} \exp\left(\frac{-\Delta H_{II}}{RT}\right) + \frac{C_{III}}{T}\, \tau^{n_{III}} \exp\left(\frac{-\Delta H_{III}}{RT}\right) \qquad (15\text{-}11)$$

The parameters in Eq. (15-11) are as follows:

$C_{II} = 1.39 \times 10^{-6}$
$C_{III} = 2.38 \times 10^{3}$
$T$ = temperature (K)
$\tau$ = shear stress (MPa)
$R$ = gas constant = $8.3 \text{ J mol}^{-1} \text{ K}^{-1}$
$n_{II}$ = Region II stress exponent = 1.96
$n_{III}$ = Region III stress exponent = 7.1
$d$ = grain size (m)
$p_{II}$ = Region II grain size exponent = 1.8

Since Eq. (15-6) was originally formulated with the grain size parameter $d$ being the mean phase intercept spacing, this measurement was also used here for $d$. It is acknowledged that this may not be the mechanistically correct grain size parameter; however, it is generally assumed that some relationship exists between the mean phase size and the size of the grains involved in the grain boundary sliding.

Deformation in thermomechanical fatigue has been shown to concentrate in a coarsened band and at intercolony regions in cast solder joints that display a colony structure.[11] In contrast, eutectic Pb–Sn specimens prepared by cold working and recrystallization to give a fine, equiaxed grain structure without colonies have been shown to deform by grain boundary sliding at the intraphase and interphase boundaries in Region II.[14,22] For this fine grain morphology, use of phase size has been shown to accurately describe plastic deformation for use in constitutive equations that include a grain size term.[15,23] As the cooling rate is increased in cast solder joints, the structure tends towards a finer morphology in which colony and lamellar structures are less apparent. Fine equiaxed grains whose appearance is similar to that of cold rolled and recrystallized eutectic solder now become more predominant.[12] These quickly cooled specimens also displayed mechanical behavior akin to the cast and recrystallized solder specimens, in that they exhibited similar stress exponents and activation energies for deformation. The solder joints used in our tests were cooled relatively quickly, resulting in a morphology more like that of quenched solder joints; however, there were variations across the specimens as seen in Fig. 15-9. Because of the strong similarities between the microstructure of our specimens and the quenched specimens of ref. 12, it was felt that it would be meaningful to include grain size in our analysis. It was necessary to do so because of the importance of microstructure in thermomechanical fatigue behavior. We did simulations only for the initial hysteresis loops obtained, before effects of inhomogeneous coarsening made disparities in grain size so large that the average initial grain size term became meaningless.

The mean value $d$ was measured after the third 24-min period cycle and was found to be 1.0 μm (standard deviation = 0.7 μm), Fig. 15-14. This value was assumed to be an average value after three cycles of 12-, 24-, and 48-min periods. It was used as input in the 12-, 24-, and 48-min period simulations. In the 300-min period tests, where signficant grain coarsening occurred

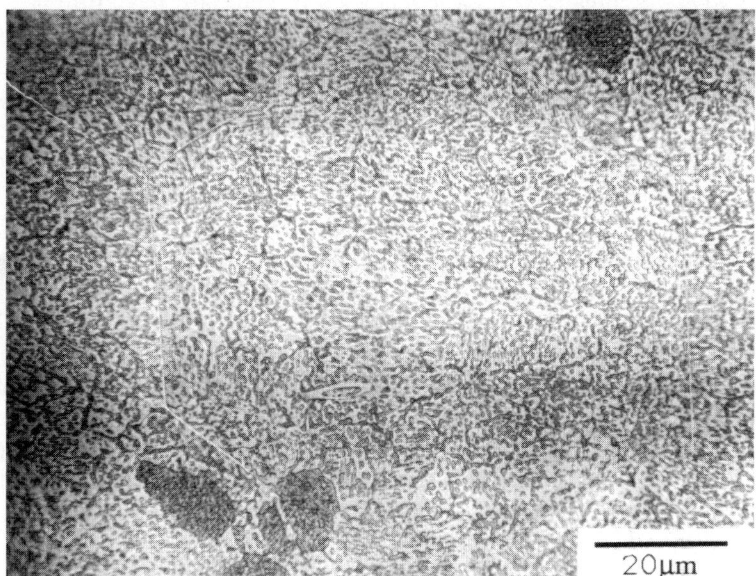


**Figure 15-14**   Optical micrograph of the microstructure after the third 24-min period cycle. Dark phase is Pb-rich; light phase is Sn-rich.

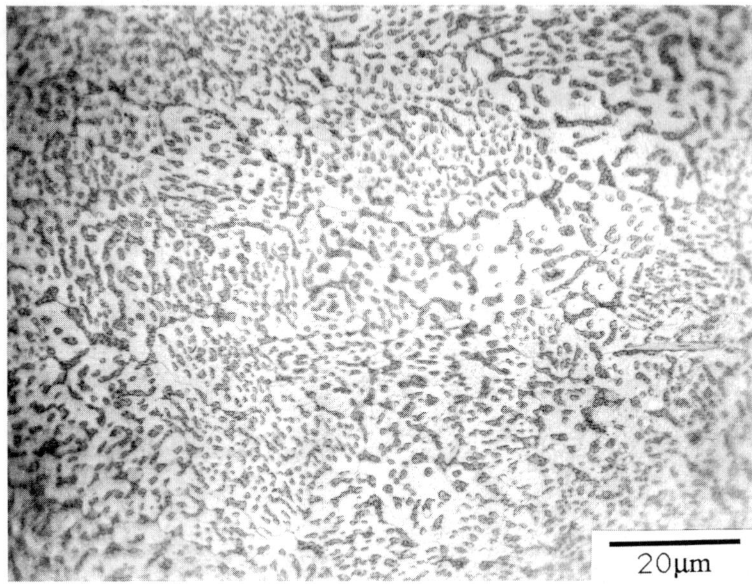


**Figure 15-15**    Optical micrograph of the microstructure after the third 300-min period cycle. Dark phase is Pb-rich; light phase is Sn-rich.

(Fig. 15-15), the mean value of $d$ was 2.4 μm (standard deviation = 2.2 μm) and was used in the modeling for this cycling period in Fig. 15-13. Grain coarsening reduced the contribution of grain boundary sliding to the plastic deformation, i.e., as the grain size increased, the fraction $\dot{\gamma}_{II}/\dot{\gamma}_{III}$ decreased. A log-plot of $\dot{\gamma}_p$ vs. $\tau$ derived using Eq. (15-11) for the temperature range and grain size employed in the present tests is given in Fig. 15-16. The intersection of the constant-stress lines (representing the maximum plus and minus shear stresses in Figs. 15-10 (1 μm grain size) and 15-13 (2.4 μm grain size) indicates that the major component of the plastic flow during the thermal cycling was superplastic flow, i.e., Region II.

The derived hysteresis loops employing the constitutive equation (15-11) with the iterative procedure described above are compared with those determined experimentally in Figs. 15-10 to 15-13. There is reasonable, though not ideal, correspondence between the iterated and the experimentally obtained loops. Differences between the two could result from a number of sources. With respect to the experimentally obtained hysteresis loops, irregularities that may alter the results include temperature gradients in the cycling frame and the solder joint. Gradients are a bigger problem at the higher cycling frequency tests. An additional consideration is that the frame consists of parts screwed together. The dimension tolerances are very stringent in this type of apparatus; any slippage between parts greatly skews

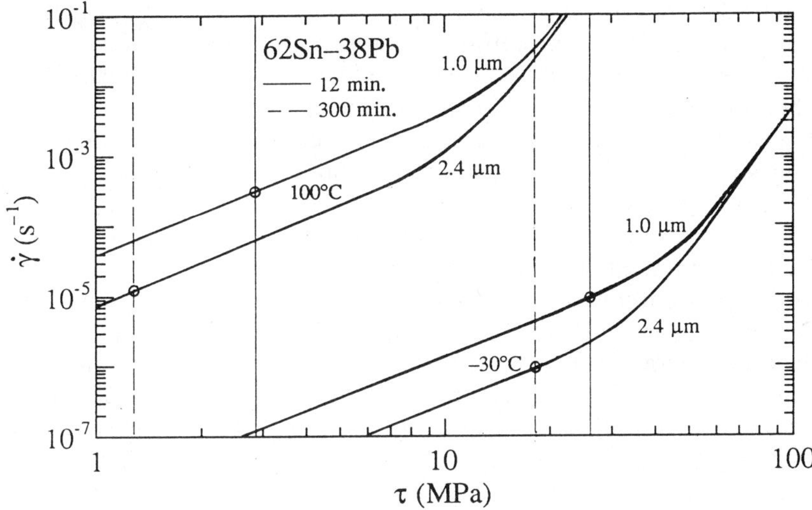



**Figure 15-16**   Log shear rate vs. log shear stress for 63Sn–37Pb alloy with $d = 1.0$ µm and 2.4-µm interphase spacing derived using Eq. (15.11) of the text. The vertical lines indicate the maximum stresses measured at the two temperature regimes for the two grain sizes.

results. By careful construction and adjustments in the apparatus, we believe that we have eliminated these problems.

Issues pertaining to the computer simulation include the question of applicability of the particular constitutive equation chosen and the inherent assumptions made in its derivation. Differences in microstructure and the slight difference in composition between the specimens used to derive Eq. (15-6)[15] and those in the present study are likely to give some variation. Further, Eq. (15-6) is based on isothermal monotonic loading, whereas in the present work it is applied to thermomechanical fatigue.

Although dwells at the temperature extremes were not included in the present tests, some mention should be made of the effect of holding period. Hall's experiments[5,19] for a temperature cycle between $-25°C$ and $125°C$ ramping at $30°C/h$ contained dwell times of 2 h at either extreme. Upon ramping to the high-temperature extreme of the cycle, significant plastic flow already occurs; therefore, little additional relaxation takes place during the hold. At the low-temperature extreme, $-25°C$, the plastic strain rate is very slow, and again, little relaxation occurs. Clech and Augis,[21] upon simulating thermomechanical fatigue cycles taken from ref. 19, showed that the dwell time at high temperature could be reduced to hasten the time required to determine the cycles to failure of a solder joint. Frear et al.[24] also found that the time spent at high temperature does not have a strong effect on solder joint lifetime. An increased coarsening of the grains and formation of intermetallics at the solder/substrate interface will occur, however. Cracks through the intermetallic at the copper–solder interface have been observed for thermomechanical tensile/compressive loading.[24]

It is well known that the damage per cycle during isothermal fatigue cycling is related to the amount of plastic deformation. The Coffin–Manson relation applied to solder fatigue data shows that a larger cyclic plastic strain range causes a reduction in the number of cycles to failure. Engelmaier[7] applied a version of the Coffin–Manson relation to results from widely varying test conditions in thermomechanical fatigue of solder joints. As in isothermal fatigue, the number of cycles to failure decreases as the shear strain range increases. During low-temperature holds, elastic strain in the package assembly is transferred to plastic strain in the solder joint, thus increasing the total cyclic plastic strain range. The deformation rate, mechanism, and the amount of damage accumulating in the solder joint during a hold at the low-temperature extreme depends on the microstructure, load, and temperature.

The present computer model was also applied to the thermomechanical hysteresis loop undergone by a solder joint in a CCC/PWB assembly obtained by Hall;[5,19] see Fig. 15-17. The parameters derived from that publication and employed as input for the present simulation are the assembly stiffness, $K = 2.8 \times 10^4$ kg/m, and the displacement per °C between the materials being joined, $\Delta\alpha \cdot L = 1.3 \times 10^{-7}$ m/°C. The only unpublished

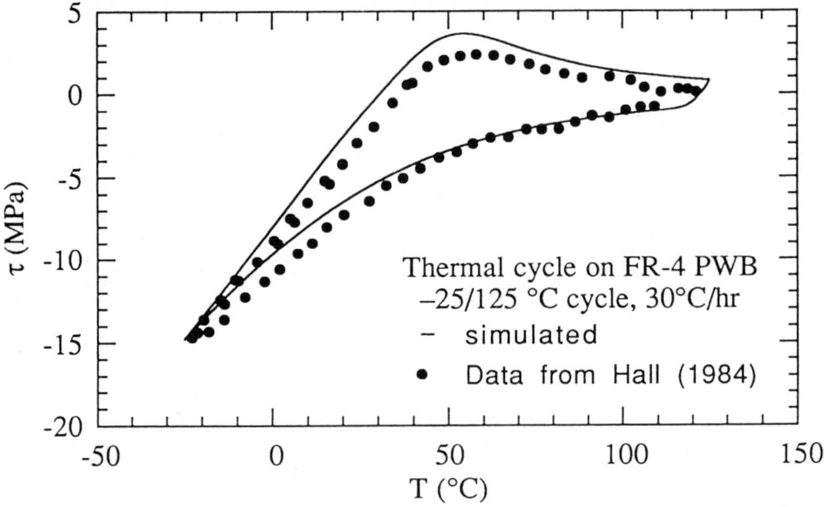


**Figure 15-17**  Shear stress vs. temperature hysteresis loop for eutectic Pb–Sn solder joint taken from Hall.[19]

parameter was the mean phase spacing, taken here as $d_{\text{eff}} = 4.8\ \mu\text{m}$, a reasonable value for the test conditions.[25] The dwells at either temperature extreme were neglected. This was believed to be reasonable because the plastic deformation undergone during the holds was minimal. Figure 15-17 shows that the present simulation gives reasonably good agreement with the experimentally obtained results if the above-mentioned assumptions are made. Modeling of Hall's data was also done by Stone et al.[26] and Wilcox et al.[27] with nearly perfect correspondence using the state variable approach.

## 15.4   CRACK INITIATION AND GROWTH

During the thermal cycling it was found that the maximum load (nominal stress) at each end of the hysteresis loop decreased with the number of cycles; Fig. 15-18. The maximum load at the low-temperature end of the loop remains relatively constant up to $\sim 50$ cycles and then begins to decrease significantly with increasing number of cycles. In contrast, the maximum load at the high-temperature end remains relatively constant to about 150 cycles and then decreases only slightly. There is a significant increase in the maximum shear strain at the low-temperature segment of the cycle, while the high-temperature strain remained constant. The load range $\Delta P$ decreases and the strain range $\Delta \gamma$ in the solder joint increases with number of thermal cycles.

The change in the load at the low-temperature extreme can be expressed by the parameter

$$\phi_{-30^\circ} = \left(1 - \frac{P_N}{P_0}\right) \tag{15-12}$$

where $P_N$ is the load after $N$ cycles and $P_0$ is that for the first cycle. It was shown by Guo et al.[28] in isothermal fatigue tests on Pb–Sn solder joints that a decrease in maximum load was directly proportional to the fraction of the specimen cross-sectional area that had fatigue-cracked, so that

$$\phi = \frac{A_c}{A_0} \tag{15-13}$$

where $A_c$ is the cracked area and $A_0$ is the original area. The cracked area was determined by immersing and breaking the fatigued specimen in liquid nitrogen. This procedure was also performed on specimens that had been thermal-cycled.[29] Typical structures observed by SEM on the fracture surface of a specimen broken in this manner are presented in Fig. 15-19 for a specimen cycled to $N = 182$ (12-min period) and $\phi = 17\%$. The

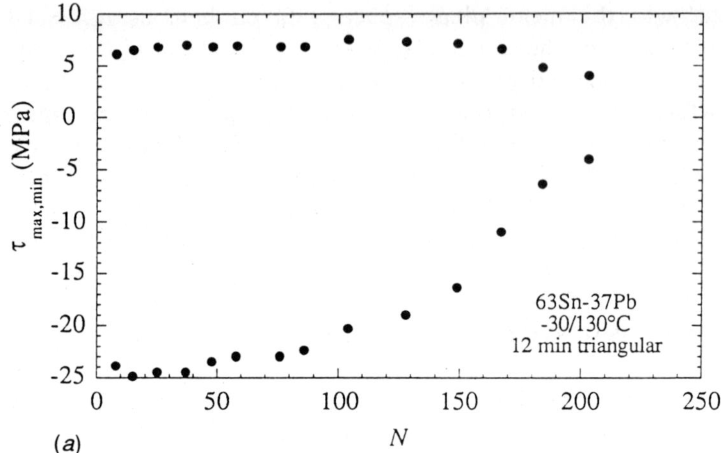


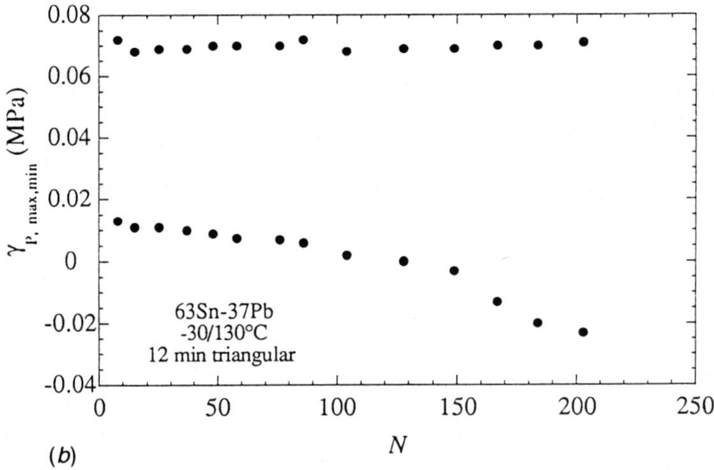


**Figure 15-18** Change in nominal shear stress and shear strain with number of thermal cycles (−30°C to 130°C, 12-min period: (a) maximum shear stress at the two temperature extremes of the hysteresis loop and (b) cyclic shear strain extremes.

low-magnification overview in Fig. 15-19 exhibits three regions: (i) a central region, (ii) a transition region, and (iii) a region near the circumference of the specimen. Higher-magnification views of the three regions are given in Figs. 15-19b, c, and d, respectively. These higher-magnification photomicrographs indicated the following regarding the three regions.

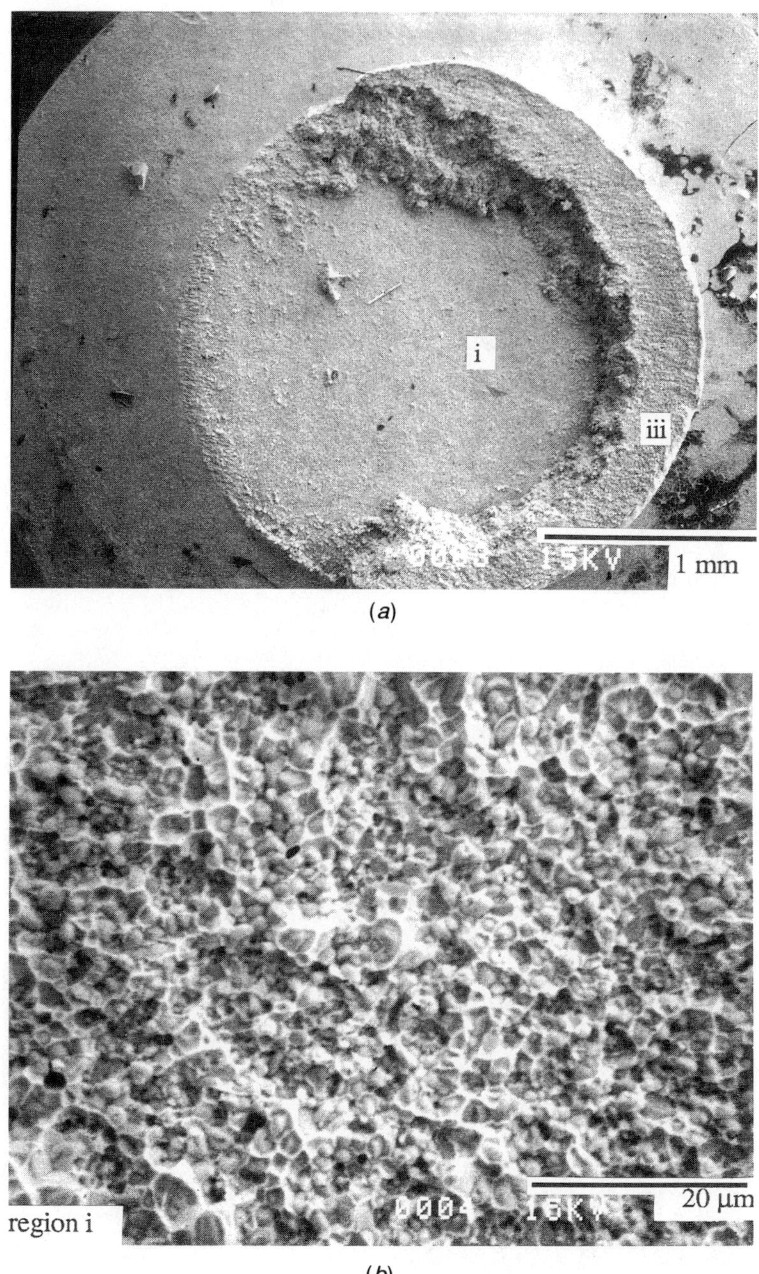



(a)

(b)

**Figure 15-19**    SEM micrographs of the fractured surface of a thermal-cycled ($-30/130°C$, 12 min) specimen to $N = 182$, with 17% drop in load. The joint was then manually broken in liquid nitrogen: (a) overall view; (b) region i. (*Continued*)

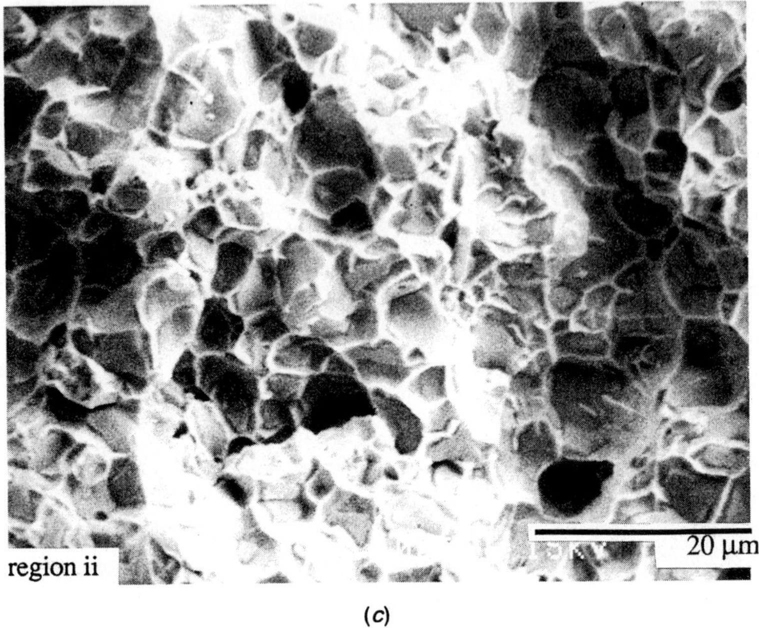



(*c*)

(*d*)

**Figure 15-19** (*continued*)   (*c*) region ii; (*d*) region iii.

1. Region $A_i$ in Fig. 15-19$b$ was identified as the solder–copper interface region, which had not been cracked during the fatigue tests. This structure was also observed in ref. 28. Energy-dispersive spectroscopy of this surface indicated the presence of a copper–tin intermetallic compound.

2. Region $A_{ii}$ in Fig. 19$c$ generally connected Regions (i) and (iii). It was also found in islands in the intermetallic region ($A_i$). The structure had the appearance of intergranular fracture. It is suggested that this region reflects grain boundary or interphase sliding that occurred during the thermomechanical fatigue tests. The sliding could create fissures along the boundaries and grain corners, resulting in their separation upon manual breaking at liquid nitrogen temperature.

3. Region $A_{iii}$ in Fig. 15-19$d$ had somewhat the appearance of having been mechanically abraded. It occurred at the perimeter of the solder joint where cracking has been observed to initiate.[11] More of region $A_{ii}$ generally occurred as the number of cycles $N$ increased. It was not observed in specimens that had not been thermal-cycled. This region was therefore identified as crack surfaces created during thermo-mechanical fatiguing.

Figure 15-20 is a plot of $\phi_{-30°}$ vs. $A_c/A_0$ where $A_{iii}$ is taken to represent the fatigue cracked area $A_c$. The data are in reasonable accord with Eq. (15-13). Figure 15-21 is a plot of load at the low-temperature extreme vs. the

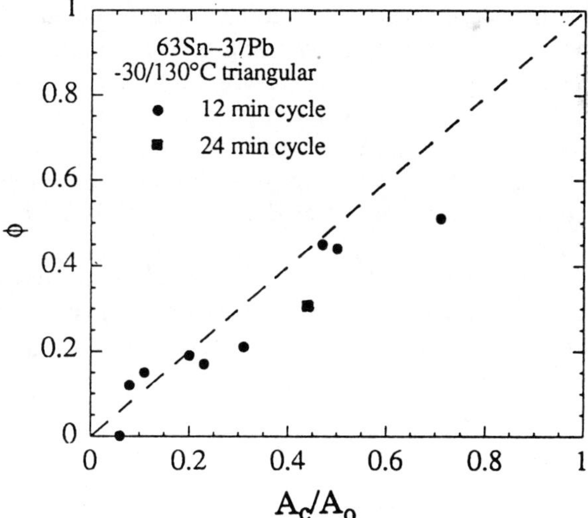


**Figure 15-20** Load drop parameter $\phi$ vs. the ratio of cracked area to original cross-sectional area, $A_c/A_0$.

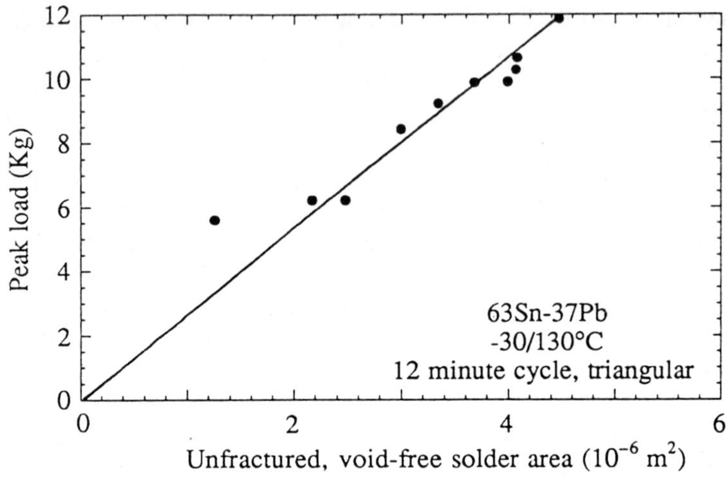



**Figure 15-21** Maximum load at the low-temperature extreme of the $\tau$–$\gamma$ hysteresis loop vs. the unfractured, void-free cross-sectional area for 12-min period cycling.

area of sound material. The sound material is the sum $A_i + A_{ii}$, i.e., the unfractured, void-free cross-sectional area. The decrease in load at the low-temperature extreme of the thermal cycle thus appears to have occurred as a result of fatigue cracking, which decreased the area available for supporting the load. The straight line in Fig. 15-21 then indicates that the true shear stress remained essentially constant throughout the fatigue test.

Crack growth in solder joints undergoing thermomechanical fatigue has been approached phenomenologically using a stress intensity factor $K$.[1,28] Although the mechanisms of crack growth in solder joints are not well understood, and they change with temperature and deformation mechanism,[5,30] $K$ nevertheless probably has some merit in initial attempts to characterize crack growth. Use of $K$ presumes that the combined effects of various damage mechanisms are related to the growth of the major crack that can be measured.

Figure 15-21 indicates that fatigue cracks had initiated early in the fatigue test and thereafter grew with increasing number of cycles. A plot of $A_c$ vs. $N$ is given in Fig. 15-22, and a log–log plot of $dA_c/dN$ vs. $A_c$ is presented in Fig. 15-23, which yields

$$\frac{dA_c}{dN} = CA_c^q \tag{15-14}$$

with $C = 1.3 \times 10^1$ m$^2$ and $q = 1.5$. Since $\tau$ was established to be approximately constant throughout the fatigue test, Eq. (15-14) can be considered

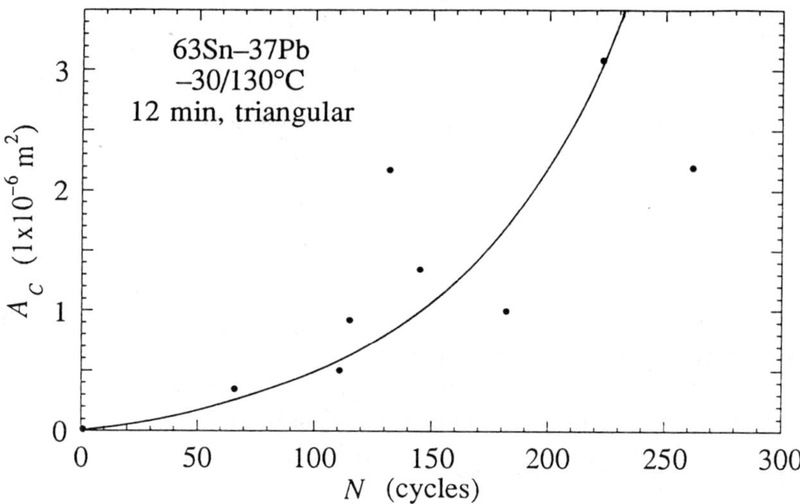



**Figure 15-22**    Fatigue-cracked area vs. number of 12-min period thermal cycles.

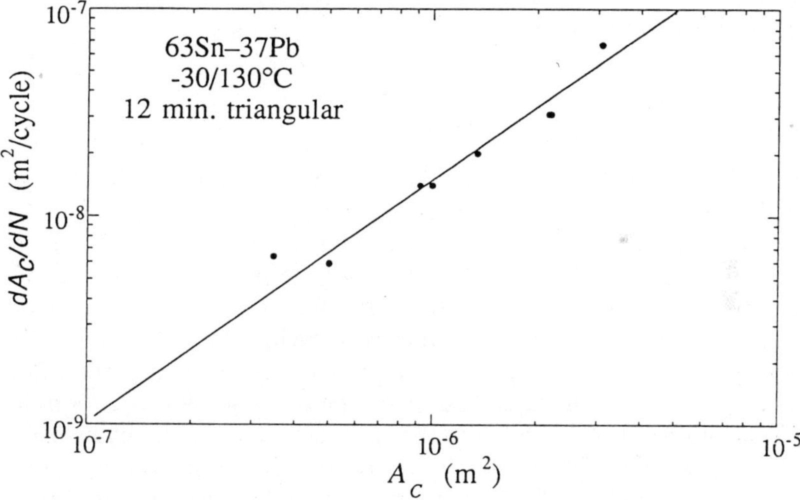



**Figure 15-23**    Log–log plot of the rate of growth of the cracked area $dA_c/dN$ vs. $A_c$ for the 12-min period thermal cycles.

to be equivalent to the Paris–Erdogan equation[31]

$$\frac{da_c}{dN} = C' \, \Delta K^{q'} \qquad (15\text{-}15)$$

where $a_c$ is the crack length and $\Delta K = \Delta\tau\sqrt{\pi a_c}$ is the stress intensity factor.

Taking

$$A_c = wa_c \tag{15-16}$$

where $w$ is the width of the crack and assuming that $w$ is relatively constant,[17] we obtain from Eq. (15-14)

$$\frac{da_c}{dN} = Cw^{q-1}a_c^q \tag{15-17}$$

Upon comparing Eq. (15-17) with Eq. (15-15), we find that $q' = 2q = 3$ and that $Cw^{q-1} = C_1(\Delta\tau\sqrt{\pi})^{q'}$. The value of $q' = 3$ obtained here for shear (presumably Mode II crack propagation) is similar to that found for metals in general tested in Mode I.[32] This may indicate that the crack growth we measured includes the tensile component of the bending moment that occurs in solder joints during a shear test. The crack growth may have occurred by mixed Modes I and II.

As with the approach of Li and coworkers,[1,2,27] the above consideration of crack growth in solder joints in shear is based on the stress intensity factor $K$. It may be that a more appropriate parameter is the $J$-integral.[9] We are presently investigating this possibility.

## 15.5 LIFE PREDICTION

The present results and those by others[1–5,7,8,26–28] indicate that the stress vs. strain hysteresis loops obtained by thermal cycling can be predicted reasonably well using constitutive equations or parameters derived from isothermal monotonic tests. The accuracy of the prediction will depend on the correctness of the chosen constitutive equation. Especially important is the initial microstructure and its change with cycling. Having established the stress–strain behavior, it appears that one may be able to predict fatigue damage by employing a Paris–Erdogan type crack growth equation determined in isothermal tests, similar to the approach of Subrahmanyan et al.,[1] or perhaps by a $J$-integral approach. Again, the degree of success will depend on how well the parameters of the equation apply to the existing microstructure.

Accurate prediction of fatigue life in thermal cycling requires more information on the detailed mechanisms responsible for the inelastic deformation and the nucleation and growth of cracks.

## 15.6 SUMMARY AND CONCLUSIONS

A testing frame was constructed to simulate the stress $\tau$ vs. strain $\gamma$ hysteresis loops in shear experienced during the thermal cycling of solder joints in microelectronic packages. The $\tau$ vs. $\gamma$ hysteresis loops generated by the device were simulated reasonably well employing a constitutive equation based on a combined theoretical–phenomenological treatment of isothermal deformation kinetics data obtained on bulk specimens in monotonic loading.

The fraction of the specimen cross-sectional area that had fatigue-cracked during testing was measured directly and found to be proportional to the decrease in load at the low-temperature extreme of the thermal cycle. This indicates that the true applied stress remained essentially constant throughout the thermal fatigue test. The crack growth rate was in reasonable accord with the Paris–Erdogan equation, giving a value of 3 for the exponent over the stress intensity range.

The present results, along with those of others, suggest that thermal cycle fatigue life can be predicted by employing appropriate and accurate constitutive equations for the inelastic deformation and a Paris–Erdogan (or $J$-integral) type equation for crack growth rate.


## ACKNOWLEDGMENT

This research was supported by the Electronics Division of E.I. Du Pont de Nemours and Company with Dr. J. Dorfman as technical monitor.


## REFERENCES


1. Subrahmanyan, R., J. R. Wilcox, and C.-Y. Li, "A Damage Integral Approach to Thermal Fatigue of Solder Joints," *IEEE Trans. Components, Hybrids, and Manufacturing Technology,* **12**(4), 1989, pp. 480–491.
2. Li, C.-Y., R. Subrahmanyan, J. R. Wilcox, and D. Stone, "A Damage Integral Methodology for Thermal and Mechanical Fatigue of Solder Joints," *Solder Joint Reliability,* J. H. Lau, ed., Van Nostrand Reinhold, New York, 1991, pp. 361–383.
3. Shine, M. C., and L. R. Fox, "Fatigue of Solder Joints in Surface Mount Devices," *Low Cycle Fatigue,* ASTM Special Technical Publication 942, 1987, pp. 588–610.
4. Knecht, S., and L. Fox, "Integrated Matrix Creep: Application to Accelerated Testing and Lifetime Prediction," ref. 2, pp. 508–544.
5. Hall, P. M., "Creep and Stress Relaxation in Solder Joints," ref. 2, pp. 306–332.
6. Solomon, H. D., "Predicting Thermal and Mechanical Fatigue Lives from Isothermal Low Cycle Data," ref. 2, pp. 406–454.
7. Engelmaier, W., "Solder Attachment Reliability, Accelerated Testing and Result Evaluation," ref. 2, pp. 545–587.
8. Clech, J. P., and J. A. Augis, "Surface Mount Attachment Reliability and Figures of Merit for Design for Reliability," ref. 2, pp. 588–613.
9. Arrowood, R., A. Mukherjee, and W. R. Jones, "Hot Deformation of Two-Phase Mixtures," *Solder Mechanics,* D. R. Frear, W. D. Jones, and K. R. Kinsman, TMS, Warrendale, PA, 1991, pp. 107–153.

10. Bird, J. E., A. K. Mukherjee, and J. E. Dorn, "Correlations Between High-Temperature Creep Behavior and Structure," *Quantitative Relation Between Properties and Microstructure*, A. Rosen, ed., Israel University Press, Jerusalem, 1969, pp. 255–342.
11. Morris, Jr., J. W., D. Tribula, T. S. E. Summers, and D. Grivas, "The Role of Microstructure on Thermal Fatigue of Pb–Sn Solder Joints," ref. 2, pp. 225–265.
12. Mei, Z., and J. W. Morris, Jr., "Fatigue Lives of 60Sn–40Pb Solder Joints Made With Different Cooling Rates," *Journal of Electronic Packaging*, **114**(2), 1992, pp. 104–108.
13. Mei, Z., D. Grivas, M. C. Shine, and J. W. Morris, Jr., "Superplastic Creep of Eutectic Tin–Lead Solder Joints," *Journal of Electronic Materials*, **19**(11), 1990, pp. 1273–1280.
14. Vastava, R. B., and T. G. Langdon, "An Investigation of Intercrystalline and Interphase Boundary Sliding in the Superplastic Pb–62% Sn Eutectic," *Acta Metallurgica*, **27**, 1979, pp. 251–257.
15. Grivas, D., K. L. Murty, and J. W. Morris, Jr., "Deformation of Pb–Sn Eutectic Alloys at Relatively High Strain Rates," *Acta Metallurgica*, **27**, 1979, pp. 731–737.
16. Guo, Z., A. F. Sprecher, and H. Conrad, "Plastic Deformation Kinetics of Eutectic Pb–Sn Solder Joints in Monotonic Loading and Low-Cycle Fatigue," *ASME Symp. Mechanics of Surface Mount Assemblies*, Atlanta, GA, December 1991. *Journal of Electronic Packaging*, **114**(2), 1992, pp. 112–117.
17. Solomon, H. D., "Low Cycle Fatigue of 60/40 Solder-Plastic Strain Limited vs. Displacement Limited Testing," *ASM 2d Electronic Packaging Materials and Processes Conference*, Bloomington, MN, 29–31 October 1985, ASM, Metals Park, OH, pp. 29–47.
18. Bae, K., A. F. Sprecher, D. Y. Jung, and H. Conrad, "Fatigue of 63Sn–37Pb Solder Used in Electronic Packaging," *International Symp. Testing and Failure Analysis*, ISTFA 1988, ASM, Metals Park, OH, 1988, pp. 53–61.
19. Hall, P. M., "Forces, Moments and Displacements During Thermal Chamber Cycling of Leadless Ceramic Chip Carriers Soldered to Printed Boards," *IEEE Trans. Components, Hybrids, and Manufacturing Technology*, **CHMT-7**(4), 1984, pp. 314–327.
20. Liljestrand, L.-G., and L.-O. Andersson, "Accelerated Thermal Fatigue Cycling of Surface Mounted PWB Assemblies in Telecom Equipment," *Proc. Seventh Annual Electronics Packaging Conference*, Boston, MA, 1987, pp. 411–424.
21. Clech, J.-P., and J. A. Augis, "Engineering Analysis of Thermal Cycling Accelerated Tests for Surface-Mount Attachment Reliability Evaluation," *Proc. 7th Annual International Electronics Packaging Conference*, Boston, MA, 1987, pp. 385–424.
22. Langdon, T. G., "The Significance of Grain Boundary Sliding in Creep and Superplasticity," *Metals Forum*, **4**, 1981, pp. 14–23.
23. Kashyap, B. P., and G. S. Murty, "Experimental Constitutive Relations for High Temperature Deformation of a Pb/Sn Eutectic Alloy," *Materials Science and Engineering*, **50**, 1981, pp. 205–213.
24. Frear, D., Grivas, D., and Morris, Jr., J. W., "Parameters Affecting Thermal Fatigue Behavior of 60Sn–40Pb Solder Joints," *Journal of Electronic Materials*, **18**(6), 1989, pp. 671–680.
25. Conversation with P. Hall, 1990.

26. Stone, D., S.-P. Hannula, and C.-Y. Li, "The Effects of Service and Material Variables on the Fatigue Behavior of Solder Joints during the Thermal Cycle," *Proc. 35th Electronic Components Conference*, IEEE, 1985, pp. 46–51.

27. Wilcox, J. R., R. Subrahmanyan, and Che-Yu Li, "Thermal Stress Cycles and Inelastic Deformation in Solder Joints," *Proc. 2d ASM International Electronic Materials and Processing Congress*, Philadelphia, PA, 1989, pp. 203–211.

28. Guo, Z., A. F. Sprecher, and H. Conrad, "Crack Initiation and Growth During Low-Cycle Fatigue of Pb–Sn Solder Joints," *41st Electronic Components and Technology Conference*, 41st ECTC, IEEE CHMT, IEEE Catalogue No. 91CH2989-2, 1991, pp. 658–666.

29. Hacke, P. L., A. F. Sprecher, and H. Conrad, "Modeling of the ThermoMechanical Fatigue of 63Sn–37Pb Alloy," *ASTM Symposium Thermomechanical Fatigue Behavior of Materials*, San Diego, CA, 1991, ASTM STP 1186, 1993.

30. Solomon, H. D., "Fatigue of 60/40 Solder," *IEEE Trans. Components, Hybrids, and Manufacturing Technology*, **CHMT-9**(4), 1986, pp. 423–432.

31. Paris, P. C., and F. Erdogan, "A Critical Analysis of Crack Propagation Laws," *Trans. ASME, Journal of Basic Engineering*, **85**, 1963, pp. 528–534.

32. Dieter, G. E., *Mechanical Metallurgy*, 3d edn, McGraw-Hill, New York, 1986, pp. 398–401.

# 16

# A Prediction of the Thermal Fatigue Life of Solder Joints Using Crack Propagation Rate and Equivalent Strain Range

*Ryohei Satoh*

## 16.1 INTRODUCTION

With high performance and high density of computer and other electronic circuits, it becomes very important to assure the reliability of smaller and smaller solder joints. In particular, more accurate prediction of thermal fatigue life is required for large thermal stress induced by power and temperature cycling. In this chapter an accurate estimation method for thermal fatigue life is described.

Usually tin–lead solders (such as Sn–37Pb, m.p. 183°C; Sn–95Pb, m.p. 314°C; and Sn–3.5Ag, m.p. 220°C) are used for solder joints of electronic circuits. The Sn–37Pb eutectic alloy is used in electronic circuit inter-connections at a maximum temperature of 100°C (373 K). This temperature is roughly $0.8T_m$ and is higher than the homologous temperatures of structural materials (stainless steel, super alloy based on Ni or Co, etc.) used in gas turbines, boilers, etc. (Fig. 16-1). Further, solder joints can undergo cyclical thermal strain corresponding to power on–off cycling of the equipment or environmental temperature fracture forces so that thermal fatigue fractures can occur. Therefore, in order to produce reliable electronic circuits, we must understand the thermal fatigue processes of solder joints at high temperatures in order to be able to take countermeasures.

The thermal fatigue behavior of Sn–Pb solder joints has been investigated by numerous authors.[1–16, 27–29, 36–38].

The thermal fatigue life, $N_f$, can be described by Coffin–Manson low-cycle

500

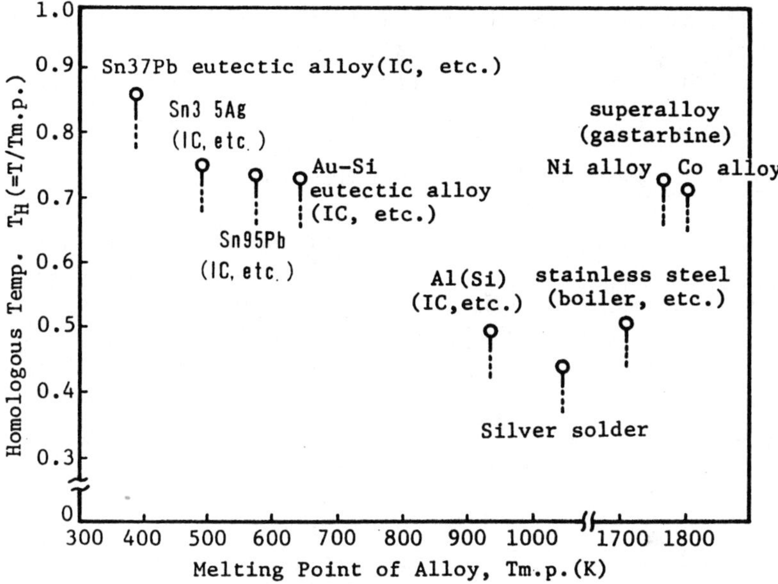


**Figure 16-1**   Relationships of actual temperatures of use for melting points on the alloys.

fatigue.[17–19] Norris and Landzberg[1] have proposed an equation that combines the effects of frequency $f$ and maximum temperature $T_{max}$:

$$N_f = Cf^{1/m}(\Delta\varepsilon_p)^{-n} \exp(Q/kT_{max}) \tag{16-1}$$

where $C$, $m$, and $n$ are material constants, $\Delta\varepsilon_p$ is the plastic strain range, $Q$ is the activation energy, and $k$ is the Boltzmann constant. This agrees with the crack propagation equation of the dynamic fracture theory of Yokobori.[20] However, there are some problems in applying this equation to actual solder joints:

1. As experimental crack propagation rates are not made clear, the joint size effect is incompletely understood.
2. This is thermal fatigue, not isothermal fatigue, so it is important to simulate the thermal strain range of solder at large plastic deformation.
3. The thermal fatigue fracture mechanism of solder alloy is not clear.

This chapter presents experimental results related to these problems. The purpose is to estimate thermal fatigue lifetimes for Sn–37Pb, Sn–95Pb, and Sn–3.5Ag alloy solder joints with different melting points. In particular, using the crack propagation model and the accurate estimation method of thermal strain range, the purpose is to offer a method of estimation of the thermal fatigue life for actual solder joints.

## 16.2 TENSILE PROPERTIES OF CAST SOLDERS

Temperature-dependent tensile properties are needed to simulate the thermal stress–strain relations of a solder joint under thermal cycle test conditions. The test samples used were JIS A class and impurities were less than 0.05 wt% Bi, 0.3 wt% Sb, 0.05 wt% Cu, 0.03 wt% Fe, and 0.03 wt% As. Solders were cast in the form shown in Fig. 16-2. Cooling rates were matched with those of joint solder reflow (about 0.5–1.0°C/sec) in order to obtain an as-solidified structure. Also, the deformation rates used in the test matched the strain rates (about $1.5 \times 10^{-4}$/sec) of controlled collapse bonding (CCB) and quad flat package (QFP) joints during thermal cycle testing, shown in Fig. 16-3. Stress–strain curves of Sn–37Pb, Sn–95Pb, and Sn–3.5Ag cast solders are shown in Figs. 16-4 and 16-5. The strength of the solder decreases as temperature rises. The yield strength of the Sn–37Pb eutectic solder increases 10 times between $+150°C$ and $-50°C$. However, these tensile elongations are smaller than those reported by Thwaites and Hampshire.[22] The Pb–Sn eutectic solder shows superplastic behavior when it is worked to recrystallization. Cast solders do not show this phenomenon; cast defects and large grain size control the superplastic deformation and limit elongation. Thus, thermal fatigue analysis must consider these solder properties.

Sn–3.5Ag and Sn–37Pb solders with the addition of gold were also examined for tensile characteristics and microstructure. The solderable area of electronic parts is frequently plated with gold because gold is unlikely to oxidize and has good wettability. However, gold dissolves into the molten solder, forming brittle intermetallic compounds with the solder that could cause fracture. Solder ingots were melted at 250°C and gold was added at 1, 2, 3, 5, 6, 7, 8, and 10 mass%. The gold content in the solder joints of

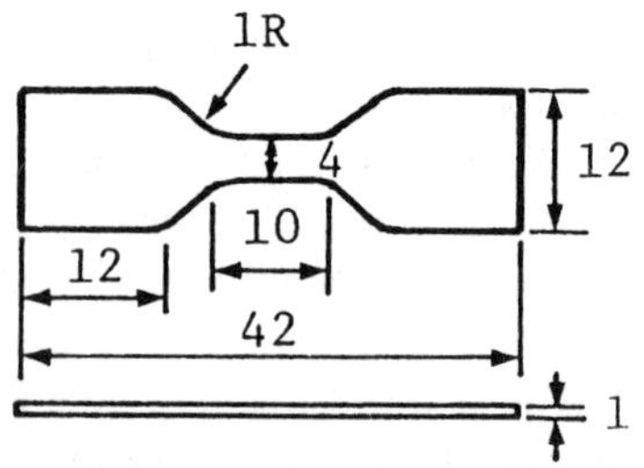


**Figure 16-2**   Shape of tensile specimen.

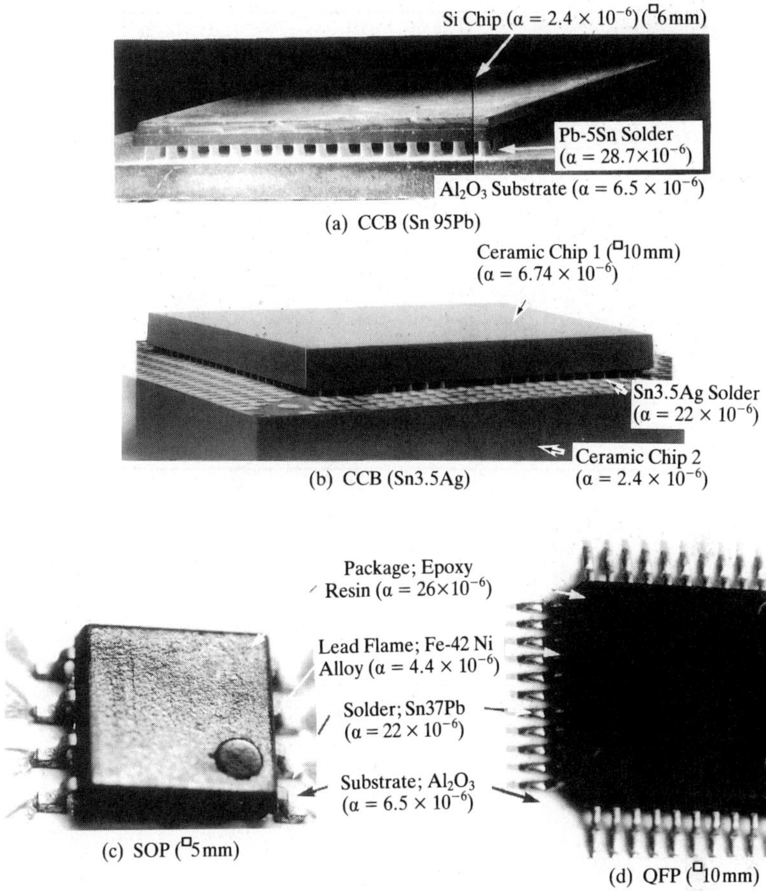

**Figure 16-3**   Microsolder joint structure.

electronic equipment is approximately 1–5 mass%. In this experiment, gold was added to make up more than 5 mass% so as to determine the limit of the gold content for reliable interconnections.

Figure 16-6 shows changes in tensile fracture elongation and tensile strength as functions of gold content. Without gold, both Sn–3.5Ag and Sn–37Pb have similar fracture elongations. Sn–37Pb with more than 3 mass% gold has a poor elongation of less than 15%, but Sn–3.5Ag with up to 6 mass% gold retains good elongation up to 40%. The tensile strength of Sn–3.5Ag increases gradually with the addition of gold. The tensile ductility and microstructure of Sn–37Pb solder as a function of gold content have been investigated by Bester,[30] who found results similar to this experiment.

Figure 16-7 shows the fracture surfaces of tensile specimens for both types

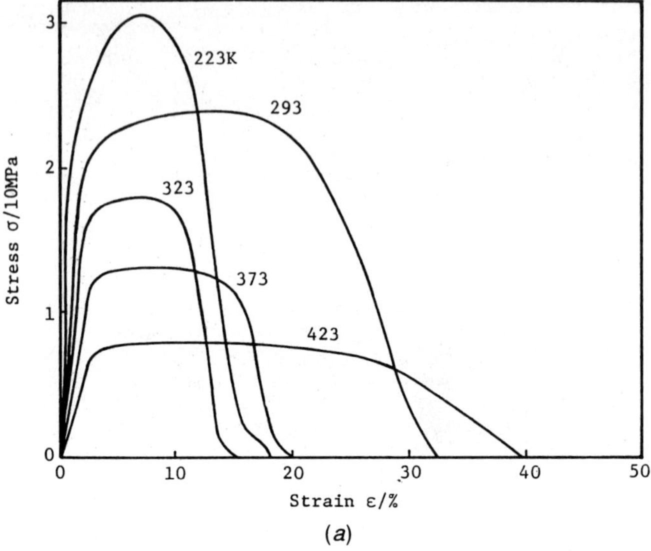


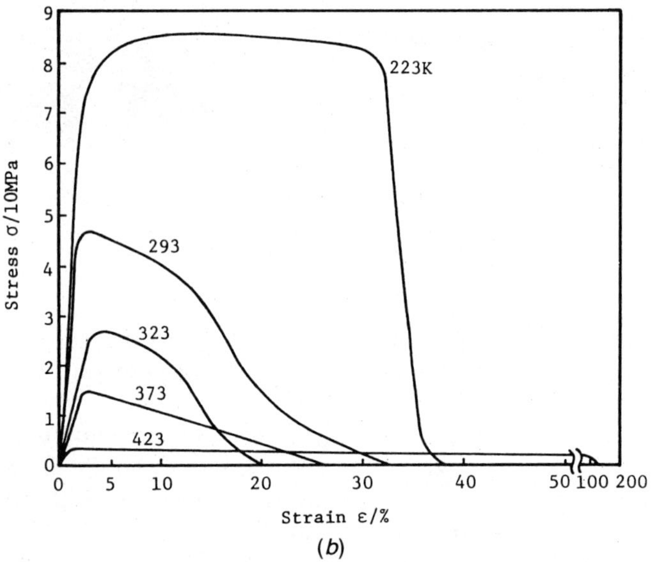


**Figure 16-4**  Tensile stress–strain curves of cast Pb–Sn alloys with temperature dependence ($\dot{\varepsilon} \approx 1.5 \times 10^{-4}$/sec). (*a*) Pb–5Sn. (*b*) Sn–37Pb.

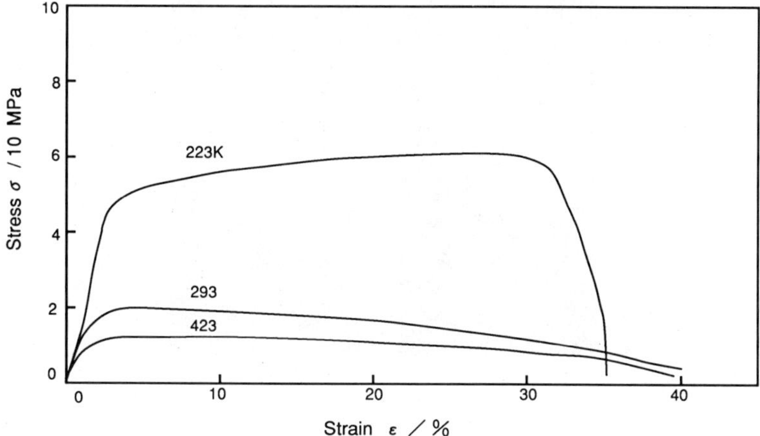



**Figure 16-5**  Tensile stress–strain curves of cast Sn–3.5Ag alloy with temperature dependence ($\dot{\varepsilon} \approx 1.5 \times 10^{-4}$/sec).

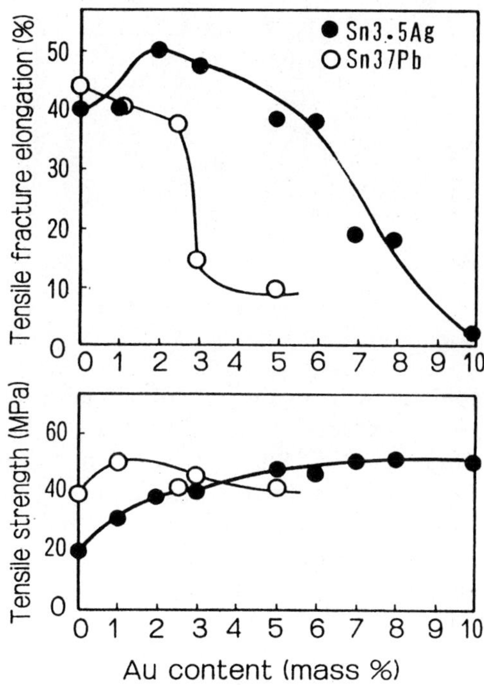



**Figure 16-6**  Effect of gold content on tensile fracture elongation and tensile strength.

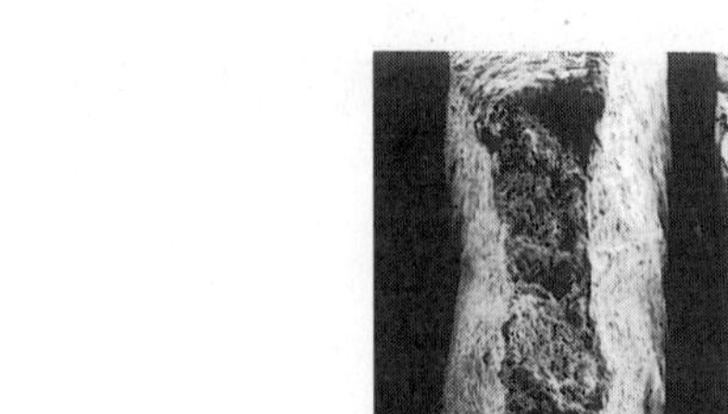
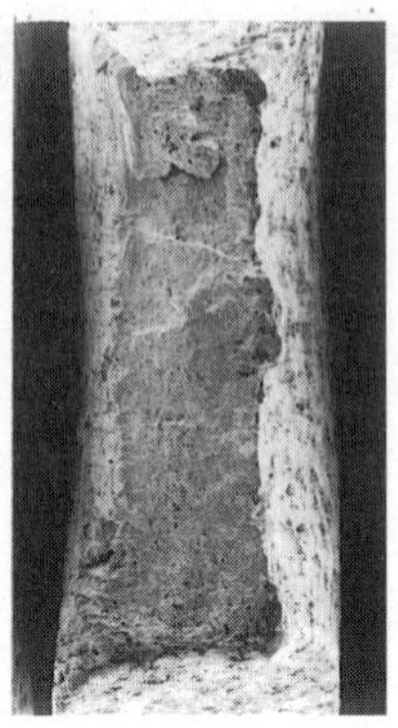
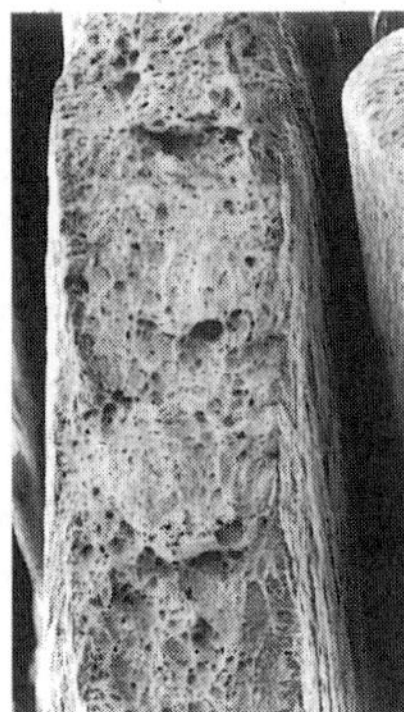
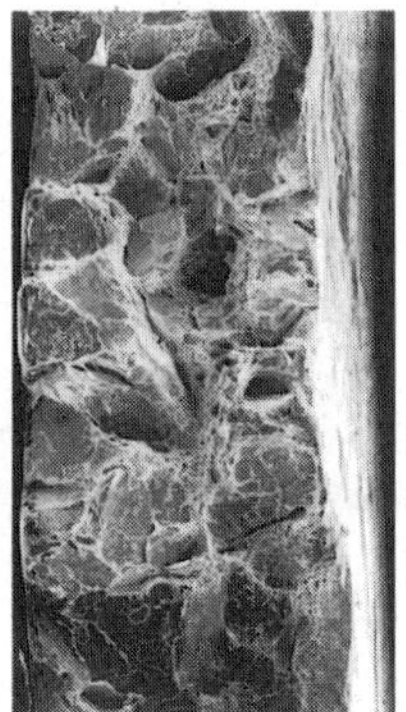

**Figure 16-7**   Fractured surfaces of tensile specimens. *Upper,* Sn–3.5Ag; *lower,* Sn–37Pb. *Left,* without Au; *right,* Au 5 mass%.

of solder, without gold and with a gold content of 5 mass%. Dimple patterns appear on the fractured surfaces of Sn–3.5Ag and Sn–37Pb without gold and indicate ductile fractures.[31–32] Sn–3.5Ag with 5 mass% gold content also fractured ductilely. However, on the fractured surface of the Sn–37Pb with 5 mass% gold, a "rock candy" pattern appears. This is an intergranular fractured surface. The specimen cleaved along grain boundaries formed by intermetallic compounds of gold and tin.

The microstructure of the specimens with 5 mass% gold is illustrated by the micrograph shown in Fig. 16-8. The size of the intermetallic compounds formed by Sn–3.5Ag is very small, while the intermetallics in Sn–37Pb are large.

**Figure 16-8** Metallographical micrograph of microstructure of tensile specimens. *Left,* Sn–3.5Ag (Au 5 mass%); *right,* Sn–37Pb (Au 5 mass%).

## 16.3 THERMAL FATIGUE FRACTURE MECHANISM

It is very difficult to substitute mechanical test properties for thermal cycling because the mechanical properties of solders change as temperature rises. Therefore, thermal cycle testing was performed using surface mounted devices. Test conditions were a cycle temperature range of $-55°C$ to $+150°C$ at a frequency of 1 cycle/hour ($2.8 \times 10^{-4}$ Hz). The temperatue profile is shown in Fig. 16-28($c$). Thermal fatigue crack propagation of the solder joints during thermal cycle testing is shown in Figs. 16-9, 16-10, and 16-11. Specimens were prepared on 6 mm × 6 mm 60 I/O CCB Si chips and conventional surface mounts of 10 mm × 10 mm 44 I/O QFPs and 5 mm × 5 mm 8 I/O small-outline plastic packages (SOP) on $Al_2O_3$ substrates, as shown in Fig. 16-3. Similarly, specimens of Sn–3.5Ag were prepared by controlled collapse bonding between $Al_2O_3$ and mullite ceramic chips (10 mm × 10 mm, 582 bumps). The mechanical and physical material properties of the CCBs and QFP and SOP assemblies are shown in Table 16-1.

Thermal fatigue crack propagation in the CCB generally occurs in the solder near the outer and inner interface between the solder joint and the Si chip or ceramic chip after about 50 cycles. The crack propagation rate depends on the bonded solder shape. The fatigue life of the hourglass-shaped CCB is about three times that of a typical barrel-shaped CCB. Crack

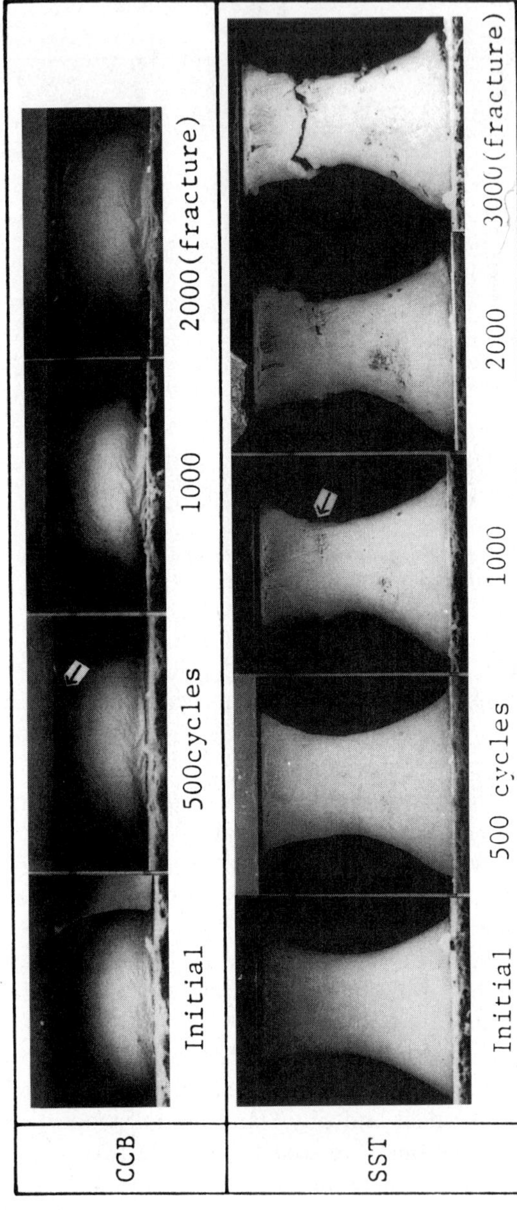


**Figure 16-9** Thermal fatigue crack propagation in CCB and SST. Test conditions: $-50°C$ to $+150°C$, 1 cycle/hour. Solder: Sn–95Pb.

508

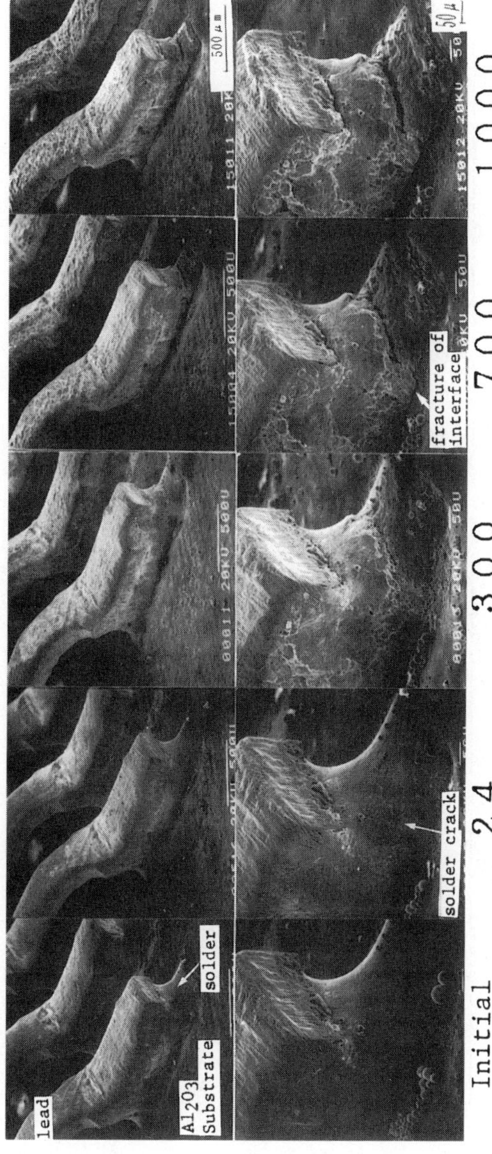



**Figure 16-10**  Thermal fatigue crack propagation in QFP assembly. Test conditions: −50°C to +150°C, 1 cycle/hour. Solder: Sn–37Pb.

**Table 16-1**  Properties of Structural Materials

| | Melting Point (K) | Thermal Expansion Coefficient ($\times 10^{-6}$/K) | Young's Modulus ($\times 10^5$ MPa) | Poisson's Ratio | Yield Stress ($\times 10$ MPa) |
|---|---|---|---|---|---|
| Sn–95Pb alloy | 587/578 | 28.7 | 0.026 | 0.3 | 1.8 |
| Sn–37Pb alloy (eutectic) | 456 | 22 | 0.032 | 0.3 | 4.0 |
| Sn–3.5Ag alloy | 493 | 22 | 0.024 | 0.3 | 3.7 |
| Si (LSI) | 1703 | 2.4 | 1.7 | 0.06 | – |
| $Al_2O_3$ (substrate) | | 6.5 | 2.6 | 0.23 | – |
| Ceramic chip 1 | – | 6.74 | 2.65 | 0.25 | – |
| Ceramic chip 2 | – | 2.4 | 1.85 | 0.2 | – |
| Epoxy resin (package) | 438 (glass soft point) | 26 | 0.2 | 0.4 | 4.0 |
| Fe–42Ni (42 alloy lead) | 1723 | 4.4 | 1.4 | 0.3 | 40.0 |

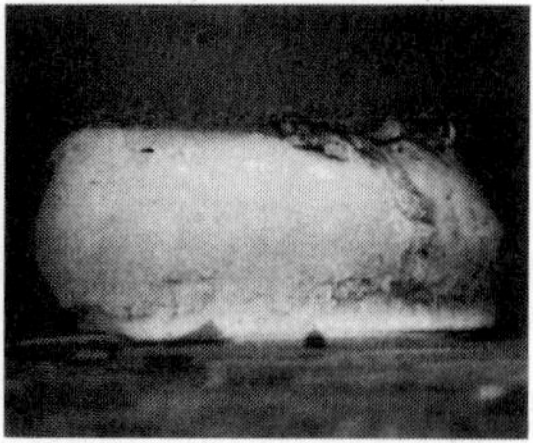
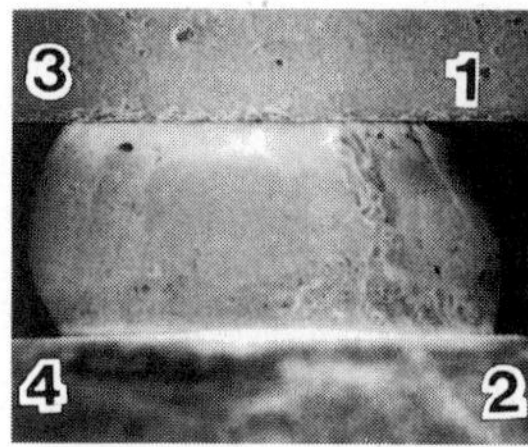
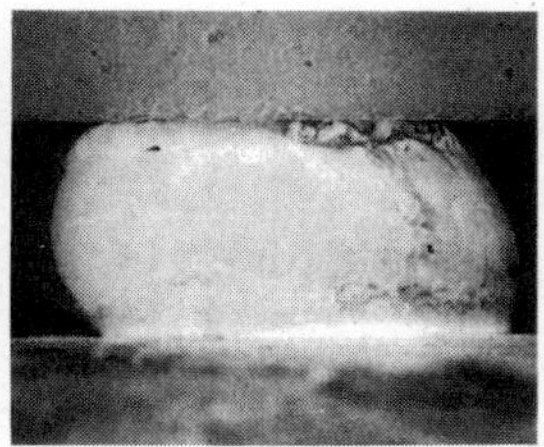

**Figure 16-11**   Thermal fatigue process of solder joint. Solder: Sn–3.5Ag. *Left*, initial; *center*, 500 cycles; *right*, 1000 cycles.

propagation in the QFP occurs similarly and finally reaches the interface between the metal and the substrate. This interface fracture depends on high stress at a low temperature of $-55°C$.

As shown in Fig. 16-4, the strength of the Sn–37Pb eutectic solder increases at low temperature. The interface fracture is related to this characteristic. For investigation of the fracture mechanism, the fracture surface was observed using high-resolution field emission SEM (Hitachi S-800) operated at an acceleration voltage of 1–5 kV in order to inspect the surface without conductive treatment. The thermal fatigue fracture surfaces of each solder joint are shown in Figs. 16-12, 16-13, 16-14, and 16-15. Very small striations (0.05–1 μm) are observed on both fracture surfaces.

The observation of solder joint striations had not been reported prior to 1988.[15] The main reason for this is the resolution of SEM. Since SEM is usually performed under conditions of high acceleration voltage (10–30 kV), to obtain high magnification, the electron beam penetrates deep into the fracture surface and small striations cannot be seen. This study made observations at a low acceleration voltage (1–5 kV) using field emission–scanning electron microscopy (FE-SEM) and could delineate striations down to 0.05 μm.

In light of these results, the fracture mechanism can be discussed. First the striations of CCB joints will be discussed. The striation of a CCB joint using Pb–5Sn solder is shown in Fig. 16-12. The striation size is about 0.15 μm and gradually becomes larger away from the joint edge. The number of striations agrees approximately with the number of temperature cycles (176 cycles). Many striplike traces are also formed in the direction of the Si chip center. The striations have a small angle to this stripe, and are connected to it. This shows that the formation of the striations is related to the slip plane of the (FCC) crystal structure of Pb. The thermal fatigue life shows that crack propagation life dominates, and that the crack generation life is very short.

The thermal fatigue fracture mechanism proposed here is that during temperature cycling, when tensile and compressive forces alternate on a

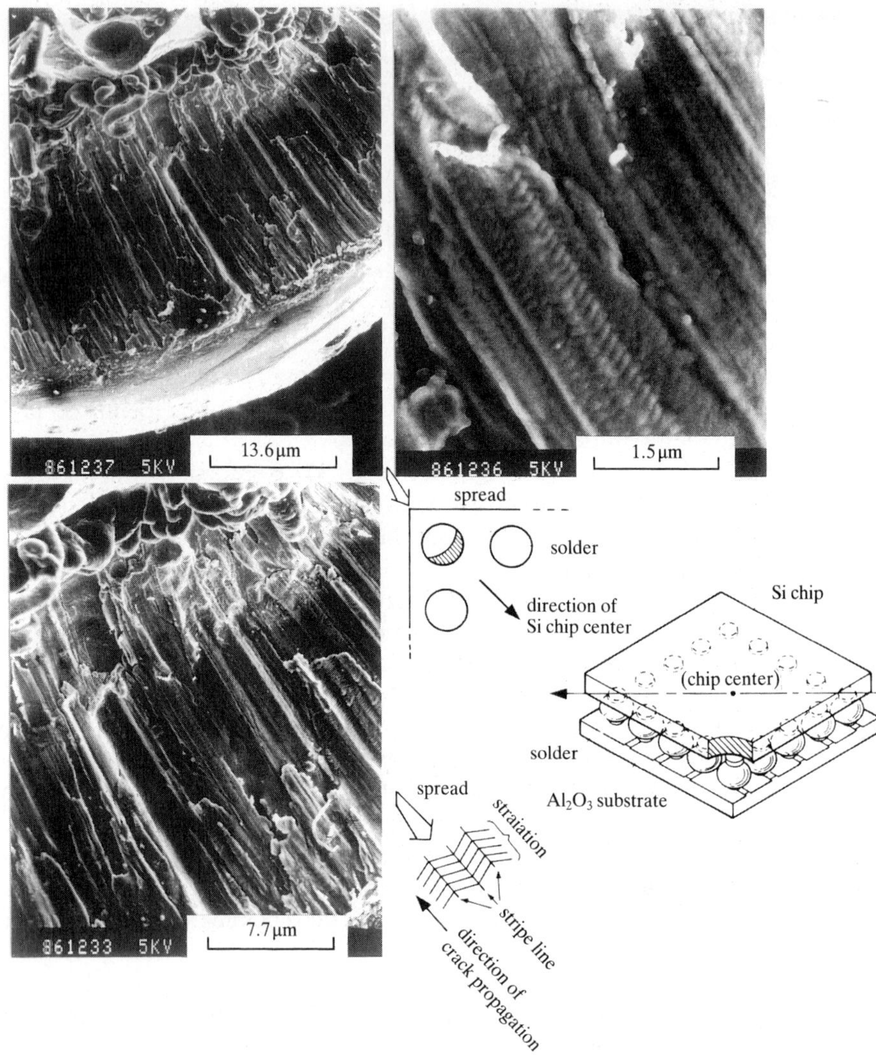



**Figure 16-12**  Typical thermal fatigue striation on a CCB fatigue fracture surface. Sn–95Pb alloy solder, after 176 cycles.

crack tip, intergranular sliding occurs on many equivalent slip planes {111}, and one striation is produced per cycle, as shown in Fig. 16-16. Accordingly, it is assumed that the plastic strain range of thermal fatigue is not the ordinary plastic strain range (i.e., shear or principal strain range) but is the equivalent strain range. This equivalent strain is von Mises's yield criterion under a multiaxial stress–strain condition. This mechanism agrees with the

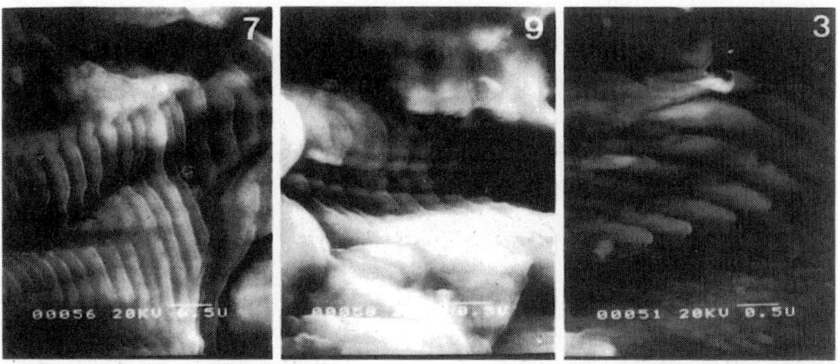

**Figure 16-13**  Typical thermal fatigue striation on the fracture surface of a QFP solder joint. *Left, da/dN* = 0.28 μm (300 cycles); *center,* 0.3 μm (500 cycles); *right,* 0.38 μm (700 cycles).

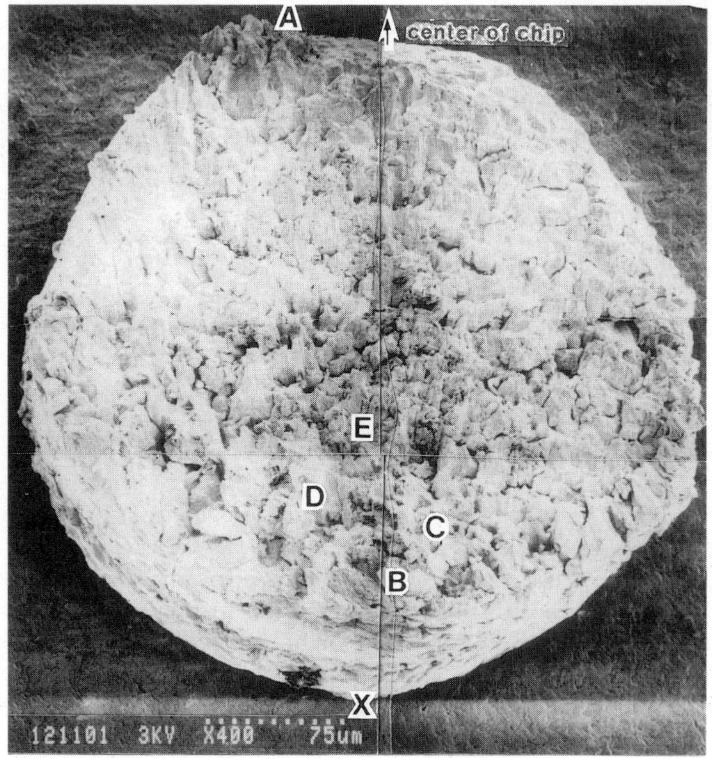


**Figure 16-14**  Fractured surface of Sn–3.5Ag solder joint.

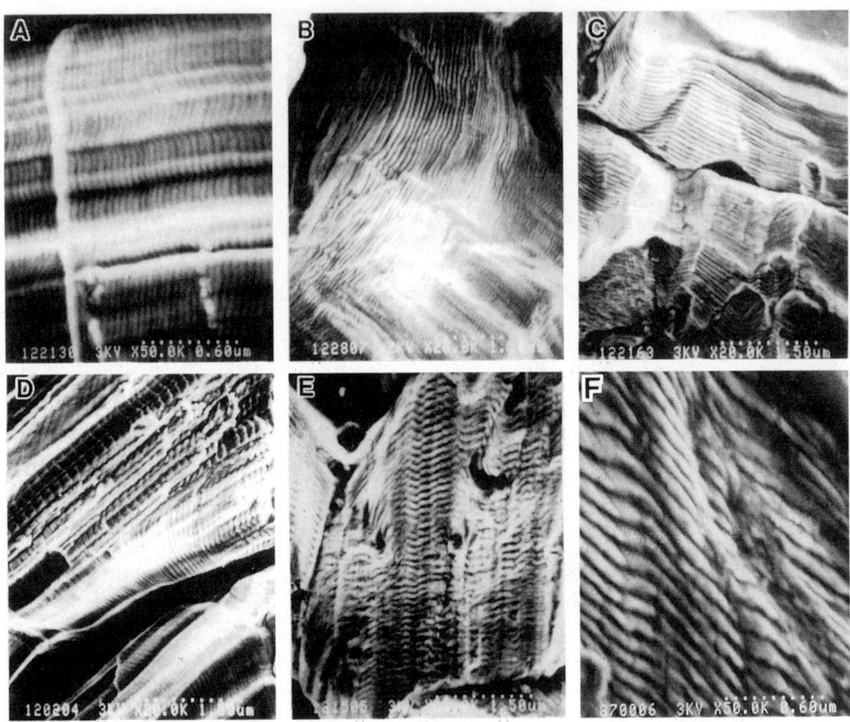

**Figure 16-15**  Scanning electron micrographs of the striations on the thermal fatigue fractured surface. A–E, Sn–3.5Ag solder joint; F, Sn–37Pb solder joint.

models of Laird[23,24] and Pelloux.[25] The striation when using Sn–Pb and eutectic solder is shown in Figs. 16-13 and 16-15. A similar striation is observed on the fracture surface.

## 16.4  THERMAL FATIGUE CRACK PROPAGATION RATE

Crack propagation can be used to estimate fatigue life. If such a rate can be measured, we can estimate fatigue life from it for the design of structures and materials. Fractographic studies in Figs. 16-9 to 16-15 show that the crack propagation occurs in the joint, and the striation sizes on the fracture surfaces shown in Figs. 16-12 to 16-15 correlate with the crack propagation rate. Figure 16-17 shows the relationship of crack propagation rate ($da/dN$) to crack length $a$ for CCB joints, while Fig. 16-18 shows the same relationship for the heel side of QFP joints. The crack propagation rate of the heel side is faster than that of the toe side, so we used the faster crack propagation rate of the heel in the QFP joints.

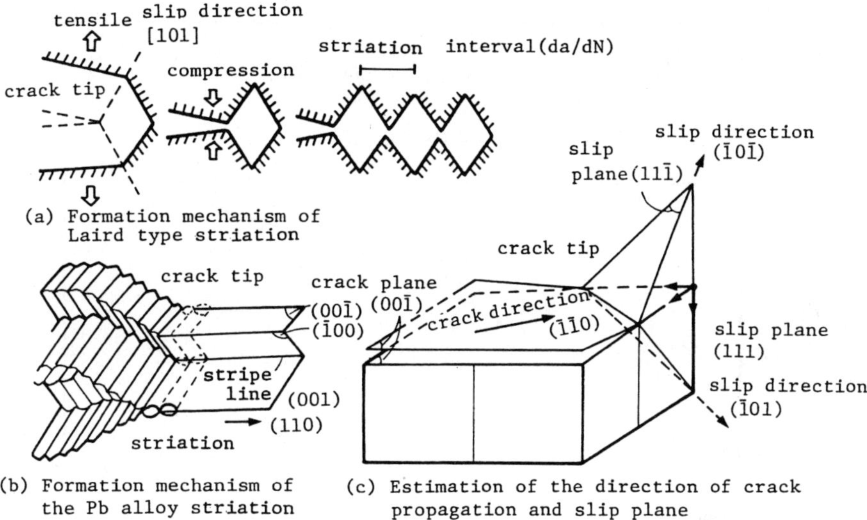



**Figure 16-16**   Schematic of the thermal fatigue fracture mechanism in Sn–95Pb alloy.

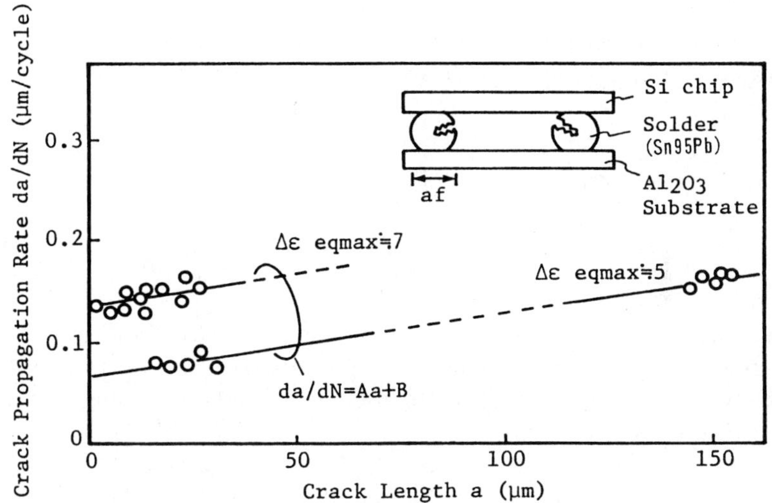



**Figure 16-17**   Thermal fatigue crack propagation rate by the striations in CCB joints.

The crack propagation rate can be approximated in terms of crack length $a$ by

$$\frac{da}{dN} = Aa + B \qquad (16\text{-}2)$$

where $A$ and $B$ are constants.[33] $A$ is a parameter of crack propagation

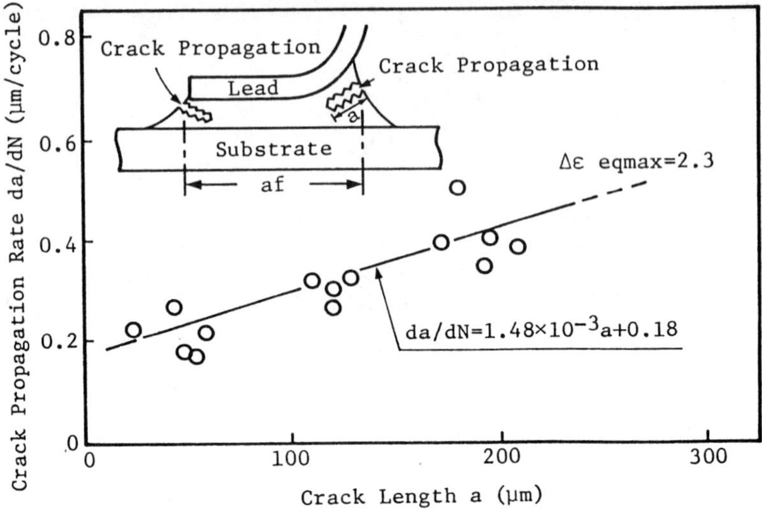



**Figure 16-18**  Thermal fatigue crack propagation rate by the striations in QFP joints.

resistance and *B* is a parameter of crack initiation. If this equation is integrated, the final fatigue life $N_f$ can be determined:

$$dN = \frac{1}{(Aa + B)}\, da$$

$$N_f = \int_{N_0=0}^{N_f} dN = \int_{a_0}^{a_f} \frac{da}{Aa+B} = \frac{1}{A}\ln\frac{|Aa_f + B|}{|Aa_0 + B|} \tag{16-3}$$

where $a_f$ is joint length and $a_0$ is an initial surface defect in the solder, e.g., a solidification defect like a dendrite, ordinarily 0.01–0.1 μm in size. From the above results we can estimate the life directly by using the crack propagation as a condition.

As shown in Fig. 16-17, crack propagation rate increases as strain range becomes larger. For similar joint shapes this means that crack propagation rate changes with strain range, $\Delta\varepsilon_{eqmax}$. This relation can be approximated by

$$\frac{da}{dN} = (Aa + B)(\Delta\varepsilon_{eqmax})^n \tag{16-4}$$

The change of $\Delta\varepsilon_{eqmax}$ with joint shapes and test conditions was investigated.

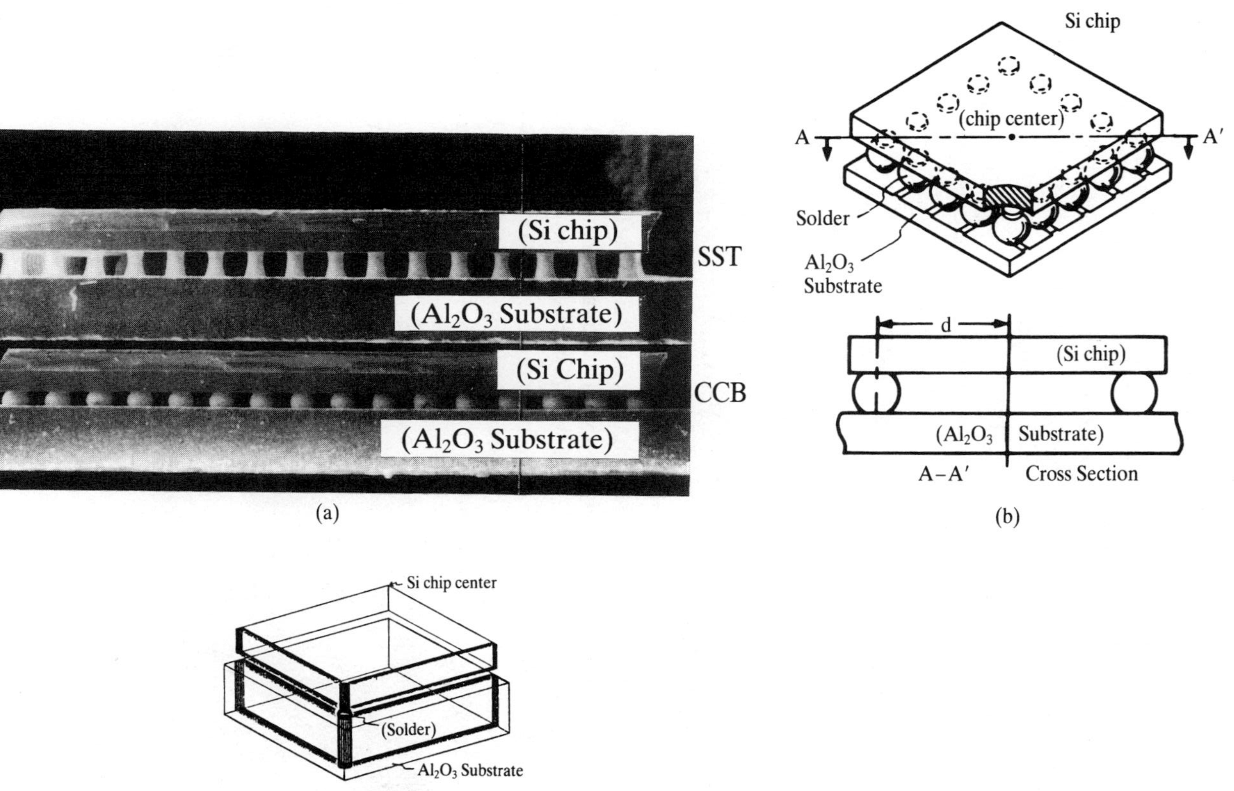



**Figure 16-19**   CCB microjoint structure and three-dimensional FEM model. (*a*) CCB/SST microjoint structure. (*b*) Model of CCB microjoint structure. (*c*) Three-dimensional FEM model. (Load condition: change of temperature 223–423 K.).

## 16.5 SIMULATION OF THERMAL FATIGUE STRAIN RANGE

It is difficult to obtain an experimental thermal fatigue strain range, so finite element method (FEM) simulation was done. In this type of simulation the solder joints are asymmetrical structures (see Figs. 16-18 and 16-20) and the mechanical properties of solders depend on temperature and follow plastic deformation. Thus, a three-dimensional thermal elasto-plastic structure analysis was conducted. The FEM program was ADINA;[21] the computers used were HITAC S810, M-280, and M-680. The Coffin–Manson equation[34,35] was applied for isothermal low-cycle fatigue with plastic deformation:

$$N_f = C \, \Delta\varepsilon_p^{-n} \tag{16-5}$$

where $C$ and $n$ are material constants and $\Delta\varepsilon_p$ is the plastic strain range. The thermal fatigue life will basically obey this equation as large-scale deformation occurs in solder joints. The study was conducted to determine an accurate strain range $\Delta\varepsilon_p$ by FEM.

A three-dimensional FEM model was used because of asymmetry of the solder joints and LSI packaging. As the mechanical properties of Sn–95Pb, Sn–37Pb, and Sn–3.5Ag solders change in a temperature range from $-50°C$ to $+150°C$, as shown in Figs. 16-4 and 16-5, a thermo-elasto-plastic model has to be used in the FEM. The ADINA FEM program is very useful in order to simulate the strain range of this model. FEM models with one solder joint for each package and chip are shown in Figs. 16-19 to 16-21.

The resulting equivalent strain[26] contours for a CCB with two different

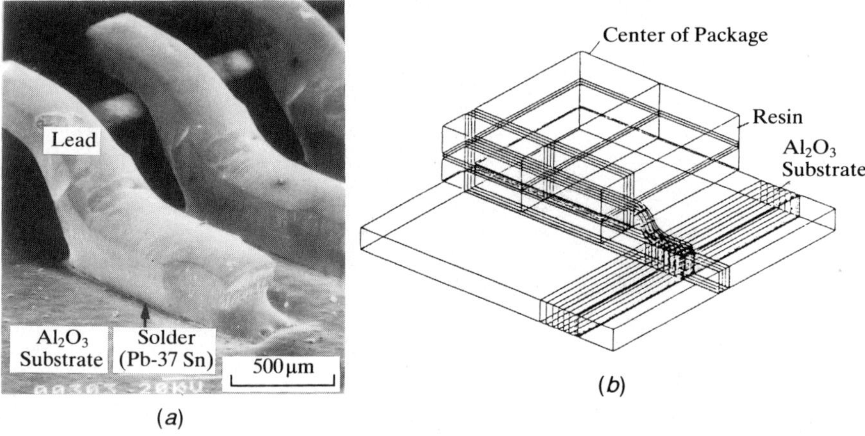


**Figure 16-20**  QFP/SOP microsolder joint structure and three-dimensional FEM model. *Left,* spread of QFP bonding solder; *right,* 3D FEM model. (Load condition: change of temperature 223–423 K.).

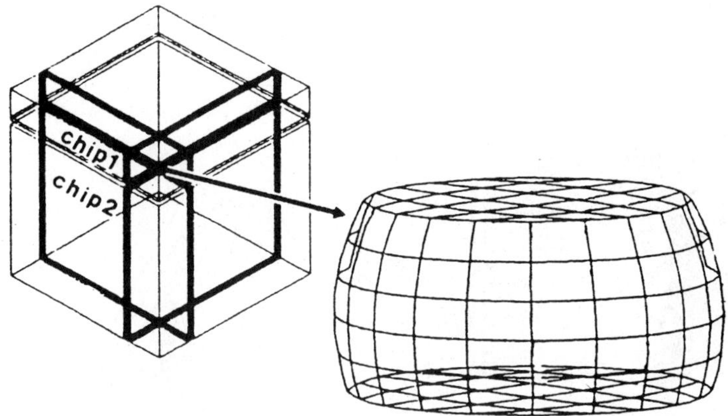


**Figure 16-21**   FEM analysis model with Sn–3.5Ag CCB joint.

joint shapes, a barrel shape and an hourglass shape, are shown in Figs. 16-22 to 16-24. Equivalent strain $\varepsilon_{eq}$ is defined as

$$\varepsilon_{eq} = \frac{\sqrt{2}}{3}[(\varepsilon_1 - \varepsilon_2)^2 + (\varepsilon_2 - \varepsilon_3)^2 + (\varepsilon_3 - \varepsilon_1)^2]^{1/2} \qquad (16\text{-}6)$$

where $\varepsilon_1$, $\varepsilon_2$, and $\varepsilon_3$ are principal strains. It can be seen that maximum strain $\Delta\varepsilon_{eqmax}$ occurs near the Si chip and ceramic chip for the barrel-shaped CCB joints, while the hourglass-shaped CCB joint has the maximum strain above the joint center. Similar strain contours for QFP solder joints are shown in Fig. 16-25. In this case it can be seen that maximum strain occurs near the tip of the outer solder fillet. In each case, the site of failure agrees with that of each value of $\Delta\varepsilon_{eqmax}$, as shown in Figs. 16-22c, 16-23c, 16-10, and 16-11 (1000 cycles). From these results, the use of these FEMs is assumed to be justified in determining the thermal fatigue of solder joints.

Next, the equivalent strain ranges illustrated in Figs. 16-26 and 16-27 are produced by this simulation. In mechanical fatigue, the strain range of low-cycle fatigue is usually the plastic strain range at constant temperature, shown in Fig. 16-28a. However, in thermal fatigue, the equivalent stress and strain curve shows a complex movement with thermal cycle temperature (Fig. 16-28c), as shown in Fig. 16-28b. The equivalent strain increases as temperature rises to $+150°C$ and decreases as temperature falls from $+150°C$ to $-50°C$. $\varepsilon_{eq}$ attains a maximum value at $+150°C$ and a minimum value at $-50°C$. As similar behavior continues in the following cycles, it can be assumed that the plastic strain range $\Delta\varepsilon_{eq}$ of the Coffin–Manson equation

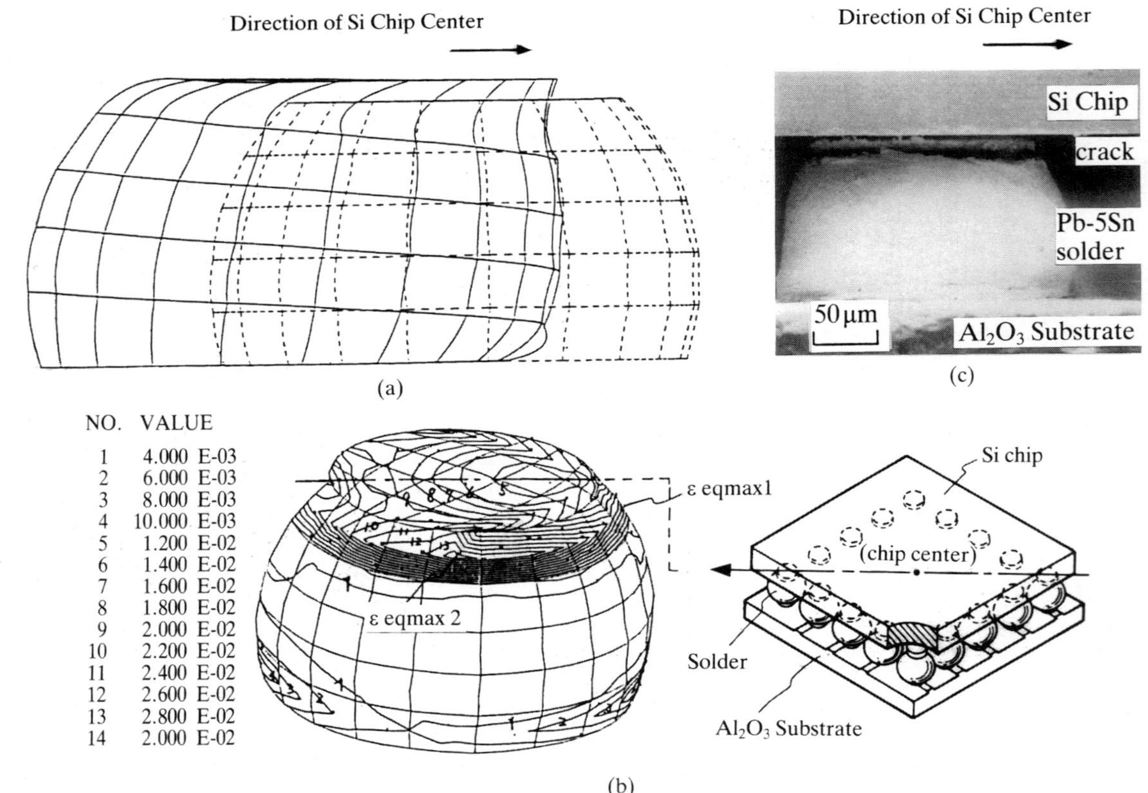



**Figure 16-22**  Deformation of a CCB microsolder joint and equivalent strain distribution by FEM simulation (at 423 K). (*a*) Deformation state. (*b*) Equivalent strain contour line. (*c*) Thermal fatigue fracture (after 300 cycles; −50°C to +150°C, 1 cycle/hour).

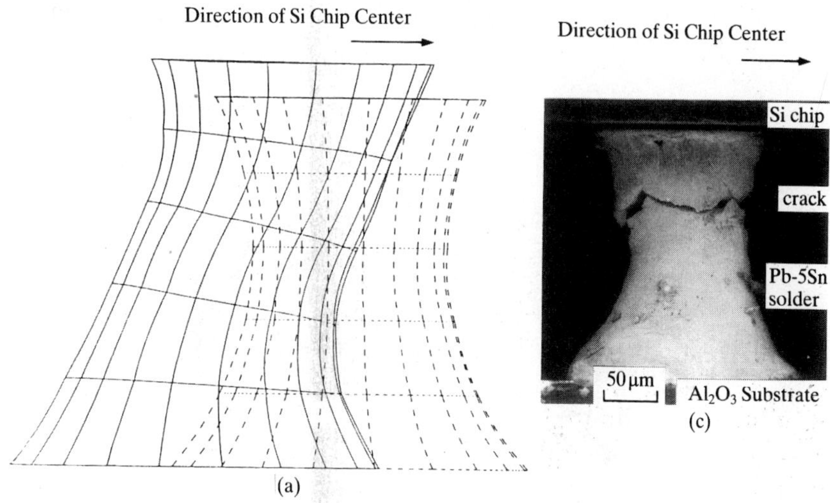


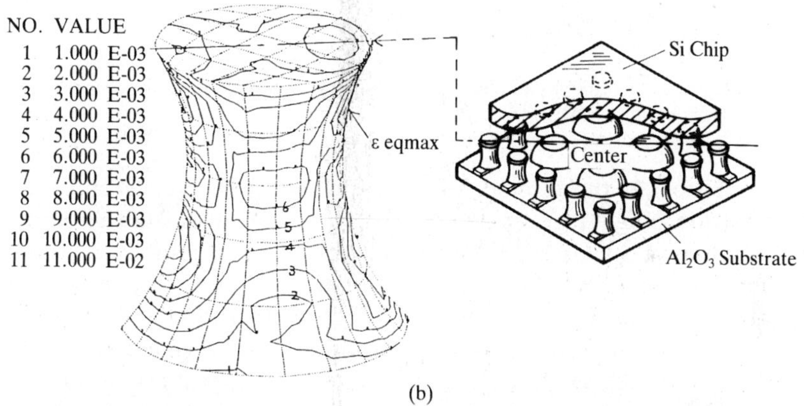


**Figure 16-23**   Deformation of a CCB (SST) microsolder joint by FEM simulation and equivalent strain distribution (at 423 K). (*a*) Deformation state. (*b*) Equivalent strain contour line. (*c*) Thermal fatigue fracture (after 3000 cycles; −50°C to +150°C, 1 cycle/hour).

is a maximum equivalent strain range in the solder joints between −50°C and +150°C.

## 16.6 THERMAL FATIGUE LIFE OF SOLDER JOINTS

The correlation of actual thermal fatigue life and $\Delta\varepsilon_{eqmax}$ was investigated. Thermal cycle tests were conducted to correlate the simulation results and to obtain master curves of thermal fatigue life.

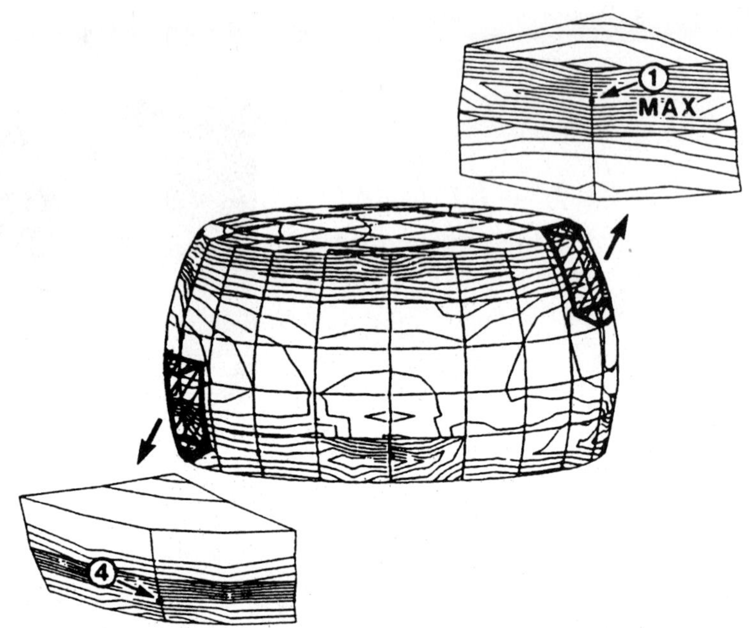


**Figure 16-24**   Results of FEM calculations showing the equivalent strain distribution at 150°C of the solder joint with $d = 6.12$ mm.

The correlation of the number of cycles to failure $N_f$ versus maximum equivalent strain range $\Delta\varepsilon_{eqmax}$ is shown in Figs. 16-29 to 16-31. The thermal fatigue life curve of a CCB with various shapes of solder joints is a straight line. This means that the Coffin–Manson equation using $\Delta\varepsilon_{eqmax}$ can be used for the thermal fatigue life of solder joints. The correlation of life $N_f$ versus maximum shear strain range $\Delta\gamma$ (ref. 1), is not linear, as shown in Fig. 16-32.

A similar correlation exists for the QFP and SOP assemblies. In this case, QFP and SOP solder joints with different bonding electrode lengths are correlated with the number of cycles to failure, $N_f$, per single joint length (normalized thermal fatigue life $N_f^*$) modified by a crack propagation rate to the maximum equivalent strain range. This shows the effects of joint length on fatigue life in actual joints.

Furthermore, if the crack propagation rate depends on the frequency, $f$ and maximum temperature $T_{max}$, as shown by Yokobori,[20] Eq. (16-4) becomes

$$\frac{da}{dN} = K(Aa + B)(\Delta\varepsilon_{eqmax})^n f^{-1/m} \exp(-Q/kT_{max}) \qquad (16\text{-}7)$$

If this equation is integrated, the final fatigue life $N_f$ considering all factors

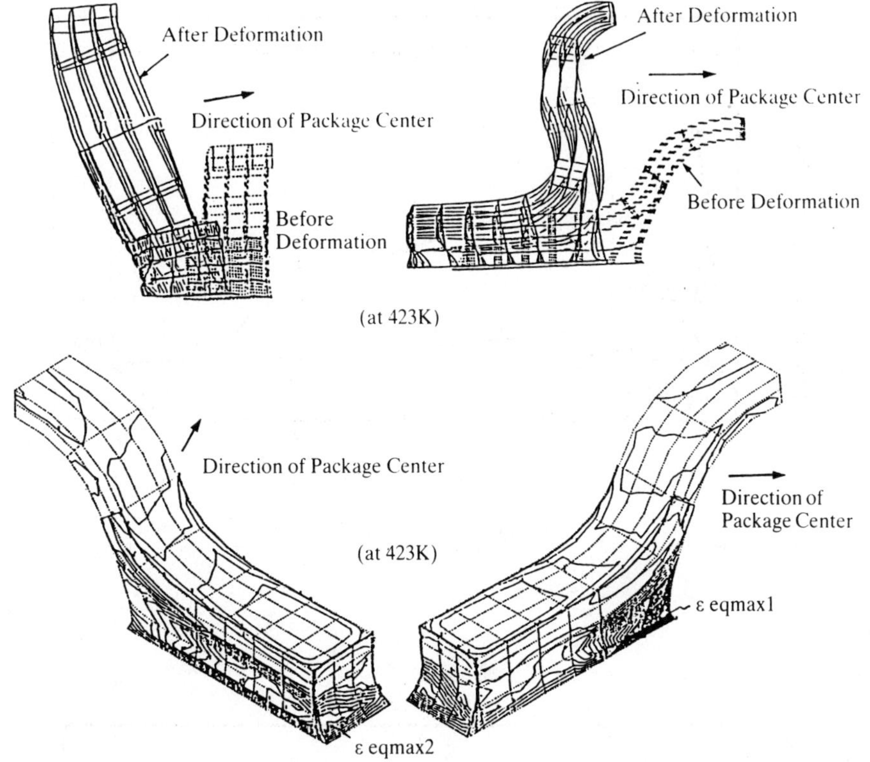


**Figure 16-25**    Deformation of QFP microbonding and equivalent strain contour line by FEM simulation.

can be arrived at:

$$N_f = K^{-1} \int_{a_0}^{a_f} \frac{da}{(Aa + B)} (\Delta\varepsilon_{\text{eqmax}})^{-n} f^{1/m} \exp(Q/kT_{\text{max}}) \qquad (16\text{-}8)$$

$$= K^{-1} A^{-1} \ln\left(\frac{|Aa_f + B|}{|Aa_0 + B|}\right)(\Delta\varepsilon_{\text{eqmax}})^{-n} f^{1/m} \exp(Q/kT_{\text{max}}) \qquad (16\text{-}9)$$

where

$$
\begin{aligned}
N_f &= \text{thermal fatigue life} \\
\Delta\varepsilon_{\text{eqmax}} &= \text{maximum equivalent strain range} \\
Q &= \text{activation energy} \\
T_{\text{max}} &= \text{maximum temperature} \\
f &= \text{frequency of temperature cycle} \\
k &= \text{Boltzmann's constant} \\
A, B, C, m, \text{ and } n &= \text{constants}
\end{aligned}
$$

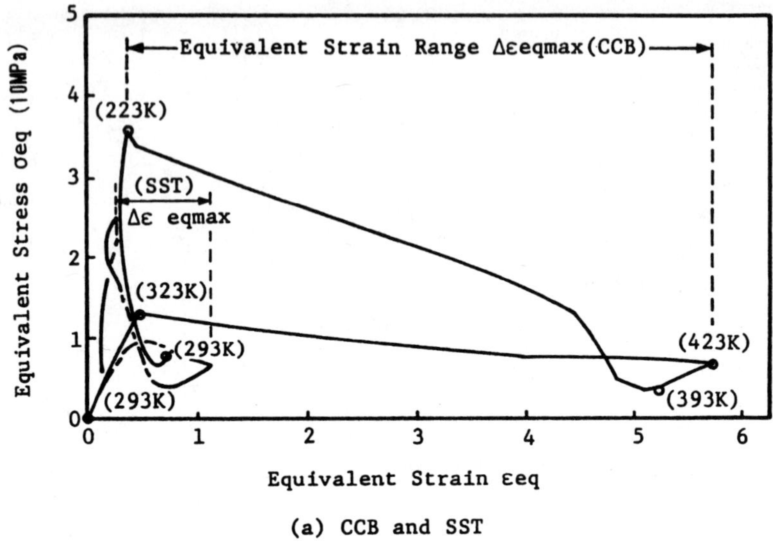



(a) CCB and SST

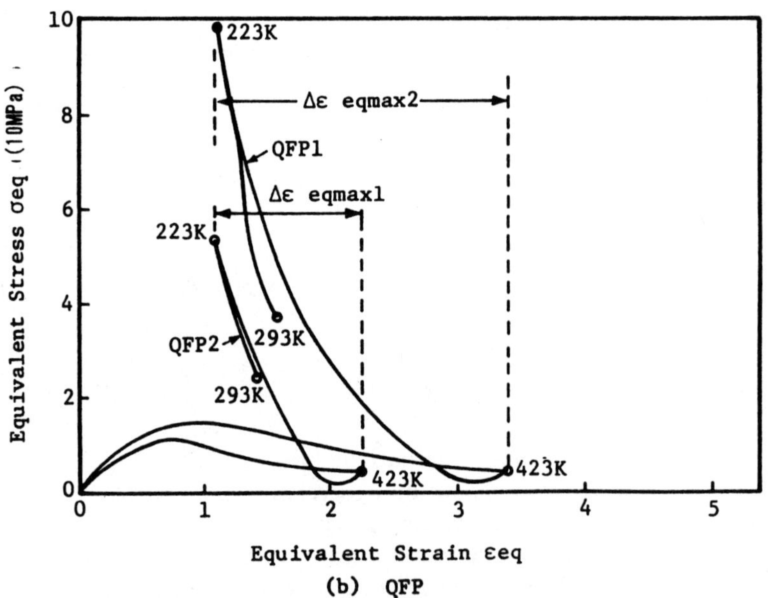



(b) QFP

**Figure 16-26**   Equivalent stress–strain curves and equivalent strain range of CCBs and QFPs under thermal cycle testing.

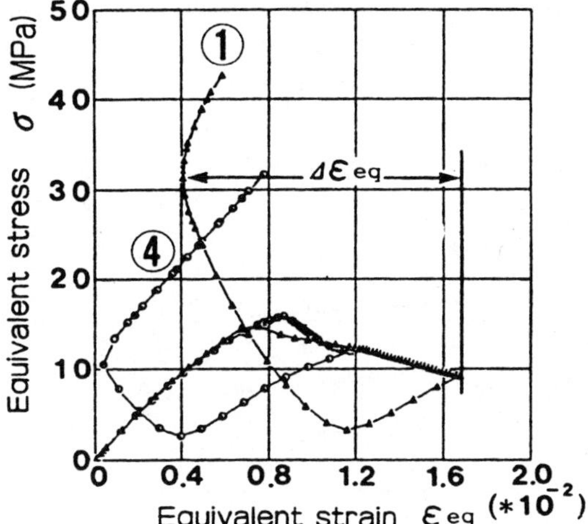



**Figure 16-27**   Relation between equivalent stress and equivalent strain.

The values of $n$ in (16-9) are 1.2 (Sn–3.5Ag, Sn–95Pb) and 1.5 (Sn–37Pb), as found from the slope of the lines in Figs. 16-29 to 16-31.

This equation considers all parameters affecting the thermal fatigue life of actual solder joints—the concentrated strain range (the effect of shape), the effect on solder thermal mechanical properties as temperature changes, the effect of joint length, and the effects of frequency and maximum temperature.

## 16.7 SUMMARY

The accurate estimation of thermal fatigue life has been investigated to enable production of SMT components with highly reliable solder joints. The conclusions of this chapter can be summarized as follows.

1. Thermal fatigue cracking occurs in solder near the chip in the barrel-shaped CCB joint and in the center of solder joints in the hourglass-shaped CCB, and propagates in both directions to the center of the chip. A similar pattern occurs in QFP and SOP assemblies.
2. The thermal fatigue fracture shows clear fatigue striation on the fracture surface. In the Pb–Sn alloy solder joint, this was observed by FE-SEM. From photographic analysis, this fracture mechanism agrees with the Laird and Pelloux type, in which cracks propagate with multislip on equivalent slip planes of the crack tip. Because the crack propagation rate was produced by this analysis, it is possible to accurately estimate fatigue life.

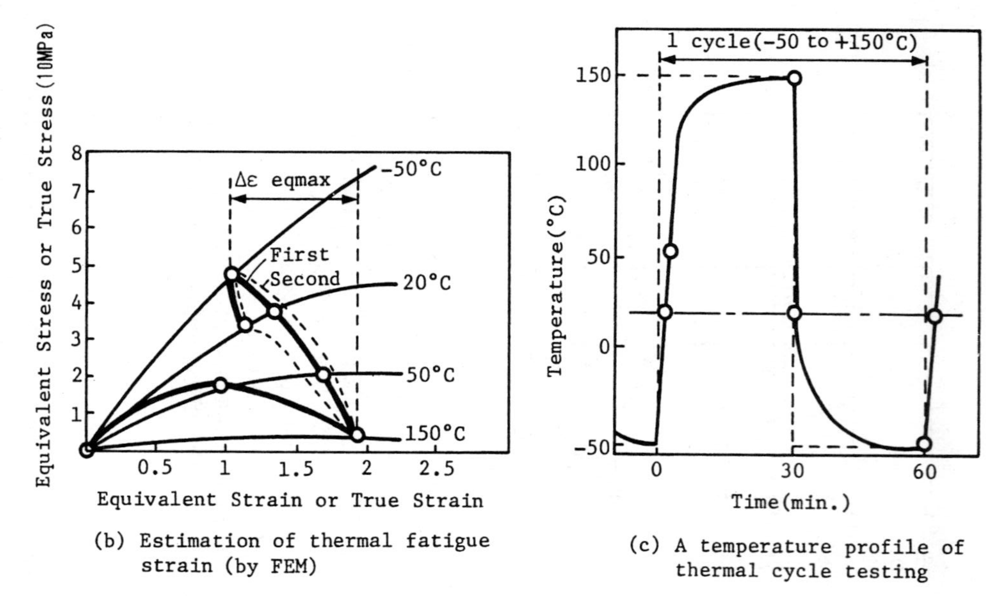


(a) Usual method of plastic
strain range

(b) Estimation of thermal fatigue
strain (by FEM)

(c) A temperature profile of
thermal cycle testing

**Figure 16-28**   Simulation method of thermal fatigue strain range.

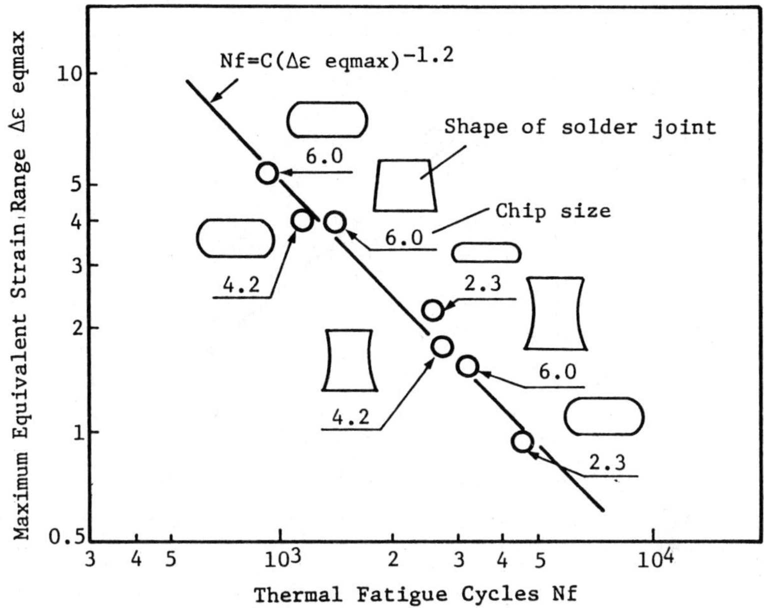


**Figure 16-29**    Thermal fatigue life of CCB microsolder joints.

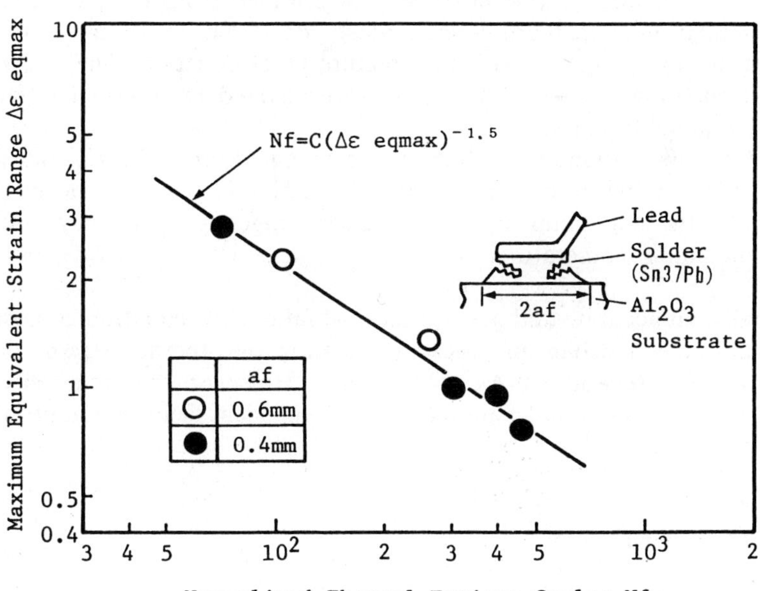


**Figure 16-30**    Thermal fatigue life of QFP and SOP microsolder joints.

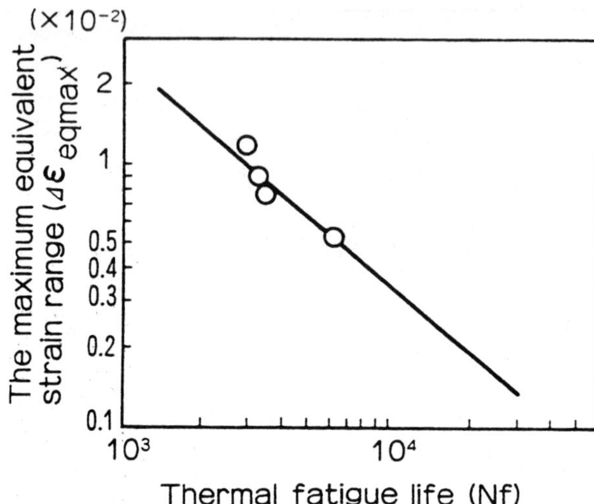


**Figure 16-31** Relation between thermal fatigue life of Sn–3.5Ag solder joint and the maximum equivalent strain range..

3. According to this mechanism, the dominant strain range for thermal fatigue is the equivalent strain range using the von Mises criterion under thermal cycling.
4. The equivalent strain range was determined from three-dimensional simulation of the thermo-elasto-plastic FEM. The maximum equivalent strain range agreed with the fracture crack location. The equivalent strain range and its simulation were justified for thermal fatigue of actual solder joints.
5. A thermal fatigue life equation was obtained for various solder joints under temperature cycling that is a function of the crack propagation rate, the maximum equivalent strain range, the frequency, and the maximum temperature.

Finally, an accurate and general thermal fatigue life equation considering thermal fatigue mechanism, crack propagation, and thermal strain range is given for Pb–Sn and Sn–Ag solder joints. However, the creep effect for thermal fatigue is not understood and time-dependent deformation properties must be explored.


## ACKNOWLEDGMENTS

The authors thank Dr. K. Watanabe and Dr. H. Kobayashi, and appreciate the support, technical discussions, and helpful suggestions of many persons

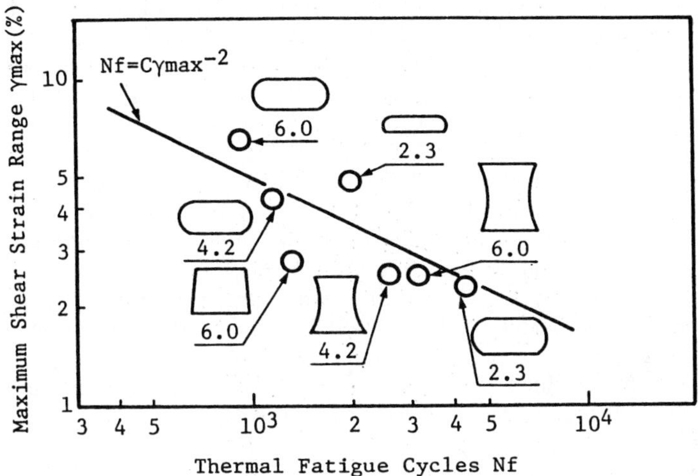



**Figure 16-32**    Thermal fatigue life of CCB microsolder joints with maximum shear strain range.

in the Hitachi Group. The authors also thank Y. Miyazaki for technical assistance.

## REFERENCES


1. Norris, K. C., and A. H. Landzberg, "Reliability of Controlled Collapse Interconnections," *IBM J. Res. Develop.*, **13**, 1969, p. 266.
2. Wild, R. N., "Fatigue Properties of Solder Joints," *Welding J.*, **51**, 1972, pp. 521S–526S.
3. Wild, R. N., "Some Fatigue Properties of Solders and Solder Joints," presented in Internepcon, Brighton, England, 1975.
4. Goldmann, L. S., "Geometric Optimization of Controlled Collapse Interconnections," *IBM J. Res. Develop.*, **13**(3), 1969, pp. 251–265.
5. Tobias, P. A., N. A. Sinclair, and A. S. Van, "The Reliability of Controlled-Collapse Solder LSI Interconnections," *ISHM Proc.*, 1976, p. 60.
6. Shah, H. J., and J. H. Kelly, "Effect of Dwell Time on Thermal Cling of the Flip Chip Joint," *ISHM Proc.*, 1970, paper 3, 4.
7. Sinclair, N. A., "Thermal Cycle Fatigue Life of LSI Solder Interconnections," *Proc. International Electrical Conference and Exposition*, Toronto, 1982, p. 56.
8. Totta, P. A., "Flip-Chip Solder Terminals," *Proc. 21st Electronics Components Conference*, Washington, DC, 1971, p. 275.
9. Engelmaier, W., "Fatigue Life of Leadless Chip Carrier Solder Joints During Power Cycling," *IEEE Trans. Components, Hybrids, and Manufacturing Technology*, **CHMT-6**, 1983, pp. 232–237.

10. Tribula, D. et al., "Observations on the Mechanisms of Fatigue in Eutectic Pb–Sn Solder Joints," *ASME Trans. J. Electron. Pack.*, **111**, June 1983, pp. 83–89.

11. Fox, L. R., J. W. Sofia, and M. C. Shine. "Investigation of Solder Fatigue Acceleration Factors," *IEEE Trans. Components, Hybrids, and Manufacturing Technology*, **CHMT-8**, 1985, pp. 275–281.

12. Solomon, H. D., "Fatigue of 60/40 Solder," *IEEE Trans. Components, Hybrids, and Manufacturing Technology*, **CHMT-9**, December 1986, pp. 423–432.

13. Lau, J. H., D. W. Rice, and P. A. Avery, "Elasto-plastic Analysis of Surface-mount Solder Joints," *IEEE Trans. Components, Hybrids, and Manufacturing Technology*, **CHMT-10**, September 1987, pp. 346–357.

14. Satoh, R., M. Ohshima, K. Hirota, and I. Ishi, "Optimum Bonding Shape Control on Micro Solder Joints of IC and LSI," *J. Japan. Inst. Metals.*, **51**(6), 1987, pp. 553–560.

15. Satoh, R., M. Ohshima, K. Arakawa, "Thermal Fatigue Life of Pb–Sn Alloy Joint on Electronics Circuits," *Proc. Conference Japan Institute of Metals*, 1988, p. 144.

16. Zubelewicz, A., et al., "Lifetime Prediction of Solder Materials," *ASME Trans. J. Electronic Packaging*, **111**, September 1989, pp. 179–182.

17. Manson, S. S., "Fatigue: A Complex Subject—Some spl Approximations," *Exp. Mech.*, **5**, 1965, pp. 193–226.

18. Coffin, L. F., "Low Cycle Fatigue: A Review," *Appl. Mech. Res.*, **1**(3), October 1962, pp. 129–141.

19. Coffin, L. F., Jr., "A Study of the Effects of Cyclic Thermal Stresses on a Ductile Metal," *Trans. ASME*, **76**, 1954, pp. 931–950.

20. Yokobori, T., *Rep. Res. Inst. Str. and Frac. Mater., Tohoku Univ.*, **5**, 1969, p. 19.

21. Bathe, K. J., "ADINA—A Finite Element Program for Automatic Dynamic Incremental Nonlinear Analysis," Rep. 82448-1, MIT, 1975.

22. Thwaites, C. J., and W. B. Hampshire, *Welding J.*, **55**, 1976, p. 323s.

23. Laird, C., and G. C. Smith, *Phil. Mag.*, **7-77**, 1962, p. 847.

24. Laird, C., *ASTM STP*, **415**, 1967, p. 131.

25. Pelloux, R. M. N., *Trans. Am. Soc. Metals*, **62**(1), 1969, p. 281.

26. Von Mises, R., *Nachr. Ges. Wiss. Gott.*, 1913, p. 582.

27. Davies, R. L., "High Strength, Low Temperature Bonding With Silver–Tin Solders," *Welding J.*, 1976, pp. 838–842.

28. Lau, J. H., and D. W. Rice, "Effects of Standoff Height on Solder Joint Fatigue," *Proc. NEPCON West*, 1986, pp. 437–454.

29. Lau, J. H., and D. W. Rice, "Solder Joint Fatigue in Surface Mount Technology: State of the Art," *Solid-State Technol.*, **28**, 1985, pp. 91–104.

30. Bester, M. H., "Metallurgical Aspects of Soldering Gold and Gold Plating," *Proc. Tech. Programme Intern.*, 1968, pp. 211–231.

31. Fellows, J. A., et al. (ed.) "Fractography and Atlas of Fractographs," *Metals Handbook*, Vol. 9, American Society for Metals, 1974.

32. Boyer, H. E., et al. (ed.), "Fracture Analysis and Prevention," *Metals Handbook*, Vol. 11, American Society for Metals, 1986.

33. Frost, N. E., and D. S. Dugdale, *J. Mech. Phys. Solids*, **6**, 1958, p. 92.

34. Manson, S. S., *Thermal Stress and Low Cycle Fatigue.* McGraw-Hill, New York, 1966.

35. Coffin, L. F., Jr., "Fatigue at High Temperature" *Fatigue at Elevated Temperatures*, ASTM STP 520, American Society for Testing and Materials, 1973, pp. 5–34.
36. "Development of Highly Reliable Soldered Joints for Printed Circuit Boards," Westinghouse Rep., no. N69-25697, 1968.
37. Clatterbaugh, G. V., and H. K. Charles, Jr., "Thermomechanical Behavior of Soldered Interconnects for Surface Mounting: A Comparison of Theory and Experiment," *Proc. 35th Electronic Components Conference*, 1985, pp. 60–72.
38. Lau, J. H., and G. Harkins, "Thermal-Stress Analysis of SOIC Packages and Interconnections," *Proc. 38th Electronic Components Conference*, 1988, pp. 23–31.

# 17

# Microstructural Evaluation of Sn–Pb Solder and Pd–Ag Thick-Film Conductor Metallization Under Thermal Cycling and Aging Conditions



*Jenq-Gong Duh, Kuo-Chuan Liu, and Bi-Shiou Chiou*



## 17.1 INTRODUCTION

Soldering is a technique for bonding different metals in which liquid solder wets the base metal surface by reducing the surface energy,[1] and thus makes a complete joint. Sn–Pb solder is the most popular one that is applied to bond many different kinds of metals in the electronics industry. The soldered joint is found in all electronic products, such as computers, radios, TV sets, etc.[2] However, the soldered joint will degrade after a period of time. The mechanical strength will decrease, the electrical resistance will increase, and the fillet's volume will shrink.[3,4] The degradation is believed to be accompanied by an interfacial interaction between the base metal and the solder that produces an intermetallic compound at the interface.[3] To prevent joint degradation, the mechanism must be understood.

The thick-film process is widely used in hybrid microelectronics. Power dissipation in chip carriers, daily and seasonal temperature variations, and power on/off introduce thermal fatigue into the solder joints. Under these circumstances, the soldered joints suffer very complex stress environments. Thus the study of thermal cycle effects in hybrid circuits is of great importance from a practical viewpoint.

Among the thick-film conductors, Pd–Ag is the one most often used. The Pd–Ag conductor is a mixture of Pd, Ag, and glass/oxide powders in the as-received state. Firing is the process that sinters the Pd and Ag powders and initiates a reaction that takes place between the glass/oxide and the substrate. There are several problems encountered with Pd–Ag thick films in practical application, such as silver migration due to humidity and voltage, and solder leaching in the soldering process. Adhesion loss is another focal point of interest. The adhesion loss of soldered Pd–Ag thick films after aging is attributed to the volume change associated with the intermetallic compound formation, which weakens the glass network,[5] or to tin diffusion into the conductor–substrate interface, affecting the silver conductivity.[6] Taylor et al.[7] introduced another factor that related to the redox reaction with bismuth oxide at the conductor grain boundary. The reaction results in a continuous drop of adhesion after the compound is formed.

With different solder–base metal combinations, a variety of intermetallic compounds are always present at the interface, such as $Cu_6Sn_5$ and $Cu_3Sn$ in the copper–tin system, $Ag_3Sn$ and $Ag_5Sn$ in the silver–tin system, and $AuSn_4$ in the gold–tin system.[8,9] The intermetallic compounds possess a different structure compared to their parent phases. $Ag_3Sn$ is orthorhombic, $Ag_5Sn$ and $Cu_6Sn_5$ are hexagonal.[10–13] In general, the intermetallic compound is brittle.[3,14] The strength decrease of the solder joints and the initiation of the fatigue cracks during thermal cycles are ascribed to the embrittlement that is the main deleterious effect on the solder joint.[3,15,16] In general, the thick-film metallization produces the same compounds with solder as the bulk base metals and has the same growth rate in the copper system.[17] However, the failure of the soldered thick film is not simply caused by the embrittlement due to intermetallic compounds, as the thick-film paste is not only composed of metals. For the Pd–Ag conductor, the $Ag_3Sn$ intermetallic compound is nonembrittling but rather ductile and soft.[18]

In real applications, the joint layout and the size of the leadless ceramic chip result in different stresses on the joints when the temperature is not uniform. A temperature difference exists between substrate and chips that may be caused by the power dissipation or local heating. In thick-film processes, the chip resistor soldered with a large solder fillet fails at the resistor–conductor–solder interface where a tensile stress exists.[4] Relatively speaking, the thick-film conductor fails easily when a tensile stress exists. Under thermal cycling, the solder joint experiences a rather complicated stress that is determined by the thermal expansion of the chip and substrate, and the geometry of the joint. Recently, a microstructural study of the thermal fatigue failures of 60Sn–40Pb solder joint was investigated by Frear and co-workers.[19] Thermal cycling tests were performed on 60Sn–40Pb joints using a $-55°C$ to $125°C$ cycle and 19% imposed shear strain. A heterogeneously coarsened region of both Pb- and Sn-rich phases developed within the solder joints. Cracks initiated in the heterogeneously coarsened Sn-rich phase at the Sn–Sn grain boundaries. The elevated temperature portion of the thermal cycle was found to be the most significant factor in the heterogeneous coarsening and failure of the solder joints.

In summary, aging degradation and thermal fatigue failure of solder joints in electronics packages is of critical concern in the microelectronics industry. To further complicate these problems, the trend in microelectronics packaging is towards larger integrated chips and therefore larger chip carriers. Thus, exacerbated failure in solder joints may be encountered due to larger strains imposed, and the aging and thermal cycling behavior of solder joints must be better understood in order to alleviate the associated problems. A microstructural analysis of the solder joints under aging and thermal cycling is the most appropriate approach to elucidating the picture. Therefore, the purpose of this paper is to describe, to document, and then to understand the microstructural development of Sn–Pb solder joints subjected to aging and thermal cycling. These observations help to characterize the effects of aging and thermal cycling on the microstructural evolution in soldered thick-film joints.

## 17.2 EXPERIMENTAL TECHNIQUES AND INSTRUMENTATION

To investigate the interfacial reaction between Sn–Pb solder and Pd–Ag thick-film conductor, the specimen is subjected to high temperature in an accelerated life test to show the degradation behavior.

The test specimens are fabricated by the following procedures.[20–22]

1. The conductor paste is printed through a 325 mesh stainless steel screen with the pattern shown in Fig. 17-1 on a 96% $Al_2O_3$ substrate, then leveled in air for 10–15 minutes, dried at 150°C for 15 minutes, and fired in a belt furnace with a 32-minute cycle having a dwell time of 850°C peak temperature about 7–9 minutes. The temperature profile is shown in Fig. 17-2a.

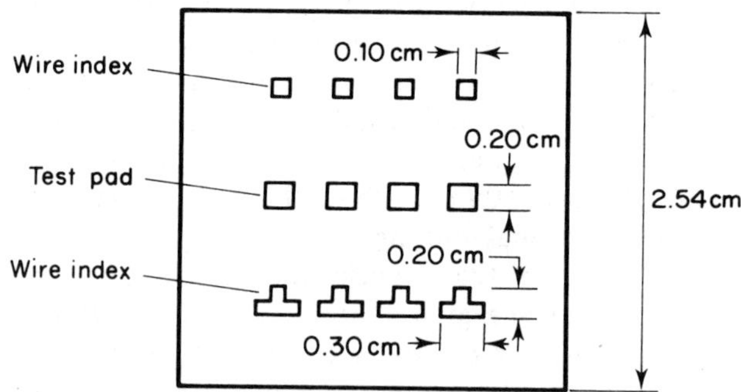


**Figure 17-1** Conductor test pattern employed in the soldered joint aging and thermal cycling test.

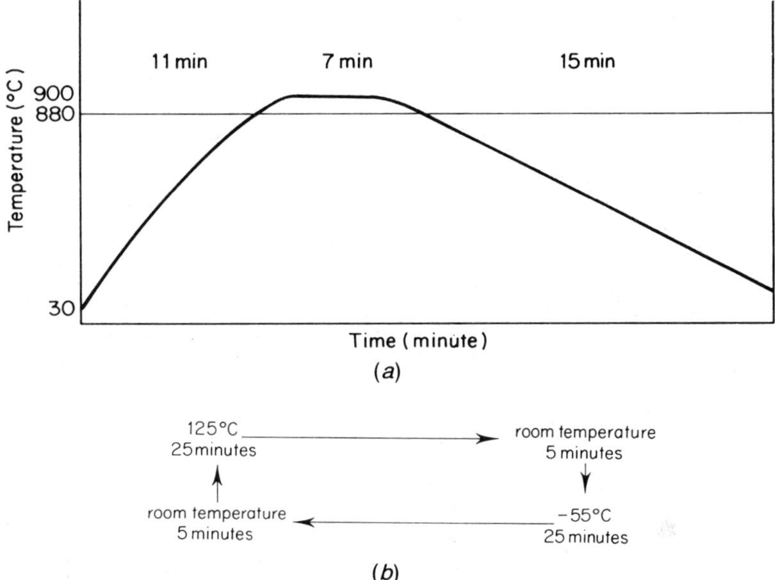



**Figure 17-2**    (a) Temperature profile of the thick-film furnace. (b) Conditions for thermal cycling.

2. Crooks are carefully formed at one end of each wire lead as shown in Figs. 17.3a to c, then the leads are slipped onto the test pattern with wire indexes to center leads over the test pads. Each lead should grip around the substrate edge firmly at its crooked end and stay in contact with the underlying pads. A fixture that makes the specimen handle easily during soldering is applied on the other end of the substrate to keep the correct alignment.

3. Specimens are degreased by a solvent in an ultrasonic cleaner for at least 15 minutes, allowed to dry, and then dipped into the RMA-type flux for a length from the crooked end of about two-thirds of the substrate. The molten solder bath is held at 230°C, and the surface is free of flux residues and dross. Substrates are vertically dipped into the bath until the adhesion test pads are fully immersed after the flux is apparently wicked up. The dwell time is 5 seconds. Soldered specimens are cleaned in three solvent-containing beakers in turn in an ultrasonic cleaner, and the total cleaning time is about 20 minutes.

4. After a 12-hour room-temperature storage period, specimens are placed in an oven at 130 ± 5°C for the aging test. Thermal cycle tests are performed with a 1-hour cyclic period from −55°C to 125°C between a freezing chamber and the oven. The conditions of the thermal cycle test are shown in Fig. 17-2b.

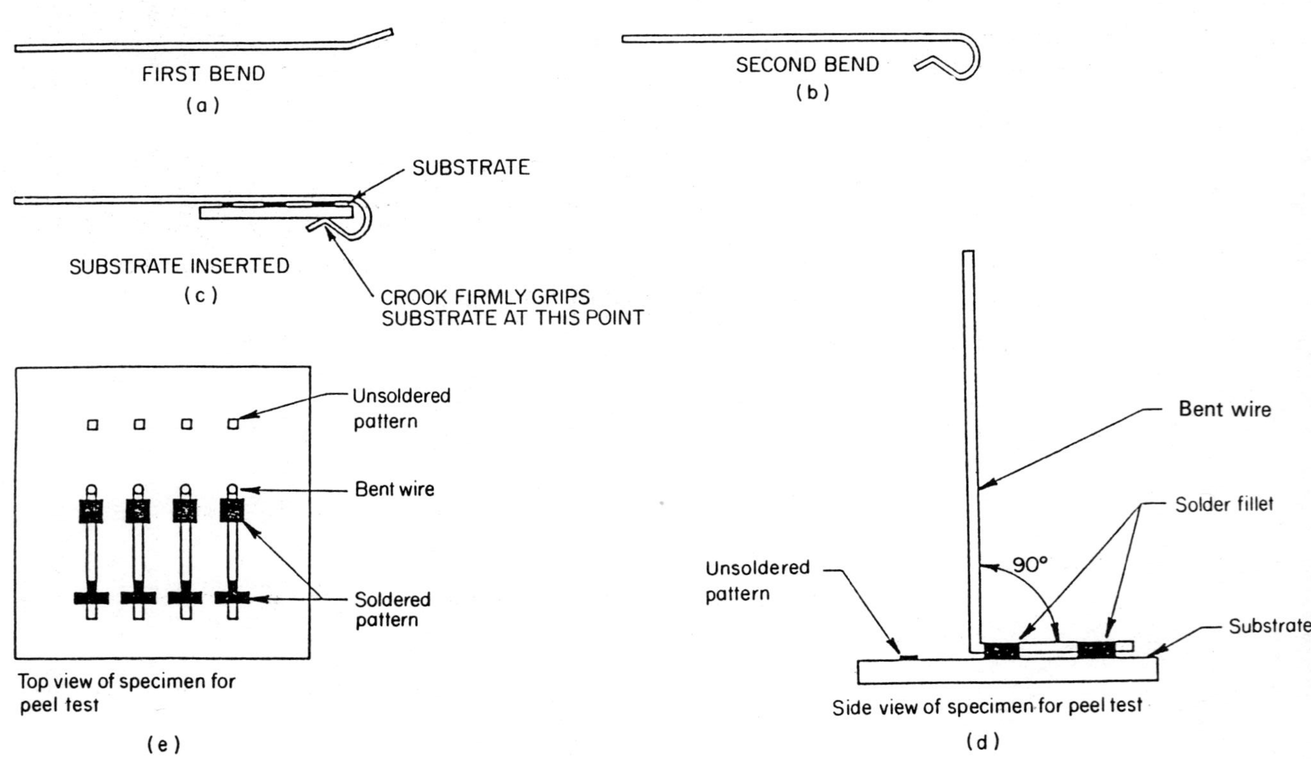


**Figure 17-3**   Schematic of the adhesion test.

5. The wires are bent in a 90° configuration, as shown in Figs. 17-3*d* and *e*, after the specimens have been taken out of the oven after a period of aging time or after a number of thermal cycles. Adhesion measurements are conducted in a universal material test machine with a cross-head speed of 26.2 or 2.28 mm/minute. The adhesion strength is measured as the maximum breaking force divided by the test pad area, 2 mm × 2 mm.
6. The fracture surfaces of the test pads are evaluated by scanning electron microscopy (SEM) and electron probe microanalysis (EPMA). The phases are identified by X-ray diffractometry (XRD). Several specimens were cut transversely with a diamond saw, and the cross-sectional view and elemental distribution were inspected by EPMA.

## 17.3 CHARACTERISTICS OF THE CONDUCTOR

The intrinsic properties of the conductor paste provide the basic knowledge for the understanding of the conductor–solder interaction, hence investigation of the composition of the conductor employed is needed. Two types of Dupont mixed bonded Pd–Ag conductors, 6134 and 6125, were employed. They were investigated by X-ray diffraction (XRD) to evaluate the palladium/silver ratio and also by atomic emission spectrometry (AES) to identify the major constituents, especially the binder.

The alloy of palladium and silver, which has a face-centered cubic crystal structure, is isomorphous as shown in the phase diagram of Fig. 17-4.[23] Fired

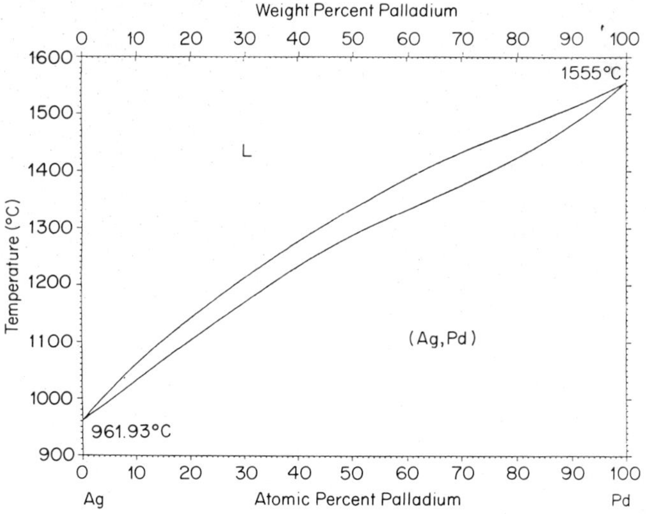


**Figure 17-4**   Phase diagram of palladium and silver.[22]

conductor pastes made from alloys of palladium and silver were investigated with step scanning by X-ray diffractometry at the (111), (200), (220), (311), and (222) planes. In order to precisely determine the lattice parameters, the $K\alpha_1$ and $K\alpha_2$ separation was obtained by computer using a specially designed program. The lattice parameter $a$ calculated from the Cu $K\alpha_1$ diffraction angle on the five planes is plotted versus the Nelson–Riley function[24,25] $[(\cos^2 \theta/\sin \theta) + (\cos^2 \theta/\theta)]$ for two conductors in Figs. 17-5$a$ and $b$. The least-squares fitted linear equations are[20]

$$\text{For 6125:} \quad a = 4.017 - 1.1218 \times 10^{-3}\left(\frac{\cos^2 \theta}{\sin \theta} + \frac{\cos^2 \theta}{\theta}\right) \tag{17-1}$$

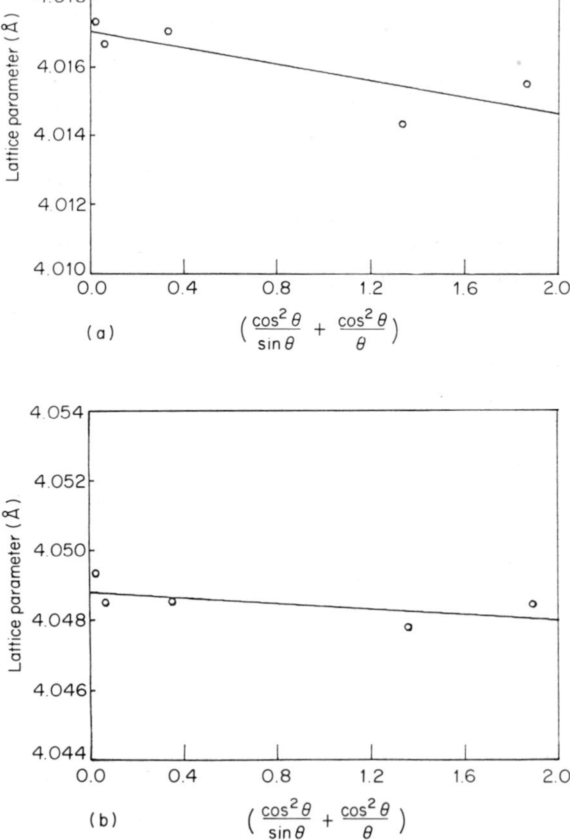

Figure 17-5   Lattice parameter vs. Nelson–Riley function for (a) 6125; (b) 6134.

For 6134:    $a = 4.049\text{--}4.060 \times 10^{-4}\left(\dfrac{\cos^2 \theta}{\sin \theta} + \dfrac{\cos^2 \theta}{\theta}\right)$    (17-2)

The lattice parameters of the alloy can be evaluated from the intercepts of the plots, which are 4.017 Å and 4.049 Å for alloys 6125 and 6134, respectively. From Fig. 17-6, a plot of lattice parameter versus alloy composition, the Pd:Ag ratios are 1:2.38 and 1:7.92 for alloys 6125 and 6134, respectively.

Quantitative analysis by X-ray diffraction is based on the fact that the intensity of the diffraction pattern of a particular phase in a mixture of phases depends on the concentration of that phase in the mixture. The relation between the diffracted intensity and concentration of a mixture containing two phases, $\gamma$ and $\alpha$, can be expressed as[24]

$$\frac{I_\gamma}{I_\alpha} = \frac{R_\gamma C_\gamma}{R_\alpha C_\alpha}$$    (17-3)

where   $I_\gamma$ and $I_\alpha$ = the integrated intensity of $\gamma$ and $\alpha$ phases, respectively
   $R_\gamma$ and $R_\alpha$ = constants depending on $\theta$, $hkl$, and the absorption
         coefficients of $\gamma$ and $\alpha$ phases
   $C_\gamma$ and $C_\alpha$ = the volume fractions of the $\gamma$ and $\alpha$ phases

The unfired conductor pastes containing the silver and palladium powders

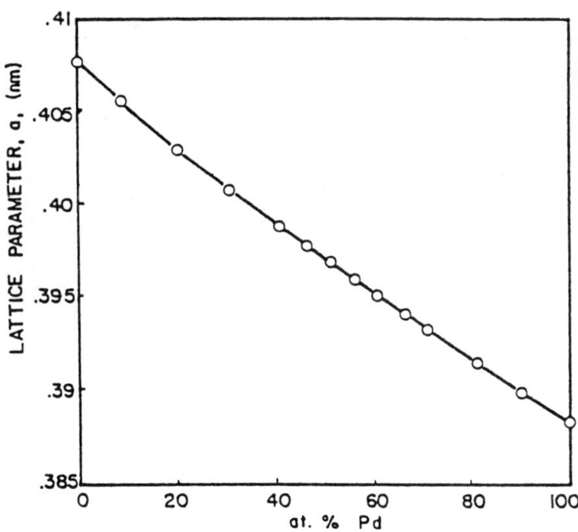


**Figure 17-6**    Effect of composition on Ag–Pd solid solution lattice parameter.[22]

are step-scanned around the diffraction peaks of the (111) plane in the X-ray diffractometer. The absorption coefficients of palladium and silver are similar, so the integrated intensities of palladium and silver are directly proportional to their concentrations. The Pd:Ag ratios obtained from integrated intensities are 1:2.68 and 1:6.35 for alloys 6125 and 6134, respectively, as shown in Figs. 17-7a and b.

In addition, the Pd:Ag ratio can be determined by atomic emission spectrometry (AES) analysis of the compositions in the pastes as listed in

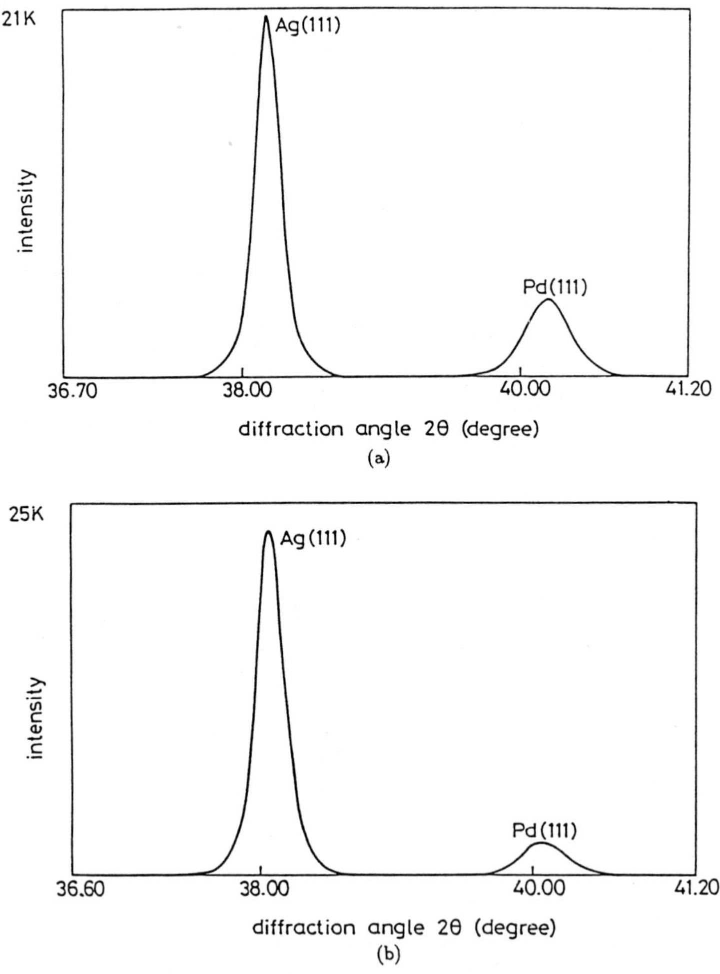


**Figure 17-7** X-ray diffraction pattern of unfired pastes for quantitative analysis: (a) 6125; (b) 6134.

**Table 17-1**   The Composition of Thick-film Pastes by AES

| | 6125 | | 6134 | |
|---|---|---|---|---|
| | p.p.m. | Normalized to (Ag + Pd) = 10 000 | p.p.m. | Normalized to (Ag + Pd) = 10 000 |
| Ag | 1519.4 ⎱ | 10 000 | 2277.8 ⎱ | 10 000 |
| Pd | 489.4 ⎰ | | 325.6 ⎰ | |
| Bi | 260.8 | 1 298.28 | 269.3 | 1 034.42 |
| Pb | 38.26 | 190.46 | 40.00 | 153.65 |
| Cu | 1.81 | 9.01 | 1.57 | 6.03 |
| Co | 0.22 | 1.095 | 0.22 | 0.845 |
| Ni | 0.09 | 0.448 | 0.09 | 0.346 |
| Si | 9.74 | 48.49 | 9.58 | 36.88 |
| Al | 3.26 | 16.23 | 3.26 | 12.52 |
| Ca | 6.33 | 31.51 | 6.77 | 26.00 |

**Table 17-2**   Summary of the Pd:Ag Ratio from Three Different Methods

| | N-R Relation | Integrated Area | AES | Average |
|---|---|---|---|---|
| 6125 | 1:2.38 | 1:2.68 | 1:3.06 | 1:2.7 |
| 6134 | 1:7.92 | 1:6.35 | 1:6.90 | 1:7.05 |

Table 17-1. Pd:Ag atomic ratios are evaluated as 1:3.06 and 1:6.90 for alloys 6125 and 6134, respectively.

The results obtained from three different methods are summarized in Table 17-2. The average values of the Pd:Ag ratio on the basis of these three methods are 2.71 and 7.05, and alloys 6125 and 6134, respectively. It appears that the Pd:Ag ratios evaluated by three different methods are fairly comparable.


## 17.4 MORPHOLOGY AND ELEMENTAL DISTRIBUTION IN THE AS-FIRED CONDUCTOR

Firing of the conductor is, in fact, a sintering process for the constituent Pd and Ag powders. During firing, reactions take place among Pd/Ag,

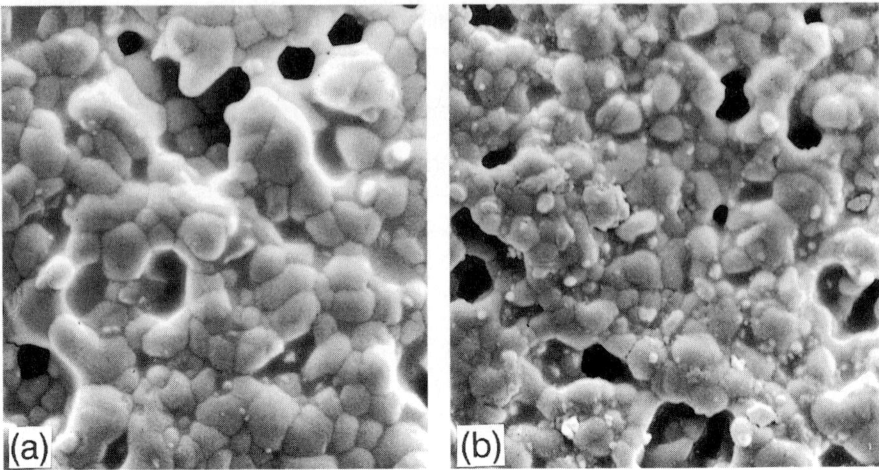

**Figure 17-8**  SEM micrograph for the as-fired conducto  surface: (a) 6125; (b) 6134.

glass/oxide, and the substrate. Investigation by electron microscopy provides a basic understanding of the interfacial phenomena.

The as-fired surface morphology and cross-sectional views of conductors are shown in Figs. 17-8 to 17-10. Near the surface, lots of pores are present, which are left by the organic content in pastes. Dupont 6134 conductor has more fine particles at the grain boundary than 6125.[20] The compositions of these particles from the EPMA qualitative analysis are Al, Si, Co, Ca, and Bi. AES results show that 6125 has a greater nonmetal/metal ratio, which means that if there is identical metal content, 6125 paste should contain more nonmetal phases. This also implies that the glass distribution in the film is different for various types of conductors.

Although the cross-sectional view does not clearly reveal the glass distribution, interaction of the conductor paste and the substrate is observed. The conductor paste penetrates into the substrate along the alumina grain boundary to about 10 μm in depth as a mechanical interlock. Another binding force may be due to reactive bonding. The oxides in the paste, such as bismuth oxide, copper oxide, and cobalt oxide, react with substrate and form the spinel structure products that result in the reactive bonding force.

The main constituent in the binder is bismuth oxide, as shown in Table 17-1. During the firing process, the molten bismuth oxide not only reacts with the alumina substrate but also penetrates into its interior. The silver and palladium powder is carried by the molten bismuth oxide into the alumina grain boundary, and subsequent sintering takes place. This enhances the mechanical interlock strength, and makes the adhesion stronger.

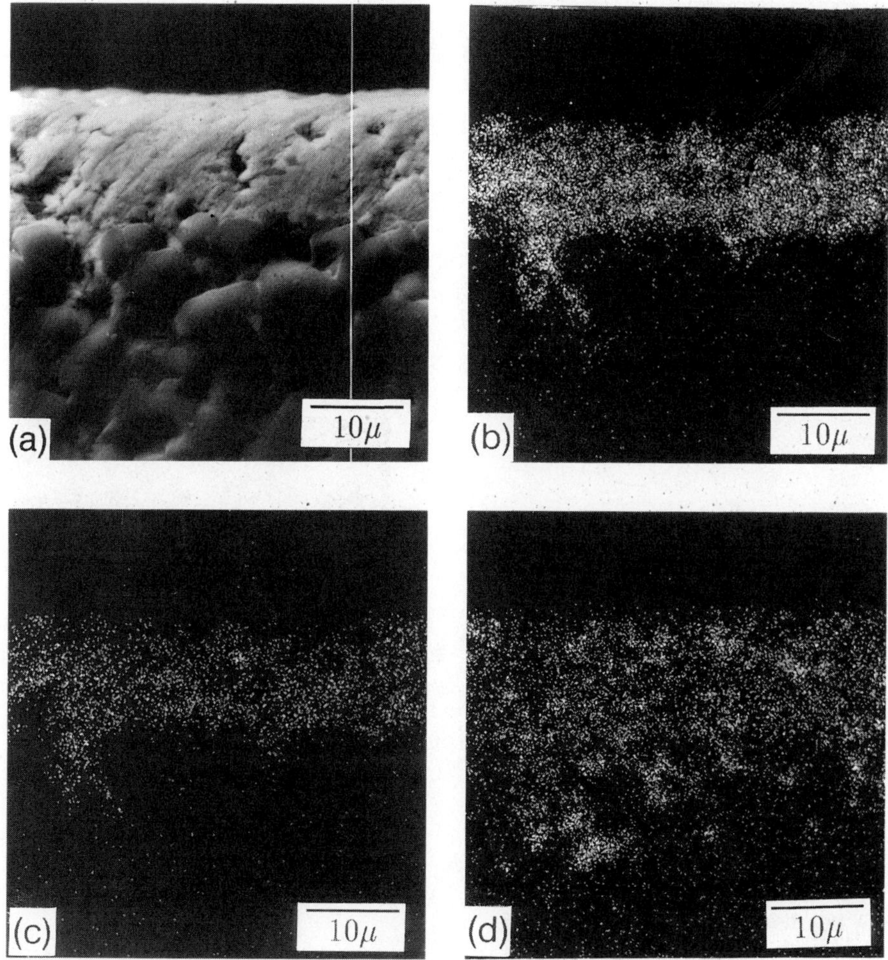

**Figure 17-9**  Cross-sectional SEM micrograph (*a*) and X-ray mapping of as-fired 6125; (*b*) Ag X-ray mapping; (*c*) Pd X-ray mapping; (*d*) Bi X-ray mapping.

## 17.5 AGING EFFECTS ON SOLDERED THICK-FILM JOINTS

### 17.5.1 Adhesion Loss Mechanism

In most cases, the soldered joint will degrade after long periods of use. The mechanical strength will decrease and the electric resistance will increase. To investigate the joint degradation, an accelerated life test is usually employed in which the specimen is put under an elevated temperature, and the

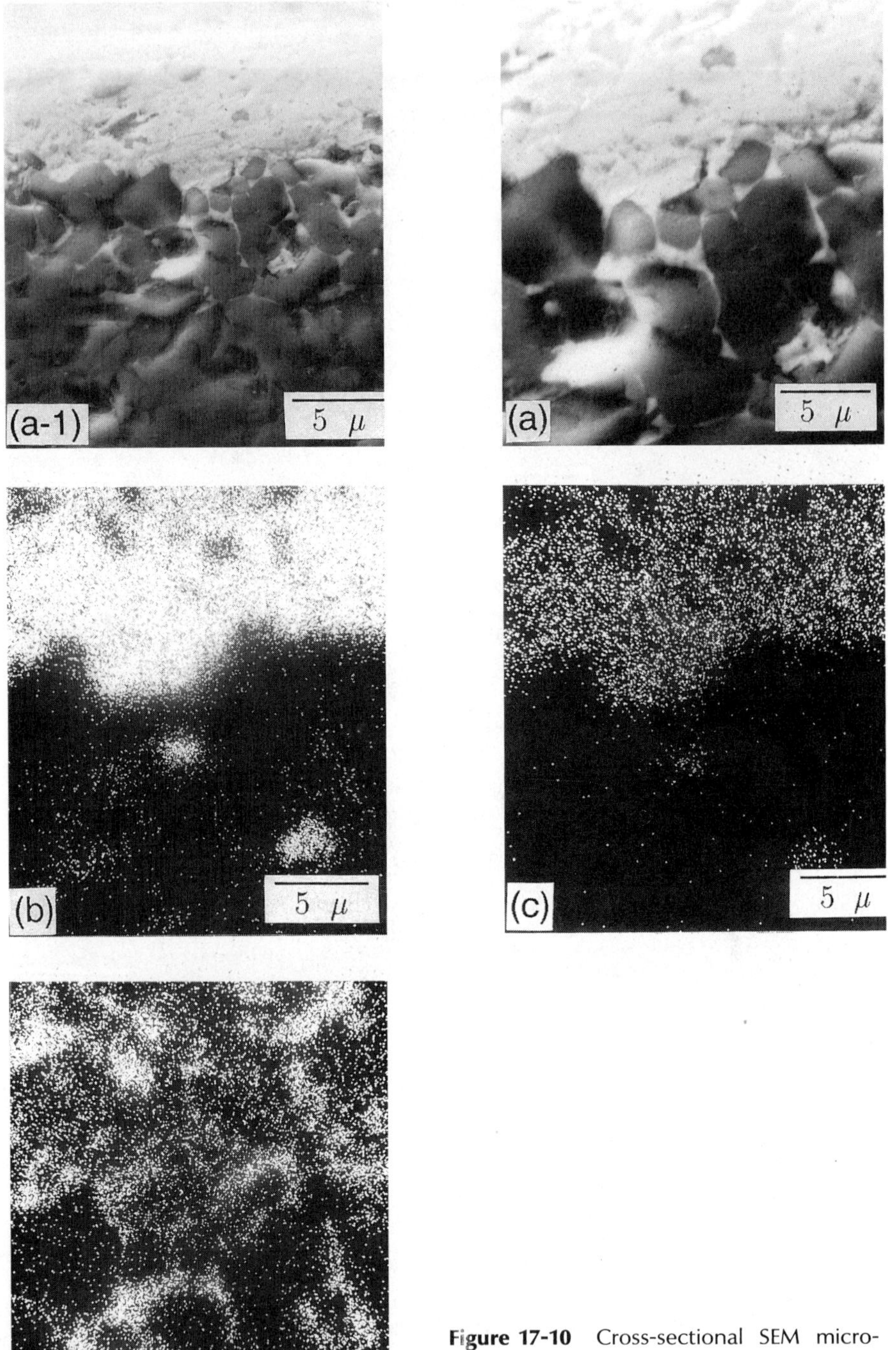

**Figure 17-10** Cross-sectional SEM micrograph (*a*) and X-ray mapping of as-fired 6134; (*b*) Ag X-ray mapping; (*c*) Pd X-ray mapping; (*d*) Bi X-ray mapping.

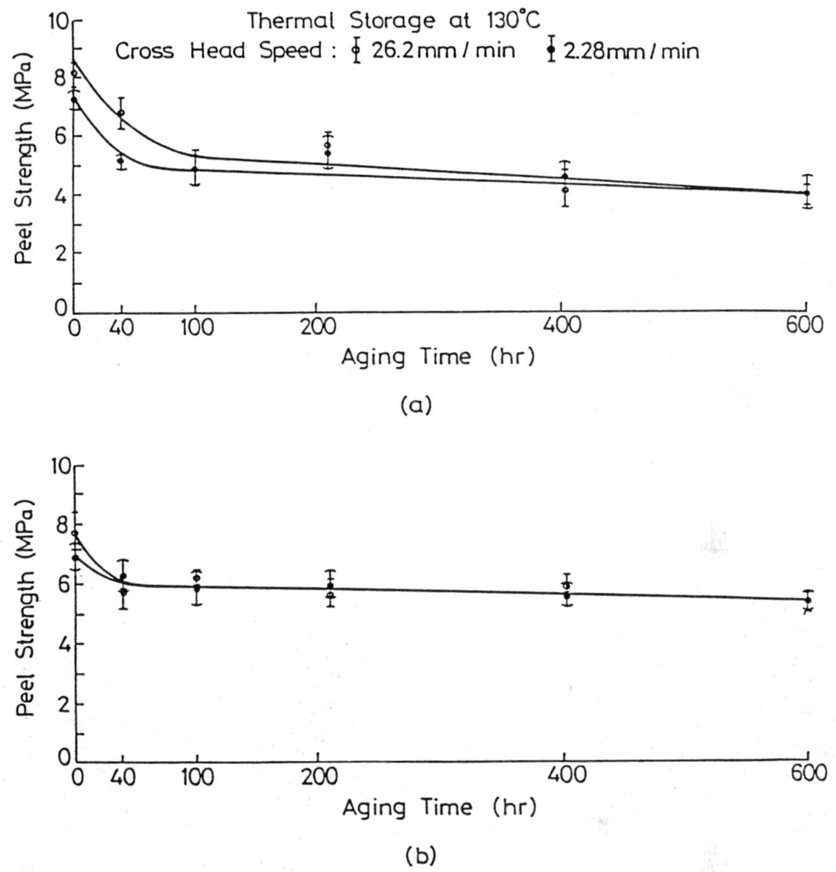


**Figure 17-11**    Peel strength as a function of aging time for (*a*) 6134, (*b*) 6125.

degradation behavior is studied. Possible mechanisms could be determined on the basis of such observations.

In Liu's study[20,21] test specimens aged at 130°C were removed after 40, 100, 210, 400, and 600 hours to test their adhesion strength and to evaluate phases and microstructure. The adhesion strength of Dupont 6125 and 6134 conductors decreases with increasing aging time, as shown in Fig. 17-11. The strength loss is relatively abrupt in the first 40 hours of aging for 6125 and in the first 100 hours of aging for 6134. After this initial period, the strength loss is rather slow. Identical trends are observed for both the faster cross-head speed of 26.2 mm/min and the slower one of 2.28 mm/min. Phases at the fracture surface of the specimen are Pd–Ag alloy for 6125, and $Ag_3Sn$, $Pd_3Pb$, and $Pd_3Sn$ for 6134 at the beginning of aging, as shown in Figs. 17-12*a* and 17-13*a*.

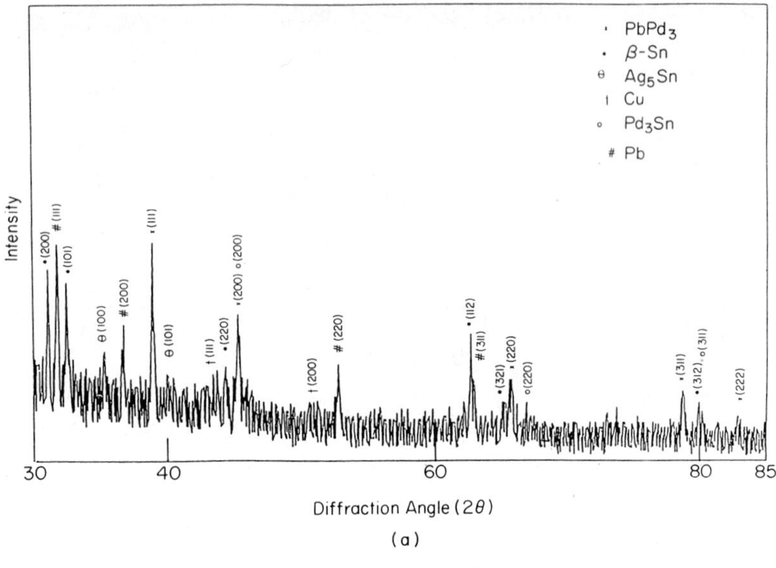


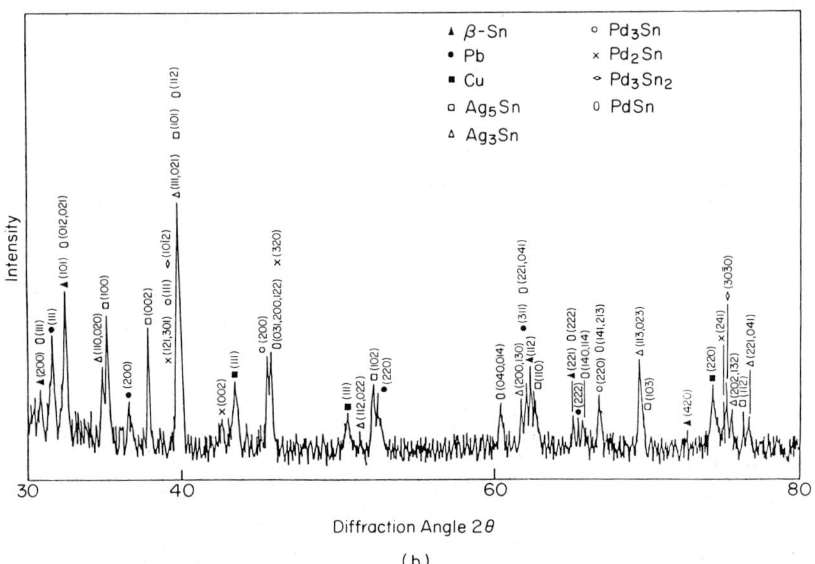


**Figure 17-12**   X-ray diffraction pattern of samples with 6134 conductor: (*a*) no aging; (*b*) 100 hours aging. (*Continued*)

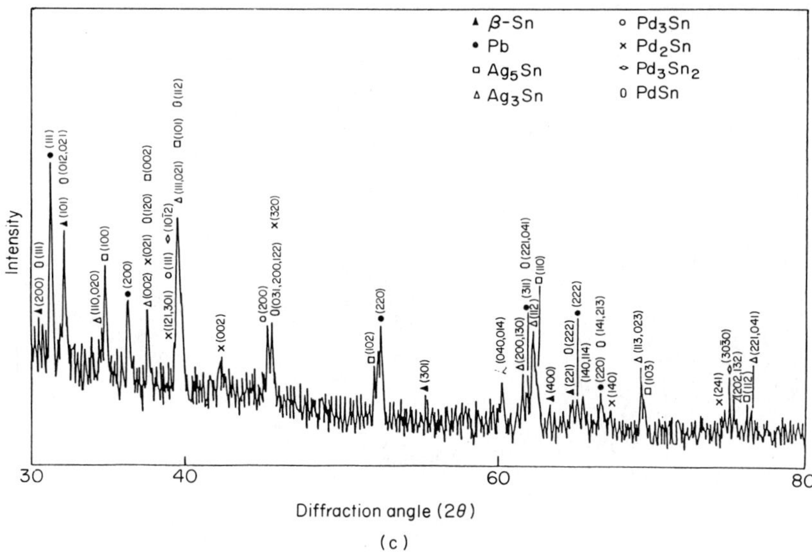



**Figure 17-12** (*continued*)   X-ray diffraction pattern of samples with 6134 conductor: (*c*) 210 hours aging.

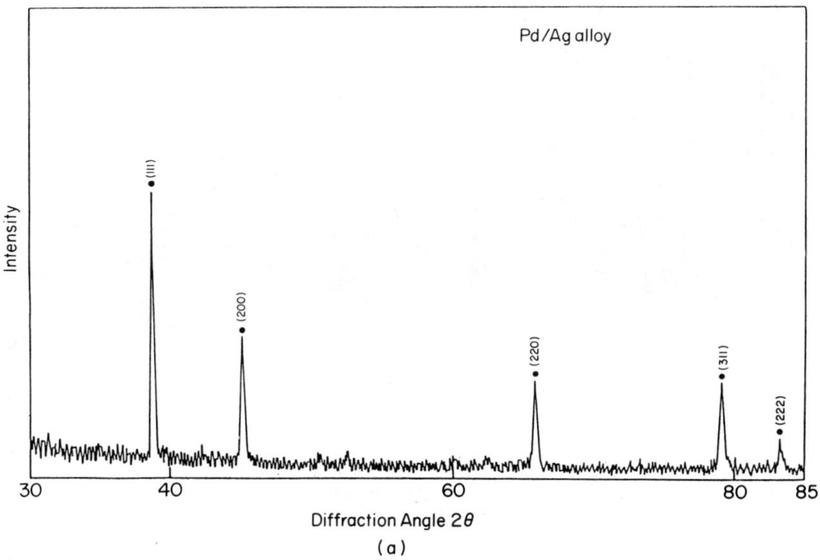



**Figure 17-13**   X-ray diffraction pattern of samples with 6125 conductor: (*a*) no aging. (*Continued*)

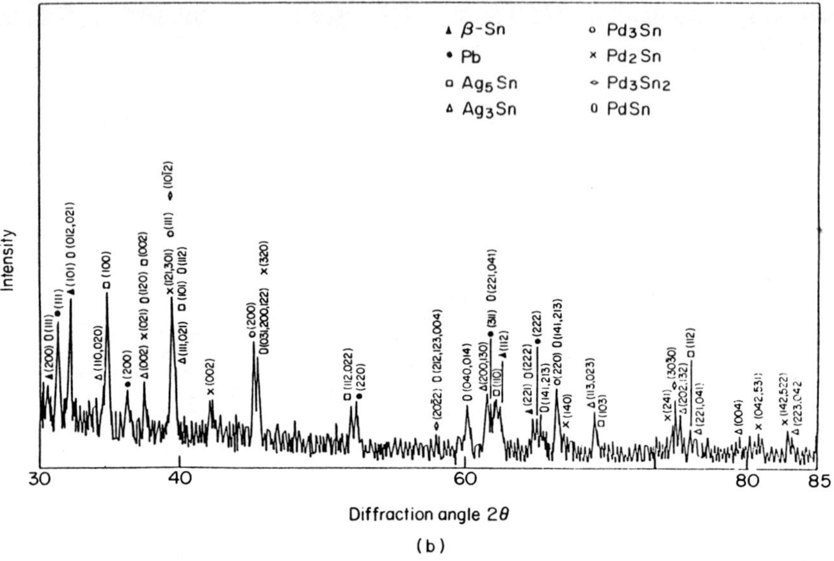

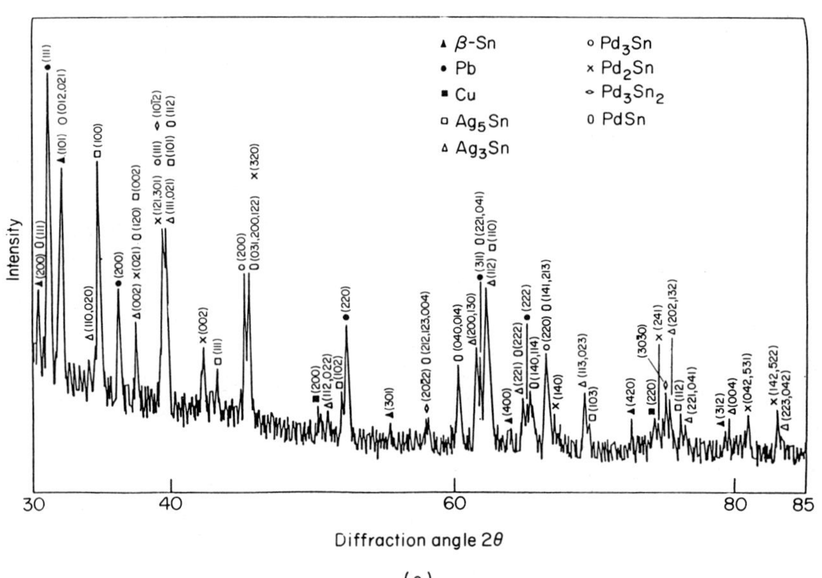

**Figure 17-13** (*continued*)   X-ray diffraction pattern of samples with 6125 conductor: (*b*) 40 hours aging; (*c*) 210 hours aging.

After longer aging time, phases of $Ag_3Sn$, $Ag_5Sn$, $Pd_3Sn$, $Pd_2Sn$, $Pd_3Sn_2$, and PdSn occur in both conductors, as shown in Figs. 17-12$b$, $c$ and 17-13$b$, $c$. Figures 17-14 to 17-25 show cross-sectional views of 6125 and 6134. The overall microstructural evolution is summarized as follows.

1. Tin diffuses into the conductor film during and after soldering. It takes about 40 hours to diffuse up to the conductor–substrate interface for 6125, and 100 hours for 6134.
2. Lead in solder segregates above the Ag-rich layer, which implies that Pb is left behind as Sn diffuses. The degree of segregation is greater for the high silver content conductor (6134) than the lower one (6125).
3. The palladium layer always has a greater thickness than the silver. It covers the Pb layer, as Pd, can form intermetallic compounds with lead on the basis of the Pd–Pb phase diagram. $Pd_3Pb$ is the one found in the X-ray spectrum. It appears that $Pd_3Pb$ segregates above the silver-rich compounds, $Ag_3Sn$ and $Ag_5Sn$, while there is random distribution of the Pd–Sn compounds.
4. After a period of aging, the intermetallic compounds exhibit grain growth and show a distinct Pd-rich and Ag-rich separated region. For the high silver content conductor, it takes longer, about 600 hours, to reveal the separated region. However, for the high-palladium conduction, the separated region is observed at about 40 hours of aging.

To characterize the interface where the fracture takes place, the fracture behavior of the adhesion test is classified into three modes. Mode A is defined as failure at the conductor–substrate interface; mode B represents a solder–conductor interface failure; and mode C is a wire–solder interface failure. Different types of fracture modes indicate different interfaces where the fracture occurs. In the real situation, joint pads are not fractured in a single mode. If the conductor–substrate interface fracture area is large, then it is designated as B mode fracture. If the conductor–solder interface area is large, then the A mode is specified. Numbers of test pads with different fracture modes versus aging time are listed in Table 17-3.[20,21]

The fracture surface at the conductor–solder interface of Dupont 6134 is shown in Figs. 17-26$a$ to $f$. The surface morphologies vary with different aging times. For specimens prior to the aging test, there are lots of bubbles and dimples on the fracture surface. For aged specimens, ductile fracture forms a dimple structure that tends to be more complete as the aging time is increased. Figure 17-27 shows the fracture surface at the conductor–substrate interface, which is filled with the bowl-type structure. The bowl-type structure is left by spalling of the conductor grain. From Table 17-3, mode A fracture occurs after 100 hours of aging, and at this time tin has diffused up to the conductor–substrate interface. The formation of

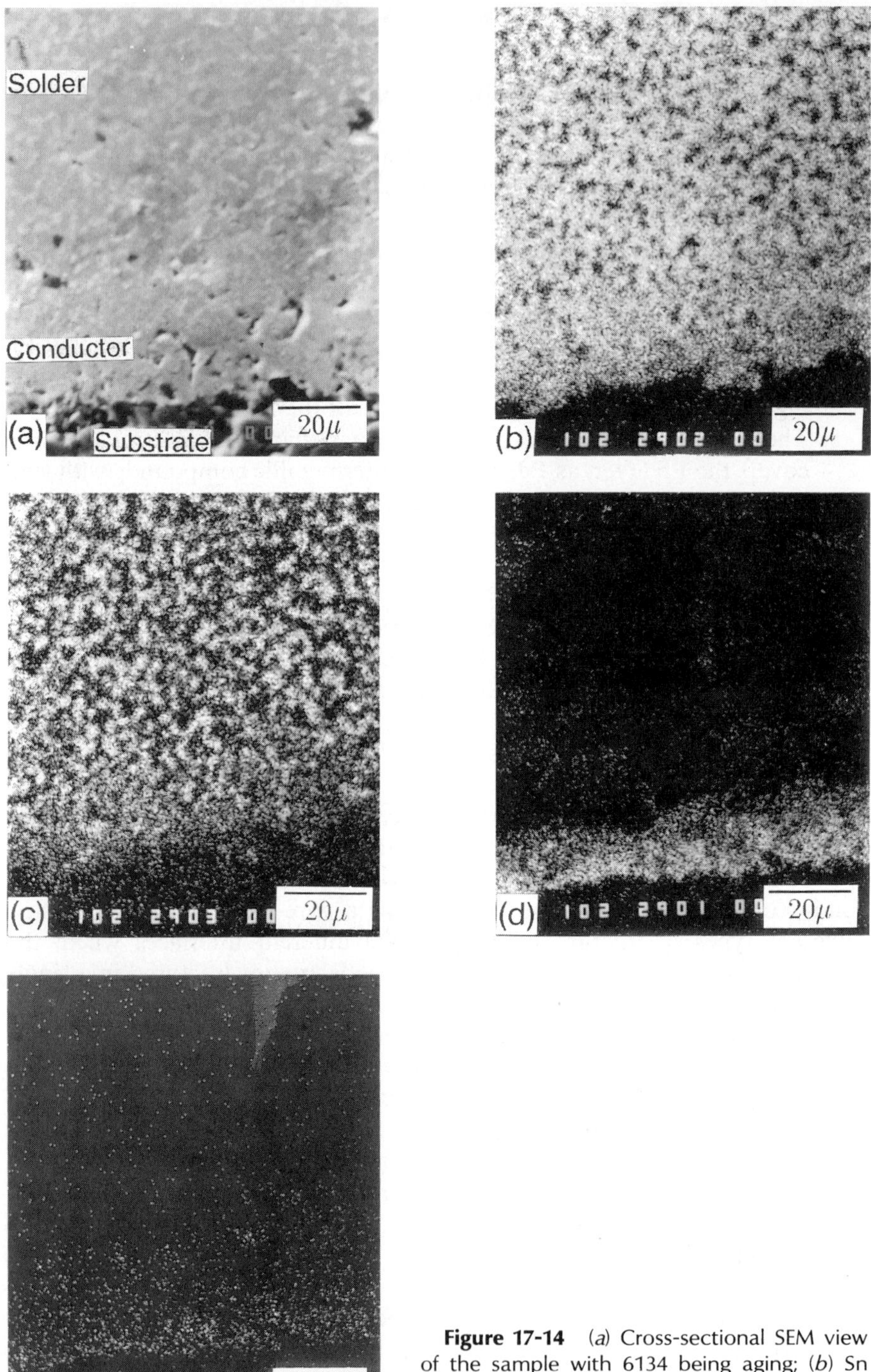



**Figure 17-14** (a) Cross-sectional SEM view of the sample with 6134 being aging; (b) Sn X-ray mapping; (c) Pb X-ray mapping; (d) Ag X-ray mapping; (e) Pd X-ray mapping.

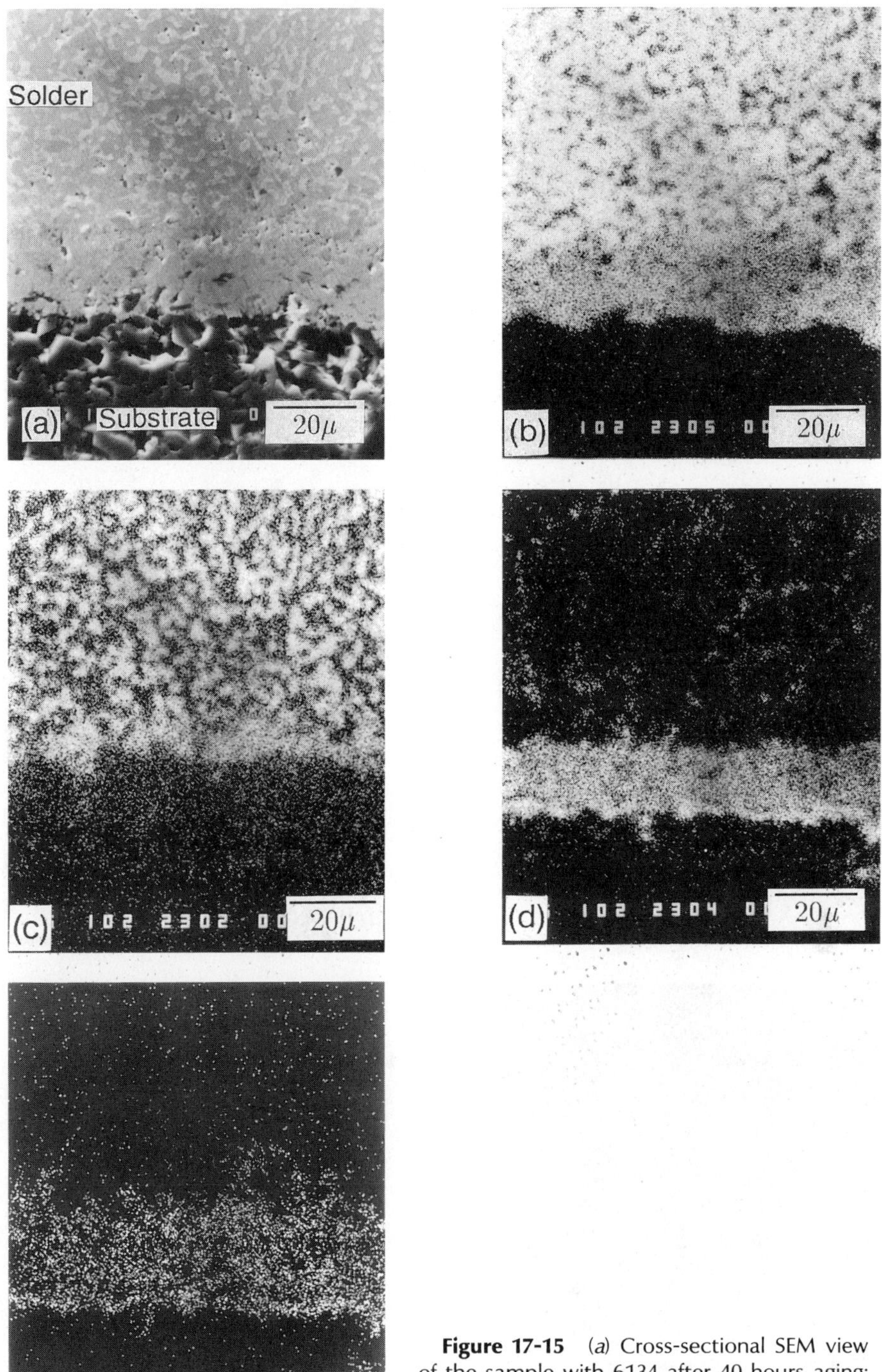

**Figure 17-15** (*a*) Cross-sectional SEM view of the sample with 6134 after 40 hours aging; (*b*) Sn X-ray mapping; (*c*) Pb X-ray mapping; (*d*) Ag X-ray mapping; (*e*) Pd X-ray mapping.

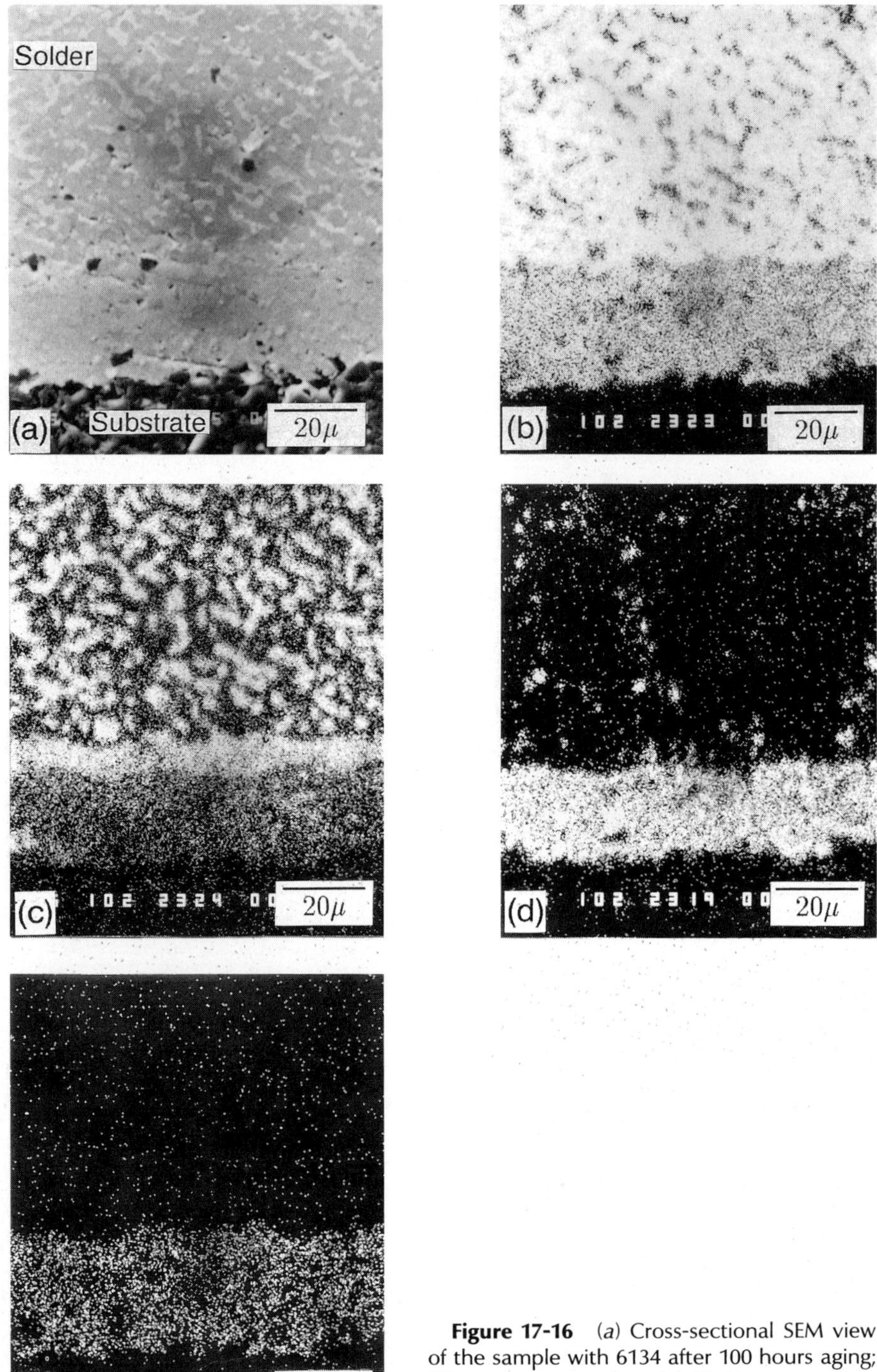

**Figure 17-16** (*a*) Cross-sectional SEM view of the sample with 6134 after 100 hours aging; (*b*) Sn X-ray mapping; (*c*) Pb X-ray mapping; (*d*) Ag X-ray mapping; (*e*) Pd X-ray mapping.

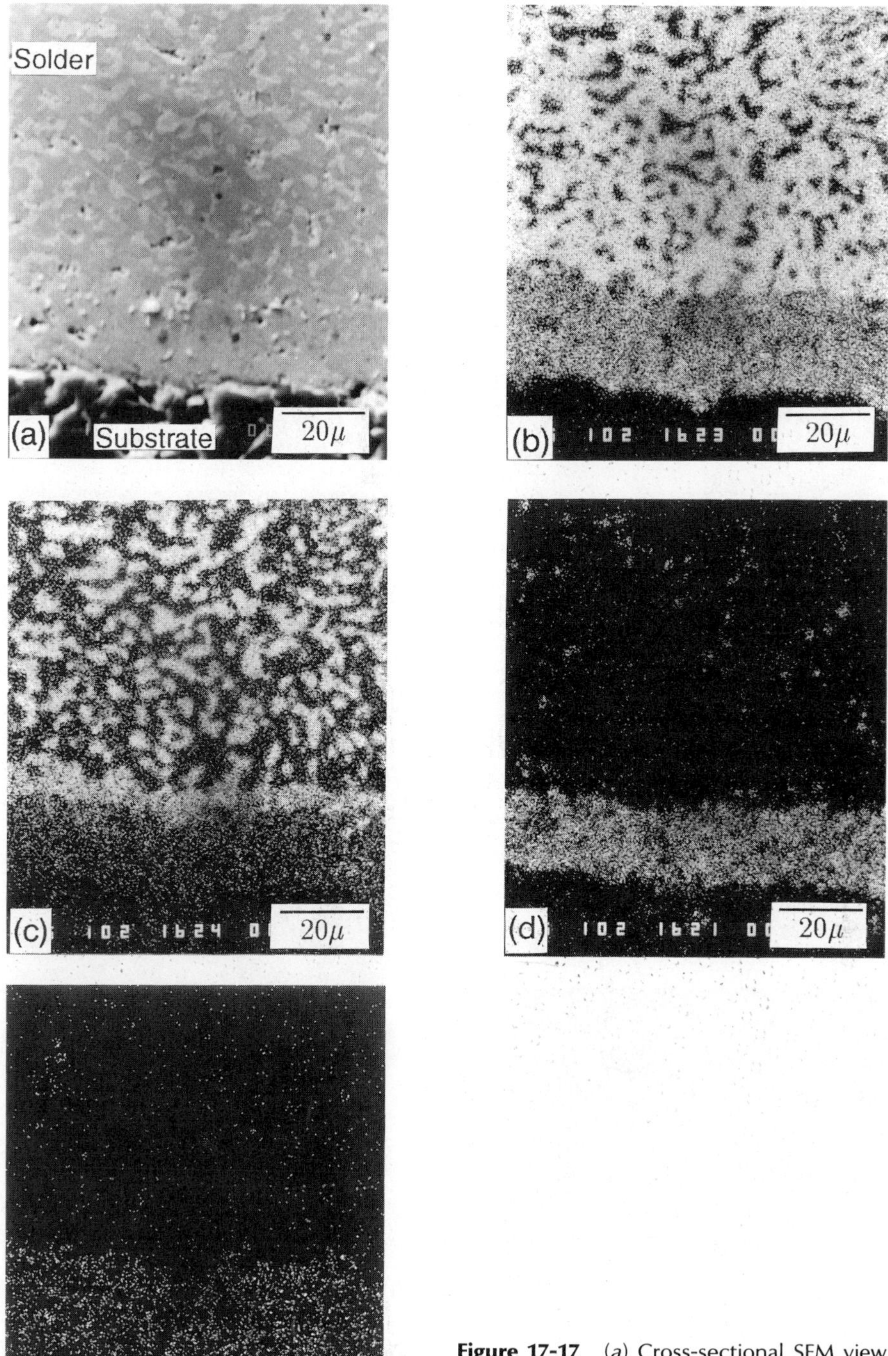

**Figure 17-17** (*a*) Cross-sectional SEM view of the sample with 6134 after 210 hours aging; (*b*) Sn X-ray mapping; (*c*) Pb X-ray mapping; (*d*) Ag X-ray mapping; (*e*) Pd X-ray mapping.

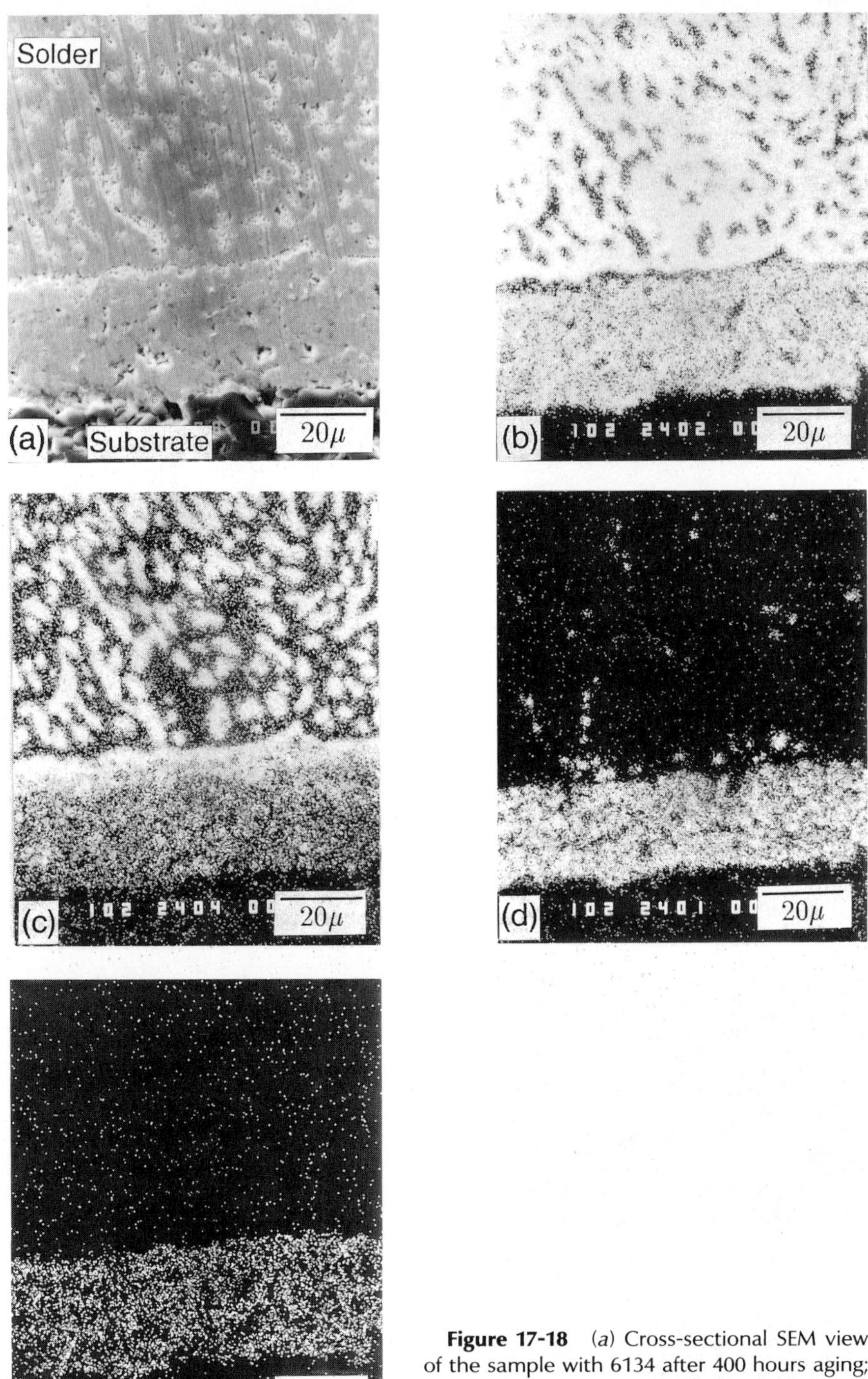

**Figure 17-18** (a) Cross-sectional SEM view of the sample with 6134 after 400 hours aging; (b) Sn X-ray mapping; (c) Pb X-ray mapping; (d) Ag X-ray mapping; (e) Pd X-ray mapping.

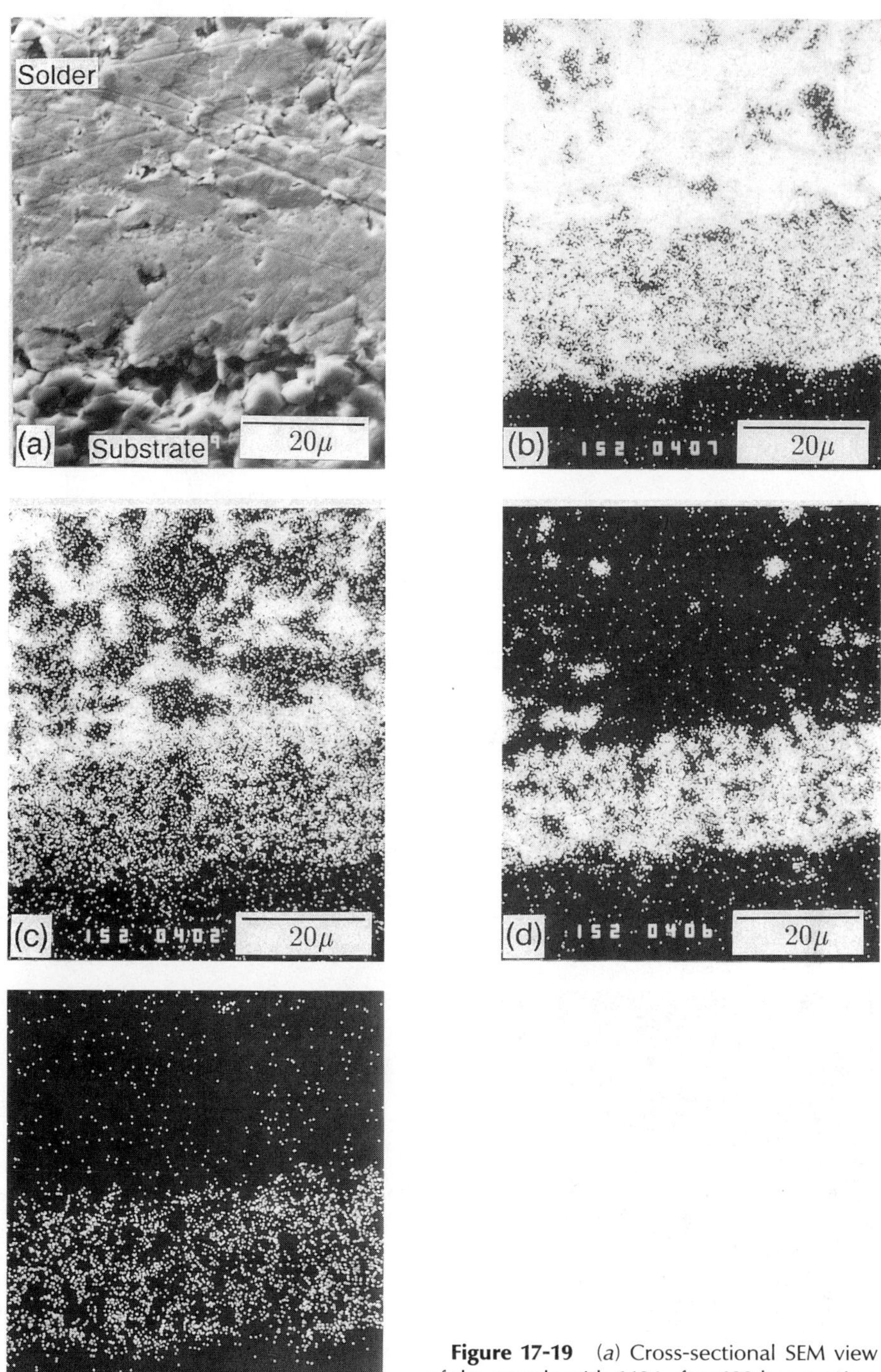

**Figure 17-19** (*a*) Cross-sectional SEM view of the sample with 6134 after 600 hours aging; (*b*) Sn X-ray mapping; (*c*) Pb X-ray mapping; (*d*) Ag X-ray mapping; (*e*) Pd X-ray mapping.

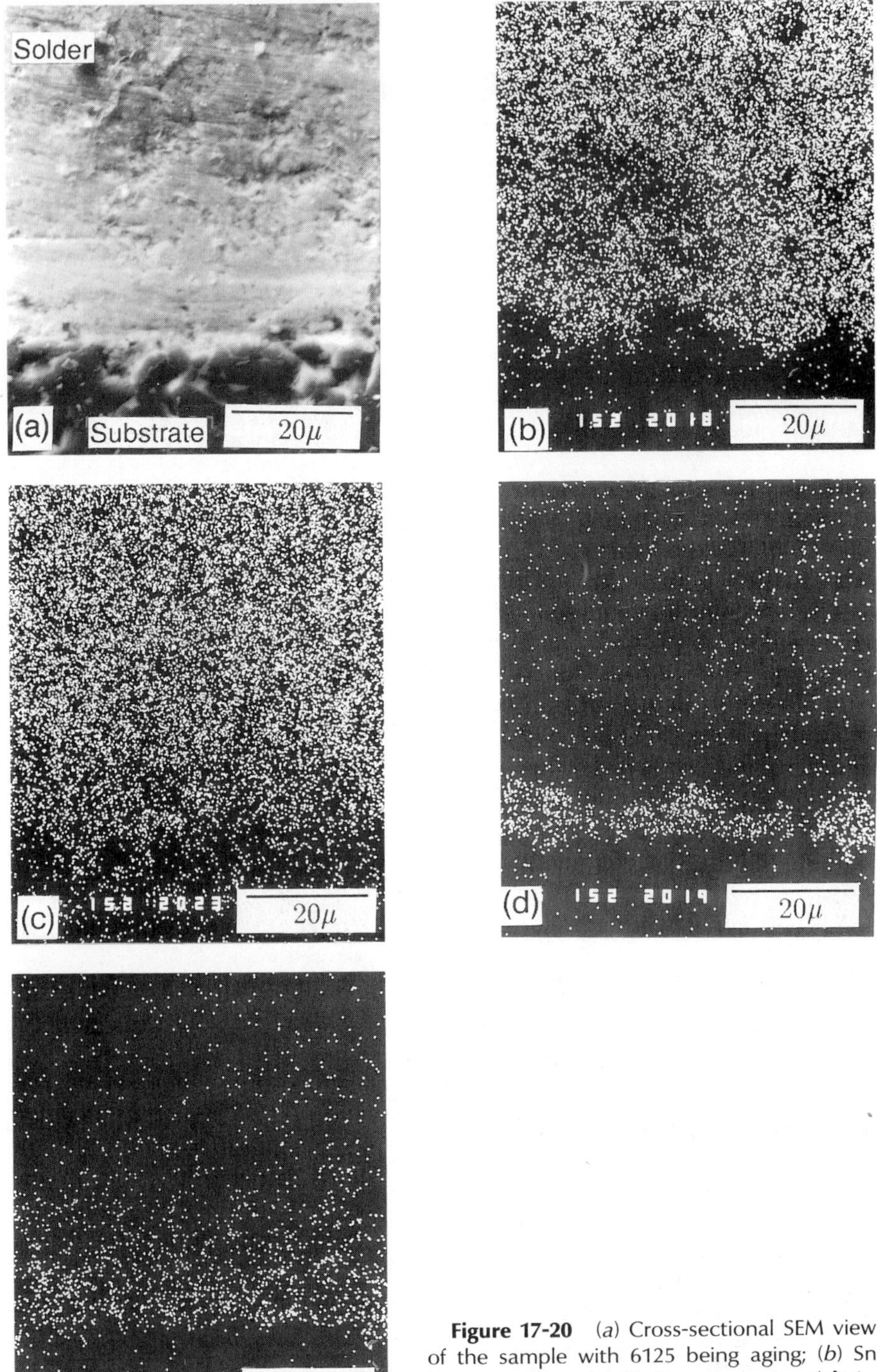



**Figure 17-20** (*a*) Cross-sectional SEM view of the sample with 6125 being aging; (*b*) Sn X-ray mapping; (*c*) Pb X-ray mapping; (*d*) Ag X-ray mapping; (*e*) Pd X-ray mapping.

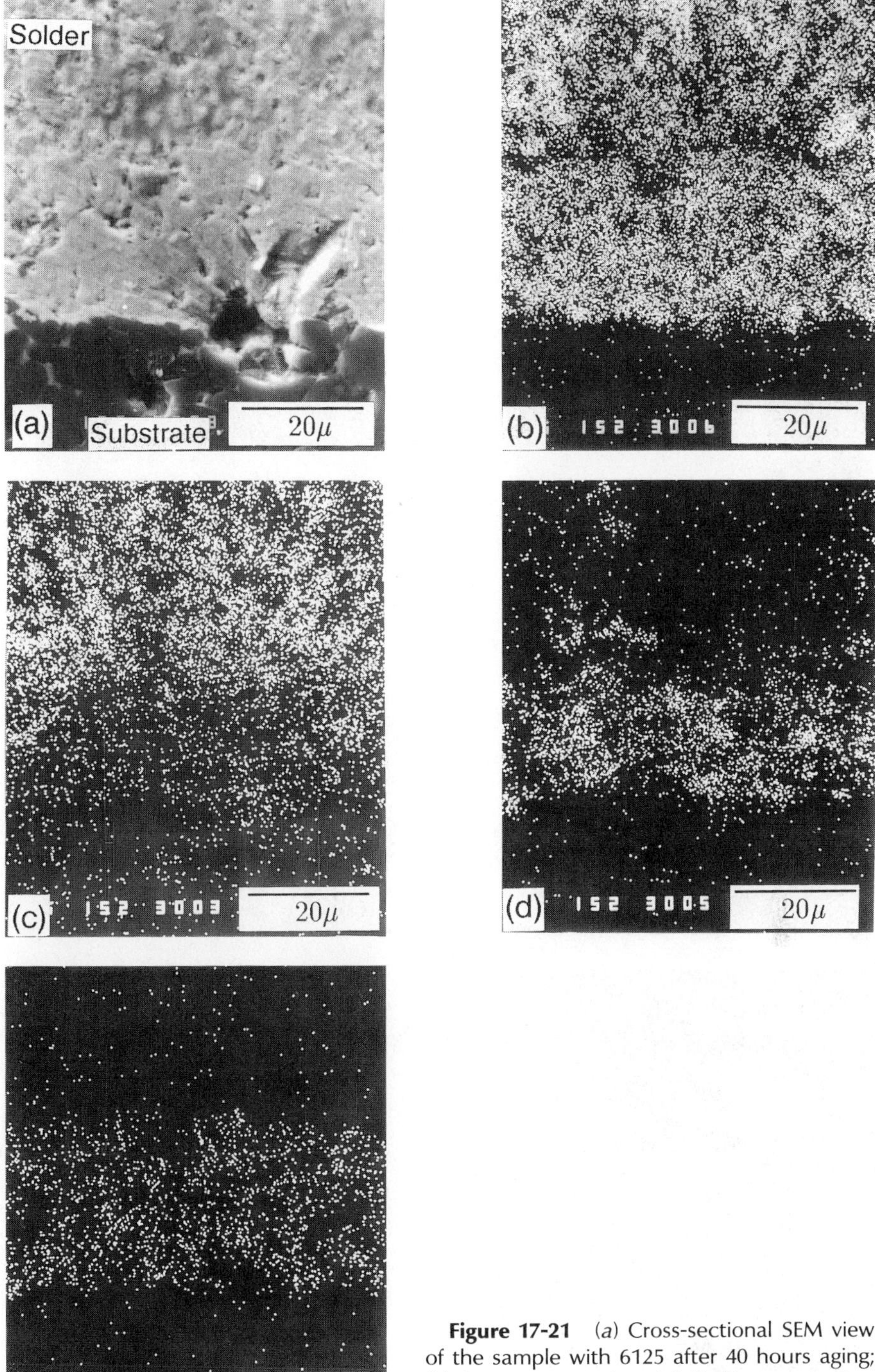

**Figure 17-21** (*a*) Cross-sectional SEM view of the sample with 6125 after 40 hours aging; (*b*) Sn X-ray mapping; (*c*) Pb X-ray mapping; (*d*) Ag X-ray mapping; (*e*) Pd X-ray mapping.

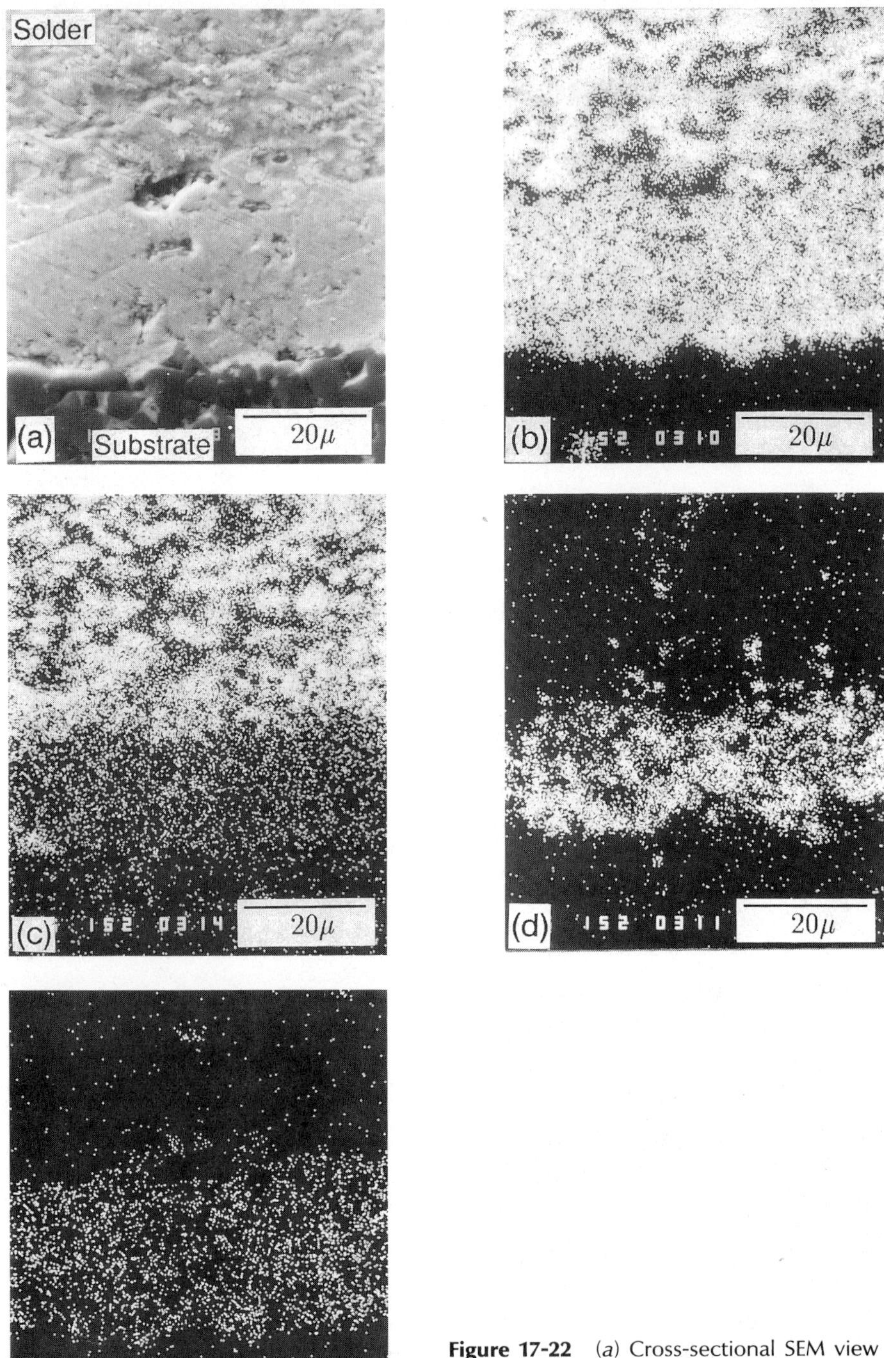

**Figure 17-22** (*a*) Cross-sectional SEM view of the sample with 6125 after 100 hours aging; (*b*) Sn X-ray mapping; (*c*) Pb X-ray mapping; (*d*) Ag X-ray mapping; (*e*) Pd X-ray mapping.

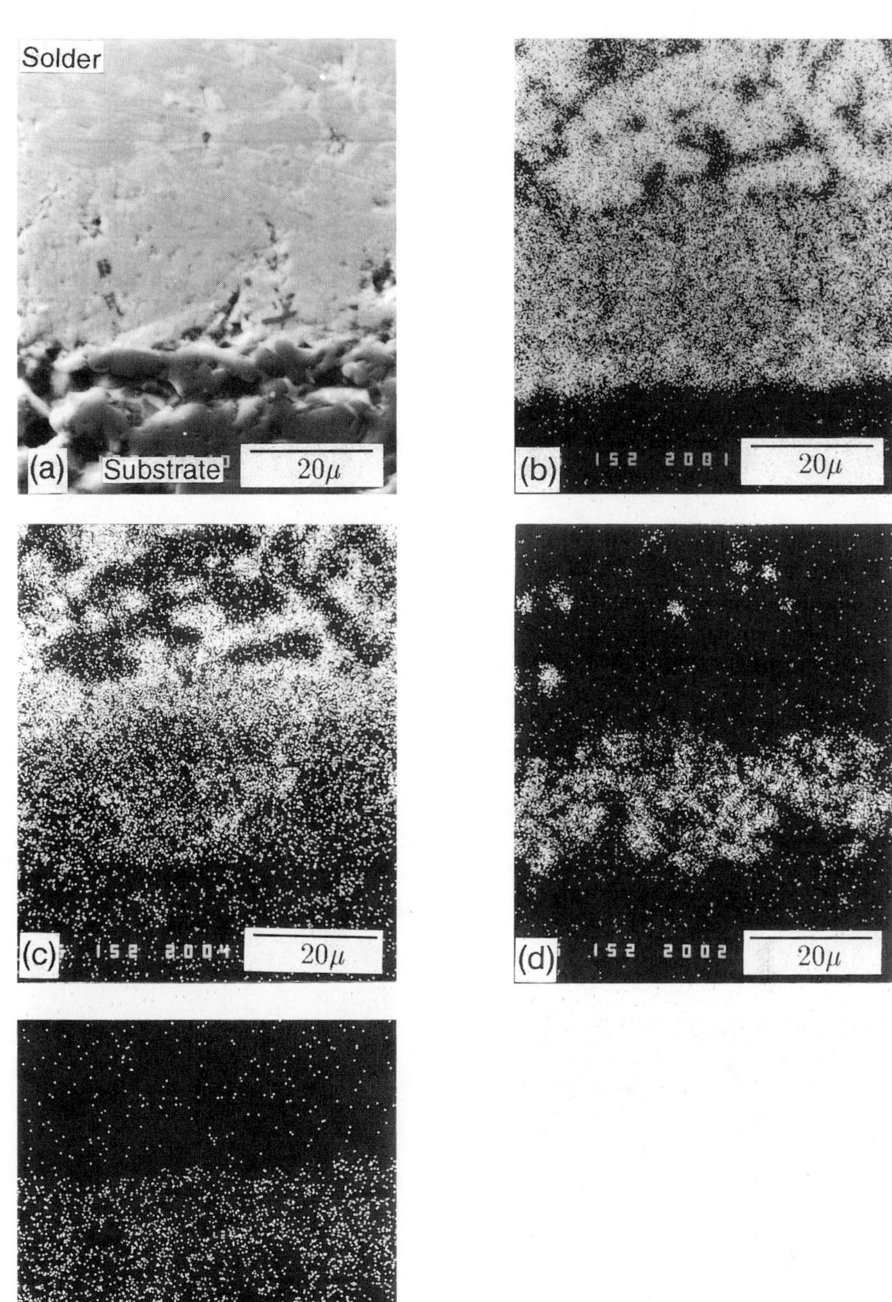

**Figure 17-23** (*a*) Cross-sectional SEM view of the sample with 6125 after 210 hours aging; (*b*) Sn X-ray mapping; (*c*) Pb X-ray mapping; (*d*) Ag X-ray mapping; (*e*) Pd X-ray mapping.

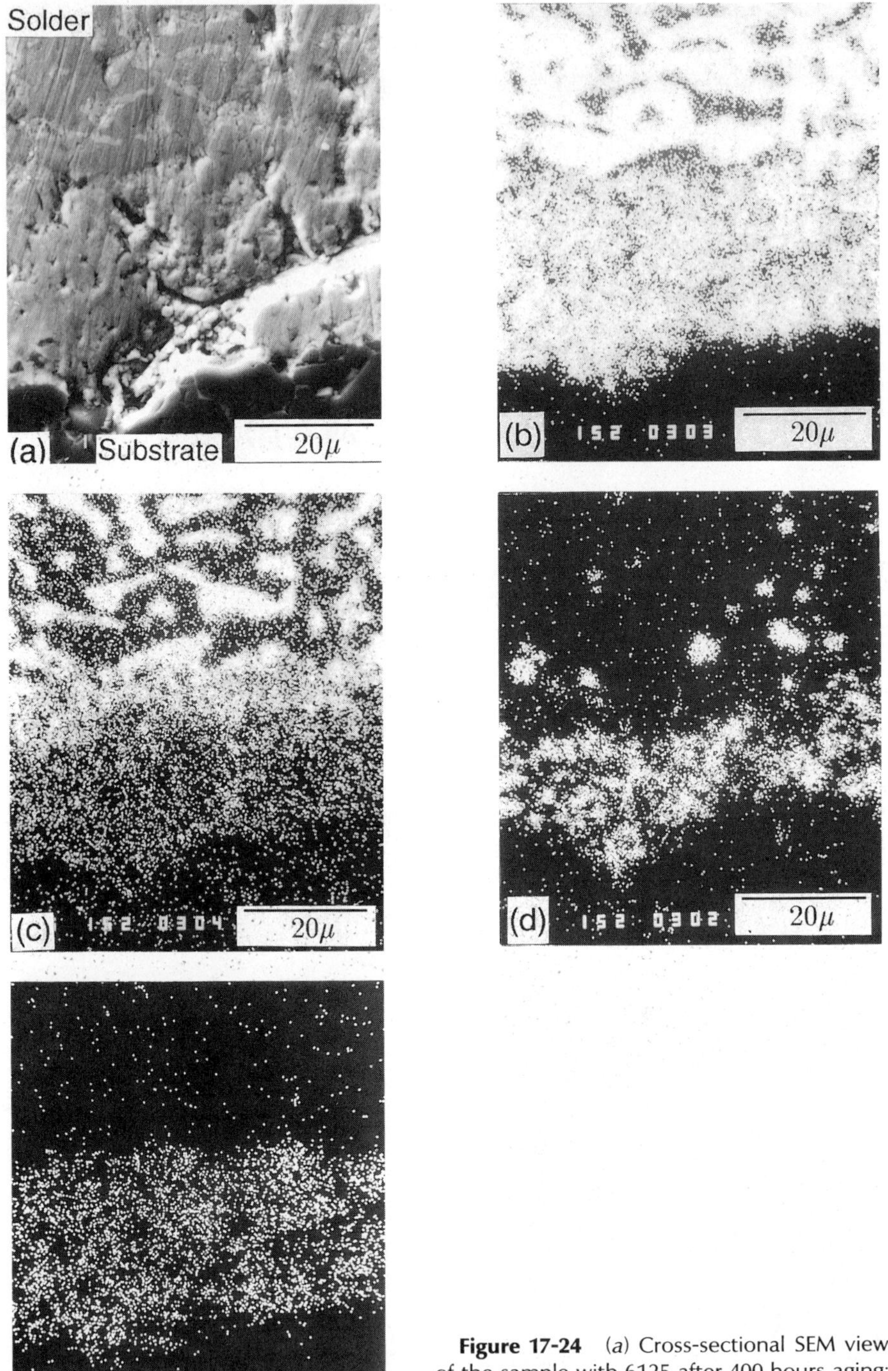

**Figure 17-24**  (*a*) Cross-sectional SEM view of the sample with 6125 after 400 hours aging; (*b*) Sn X-ray mapping; (*c*) Pb X-ray mapping; (*d*) Ag X-ray mapping; (*e*) Pd X-ray mapping.

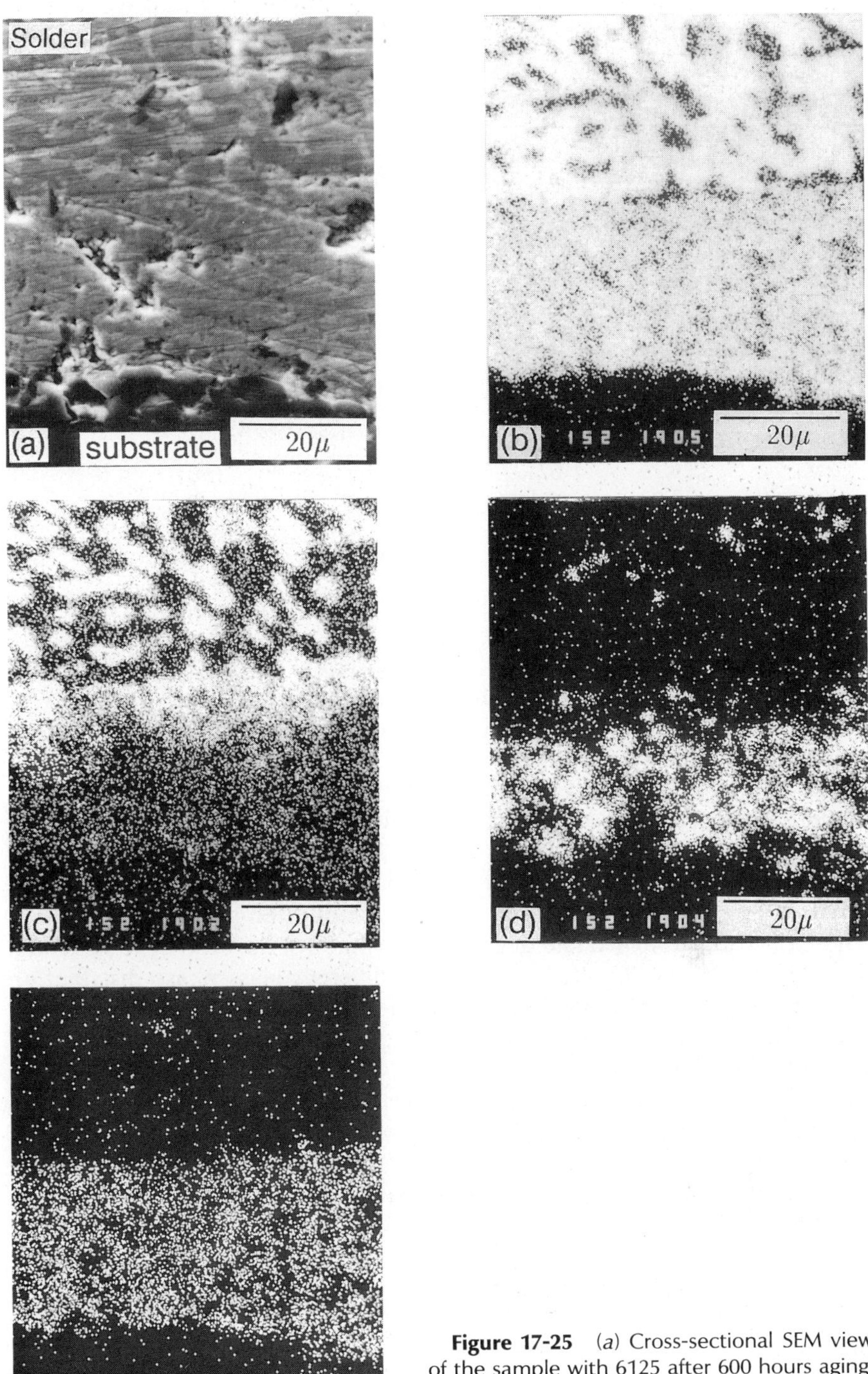



**Figure 17-25** (*a*) Cross-sectional SEM view of the sample with 6125 after 600 hours aging; (*b*) Sn X-ray mapping; (*c*) Pb X-ray mapping; (*d*) Ag X-ray mapping; (*e*) Pd X-ray mapping.

**Table 17-3**   Failure Mode of Conductor After Peel Test

| Aging Time (h) at 130°C | Strain Rate (mm/min) | 6134 | | | 6125 | | |
| --- | --- | --- | --- | --- | --- | --- | --- |
| | | Fracture Mode | | | Fracture Mode | | |
| | | A | B | C | A | B | C |
| 0 | 26.2 | | 12/12* | | 8/8 | | |
| | 2.28 | | 9/12 | 3/12 | 8/8 | | |
| 40 | 26.2 | | 8/8 | | 8/8 | | |
| | 2.28 | | 8/8 | | 6/8 | 2/8 | |
| 100 | 26.2 | 5/8 | 3/8 | | 6/8 | 2/8 | |
| | 2.28 | 4/8 | 4/8 | | 2/8 | 3/8 | 3/8 |
| 210 | 26.2 | 6/12 | 6/12 | | 6/8 | 1/8 | 1/8 |
| | 2.28 | | 8/8 | | 4/8 | 4/8 | |
| 402 | 26.2 | 7/8 | 1/8 | | 4/8 | 1/8 | 3/8 |
| | 2.28 | 4/8 | 4/8 | | 3/8 | 4/8 | 1/8 |
| 600 | 26.2 | | 8/8 | | 4/8 | 2/8 | 2/8 |
| | 2.28 | | 8/8 | | 2/8 | 4/8 | 2/8 |

* The value in the denominator indicates the total number of test samples, while that in the numerator is the number of samples failed as the specified mode.

intermetallic compounds is accompanied by a volume change. The adhesion of the conductor grain and reacted alumina substrate is destroyed by the volume change. As a result, the bowl-type structure is left after the peel test.

Figure 17-28 shows the fracture surface at the conductor–substrate interface for 6125, which exhibits a more complicated structure than 6134. The surface looks rugged and does not have much of the bowl-type structure. This difference may be attributed to the different glass phase distribution between 6125 and 6134 as discussed earlier. It should be pointed out that the glass phase is more extensive at the conductor–substrate interface for 6125 than for 6134.

Several investigators proposed the mechanism of adhesion loss. Loasby et al.[5] found that silver conductor film was swelled by the formation of intermetallic compound after tin diffused through it. Milgram[6] observed that tin diffusing to the conductor–substrate interface replaced the silver. In addition, a redox reaction

$$2Bi_2O_3(s) + 3Sn(s) \rightarrow 3SnO_2(s) + 4Bi(s) \qquad (17-4)$$

at the conductor grain boundary was reported by Taylor et al.[7] Recently, Liu[21] evaluated the binding energy of bismuth $4f_{5/2}$ electrons at the fracture surface of Dupont 6125 conductor specimens, unaged and aged for 40 hours,

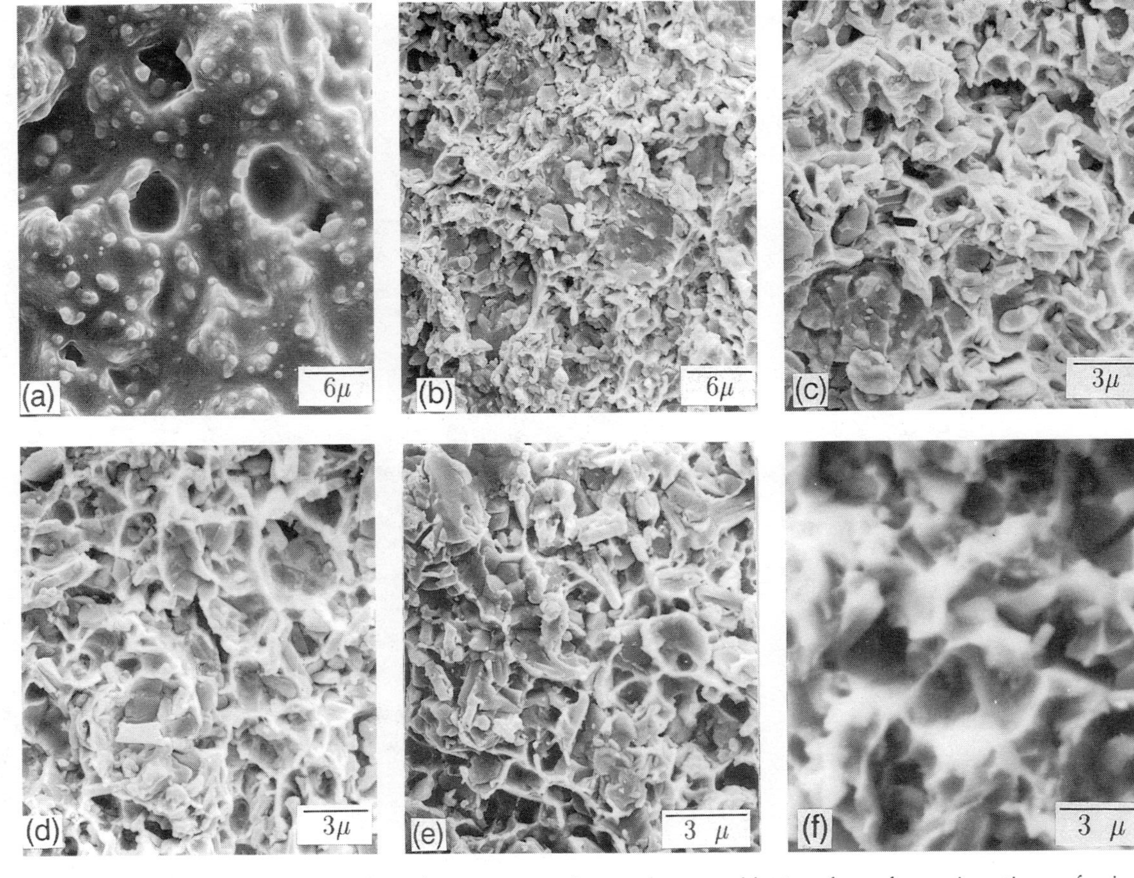

**Figure 17-26** Fracture surface of conductor 6134 at the conductor–solder interface after various times of aging at 130°C: (*a*) 0 hours; (*b*) 40 hours; (*c*) 100 hours; (*d*) 210 hours; (*e*) 400 hours; (*f*) 600 hours.

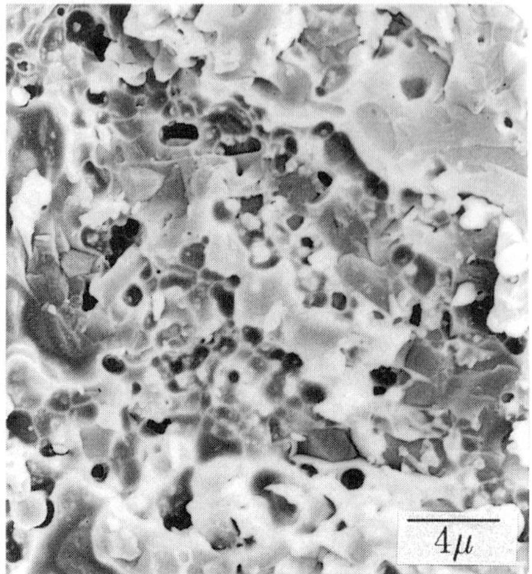


**Figure 17-27**  SEM micrograph of the fracture surface at the conductor–substrate interface for 6134.

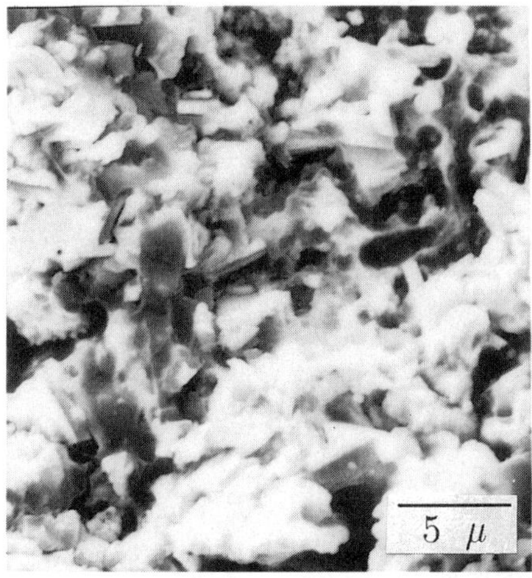


**Figure 17-28**  SEM micrograph of the fracture surface at the conductor–substrate interface for 6125.

using ESCA. The experimental values are 164.8 eV for both the unaged and aged samples after normalization to the C $1s$ line at 285 eV. The literature binding energy of the $4f_{5/2}$ electrons in bismuth oxide is 164.9 eV[26] and 163.8 eV,[27] where the former is normalized to the C $1s$ line but the latter is not. The literature value from ref. 26 is identical to the experimental result of Liu. This implies that bismuth retains the same state as the bismuth oxide after 40 hours of aging, and it appears that the initial adhesion loss does not result from the redox reaction.

The initial conductor film is face-centered cubic with lattice parameters about 4.017 Å and 4.049 Å for 6125 and 6134, respectively. The compounds possess different structures and lattice constants as listed in Table 17-4.[28] Swelling in the conductor film is inevitable, but another factor may be introduced in the Pd–Ag conductor: incoherency in the conductor film might be attributed to the formation of miscellaneous compounds. These two factors cause adhesion loss in the Pd–Ag conductor.

Chiou and co-workers proposed a mechanism of adhesion loss for Dupont 6134 conductor.[20] In the initial stage, the surface nonmetal constituents retained at the conductor–solder interface after soldering make fracture occur. The fracture mode is pure B mode, and the fracture interface is where the $Pd_3Pb$ exists, which is rather near the solder. After tin has diffused up to the conductor–substrate interface, the adhesion strength drops to a minimum value. Mode A fracture occurs after 100 hours of aging. Afterwards, both A mode and B mode fractures take place and the adhesion loss is rather slow. At 600 hours, the Ag-rich and Pd-rich separated regions are observed, and the fracture mode becomes pure B mode. The phases at the fracture surface are the same as those found before 600 hours of aging, and the adhesion loss takes place among the intermetallic compounds.

The mechanism of adhesion loss for Dupont 6125 conductor differs a little from that for Dupont 6134.[20] Initially, the fracture mode is pure A, and the phases analyzed by XRD show the Pd–Ag alloy. This implies that the initial strength is controlled by the glass phase at the conductor–substrate interface. After 40 hours of aging, the phases at the fracture surface are $Ag_3Sn$, $Ag_5Sn$, $Pd_3Sn$, $Pd_2Sn$, $Pd_3Sn_2$, and $PdSn$. The fracture mode becomes a different type, i.e., B mode, and the strength decreases rather fast. At this time, the

**Table 17-4**  Lattice Constants of Intermetallic Compounds[28]

| Compound | Structure | $a_0$ (Å) | $b_0$ (Å) | $c_0$ (Å) | $V_0$ |
|---|---|---|---|---|---|
| $Ag_3Sn$ | Orthorhombic | 2.995 | 5.159 | 4.781 | 73.87 |
| $Ag_5Sn$ | Hexagonal | 2.966 | | 4.782 | 34.63 |
| $Pd_3Sn$ | Cubic (FCC) | 3.97 | | | 62.57 |
| $Pd_2Sn$ | Orthorhombic | 8.11 | 5.662 | 4.234 | 194.42 |
| $Pd_3Sn_2$ | Hexagonal | 4.39 | | 5.655 | 94.38 |
| $PdSn$ | Orthorhombic | 3.86 | 6.12 | 6.31 | 149.06 |
| $PbPd_3$ | Cubic (FCC) | 4.024 | | | 65.16 |

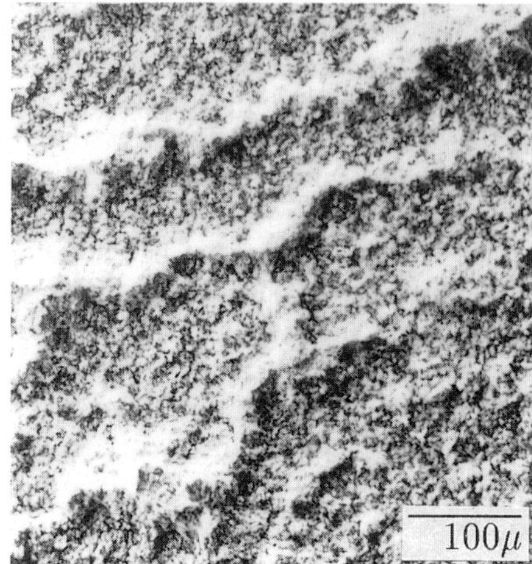

**Figure 17-29**   SEM micrograph of the fracture surface at the conductor–substrate interface for 6125 with the "river pattern."

**Figure 17-30**   SEM micrograph of the fracture surface at the conductor–substrate interface for 6134 without the "river pattern."

fracture surface exhibits a river pattern that is different from that for Dupont 6134, as shown in Figs. 17-29 and 17-30. It is argued that the incoherency between the compounds, i.e., Ag-rich and Pd-rich separated regions, is relatively large at this time, and the cracks propagate through the compounds during the peel test and leave the river pattern. When the aging time is extended, the grain growth of the compound, which is a time-related process, makes the incoherency more evident. The river patterns are always found at the fracture surface, and no dominating fracture type is observed, as shown in Table 17.3. The slow adhesion loss with longer aging may be attributed to the incoherency-enhanced process that results from the grain growth of compounds.

The adhesion loss of the thick-film conductor is initially abrupt, and is caused by the volume change accompanying the compound formation. However, it tends to be slow at longer times. The reason for the slow loss is not yet well understood. Although the redox reaction indicated in Eq. (17-4) was proposed,[5] the enhanced incoherency between the compounds seems to be the major effect.[21] Even for conductors with different Pd:Ag ratios, this effect is evident. The only difference is the variation in transition time for the incoherency to be observed.

### 17.5.2 Influence of the Phase Distribution in the Solder Joints

In real applications, the solder joint acts as a conductive path between the chips and the printed circuit boards (PCBs). The resistivity increase after aging does not favor the performance of equipment. For example, the power dissipation results in waste of energy, and the heat evolved makes the joint degradation more rapid. The resistivities of the intermetallic compounds formed between Sn–Pb solder and Pd–Ag thick film and of the raw metals are listed in Table 17-5.[29] The intermetallic compound always has greater resistivity than that of the conductive metals. In particular, the $Pd_3Pb$ layer, a Pb-rich layer formed across the conductive path, increases the resistivity drastically.

**Table 17-5**  Resistivities of the Intermetallic Compounds and Raw Metals[29]

| Compounds | Resistivity at 20°C ($\mu\Omega$-cm) |
|---|---|
| $Pd_3Sn$ | 32.4 ± 4.7 |
| $Pd_3Sn_2$ | 32.4 ± 4.7 |
| PdSn | 32.4 ± 4.7 |
| $Pd_3Pb$ | 32.4 ± 4.7 |
| Solder (63/67) | 14.6 |
| Pd | 10.8 |
| Ag | 1.59 |

## 17.6 THERMAL CYCLE EFFECT ON SOLDERED THICK FILM

In practical usage a temperature transient caused by the power dissipation or local heating would build up between the substrate and the conductor. The thermal deformation of the assembly could, in turn, affect the integrity of the solder joint. A simulated thermal cycle experiment is usually conducted to investigate the thermal effect.

The adhesion strength of soldered test pads with Dupont 6134 thick-film conductor dropped nearly to zero after 63 thermal cycles.[30] The thermal cycle was performed between $-55°C$ and $125°C$ with 25 minutes dwell time. Figures 17-31$a$ to $e$ show the elemental distribution of the cross-sectional view. Except for the transverse crack, the morphology is almost identical to that in the aging test as described in the last section. Tin diffuses into the conductor, and the bismuth-rich phase penetrates into the substrate to about $10\,\mu m$ depth. The crack resulting in the adhesion strength loss to zero propagates across the soldered thick film from the conductor–substrate interface up to the solder–conductor interface. The phases at the fracture surface are as complex as those in the aging test, as indicated in Fig. 17-32. $Pd_3Pb$, a face-centered cubic structure with lattice constant of about 4.204 Å, is found. As discussed earlier, $Pd_3Pb$ is the phase located at the interface near the solder. The crack propagates through $Pd_3Pb$ when it reaches the solder–conductor interface.

A typical fracture surface of the substrate side, as shown in Fig. 17-33, is divided into two distinct regions including the substrate exposed region and the conductor region. At the conductor region, the surface morphology is almost the same as the fracture surface of the aging test specimen at the solder–conductor interface. At the substrate exposed region, there exists the reacted alumina, similar to the type of fractured structure in the aging test at the conductor–substrate interface. However, in some areas, grains of alumina substrate can be observed as shown in Fig. 17-34$a$, along with microcracks at the grain boundary. Cracks exist at both the reacted alumina region and the interfaces between reacted and unreacted grains, as indicated in Figs. 17-34$b$, $c$, and $d$. The bismuth-rich phase is segregated at the alumina grain boundaries. It is argued that during the thermal cycles the thermal expansion coefficient mismatch between the alumina grain and the bismuth-rich phase produces the cracks. The bismuth-penetrated region in the thick-film conductor fails to bear the cyclic thermal shock between $-55°C$ and $125°C$.

Differences of the thermal expansion coefficients are always cited as the reason for the joint failure under thermal cycling. To simulate the performance of the solder and alumina under the temperature ramping part of the thermal cycle, a thermal dilatometer was used to measure the extent of the solder and alumina expansion with a fast temperature ramping rate.

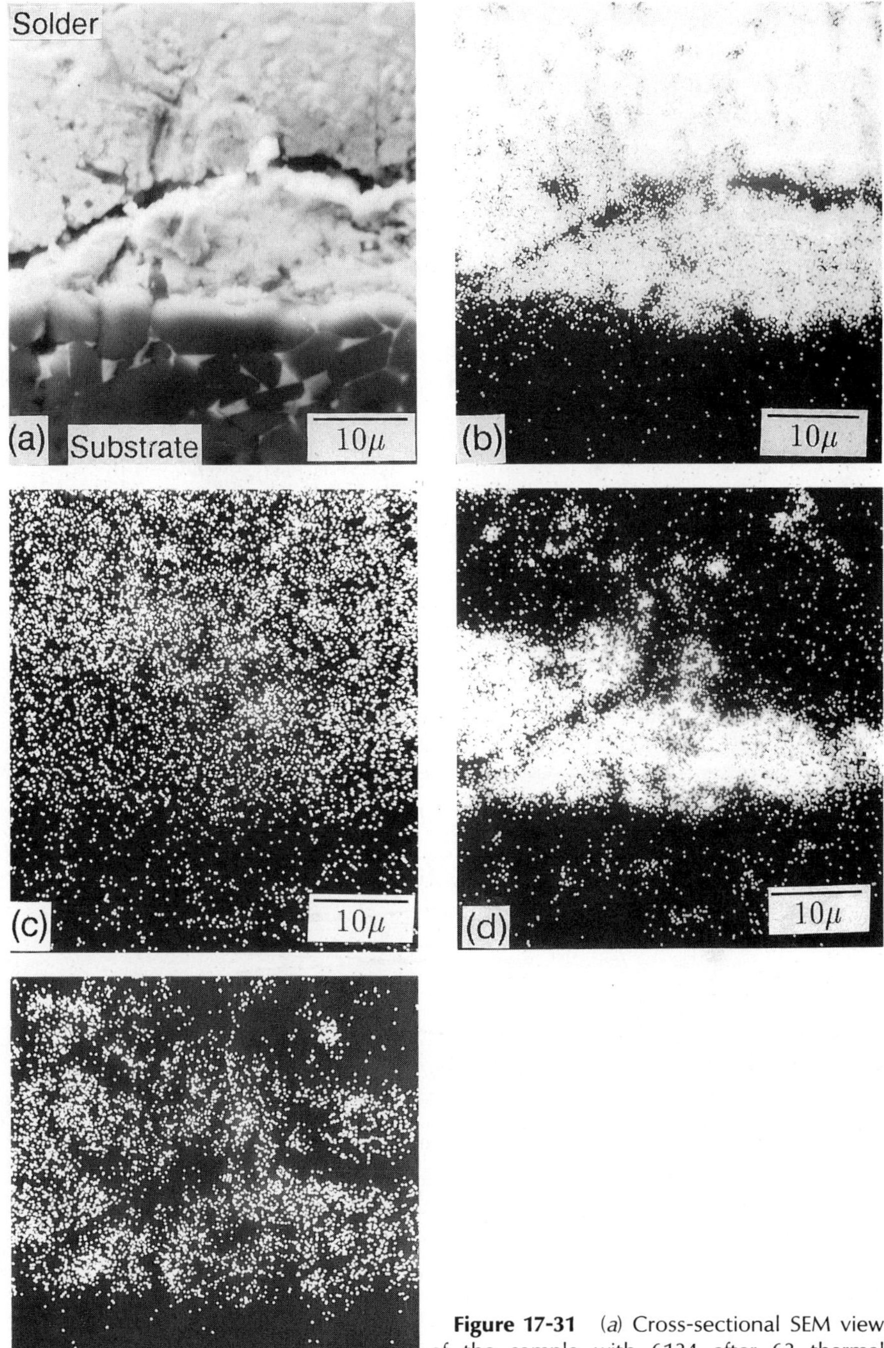

**Figure 17-31** (*a*) Cross-sectional SEM view of the sample with 6134 after 63 thermal cycles: (*b*) Pb X-ray mapping; (*c*) Pb X-ray mapping; (*d*) Ag X-ray mapping; (*e*) Pd X-ray mapping.

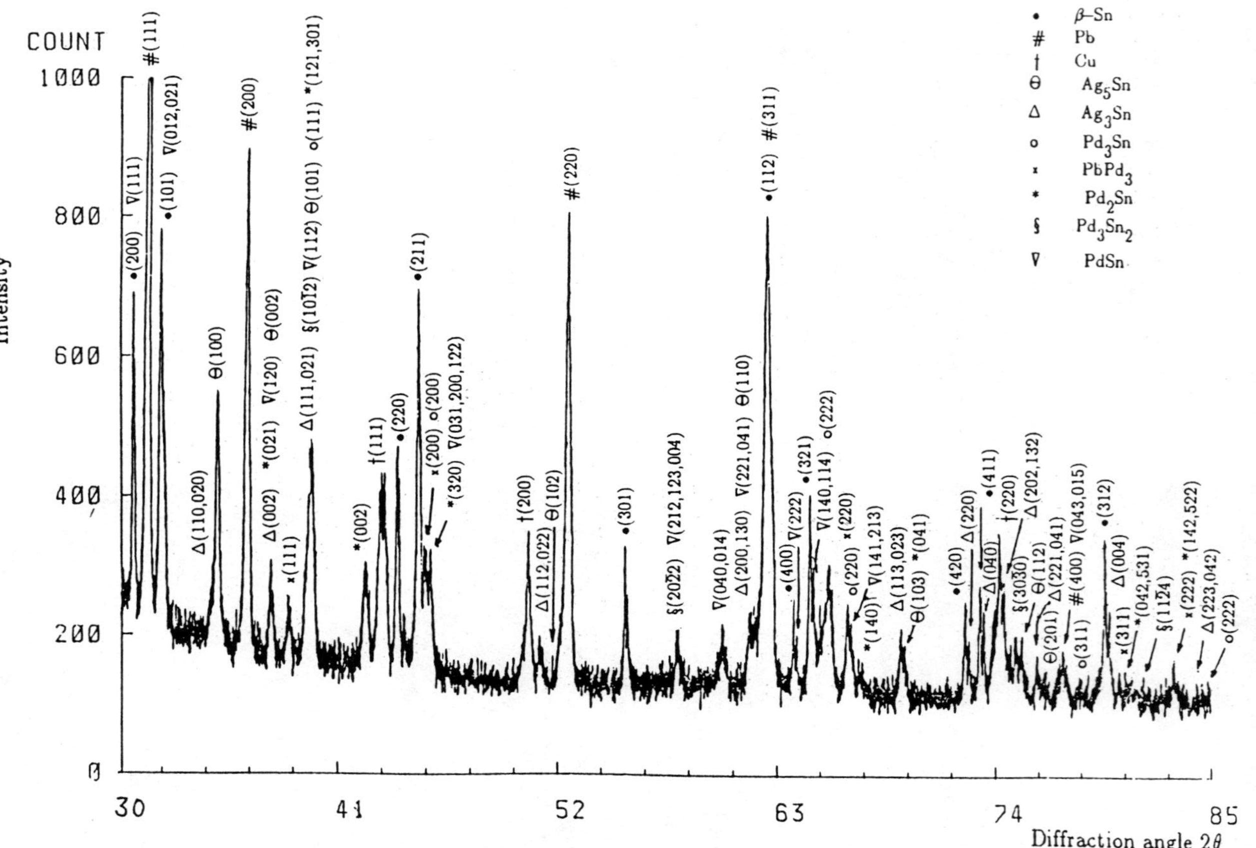


**Figure 17-32**  X-ray diffraction pattern of fracture surface for 6134 after 63 thermal cycles.

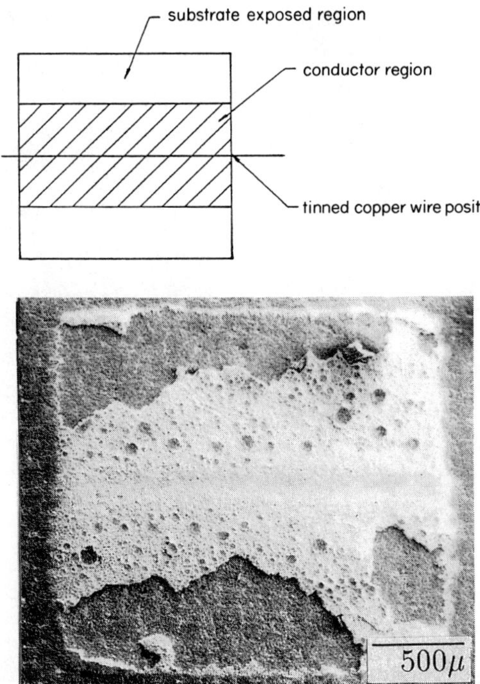


**Figure 17-33**    A typical view of fracture surface for the failed joint with 63 thermal cycles.

A 12.85-mm solder bar was tested at furnace temperature ramp rates of 4°C/sec and 15°C/sec, from room temperature to 120°C. A 13.55-mm alumina slice was tested with the same rates, up to 130°C.[30] The solder, with a larger thermal expansion coefficient and thermal conductivity as listed in Table 17-6, should have a faster temperature rise and greater thermal strain. Figures 35*a* and *b* indicate typical curves for temperature rise and expansion

**Table 17-6**    Thermal expansion coefficients of selected materials[18,31,32]

| Materials | Thermal Expansion Coefficient ($\mu/°C$) |
|---|---|
| Solder (63/37) | $24.7 \times 10^{-6}$ |
| $Al_2O_3$ | $8.8 \times 10^{-6}$ |
| Cu | $16.6 \times 10^{-6}$ |
| Pd | $11.8 \times 10^{-6}$ |
| Ag | $18.9 \times 10^{-6}$ |

**Figure 17-34** SEM micrograph at the substrate exposed region for the failed joint with 63 thermal cycles: (*a*) microcracks between exposed alumina grain; (*b*) cracks exist at the reacted alumina substrate; (*c*) cracks between the alumina grain and reacted alumina substrate; (*d*) cracks between the alumina grain and reacted alumina substrate with Bi-rich phase.

versus time. For the furnace temperature ramp rate of 4°C/sec, after a finite delay time, the solder exhibits an initial strain rate of $47.54 \times 10^{-6}$ μm/μm-sec and a temperature rise rate of 8.11°C/sec. For alumina, the initial strain rate and temperature rise rate are $11.04 \times 10^{-6}$ μm/μm-sec and 4.44°C/sec, respectively. For the temperature ramp rate of 15°C/sec, the alumina strain rate becomes $23.98 \times 10^{-6}$ μm/μm-sec, while it is $169.3 \times 10^{-6}$ μm/μm-sec for solder. This implies that if the environmental temperature changes very

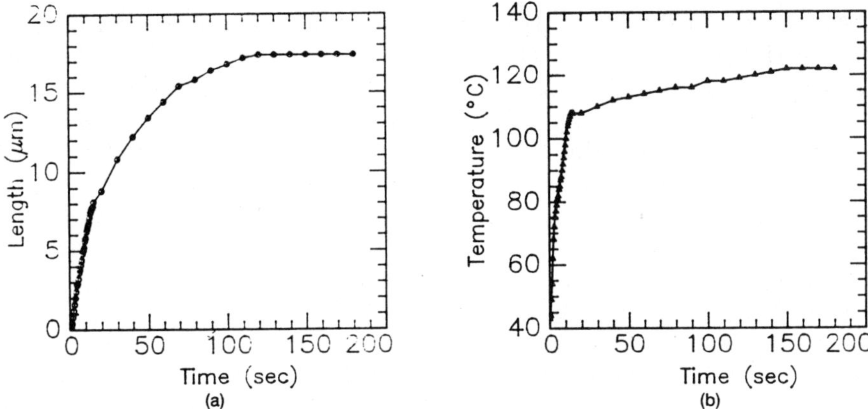

**Figure 17-35**   (*a*) Solder expansion in the thermal dilatometer with a furnace temperature ramp rate of 4°C/sec. (*b*) Solder temperature in the thermal dilatometer with a furnace temperature ramp rate of 4°C/sec.

fast, as in the ramping part and the falling part during the thermal cycle, the solder and alumina will expand and shrink at different rates. The more rapid the environmental temperature changes, the more enhanced is the strain rate difference.

Joint geometry is another factor that affects the joint's failure. The difference in thermal expansion coefficients result in stress, and the joint geometry defines where the stress is created. The inclined surface of the solder fillet at both sides of the test pads creates a stress with an elevation angle $\theta$, as shown in Fig. 17-36, which is resolved into tensile and shear components. Near the copper wire, the stress field becomes simply shear due to the thermal expansion coefficient misfit between the copper wire and the substrate.

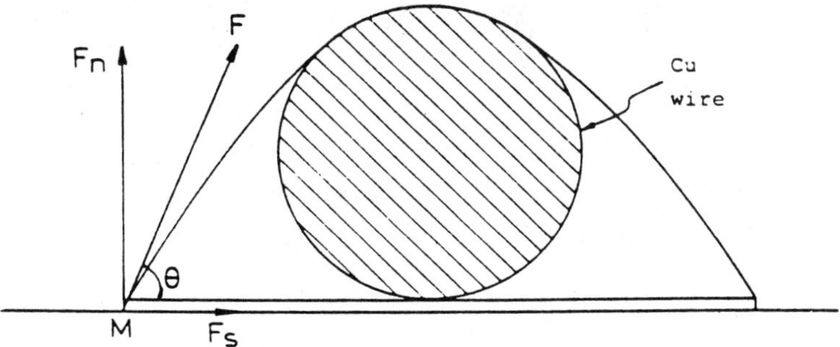


**Figure 17-36**   Schematic showing the solder shrinkage force, **F**, applied on the conductor–substrate interface. The force **F** can be resolved into the tensile component, **F**$_n$, and shear component **F**$_s$.

A top view of the test pads shows two distinct regions that correspond with various stress fields. The characteristics of the two regions, the conductor–substrate interface and the conductor–solder interface, have been discussed in an earlier section. A mechanism for the transverse crack initiation and propagation has been proposed.[30] The transverse crack initiates at the conductor–substrate interface where microcracks exist, and the tensile stress drives the crack to propagate along the conductor–substrate interface. When the tensile stress is reduced, the overall stress field is dominated by the shear stress, and the crack propagates up to the solder–conductor interface, as illustrated in Fig. 17-37.

Occasionally, the crack propagates down to the alumina substrate, as shown in Fig. 17-38. This is possible in about 10 percent of the total joints.[21] The occurrence of this condition depends on the accidental meeting of the transverse crack and the downward microcracks at the alumina grain boundary.

The strength of the interface, joint geometry, and the materials system determine the location where the failure takes place. For thick-film conductors, the interface strength is not strong enough to bear the cyclic stress, and joint failure occurs before solder fatigue is critical. In practical applications, approaches to designing an appropriate joint geometry, to improving the interface strength, and to selecting the materials system with a negligible thermal expansion coefficient mismatch will be crucial to increasing the lifetime of solder joints in thick-film electronics packages.

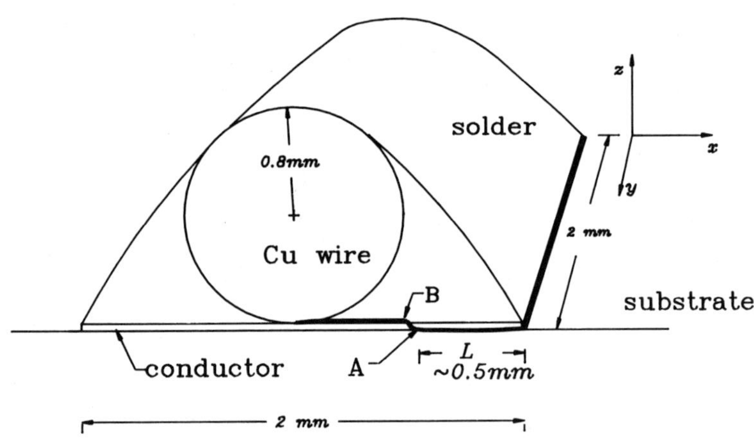


**Figure 17-37**   Schematic illustrating the crack configuration and the shear stress effect on the solder–conductor interface. Point A indicates the breaking point of the crack propagation at the conductor–substrate interface. Point B indicates the beginning of the crack propagation at the solder–conductor interface.

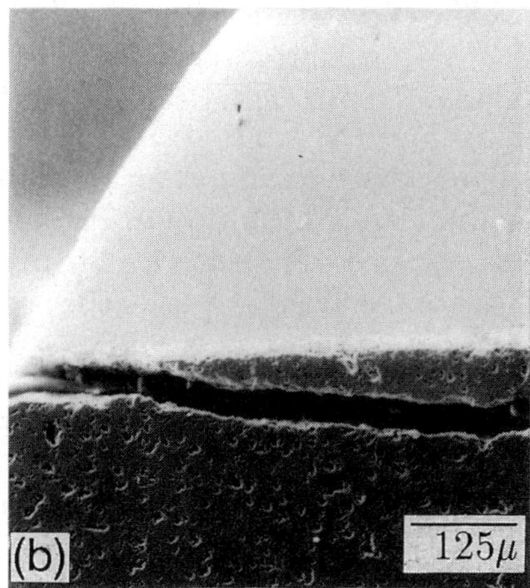

**Figure 17-38**   Transverse crack results in joint failure: (*a*) crack propagates along the conductor–substrate interface, then up to the solder–conductor interface; (*b*) crack propagates down to the substrate.

## 17.7 SUMMARY

In a simulated microelectronic hybrid circuit of Sn–Pb solder and Pd–Ag thick-film conductor metallization, microstructural evolution of the interfacial morphology, and elemental and phase distribution in the solder–conductor assembly were probed with the aid of electron microscopy and X-ray diffraction. There exists mechanical interlocking between conductor and substrate. Penetration of Bi from the oxide phase into the substrate is observed. Decrease in adhesion strength occurs when the sample is aged at 130°C for long periods. Microstructural analysis reveals the formation of intermetallic compounds $Pd_3Sn$, $Pd_2Sn$, $Pd_3Sn_2$, $PdSn$, $Pd_3Pb$, $Ag_5Sn$, and $Ag_3Sn$. Grain growth of the compounds after aging is also observed. A mechanism is proposed to explain the continuous adhesion loss after compounds are formed. It is argued that volume change of the film resulting from intermetallic formation and incoherency between the compounds due to grain growth are major factors in the degradation of the peel strength. Compound formation also occurs after 63 thermal cycles from −55°C to 125°C. A transverse crack propagates across the joint from the conductor–substrate interface to the conductor–solder interface, which makes the joint fail. The microstructural investigation and stress analysis reveal that the transverse crack is initiated by microcracks at the glass-penetrated region of the substrate and propagates under a tensile stress that results from solder shrinkage.

## ACKNOWLEDGMENT

This work was initiated at Delco Electronics, General Motor Corporation, Kokomo, IN, while one of the authors, B. S. Chiou, was on leave. She appreciates the support and friendship received from the Delco Group. The authors are grateful to Dr. P. Samy Palanisamy for his kindness in offering the research opportunity. Part of the work was supported by the National Science Council, Taiwan, under contract no. NSC 79-0404-E009-53.

## REFERENCES

1. Smith, G. C., and C. Lea, "Wetting and Spreading of Liquid Metals: The Role of Surface Composition," *Surface and Interface Analysis*, **9**, 1986, pp. 145–150.
2. Short, R. H., and G. P. Evans, "Soldering as a Forethought," *Assembly Eng.*, **3**, 1988, pp. 44–47.
3. Chadwick, R., "The Influence of Surface Alloying on the Strength of Soft Soldered Joint," *J. Inst. Metals*, **62**, 1938, pp. 277–295.
4. Hamer, D. W., and G. R. Sellers, "Chip Resistor Failures Associated with Solder Joint Degradation," *Proc. 1985 International Symposium on Microelectronics*, 1985, pp. 31–42.
5. Loasby, R. G., N. Davey, and H. Barlow, "Enhanced Property Thick-Film Conductor Pastes," *Solid State Technol.*, May 1972, pp. 46–50, 72.
6. Milgram, A. A., "Influence of Metallic Diffusion on the Adhesion of Screen Printed Silver Films," *Metall. Trans.*, **1**, 1970, pp. 695–700.
7. Taylor, B. E., J. J. Felten, and J. R. Larry, "Progress in and Technology of Low-Cost Silver Containing Thick-Film Conductors," *IEEE Trans. Comp. Hybrids, Manuf. Technol.*, **CHMT-3**(1), 1980, pp. 504–517.
8. Ohriner, E. K., "Intermetallic Formation in Soldered Copper-based Alloy at 150°C to 250°C," *Welding Res. Suppl.*, July 1987, pp. 191–202s.
9. Brothers, E. W., "Intermetallic Compound Formation in Soft Solders," *The Western Electric Engineer*, Spring/Summer 1981, pp. 49–63.
10. Gangulee, A., G. C. Das, and M. B. Bever, "An X-ray Diffraction and Calorimetric Investigation of the Compound $Cu_6Sn_5$," *Metall. Trans.*, **4**, 1973, pp. 2063–2066.
11. Nial, O., A. Almin, and A. Westgren, "Röntgenanalyse der Systeme Gold–Antimon und Silber–Zinn," *Z. Physikal. Chem. Abt. B.*, **138**, 1931, pp. 81–90.
12. King, H. W., and T. B. Massalski, "Lattice Spacing Relationships and the Electronic Structure of H.C.P. $\zeta$ Phases Based on Silver," *Phil. Mag.*, **6**, 1960, pp. 669–682.
13. Harris, I. R., and M. Cordey-Hayes, "A Study of Some Palladium–Tin, Silver–Tin and Palladium–Silver–Tin Alloy," *J. Less-Common Metals*, **16**, 1968, pp. 223–232.
14. Blum, P. L., J. Pelissier, and G. Silvestre, "An Investigation of Soldered Copper–tin Bond Brittleness by Electron Microscopy," *Solid State Technol.*, March 1973, pp. 55–58.

15. Duckett, R., and M. L. Ackroyd, "The Influence of Solder Composition on the Embrittlement of Soft-Soldered Joints on Gold Coatings," *Electroplating and Metal Finishing*, May 1976, pp. 13–20.

16. Waine, C. A., C. J. Brierley, and D. J. Pedder, "Thermal Fatigue Failure of Solder Joints in Printed Circuit Assemblies," *Reliability in Electrical and Electronic Components and Systems*, E. Lauger and J. Møltoft, eds., North-Holland, Amsterdam, 1982, pp. 231–235.

17. Muckett, S. J., M. E. Warwick, and P. E. Davis, "Thermal Aging Effects Between Thick-Film Metallization and Reflowed Solder Creams," *Plating and Surface Finishing*, January 1986, pp. 44–50.

18. Steen, H. A. H., and G. Becker, "The Effect of Impurity Elements on the Soldering Properties of Eutectic and Near-eutectic Tin–lead Solder," *Brazing & Soldering*, Autumn [11], 1986, pp. 4–11.

19. Frear, D., Grivas, D., and Morris, Jr. J. W., "A Microstructural Study of the Thermal Fatigue Failures of 60Sn–40Pb Solder Joints," *J. Electronic Materials*, **17**, 1988, pp. 171–180.

20. Chiou, B. S., K. C. Liu, J. G. Duh, and P. Samy Palanisamy, "Intermetallic Formation on the Fracture of Sn/Pb Solder and Pd/Ag Conductor Interfaces," *IEEE Trans. Comp., Hybrids, Manuf. Technol.*, **CHMT-13**(2), June 1990, pp. 267–274.

21. Liu, K. C., "Solder–Conductor Interaction and Dielectric–Substrate Interfacial Phenomenon in Hybrid Circuit," Master's thesis, National Tsing Hua University, Hsinchu, Taiwan, May 1989.

22. Duh, J. G., and K. C. Liu, "Microstructural Evolution in Sn/Pb Solder and Pd/Ag Thick Film Conductor Metallization," *41st Electronic Components and Technology Conference*, Atlanta, GA, 1991, pp. 653–657.

23. Karakaya, I., and W. T. Thompson, "The Ag–Pd (Silver–Palladium) System," *Bull. of Alloy Phase Diagrams*, **9**(3), 1988, pp. 237–243.

24. Cullity, B. D., *Elements of X-ray Diffraction*, 2d edn., Addison-Wesley, London, 1978, pp. 356, 412.

25. Klug, H. P., and L. E. Alexander, *X-ray Diffraction Procedures*, 2d edn., Wiley, New York, 1974, pp. 594.

26. Morgan, W. E., W. J. Stec, and John R. Van Wazer, "Inner-Orbital Binding-Energy Shifts of Antimony and Bismuth Compounds," *Inorg. Chem.*, **12**(4), 1973, pp. 953–955.

27. Debies, T. P., and J. W. Rabalais, "X-ray Photoelectron Spectra and Electronic Structure of $Bi_2X_3$ (X = O, S, Se, Te)," *Chem. Phys.*, **20**, 1977, pp. 277–283.

28. *A Handbook of Lattice Spacings and Structure of Metals and Alloys*, Vol. 2, W. B. Pearson, ed., Pergamon Press, Oxford, 1967.

29. Wopersnow, W., and C. J. Raub, "Eigenschaften Einiger Binärer Intermetallischer Phasen des Palladiums und Rutheniums mit Anderen Metallen," *Metall: Internationale Zeitschrift fuer Technik und Wirtschaft*, **33**(7), 1979, pp. 736–740.

30. Chiou, B. S., K. C. Liu, J. G. Duh, and P. Samy Palanisamy, "Temperature Cycling Effects Between Sn/Pb Solder and Thick Film Pd/Ag Conductor

Metallization," *IEEE Trans. Comp., Hybrids, Manuf. Technol.*, **CHMT-14**(1), March 1991, pp. 233–237.

31. Kingery, W. D., H. K. Bowen, and D. R. Uhlmann, *Introduction to Ceramics*, 2d edn., Wiley, New York, 1976, p. 595.

32. *CRC Handbook of Chemistry and Physics*, R. C. Weast, ed., CRC Press, Boca Raton, FL, 1980.

# 18

# Solder Joint Reliability of Leadless Chip Carriers

*Jim Lynch and Alberto Boetti*

## 18.1 INTRODUCTION

Ever since the introduction of the integrated circuit in the late 1950s there has been a continuous growth in the levels of complexity and density of functions in electronic systems. This has progressed in parallel with the density of function within such devices and with the degree of interconnection to other system levels. The two prime technology bases enabling interconnection between devices are the printed wiring (or printed circuit) board and the thick-film hybrid. The historical difference between printed wiring boards and hybrid technology was the method used to attach the integrated circuit and the method of protecting it from mechanical and environmental damage.

On traditional printed wiring boards the integrated circuit is packaged in a dual-in-line package and solder-attached into holes in the board. The advent of LSI and VLSI devices meant that the dual-in-line package was no longer appropriate as the packaging medium.

On the alumina substrate of a thick-film hybrid it has been normal practice to attach naked integrated circuit chips to the substrate and wire bond out to the screen-printed interconnections on the alumina. The whole circuit would then be hermetically sealed. This technology suffers from several disadvantages, one of which was the difficulty in procuring fully tested chips. Another is the reliability of interconnection, which was always suspect and became more so as the number of wirebonds on a hybrid increased.

The advent of surface mount technology and, particularly, the leadless chip carrier offered a potential solution to the problems of both the printed wiring board assembler and the thick-film hybrid manufacturer. The chip carrier has contributed a number of advantages. It allows pretestability, active device hermeticity, and good device interconnection without obviously sacrificing flexibility and without incurring extra investment or working costs.

Although industry was quick to seize upon the solution to tomorrow's interconnection problems, it still felt uneasy about one particular aspect of surface mount technology. The link between the device and its substrate is that all-important yet vulnerable constituent, the solder joint. Unlike conventional through-hole attachment where the device has a separate continuous lead of, say, Kovar that is only coated with the solder, surface mount technology places total reliance on a column of solder. This column acts as the medium for attachment (and flexibility), for electrical interconnection, and, in many cases, as a dissipator of heat from the device.

The first question to ask is what is the short-term reliability of the surface mount technology? In other words will the solder joints survive the initial mechanical and thermal excursions required of the printed circuit board or thick-film hybrid microcircuit. The answer to that question is relatively easily obtained. Of far more concern is the question of long-term reliability. This is a requirement not only of aerospace and military equipment but also of automotive and household goods.

A surface mount device in a satellite may see millions of thermal excursions over a period of years in space as the device powers up and down and as the ambient temperature is affected by orientation with respect to the sun. Military equipment using surface mount can be stored for long periods in desert or arctic conditions (or alternate between the two), can be subject to severe mechanical stresses during field commissioning, and then be used in earnest in the battlefield scenario, again under the most environmentally and mechanically taxing conditions.

Electronic circuitry has been used in household goods for some years and many of these goods now contain surface mount technology. Consumer awareness and ferocious competition between suppliers mean that manufacturers of such items as televisions and washing machines are now being forced to offer ever-increasing warranty periods on goods that are the subject of almost constant use, stringent duty cycles, and minimal respect from the user.

The surface mounted component is now in general use in the automotive industry. The environment created by the automobile and its user is extremely severe and, again, competition means that the solder joint must remain reliable for periods of several years while being subjected to a variety of combinations of stress such as heat, cold, shock, vibration, noxious chemicals, water, snow, or sand.

The ability of the supplier of substrates utilizing surface mount technology to predict long life for solder joints under such taxing conditions has been the subject of much research and not a little speculation. This chapter gives details of work carried out on various aspects of the solder joint reliability of surface mount technology. It looks primarily at leadless chip carriers mounted upon thick-film alumina substrates, but also makes comparison with leadless chip carriers mounted upon printed wiring boards and considers the use of leaded chip carriers.

## 18.2 INTRINSIC RELIABILITY OF THE CHIP CARRIER

### 18.2.1 Hermetic Leadless Chip Carriers

Early work on the reliability of the chip carrier needed to consider two quite distinct questions. First, was the chip carrier an adequate package for chip components, such as integrated circuits, and second, could it be reliably connected to motherboards, such as printed wiring boards or alumina hybrid substrates? As an example, the first question was answered[1] by subjecting integrated circuits packaged in three-layer ceramic leadless chip carriers to a test program based as closely as possible on an existing European Space Agency specification.[2] This program was derived originally for the qualification, endurance, and environmental testing of integrated circuits packaged in "conventional" packages, such as dual-in-line.

Seventy devices packaged in chip carriers were subjected to a series of subgroups of sequential testing. The tests included shock, vibration, acceleration, hermeticity, temperature cycling, thermal shock, moisture resistance, salt atmosphere, resistance to solvents, high temperature storage, electrical endurance, and solderability. The samples met the requirements of the test program and it was concluded that the test sequence under consideration was appropriate as a vehicle for testing leadless chip carriers and that the ceramic leadless chip carrier was inherently reliable as a packaging medium.

The information given above provides little more than a background to the overall purpose of this chapter, but one of the tests—solderability—is of more relevance than the others. Interconnection between a leadless chip carrier and its motherboard depends upon a solder pillar and the solderability of the surfaces at either end. The solderability of the leadless chip carrier is, therefore, an intrinsic consideration in the question of solder joint reliability.

Terminations of chip carriers are metallized with tungsten-based material finished with nickel- and gold-plated films. They should be solderable for long periods since the outer layer of gold acts as a protection against the environment. In order to ensure that solderability has been maintained, and to facilitate rapid wetting at assembly, leadless chip carriers are normally solder-dipped immediately before use. It is important to establish that this solder dipping process prior to attachment is controlled. Fears of brittle intermetallic compound formation in the joint at high gold concentrations can be minimized by leaching most of the gold with the large reservoir of molten solder in the bath used. Regular analysis of the bath will establish when the gold concentration becomes too high. Certain considerations are important. The relationship between the solder reservoir temperature and dipping time has been found to be important, since the thermally activated process of gold dissolution into the solder does not appear to be instantaneous.[3] Some chip carrier assemblers demand two reservoirs for the solder dipping process in order to ensure that the second dip removes any vestigial remnants of gold on the chip carrier's external termination.

The solderability tests performed on chip carriers are well defined by national and international standards and these need to be performed together with resistance to solder heat tests. However, it should be borne in mind that these tests need to be tailored to determine whether the chip carrier is able to withstand the individual pretinning and solder attachment processes that the chip carrier assembler is using.

### 18.2.2 Hermetic Leaded Chip Carriers

The use of leaded surface mountable devices is a compromise between through-hole leaded assemblies and the leadless chip carriers. By interposing a compliant lead between the chip carrier and the substrate, the thermal mismatch of package and board is no longer such an important consideration. Two styles of lead tend to predominate. Gull-wing leads spread out from the chip carrier, give a low board profile but take up more board space. J-leads curl under the package and thus take-up less board space but sit higher when mounted. These leaded versions have been shown to be perfectly adequate as packaging mediums and the only problem is that of ensuring that the leads do not get bent. The high pin counts and low space demanded by VLSI result in the larger packages having extremely flimsy leads that can be damaged at any time during manufacture or board assembly. There is no evidence to suggest that, once properly mounted, the leads of leaded chip carriers introduce any form of reliability hazard. Leaded chip carriers are normally found on printed wiring boards. However, as termination counts approach 200 there is a lack of suitable leadless chip carriers, and for this reason leaded chip carriers are found on the alumina substrates of thick-film hybrids.

### 18.2.3 Plastic Surface Mountable Packaging

The plastic integrated circuit package has faced a long uphill struggle to gain respectability within the electronics industry. There has always been concern over moisture-related problems and plastic surface mountable packages, the majority of which are leaded, share this stigma. One potential reliability hazard is specific to surface-mountable plastic devices.[4] Plastic-encapsulated devices, once molded and cured, may be stored for long periods of time under uncontrolled conditions. During this storage period the plastic can become saturated with moisture vapour. In itself this rarely does any harm to the device since most chips are well passivated before encapsulation. It is during the surface mount attachment process that the problem arises. At the point of reflow the package is suddenly subjected to temperatures in excess of 200°C with a steep profile. The entrapped moisture is unable to escape and the result is a severe pressure build-up that can cause rupture of the plastic bonds, cavity formation, chip bowing, and deformation of the copper lead frame on which the chip sits. Various secondary mechanisms can occur either immediately or much later, resulting in either yield problems or, worse, a reliability hazard.

This problem has been addressed by recommendations for modifications in package design[5] and standards to assess the susceptibility of plastic surface mountable components to the moisture-induced damage.[6]

## 18.3 THE RELIABILITY OF CHIP CARRIER SOLDER ATTACHMENT

### 18.3.1 The Philosophy of a Reliability Program

The quality of a solder joint can be tested in a number of ways. The reliability of a joint is much more difficult to assess. If quality is considered to be reliability at time zero, then the value of a quality-related test in assessing reliability is a question of whether an acceleration factor can be assigned. In most standard tests this is not possible. Cyclic humidity tests such as MIL STD 883, Method 1004 define one cycle and give a choice of the number of cycles. A few cycles can give an indication of quality, whereas a few tens of cycles can give some confidence of reliability provided the test actually does tend to cause degradation. Similarly, the shock, vibration, and bump tests of MIL STD 883 are intended to subject the device to the full range of conditions that it will meet during handling, transport, and use; they are not intended to simulate long-term mechanical exposure unless extended durations or numbers of cycles are prescribed.

The use of tests involving heat is more readily interpreted. Once a failure mechanism has been established and its activation energy determined, an acceleration factor can be established. This will allow a lifetime at elevated temperature to be calculated and then extrapolated to produce a lifetime under normal operating conditions. Unfortunately, one of the principal failure modes of joints on surface mountable devices occurs during temperature cycling or power cycling, and the acceleration factor is much more difficult to determine.

In order to investigate all aspects of the reliability of chip carrier solder joints on ceramic substrates it was felt necessary to consider a range of tests, environmental, mechanical, as well as thermal. The baseline for the program was to consider the chip carrier as a solder-attachable thick-film hybrid component. This allowed a series tests to be chosen from an existing European Space Agency Standard[7] designed to test the space-worthiness of components solder-mounted to thick-film hybrid substrates. Full details of the assessment have been reported elsewhere[1] but a summary is given below.

### 18.3.2 Test Vehicles for Chip Carriers Attached to Thick-Film Alumina Substrates

Two test vehicles were used. The first consisted of a 2 in. × 2 in. thick-film substrate containing six 24-lead, 0.050 in. pitch, three-layer ceramic chip carriers (Fig. 18-1). The second substrate (also 2 in. × 2 in. thick-film

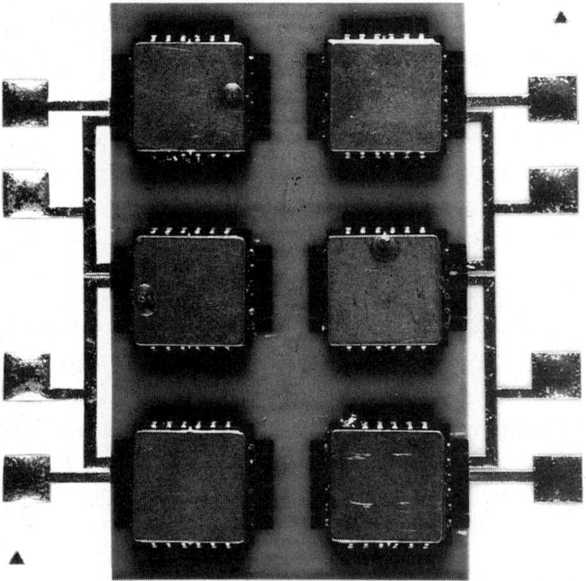

**Figure 18-1**    24-Lead chip carriers on substrate.

**Figure 18-2**   64-Lead chip carriers on substrate.

alumina) contained a single 64-lead, 0.040-in. pitch, three-layer ceramic chip carrier (Fig. 18-2). The chip carriers were mounted on the substrates so that the resistance of the series connection of all the terminals could be measured. This was done by alternately connecting together the internal terminals by a gold wire daisy chain. None of the chip carriers contained a chip. The 2 in. × 2 in. substrates were processed by standard thick-film procedures with platinum–gold paste and the chip carriers were attached by 62/36/2 Pb/Sn/Ag screen-printed solder cream followed by reflow.

### 18.3.3 Test Program and Results

**Subgroup I: Control**   No environmental stresses were placed upon the chip carrier hybrids, but half of the chip carriers were subjected to a 10 kg shear and then all chip carriers were torque-wrenched from the substrates. This test, in conjunction with resistive continuity electrical measurement and visual inspection, was used as a post-test end-point for subgroups II–V. A typical solder joint is shown in Fig. 18-3.

**Subgroup II: High-Temperature Storage**   Substrates were subjected to 1000 hours storage at 150°C without bias. At the end of the test, electrical

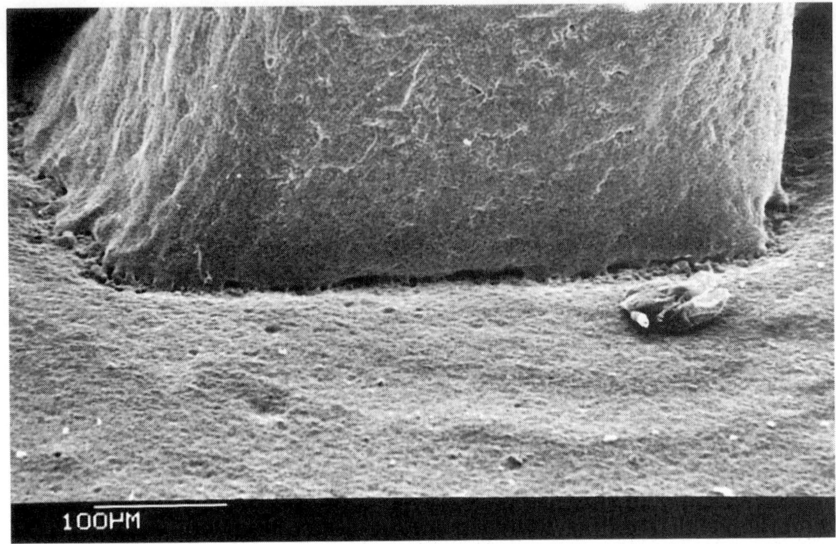

**Figure 18-3**   Control.

integrity was seen to have been maintained but visual inspection, SEM, and microsection showed marked evidence of microstructural coarsening in the chip carrier solder joints (see Fig. 18-4) and intermetallic formation between the solder and the gold and nickel layers of the chip carrier solder pads. There was no failure at shear test, but torque wrenching to destruction gave values consistently 35% lower than in the Control subgroup.

*Subgroup III: Humidity Steady State*   Substrates were subjected to 540 hours of steady-state humidity at 40°C and 90–95% RH without bias. The test procedures caused no deterioration compared with the Control subgroup.

*Subgroup IV: Temperature Cycling*   The test vehicles were temperature-cycled between −65°C and 150°C, 1000 times, with a ramp of 5 minutes and a dwell of 10 minutes. The post-test examinations showed no electrical degradation but some microcracking at the base of the solder joints. An example is shown in Fig. 18-5. There were no failures at shear test and torque test results were consistently 20% lower than in the Control subgroup.

*Subgroup V: Humidity Cycling*   The chip carriers on hybrids were temperature-cycled between 25°C and 65°C over a 24-hour cycle with humidity at approximately 90% RH. The cycle was repeated 10 times, with

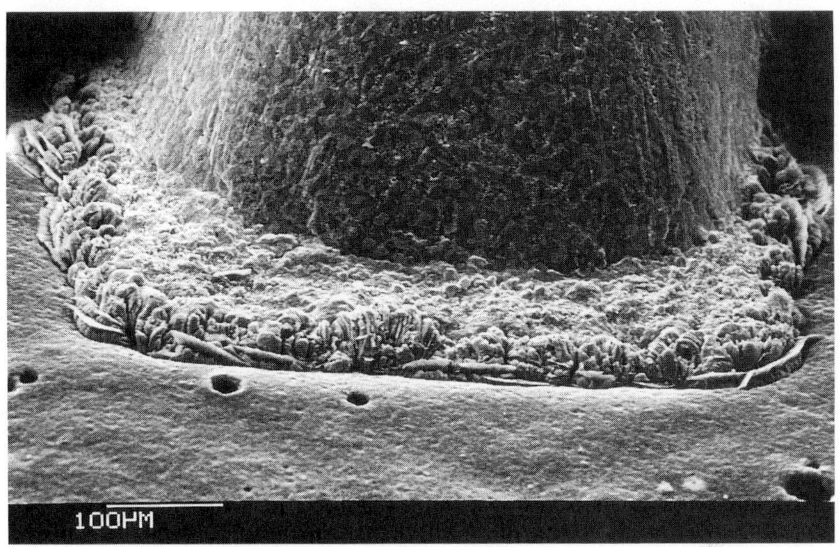

**Figure 18-4**    High-temperature storage.

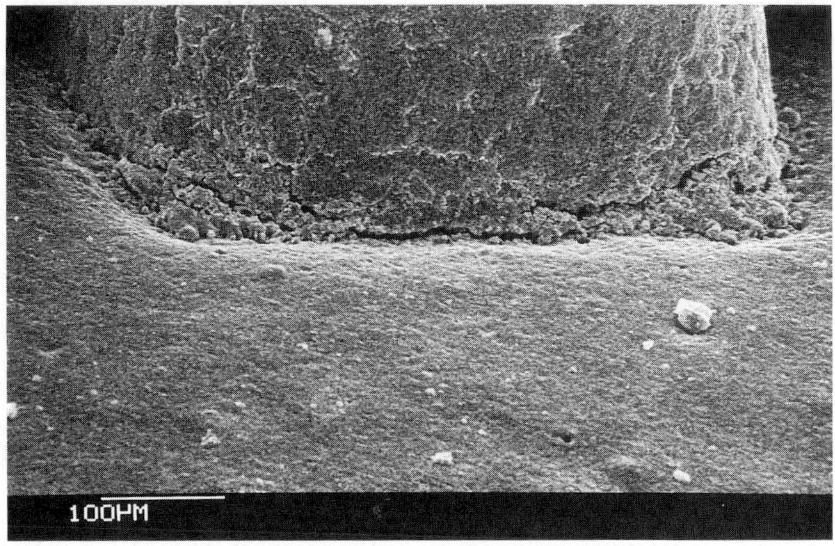

**Figure 18-5**    Thermal cycling.

3-hour excursions to $-10°C$ on any 5 of the 10 cycles. No deterioration compared with the Control subgroup was observed.


### 18.3.4 Interpretation of Results

All of the chip carriers attached to the thick-film substrates survived all of the various tests without change in electrical resistance.

There was some change in the chip carrier adherence as measured by torque testing in the two temperature-related tests but it can be concluded that exposure to humidity, whether under steady-state conditions or coupled with mild temperature cycling, has no effect. Storage at 150°C and temperature cycling between $-65°C$ and $+150°C$ both cause a deterioration of the torque strength that can be analyzed by the use of standard structural analysis techniques. The failure mechanism can be attributed, in the main, to a decrease in solder bond strength caused by grain growth, cracking at the castellation interface, and by intermetallic formation between the tin of the solder and gold of the substrate and chip carrier solder pads. As a crude approximation, when comparing the solder joints after temperature cycling and high-temperature storage, the strengths appear to depend upon the total time at high temperature rather than temperature variation.

The torque test results show a quantitative agreement with shear test results performed on chip carriers solder-attached to thick-film hybrids (e.g., ref. 8). The results are also of the correct order when compared with theoretical calculations of the torque strength based on the strength of the solder itself.[9]

The overall conclusion of this particular set of tests indicates that three-layer chip carriers can be reliably attached to thick-film hybrid substrates and that the test plan used forms a good basis for specifying chip hybrid attachment for high-reliability applications.


### 18.3.5 Comparison with Chip Carriers Solder-Attached to Printed Wiring Boards

Much work has been done on the subject of chip carrier attachment to printed wiring boards, the main objective being to discover whether the solder joint can reliably sustain the thermal mismatch between the ceramic carrier and normal glass-fiber-reinforced polymer (FR4). Brierley et al.[10] have assessed previous results and compared them with their own.

The conclusion is that small chip carriers (up to 24 leads) on FR4 can withstand temperature cycling in general accordance with MIL STD 883, Method 1010 to the same degree as those on the ceramic thick-film hybrids, detailed above. However, as the termination count on the chip carrier increases, the solder rapidly becomes less able to sustain the thermal mismatch between the FR4 and the larger area of chip carrier. Thus, Brierley shows that a 16-lead chip carrier on FR4 can withstand almost 1000 cycles between $-55°C$ and $+125°C$, whereas a 68-lead chip carrier's solder joints fail soon after 100 cycles.

As Brierley points out, the mechanical stresses created by the temperature coefficients of expansion between the alumina of the chip carrier and the polymer of the printed wiring board are virtually all relieved by the plastic strain in the eutectic tin–lead solder. These stresses are highly destructive and the solder joint succumbs to a mechanism of low-cycle fatigue failure in accordance with the Coffin–Manson equation.[11]

Brierley then goes on to consider a number of alternative substrates. Polyimide Kevlar material was rejected on the basis that microcracks, formed in the laminate during thermal cycling, propagated to the surface and resulted in copper track breakage. An otherwise conventional FR4 coated with elastomer prior to copper foil cladding proved cumbersome in application and gave little reliability improvement. However, both polyimide quartz and copper–invar–copper cored materials gave a high degree of reliability, with the metal-cored laminate being preferred because of its better rigidity and improved thermal conductance. There is, however, a cost penalty to be paid since copper-clad invar is significantly more expensive than FR4.

## 18.4 POWER CYCLING

### 18.4.1 The Value of Power Cycling

The tests described above showed that temperature cycling and high-temperature storage caused degradation of the solder joints of ceramic chip carriers. For chip carriers attached to ceramic hybrid thick-film substrates this was significant but not catastrophic under the conditions chosen. Temperature cycling has for many years been considered one of the most important environmental tests in the assessment of electronic components. It is normally performed between the maximum and minimum temperature of operation of the component. Under the chamber method the whole environment is at one temperature at any one time and, thus, so are the jointed components under assessment. For a ceramic chip carrier on a ceramic substrate the temperature cycle is testing the ability of the compliant solder joint to withstand the marginal thermal mismatch between two pieces of very similar ceramic.

As integrated circuits become bigger and geometries become smaller (and more densely packed) the heat output can become considerable. A chip carrier with external dimensions in excess of 1 in. (25 mm) square with over one hundred solderable interconnection pads may be required to dissipate several watts of power. Under these circumstances there is a temperature gradient from the package to the soldered castellations and from the castellations to the substrate. This increases the effective thermal mismatch between chip carrier and substrate, and if the device is switched on and off there is the potential for induced temperature cycling to take place. This is called power cycling, and as the device is switched the solder joints can experience thermal fatigue cycles that are much more severe than those generated by conventional chamber methods.

The importance of power cycling and its simulation as a test method was pointed out by Howard et al.[12] and Engelmaier[13] who calculated that, unless precautions were taken, it was possible for failure of a solder joint to take place within a finite number of power cycles.

The following sections show the development of a test program designed not only to evaluate the importance of power cycling but also to make a direct comparison with temperature cycling and high-temperature storage performed in chambers. The work is described in more detail elsewhere.[14]

### 18.4.2 Derivation of Test Program

The determination of test conditions considered primarily the solder joint linking the chip carrier to the substrate. For high-temperature storage in a chamber, the whole assembly and, therefore, the solder joints were held at a temperature of 150°C for 1000 hours. Accordingly, it was decided to generate heat within the chip carrier so that the solder joints rose to a temperature of 150°C. Similarly, temperature cycling generates a temperature of 150°C for 10 minutes, a fall to −65°C over 5 minutes, dwell at the lower temperature for 10 minutes, and then return to 150°C over 5 minutes. This process was repeated 1000 times in the previous series of tests. Again, for the power-induced study the upper limit was set at 150°C, measured at the solder joint. No attempt was made, however, to repeat a lower temperature of −65°C at the solder joint. Instead it was considered sufficient to define the laboratory ambient of 20°C that would be realized in a power-off situation. The dwell time at temperature extremes was kept at 10 minutes and time for changeover maintained at 5 minutes. Although it proved possible to raise the temperature of the solder joint to 150°C within 5 minutes by using the power available within the chip carrier, natural cooling failed to return the temperatures to laboratory ambient once the power had been switched off. It therefore proved necessary to use forced convection, and a cooling fan was linked into the circuitry.

The apparatus required for high-temperature storage consisted simply of the requisite number of power supplies with leads soldered to the input pads on the thick-film substrates. A custom-designed circuit was used to produce the temperature cycle. The circuit is shown in Fig. 18-6. Oscillator IC1 produces a square wave of 1.1 Hz which is reduced in frequency by a factor of 1024 by the counter IC2, thus changing the state of its output every $1024/(1.1 \times 60)$ min., i.e., every 15 minutes. This output is amplified by IC3 and used to drive the relay S1 which in turn switches off the power supply when energized or the cooling fan when not energized. The arrangement gave a heat-up period of about 4 minutes and cooling to ambient within 3 minutes.

The substrates and chip carriers used for the investigation were exactly the same as had been used for the chamber work above (viz., $6 \times 24$-lead chip carrier on a 2 in. $\times$ 2 in. substrate and $1 \times 64$-lead chip carrier, also on a 2 in. $\times$ 2 in. substrate). The heating elements used in the chip carriers were specially prepared thick-film resistors on alumina. In the case of the 24-lead chip carrier a single 5-ohm resistor was found to be adequate, whereas for the 64-lead chip carrier a small array of two pairs of 5-ohm resistors in series–parallel (i.e., a net of 5 ohms) was required to provide the necessary power handling.

The relationship between power output and joint temperature was investigated in a series of preliminary experiments. Thermocouples were placed in the solder joint prior to initial reflow. A series of measurements was taken of electrical input and joint temperature and the overall thermal stability of the system was established; no significant drifting was observed. Measurements of temperature were also taken at various points on the substrate and the temperature attained by the thick-film resistors was evaluated from the accurately known temperature coefficient of resistance of the thick-film resistor ink compared with resistance values obtained by heating the resistors in an oven. Each substrate used in the main test program was calibrated individually to ensure that the correct power input was used to produce the required joint temperature.

### 18.4.3 Test Plan and Results

The effects of high-temperature storage and temperature cycling produced by power heating were assessed as follows.

*Subgroup I (Six Substrates of Each Type)*
Step 1:    *Control*: (No conditioning)
Step 2:    *Visual inspection*: Examination to specified hybrid visual standards.

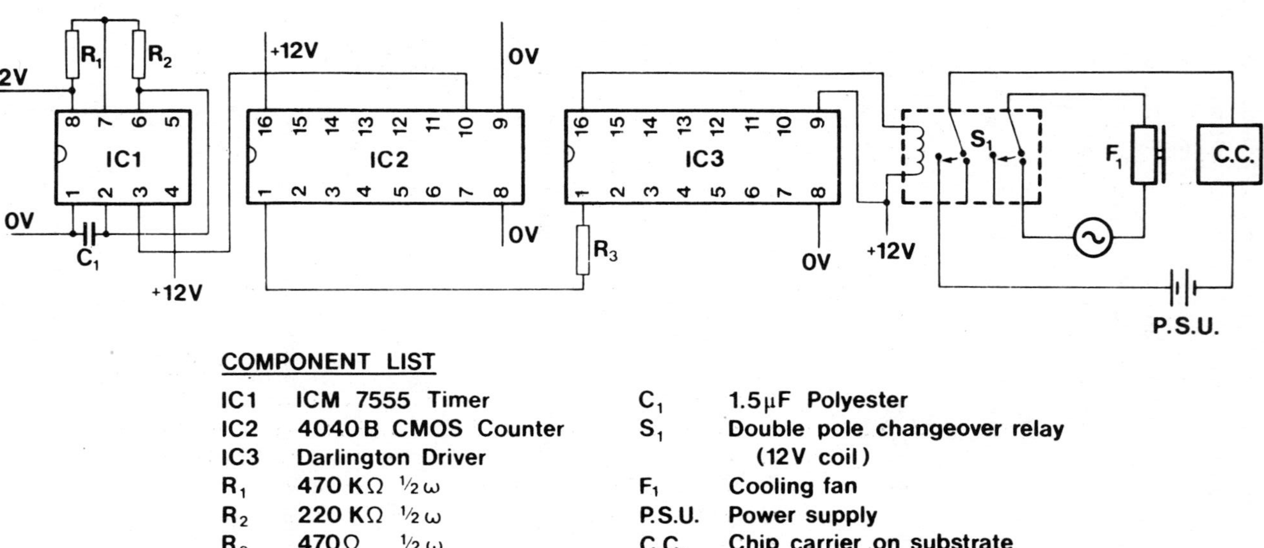


## COMPONENT LIST

| | | | |
|---|---|---|---|
| IC1 | ICM 7555 Timer | $C_1$ | 1.5 μF Polyester |
| IC2 | 4040 B CMOS Counter | $S_1$ | Double pole changeover relay (12V coil) |
| IC3 | Darlington Driver | | |
| $R_1$ | 470 KΩ ½ ω | $F_1$ | Cooling fan |
| $R_2$ | 220 KΩ ½ ω | P.S.U. | Power supply |
| $R_3$ | 470 Ω ½ ω | C.C. | Chip carrier on substrate |

**Figure 18-6**  Circuit for power cycling.

Step 3:   *Chip carrier adherence*: All chip carriers of all but one substrate of each type subjected to torque test to destruction.

Step 4:   *SEM and microsectioning*: One substrate of each type examined by SEM and then by microsectioning.

*Subgroup II (Six Substrates of Each Type)*

Step 1:   *High-temperature storage*: 150°C (chip carrier-to-substrate solder joint temperature) power-on heating for 1000 h with substrates in free air at ambient of 20°C.

Step 2:   *Visual inspection, Chip carrier adherence, SEM and microsectioning*: As in subgroup I.

*Subgroup III (Six Substrates of Each Type)*

Step 1:   *Temperature cycling*: 20°C for 10 min, heat to 150°C over 5 min, maintain for 10 min, return to 20°C over 5 min. Repeat 1000 cycles. All temperatures refer to chip carrier-to-substrate solder joints power heating with substrates in free air at ambient of 20°C.

Step 2:   *Visual inspection, Chip carrier adherence, SEM and microsectioning*: As in subgroup I.

Chip carrier adherence, in all cases, was assessed by torque testing. The ceramic substrates were rigidly clamped and the chip carriers were torque-tested to destruction as previously.

Chip carriers in subgroup I, the control, separated at an average torque of 3.3 N m for the 24-lead and 16 N m for the 64-lead, which compares well with the controls used in the previous chamber tests.

In subgroup II, following 1000 hours of power heating at a solder joint temperature of 150°C, all 12 substrates were still electrically operational. Visual and SEM examination showed that the joints resembled those heated to 150°C for 1000 hours in the chamber tests. Torque test results showed a drop in strength to 1.9 N m for the 24-lead and 9 N m for the 64-lead, in both cases a deterioration of about 40% compared with control. A similar deterioration was observed with the chamber tests.

On completing 500 of the 1000 power cycles of subgroup III it became apparent that the solder joints were showing clear signs of damage, even to the naked eye. Electrical probing showed an open-circuit condition on two of the 64-lead chip carriers and these were removed from test. All other substrates continued to 1000 cycles with no electrical failures noted. The early failures were subjected to structural analysis by SEM and microprobe. This examination showed severe surface cracking and some of the cracks had extended inwards sufficiently to create an open-circuit condition. The joints of the early failures were found to be typical of all joints on substrates of this subgroup showing extensive cracking at the lower portion of the joint

**Figure 18-7**  Power cycling.

and a degree of thermal etching indicative of deformation. An example is shown in Fig. 18-7 and a metallographic section, showing a fully propagated crack, is shown in Fig. 18-8.

The torque test results gave an average of 1.7 N m for the 64-lead chip carriers, a drop in strength of 90% compared with control, and 1.1 N m for the 24-lead chip carrier, about 70% deterioration compared with control.


### 18.4.4 Interpretation of Power Cycling Results

The 1000-hour high-temperature storage with the solder joints held at 150°C in a laboratory ambient of 20°C gave similar results for both the 24-lead and the 64-lead chip carriers. The torque strength was 40% lower than controls and examinations by cross-section and SEM showed microstructural coarsening caused by the high-temperature exposure. This coarsening probably resulted in the lower torque strength. The results and interpretation are very similar to those obtained from the high-temperature storage using the chamber method. It seems reasonable to conclude that at the constant elevated temperature of 150°C the joint strength deterioration is primarily a function of its temperature and dwell time. It is worth noting at this point that even after such a severe test the joints are still acceptably strong.

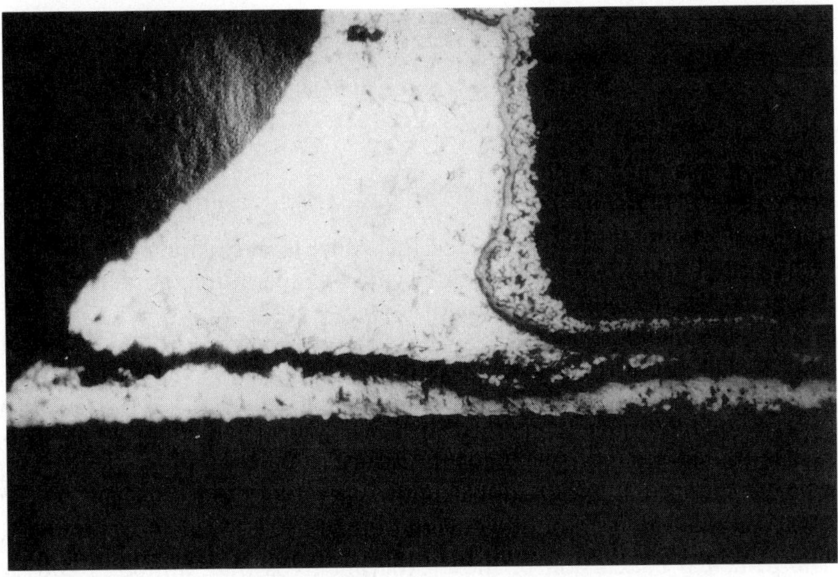

**Figure 18-8**    Section of solder joint.

Any resemblance between power heating and chamber heating ends abruptly when the cycling subgroups are compared. The lowering of torque strength compared with controls was very much greater for the power cycling than for the chamber cycling, and in the case of two of the 64-lead chip carriers the joints went open-circuit before the completion of the test. The reason for this significant difference between chamber cycling and power cycling is the direct result of the thermal gradient between the chip carrier and the substrate through the solder joint. If we consider the first-order equation of Engelmaier,[13] the thermal expansion mismatch $\Delta\varepsilon$ between the chip carrier and substrate is given by

$$\Delta\varepsilon = \alpha_c(T_c - T_o) - \alpha_s(T_s - T_o)$$
$$= (\alpha_c - \alpha_s)(T_c - T_o) + \alpha_s(T_c - T_s)$$

where  $\alpha_c, \alpha_s$ = coefficients of linear thermal expansion for chip carrier and substrate, respectively

$T_c, T_s$ = temperature of chip carrier and substrate, respectively

$T_o$ = power off steady-state temperature.

For the case of a ceramic chip carrier on a ceramic substrate temperature-cycled in an oven,

$$\alpha_c \cong \alpha_s \qquad \text{and} \qquad T_c = T_s$$

and so

$$\Delta\varepsilon \cong 0$$

As discussed above, the net effect on the solder joint can then be considered as the time that it actually spends at elevated temperature and deterioration in strength is caused by grain growth.

When the solder joint is power-cycled, $T_c \neq T_s$ and $\Delta\varepsilon \neq 0$. This results in a tendency for joint displacement, which is maximized at the corners of the chip carrier and is worse the larger the physical size of the chip carrier. At 150°C the solder alloy is at approximately 0.9 of its absolute melting point with a low yield stress. The thermal mismatch is thus relieved by plastic deformation of the joint and this results in the phenomenon of low-cycle fatigue based on the Coffin–Manson equation.[11]

The principal conclusion of these experiments is that, although conventional chamber cycling is an important technique, the use of power cycling can give much more information. The value of the conditions used in the power cycling experiments does, however, need examination. Although components used in chip carriers may generate several watts of energy, it is unlikely that the designer will permit the internal temperature of the chip carrier to rise above 150°C, which is generally considered to be the maximum junction temperature of silicon integrated circuits. The relationship between output power of the chip carrier and joint temperature was reinvestigated and it was found that an internal chip carrier temperature of 150°C provided a joint temperature on the 64-lead chip carrier of approximately 100°C. This provided the basis for a supplementary set of experiments.

The samples chosen consisted of seven standard 2 in. × 2 in. substrates with a 64-lead chip carrier, five of which were power-cycled between laboratory ambient of 20°C and a power-on temperature of 100°C measured on the solder joint. The remaining two substrates were kept as controls. Following 1000 cycles all seven substrates were visually examined and torque-tested to destruction. No serious deterioration in the visual appearance of any of the solder joints was observed apart from some microcracking at the substrate-to-solder interface in some of the joints that had been cycled. The power-cycled substrates gave an average torque value of 11.5 N m compared with 16 N m for the controls—a deterioration of about 30%.

This result shows that by reducing the power cycle from 150°C and 20°C to 100°C and 20°C there is a dramatic improvement in solder joints with respect to both visual appearance and torque strength. It may also be tentatively concluded that 1000 power cycles between ambient and a joint temperature of 100°C can prove an adequate test for ceramic chip carrier solder attached to thick-film alumina substrates.

### 18.4.5 Power Cycling on Printed Wiring Boards

The Engelmaier equation[13] used to determine the thermal expansion mismatch between chip carrier and substrate is given, as above, by

$$\Delta\varepsilon = (\alpha_c - \alpha_s)(T_c - T_o) + \alpha_s(T_c - T_o)$$

It was shown that for ceramic chip carriers on alumina, temperature cycled in a chamber, $\alpha_c \cong \alpha_s$, $T_c \cong T_s$, and so $\Delta\varepsilon = 0$. During power cycling of chip carriers on alumina $\alpha_c$ is still approximately equal to $\alpha_s$ but there is a gradient down the solder joint and $T_c$ is not equal to $T_s$. $\Delta\varepsilon$ is now finite and this explains why power cycling is potentially more damaging than chamber cycling. When ceramic chip carriers are attached to printed wiring boards such as FR4, where there is a significant thermal mismatch, $\alpha_c - \alpha_s$ is no longer zero, and all three components of the equation have a finite value. This explains, albeit simplistically, why power cycling has such a destructive effect on FR4 printed wiring boards unless the value of $(T_c - T_o)$ is kept as low as possible. In practice this means that the upper temperature must be kept low.

The work described above on ceramic chip carriers on alumina suggested that the upper temperature of the solder joint should be kept below 100°C, which corresponded to a junction temperature within the chip carrier of 150°C.

Brierley et al.[10] decided that for ceramic chip carriers on FR4 the package case temperature should be set at 105°C, thus ensuring that the solder joint temperature is well below 100°C. Under these circumstances they were able to power-cycle leadless chip carriers with up to 68 leads beyond 1500 cycles without any joint failure. This demonstrates that chip carriers having a significant thermal mismatch with the printed wiring board on which they are soldered can be reliable during power cycling if attention is paid to the temperature of the solder joint.

## 18.5 OTHER RELIABILITY CONSIDERATIONS

### 18.5.1 The Effect of Flux on Surface Mount Solder Joint Reliability

The attachment of leadless components, such as the chip carrier, is generally carried out by solder reflow, which involves the use of a flux. Concern has been expressed about the corrosive effect of such fluxes. Dunn and Chandler[15] have documented several problems associated with the corrosive nature of liquid soldering fluxes and their residues. In particular, they have noted a relationship with the phenomenon of stress corrosion resulting in the fracture of leads, which, they feel, is attributable to flux residues.

To assess the problem associated with chip carrier attachment and RMA (rosin, mildly activated) flux residues, a series of experiments was set up[16] examining

1.  Methods of flux removal and their measurement
2.  The effect of residual flux on thick-film resistors
3.  The effect of residual flux on other thick-film hybrid components

Most hybrid manufacturers have their own method of flux removal after solder reflow of leadless components such as chip carriers. The single important factor seems to be removal of flux very shortly after reflow while the substrate is still hot. Flux residues that are left tend to solidify, polymerize, harden, and become extremely difficult to remove.

Two cleaning techniques were compared (rinses with trichloroethylene and acetone, and a commercial flux remover). The efficiency of these techniques was examined both by Auger electron spectroscopy (AES) and by dynamic conductivity monitoring (DCM). Comparison with controls showed that both cleaning techniques were effective, as were both analytical techniques. DCM was favored as a routine analytical tool because it has a sensitivity comparable to AES, it is nondestructive, and it has an overall superiority with respect to price and convenience within the typical hybrid or printed wiring board manufacturing area. The levels of flux found beneath leadless chip carriers, regardless of cleaning technique or method of determination, were very low.

In order to assess the deleterious effect of flux residues on thick-film hybrids containing leadless chip carriers, substrates were subjected to environmental conditioning designed to promote the corrosive effect of any flux residues. In the first set of experiments substrates with thick-film resistors beneath the chip carrier footprint were subjected to normal surface mount soldering procedures, including cleaning. The assemblies were then subjected to high-temperature storage (150°C) unpowered for 1500 hours, or 85°C/85% RH humidity testing for 1500 hours, either unpowered or with the underlying resistors powered at $84\,mW/cm^2$ (equivalent to $200\,mW/mm^2$ at 25°C, derated). Except in one specific case (unprotected, powered resistors under 85°C/85% RH) where some resistor drift was noted, the resistors remained unaffected and no corrosion was detected in the solder joints of the leadless chip carriers.

The second set of experiments utilized an artificially high concentration of flux residues. AES and DCM examination had shown that flux residues remaining after normal attachment and cleaning processes were similar to those measured after coating the substrates with 0.1% RMA flux dissolved in isopropyl alcohol. The substrates used in these experiments were coated with ten times that value (i.e., 1%) and then dried at 85°C for 30 minutes. The components used on the substrates were thick-film conductors (fine lines in parallel) and chip resistors and chip capacitors, both solder-attached. The substrates were then subjected to 1000 hours at 150°C or 1000 hours at 85°C/85% RH, both with and without bias. Again, no deterioration of electrical performance or visual quality attributable to flux residue-induced corrosion was found compared with controls.

The two sets of experiments on the effect of flux residues show that if reasonable care is taken with the cleaning of thick-film hybrid or printed wiring assemblies after surface mount reflow, no reliability hazards attributable to flux residues are likely on the surface mounted or other components.

### 18.5.2 The Effect of Thick-Film Multilayers and Leaded Components on Chip Carrier Reliability

The work on temperature and power cycling described above considered the attachment of chip carriers directly to the ceramic substrate via a single conductor layer. Circuits designed for practical applications may place a priority on compactness and it is likely that the chip carriers will be mounted on to multilayer substrates.

Problems with chip carrier attachment to multilayer thick-film hybrid substrates have been reported by many workers and so this work was preceded by a thorough investigation of materials and footprint geometry. It was concluded that material selection and combination was extremely important. The theoretical work on the geometry of component footprint concluded that significant improvements in the long-term reliability of chip carriers attached to thick film, by solder, could be made by modification of the footprint extension. An extensive set of experiments was carried out to optimize both footprint geometry and also thick-film material combinations.

The experimental assessment of the reliability of solder attachment of chip carriers to thick-film multilayers was performed using similar techniques to those already described.

The substrate was of multilayer construction, comprising four conductive layers separated by dielectric. A second substrate circuit having the same layout but fabricated using only a single conductive layer of thick film was produced as a control and included in each aspect of the test program.

Among the components solder-attached to the substrate were leadless ceramic chip carriers with 68 and 84 terminations. The work was extended by including a J-leaded chip carrier with 84 terminations. The substrate containing the three chip carriers is shown in Fig. 18-9. As in previous work the substrates were subjected to high-temperature chamber storage at 150°C for 1000 hours, 1000 chamber temperature cycles between −65°C and +150°C, and power cycling between laboratory ambient of 20°C and 100°C. Once again, the power cycling was performed using internal resistive heating and cooling was assisted by an external fan built into the circuitry. Altogether 42 substrates were used in the assessment.

The results cannot be compared directly with the previous work described above since earlier the larger chip carrier had 64 terminations with 0.040-in. pitch and in this case the two leadless chip carriers with 68 and 84 terminations both had a 0.050 in. pitch. The control samples of the 68-lead chip carrier had an average torque strength slightly in excess of 20 N m for both single-layer and multilayer substrates. That for the 84-lead chip carrier could not be assessed since, in all but one case, the torque strength was above 25 N m, the maximum reading on the torque meter. After high-

**Figure 18-9**  High-termination chip carriers on substrate.

temperature storage some degradation was noted, but for the 68-lead carrier on multilayer it was only about 25%. Although some torque strengths of the 84-lead chip carrier remained above the measurable maximum after 1000 hours storage at 150°C, the degradation in torque strength compared with control seemed to follow a comparable pattern, both for single-layer and multilayer. Similar results were obtained after chamber temperature cycling and power cycling. Some degradation compared with control was observed, none of it catastrophic, and in all cases the multilayer substrates performed at least as well as the single-layer.

The J-leaded chip carrier was included on the substrates to make a direct comparison between two chip carriers with 84 terminations and 0.050-in. pitch—one leaded and one leadless. Comparison between control, chamber heating, chamber temperature cycling, and power cycling again showed that the torque strength of the leaded chip carrier was not affected by whether the thick film was single-layer or multilayer. This result agreed qualitatively with the results obtained for the leadless chip carriers on the same substrates. Quantitatively, the leaded chip carrier performed worse than its leadless counterpart under all conditions. Whereas it proved difficult to torque off the 84-termination leadless chip carrier below 25 N m, the average failure value for the leaded chip carrier was 18 N m. After high-temperature storage the torque strength of the 84-termination leadless chip carrier remained high (in excess of 25 N m for some of the samples) whilst its J-leaded counterpart showed an average value of less than 8 N m, at least 60% worse. Comparison of the two chip carriers after 1000 temperature cycles showed torque strength averages of 20 N m for the leadless chip carrier and 8.3 N m for the leaded chip carrier (again 60% worse). Power cycling under the prescribed conditions had the least effect of all three thermal tests. None of the 84-termination leadless chip carrier that were torque-tested failed below 25 N m and the average value for the 84-termination J-leaded chip carrier was slightly under 17 N m, at least 30% worse.

The first conclusion to be drawn from this set of experiments is that, provided care is taken over footprint geometry and material combinations, both leaded and leadless chip carrier can be attached as reliably to multilayer thick-film substrates as they can to single-layer thick-film substrates.

The second conclusion is at first sight more surprising. The experiments showed quite conclusively that 84-termination J-leaded chip carriers have a significantly lower torque strength compared with 84-termination leadless chip carriers when subjected to high-temperature storage, temperature cycling, and power cycling. The leaded chip carrier is supposedly more reliable than the leadless chip carrier because its compliant leads are more able to absorb thermal mismatch. This is undoubtedly true when there is a significant thermal expansion mismatch between chip carrier and substrate, such as there is between ceramic chip carriers and FR4 printed wiring boards. However, on a ceramic substrate the mismatch is much less important. It may be postulated that the relatively adverse performance of the J-leaded package is due to the lower mass of solder forming the bond compared with the leadless chip carrier. The formation of intermetallic compounds in solder joints held at temperatures above ambient has been extensively reported. The presence of such compounds brings about joint embrittlement and the small volume of solder in the J-lead joint enhances the effect.

Confirmation of the lower strength of J-leaded chip carriers compared with leadless chip carriers is not readily found in the work of other authors. This is primarily because, although leaded chip carriers are used extensively where there is a large thermal mismatch between chip carrier and substrate materials, they are rarely found on ceramic thick-film substrates. The exception is for chip carriers that have a large termination count and lower pitch than was used in this work. Such small geometries result in an increase in lead compliance that can increase the reliability of the joint.[17]


### 18.5.3 The Effect of Cyclic Mechanical Stresses on Flexible Surface Mount Assemblies

Although much work has been done on the effect of thermally driven cyclic stresses on chip carrier assembly structures, comparatively little information has been published on mechanical stresses at ambient temperature. Such stresses can readily occur during board insertion and removal or as a result of shock and vibration during normal service.

Brierley and McCarthy[18] describe a custom-designed mechanical flexure apparatus and a series of experiments designed to test populated printed wiring boards. They show that under worst-case conditions (high amplitude coupled with low cycle frequency) damage does occur but it tends to begin after 5–10 times as many cycles as are required for thermal cycling testing on equivalent structures.

The work pointed out, nevertheless, the vulnerability of surface mounted structures to mechanical cycle damage if rigorous design and process control are not maintained. The importance of the resistance of surface mounted devices to bending of the substrate is emphasized by a proposal from IEC (International Electrotechnical Commission). This recommends the inclusion of a new test[19] entitled "Substrate bending test for SMD-Semiconductors" to be included in existing documentation on terminal robustness. The proposal defines a test vehicle, bending equipment, choice of amplitudes, and frequency of bend. After testing, the substrates are examined visually and then subjected to climatic testing to discover hidden mechanical failures.

## 18.6  CONCLUSIONS

The reliability of the solder joints interconnecting chip carriers to their substrate can be assessed in a number of ways. Ceramic leadless chip carriers solder-attached to thick-film hybrid ceramic substrates were subjected to a number of environmental test conditions. Under conditions of damp heat, both steady-state and cyclic, there was no deterioration of the solder joints, either electrically, visually, or mechanically. Storage for 1000 hours under a steady-state condition of 150°C and 1000 temperature cycles between −65°C and +150°C both produced a significant although not catastrophic deterioration in chip carrier adherence. The deterioration caused by thermal cycling and high-temperature storage seems well explained by mechanical analysis. The important factor in the decrease in bond strength appears to be intermetallic formation and depletion, and this is dependent on total dwell time at elevated temperature rather than temperature variation. This conclusion is reasonable where the material of the chip carrier and the substrate to which it is attached are similar. However, where there is a large thermal mismatch between chip carrier and substrate the ability to withstand temperature cycling is much more dependent on temperature variation. Small leadless chip carriers (e.g., 16-termination) are able to withstand 1000 cycles between −55°C and +125°C, but larger chip carriers tend to fail soon after 100 such cycles. The accepted solution is to use leaded chip carriers for attachment to substrates such as FR4. These combinations allow the successful completion of extended temperature-cycling sequences, even when the chip carrier termination count is high.

By careful choice of thick-film substrate materials and the geometry of the chip carrier footprint it has been shown that leadless chip carriers can be attached as reliably to thick-film multilayer hybrids as they can to single-layer substrates. This observation is important since practical requirements are at least as likely to see thick-film hybrid circuits designed on multilayer as they are on single-layer.

The use of high-temperature storage as a guide to reliability is scientifically sound since, provided the failure mode at high temperature is the same as the failure mode at ambient temperature, a thermally activated process can be assigned an acceleration factor. It is more difficult to equate temperature-cycling experiments with real life since the test conditions differ markedly from service conditions and the acceleration factor is harder to calculate. It is accepted, however, that chamber cycling can produce satisfactory first-order reliability information quickly and easily. Calculations involving the use of the Coffin–Manson equations for the low-cycle fatigue[11] can be used.

Refinements to the universally used chamber temperature-cycling test led to the derivation of power-cycling tests.[12,13] These can be much more realistic since chip carrier assemblies may well undergo thermal excursions in service that are brought about by heat from a chip in the chip carrier as the assembly is switched on and off. A comparison between high-temperature storage of chip carrier assemblies in a chamber at 150°C and chip carrier assemblies internally power-heated to a solder temperature of 150°C showed great similarity after 1000 hours. Power cycling up to 150°C joint temperature for 1000 cycles was much more damaging to the joint than chamber temperature cycling up to 150°C for 1000 cycles. This is simply explained by the effect of the thermal gradient along the solder joint between chip carrier and substrate. The net thermal expansion mismatch thus produces joint failure through low-cycle fatigue. Further testing showed that the same chip carrier assemblies could more readily sustain 1000 power cycles if the joint temperature was not allowed to rise above 100°C (instead of 150°C). For the samples under test this actually corresponded to a chip temperature of 150°C. It has also been shown that leadless chip carriers can withstand prolonged power cycling when attached to printed wiring boards, such as FR4, provided that the solder joint's upper temperature is carefully controlled.

The use of mechanical cycling has been examined and has been shown to be a relevant reliability tool. By subjecting the substrate (generally only applicable to semiflexible printed wiring boards such as FR4) to a bending cycle it is possible to cause deterioration to the solder joints but, typically, only after 5–10 times as many cycles as are required by thermal cycling regimes.

The high-temperature storage, temperature cycling, power cycling, and mechanical cycling experiments described in this chapter have all shown that it is possible to derive test conditions that bring about solder joint degradation. As a first-order assessment procedure, each can give an indication of the solder joint reliability. By deriving standard tests from the experiments it is possible to make a clear decision as to whether a solder joint passes or fails and, albeit arbitrarily, whether a particular solder-attachment process is reliable. In order to refine the test method it needs to be more application-specific. For example, power-cycling or temperature-cycling tests should be based on the thermal excursions that will be experienced in service. This will enable the elimination of some unknowns during the modeling of fatigue-life behavior and permit the accelerated testing to be more realistic.

The only way to be sure that a leadless chip carrier can be reliably solder-attached to its substrate is to perform real-time testing. For most requirements this requires the assembly to have been in service for at least 10, possibly 20 years. Many organizations have now been involved in leadless chip carrier assembly for upwards of 10 years. Some have been in full-scale production longer than 10 years and are able to prove real-time reliability with hard data[20] and authenticated lifetimes up to 15 years with no joint degradation.

It may be taken, as a final conclusion, that solder-attachment of leadless chip carriers can be a reliable interconnection procedure. The work described above has demonstrated this and gives adequate guidelines for the derivation of a number of test methods that will enable anyone using chip carrier solder-assembly techniques to verify their own process.

## ACKNOWLEDGMENT

Parts of this work were performed under contract to and with the technical direction of the European Space Agency.

## REFERENCES

1. Lynch, J. T., G. M. Brydon, M. R. Hepher, J. P. McCarthy, and A. Boetti, "Environmental Assessment of Ceramic Chip Carriers Solder-attached to Thick Film Alumina Substrates," *Proc. 32nd Electronic Components Conference*, San Diego, 1982, p. 385.
2. ESA/SCC Generic Specification No. 9000 (Integrated Circuits, Monolithic).
3. Hong, J., and M. Chason, "Gold Removal from Component Terminations via Solder Dipping," *Proc. 39th Electronic Components Conference*, Houston, 1989, p. 351.
4. Fukusawa, I., S. Ishiguro, and S. Nanbu, "Moisture Resistance Degradation of Plastic LSIs by Reflow Soldering," *Proc. International Reliability Physics Symposium*, Orlando, 1985, p. 192.
5. Nishioka, T., M. Nagasawa, K. Igarashi, and M. Kohmoto, "Special Properties of Molding Compound for Surface Mounting Devices," *Proc. 40th Electronic Components and Technology Conference*, Las Vegas, 1990, p. 625.
6. "Test Method of Resistance to Soldering Heat of Surface Mounting Devices for Integrated Circuits," Provisional Standard of Electronic Industries Association of Japan, EIAJ EDX-4701, 1990.
7. "The Evaluation and Qualification of Thick Film Hybrid Microcircuits," European Space Agency Product Assurance Document PSS-01-606.
8. Shelton, T. M., "Testing of Properties for Soldered Leadless Chip Carrier Assemblies," *Proc. 30th Electronic Components Conference*, San Francisco, 1980, p. 452.
9. Taylor, J. R., and D. J. Pedder, "Joint Strength and Thermal Fatigue in Chip Carrier Assembly," *International Hybrid Microelectronics*, **5**(2), 1982, pp. 209–214.
10. Brierley, C. J., D. J. Pedder, and J. P. McCarthy, "The Characterization of Novel PWB Substrate Materials for Leadless Ceramic Chip Carrier Attachment," *Proc. 5th International Electronics Packaging Conference*, Orlando, 1985, p. 586.
11. Manson, S. S., *Thermal Stress and Low Cycle Fatigue*, McGraw-Hill, New York, 1966.

12. Howard, R. T., C. C. Sanetra, and S. W. Sobeck, "A New Package-related Failure Mechanism for Leadless Ceramic Chip Carriers (LC-3's) Solder Attached to Alumina Substrates," presented at a VLSI Packaging Workshop, NBS, Gaithersburg, MD, September 13, 1982.
13. Engelmaier, W., "Effects of Power Cycling on Leadless Chip Carrier Mounting Reliability and Technology," *Proc. 2nd Annual International Electronics Packaging Conference*, San Francisco, 1982, p. 15.
14. Lynch, J. T., M. R. Ford, and A. Boetti, "The Effect of High Dissipation Components on the Solder Joints of Ceramic Chip Carriers Attached to Thick Film Alumina Substrates," *IEEE Trans. Components, Hybrids, and Manufacturing Technology*, **CHMT-6**(3), 1983, p. 237.
15. Dunn, B. D., and C. Chandler, "The Corrosive Effect of Soldering Fluxes and Handling on Some Electronic Components," *Welding Journal*, **59**(3), 1980, p. 289.
16. Lynch, J. T., R. T. Bilson, N. R. Matthews, and A. Boetti, "The Measurement of Flux Residues from Chip Carrier Attachment and Their Effect on Other Thick Film Hybrid Components," *IEEE Trans. Components, Hybrids, and Manufacturing Technology*, **CHMT-7**(4), 1984, p. 336.
17. Kotlowitz, R. W., "Compliance Metrics for Surface Mount Component Lead Design," *Proc. 40th Electronic Component and Technology Conference*, Las Vegas, 1990, p. 1054.
18. Brierley, C. J., and J. P. McCarthy, "The Reliability of Surface Mounted Solder Joints Under PWB Cyclic Mechanical Stresses," *Proc. 5th International Electronics Packaging Conference*, Orlando, 1985, p. 312.
19. IEC Technical Committee 47, Reference Number 47 (Secretariat) 1228, proposal from Working Group 10, "Substrate Bending Test for SMD Semiconductors," October 1991.
20. Scher, P., and H. F. Inacker, "Experiences Utilizing LCC Module Design on Military Systems," *Proc. 35th Electronics Components Conference*, Washington, 1985, p. 174.

# 19

# Solder Creep–Fatigue Interactions with Flexible Leaded Surface Mount Components

*R. G. Ross, Jr. and L.-C. Wen*

## 19.1 INTRODUCTION

In most electronics packaging applications it is not a single high-stress event that breaks a component solder joint; rather it is repeated or prolonged load applications that result in fatigue or creep failure of the solder. The principal strain in solder joints is caused by differential expansion between the part and its mounting environment due to changes in temperature (thermal cycles) and/or due to temperature gradients between the part and the board.

The function of strain-relief elements—such as the flexible metal leads of the electronic components—is to lower the differential expansion-induced loads (stresses) on the solder joints to levels well below the solder yield strength, thus significantly reducing the generation of plastic strain. By reducing the developed strain, flexible leads greatly enhance solder joint reliability. Figure 19-1 illustrates a variety of flexible leads common to surface mount electronic parts.

While flexible leads invariably increase the reliability of soldered electrical connections, the level or amount of improvement varies widely with the specific application. This variability stems from the fact that the elastic spring properties of the leads that reduce the load also prolong the load on the solder. Prolonging the load over long time spans raises the specter of time-dependent failure mechanisms involving phenomena such as creep, corrosion, and long-term metallurgical changes. Creep strain is probably the most important time-dependent damage accrual mechanism; under typical multi-hour loading conditions the solder joints of flexible leaded parts can

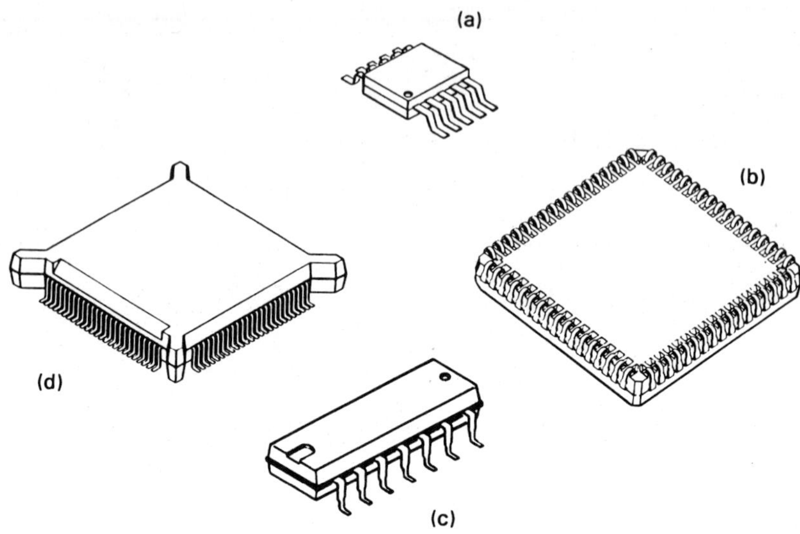


**Figure 19-1** Typical flexible lead geometries found on surface mount electronic components: (*a*) flat-pak with gull-wing leads; (*b*) plastic leaded chip carrier (PLCC) with J-leads; (*c*) DIP with gull-wing leads; (*d*) quad flat-pak with fine-pitch gull-wing leads.

be expected to undergo significant levels of strain due to creep of the solder in response to the applied elastic forces from the strain-relief elements. This creep-induced strain has the same damaging effect as more immediately induced plastic strain, and must be summed with the plastic strain to achieve the total plastic strain that is correlated with fatigue life.

Because of creep effects, the total strain range during any given loading cycle can be a strong function of solder temperature, the loading time per cycle, the applied solder stress, and the spring constant of the strain relief elements. The implication is that creep-related strain—and in particular the time, temperature, and stress dependency of creep—must be carefully factored into the design and testing of flexible leaded parts.

Although extensive research in recent years has led to an increasingly mature understanding of the creep–fatigue properties of solder as an engineering material, much less headway has been made in quantifying solder creep–fatigue behavior at the electronics package systems level. The objective of this chapter is to explore the complex systems-level creep–fatigue interactions involved in electronic part solder joints, and in particular to illustrate the key solder-joint structural configuration and environmental stress dependencies. Important issues include the effects of component lead flexibility, operating temperature, thermal cycling depth, and cyclic frequency of loading. Understanding these issues is important both to achieving robust electronic packaging designs that meet the end-use requirements, and to defining and interpreting appropriate accelerated testing procedures for hardware qualification.

## Chapter Organization

The chapter begins by first reviewing the Coffin–Manson fatigue life relationships that highlight the importance of repeated plastic strain—including creep strain —in determining the useful life of solder joints. Next, the fundamental parameters controlling creep in solder are examined together with the constitutive models that capture the parameter dependencies in mathematical terms. Combining these constitutive models with nonlinear finite element modeling techniques provides one of the best means of computing and studying the creep–fatigue behavior of solder in complex flexible lead loading conditions. One such model, developed by the authors,[1,2] will be used extensively throughout the chapter to explore the role of lead flexibility in both isothermal mechanical cycling and thermal cycling environments. Mechanical cycling with flexible leaded parts will be shown to introduce strong dependencies on the rate of cycling and the temperature of the solder. Thermal cycling will be shown to additionally introduce unidirectional creep-ratcheting that can cause parts to climb out of the solder or into the solder normal to the loading direction; failure models based on the Coffin–Manson fatigue life relationships will be examined and used to interpret the relative importance of combined creep–fatigue and creep-ratcheting mechanisms. The chapter ends with a summary of the diverse issues surrounding creep–fatigue interactions with flexible leaded parts.

## 19.2 MODELING OF SOLDER CREEP–FATIGUE INTERACTIONS

The most important cause of solder joint failure is repeated straining during temperature cycling as a result of mismatch of the coefficients of thermal expansions (CTE) of the electronic part–lead assembly and the substrate to which it is soldered.

The fatigue-dominated process, often referred to as low-cycle fatigue, is characterized by a large plastic strain range and is generally modeled by a Coffin–Manson type relationship modified to account for any creep strain and its dependence on cycle frequency and temperature. Lau[3] and Frear[4] provide extensive reviews of the literature in this area.

Figure 19-2 shows representative fatigue life data for 63–37 Sn–Pb near-eutectic solder.[5–7] The plot illustrates the typical logarithmic dependency

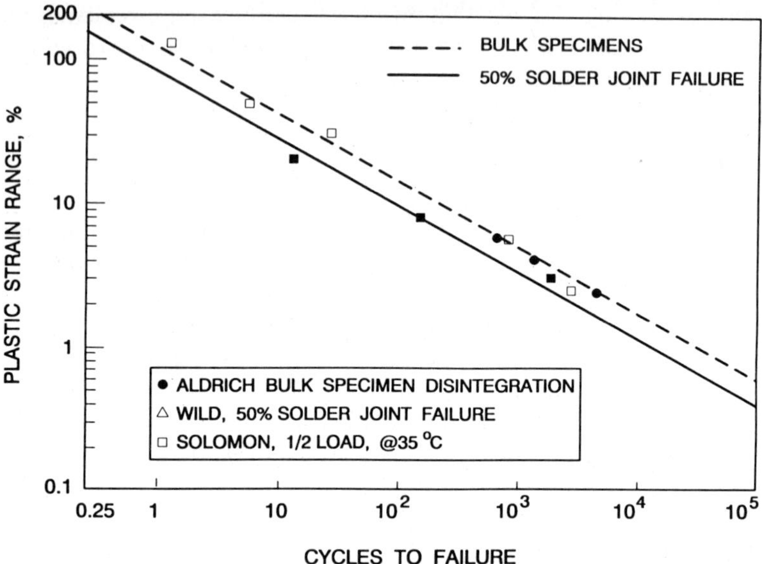


**Figure 19-2**   Coffin–Manson curves fitted to typical eutectic Sn–Pb solder fatigue data from Aldrich,[5] Wild,[6] and Solomon.[7]

between cycles-to-failure and the level of plastic strain range. The data can be fitted to the empirical Coffin–Manson relationship[8,9]

$$N_f^\beta \, \Delta\varepsilon = C \qquad\qquad (19\text{-}1)$$

where $N_f$ = number of cycles to failure
$\Delta\varepsilon$ = cyclic plastic strain range
$\beta \approx 0.40$ (from curve slope in Fig. 19-2)
$C \approx 0.80$ (for 50% solder failure limit in Fig. 19-2)

Although the intercept $C$ varies somewhat for various alloys, the slope $\beta$ has been found to be around 0.5 for most ductile metals including solder, aluminum, copper, and stainless steel.[5] However, $\beta$ has been found to vary modestly with the degree of damage used to define failure.[10,11] Lower levels of damage, such as "cycles to 50% load reduction for a constant strain," tend to yield steeper Coffin–Manson slopes; on the other hand, higher levels of damage, such as complete separation of the solder joint, tend toward lower $\beta$ values. This trend is illustrated in Fig. 19-3.

A flatter slope ($\beta \approx 0.40$) is used in this chapter as more representative of the dependence of the number of cycles required to achieve total joint separation, and thus an electrical open circuit.

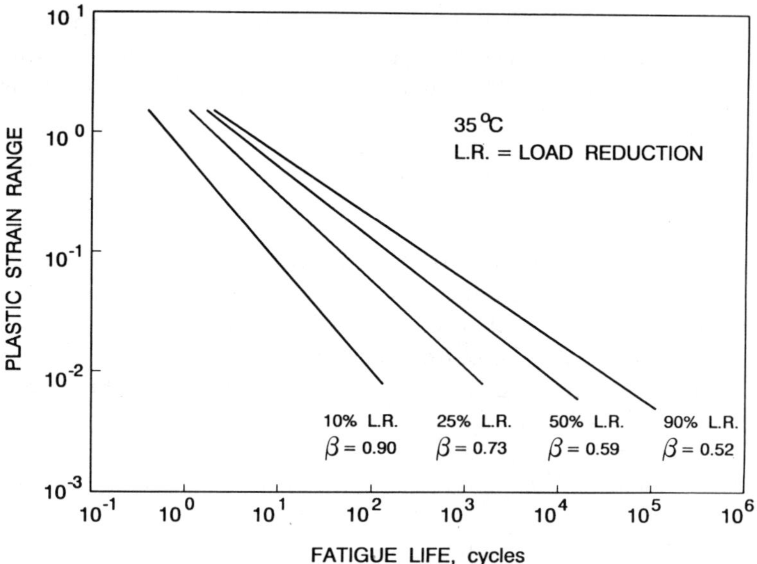


**Figure 19-3**   Effect of definition-of-failure on slope of Coffin–Manson fatigue curves. (From Solomon.[11])

### 19.2.1 Use of Coffin–Manson Relationships to Account for Creep Strain

The Coffin–Manson relationship (Eq. (19-1)) can be used for life prediction with flexible leaded parts only if the strain used is the total plastic strain including creep strain. This requires that the creep strain be accurately determined and added to any immediately produced plastic strain.

In practice, experimentally or analytically determining the creep strain can be exceedingly difficult; this has led to empirical modification of the Coffin–Manson equations to indirectly include the contributon of creep strain likely to occur, and its dependence on cycle rate and temperature. The simplest of the frequency-modified Coffin–Manson relationships[12,13] is of the form

$$N_f^\beta \, \Delta\varepsilon \, f^m = C \tag{19-2}$$

where $f$ is the frequency of applied loading and $m$ is a design-specific, temperature-dependent material parameter. Solving explicitly for cycle life gives

$$N_f = \left( \frac{C}{\Delta\varepsilon f^m} \right)^{1/\beta} \tag{19-3}$$

To obtain a quantitative feel for the value of the frequency exponent $m$, it is useful to compare Eq. (19-3) with representative frequency-dependence data. Fitting Eq. (19-3) to the low-frequency portion of the data shown in Fig. 19-4[10,14,15] gives

$$N_f \propto f^{-m/\beta} \approx f$$

This corresponds to $m = -\beta \approx -0.4$. In contrast, at frequencies above 0.1 cycle/min, fatigue life for this loading system is seen to become much less sensitive to frequency as creep strain becomes small relative to other time-independent strain mechanisms. Thus the value of $m$ varies from around $-0.4$ to zero over the range of frequencies tested.

Although Eq. (19-2) recognizes the strong time dependence of creep strain, its simple form is unable to capture the important interrelationships between frequency, temperature, lead stiffness, and other parameters known also to have strong influences on generated strain. This limits its applicability to a specific design for which the equation has been calibrated. The $m = -0.4$ value derived above may therefore be of limited use for the prediction of performance of solder-joint systems significantly different from those used to gather the data in Fig. 19-4.

To help correct for this important weakness, Engelmaier[16–18] has developed a more complete fatigue life relationship that explicitly accounts

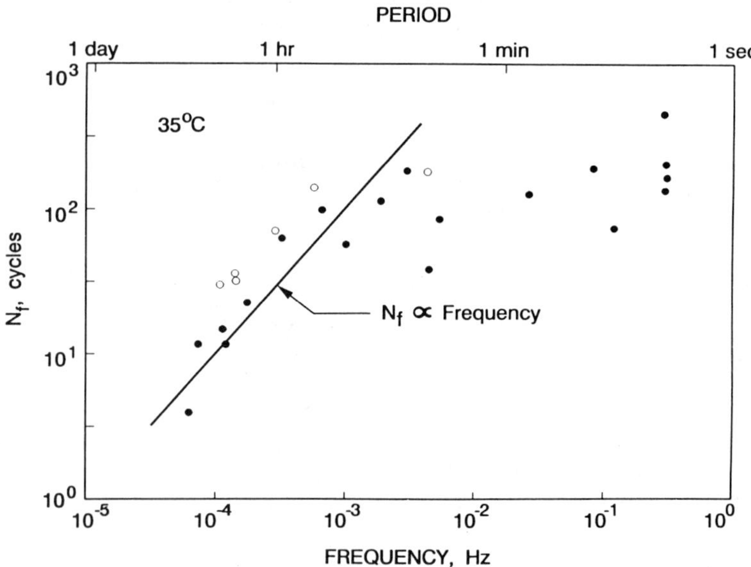


**Figure 19-4**   Dependence of fatigue life of 60–40 Sn–Pb solder on loading frequency as measured by various investigators.[10,14,15]

for lead stiffness and other parameters such as cyclic loading rate and solder and substrate temperature. Engelmaier's modified version of the Coffin–Manson equation is of the form

$$N_f = \left( \frac{CAh}{K\,\Delta x^2} \right)^{1/\beta} \tag{19-4}$$

where   $C$ = constant
$N_f$ = mean cycles to fatigue failure
$K$ = stiffness of part lead in direction of loading (lb/in.)
$A$ = effective solder joint loading area (in.$^2$)
$\Delta x$ = lead displacement range = $x(\alpha_p\,\Delta T_p - \alpha_b\,\Delta T_b)$ (in.)
$x$ = distance from part center to corner lead (in.)
$\alpha_p$ = CTE of part-lead ($^\circ$C$^{-1}$)
$\alpha_b$ = CTE of substrate board ($^\circ$C$^{-1}$)
$\Delta T_p$ = temperature change of part ($^\circ$C)
$\Delta T_b$ = temperature change of board ($^\circ$C)
$h$ = solder joint effective height (in.)
$\beta = 0.442 + 6.0 \times 10^{-4}T_m - 0.0174\ln(1 + 360/\tau)$
$T_m$ = mean solder temperature ($^\circ$C)
$\tau$ = half-cycle dwell time (min)

Equation (19-4) is equivalent to saying that the strain range in the Coffin–Manson relationship is given by

$$\Delta\varepsilon \propto \left( \frac{K\,\Delta x^2}{Ah} \right) \tag{19-5}$$

Equation (19-5) suggests an approximately linear relationship between strain range and lead stiffness ($K$), and a squared dependency on lead deflection range ($\Delta x$). Note that cyclic frequency of loading is not explicitly dealt with; rather half-cycle dwell time ($\tau$), which is the more relevant parameter for creep, is incorporated as a variation in the Coffin–Manson slope $\beta$. Varying $\tau$ from 1 min to 10 min causes $\beta$ to vary from 0.354 to 0.394. To understand this change consider a system exhibiting $N_1 = 1000$ cycles to failure with $\tau = 1$ min. If all other parameters remain constant, the reduction ($N_2/N_1$) in the cycles-to-failure when $\tau$ is increased to 10 min can be computed by noting that

$$N_1^{\beta_1} = N_2^{\beta_2}$$

thus

$$\frac{N_2}{N_1} = N_1^{(\beta_1/\beta_2 - 1)} \tag{19-6}$$

This change in $\beta$ can also be converted to an equivalent change in strain range by noting that

$$N_1^\beta \, \Delta\varepsilon_1 = N_2^\beta \, \Delta\varepsilon_2$$

thus

$$\frac{\Delta\varepsilon_2}{\Delta\varepsilon_1} = \left(\frac{N_2}{N_1}\right)^{-\beta} \qquad (19\text{-}7)$$

For our example of a tenfold change in dwell time, Eqs. (19-6) and (19-7) give

$$\frac{N_2}{N_1} = (1000)^{(0.354/0.394-1)} = 0.496$$

and

$$\frac{\Delta\varepsilon_2}{\Delta\varepsilon_1} = (0.496)^{-0.39} = 1.31$$

Thus the cycles to failure are reduced from 1000 to 496 by increasing the dwell time from 1 min to 10 min. This is seen to also correspond to roughly a 30% increase in the strain range. To compare these results to the predictions of Eq. (19-3), one can assume that the tenfold increase in dwell time corresponds to a tenfold reduction in frequency; thus, from Eq. (19-3),

$$\frac{N_2}{N_1} = \left(\frac{f_2}{f_1}\right)^{-m/\beta} = (0.1)^{1.0} = 0.1$$

This is a considerably larger change than predicted by Engelmaier's relationship, and is indicative of the difficulty in extrapolating data from one set of test conditions to another. Understanding the issues underlying these difficulties is an important objective of the remaining sections of this chapter.

As a final step in examining the solder strain dependencies implied by Engelmaier's equation, one can examine the sensitivity of fatigue life to solder temperature. Like loading dwell time, mean solder temperature ($T_m$) is incorporated into Eq. (19-4) as a variation in the Coffin–Manson slope $\beta$. For 20°C and 30°C solder temperatures and a 10-min dwell time, we obtain $\beta$ values of 0.391 17 and 0.397 17, respectively. Substituting these into Eqs. (19-6) and (19-7) gives

$$\frac{N_2}{N_1} = (1000)^{(0.39117/0.39717-1)} = 0.90$$

and therefore

$$\frac{\Delta\varepsilon_2}{\Delta\varepsilon_1} = \left(\frac{N_2}{N_1}\right)^{-\beta} = (0.9)^{-0.39} = 1.04$$

This implies that a 10°C increase in temperature causes a 4% increase in strain range and a corresponding 10% reduction in fatigue life.

Although Engelmaier's equation has considerably greater applicability than the original frequency-compensated Coffin–Manson equation, a remaining weakness in Eq. (19-4) is the fact that the influences of temperature, lead stiffness, dwell time, and deflection range are treated as independent of one another. We will see later in this chapter that the interdependence is, in fact, very strong; for example, the influence of temperature on solder creep strain with highly flexible leads is very different from the temperature dependence with stiff leads (LCC parts). To explore these issues in greater depth we must turn to more elaborate and comprehensive modeling approaches.

### 19.2.2 Solder Constitutive Relationships

As an alternative to the modified Coffin–Manson equations of Engelmaier and others, recent research has turned toward computing the plastic and creep strain dependencies on applied loads and temperatures using special nonlinear finite element creep simulation models.[1,2,19] The finite element programs make use of the fundamental constitutive equations that model the dependency of solder strain rate on applied stress, temperature, and metallurgical properties. This has the advantage of accurately including the stiffness and geometry of the entire part–lead–substrate system, spatial detail within the solder joint itself, as well as the measured metallurgical properties of solder.

Figures 19-5 and 19-6 summarize representative data for the creep properties of solder;[20–26] an extensive review of these data is presented by Arrowood et al.[27] Figures 19-5 and 19-6 highlight the strong sensitivity of strain rate to solder age and metallurgical history and the strong Arrhenius dependency on temperature. Notice that solder exhibits three distinct strain-rate slope behaviors depending on the stress and temperature level. Regions I and III are regions of classic metal plastic deformation, whereas region II is a so-called region of superplastic behavior. In this superplastic region the solder is able to withstand much larger levels of deformation without rupture. Within each region the creep behavior of solder is well

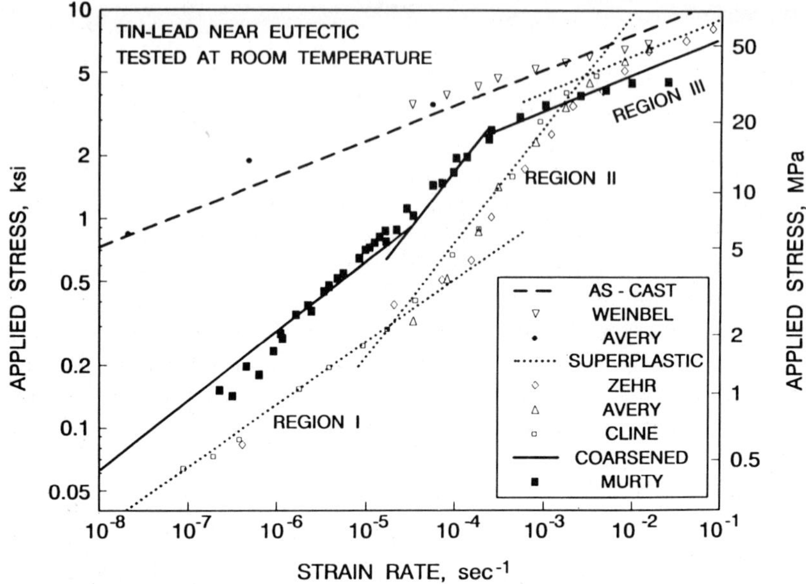


**Figure 19-5**  Dependence of strain rate on applied stress and metallurgical condition of room-temperature eutectic Sn–Pb solder.

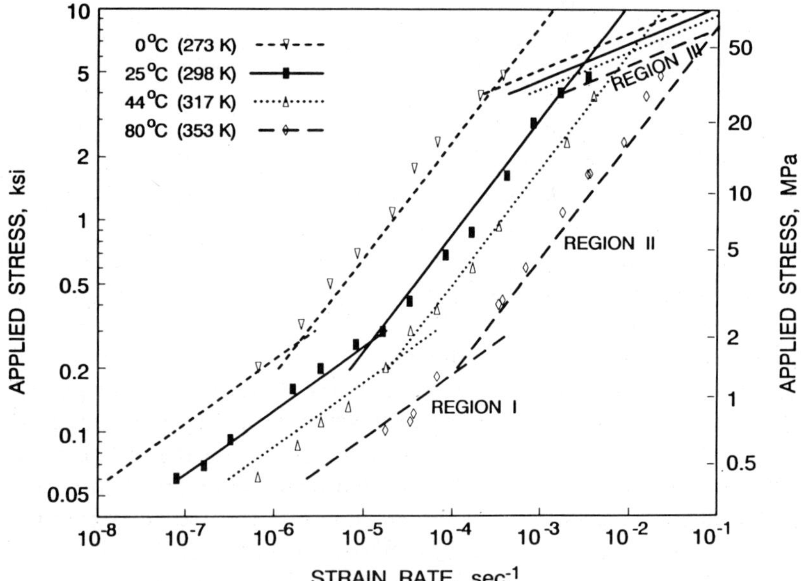


**Figure 19-6**  Dependence of strain rate on applied stress and temperature for eutectic Sn–Pb solder. (From Cline and Alden.[20])

approximated by[27]

$$\dot{\varepsilon} \propto \sigma^n g^{-p} \exp(-Q/kT) \cong \sigma^n g^{-p} \xi^{(T/10)} \qquad (19\text{-}8)$$

where  $\dot{\varepsilon}$ = solder strain rate (sec$^{-1}$)
$\sigma$ = solder stress
$n$ = creep exponent (typically $n$ = 2 to 3)
$g$ = grain size
$p$ = grain size exponent (typically $p$ = 1.6 to 2.3)
$Q$ = thermal activation energy of creep
$\xi$ = temperature dependence per 10 K (typically $\xi$ = 1.5 to 2)
$T$ = solder temperature (K)
$k$ = Boltzmann constant

Note in Fig. 19-5 that "as-cast" solder has been found to have extremely high creep resistance when compared to fine-grain "superplastic" solder; similarly, larger-grained "coarsened" solder, generally the result of prolonged aging or thermal processing, has been found to be intermediate in creep resistance. With respect to the "as-cast" solder, Arrowood et al.[27] have observed that the much greater creep resistance reported for this solder may be caused by the highly banded Pb–Sn lamellar structure often encountered with slowly cooled cast solder joints. If so, this high level of creep resistance may not be relevant to small rapidly cooled electronic component solder joints, which rarely exhibit banded solder structures.

In contrast to the anomalously low creep rate of "as-cast" solder, "superplastic" solder exhibits a much higher creep rate and tolerance to plastic deformation; this superplastic behavior is often associated with fine uniformly distributed grain structures similar to those found in electronic solder joints that are relatively small and have solidified rapidly from the melt (see Fig. 19-7). Because high creep rate correlates with increased strain with flexible leaded parts, prudence demands that the high-creep-rate behavior of superplastic solder be carefully considered when dealing with flexible leaded parts.

When solder undergoes long-term aging of a year or more, or is thermally processed, the grain size of the solder grows as shown in Fig. 19-7.[2,28] We refer to this solder as "coarsened solder." Of the three solder structures, coarsened solder is probably the most representative of aged solder joints in real electronics applications. The increased grain size of coarsened solder restricts creep strain and thus makes it more resistant to failure under creep loading conditions.[25,26] Because of the significant variability between "superplastic" solder and "coarsened" (larger-grain) solder, the properties of both will be treated parametrically in later sections of this chapter.

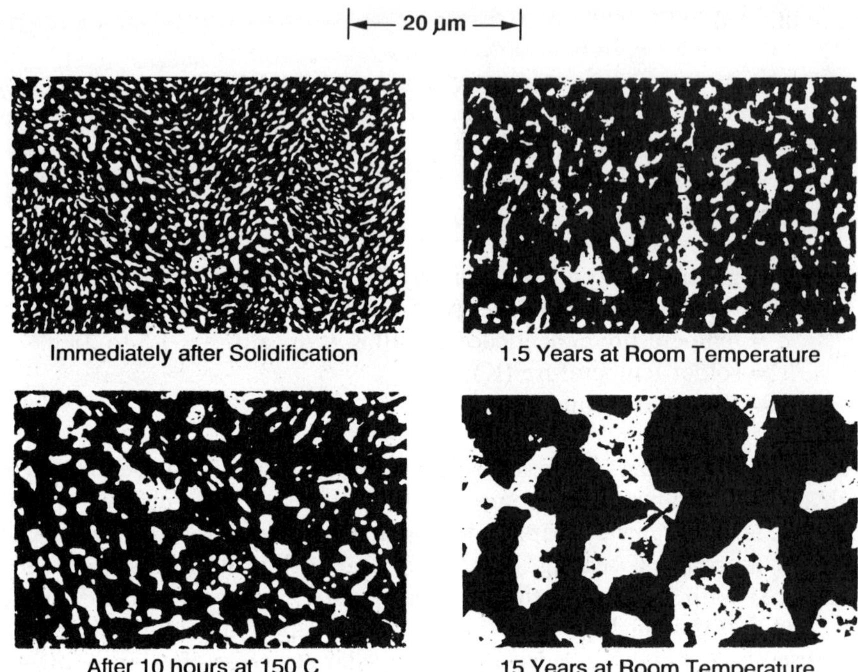


**Figure 19-7**   SEM back-scatter photographs (tin is dark regions) showing cross-sections of typical gull-wing flat-pak solder joints taken from spacecraft electronics with the indicated aging histories. (From Ross et al.[2])

### 19.2.3  Finite Element Creep–Fatigue Modeling

Given a time history of environmental loading (e.g., electronic package temperature vs. time), a finite element creep simulation program uses the constitutive properties of solder, such as those in Fig. 19-6, to compute the resulting strain in each element of the solder joint as a function of the applied stress over time. Varying the system geometry, cycle rate, thermal cycle depth, or any of a wide variety of parameters, allows the resulting effects on solder strain history to be observed directly.

In operation, a finite element creep simulation program divides the time response of the solder joint into thousands of tiny time increments; for each time increment the solder's incremental plastic strain response is computed, based on the stress existing in the solder and the solder's temperature. One common approach is to start each time increment by computing the instantaneous stress in each solder element using a classical finite element elastic structural model of the complete part–lead–solder–PWB system. The forcing function for the stresses is the instantaneous geometry of the package system, including previously accrued plastic deformation of the solder elements together with lead deflections applied externally by thermal cycle differential expansions or mechanical (isothermal) cycling. Following computation of the stress, the above-described constitutive equations for solder are used together with the current solder temperature to compute the plastic strain rate in each solder element, and thus the solder incremental strain during the time interval.

Some programs continuously adjust the size of the time increment to insure near constant strain rate during any given time increment. This is desirable to prevent strain integration errors and model instabilities caused by excessive stress relaxation during any time interval. In general, simulations of multi-hour loading cycles representative of typical electronics applications involve time steps ranging from 0.001 sec to 1 min, thus leading to as many as 1 million strain increment solutions per simulation run. The shortest time increments are generally required during periods of elevated temperature where stress relaxation occurs rapidly.

Because of the heavy computational load presented by creep simulation algorithms, the finite element description of the solder system often has to be carefully simplified to preserve the important structural and constitutive features of the solder system while sacrificing noncritical detail. Example finite element creep fatigue simulation runs are used throughout the remainder of this chapter to further explore the nature of flexible leaded parts and to contrast the results with the modified Coffin–Manson equations described earlier.

## 19.3 ROLE OF LEAD STIFFNESS IN ISOTHERMAL LOADING

It is clear from the constitutive relationships in Figs. 19-5 and 19-6 that applied stress, solder metallurgical structure, and temperature strongly influence the strain developed in solder during an application loading condition. As previously pointed out, a key function of flexible leads is to reduce the stress on the solder and thereby to reduce the damaging accrual of solder strain. Unfortunately, a second manifestation of lead stiffness is to prolong the application of stress during periods of loading dwell. If the dwell period is sufficiently long, the elastic stress in the flexible leads will fully relax through creep of the solder joint, and no strain reduction will be realized. The time required for significant stress relaxation to occur is largely controlled by the stiffness of the part–lead–board system, which sets the initial solder stress for a fixed load displacement and governs the rate at which the stress decreases as the solder strains. For loading conditions involving highly flexible leads, stress decreases slowly with increasing strain;

this provides the conditions for the accrual of large solder strains over prolonged time periods.


### 19.3.1 Creep Strain During Stress Relaxation

Because creep strain development during stress relaxation is such an important part of solder joint performance with flexible leaded parts, it is useful to first carefully examined the dynamics of simple elastic stress relaxation before proceeding to more complex cyclic loading conditions. An understanding of stress relaxation is also germane to the issue of long-term creep rupture of solder-joint systems that have accumulated built-in residual stresses following the soldering operation. Such stress can occur due to differential expansion upon assembly cool-down following machine soldering, or due to elastic deflections applied to displace a misaligned lead during hand soldering.

Figure 19-8 presents a schematic representation of the elastic stress relaxation problem where the spring stiffness $K$ is the effective stiffness of the combined solder–lead system; for the simple system shown in Fig. 19-8, this effective stiffness is given by

$$K = \left( \frac{1}{K_L} + \frac{1}{K_S} \right)^{-1} \tag{19-9}$$

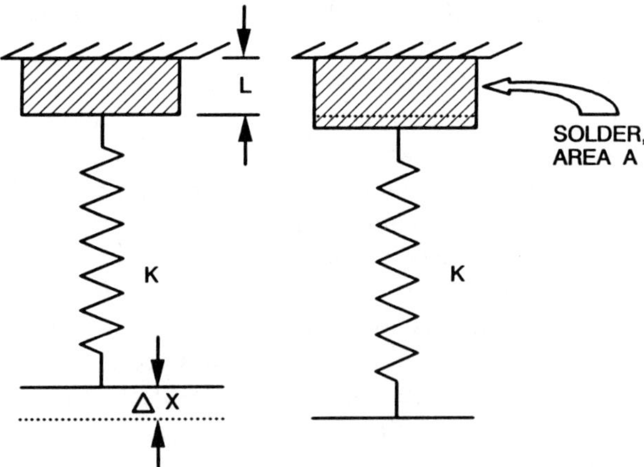


**Figure 19-8**   Schematic of single-node solder element with series spring representing the combined elastic flexibility of the solder and electronic component lead.

where $K_L$ = lead stiffness (lb/in.)
  $K_S$ = solder stiffness = $AE/L$ (lb/in.)
  $L$ = solder length (in.)
  $A$ = solder cross-sectional area (in$^2$)
  $E$ = modulus of elasticity of solder ($\approx 4 \times 10^6$ psi)

The stress relaxation problem is initiated by instantaneously elongating the outboard end of the spring by amount $\Delta x$ and then holding it at this position; this applies an initial stress ($\sigma_0 = K \, \Delta x/A$) to the solder, where $A$ is the cross-sectional area of the solder. As the solder creeps under the applied stress, the spring load gradually decreases and the creep rate slows. Eventually the solder creeps the total distance $\Delta x$, and the spring load and solder stress reduce to zero.

To obtain a quantitative feel for the stress relaxation behavior of solder, it is useful to examine two cases: one with a displacement ($\Delta x = \sigma_0 A/K$) corresponding to an initial solder stress of $\sigma_0 = 1$ ksi, and the second with a displacement ($\Delta x = L/10$) corresponding to a 10% solder strain upon complete stress relaxation. The 1-ksi initial-stress scenario represents typical conditions immediately following a large deflection that exceeds the solder yield strength; in this instance the solder starts with a stress close to the yield stress independently of lead stiffness. The second case represents a fixed lead offset that does not exceed the solder yield strength. Note that yield stress for our purposes is the stress in Fig. 19-5 at which the strain rate leads to near instantaneous straining (e.g. $\dot{\varepsilon} > 10^{-2}$/sec).

Figures 19-9 and 19-10 present stress relaxation plots computed for these two cases using the authors' finite element creep simulation computer program. The computations are based on the room-temperature solder constitutive properties of "superplastic" and "coarsened" solder as displayed in Figs. 19-5 and 19-6. The results are presented parametrically as a function of spring stiffness ratio $\kappa$, where $\kappa$ is the ratio of the stiffness of the combined solder–lead system (the spring in Fig. 19-8) to the stiffness of the solder element by itself, i.e.,

$$\kappa = \frac{K}{K_S} = \frac{KL}{AE} \tag{19-10}$$

Because $K$ is the effective stiffness of the combined solder–lead system as defined by Eq. (19-9), $\kappa = 1$ corresponds to solder with no additional lead flexibility, and $\kappa = 0.1$ and $0.01$ represent systems 10 and 100 times more flexible than the solder itself, respectively.

From Fig. 19-9 it can be seen that without the addition of system flexibility the maximum *stress-relaxation strain* that can occur in solder by itself is a negligible 0.025%. This strain equals the initial elastic strain ($\varepsilon = \sigma_0/E$) in

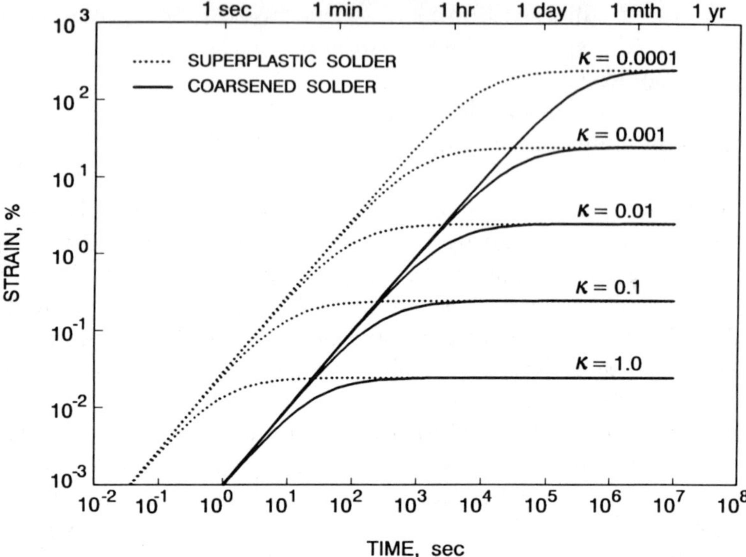


**Figure 19-9**   Creep-relaxation response of solder as a function of relative spring stiffness (κ) for a fixed initial loading stress of 1 ksi.

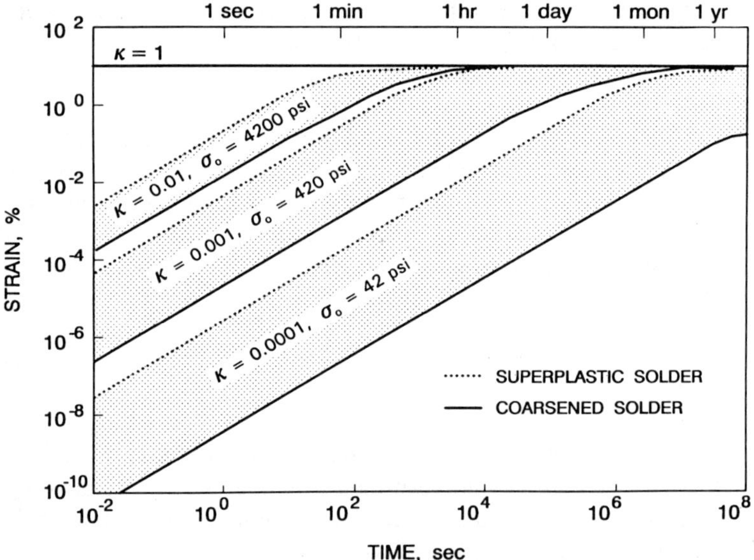


**Figure 19-10**   Creep-relaxation response of solder as a function of relative spring stiffness (κ) for a fixed spring-tip displacement (Δx) corresponding to a 10% strain upon complete stress relaxation.

the solder at 1 ksi, and is seen from Fig. 19-2 to correspond to a fatigue life of tens of millions of cycles. This is important because it confirms that creep and stress relaxation are not important unless significant additional system flexibility is present.

In contrast, notice that for the same initial 1-ksi solder stress, flexible leads can result in very damaging levels of stress-relaxation strains. Making the leads more flexible clearly worsens the problem when the loading condition is a modestly high fixed initial stress. This is apparent by noting that the final creep strain ($\varepsilon_f$) is defined by

$$\varepsilon_f = \frac{\Delta x}{L} = \frac{\sigma_0 A}{KL} = \frac{\sigma_0}{\kappa E} \tag{19-11}$$

To prevent stress-relaxation strain from nulling the effectiveness of flexible strain-relief elements over multi-hour loading cycles, it is necessary to limit the maximum creep strain rate to well below $10^{-6}$/sec; this corresponds to approximately a 1% strain over a 3-hour period. From Fig. 19-5 it can be seen that this requires that the maximum solder stress be less than 100 psi at room temperature—a small fraction of the solder yield stress.

This requirement for low initial stress ($\sigma_0 < 100$ psi) is also apparent from the case of a fixed initial lead deflection as shown in Fig. 19-10. Because Eq. (19-11) defines a fixed relationship between initial stress ($\sigma_0$), strain displacement ($\varepsilon_f$), and lead stiffness ($\kappa$), Fig. 19-10 is marked with the initial stress corresponding to each of the $\kappa$ values. In this case only leads with a relative stiffness ($\kappa$) less than around 0.001 have low initial stress and limit stress relaxation during a typical multi-hour loading cycle. Notice also the important difference made by the solder metallurgical properties ("super-plastic" versus "coarsened" as defined in Fig. 19-5).

By noting the similarity between Fig. 19-9 and 19-10, it may be suspected that a single relationship can be used to describe both; indeed this is true, and Fig. 19-11 provides this relationship as a so-called master curve for creep relaxation. With it, the creep relaxation behavior with any initial stress or spring-tip displacement can easily be determined.

### 19.3.2 Strain Range During Isothermal Mechanical Cycling

As the next step in systematically understanding the role of lead flexibility on solder-joint life, it is useful to extend the single-node solder model presented in Fig. 19-8 by imposing a cyclic mechanical displacement to the end of the spring instead of the fixed-displacement used in the stress-relaxation computations. Figures 19-12 and 19-13 describe the influence of

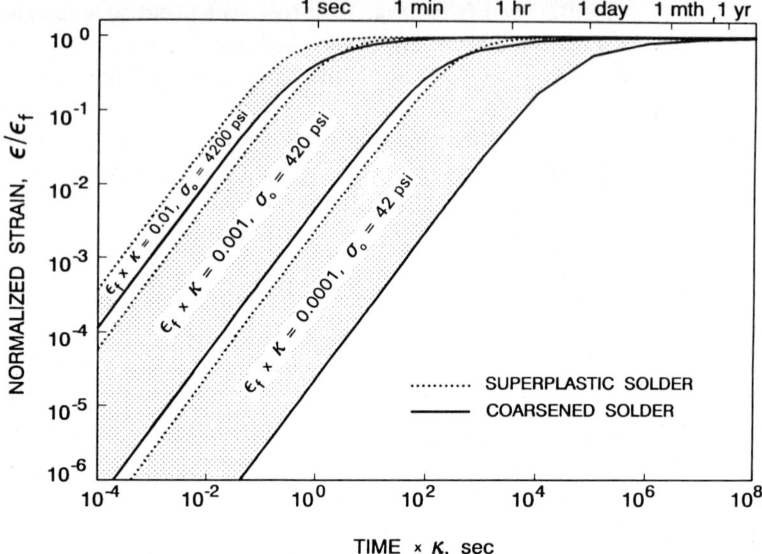


**Figure 19-11** Creep-relaxation master curves that describe creep-relaxation response of solder as a function of relative spring stiffness ($\kappa$) for initial stress ($\sigma_0$) or a spring-tip displacement ($\Delta x$) corresponding to a final strain ($\varepsilon_f$) upon complete stress relaxation.

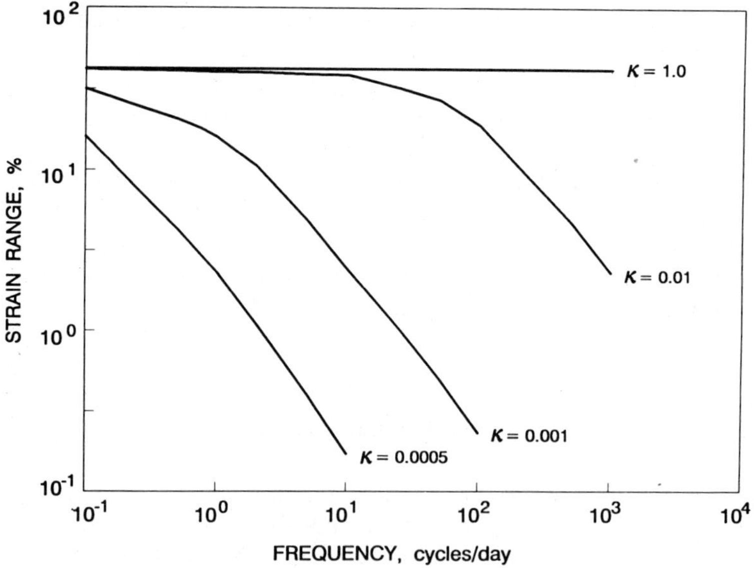


**Figure 19-12** Effect of cyclic frequency of loading and relative spring stiffness ($\kappa$) on computed solder strain range.

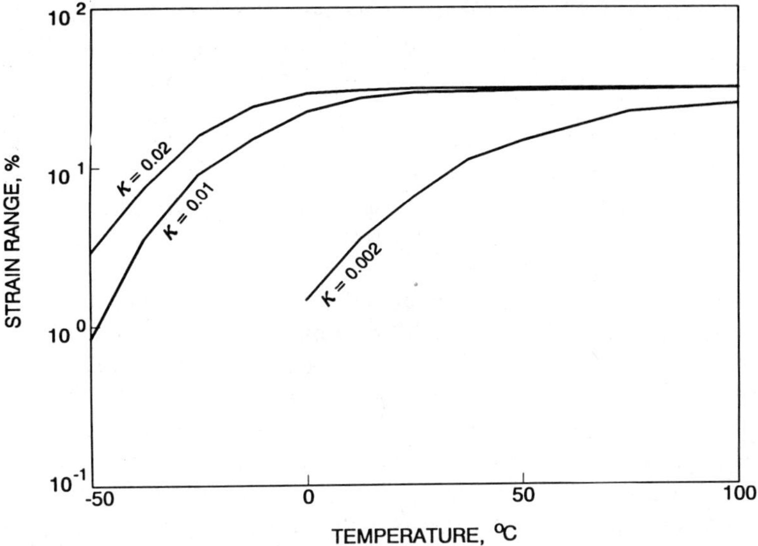

**Figure 19-13**   Effect of solder temperature and relative spring stiffness ($\kappa$) on computed solder strain range.

lead stiffness, cyclic frequency, and solder temperature on the computed strain response to the cyclic displacement profile noted in Fig. 19-14. Because the amplitude of the profile was chosen to correspond to a maximum solder strain of 42% during circumstances of complete stress relaxation, each of the curves in Figs. 19-12 and 19-13 approaches this 42% strain value asymptotically.

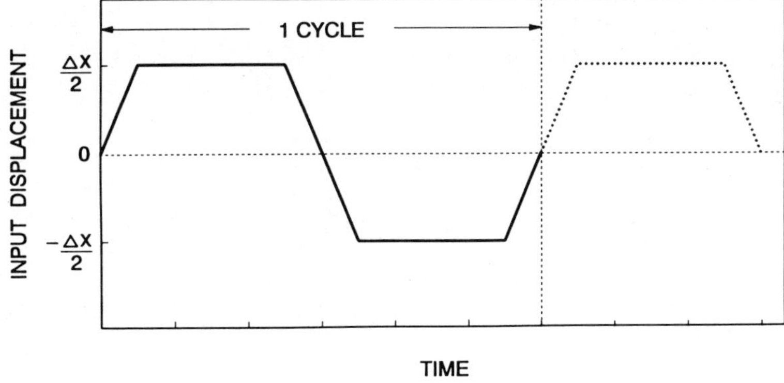

**Figure 19-14**   Cyclic displacement profile ($\Delta x$ versus time) used for isothermal strain studies with single-node spring-loaded solder element.

Looking first at Fig. 19-12, it can be seen that very flexible leads with $\kappa < 0.001$ significantly reduce the generated strain range from cyclic loading. In addition, for highly flexible leads, the strain range amplitude is inversely proportional to frequency—thus proportional to dwell time; this is expected for constant applied stress (minimal stress relaxation), which causes constant creep strain rate during each half-cycle. For very low frequencies, or very stiff leads, the solder creeps to its asymptotic equilibrium value; under such conditions solder strain is determined by the lead deflection and is independent of frequency.

Turning next to solder temperature effects, Fig. 19-13, which is similar to Fig. 19-12, provides the computed response of cyclic loading to solder temperature. The results for highly flexible leads are again predictable, and reflect a factor of 1.5 to 2.0 increase in strain range for each 10°C increase in temperature. This is in agreement with the classical creep rate dependence on temperature[26] shown in Fig. 19-6.

Lastly, Fig. 19-15 describes the effect of variable loading displacement ($\Delta x$ as defined in Fig. 19-14) on solder strain range; in this case the solder temperature is held fixed at 25°C and cyclic frequency is held at 1 cycle per 3 hours. Note that for highly flexible leads the strain range varies approximately as the square of the deflection range ($\Delta x$), and for very stiff leads the curves asymptotically approach the linear ($\Delta \varepsilon = \Delta x/L$) dependency

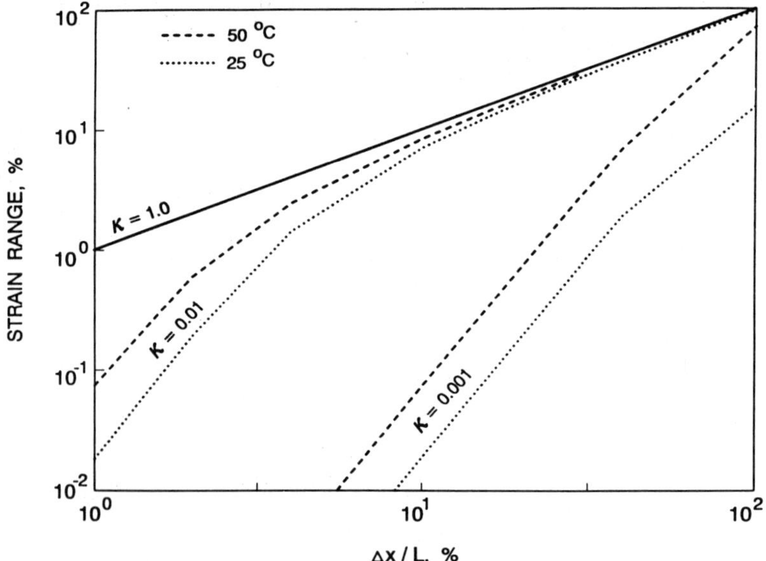


**Figure 19-15**   Effect of input displacement amplitude ($\Delta x$) and relative spring stiffness ($\kappa$) on solder cyclic strain range ($\Delta \varepsilon$).

defined by an infinitely stiff lead (i.e., $\kappa = 1$). Also note that for a fixed $\Delta x$, the strain range for highly flexible leads varies approximately as the square of the relative spring stiffness ($\kappa$).

This stiffness and deflection-range dependence for highly flexible leads follows directly from the constitutive properties of solder as presented earlier in Eq. (19-8). For the case of minimal stress relaxation (i.e., constant stress during each loading half-cycle), the strain rate $\dot{\varepsilon}$ is approximately constant; thus the total strain range for each cycle is

$$\Delta\varepsilon = 2 \int \dot{\varepsilon}\, dt = 2\dot{\varepsilon}\tau \tag{19-12}$$

where $\tau$ is the half-cycle dwell time. Combining Eq. (19-12) with the constitutive properties defined by Eq. (19-8) gives

$$\Delta\varepsilon \propto \tau\sigma^n g^{-p}\xi^{(T/10)} \tag{19-13}$$

where $\sigma$ = solder stress
       $n$ = creep exponent (typically $n = 2$ to 3)
       $g$ = grain size
       $p$ = grain size exponent (typically $p = 1.6$ to 2.3)
       $\xi$ = temperature dependence for 10°C (typically $\xi = 1.5$ to 2)
       $T$ = solder temperature (°C)
       $\tau$ = half-cycle dwell time

Noting that the solder stress is determined by $\sigma = K\,\Delta x/A$ gives

$$\Delta\varepsilon \propto \frac{\tau K^n\, \Delta x^n}{A^n g^p}\, \xi^{(T/10)} \tag{19-14}$$

where $A$ is the cross-sectional area of the solder under load. This expression for very flexible leads notes that the strain range is proportional to roughly the square of the lead stiffness, the square of the lead displacement, and the inverse square of the grain size as defined by the values of $n$ and $p$ associated with Fig. 19-5 and Eq. (19-8). This agrees with, and provides physical insight into, the numerical simulation results presented in Fig. 19-15. For the limiting case of an infinitely rigid lead with no creep, Eq. (19-14) reduces to $\Delta\varepsilon \propto \Delta x$.

The important implication of Eq. (19-14) is that, for very flexible leads, the dependency of the strain range of operational parameters is fundamentally determined by the constitutive properties of solder. If these constitutive properties change for different solder types or aging conditions, the power-law dependencies will also change.

### 19.3.3  Complex Strain Behavior with Leaded Parts

The preceding discussion has highlighted the important parameters influencing the development of creep strain in elastically loaded solder joints. Because creep rate is strongly dependent on the applied stress, flexible leads that lower the stress can greatly diminish the developed strain. However, with flexible leaded parts, such as a gull-wing flat-pak shown in Fig. 19-16, the solder is subjected to a much more complex loading environment than the previously discussed single-node model. There are three principal contributors to this increased complexity:

1. The solder joint is spatially distributed and attached to the part lead at multiple locations. For clarity, it is often useful to consider the joint as made up of numerous independent solder elements attached in parallel between the part lead and the substrate.
2. The complex lead geometry causes the individual solder elements to be loaded by different stress levels in different directions. This causes some solder elements to be primarily loaded normal to the board in tension–compression, while others are primarily loaded parallel to the board in shear. The lead itself undergoes complex motion involving rotation as well as translation.
3. Because the lead is flexible, the load on the individual solder elements is highly interactive and often timewise out of phase. For example, on a gull-wing lead, the solder attaching the toe of the lead cannot be significantly loaded until the heel has deflected. This causes a phase shift between the times when the heel and toe undergo maximum straining.

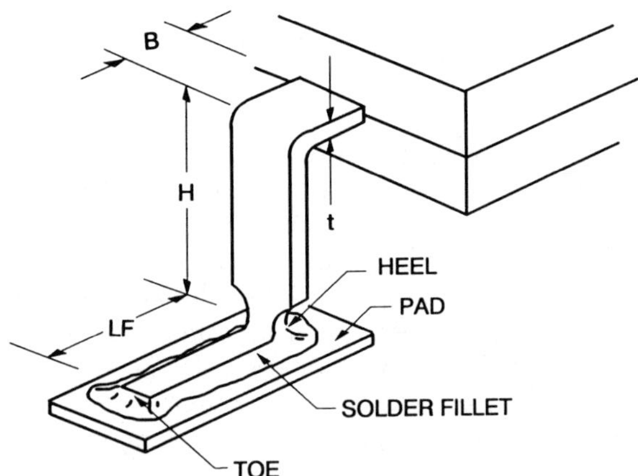


**Figure 19-16**  Solder-joint nomenclature and dimension definitions for flat-pak parts mounted with gull-wing leads.

This lack of simultaneity is important because it implies that there is no single point in time when each element of the solder is at its maximum strain level; thus no single snapshot in time can be used to evaluate the strain range occurring in different elements of the solder joint.

One of the only practical means of analyzing solder joints with flexible leads is the nonlinear finite element creep simulation approach discussed earlier. Figure 19-19 illustrates the solder strain response computed for a flat-pak part using the model shown in Fig. 19-17 and the lead deflection profile shown in Fig. 19-18; the lead deflection is noted with sign as $+X$ in Fig. 19-17. The lead deflection profile shown in Fig. 19-18 ($\pm 0.0003$ in.) is the same magnitude as would be seen by a ceramic chip mounted to an FR-4 board and cycled between $-35°C$ and $+100°C$.

The results presented in Fig. 19-19 correspond to a gull-wing lead with a 20 mil height, a constant 32.5°C operating temperature, and a 3-hour loading cycle. The structure of the finite element model implicitly accounts for the important geometry and distributed flexibility of the lead, and the distributed nature of the solder joint.

In the results shown in Fig. 19-19, the heel and toe rock up and down, applying normal-to-the-board tension and compression to the solder. Note that the heel and toe strains are out of phase with respect to one another, and are substantially different in amplitude. Although some asymmetry occurs at the start of loading, the strain eventually becomes symmetrical about the zero-strain axis.

Figure 19-20 expands on the results of Fig. 19-19 by describing the dependence of the gull-wing solder strain range on lead height, for the same 3-hour cycle and lead-deflection profile shown in Fig. 19-18. Notice that

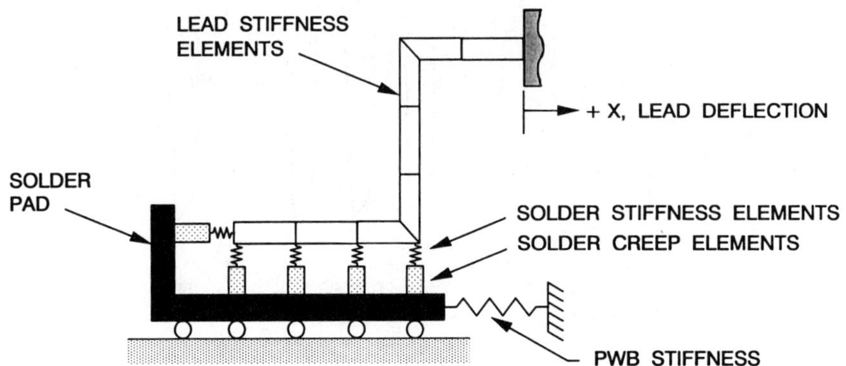


**Figure 19-17**   Schematic of finite element elastic–plastic creep model of solder joint with gull-wing lead.

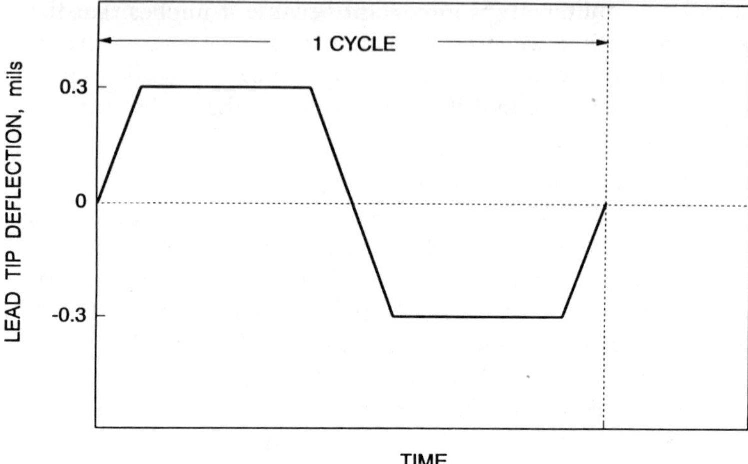


**Figure 19-18** Cyclic displacement profile ($\Delta x$ versus time) used for isothermal strain response studies with gull-wing flat-pak.

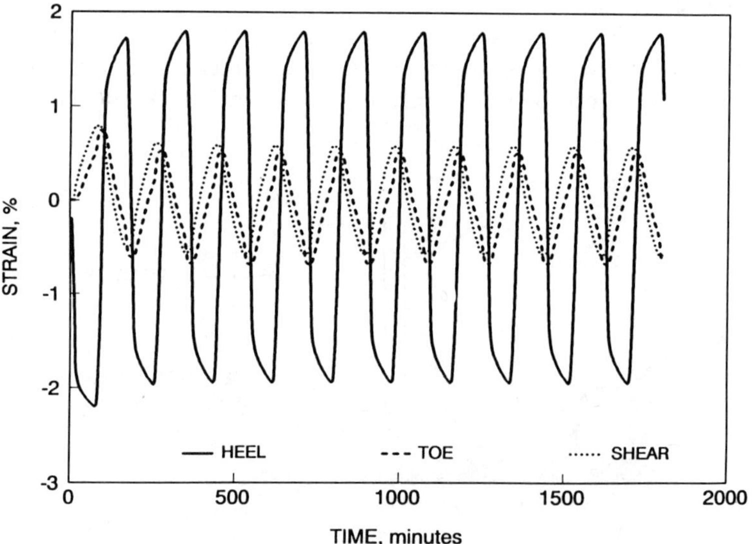


**Figure 19-19** Typical heel, toe, and shear strain response of mechanically cycled gull-wing lead (20-mil high flat-pak).

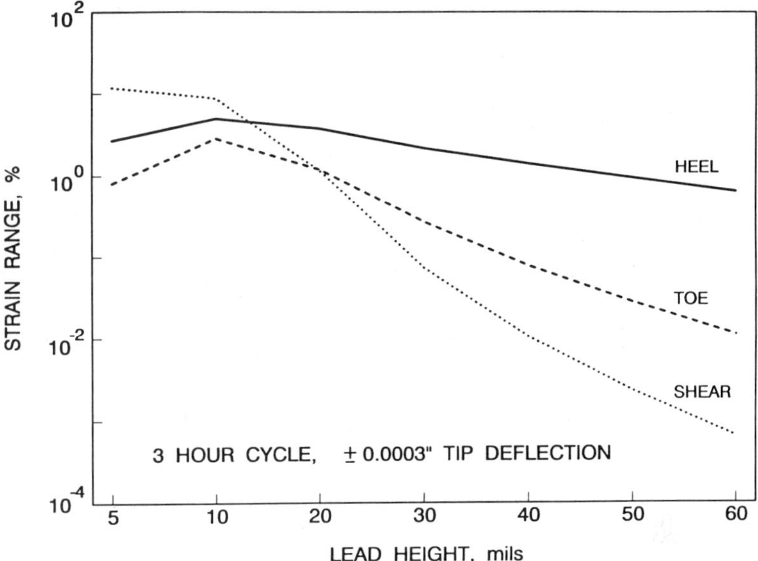



**Figure 19-20**   Computed dependence of strain range on lead height for flat-pak parts with gull-wing leads.

horizontal shear strain, which is negligible with the taller lead heights, becomes dominant when the lead height is reduced below 20 mils. Notice also that the strain at the toe decreases much more rapidly with increasing lead flexibility than does the strain at the heel. These results graphically illustrate the complex interdependency between shear strain and heel–toe tension–compression strain with gull-wing or J-lead type leads. This inter-dependency reflects the fact that the effective stiffness raio ($\kappa_s$) in the horizontal shear direction drops off much faster with increasing lead height than does the effective stiffness ratio ($\kappa_t$) in the normal-to-the-board tension–compression direction. Table 19-1 presents lead dimensions and approximate $\kappa$ values for the popular component lead types illustrated earlier in Fig. 19-1. Nomenclature is presented in Fig. 19-16.

Because of these complex lead stiffness interdependencies, it is instructive to revisit the sensitivity of strain range to lead deflection range $\Delta x$, solder temperature, and cyclic frequency noted in the last section for a single-node solder joint. Figure 19-21 combines the effects of solder temperature, lead stiffness, and lead deflection range into a single plot. Notice that the 20-mil flat-pak lead is quite stiff in that it exhibits little strain range reduction, near linear strain dependency on $\Delta x$, and very little temperature dependency. This is indicative of near complete creep relaxation. In contrast, the quad-pak with much more flexible fine-pitch leads (see Table 19-1) significantly

**Table 19-1** Approximate Dimensions and Relative Stiffnesses for Popular Electronic Component Leads*

|  | Typical Flatpack | Gull-wing Dip | Gull-wing Quad | J-Lead Quad |
|---|---|---|---|---|
| $LF$ (mils) | 50 | 50 | 40 | 45 |
| $H$ (mils) | 20 | 115 | 100 | 115 |
| $B$ (mils) | 35 | 20 | 17 | 30 |
| $t$ (mils) | 7 | 11 | 6.5 | 9 |
| $EI_{LF}$ (lb-mil$^2$) | 8 500 | 34 000 | 3 910 | 19 550 |
| $EI_H$ (lb-mil$^2$) | 8 500 | 113 050 | 3 910 | 29 750 |
| $EI_B$ (lb-mil$^2$) | 8 500 | 113 050 | 3 910 | 20 400 |
| $\kappa_t$, heel | 0.009 | 0.0015 | 0.000 08 | 0.0004 |
| $\kappa_s$, shear | 0.003 | 0.0005 | 0.000 03 | 0.0001 |

* For notation see Fig. 19-16.

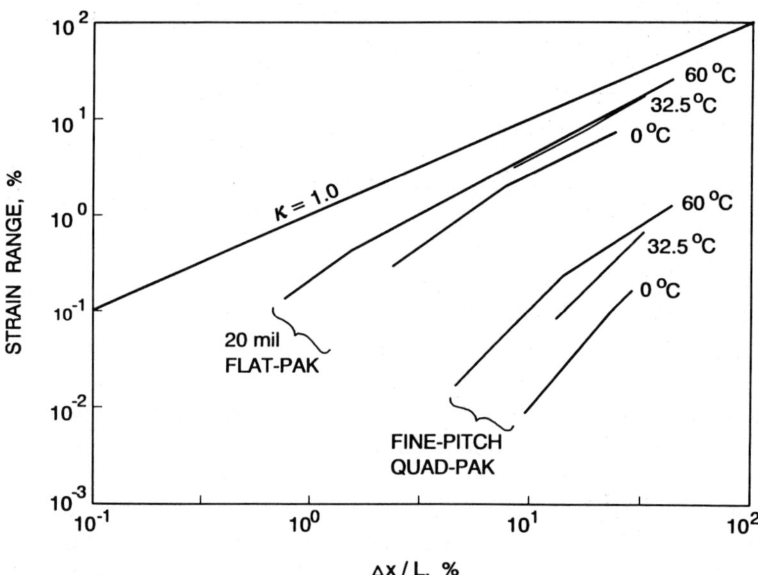


**Figure 19-21** Dependence of solder strain range on lead deflection ranges ($\Delta x$) and temperature for gull-wing leads with different stiffnesses.

reduces the strain range; as a consequence, it exhibits the strong ($\Delta x^n$) strain dependency and Arrhenius temperature dependency associated with the creep properties of solder in Figs. 19-5 and 19-6 and Eq. (19-14). Figure 19-22 amplifies on the temperature dependency of strain range for these two lead configurations and graphically illustrates that parallel-to-the-board shear strain has a much stronger temperature dependency than does the normal-to-the-board tension–compression strain at the lead's heel. This is cause by the relative differences in $\kappa$ for the shear and tension–compression loading directions, and by the complex interplay associated with simultaneous straining in these two orthogonal directions. This complex strain interdependency will become even more important in the thermal-cycling environments discussed in the next section.

Lastly, Fig. 19-23 displays the computed dependence of strain range on loading frequency for gull-wing leads with a constant $\pm 0.0003$ in. lead deflection amplitude, as shown in Fig. 19-18, and 32.5°C solder temperature. Notice that the frequency dependence of the more highly flexible fine-pitch quad-pak lead is very similar to the 1/frequency dependence previously noted in Fig. 19-12 and Eq. (19-14). In contrast, the 20-mil flat-pak lead is seen to exhibit a significantly lower frequency dependence, consistent with its higher stiffness.

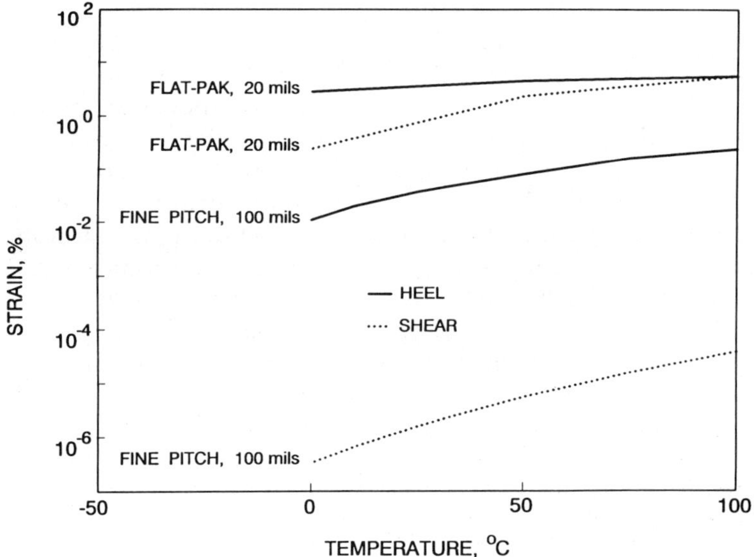


**Figure 19-22**  Temperature dependence of solder heel and shear strain range for gull-wing leads of different stiffnesses.

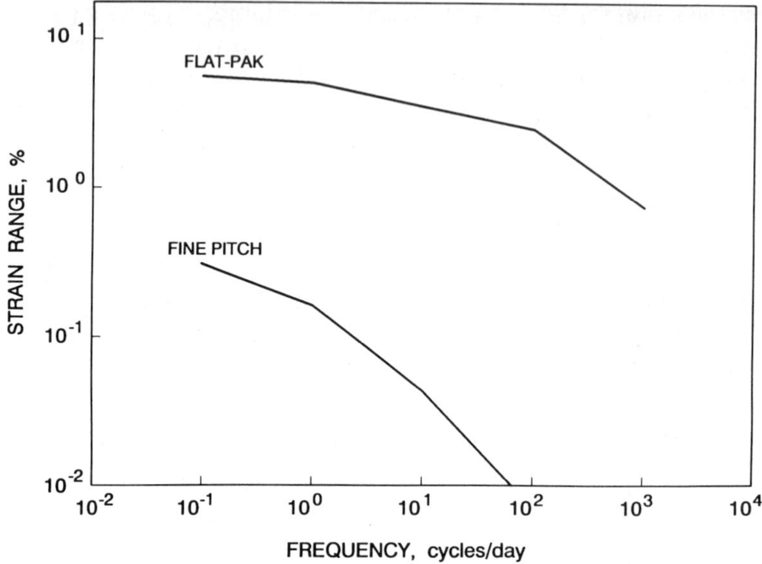


**Figure 19-23**  Dependence of solder strain range on cyclic frequency for gull-wing leads of different stiffnesses.

## 19.4 ROLE OF LEAD STIFFNESS WITH THERMAL CYCLING

Thermal cycling, or thermomechanical loading as it is often referred to, generates similar displacement loading profiles to those noted in the last section. However, the displacements generally arise from differential thermal expansions and contractions of the electronic part relative to those of the substrate to which it is mounted. Although the resultant lead deflection may be similar, a fundamental difference between isothermal mechanical cycling and thermal cycling is the fact that the solder is at different temperatures at different points in the loading cycle. This alters the creep strain contributions from various parts of the cycle, and introduces important asymmetries in the resulting strains. With shallow cycle depths the differences between isothermal cycling and mechanical cycling can be quite small and accurately ignored. However, with increasing depth, the difference can become very pronounced and must be carefully considered.

Figure 19-24 displays the computed strain response during a representative field-application thermal cycle with 15°C temperature range (25°C to 40°C) and 3-hour cycle period. The input horizontal displacement ($\Delta x$) corresponds to the thermal mismatch between a ceramic flat-pak ($x = 1.23$ cm, $\alpha_p = 6$ p.p.m./°C) and an FR-4 PWB bonded to an aluminum substrate ($\alpha_b = 20$ p.p.m./°C). The mean temperature is 32.5°C in this case, and the

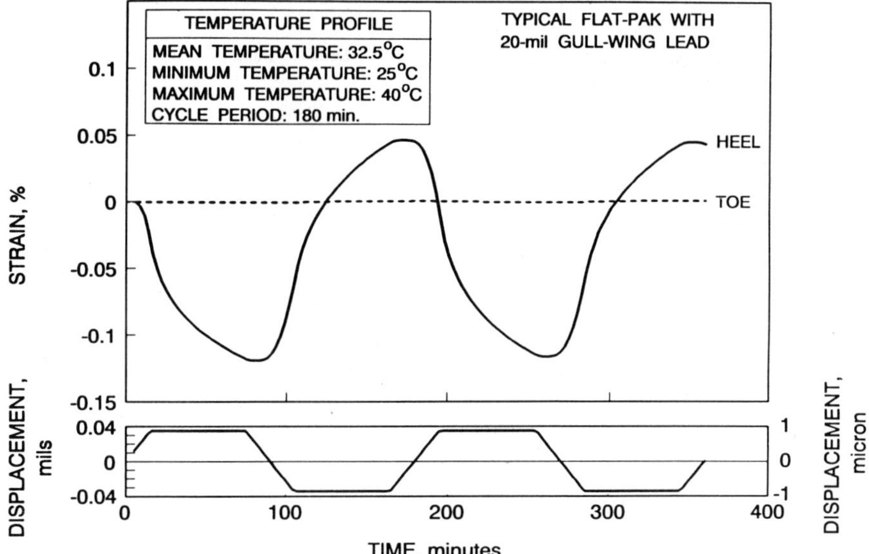



**Figure 19-24**   Example flat-pak lead displacement and solder strain response to 15°C thermal cycle with 3-hour period.

displacement variation, shown at the bottom of Fig. 19-24, corresponds to a $\pm 7.5°C$ temperature swing over the 3-hour cycle. The first two cycles of computed strain variation at the solder joint heel and toe are shown in the top of Fig. 19-24 for a typical 20-mil lead height and 7-mil lead thickness. The maximum strain range (about 0.17%) occurs as a normal-to-the-board tension–compression at the heel, while the toe undergoes much less strain.

### 19.4.1 Creep Ratcheting

Perhaps the most important difference between isothermal mechanical cycling and thermal cycling is the existence of progressive strain drift or "creep ratcheting" that accompanies the alternating fatigue component in long-term thermal cycling. As an example, Fig. 19-25 displays the long-term strain-cycling behavior of the flat-pak part used in Fig. 19-24, together with the long-term behavior of parts with both taller (30-mil) and shorter (10-mil) lead heights. Although the computed cyclic strain ranges (0.62% and 0.051%, respectively) vary as expected with the increasing lead height, the solder also undergoes a lead-height-dependent long-term strain drift. In this example, this progressive strain drift is 2.5%, 0.5%, and 0.04%, respectively, after 1000 thermal cycles.

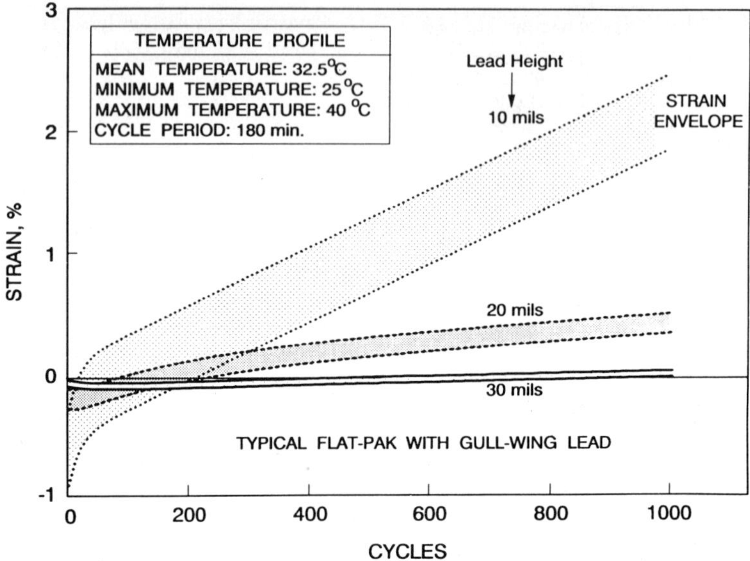


**Figure 19-25**   Thermal cycle strain response showing dependence of creep ratcheting on lead height.

This long-term drift is referred to as "creep ratcheting" in recognition of its ratchet-like directional dependence, which stems from the strong temperature dependence of creep during different phases of the loading cycle. This creep-ratcheting phenomenon has only recently been understood through the use of finite element strain simulation analyses.[1,2,19]

To further illustrate the observed creep ratcheting phenomenon, a pair of simulations is presented in Fig. 19-26 based on the same 20-mil flat-pak used in Fig. 19-24 but with the displacements provided in one case by the 25°C to 40°C thermal cycle and in the other case by an isothermal mechanical cycle with the same lead displacement and 32.5°C mean temperature. The left side of the figure compares the strain variations in the first five cycles. It can be seen that the shapes and amplitudes of the two cases are very similar. However, after 1000 cycles, the mechanically cycled straining remains essentially unchanged and is symmetrical about the neutral position. The thermally-cycled straining, on the other hand, has ratcheted—in essence, drifted—an additional 0.6% over and above the mechanical cycling component.

The mechanics of creep ratcheting can be better understood by studying the exaggerated movements occurring in deep temperature cycles ($-35°C$ to $100°C$); these are typical of qualification test levels for electronic circuit boards.[29]

Figure 19-27 displays the tension–compression strain variations occurring between the 90th and 100th deep thermal cycles (range = 135°C, mean

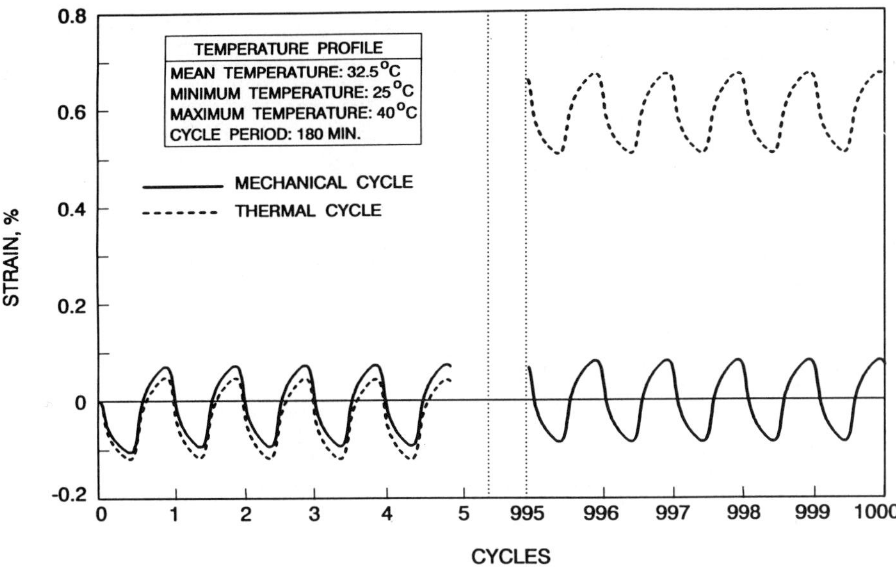



**Figure 19-26**   Creep ratchet visible in thermal cycle strain response as compared with isothermal mechanical cycling.

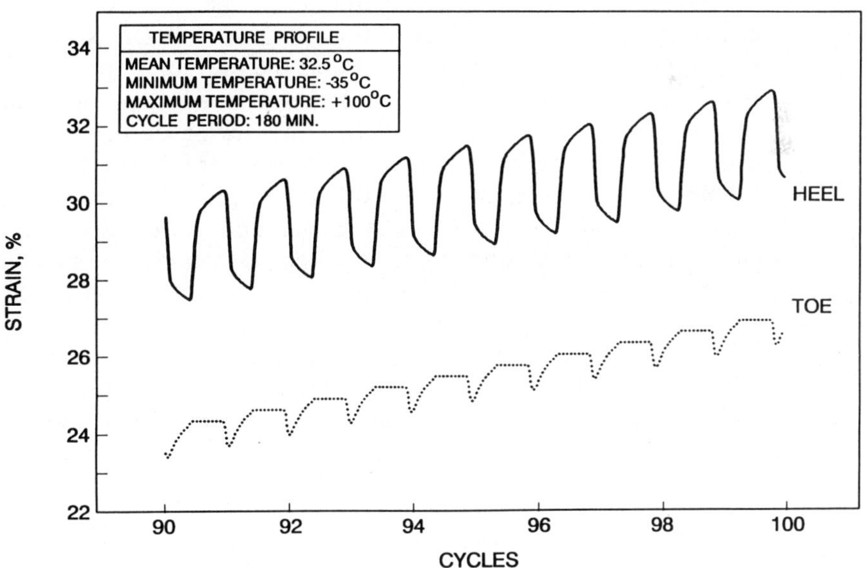



**Figure 19-27**   Creep ratcheting apparent in heel and toe strain response to 135°C thermal cycling.

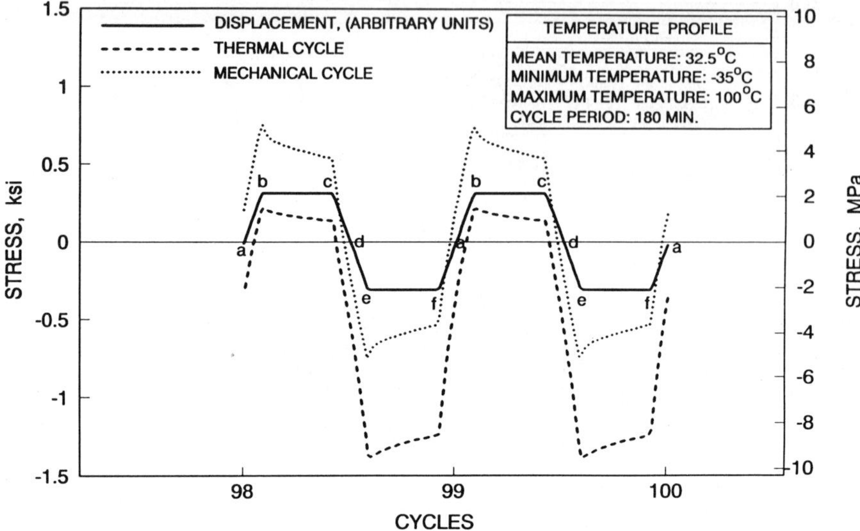


**Figure 19-28**  Comparison of steady-state shear stress vs. time for isothermal mechanical cycling and thermal cycling having the identical (shown) displacement.

temperature = 32.5°C) as computed at the heel and at the toe for the same lead–solder system used in Fig. 19-25. The rate of creep ratcheting is considerably larger for the 135°C cycle depth than for the 15°C cycle depth; this can be seen by comparing the data of Figs. 19-25 and 19-27. Note that the tension–compression strains at the heel and toe are out of phase and that the toe strains have a very different time response than during isothermal cycling (Fig. 19-19). The nature of the heel-to-toe rocking during thermal cycling causes the creep ratcheting that causes a bias "drift" in a direction normal to the board—away from the board if $\alpha_b > \alpha_p$, toward the board if $\alpha_b < \alpha_p$; recall that $\alpha_p$ and $\alpha_b$ are the CTEs of the part and board, respectively.

To explore the fundamentals underlying the creep-ratcheting phenomenon, it is useful to examine closely the 135°C thermal cycling simulation results. Figure 19-28 shows the computed shear stress variations for this thermal cycle (for $\alpha_b > \alpha_p$) overlaid on the corresponding data for isothermal mechanical cycling, i.e., with identical part displacement profiles. It is helpful to first examine the mechanics of the mechanical cycling, and then to contrast them with the observed differences for thermal cycling.

### 19.4.2 Strain Development During Isothermal Mechanical Cycling

During the expansion phase of the isothermal mechanical cycle, there is a positive displacement of the part lead in the $+x$ direction (defined in

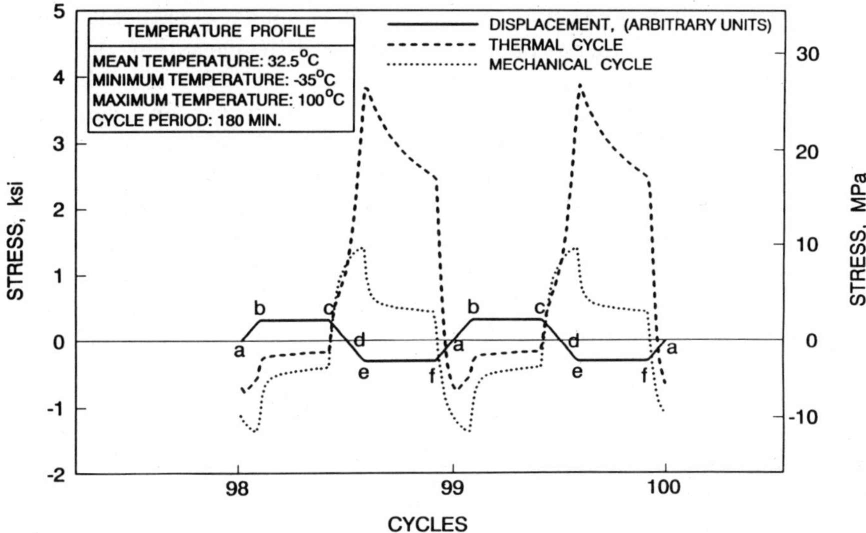


**Figure 19-29**    Comparison of steady-state heel stress vs. time for isothermal mechanical cycling and thermal cycling having the identical (shown) displacement.

Fig. 19-17), and a resulting horizontal force applied to the solder fillet in the $+x$ direction. This force generates both horizontal shear stresses and vertical tension–compression stresses within the solder. The corresponding shear stress in the solder increases to a maximum as the displacement proceeds from point a to point b (Fig. 19-28). During the dwell time (b to c), the stress relaxes due to creep until the displacement direction reverses (c to d) and ultimately achieves a maximum displacement in the $-x$ direction (point e) just before the stress again undergoes creep relaxation during the dwell period. In essence, once a steady-state condition is achieved, the shear stress variation during isothermal mechanical cycling is symmetrical relative to the neutral position and it closely tracks the displacement profile.

The corresponding tension–compression heel stress variation in isothermal mechanical cycling is also symmetrical about the zero strain axis, as shown in Fig. 19-29. The heel is under compression, while the toe is under tension when the lead displacement is positive. These stresses relax during the dwell time and reverse their sense when the lead displacement changes direction.

### 19.4.3 Strain Development During Thermal Cycling

If the part displacements arise from differential thermal expansion–contraction during temperature cycling, the dynamics are quite different from those for isothermal mechanical cycling; this is because solder creeps much more rapidly during the higher-temperature portion of the cycle than during the lower-temperature portion, in accordance with Fig. 19-6. The difference between isothermal mechanical cycling and temperature cycling manifests in the very first cycle. If the thermal cycle starts with cooling, a net initial displacement in the $-x$ direction, the solder is quite creep-resistant because of the low temperature; the resulting shear strain is therefore very limited and most of the would-be solder displacement is absorbed by lead deflection. As the solder temperature increases during the warm-up phase of the cycle, it more readily deforms in response to the force exerted by the deflected lead; this produces a large shear strain in the $+x$ direction (almost 3% in the first cycle as shown in Fig. 19-30). In the subsequent cooling phase the solder again loses its ductility and yields only slightly in the reverse $(-x)$ direction. After a number of cycles the lead achieves a deflection bias in the $+x$ direction that provides a greater force in the yielding direction during the cold half-cycle, and a reduced force during the hot half-cycle. This equilibrium force offset is that required to provide equal horizontal creep strain in the plus and minus shear directions during each cycle; at this point the shear strain cycling becomes symmetrical, and the shear creep ratcheting equilibrates at a constant value. The number of cycles to achieve this symmetry is a function of the lead stiffness and lead–solder properties.

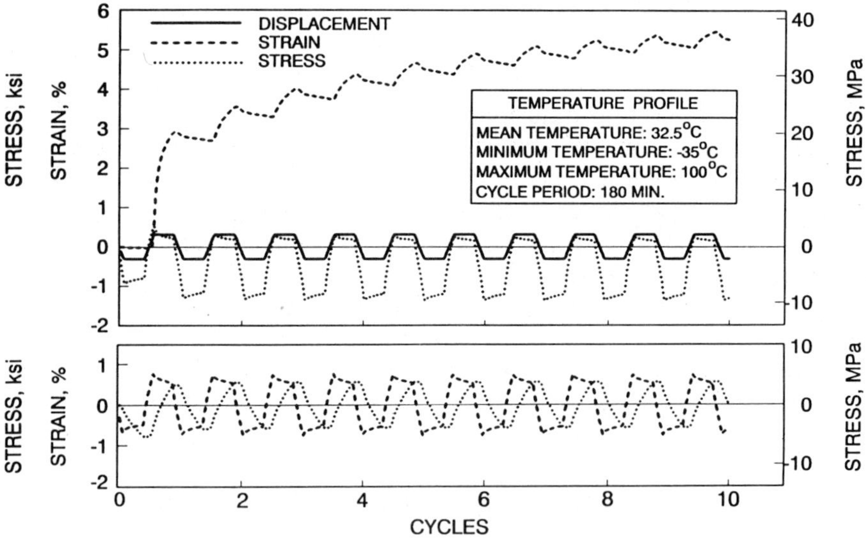


**Figure 19-30**  Transient shear stress and strain response to isothermal mechanical cycling (bottom) and thermal cycling (top) with identical displacement.

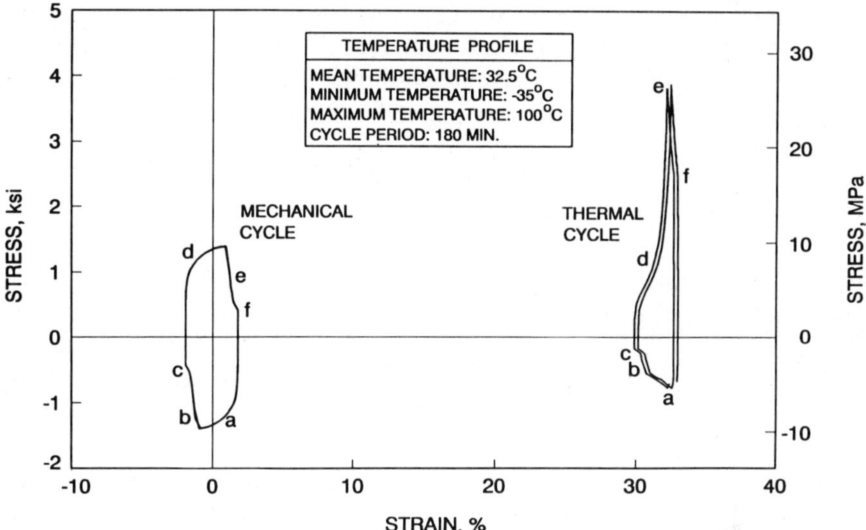


**Figure 19-31**   Heel stress–strain hysteresis loops for thermally and mechanically strained solder joints after 900 cycles.

In response to this horizontal lead deflection offset that develops to maintain shear strain symmetry, out-of-phase tension–compression forces are developed normal to the board, favoring tension–creep in the solder joint for the case $\alpha_b > \alpha_p$, and compression–creep in the solder joint for the case $\alpha_b < \alpha_p$. As a result, the tension–compression stress at the heel during the cooldown (c–d–e) is significantly larger than the same stress during the warmup (f–a–b), as shown in Figs. 19-29 and 19-31. Consequently, a resultant creep-ratchet strain in the tension–compression direction is generated each cycle. Since normal-to-the-board tension drift has no means of generating a self-arresting force, it continues ratcheting at a constant rate as illustrated in Fig. 19-25. For the opposite case ($\alpha_b < \alpha_p$), where normal-to-the-board ratcheting occurs in the compression direction, ratcheting will self-arrest when the lead bottoms out against the solder pad.

### 19.4.4 Damage Prediction with Combined Creep Ratcheting and Fatigue

Before further examining the implications of the thermal cycling strain simulation results it is useful to develop an analysis structure for comparing the damage caused by creep-ratcheting strain relative to that caused by cyclic fatigue strain.

Although there are many possible approaches to assessing the interactive damage contributions of creep rupture and fatigue, the most straightforward technique is to use Miner's rule, which additively accumulates the individual fractional damage rates as follows:

$$\phi = \eta \frac{N}{\Gamma} + \frac{N}{N_f} = N\left(\frac{\eta}{\Gamma} + \frac{1}{N_f}\right) \tag{19-15}$$

where  $\phi$ = fraction of useful service life consumed
$\eta$ = creep-ratchet strain per cycle
$\Gamma$ = strain limit for creep rupture
$N$ = number of cycles of applied stress
$N_f$ = number of cycles to fatigue failure

Following our earlier lead, we use Manson's[9] fatigue relationship as embodied in Fig. 19-2 to predict $N_f$, thus

$$N_f^{\beta}\, \Delta\varepsilon = C \tag{19-16}$$

where $N_f$ = number of cycles to fatigue failure
$\Delta\varepsilon$ = cyclic plastic strain range
$\beta \approx 0.40$
$C \approx 0.80$

To estimate the strain rupture limit ($\Gamma$), we again follow the lead of Manson, who in the development of his strain-cycle fatigue relationship (Eq. 19-16), viewed the tensile test as a cyclic fatigue failure at 1/4-cycle. In a similar fashion, the strain range at the 1/4-cycle point on the Coffin–Manson plot can be taken as a rough estimate of the creep extension limit, i.e.,

$$\Gamma \approx \frac{C}{(0.25)^{\beta}} \tag{19-17}$$

Combining Eqs. (19-15), (19-16), and (19-17) gives

$$\phi = N\left[\frac{\eta(0.25)^{\beta}}{C} + \left(\frac{\Delta\varepsilon}{C}\right)^{1/\beta}\right] \tag{19-18}$$

Equation (19-18) can also be used to describe the acceleration of a particular test condition (1) over that of another condition (2) by noting that the acceleration ratio $R$ is given by the ratio $N_2/N_1$, where $N_1$ is the number of environmental stress cycles of condition (1) required to yield the same damage as $N_2$ cycles of environmental stress condition (2), i.e., $R = N_2/N_1$ for $\phi_1 = \phi_2$. Thus,

$$R = \frac{[\eta_1(0.25)^\beta/C + (\Delta\varepsilon_1/C)^{1/\beta}]}{[\eta_2(0.25)^\beta/C + (\Delta\varepsilon_2/C)^{1/\beta}]} \qquad (19\text{-}19)$$

Returning to the above examples, Eq. (19-18) can now be used to obtain an estimate of the relative importance of the observed strain ratcheting and cyclic fatigue.

With the modest 15°C-cycle depth used in Fig. 19-25, recall that the creep ratcheting drifts were 2.5%, 0.5%, and 0.04% after 1000 cycles for the 10-, 20-, and 30-mil lead heights, respectively, and the corresponding cyclic fatigue strain range per cycle was 0.62%, 0.16%, and 0.051%, respectively. When these data and the parameters from Fig. 19-2 are entered into Eq. (19-18), the creep-ratchet and fatigue damage are found to be about equal in level.

Continuing this analysis for the simulation results of the deep 135°C thermal cycles provides the data assembled in Table 19-2. In the cases presented in Table 19-2, the dominating failure mechanisms are extension–compression low-cycle fatigue and creep at the heel (or toe) of the solder joints. Note that creep-ratcheting represents over 50% of the damage in most of the cases.

**Table 19-2**  Comparison of Computed Cyclic and Creep-Ratchet Damage to Solder Joints on Flak-Pak Parts with Different Lead Heights Exposed to 15°C and 135°C Thermal Cycle Depths

| Lead Height (mils) [mm] | Cycle Depth (°C) | Cycle Strain Range (%) | Percent Creep Ratchet Strain per 1000 Cycles | Number of Cycles to Failure | Percent Damage due to Creep Ratchet | Acceleration, Factor, 135°C to 15°C |
|---|---|---|---|---|---|---|
| 10 | 15 | 0.512 | 2.50 | $2.9 \times 10^4$ | 49 | 1 |
| [0.25] | 135 | 4.630 | 683 | 152 | 69 | 195 |
| 20 | 15 | 0.160 | 0.50 | $2.1 \times 10^5$ | 71 | 1 |
| [0.51] | 135 | 2.830 | 330 | 347 | 76 | 618 |
| 30 | 15 | 0.051 | 0.04 | $2.6 \times 10^6$ | 70 | 1 |
| [0.76] | 135 | 1.470 | 9.13 | 4370 | 26 | 607 |

## 19.5 DISCUSSION AND CONCLUDING SUMMARY

It has been shown that the use of highly flexible leads can greatly reduce the strain developed in solder joints during cyclic loading conditions. The key requirement on the leads is to limit the maximum stress that is applied to the solder joint during peak loading conditions so as to minimize the generation of creep strain. For multi-hour loading dwells, such as are common for electronic assemblies that are power-cycled daily (cars, computers, TVs), stress levels must be reduced significantly below the solder yield stress. Depending on the lead displacement applied by the part–board system, lead stiffnesses must generally be at least 1000 times more flexible than the solder itself (i.e. $\kappa < 10^{-3}$, as defined in Eq. (19-10)). Failure to use a sufficiently flexible lead may result in stresses that cause excessive creep-strain generation.

The rate at which solder creeps is a strong function of not only the applied stress but also the solder temperature and metallurgical structure. As shown in Figs. 19-5 and 19-6, temperature and solder aging conditions can have order-of-magnitude influences on creep-strain generation. The parametric studies presented provide an overview of the expected sensitivities to cycle rate, temperature, lead stiffness, and thermal cycle depth. When complex lead geometries such as gull-wing leads are used, the type of loading (lateral shear versus normal-to-the-board tension–compression) is strongly dependent on the lead flexibility. Solder joints with leads having different compliance may result in different damage mechanisms in the same environment. For highly compliant leads, solder-joint failure is generally caused by tension–compression cyclic fatigue at the heel. However, for a very stiff lead, the dominant damage mechanism can be cyclic shear fatigue. When thermal cycling is introduced, even more complex behavior results from the strong creep-rate dependency on temperature. One such complexity is the development of creep-ratcheting strains both parallel to and normal to the board. This type of strain, which is not present in isothermal mechanical cycling, may represent 50% of the total damage associated with thermal cycling.

Because of unit-to-unit variabilities in solder joint geometry and aging condition it is difficult to expect analytical simulation results to provide more than a qualitative feel for the important solder-joint design parameters and their sensitivities. The ultimate verification of any given flexible lead design must be determined by test under conditions that accurately replicate the exact manufacturing processes and applicable loading conditions.

When viewing test results, it is important to note that the relative importance of each damage mechanism can change as time progresses. Initial shear creep ratcheting, if it does not exceed the rupture limit, is self-arresting after a few cycles. This damage mechanism can thus produce an early appearance of cracks, but does not necessarily result in crack propagation.

Instead, as more and more cycles accumulate, the dominant damage mechanism can become tensile creep ratcheting and fatigue, eventually resulting in failure. An important observation is that conclusions based upon early test inspection, when one failure mechanism is predominant, may lead to erroneous projections of failure at later times; this is because damage accumulation rates can change as different mechanisms become predominant.

Another important attribute of different lead configurations is their ability to fail gracefully with a large spread between the number of cycles for crack initiation and the number of cycles to electrical open circuit. The low toe stress with flexible gull-wing leads provides this desirable attribute and results in slow propagation of fatigue cracking down the solder side fillets. Measuring the length of cracking allows a useful quantification of the damage level accumulated to date, and the percentage of useful life remaining in the joints.

Finally, where flexible leads are used for strain relief, it is critically important that the lead flexibility not be compromised (significantly stiffened) by subsequent manufacturing and assembly processes. Two such areas to watch carefully are: (1) the application of conformal coating and heat-sink materials that can bridge the part–board gap and generate large normal-to-the-board stresses;[30,31] and (2) the wicking of solder up into the lead system such that it significantly changes the bending stiffness of the leads. In a related problem, some ceramic J-lead parts that have body metallization very close to the "J" of the lead are vulnerable to soldering of the flexible end of the lead to the bottom of the part. This "shorts out" the lead flexibility completely and can result in almost immediate solder-joint failure.

With properly designed flexible lead systems, solder strains will be substantially lower than with leadless devices, and completed boards should be capable of surviving hundreds of deep cyclic loads.

## ACKNOWLEDGMENTS

The authors' contributions in this chapter were partially supported by the Jet Propulsion Laboratory, California Institute of Technology, under contract with the National Aeronautics and Space Administration. The authors would like especially to thank Ms. Elizabeth Jetter who conducted many of the computer simulations and generated most of the figures used in this chapter.

## REFERENCES

1. Ross, R. G., Jr., et al., "Solder Creep–Fatigue Interactions with Flexible Leaded Parts," *ASME J. Electronic Packaging*, **114**, 1992, pp. 185–192.

2. Ross, R. G., Jr., et al., "Creep-Fatigue Behavior of Micro-electronic Solder Joints," Proceedings of the 1991 MRS Spring Meeting, Anaheim CA, April–May 1991.

3. Lau, J. H., *Solder Joint Reliability: Theory and Application*, Van Nostrand Reinhold, New York, 1991.

4. Frear, D. R., W. B. Jones, and K. R. Kinsman, *Solder Mechanics: A State of the Art Assessment*, Minerals, Metals and Materials Society, Warrendale, PA, 1991.

5. Aldrich, J. W., and D. H. Avery, "Alternating Strain Behavior of a Superplastic Metal," in *Ultrafine Grain Metals*, Proceedings of the 16th Sagamore Army Material Research Conference, August 1969, Syracuse University Press, 1970, pp. 397–416.

6. Wild, R. N., "Some Fatigue Properties of Solders and Solder Joints," IBM Report No. 7AZ000481, IBM Federal Systems Division, New York, 1975.

7. Solomon, H. D., "Influence of Temperature on the Fatigue of CC/PWB Joints," *Journal of the IES*, January/February 1990, pp. 17–25.

8. Manson, S. S., "Effect of Mean Stress and Strain on Cyclic Life," *Machine Design*, August 1960, pp. 129–135.

9. Manson, S. S., "Fatigue: A Complex Subject—Some Simple Approximations," *Engineering Mechanics*, **5**, 1965, pp. 193–226.

10. Solomon, H. D., "Predicting Thermal and Mechanical Fatigue Lifes from Isothermal Low-cycle Data," *Solder Joint Reliability: Theory and Application*, Van Nostrand Reinhold, New York, 1991, pp. 406–454.

11. Solomon, H. D., *ASME J. Electronic Packaging*, **111**, 1989, pp. 75–82.

12. Vaynman, S., M. E. Fine, and D. A. Jeannotte, "Low-cycle Isothermal Fatigue Life of Soldered Materials," *Solder Mechanics*, D. R. Frear et al., eds., Minerals, Metals and Materials Society, Warrendale, PA, 1991, pp. 333–360.

13. Gohn, G. R., and W. C. Ellis, *Proc. ASTM*, **51**, 1951, pp. 721–740.

14. Solomon, H. D., *Electronic Packaging Materials and Processes*, J. A. Sortell, ed., ASM, 1985, pp. 29–49.

15. Solomon, H. D., *IEEE 38th Electronic Components Conference*, May 1988, pp. 7–13.

16. Engelmaier, W., "Functional Cycling and Surface Mounting Attachment Reliability," ISHM Technical Monographic Series 6984-002, International Society for Hybrid Microelectronics, 1984, pp. 87–114.

17. Engelmaier, W., *Circuit World*, **11**(3), 1985, pp. 61–67.

18. Engelmaier, W., "Solder Attachment Reliability, Accelerated Testing, and Result Evaluation," *Solder Joint Reliability: Theory and Application*, Van Nostrand Reinhold, New York, 1991, pp. 545–587.

19. Pan, T-Y., "Thermal Cycling Induced Plastic Deformation in Solder Joints—Part 1: Accumulated Deformation in Surface Mount Joints," *ASME J. Electronic Packaging*, **113**, March 1991, pp. 8–15.

20. Cline, H. E., and T. H. Alden, "Rate Sensitive Deformation in Tin–Lead Alloy," *Trans. AIME*, **239**, 1967, pp. 710–714.

21. Zehr, S. W. and W. A. Backofen, "Superplasticity in Lead–Tin Alloys," *Trans. ASM*, **61**, 1968, pp. 300–312.

22. Guo, Z., A. F. Sprecher, and H. Conrad, "Plastic Deformation Kinetics of Eutectic Pb–Sn Solder Joints in Monotonic Loading and Low-cycle Fatigue,"

ASME Paper No. 91-WA-EEP-48 presented at the 1991 ASME WAM, Atlanta, GA, December 1991.

23. Weinbel, R. C., J. K. Tien, R. A. Pollak, and S. K. Kang, "Creep–Fatigue Interaction in Eutectic Lead–Tin Solder Alloy," *J. Materials Science Letters*, **6**, 1987, pp. 3091–3096.

24. Avery, D. H., and W. A. Backofen, "A Structural Basis for Superplasticity," *Trans. ASM*, **58**, 1965, pp. 551–562.

25. Murty, G. S., "Stress Relaxation in Superplastic Materials," *J. Materials Science*, **8**, 1973, pp. 611–614.

26. Kashyap, B. P., and G. S. Murty, "Experimental Constitutive Relations for the High Temperature Deformation of a Pb–Sn Eutectic Alloy," *Materials Science and Engineering*, **50**, 1981, pp. 205–213.

27. Arrowood, R., A. Mukherjee, and W. B. Jones, "Hot Deformation of Two-phase Mixtures," *Solder Mechanics: A State of the Art Assessment*, Minerals, Metals and Materials Society, Warrendale, PA, 1991, pp. 107–153.

28. Lampe, B. T., "Room Temperature Aging Properties of Some Solder Alloys," *Welding Research Supplement*, October 1976, pp. 330–340.

29. NASA, "Requirements for Soldered Electrical Connections," NHB 5300.4(3A-1), revalidation date June 1986.

30. Ross, R. G., Jr., *Magellan/Galileo Solder Joint Failure Analysis and Recommendations*, JPL Publication 89-35, Jet Propulsion Laboratory, Pasadena, CA, September 1989.

31. Ross, R. G., Jr., "A Systems Approach to Solder Joint Fatigue in Spacecraft Electronic Packaging," *ASME J. Electronic Packaging*, **113**, June 1991, pp. 121–128.

# 20

# Thermal Stress Issues in Plated-Through-Hole Reliability


*Donald B. Barker and Abhijit Dasgupta*


## 20.1 INTRODUCTION

Failure of plated-through-holes (PTHs) due to thermomechanical stresses is a well-established cause of failure of multilayer printed wiring boards (MLBs). This chapter uses the finite element method (FEM) to examine the nature of the stress distribution within the PTH structure when the MLB is subjected to uniform thermal loads as seen in operational environments as well as to examine the thermal transients seen in wave soldering. Guidelines are laid out for realistic modeling of material properties, boundary conditions, and stress averaging in the FEM model. Parametric studies are conducted on the critical stresses in the PTH to study the qualitative effect of several geometric parameters, including the effect of filling the PTH with solder. The purpose here is to provide an insight into the stresses that cause PTH failures and to establish guidelines for the reliable design of PTHs. Manufacturing-related problems such as defects and flaws in the plating material and actual life predictions are not addressed at this time.

A plated-through-hole is defined in ref. 1 as "A hole in which electrical connection is made between internal or external conductive patterns, or both, by plating of metal on the wall of the hole." In the past, PTHs had large diameters and served as receptacles for component leads. They almost always contained solder. More recently, PTH diameters have decreased, and more often than not they do not contain solder. This chapter deals with PTHs in multilayer fabric-reinforced rigid MLBs, although the methods suggested here can be applied to other vias and board types.

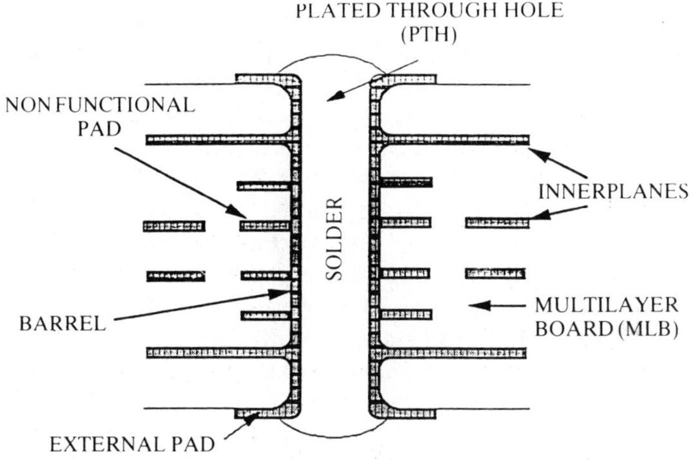


**Figure 20-1**   Cross-sectional geometry of a plated-through-hole (PTH).

Figure 20-1 shows the cross-sectional geometry of a typical PTH. Failure of a PTH constitutes an electrical discontinuity that can be caused by circumferential barrel cracks, land corner cracks, separation of lands from the MLB surface, and/or shearing of internal traces. The failures result from the thermomechanical stresses caused by the large differences in the coefficients of thermal expansion (CTE) between the PTH plating materials and the MLB in the thickness direction of the MLB. The principal failure mechanisms are ductile fracture of the plating caused by thermal shock (during soldering), fatigue fracture of the plating caused by cyclic temperature variations (operational and environmental loads), and interfacial deadhesion between the plating and the surrounding MLB caused by low temperatures.

The majority of PTH research has been in the experimental detection of PTH failures, inspection methods, metallurgical studies, and round-robin evaluations of a wide variety of commercial MLBs.[2-10] Models developed to simulate thermomechanical behavior of PTHs have followed two separate approaches. The analytical approach utilizes closed-form expressions derived from a simplified model for the critical values of stress/strain in the PTH.[11-17] The numerical approach uses finite elements to obtain local stress/strain fields in and around the PTH.[10,18-23]

In this chapter a review is presented of recent research conducted by the authors in the use of the finite element method to examine the critical thermomechanical stresses within the PTH.[24-26] The advantages of detailed numerical modeling are that material anisotropy and complex geometries can be easily handled and critical stresses can be obtained at any point within the MLB. Manufacturing defects, such as surface irregularities, variations in thickness, excessive etchback, and other flaws[17] are not addressed here. In spite of the modeling assumptions, this study reveals critical information on the effect of geometry and material properties on stresses in PTHs, and the results are of intrinsic value to the PTH designer.

The chapter is divided into four sections. The first section is more general and examines in detail the three-dimensional and axisymmetric FEM models of the PTH. Guidelines are laid out for realistic modeling of anisotropic MLB material properties and boundary conditions and also for averaging of stresses across dissimilar material interfaces. The critical stresses that cause the various failure mechanisms are also identified. Comparisons are made between the three-dimensional and axisymmetric models. In the second section, the axisymmetric model is then used for carrying out parametric studies on the effect of various geometric parameters on the critical stresses within the PTH. The study is conducted both for standard E-glass epoxy (FR-4) and for highly anisotropic Kevlar-polyimide MLBs that have a low in-plane CTE tailored for surface mount devices. The third section looks at the differences in the stress distribution in a solder-filled PTH and the effects of voids in the solder. Finally, the fourth section looks at the transient thermal stresses in a non-solder-filled PTH subjected to a typical wave soldering process.

## 20.2 FINITE ELEMENT MODELING OF A PTH

The following subsections present a methodology and guideline for axisymmetric and three-dimensional (3D) finite element methods in studying critical stresses within a PTH. Accurate boundary conditions and stress averaging are important in order to get reliable stress predictions. The overall goals are to provide modeling methods that will provide insights for reliable design of PTHs.

### 20.2.1 Boundary Conditions

The in-plane dimensions of an MLB are many orders of magnitude larger than the cross-sectional dimensions of a PTH. Therefore an appropriate "unit cell" containing only one PTH is isolated from the surrounding MLB and discretized for analysis. Suitable "periodic" boundary conditions (described later) are imposed on the unit cell to simulate the presence of surrounding MLB material. The unit cell extends half-way to the neighboring PTHs and assumes a regular rectangular arrangement of equally spaced PTHs on the board as depicted in Fig. 20-2.

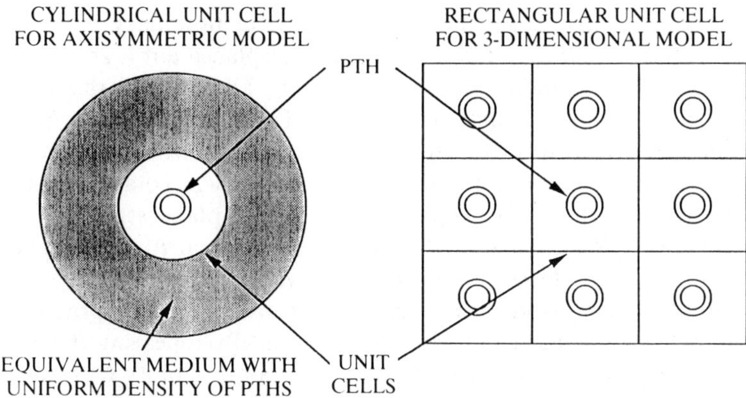


**Figure 20-2**   Unit cell for axisymmetric and 3D finite element models.

The axisymmetric and 3D models (unit cells) of the top half of a PTH structure (above the MLB mid-plane) are shown in Fig. 20-3. While the axisymmetric computational model uses 4-noded axisymmetric finite elements, the 3D model uses 8-noded isoparametric brick elements. The dimensions, mesh geometry, and mesh density are similar for the two models for comparison purposes. The difference between the axisymmetric and 3D models lies in their capability of handling periodic boundary conditions and orthotropic material properties.

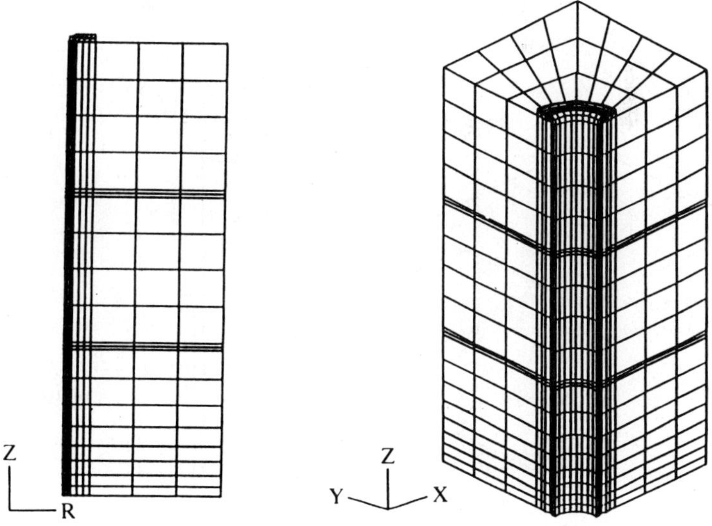


**Figure 20-3**   Finite element meshes of a PTH. *Left,* axisymmetric model. *Right,* three-dimensional model.

In the axisymmetric case, the outer cylindrical boundary either can be constrained in the radial direction (i.e., radial displacement is zero), or it can be given a periodic boundary condition whereby all nodes on the cylindrical boundary have the same radial deformation. Totally constraining the radial deformation is over-restrictive and unrealistic except when the ends of the MLB are physically constrained relative to each other. Totally constraining the radial deformation will exaggerate the out-of-plane stresses and land rotations. The periodic boundary conditions, shown in Fig. 20-4, allow the PTH and the MLB material in the immediate vicinity to deform as a cylinder. These boundary conditions are in accordance with common practice when modeling heterogeneous structures with periodically repeating inclusions.[27] However, it violates rectilinear periodicity and compatibility with surrounding unit cells. Instead it considers each PTH and a representative amount of surrounding board material to be embedded in an equivalent homogeneous medium that contains a similar density of uniformly distributed PTHs.

For the 3D rectangular unit cell model, the periodic boundary conditions permit the faces to deform but constrain them to remain plane and parallel to the original faces as shown in Fig. 20-5. These boundary conditions are more realistic than the axisymmetric boundary conditions because the rectangular unit cell deformations are compatible with self-similar periodic deformations of similar surrounding unit cells. Due to the quarter-plane symmetry of the rectangular unit cell, only one quarter of the top half is modeled.

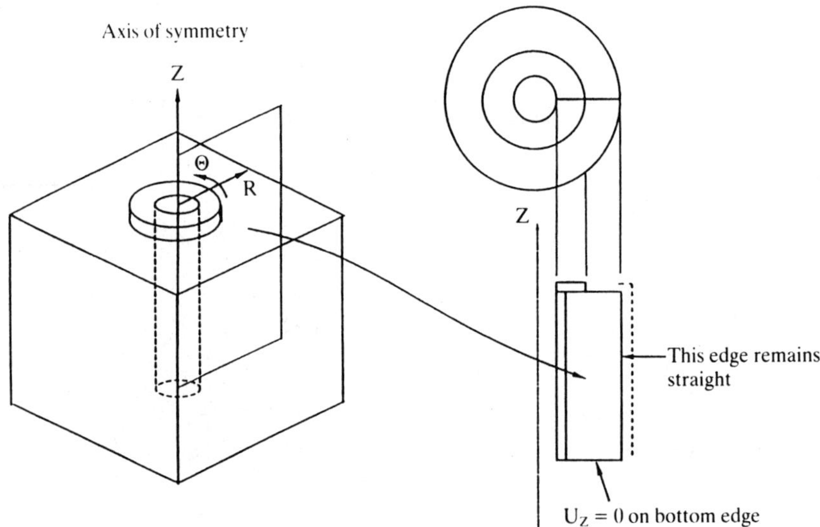


**Figure 20-4**   Boundary conditions for the axisymmetric model.

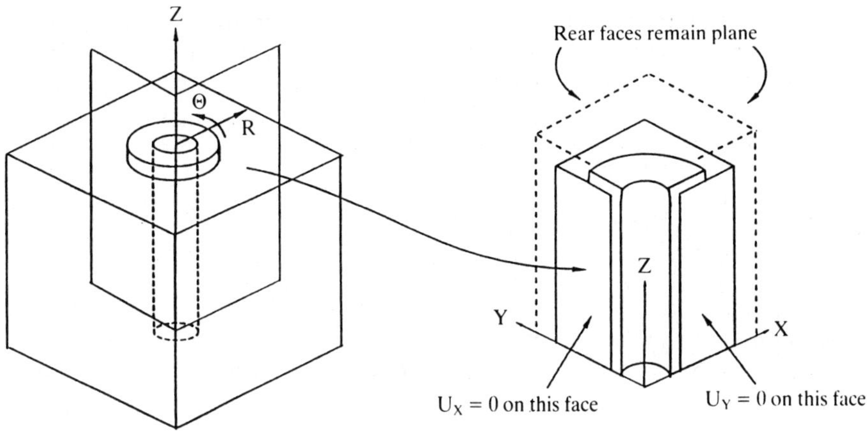


**Figure 20-5**    Boundary conditions for the 3D model.

### 20.2.2 Material Properties

Most MLBs are constructed from fabric-reinforced composites that are orthotropic materials, with two principal material directions running along the fabric fill and warp directions and a third along the out-of-plane normal. The axisymmetric model is incapable of modeling such a fully orthotropic material and is forced to model transverse isotropy in the plane of the MLB ($r$–$\theta$ plane). This introduces errors since the in-plane shear modulus $G$ for a transversely isotropic material is forced to take the value $G = E/[2(1 + v)]$ ($E$ and $v$ are the in-plane elastic modulus and Poisson's ratio, respectively). This dependency of $G$ on $E$ and $v$ is generally not true for an orthotropic material. The 3D model, on the other hand, is capable of modeling the full orthotropy of the MLB with nine independent elastic constants and three thermal coefficients of expansion and therefore represents the most accurate modeling option.

A major difficulty in full 3D modeling lies in obtaining realistic values for the material properties. While properties for plated copper and solder are widely available in the literature, the properties for copper can vary widely within a PTH,[16,17] and the properties of solder are highly temperature and rate dependent.[28] The complete set of properties for the orthotropic MLB material is also hard to obtain. Published values are available for some material constants such as the CTEs and the in-plane elastic moduli. This is partly due to the difficulty in experimental characterization of the three-dimensional behavior of laminated composites. References 25 and 32

**Table 20-1**   Material Properties of FR-4, Kevlar-49 Polyimide, and Copper Plating

| PWB Material Properties | | Copper Plating Properties |
|---|---|---|
| FR-4 | Kevlar-49 Polyimide | |
| $E_x = 17.0$ GPa | $E_x = 25.7$ GPa | $E = 130.0$ GPa |
| $E_y = E_x$ | $E_y = E_x$ | $v = 0.35$ |
| $E_z = 7.45$ GPa | $E_z = 4.58$ GPa | $\alpha = 17.0$ p.p.m./°C |
| $v_{xy} = 0.13$ | $v_{xy} = 0.04$ | |
| $v_{xz} = 0.42$ | $v_{xz} = 0.62$ | |
| $v_{yz} = v_{xz}$ | $v_{yz} = v_{xy}$ | |
| $G_{xy} = 3.0$ GPa | $G_{xy} = 2.70$ GPa | |
| $G_{xz} = 2.4$ GPa | $G_{xz} = 1.76$ GPa | |
| $G_{yz} = G_{xz}$ | $G_{yz} = G_{xz}$ | |
| $\alpha_x = 18.23 \times 10^{-6}/°C$ | $\alpha_x = 2.0 \times 10^{-6}/°C$ | |
| $\alpha_y = \alpha_x$ | $\alpha_y = \alpha_x$ | |
| $\alpha_z = 58.72 \times 10^{-6}/°C$ | $\alpha_z = 69.0 \times 10^{-6}/°C$ | |

describe a numerical technique for computing these properties by modeling the microstructure of the fabric-reinforced composite. The board properties used in this chapter are obtained from such a numerical simulation. Table 20-1 lists the material properties for copper and MLB materials (FR-4 and Kevlar-49 polyimide) used for the axisymmetric and 3D models.

### 20.2.3 Stress Averaging

It is important to realize that most commercial FEM packages average all components of stress across material interfaces during post-processing unless special precautions are taken. Improper averaging of stresses leads to significant errors in computing the maximum stress values that often occur at material interfaces.

Since the model has different materials, not all components of stresses are continuous across the material interfaces. Care is taken while averaging the stresses to maintain traction continuity across interfaces between dissimilar materials. When computing the stresses in the barrel, for example, the $\sigma_{rr}$, $\sigma_{rz}$, and $\sigma_{r\theta}$ stresses are continuous in the interfacial elements and are averaged across the elements. The other stress components, however, are not averaged. In the pads, lands, and innerplanes, the $\sigma_{zz}$, $\sigma_{zr}$, and $\sigma_{z\theta}$ stresses are continuous and are averaged.

### 20.2.4  Critical Stresses

The critical stresses for the various PTH failure modes are listed in Table 20-2. The interfacial stresses are responsible for the de-adhesion type failures while the Von Mises equivalent stresses in the barrel cause barrel cracking. It should be noted here that failure could result from a combination of failure modes, such as hole-wall pull-away during thermal cycling that weakens the barrel, followed by a circumferential barrel crack in the next cycle. In such cases, a single value of stress cannot be used as a failure indicator.

**Table 20-2**   Critical Stress for Various PTH Failure Modes

| Failure Mode | Critical Stresses |
| --- | --- |
| Circumferential barrel crack | Von Mises stress |
| Land corner crack | Von Mises stress |
| Land separation | $\sigma_{zz}$, $\sigma_{rz}$, $\sigma_{z\theta}$ |
| Hole-wall pull-away | $\sigma_{rr}$, $\sigma_{rz}$, $\sigma_{r\theta}$ |

### 20.2.5  Comparison Between 3D and Axisymmetric Analysis

For comparative purposes, a linear-elastic analysis is performed on both models for a unit rise in temperature. The general pattern of stresses agrees in both cases, except that the 3D model shows a circumferential variation of stress due to material orthotropy. The locations of maximum stress match in the two models. The $\sigma_{r\theta}$ and $\sigma_{z\theta}$ stresses are zero in the axisymmetric analysis by definition. This implies that the 3D model should be used for interfacial de-adhesion failure modeling to properly account for the shear stresses as well as the barrel normal stresses.

Figure 20-6 shows a comparison of the maximum Von Mises barrel stress at the MLB mid-plane (bottom of the model) obtained in the axisymmetric and the 3D models for various PTH separations. The maximum difference in the Von Mises stresses between the 3D orthotropic and the axisymmetric models is less than 7% (for both E-glass and Kevlar) for the geometry considered in the analysis. The 3D model predicts slightly higher stresses.

The 3D model, while more accurate, is computationally too intensive to be used for parametric studies. It is ideal for analyzing the details of a given PTH geometry or for detailed failure analysis. The axisymmetric model, on the other hand, is easier to use for parametric studies. The 5% inaccuracy in Von Mises stresses is small and is outweighed by the

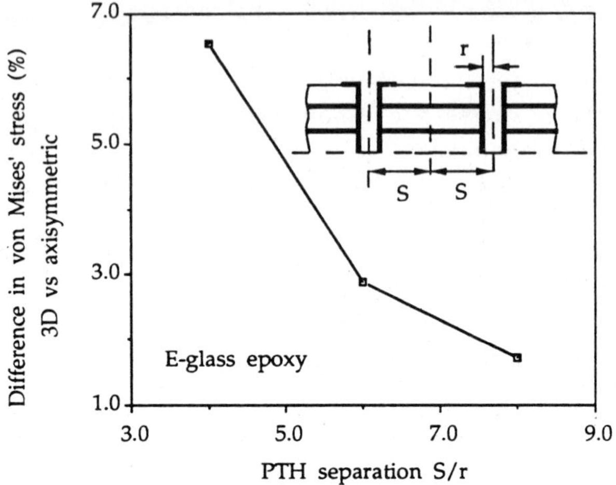


**Figure 20-6**   Comparison of maximum Von Mises barrel stress in the axisymmetric and 3D models.

assumptions and simplifications regarding geometry and material properties that are routinely made during parametric studies. Therefore the axisymmetric model is used in the following section to study the effect of various geometrical parameters.

## 20.3 PARAMETRIC STUDIES OF PTH GEOMETRY

As explained in the previous section, the axisymmetric model is used to study the effect of geometric parameters such as PTH spacing, presence and location of innerplanes, nonfunctional internal pads, and aspect ratio. It is assumed in the analysis that there are no flaws and irregularities in the plating and that the plating is perfectly bonded to the substrate. The data in this section is presented in a normalized fashion to emphasize qualitative information and general trends rather than raw numbers. These trends are of use to the PTH designer. Furthermore, no attempt is made in this chapter to compute absolute values of stresses and life, which would require a nonlinear analysis and would have to consider manufacturing defects.[17] Such analysis is currently under way and forms the topic of a forthcoming paper. The following additional considerations and assumptions are made:

1. The only plating material used is copper with no undercoats or overcoats. Additive nickel layers are sometimes used in practice but are not considered in this chapter.

2. The maximum Von Mises stress in the barrel is considered the critical stress, implying that circumferential barrel cracking is the principal failure mechanism. Land corner cracking is not considered because land corner stresses are highly dependent on the radius and local geometry at the root of the land, which is process-dependent. Land corner cracking also requires a very refined FEM mesh for correct modeling of the stress concentrations.

3. A linear-elastic analysis is carried out for a uniform positive temperature change. The results are considered good comparative figures of merit under actual conditions.

4. A hole diameter of 10 mils, aspect ratio of 12:1 (MLB thickness to hole diameter), plating thickness of 0.75 mil, and a land diameter of 15 mils are used as the starting point for each parametric study. The number of innerplanes and pads are indicated in the figures corresponding to each analysis.

5. The axisymmetric mesh used in the parametric studies uses 8-noded axisymmetric elements and is much more refined than the mesh shown in Fig. 20-3. Mesh refinements are made wherever necessary according to the geometrical parameter being varied.

### 20.3.1 PTH Spacing

The critical stresses within a PTH are affected by the presence of other PTHs around it. This is clearly shown in Fig. 20-7, which is a plot of the maximum Von Mises stresses in the PTH barrel versus the distance separating adjacent PTHs. In the figure the barrel stresses have been normalized with respect to the barrel stress for a solitary PTH with no neighboring PTHs. As expected the stresses drop as the PTH spacing is decreased. The stress drop is due to the larger number of PTHs per unit MLB area constraining the out-of-plane thermal expansion of the MLB. External land radius does not directly affect the PTH barrel stress, except for the fact that a smaller land radius allows closer PTH spacing. The PTH barrel stress becomes insensitive to the presence of neighboring PTHs when the separation distance exceeds approximately 8.5 times the inner radius $r_i$. The results are similar for both FR-4 and Kevlar-polyimide boards.

### 20.3.2 Innerplanes

The effect of the presence and location of solid copper innerplanes (power/ground) is shown in Fig. 20-8. There is a marked difference between FR-4 and Kevlar-polyimide MLBs. For FR-4 there is a lowering of

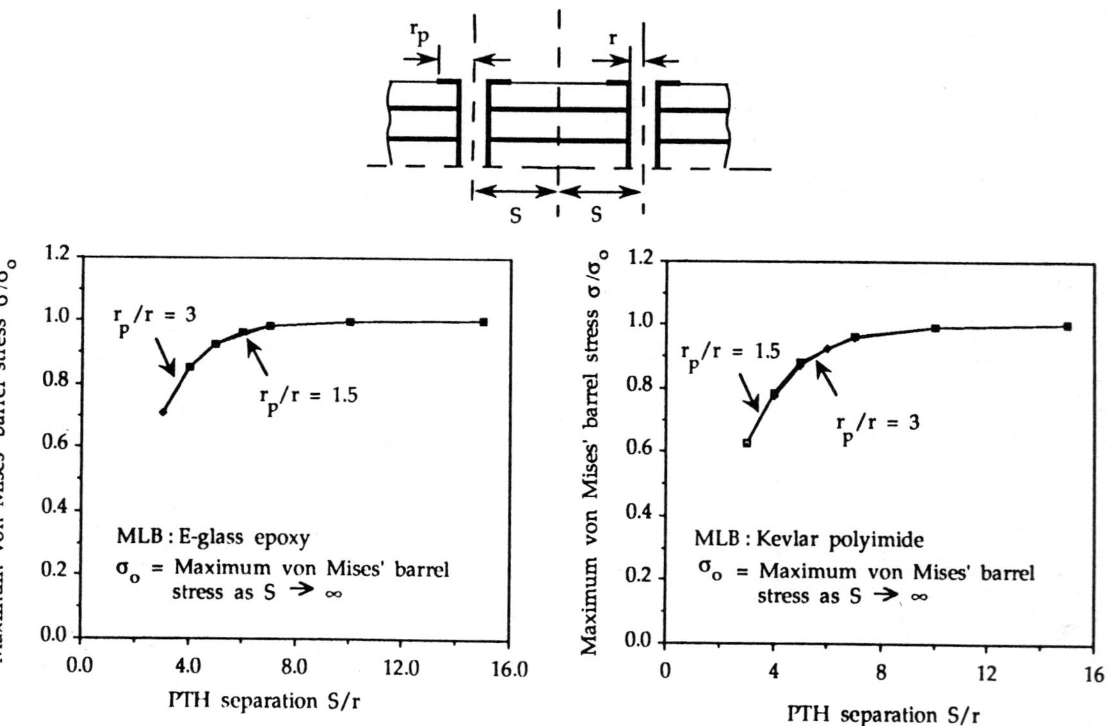


**Figure 20-7**   Effect of PTH separation on Von Mises barrel stress.

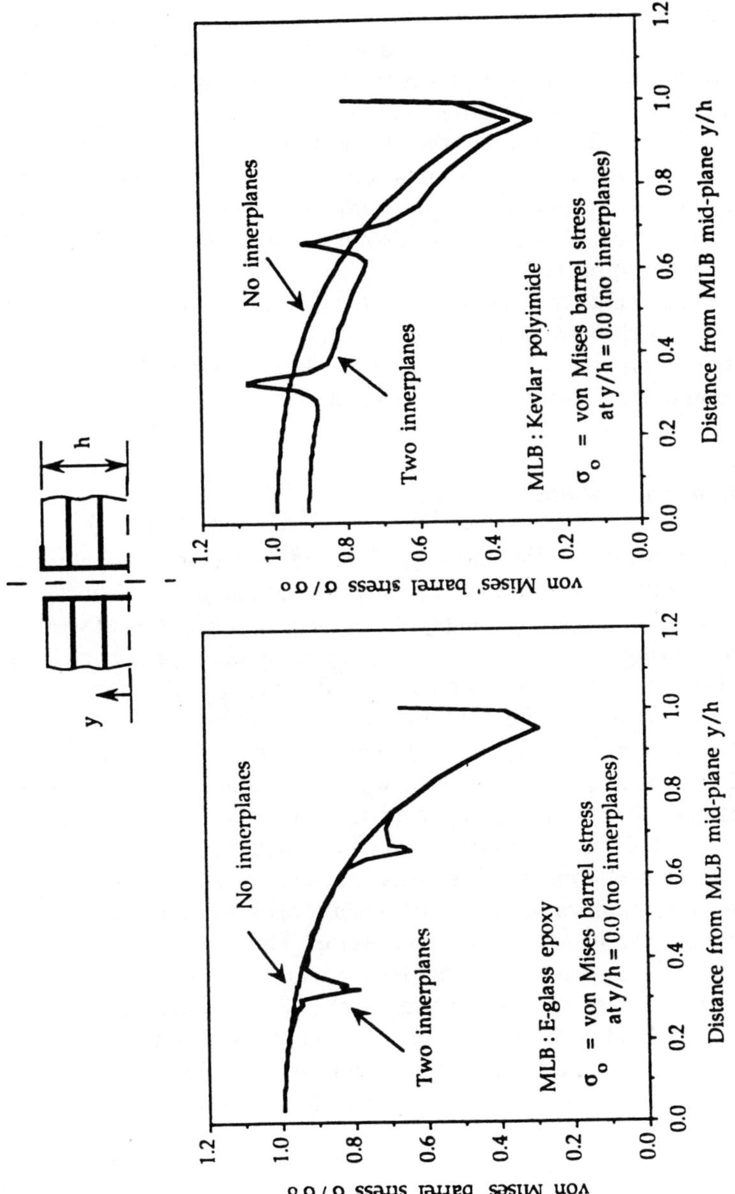


**Figure 20-8**  Effect of inner planes on Von Mises barrel stress.

Von Mises barrel stress at the innerplane and the effect is local, i.e., confined to the region of the barrel near the innerplanes. The effect is reversed for Kevlar-polyimide MLBs, where there is a local increase of stress at each innerplane accompanied by an overall lowering of stress along the entire length of the barrel. More innerplanes imply a lower overall stress. The local stress spike may exceed the stress at the mid-plane of the barrel as the innerplane is moved closer to the MLB mid-plane. While the relative magnitudes of the spikes are dependent on other geometric parameters such as plating thickness, aspect ratio, land radii, and PTH separation, it is seen that innerplanes located at a suitable distance away from the MLB mid-plane are beneficial for improved PTH life.

The difference between FR-4 and Kevlar-polyimide MLBs arises because there is a mismatch between in-plane expansion of Kevlar-polyimide MLBs and the copper innerplane. For FR-4 MLBs, the in-plane expansion between the copper and the FR-4 is closely matched.

### 20.3.3 Nonfunctional Internal Pads

It has been reported in the industry that the inclusion of nonfunctional internal pads within the MLB is beneficial for extended PTH life. This may be due to a combination of geometry-related factors such as local stress relief and manufacturing-related factors such as better anchoring of barrel plating to the substrate due to the presence of the pads. The effect on stress distributions is shown in Fig. 20-9 for internal nonfunctional pads that are of the same radii as the external pads. Perfect bonding is assumed between the barrel plating and MLB irrespective of the presence or absence of pads. The innerplanes have been removed in this case. The effect in the case of glass–epoxy MLBs is the same as for the innerplanes discussed above. However, for Kevlar-polyimide boards the stresses are dependent on the land radius. For small-radius pads there is a local lowering of stress with an upward stress spike just beginning to develop. There is no overall lowering of stress along the length of the barrel. As the pad radius increases there is an increase in the upward stress spike. This indicates that beyond a particular threshold pad radius the in-plane CTE mismatch begins to override the out-of-plane mismatch for Kevlar-polyimide boards. Again, the exact threshold radius where this happens will depend on several geometric parameters and material properties. However, a large number of internal pads of suitable diameter, located at a sufficient distance from the MLB mid-plane are beneficial for improved PTH life.

It should be noted that for both the innerplanes and nonfunctional pads there were no epoxy-rich regions modeled in the MLB. Epoxy-rich regions, especially near the center of the PTH, act as stress raisers. In such cases, the

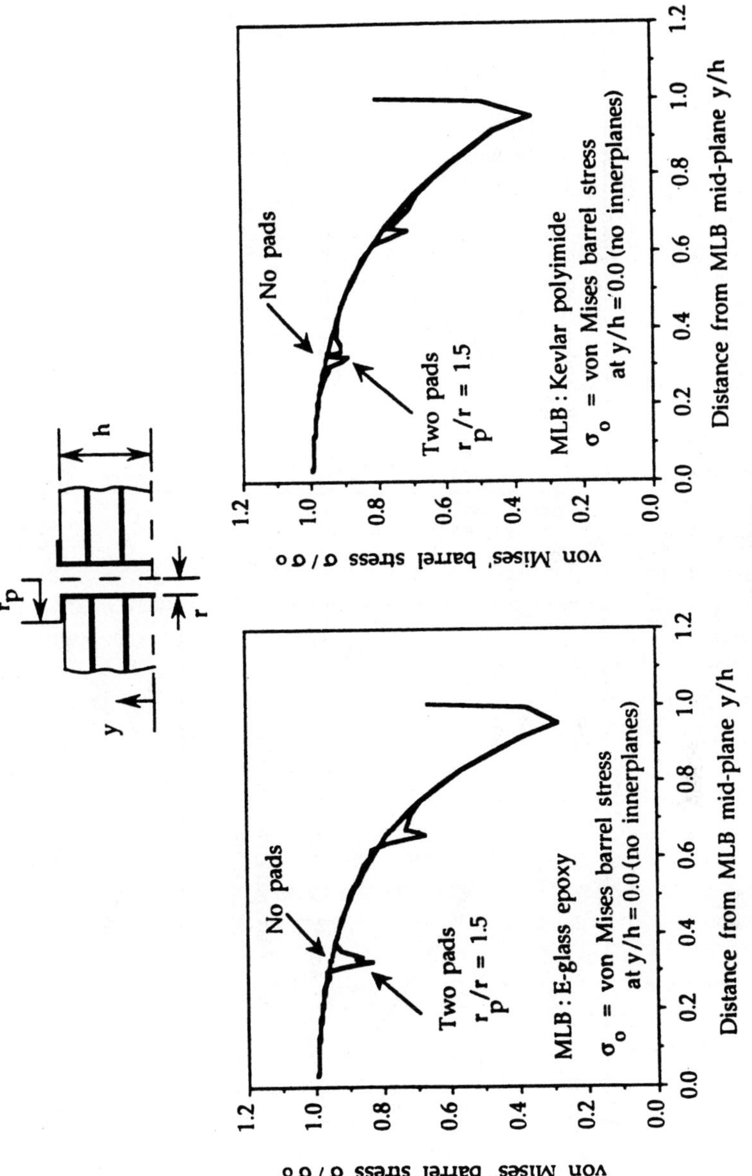


**Figure 20-9** Effect of internal nonfunctional pads on Von Mises barrel stress.

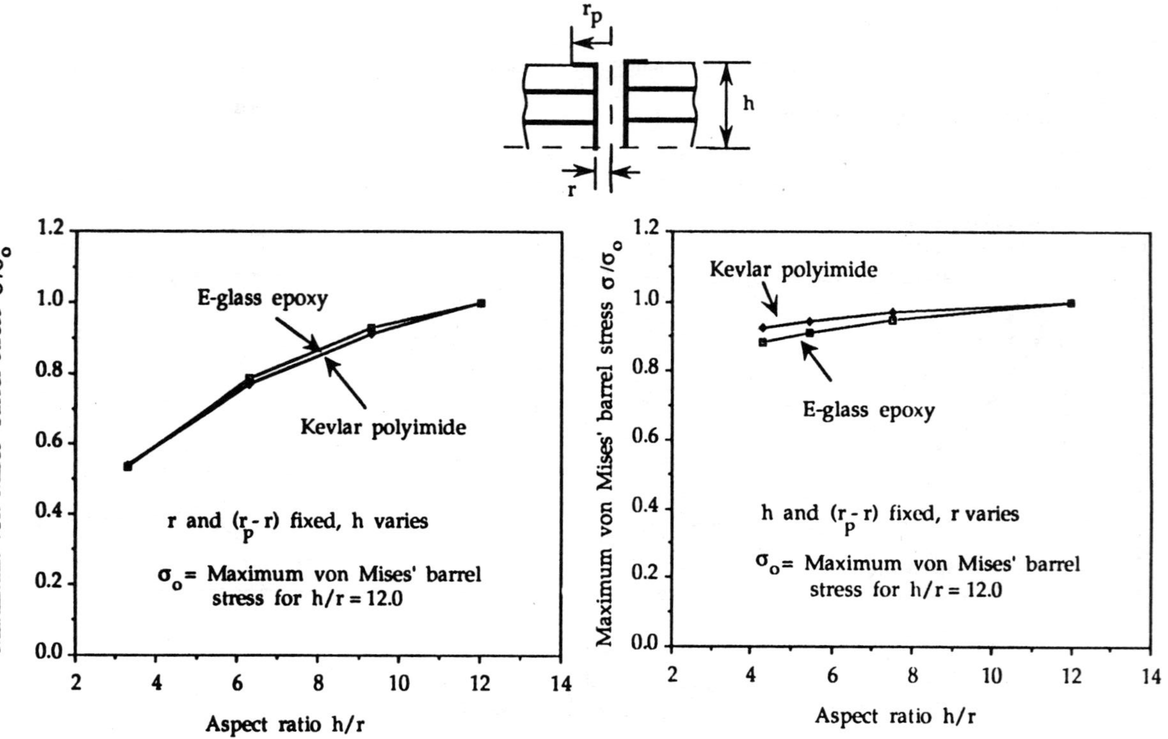


**Figure 20-10**  Effect of aspect ratio on Von Mises barrel stress.

presence of innerplanes and pads that eliminate the epoxy-rich region are beneficial for the PTH. This is true even for Kevlar-polyimide MLBs since the CTE mismatch between pure polyimide and copper is much greater than in-plane CTE mismatch between Kevlar-polyimide and copper.

### 20.3.4 Aspect Ratio

The aspect ratio (MLB thickness/hole diameter) can be varied by changing either the board thickness or the hole diameter. In this study the plating thickness is held constant. Figure 20-10 shows the expected trend of higher barrel stresses as the aspect ratio is increased, as it is well known that high aspect ratios are detrimental to PTH life.[6,7] Decreasing the MLB thickness is seen to be more effective in lowering critical stresses than increasing hole diameter for a given thickness of copper plating and land pad size ($r_o - r_i$). The results are relatively insensitive to the choice of MLB material.

### 20.3.5 Plating Thickness

The PTH plating thickness is one of the most important factors affecting PTH stresses. It is seen in Fig. 20-11 that increasing the plating thickness is an effective way of lowering the critical stresses in the PTH. The aspect ratio

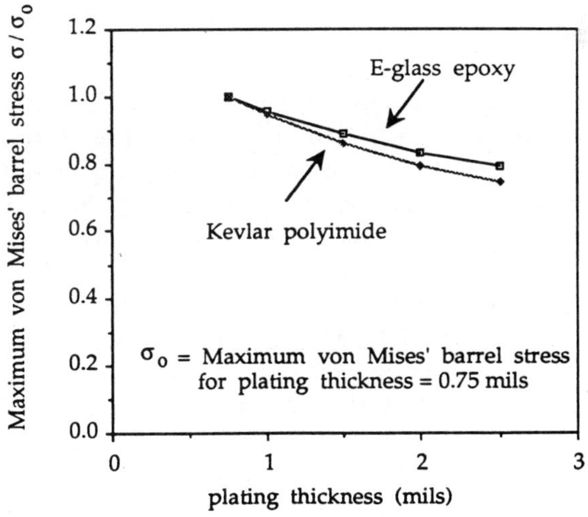


**Figure 20-11**    3D finite element model of a solder-filled PTH.

and the pad diameters were held constant in this study. The trend shown is rather obvious and has been observed in many experimental tests.[6,7]

## 20.4 SOLDER-FILLED PTH AND VOIDS IN SOLDER-FILLED PTH

The issue of filling a PTH with solder as opposed to leaving it unfilled to enhance fatigue life has been open to considerable debate. This section compares the critical thermomechanical stresses in filled and unfilled PTHs using orthotropic 3D finite element models for simulation. The effect of the presence of voids within the solder on the critical stresses in the PTH is also examined.

### 20.4.1 Modeling

Using the methodology of Section 20-2, one-quarter of a PTH together with surrounding FR-4 board material was modeled using 8-noded isoparametric brick elements. The aspect ratio of the PTH was 1:12 (hole diameter to board thickness). The copper plating thickness was 1 mil and the hole diameter 15 mils. The orthotropic MLB material properties were obtained from micromodeling the structure of the fabric-reinforced composite.[29] A solder void or gap at the MLB mid-plane, extending throughout the PTH cross-section was also modeled as shown in Fig. 20-12. The mesh density is increased near the center of the MLB in expectation of a higher stress concentration region.

### 20.4.2 Results

A linear elastic analysis was performed and the Von Mises stresses are presented in terms of a 1°C rise in temperature. Stresses for higher temperature increments can be obtained by scaling, but the results are primarily intended to be used in a comparison fashion. The changes in mechanical and thermal properties of the materials are not taken into account.

Figure 20-13, shows the Von Mises contour plot of the copper barrel stresses for an unfilled PTH. The Von Mises stresses are maximum at the MLB mid-plane, or bottom of the FEM model. Also seen in the figure and as pointed out in Section 20-3, the internal lands have the effect of lowering the barrel stresses in the vicinity of the land–barrel junction.

The effect of filling the PTH with solder, and also the effect of the presence of voids of various sizes and locations in the solder, are summarized in Table 20-3. For all the MLB materials listed in Table 20-1, the plating

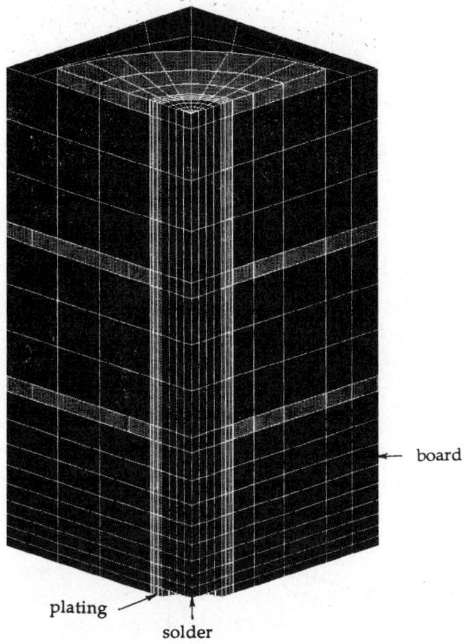


**Figure 20-12**   3D finite element model of a solder-filled PTH.

material is copper. The properties of the copper plating and solder used in the analysis are listed in Table 20-4.

Completely filling the PTH with solder does result in a small decrease in the maximum Von Mises stress in the PTH barrel. In general, solder voids have an adverse effect on the critical stresses in the PTH barrel, but the local increase in barrel stresses is rather small. A void throughout the complete

**Table 20-3**   Effect of Solder Filling and Solder Voids on PTH Critical Stresses

| Board Properties | | Normalized Von Mises Stress* | | |
|---|---|---|---|---|
| $E_z$ (GPa) | $\alpha_z$ $(10^{-6}/°C)$ | Filled $S_1$ | Full Void $S_2$ | $\sigma_{\text{unfilled}}$ (GPa) $\Delta T = 1°C$ |
| 7.731 | 77.320 | 0.977 | 1.096 | 5.21 |
| 7.889 | 78.920 | 0.977 | 1.098 | 5.18 |
| 8.327 | 71.860 | 0.978 | 1.068 | 4.59 |
| 8.673 | 82.270 | 0.978 | 1.072 | 5.55 |
| 9.008 | 85.660 | 0.975 | 1.076 | 5.95 |

* $S_1 = \sigma_{\text{filled}}/\sigma_{\text{unfilled}}$; $S_2 = \sigma_{\text{full void}}/\sigma_{\text{unfilled}}$.

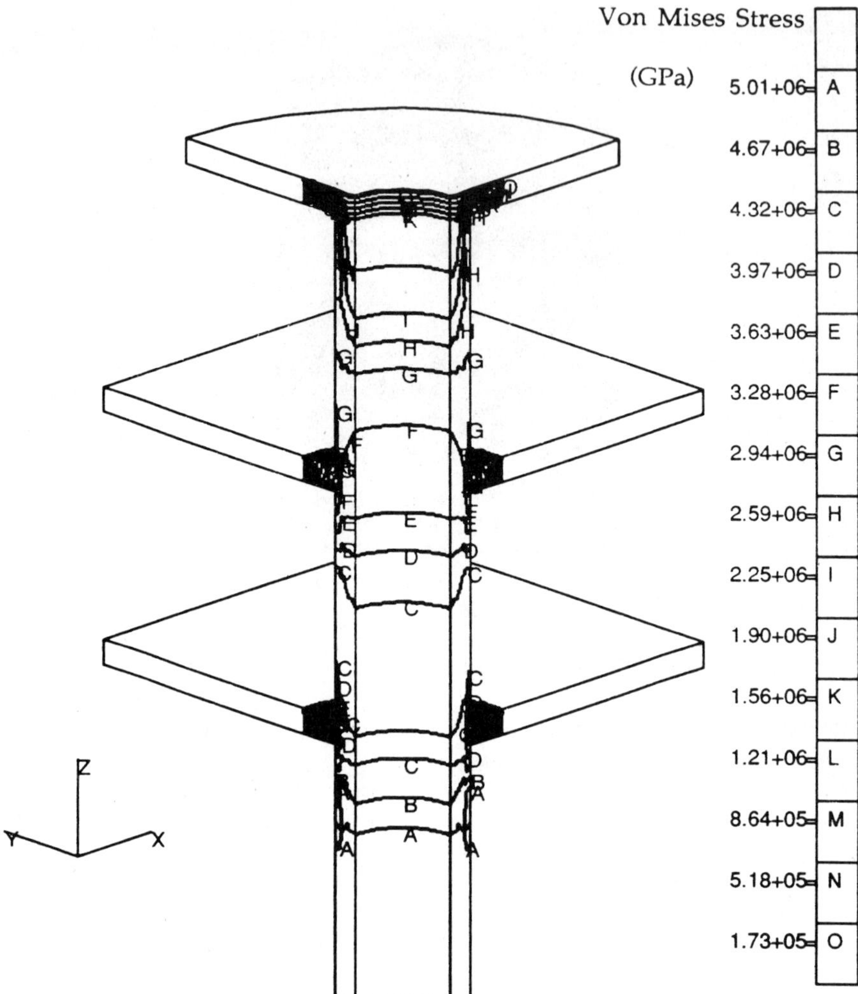


**Figure 20-13**    Von Mises stress in copper plating for a uniform $\Delta T$ of 1°C.

cross-section of the PTH (full void), located at the MLB mid-plane as shown in Fig. 20-12, is the most detrimental to the PTH as compared to other voids (partial and full) at various other locations through the PTH. Partial voids at the MLB mid-plane and voids located away from the mid-plane do locally affect the stress distribution, but the maximum barrel stress is still located at the MLB mid-plane. These local variations in stress are extremely small and are not considered to be very important.

The MLB mechanical stiffness, in addition to the CTE, are found to have an effect on the magnitude of the maximum barrel stress. Comparisons

**Table 20-4**   Material Properties Used in Solder-Filled PTH Models

| FR-4 PWB<br>Material Properties | Solder Properties | Copper Plating<br>Properties |
| --- | --- | --- |
| $E_x = 18.66$ GPa | $E = 10.34$ Gpa | $E = 130$ Gpa |
| $E_y = E_x$ | $v = 0.4$ | $v = 0.35$ |
| $E_z = 7.73$ GPa | $\alpha = 21.0 \times 10^{-6}/°C$ | $\alpha = 17.0 \times 10^{-6}/°C$ |
| $v_{xy} = 0.162$ | | |
| $v_{xz} = 0.421$ | | |
| $v_{yz} = v_{xz}$ | | |
| $G_{xy} = 5.07$ GPa | | |
| $G_{xz} = 2.17$ GPa | | |
| $G_{yz} = G_{xz}$ | | |
| $\alpha_x = 19.11 \times 10^{-6}/°C$ | | |
| $\alpha_y = \alpha_x$ | | |
| $\alpha_z = 77.32 \times 10^{-6}/°C$ | | |

between different MLB materials regarding thermomechanical PTH stresses should therefore be based not only on the CTE but also on the stiffness properties. The copper barrel stress drop due to PTH solder filling, approximately 3%, and the barrel stress rise due to the presence of mid-plane voids, approximately 9%, is the same regardless of MLB material.


## 20.5 TRANSIENT THERMAL STRESS IN A PTH SUBJECTED TO WAVE SOLDERING

This section presents some recent results of investigation into the transient thermal stress and strain distribution surrounding a PTH in an MLB subjected to a typical wave soldering operation.[24] The objective is to study the effect of wave soldering on the PTH fatigue life, with due consideration to the orthotropic and nonlinear material properties involved. This analysis does not consider discontinuities or irregularities in the copper plating, which can provide stress concentration sites for the initiation of fractures, but addresses the stresses in an ideal PTH. Furthermore, this analysis assumes that the PTH is masked and not filled with solder. Due to the complex material properties associated with the solidification and cooling of the solder, this last assumption simplifies an already complex problem to a manageable magnitude.

The wave soldering process is simulated in the general-purpose finite element code MARC.[30] A transient analysis is used because the plastic strains that result in the PTH during wave soldering are load-path-dependent and cannot be simulated with a quasi-static elastic-plastic nonlinear analysis with an assumed uniform $\Delta T$. The temperature, stress, and strain history plots obtained from the wave soldering simulation emphasize the significance of the transient analysis.

As in the previous sections, the assumed critical failure mechanism is barrel cracking and the critical parameter is the Von Mises stress in the copper barrel. In reality, during manufacturing even a significant barrel crack that is not completely around the circumference can be difficult to detect. Such a crack will only become functionally (electrically) manifested after temperature cycling. This creates the appearance that eventual failure resulted from thermal cycling during operation, even though the original fracture could have been initiated during the first soldering thermal transient. If a PTH barrel remains uncracked during initial soldering, hundreds to thousands of operational thermal cycles may be required to produce cyclic fatigue cracking of the barrel.[12]

### 20.5.1 Materials

FR-4 is the assumed MLB material, whose thermal and mechanical properties exhibit a complex dependence on temperature. Below the glass transition temperature $(T_g)$, the FR-4 is in a hard, glassy state, turning into a soft rubbery state above $T_g$. The glass transition typically occurs gradually over a temperature range of 110–140°C. This transition results in dramatic changes in the thermal and mechanical properties. For example, the out-of-plane CTE, which is the slope of the curve in Fig. 20-14, jumps from 75 p.p.m./°C to 400 p.p.m./°C above $T_g$. In comparison the CTE of copper remains nearly constant at 17 p.p.m./°C. Beside the dramatic changes in the CTE around $T_g$, the $z$-direction modulus of elasticity drops to a third of its value, as shown in Fig. 20-15. The plating and innerplane copper has nearly constant properties over the entire temperature range of $-65°C$ to 270°C and is modeled as a bilinear isotropic metal with linear work hardening, as shown in Fig. 20-16. The nonlinear orthotropic FR-4 properties are modeled as piecewise linear, with five data points describing the changes in each material constant with temperature, as listed in Table 20-5.

### 20.5.2 Transient Finite Element Model

Figure 20:17 is a 3D finite element model of the PTH in an MLB. Due to symmetry at the half-plane ($xz$ plane), only half of the PTH is modeled. The model consists of the board, the PTH barrel, lands, and innerplanes. A

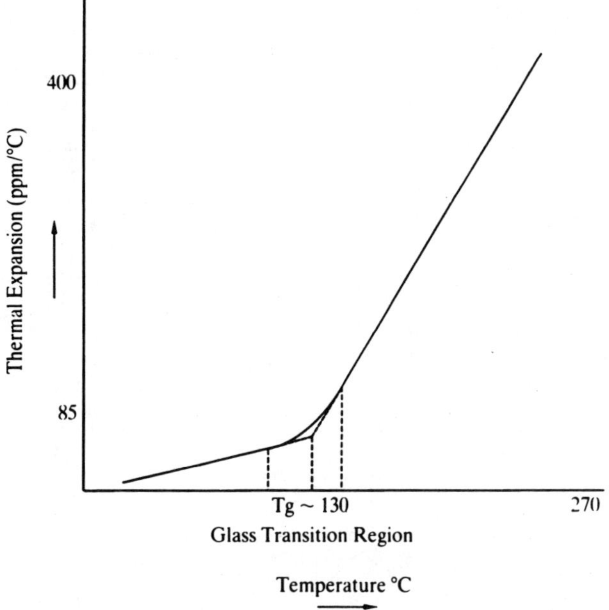


**Figure 20-14**   Thermal expansion (*x*-direction) for epoxy–fiberglass.

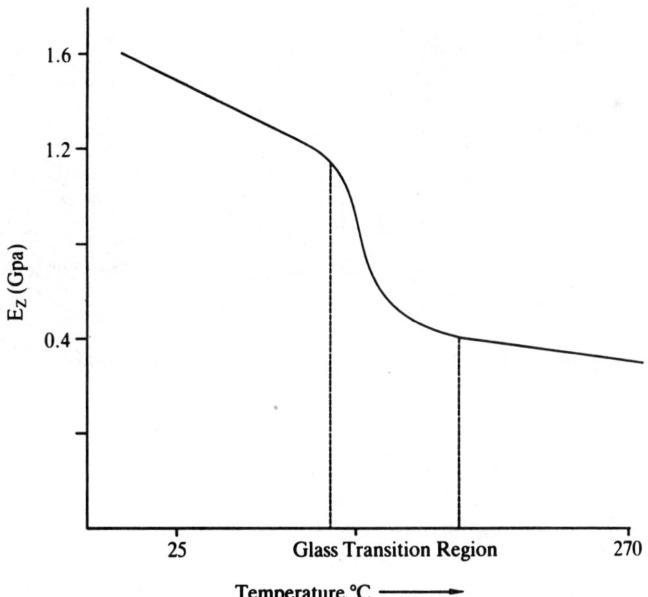


**Figure 20-15**   Variation with temperature of the modulus of elasticity (*x*-direction) for epoxy–fiberglass.

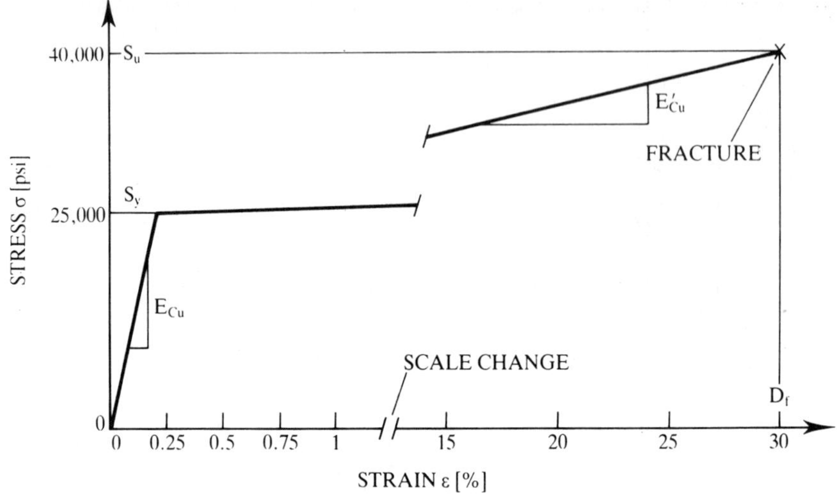


**Figure 20-16**  Linearized stress–strain diagram for electrodeposited copper[11] $\sigma_u = 40\,000$ psi, $\sigma_y = 25\,000$ psi, $E_{Cu} = 12 \times 10^6$ psi, $E'_{Cu} = 0.1 \times 10^6$ psi, $D_f = 30\%$).

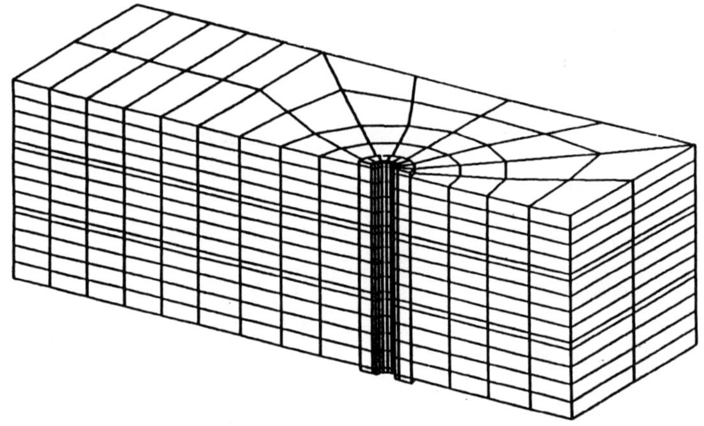

**Figure 20-17**  3D finite element model of PWB assembly.

copper plating thickness of 1.25 mils is assumed for the PTH barrel and lands. The aspect ratio (MLB thickness/PTH diameter) is 10:1, with a 60-mil thick MLB. A copper innerplane thickness of 2 mils is assumed with the innerplanes equidistant from the MLB mid-plane. The model takes into consideration the difference in thermal diffusivities between the FR-4 and the copper as well as the temperature dependence of the FR-4 orthotropic mechanical properties.

**Table 20-5**   Thermal and Mechanical Properties for Transient Nonlinear Analysis

| | Epoxy Fiber Glass | | | | | |
| | Temperature (°C) | | | | | |
| Property | 30 | 95 | 125 | 150 | 270 | Copper |
|---|---|---|---|---|---|---|
| $k$ (W/m-°C) | 0.157 | 0.157 | 0.157 | 0.157 | 0.157 | 398 |
| $\rho$ (kg/m$^3$) | 220 | 220 | 220 | 220 | 220 | 8940 |
| $C_p$ (J/kg-°C) | 840 | 840 | 840 | 840 | 840 | 384 |
| $E_x$ (Gpa) | 22.4 | 20.68 | 19.3 | 17.92 | 16.0 | 103.42 |
| $E_y$ (Gpa) | 22.4 | 20.68 | 19.3 | 17.92 | 16.0 | 103.42 |
| $E_z$ (Gpa) | 1.6 | 1.2 | 1.0 | 0.60 | 0.45 | 103.42 |
| $\nu_{xy}$ | 0.02 | 0.02 | 0.02 | 0.02 | 0.02 | 0.3 |
| $\nu_{xz}$ | 0.1425 | 0.1425 | 0.1425 | 0.1425 | 0.1425 | 0.3 |
| $\nu_{yz}$ | 0.1425 | 0.1425 | 0.1425 | 0.1425 | 0.1425 | 0.3 |
| $G_{xy}$ Gpa) | 0.63 | 0.60 | 0.50 | 0.450 | 0.441 | 39.71 |
| $G_{xz}$ (Gpa) | 0.199 | 0.189 | 0.157 | 0.142 | 0.1393 | 39.71 |
| $G_{yz}$ (Gpa) | 0.199 | 0.189 | 0.157 | 0.142 | 0.1393 | 39.71 |
| $\alpha_x$ ($10^{-6}$/°C) | 20.0 | 20.0 | 20.0 | 20.0 | 20.0 | 17.00 |
| $\alpha_y$ ($10^{-6}$/°C) | 20.0 | 20.0 | 20.0 | 20.0 | 20.0 | 17.00 |
| $\alpha_z$ ($10^{-6}$/°C) | 86.50 | 86.50 | 400.0 | 400.0 | 400.0 | 17.00 |

The following assumptions and boundary conditions are used for the thermal analysis. The temperature of the MLB bottom surface is raised to the solder wave temperature as a step function because the time required for the solder wave to travel from one end of the model to the other is much smaller than that required for the presence of the solder wave to be felt in the board. The thermal contact resistance between the solder wave and the MLB is assumed to be negligibly small. Due to symmetry at the half-plane ($xz$ plane) of the PTH, this surface is treated as insulated and displacement-constrained in the $y$-direction. The remaining board edges are also treated as insulated. Forced convective boundary conditions are applied on the top surface to model board motion through still air (2.4429 W/m$^2$-K). The material of the board is assumed to be orthotropic but homogeneous. The MLB assembly is assumed to be at 35°C at the start of the analysis and subjected to the constant 260°C solder wave for 3 seconds before removing the solder wave from the bottom of the MLB.

Besides the orthotropic 3D model, two different axisymmetric models were used, one with copper innerplanes and the other without. The axisymmetric model dimensions were equivalent to those of the 3D model. The thermal and mechanical properties are given in Table 20-5 with the limitations as explained in Section 20.2.2.

The governing heat transfer equation is

$$C(\mathbf{T})\dot{\mathbf{T}} + K(\mathbf{T})\mathbf{T} = \mathbf{Q} \qquad (20\text{-}1)$$

where $C(\mathbf{T})$ is the heat capacity and $K(\mathbf{T})$ the thermal conductivity matrix, respectively. $\mathbf{T}$ is the nodal temperature vector, and $\dot{\mathbf{T}}$ is the time derivative of the temperature vector, and $\mathbf{Q}$ is the heat flux vector. The discretization of the time variable in the above equation yields the expression

$$\left[\frac{1}{\Delta t} C(\mathbf{T}) + K(\mathbf{T})\right]\mathbf{T}_n = \mathbf{Q}_n + \frac{1}{\Delta t} C(\mathbf{T})\mathbf{T}_{n-1} \qquad (20\text{-}2)$$

where $\mathbf{T}_n$ is the temperature vector at the end of time increment $n$, $\mathbf{T}_{n-1}$ is the temperature vector at increment $n - 1$, and $\mathbf{Q}_n$ is the heat flux for the $n$th increment. The above equation computes the nodal temperatures for each time increment $\Delta t$.

For the evaluation of the temperature-dependent matrices, the temperatures at the two previous steps provide a linear (extrapolated) temperature description over the desired interval:

$$T(\tau) = T(t - \Delta t) + \frac{\tau}{\Delta t}(T(t - \Delta t) - T(t - 2\Delta t)) \qquad (20\text{-}3)$$

This temperature is then used to obtain an average property, $P$, of the material over the interval:

$$P = \frac{1}{\Delta t} \int_{t-\Delta t}^{t} P(T(t))\, dt \qquad (20\text{-}4)$$

The temperature distribution profiles generated from the transient thermal analysis are used as an input for an elastic–plastic stress analysis to obtain the stress distribution in the board and the PTH.

The MLB assembly, which was assumed to be initially at 35°C, was exposed to the solder wave for 3 seconds. At the end of this 3-second heating period, the transient thermal solution in the MLB approaches a steady-state profile. The plastic strains obtained from the transient solution and a quasi-static solution assuming a uniform $\Delta T$ are not the same because the plastic strains are load-path-dependent. A transient analysis is, therefore, the only approach to estimating the plastic strains developed in a PTH during wave soldering.

### 20.5.3 Results and Discussion

A 3D plot of the deformed shape of the finite element model is shown in Fig. 20-18 after 3 seconds exposure to the solder wave. The tension in the PTH barrel and the bond pad deformations are due to the mismatch in the $z$-direction thermal expansions of the FR-4 board and the PTH. As explained earlier in Section 20.2.4 the critical stress for circumferential barrel cracking is the Von Mises stress. Even though the Von Mises stress was found to be high at the pads and innerplanes, the remainder of the paper discusses the stress in the PTH barrel at the MLB mid-plane, since it was always found to be higher.

The PTH barrel Von Mises stress history plot is shown in Fig. 20-19. For sufficiently high temperatures, as the MLB temperature passes the FR-4 glass transition temperature, the PTH barrel is deformed to a point where yielding takes place. Figures 20-20 and 20-21 present the equivalent plastic strain and the temperature history for the PTH barrel mid-plane for the heating period. Figure 20-22 shows the corresponding Von Mises stress versus equivalent plastic strain plot for the 3D and axisymmetric models. A comparison of results from Figs. 20-19, 20-20, and 20-22 for the two axisymmetric models with and without the innerplanes shows that there is a negligible difference between the value of the equivalent plastic strain at the MLB mid-plane from the two models. (Similar results were discussed in Section 20.3.2 for FR-4 under a uniform temperature change.) The presence of the innerplanes does not affect the maximum level of plastic strain generated in the PTH

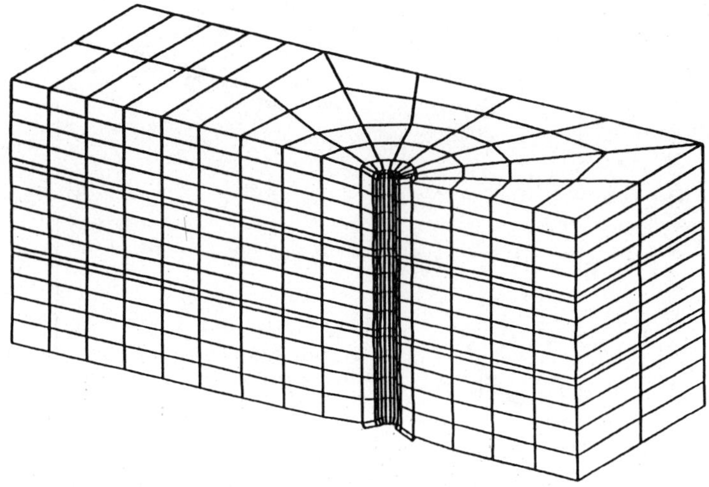

**Figure 20-18**   Deformed shape of the 3D model after heating simulation.

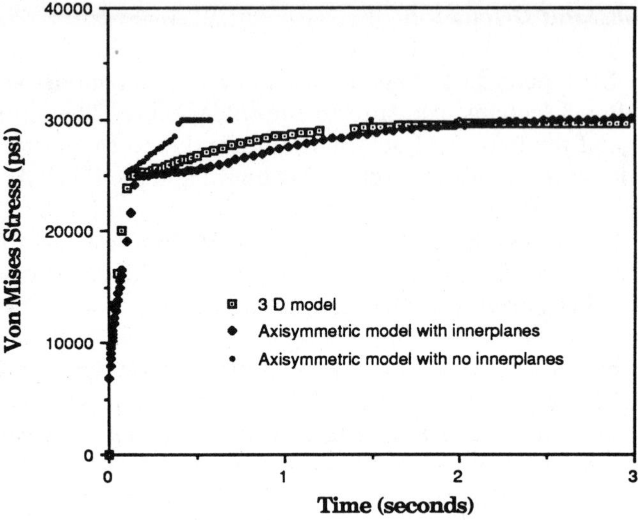



**Figure 20-19**   Von Mises stress history for PTH barrel mid-plane.

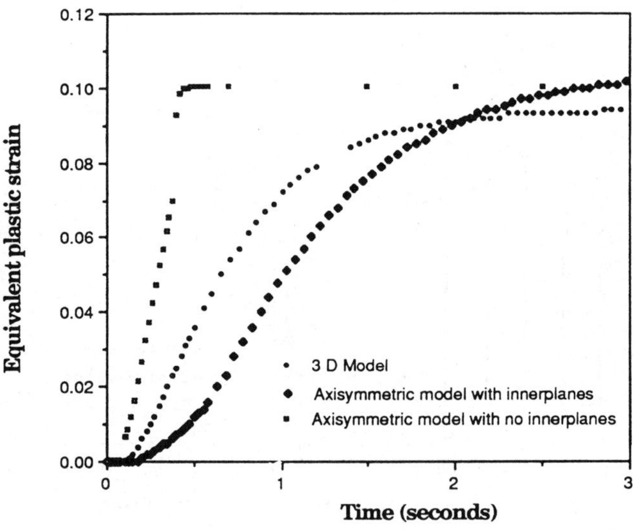



**Figure 20-20**   Strain history for PTH barrel mid-plane.

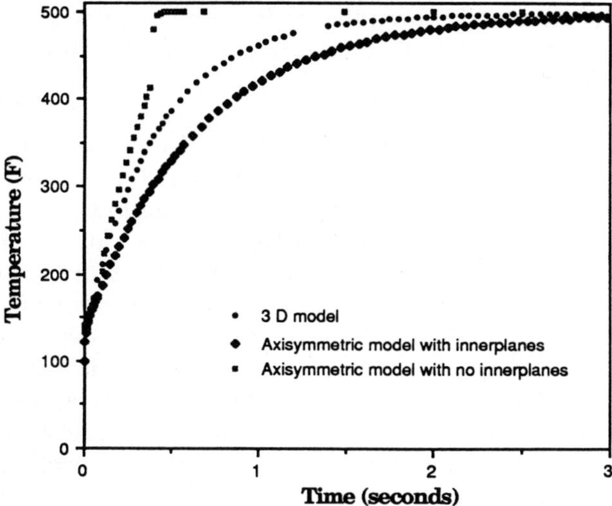


**Figure 20-21**   Temperature history for PTH barrel mid-point.

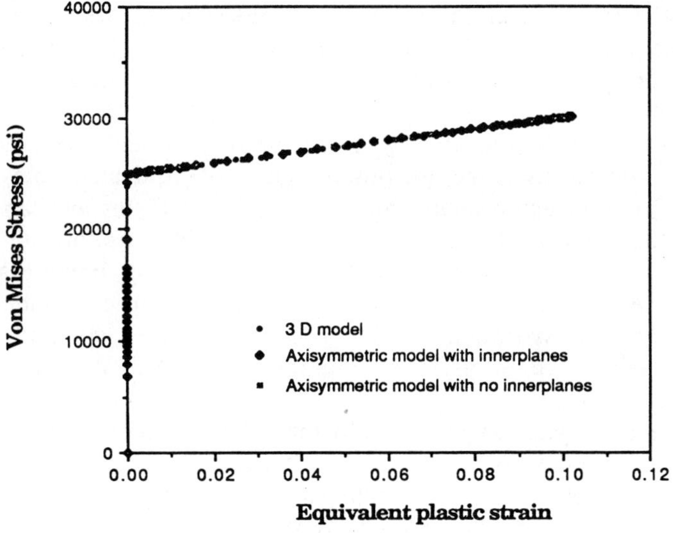


**Figure 20-22**   Von Mises stress vs. equivalent plastic strain for PTH mid-plane.

during wave soldering. However, the model without innerplanes experiences the most thermomechanical shock as evidenced by the strain rate and rate of temperature change during the early phase of the heating.

Figures 20-19 and 20-20 also point out the importance of using a transient elastic-plastic thermal stress analysis instead of a quasi-static analysis in which the temperature of the entire MLB assembly is raised uniformly from the initial temperature to the solder wave temperature. For example, the temperature of the PTH rises much faster in the case of the axisymmetric model with no innerplanes compared to the axisymmetric model with innerplanes. This difference is caused by the fact that the PTH in the case of the model with no innerplanes conducts heat radially only to the surrounding FR-4 board, whereas in the case of the model with innerplanes heat is conducted to the board surrounding the PTH as well as to the innerplanes. This difference in temperature and stress and strain history is of importance if creep is considered and strain rate-dependent properties are used. For this study, however, creep was ignored and strain rate-dependent properties were not used.

The 10% PTH barrel strains observed at the end of the heating period are far below the ductile rupture strain for electrodeposited copper, which is about 30% (Figs. 19-16, 19-20, and 19-22). Thus, the thermomechanical strains that are generated in the PTH barrel during the wave soldering are not enough to cause ductile rupture of the PTH barrel. As explained earlier, the models considered in this study are idealizations that ignore the effect of manufacturing flaws like plating thickness variations, and the presence of stress concentrations and interfacial delamination. Failure or initiation of a crack in the PTH barrel during the wave soldering operation may still arise due to the presence of such flaws, but PTHs that come close to the idealized model with no flaws should pass the wave soldering process without barrel cracking.

Comparing the results in Fig. 20-22 for the 3D and the axisymmetric analyses showed that the maximum value of equivalent plastic strain obtained from the axisymmetric analysis was about 8% higher than the result from the 3D analysis. (For the elastic analysis with a uniform temperature rise, the axisymmetric analysis yielded results about 5% lower than the 3D analysis, Section 20.2.5.) Thus the axisymmetric transient models yielded slightly conservative results, with significantly reduced computation time. This significant reduction in computation time provides an acceptable trade-off.

The trade-off was the justification for using an axisymmetric model to simulate the cooling down of the MLB assembly after the soldering. In cooling, the problems of PTH buckling and debonding of the copper surfaces from the epoxy are not considered. As shown in Fig. 20-23, the cooling of the MLB after exposure to the solder wave is much slower than the heating.

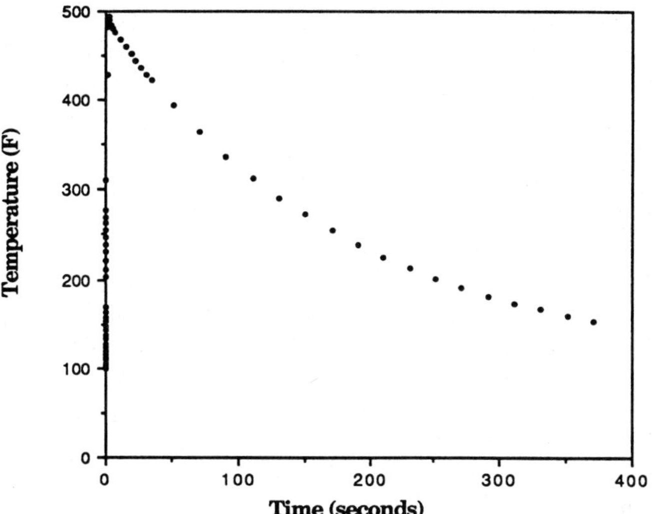

**Figure 20-23**  Temperature history for PTH barrel mid-point for axisymmetric model with two innerplanes.

The use of the 3D model to simulate such cooling of the MLB assembly would have required very long computation times, without any significant gains in accuracy.

Figure 20-24 shows the Von Mises stress history for the PTH barrel mid-plane for a heating and cooling simulation with an axisymmetric model. As the MLB assembly cools down from 260°C to a temperature of 35°C, the PTH barrel yields under axial compression. The negative sign for the Von Mises stress is artificially added to indicate this general reversal in barrel stress from tension to compression. Figure 20-25 shows the Von Mises stress versus equivalent plastic strain plot for this simulation. From the figure, the residual compressive stress remaining at the end of the heating and cooling simulation is about 34 000 psi. The presence of a compressive residual stress means an enhanced fatigue life.

The Coffin–Manson equation is used to estimate the effect of presence of the residual stresses in the PTH,

$$\frac{\varepsilon_t}{2} = \frac{\sigma_f}{E}(2N_f)^b + \varepsilon_f(2N_f)^c \tag{20-5}$$

where $\varepsilon_t/2$ is half of the plastic strain range, and for copper $\sigma_f/E = 0.016$, $\varepsilon_f = 0.398$, $b = -0.115$, and $c = -0.6$.[31]

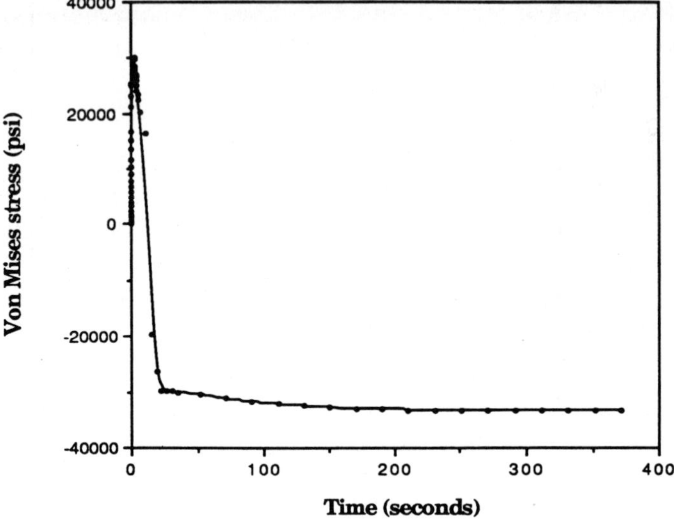

**Figure 20-24** Von Mises stress history for PTH barrel mid-point for axisymmetric model with two innerplanes.

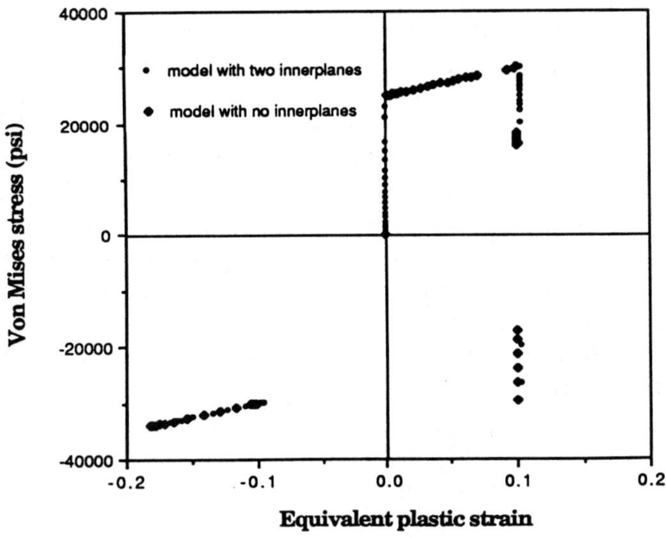



**Figure 20-25** Von Mises stress vs. equivalent plastic strain for PTH barrel mid-point for axisymmetric model with two innerplanes.

For the presence of residual stresses ($\sigma_r$), the Coffin–Manson equation is

$$\frac{\varepsilon_t}{2} = \frac{\sigma_f - \sigma_r}{E} (2N_f)^b + \varepsilon_f (2N_f)^c \tag{20-6}$$

A compressive residual stress of 34 000 psi, with $E_{Cu} = 12 \times 10^6$ psi implies that $(\sigma_f - \sigma_r)/E = 0.018\,8333$. The Coffin–Manson equation is thus

$$\frac{\varepsilon_t}{2} = 0.018\,8333(2N_f)^{-0.155} + 0.398(2N_f)^{-0.6} \tag{20-7}$$

As a case example, if the life of a plated-through-hole subjected to a COM-T-CYCLE[7] is 1200 cycles, then from Eq. (20-5), $\varepsilon_t/2 = 0.010\,98$. Substituting this number in Eq. (20-7) yields a life of 1600 cycles, representing a 33% increase.

An estimate of the fatigue damage to the PTH during wave soldering can be made by utilizing Miner's relationship,[32] where the ratio $R$ (of cycles experienced, $n_i$, to the fatigue life, $N_i$, at a specific $i$th strain level) is used to define the percentage of the fatigue life that is exhausted:

$$R = \sum \frac{n_i}{N_i} \tag{20-8}$$

(Failure is assumed to occur when $R = 1$, but commonly failure for electronic equipment is assumed to occur at a more conservative value such as 0.7, or lower when loss of life is at stake.[33] For the heating and cooling simulation represented in Fig. 20-25, Miner's equation is

$$\frac{0.7}{N_{fws}} + \frac{n_{op}}{N_{op}} = 1 \tag{20-9}$$

where $N_{fws}$ is the life for the wave soldering operation, $N_{op}$ is the operational life, $n_{op}$ is the life remaining after the wave soldering operation, and 0.7 is the ratio of area under the stress–strain curve in Fig. 20-25 to the area under one complete cycle (see Fig. 20-26). For the wave soldering transient, half of the plastic strain range is $\varepsilon_t/2 = 0.14$ in./in. Thus, from Eq. (20-5), $N_{fws} = 3$ cycles. For the previous example of the PTH worked out above, if the operational life is $N_{op} = 1200$ cycles, then from Eq. (20-9), $n_{op} = 920$ cycles. This implies a 25% reduction in the fatigue life of the plated-through-hole.

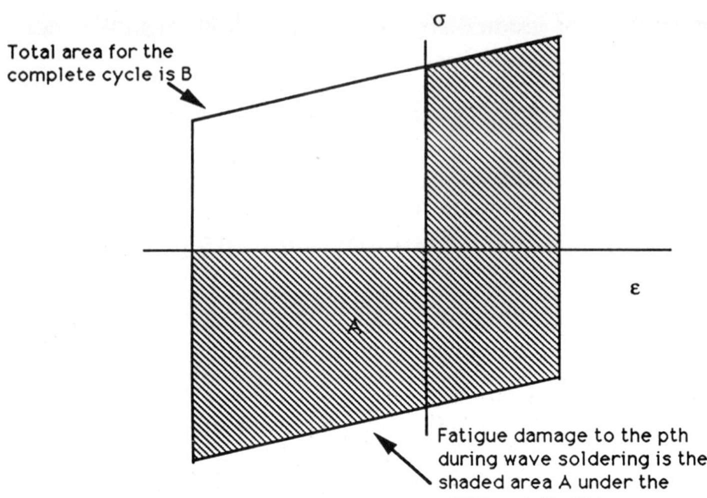


**Figure 20-26**  Fatigue damage to the PTH during the wave soldering transient.


## 20.6  SUMMARY

A methodology and guideline have been laid out for the use of axisymmetric and 3D finite element methods in studying critical stresses within a PTH. Accurate boundary conditions and stress averaging are important in order to get reliable stress predictions.

Parametric studies have been conducted for both standard FR-4 MLBs and for highly anisotropic Kevlar-49 polyimide MLBs. The following general trends were noticed:

1. As expected, decreasing PTH spacing improves mechanical reliability.
2. Decreasing PTH aspect ratio, either by decreasing MLB thickness or by increasing PTH diameter, is beneficial for the PTH, with MLB thickness having a more pronounced effect.
3. For FR-4 MLBs, the presence of solid innerplanes, nonfunctional internal pads and functional signal layers result in local stress relief in the vicinity of the innerplane/pad/layer. For Kevlar-polyimide MLBs, however, there is a local stress spike (increase) accompanied by an overall lowering of stress along the entire length of the barrel.
4. The effect of internal pads on PTH critical stresses is dependent on pad radius, location of the pad along the length of the barrel, and the in-plane CTE mismatch between the MLB and pad material.

Furthermore, it was found that filling the PTH with solder, either completely or allowing some solder voids to occur, does not significantly affect the critical barrel stresses. These results were found to be relatively independent of MLB material.

A transient nonlinear thermal stress analysis was conducted on a non-solder-filled PTH in an FR-4 MLB to determine the effect on the PTH fatigue life due to the wave soldering process. The transient analysis showed that the temperature of a PTH in an MLB without innerplanes rises much faster than a PTH in an MLB with innerplanes. This imparts a greater thermal shock to the PTH assembly and is important when accounting for creep and strain rate dependence effects. The presence of innerplanes, however, was not found to affect the maximum level of plastic strains developed in the PTH barrel during wave soldering.

The simulation of the wave soldering process for the particular PTH considered did not produce barrel stresses and strains high enough to cause circumferential cracking or ductile rupture, but the wave soldering operation did have two opposing effects on the life of the PTH. Upon cooling, the wave soldering operation results in compressive residual stresses in the PTH barrel. The presence of compressive residual stresses improves the fatigue life of the PTH. On the other hand, the soldering transient itself consumes a significant portion of the fatigue life of the PTH. For the example shown, the life of the PTH improved by 33% due to the presence of the residual compressive stress, but the fatigue damage due to the wave soldering resulted in a 25% loss in life. More detailed studies with due consideration of the transient effects such as creep and strain rate dependence are required to determine more precisely the effect of wave soldering on the life and reliability of a PTH.

The purpose here has been to provide insights for reliable design of PTHs. The models considered for this study are idealizations that ignore the effect of manufacturing flaws like plating thickness variations, presence of stress concentrations, and interfacial delamination. Studying the effect of manufacturing defects and flaws requires a more comprehensive 3D finite element model than used in the studies here.

## ACKNOWLEDGMENT

The authors acknowledge support from the University of Maryland CALCE Electronics Packaging Research Center, Engineering Research Center, Mechanical Engineering Department, and the Systems Research Center.

## REFERENCES

1. IPC-T-50, "Terms and Definitions for Interconnecting and Packaging Electronic Circuits," The Institute for Interconnecting and Packaging Electronic Circuits, Lincolnwood, IL, 1988.

2. Rudy, D. A., "The Detection of Cracks in Plated Through Holes Using Four Point Resistance Measurements," *14th Annu. Proc., IEEE Reliability Physics Symposium*, 1976, pp. 135–140.

3. Ammann, H. H., and R. W. Jocher, "Measurement of Thermomechanical Strains in Plated-Through Holes," *14th Annu. Proc. IEEE Reliability Physics Symposium*, 1976, pp. 118–120.

4. Jellison, J., "Evaluation of Multilayer Printed Wiring Boards by Metallographic Techniques," NASA Publication 1161, May 1986.

5. Tossell, D. A., and K. H. G. Ashbee, "High Resolution Optical Interference Investigation of Deformation Due to Thermal Expansion Mismatch Around Plated Through Holes in Multilayer Circuit Boards," *J. Electronic Materials*, **18**(2), 1989, pp. 275–286.

6. Gray, F., and M. Elkins, "Reliability, Thermal and Thermomechanical Characteristics of Polymer-on-metal Multilayer Boards," *Circuit World*, **14**(3), 1988, pp. 12–21.

7. "Round Robin Evaluation of Small Diameter Plated Through Holes in Printed Wiring Boards," Publication IPC-TR-579, The Institute for Interconnecting and Packaging Electronic Circuits, Lincolnwood, IL, 1988.

8. "Evaluation of Multilayer Printed Wiring Boards by Metallographic Techniques," NASA Ref. Publ. 1161, May 1986.

9. Tossell, D. A., and K. H. G. Ashbee, "High Resolution Optical Interference Investigation of Deformation Due to Thermal Expansion Mismatch Around Plated Through Holes in Multilayer Circuit Boards," *J. Electronics Materials*, **18**(2), 1989, pp. 275–286.

10. Vecchio, K. S., and R. W. Hertzberg, "Analysis of Long Term Reliability of Plated-Through Holes in Multilayer Interconnection Boards," *Microelectronics and Reliability*, **26**, 1986, pp. 715–732.

11. Olien, M. A., "A Simple Model for the Thermo-Mechanical Deformations of Plated-Through-Holes in Multilayer Printed Wiring Boards," *14th Annu. Proc. IEEE Reliability Physics Symposium*, 1976, pp. 121–128.

12. Olien, M. A., "Methods for Evaluating Plated Through Hole Reliability," *14th Annu. Proc. IEEE Reliability Physics Symposium*, 1976, pp. 129–131.

13. Kurosawa, K., Y. Takeda, K. Takagi, and H. Kawamata, "An Investigation of the Reliability Behaviour of Plated-Through Holes in Printed Wiring Boards," Technical Paper IPC-TP-385, IPC Fall Meeting, September 1981.

14. Nankey, R. A., "Thermally Induced Strain in Plated-Through-Holes in Multilayer Circuit Boards," Technical Paper IPC-TP-21, IPC 1976 Fall Meeting, September 1976.

15. Mirman, B. A., "Mathematical Model of a Plated-Through Hole Under a Load Induced by Thermal Mismatch," *IEEE Trans. Components, Hybrids, and Manufacturing Technology*, **11**(4), 1988, pp. 506–511.

16. Engelmaier, W., "Round Robin Reliability Evaluation of Small Diameter Plated-Through-Holes in Printed Wiring Boards," IPC Technical Report IPC-TR-579, The Institute for Interconnecting and Packaging Electronic Circuits, Lincolnwood, IL, September 1988, pp. 40–46.

17. Engelmaier, W., "Environmental Stress Screening and Use Environments—Their Impact on Surface Mount Solder Joint and Plated-Through-Hole Reliability,"

*Proc. Electronic Packaging Conference (IEPS)*, Marlborough, MA, September 1990, pp. 388–397.

18. Lee, L. C., V. S. Darekar, and C. K. Lim, "Micromechanics of Multilayer Printed Circuit Boards," *IBM J. Research and Development*, **28**, November 1984, pp. 711–718.

19. Chen, W. T., L. C. Lee, C. K. Lim, and D. P. Seraphim, "Mechanical Modeling for Printed Circuit Boards," Technical Paper WCIII-41, Printed Circuit World Convention III, May 1984.

20. Vecchio, K. S., and R. W. Hertzberg, "Analysis of Long Term Reliability of Plated Through Holes in Multilayer Interconnection Boards: Stress Analysis and Material Characterization," *Microelectronics and Reliabilities*, **26**(4), 1986, pp. 733–751.

21. Royce, Barrie S. H., "Differential Thermal Expansion in Microelectronic Systems," *IEEE Trans. Components, Hybrids, and Manufacturing Technology*, **11**(4), 1988, pp. 454–463.

22. Iannuzzelli, R., "Predicting Plated-Through-Hole Reliability in High Temperature Manufacturing Processes," presented at 33d IPC Annual Meeting, Boston, MA, April 1990.

23. Ozmat, B., H. Walker, and M. Elkins, "A Nonlinear Thermal Stress Analysis of the Plated-Through-Holes of Printed Wiring Boards," *Proc. Electronic Packaging Conference (IEPS)*, Marlborough, MA, September 1990, pp. 359–381.

24. Barker, D. B., S. Naqvi, M. Pecht, and A. Dasgupta, "Transient Thermal Stress Analysis of a Plated Through Hole Subjected to Wave Soldering," *J. Electronic Packaging*, **113**(2), 1991, pp. 149–155.

25. Bhandarkar, S., A. Dasgupta, D. Barker, and M. Pecht, "Effect of Voids in Solder-Filled Plated Through Holes," IPC Technical paper TP-863, 33d IPC Annual Meeting, Boston, MA, April 1990.

26. Bandarkar, S. M., A. Dasgupta, D. B. Barker, M. Pecht, and W. Engelmaier, "Influence of Selected Design Variables on the Thermo-Mechanical Stress Distributions in Plated-Through-Hole Structures," *J. Electronic Packaging*, **114**(1), March 1992, pp. 8–13.

27. Christensen, R. M., *Mechanics of Composite Materials*, Krieger Publishing Co., Malabar, FL, 1991, pp. 52–58.

28. Engelmaier, W., "Performance Considerations: Thermal-Mechanical Effects," *Technology, Electronic Materials Handbook, Vol. 1, Packaging*, ASM International, Materials Park, OH, 1989, Section 6, p. 740.

29. Dasgupta, A., S. M. Bhandarkar, D. Barker, and M. Pecht, "Thermoelastic Properties of Woven-Fabric Composites Using Homogenization Techniques," *5th Technical Conference of the American Society for Composites*, June 1990.

30. *MARC Analysis Software*, MARC Analysis Corporation, Palo Alto, CA, 1989.

31. Engelmaier, W., *Designing High Density Surface Mount PWB Assemblies for Reliability and Manufacturability*, AT&T Bell Laboratories, September 1988.

32. Shigley, Joseph E., and Larry D. Mitchell, *Mechanical Engineering Design*, 4th edn., McGraw-Hill, New York, 1983, p. 337.

33. Steinberg, Dave S., *Vibration Analysis for Electronic Equipment*, Wiley, New York, 1988.

# 21

# Nonlinear Analysis of a Ceramic Pin Grid Array (PGA) Soldered to an Orthotropic Epoxy Substrate

*John Lau, Ravi Subrahmanyan, Steve Erasmus, Sherman Leung, and Che-Yu Li*

## 21.1 INTRODUCTION

The pin grid array (PGA) is one form of first-level package.[1-4] It has had a predominant role in high-density packaging for many years and is commonly used in high-performance computers. The advantages of the PGA are: (1) excellent electrical performance (controlled impedance); (2) the I/O pins are spread over the area of substrate, making it possible to have a very large number of I/Os; and (3) the I/O pins are placed in a very small area, making it possible to have a small package with short circuit lines.

There are two ways to construct a PGA package. One is to place all the pins on the opposite side to the chip cavity (cavity up), and the other is to place all the pins on the same side as the chip cavity (cavity down). One of the important advantages of the cavity-up PGA is to have a maximum number of I/Os (the pins can be spread over the whole surface). On the other hand, one of the important advantages of the cavity-down PGA is that it enables better thermal management (the heat sink can be attached to the opposite surface of the chip cavity).

There are at least two different materials for a PGA body, namely, plastic and ceramic. While the plastic PGA is cheaper than the ceramic PGA, the ceramic PGA has better wire bond yield and hermeticity. In this study, only the 408-pin cavity-down ceramic PGA will be considered (Fig. 21-1).

There are two common ways to assemble the PGA to a printed circuit board (PCB), namely, wave soldering and solder reflow. With the increased

**Figure 21-1**   408-Pin cavity-down ceramic PGA.

usage of both sides of the PCB, the mass solder reflow methods have become more popular. Figure 21-2 shows an example of an assembled PCB with four 408-pin cavity-down ceramic PGAs.

Figure 21-3 schematically shows a PGA interconnect assembly. It consists of three major parts, namely, the ceramic PGA with Alloy-42 pins, the FR-4 epoxy–glass PCB with plated copper barrels and pads, and the 60wt%Sn–40wt%Pb or 63wt%Sn–37wt%Pb solder joints. Because of the geometry, material construction, and thermal expansion mismatch of different parts of the PGA interconnect assembly, thermal stresses and strains can occur in the assembly during manufacture and service. These stresses and strains produce the driving force for PGA interconnection failures.[1–43]

In this chapter, our focus is to determine the thermal stresses and strains

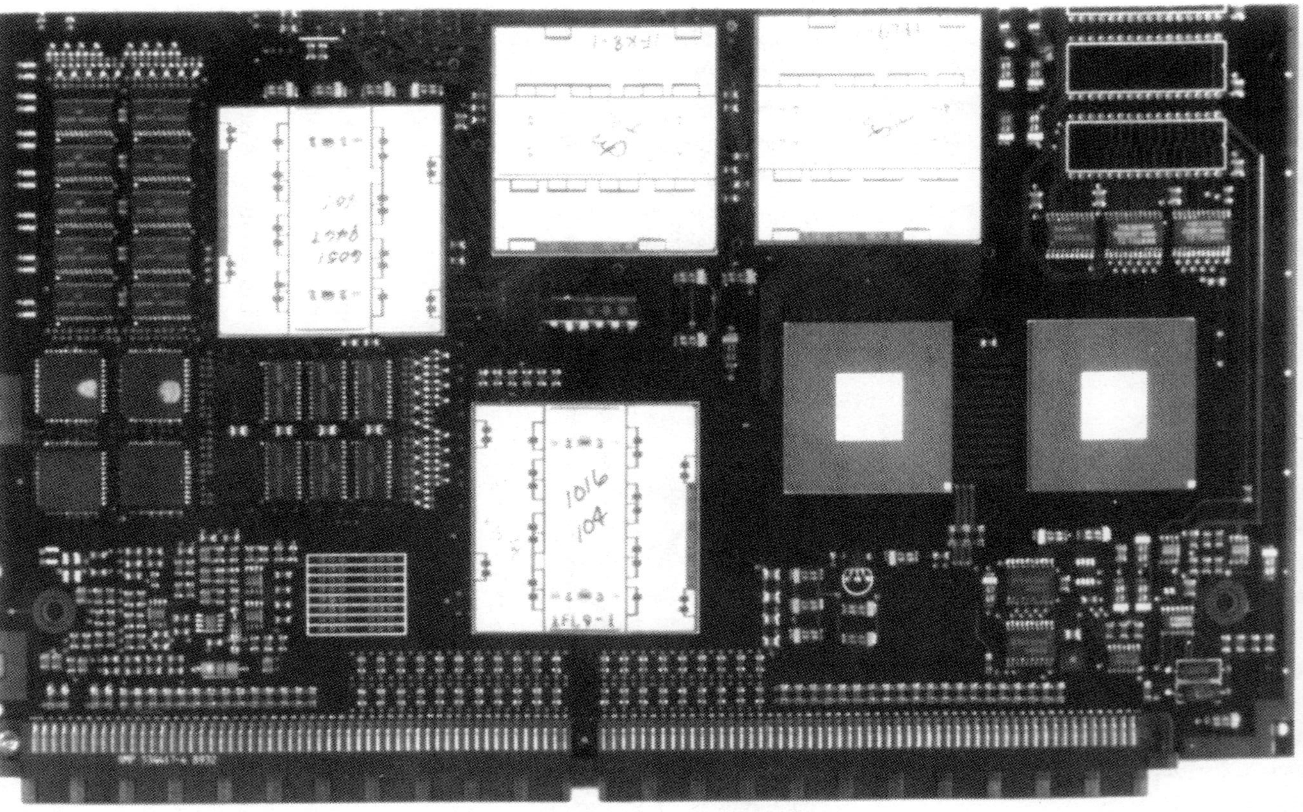

**Figure 21-2** PCB with PGA components.

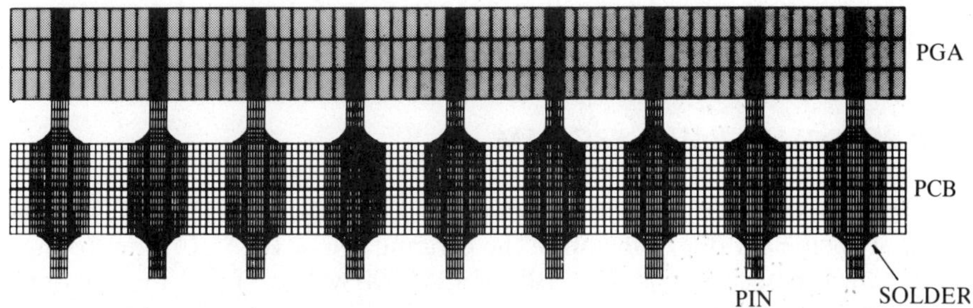


**Figure 21-3**   Pin grid array (PGA) assembly. PCB thickness = 2.36 mm; PGA thickness = 2.40 mm; stand-off height = 1.27 mm; center-to-center = 2.54 mm; hole size = 0.71 mm; pin size = 0.43 mm.

in the copper pads/barrels and solder joints of the PGA interconnect under temperature cycling conditions. Since the life of these structural components is nonlinearly related to the inverse of the stresses and strains, the reliability of these components is addressed.

The chapter has two parts. In Part I, "Reliability of Pin Grid Array Interconnects," the thermal stresses and strains in the solder joints and plated-through-hole (PTH) copper pads/barrels of a PGA assembly under thermal cycling conditions are presented. There are two major systems of thermal stresses/strains acting at the solder joint and copper. One is the transverse shear and vertical normal stress/strain due to the local thermal expansion mismatch between the pin, solder, copper, and FR-4. The other is the horizontal normal stress/strain due to the global thermal expansion mismatch between the ceramic PGA and the FR-4 PCB. The effects of the local thermal expansion mismatch on the reliability of the solder joint and PTH copper have been determined using a 3D orthotropic-elasto-plastic finite element method. The effects of the global thermal expansion mismatch on the reliability of the solder joint and PTH copper have been determined by fatigue experiments. Fatigue life of the solder joint and PTH copper was then estimated based on the calculated strains and the fatigue data on solders and coppers.

In Part II, "Effect of Rework on the Reliability of Pin Grid Array Interconnects," the reliability of solder joints and plated-through hole copper pads/barrels of the pin grid array assemblies during rework are determined by fatigue experiments. The cross-sections of the reworked PGA assemblies (before and after fatigue tests) are also provided for a better understanding of the failure mechanisms of the composite structure. Furthermore, the load-drop curves of the PGA interconnects for up to three reworks are provided for a better estimate of their fatigue life.

# Part I
# Reliability of Pin Grid Array Interconnects

## 21.2 BOUNDARY-VALUE PROBLEM

The PGA component is made of ceramic and is 0.094 in. (2.4 mm) thick. The thermal coefficient of expansion of the ceramic is $6 \times 10^{-6}/°C$. The Young's modulus of the ceramic is $37 \times 10^6$ psi ($255\,000$ MN/m$^2$) and the Poisson's ratio is 0.3 (Table 21-1).

The PGA component is 2.3 in. $\times$ 2.3 in. $\times$ 0.094 in. (5.84 cm $\times$ 5.84 cm $\times$ 0.24 cm) and has 408 pins at 0.1-in. (2.54-mm) pitch (center-to-center). The diameter of each pin is 0.017 in. (0.43 mm). The standoff height of the PGA is 0.05 in. (1.27 mm) from the PCB. The pin is made of Alloy-42 and is plated with a very thin layer of gold. The thermal coefficient of expansion of the pin is $5 \times 10^{-6}/°C$. The Young's modulus of the Alloy-42 pin is $21.5 \times 10^6$ psi ($148\,000$ MN/m$^2$) and the Poisson's ratio is 0.3.

The thickness of the PCB is 0.093 in. (2.36 mm) and it is made of FR-4 epoxy–glass. The thermal coefficient of linear expansion of the FR-4 material is $15 \times 10^{-6}/°C$ in the $X$- and $Y$- (i.e., horizontal) directions and is $85 \times 10^{-6}/°C$ in the $Z$- (i.e., vertical) direction. The Young's modulus of the PCB is $1.6 \times 10^6$ psi ($11\,000$ MN/m$^2$) and the Poisson's ratio is 0.28 (Table 21-1).

The electrodeposited copper thickness is 0.0012 in. (0.03 mm) for the PTH barrels and is 0.0016 in. (0.04 mm) for the PTH pads. The thermal coefficient of linear expansion of the copper is $17 \times 10^{-6}/°C$. The electrodeposited

**Table 21-1**   Material Properties of PGA Assemblies

| | Young's Modulus $\times 10^6$ (psi) | Poisson's Ratio | Thermal Coefficient of Linear Expansion $\times 10^{-6}$ (in./in.-°C) |
|---|---|---|---|
| Solder | 1.5 | 0.40 | 21 |
| Copper | 17.5 | 0.35 | 17 |
| Alloy-42 | 21.5 | 0.30 | 5 |
| Ceramic | 37.0 | 0.30 | 6 |
| FR-4 | | | |
| $XY$ | 1.6 | 0.28 | 15 |
| $Z$ | 1.6 | 0.28 | 85 |

copper is assumed to be an elasto-plastic material (yield stress = 9000 psi or 62 MN/m$^2$, yield strain = 0.0005, elongation = 0.12, and strain hardening parameter = 0.0075). The Young's modulus of the copper is 17.5 × 10$^6$ psi (121 000 MN/m$^2$) and the Poisson's ratio is 0.35.

Due to the low yield strength and high ductility of the Sn–Pb solder,[44–67] it is also assumed to be an elasto-plastic material (yield stress = 1200 psi or 8.3 MN/m$^2$; yield strain = 0.0008; strain hardening parameter = 0.1). The Young's modulus of the solder is 1.5 × 10$^6$ psi (10 000 MN/m$^2$) and the Poisson's ratio is 0.4. The thermal coefficient of linear expansion of the solder is 21 × 10$^{-6}$/°C. It should be emphasized that the mechanical properties of solders are strongly temperature-, frequency-, rate-, and time-dependent.[44–67] These variables have to be included in the analysis to produce an accurate estimation of the strains, and consequently the number of cycles-to-failure of the solder joint. However, for the sake of simplicity and due to the lack of a solder constitutive equation that includes all the important variables, all the material properties are assumed to be constant. Also, we assume that the small amount of gold in the solder joint does not have significant effect.[48] Furthermore, in order to have a conservative estimate of the fatigue life (based on the plastic strain calculation), the yield strength at 120°C has been used (Fig. 21-1).

The determination of thermal stresses and strains in electronics packaging is not an easy task. Closed-form solutions such as those given in refs. 68–77 are useful, but are limited to very simple geometries and loading conditions. The finite element method is one of the best candidates for obtaining numerical results for the thermal stresses and strains in electronics packages[78–87] and is used in the present study. The boundary-value problem is to calculate the thermal stresses and strains generated in the PGA assembly (Fig. 21-3) when it is subjected to a temperature cycle between 20°C and 120°C, i.e., a temperature change of 100°C. The local thermal expansion mismatch will be addressed using the finite element method, and fatigue experiments will be used to investigate the global thermal expansion mismatch.

## 21.3 LOCAL THERMAL EXPANSION MISMATCH BY FINITE ELEMENT ANALYSIS

The finite element modeling and results of the PGA assembly due to local thermal expansion mismatch will be discussed in the following sections. Since the thermal coefficient of expansion in the vertical direction is more than five times larger than that in the horizontal direction for the FR-4 epoxy–glass, the material is modeled as orthotropic.

### 21.3.1 Finite Element Modeling

Figure 21-4 shows a quarter of the solid model of a PGA assembly. It consists of the FR-4, copper, the solder joint, and the Alloy-42 pin. Since the solder joint and PTH copper are the focus of the present study, they are isolated and shown in Figs. 21-5a and 21-5b, respectively. It can be seen that, in addition to the upper and lower pads, a middle copper pad is also considered. Figure 21-6 shows the finite element model for the analysis of the PGA assembly. Due to symmetry, only the upper half of the PGA assembly is analyzed. 3D axisymmetric solid elements (2588) are used for the construction of this model: 1092 for the solder joint, 440 for the Alloy-42 pin, 542 for the FR-4, and 514 for the electrodeposited copper. Each element has eight nodal points, each with two degrees of freedom.

One of the strengths of the finite element method is its ability to handle a variable mesh. This allows one to increase the number of nodal points in stress concentration areas and relax the number of nodal points in the low-stress zones or in the areas that are of no interest. The net effect is that the overall accuracy of the method is maintained while the matrix equation to be solved is minimized. This feature of the finite element method has been used for constructing the models throughout this study.

**Figure 21-4**   3D quarter solid model of the PGA assembly.

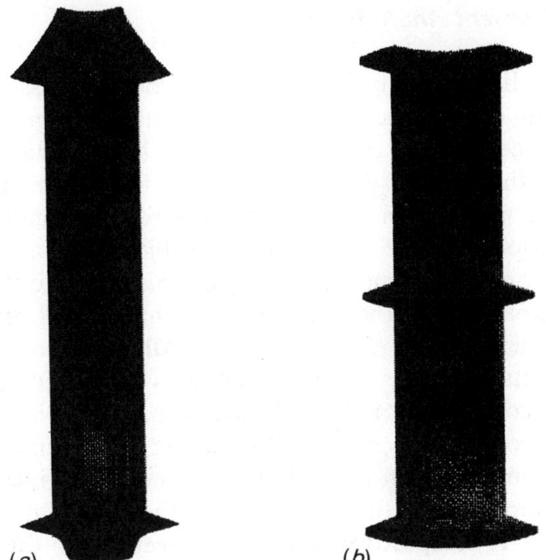

**Figure 21-5**   (*a*) 3D solid model of the PGA solder joint. (*b*) 3D solid model of the PTH copper.

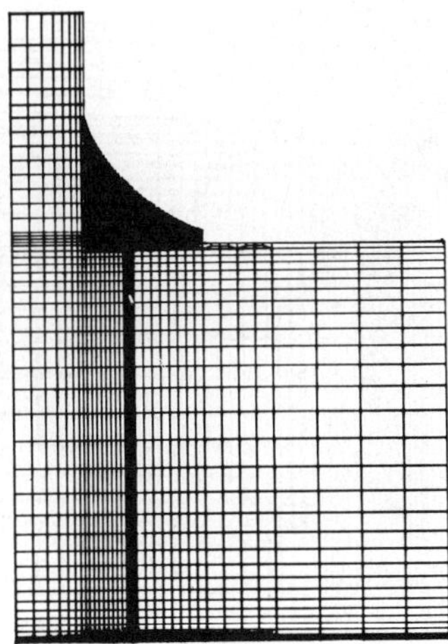

**Figure 21-6**   Finite element model of $\frac{1}{4}$ of the PGA assembly.

### 21.3.2 Finite Element Analysis and Results

The whole-field displacements (deflections) of the PGA assembly are shown in Fig. 21-7. The dotted lines are for the original mesh and the solid lines are for the displaced mesh at $\Delta T = 100°C$. It can be seen that the deformation of the FR-4 in the vertical direction is larger than that in the horizontal directions, the upper copper pad moves upward and rotates in a counterclockwise direction, and the solder joint is subjected to a combined action of shear, tension, compression, and rotation. This is due to the large thermal coefficient of expansion of the FR-4 PCB in the vertical direction and the small thermal coefficient of expansion of the Alloy-42 pin.

The incremental plastic strain components ($d\varepsilon_r^p$ = incremental plastic normal strain component in the $R$-direction, $d\varepsilon_z^p$ = incremental plastic normal strain component in the $Z$-direction, $d\varepsilon_\theta^p$ = incremental plastic normal strain component in the $\theta$-direction, and $d\gamma_{rz}^p$ = incremental plastic shear strain component in the $RZ$-plane) in the solder are shown in Figs. 21-8 through 21-11, respectively. It can be seen from Fig. 21-8 that the maximum incremental plastic normal strain component in the radial direction ($d\varepsilon_r^p$) occurs near the tip of the solder fillet and at the corner of the solder joint, with a value of 0.0047. The maximum incremental plastic normal strain component in the vertical direction ($d\varepsilon_z^p$) occurs at the corner of the solder joint (0.0058) (Fig. 21-9). This is due to the large vertical expansion of the FR-4.

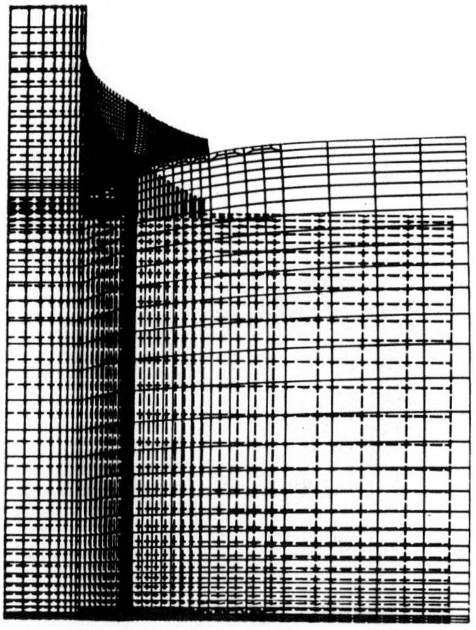

**Figure 21-7**  Displacement of $\frac{1}{4}$ of the PGA assembly.

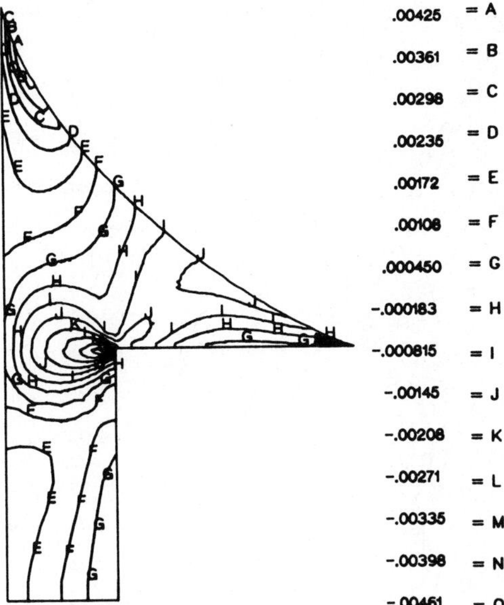


**Figure 21-8**   Solder joint incremental plastic strain in the radial direction.

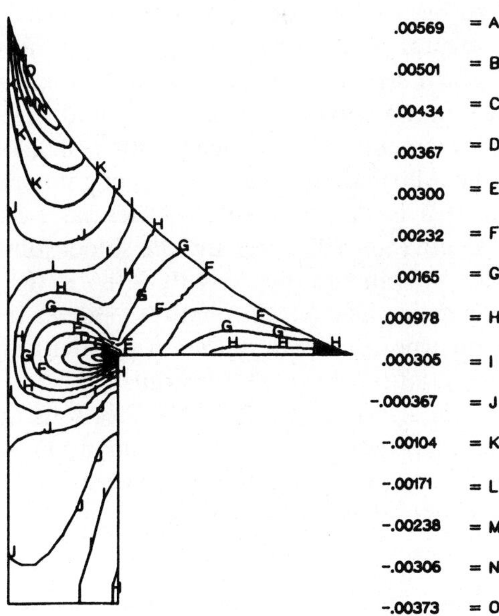


**Figure 21-9**   Solder joint incremental plastic strain in the vertical direction.

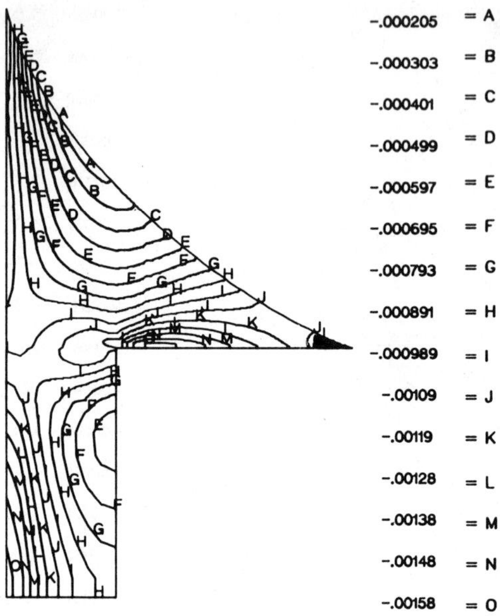


**Figure 21-10**    Solder joint incremental plastic strain in the circumferential direction.

Figure 21-10 shows the incremental plastic normal strain component in the circumferential direction $(d\varepsilon_\theta^p)$, and the maximum value (0.0016) is small compared with the other two plastic strain components.

The largest incremental plastic strain component in the solder joint is the incremental plastic shear strain component acting in the transverse direction $(d\gamma_{rz}^p)$, and its contours are shown in Figs. 21-11 and 21-12. It can be seen that the maximum incremental plastic shear strain is 0.024 and occurs at the interface between the Alloy-42 pin and the solder joint near the tip of the solder fillet. Again, this is due to the large thermal expansion mismatch between the Alloy-42 pin ($5 \times 10^{-6}/°C$) and the solder joint ($21 \times 10^{-6}/°C$), and between the Alloy-42 pin and the FR-4 PCB ($85 \times 10^{-6}/°C$). The values of the incremental plastic shear strain contours shown in Figs. 21-11 and 21-12 will be used, incorporated with Solomon's solder fatigue data,[57–61] to estimate the fatigue life of the PGA solder joint.

The plastic and total strain contours in all directions acting in the PTH copper are not presented here. However, the maximum values of the total strains are tabulated in the second column of Table 21-2.

The Von Mises stress $(\bar{\sigma})$ acting in the FR-4 epoxy–glass is shown in Fig. 21-13 and is defined as follows:

$$\bar{\sigma} = \frac{\sqrt{2}}{2} \sqrt{(\sigma_r - \sigma_z)^2 + (\sigma_z - \sigma_\theta)^2 + (\sigma_\theta - \sigma_r)^2 + 6\tau_{rz}^2}$$

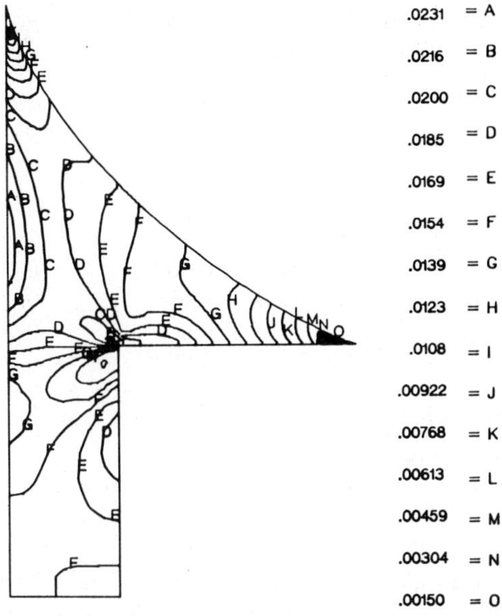


**Figure 21-11**   Solder joint incremental plastic shear strain in the transverse direction.

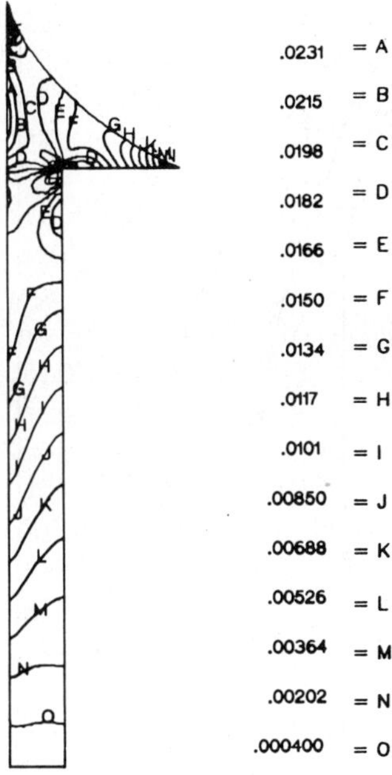


**Figure 21-12**   Solder joint incremental plastic shear strain in the transverse direction.

**Table 21-2**  Effects of Solder Geometry on the Total Strains of PTH Copper (PGA)*

| Maximum Total Strain in Copper | Solder Geometry | | |
|---|---|---|---|
| | Typical | No Fillet | Half-Filled |
| $\varepsilon_r^t$ | 0.0030 | 0.0039 | 0.0037 |
| $\varepsilon_z^t$ | 0.0051 | 0.0062 | 0.0086 |
| $\varepsilon_\theta^t$ | 0.0005 | 0.0028 | 0.0020 |
| $\gamma_{rz}^t$ | 0.0042 | 0.0035 | 0.0029 |

* Hole size = 0.028 in.; board thickness = 0.093 in.

In the above equation, $\sigma_r$ = normal stress component in the $R$-direction, $\sigma_z$ = normal stress component in the $Z$-direction, $\sigma_\theta$ = normal stress component in the $\theta$-direction, and $\tau_{rz}$ = shear stress in the $RZ$-plane. As expected, the maximum stress (13 600 psi or 94 MN/m²) occurs at the corner right underneath the intersection between the copper barrel and upper pad and decreases rapidly as soon as it is away from the corner. Since the stress

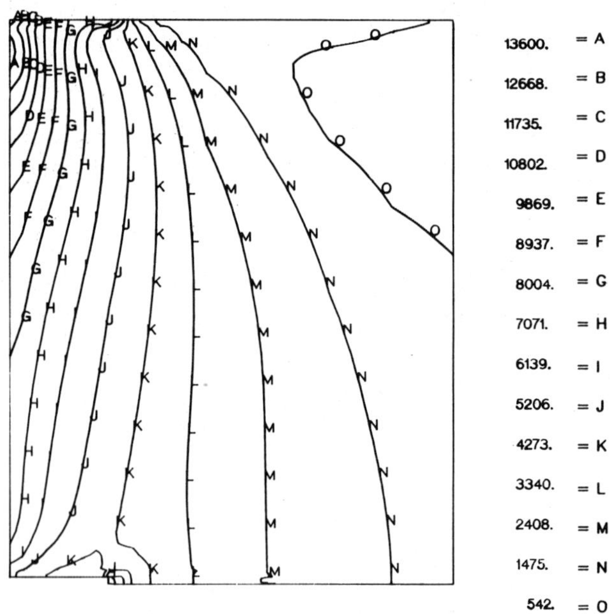


**Figure 21-13**  Von Mises stress (psi) (1 psi = 6890 Pa) in the FR-4 PCB.

value in the FR-4 PCB is smaller than the failure strength (stress in tension/compression = 35 000 psi or 240 MN/m$^2$; stress in shear = 14 000 psi or 97 MN/m$^2$) of the FR-4,[88] reliability of the PCB epoxy–glass is not of concern.

## 21.4 GLOBAL THERMAL EXPANSION MISMATCH BY FATIGUE TESTING

The horizontal dimensions of the extreme pins around the periphery of the 408-pin PGA are 2.2 in. × 2.2 in. (5.59 cm × 5.59 cm). Consequently, for the thermal condition under consideration ($\Delta T = 100°$C), the maximum thermal expansion mismatch between the FR-4 PCB ($15 \times 10^{-6}/°$C) and the ceramic PGA ($6 \times 10^{-6}/°$C) is about 0.001 in. or 0.025 mm. In this section, a displacement boundary condition of $\pm 0.001\,15$ in. (0.029 mm) has been imposed on the solder joint. The specimen preparation, test setup, and experimental results will be briefly stated in the following sections.

### 21.4.1 Specimen Preparation and Test Setup

The 408-pin PGA assembly is cut into 3-by-3 arrays (0.3 in. × 0.3 in. × 0.24 in. or 7.62 mm × 7.62 mm × 6.1 mm) (Fig. 21-14a). The pins are then cut to within 0.005 in. (0.127 mm) from the ceramic PGA using a low-speed diamond saw. During sawing, both the ceramic PGA and the FR-4 PCB are gripped such that the load supported by the pins is minimized. All the pins except the central pin are then removed; the test vehicle consisted of a single pin attached to the FR-4 PCB (Fig. 21-14b). The length of the pin from its free end to the PCB is about 0.045 in. (1.14 mm). This specimen is then placed in a fatigue machine, and a cross-section of the test setup is schematically shown in Fig. 21-15.

The fatigue machine is a custom-built micromechanical testing machine detailed in ref. 62 (Fig. 21-16). The displacements are applied through a micrometer-driven translation stage. The upper load train consists of the load cell, a coupling, and the upper sample grips. The lower load train is made up of the motor, motor coupling, a micrometer-driven translation stage, and the lower sample grips.

A variable-speed motor is used to move the translation stage. The motor speed is controlled by analog feedback using the output of a tachometer coupled to the motor shaft. The motor speed and reversal limits are determined by the software interface devised specifically for this machine.[62] The data is taken continuously using a special-purpose data acquisition program.[62]

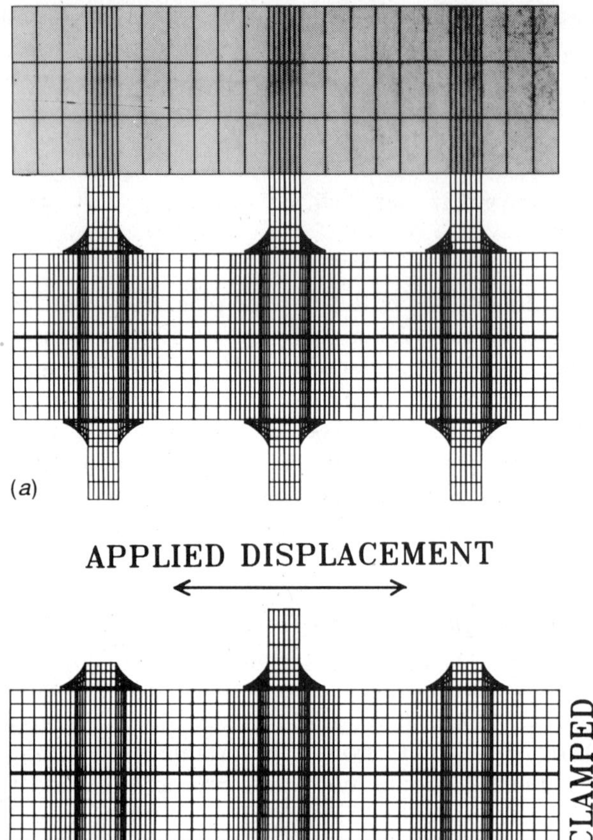



**Figure 21-14**   (a) 3-by-3 array PGA specimen. (b) Single-pin PGA fatigue specimen.

The displacements are measured using capacitance gages. The load is measured using a tension/compression resistance load cell ($\pm$ 444 N or $\pm$ 100 lb capacity). The linearity of motion of the translation stage is within 25 μm (0.001 in.) over 250 μm (0.01 in.) displacement range under typical loading conditions. The effective machine stiffness has been measured to be 1.75 MN/m (10 klb/in.). The specimen grip stiffness is 62 MN/m (354 klb/in.).[62] The minimum displacement resolution (25 nm or 0.000 001 in.) and load resolution ($6.67 \times 10^{-3}$ N or $1.5 \times 10^{-3}$ lb) are determined by the capacitance gage resolution and the voltmeter resolution ($5\frac{1}{2}$-digit

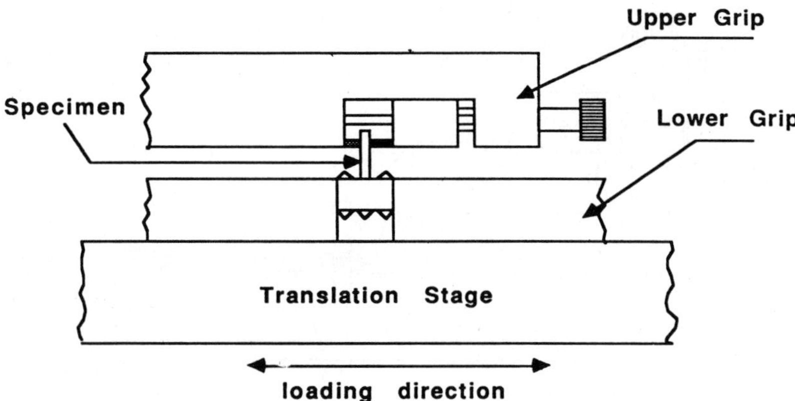


**Figure 21-15**   A cross-section view of the fatigue test setup.

resolution), respectively. The maximum displacement limit (250 μm or 0.01 in.) is due to the range of the capacitance gage.

The specimen is mounted between the two grips, one attached to the upper load train and the other attached to the translation stage (Figs. 21-15 and 21-16). The pin is attached to the upper grip while the PCB is gripped on the lower grip. The pin is loaded (in bending, shear, and axial force) with a $\pm 0.001\,15$ in. (0.029 mm) displacement boundary condition at 150 cycles/hour. The test temperature is 25°C and the ambient humidity is $65 \pm 5\%$.

### 21.4.2 Experimental Results and Discussion

The loads and displacements are continuously monitored as a function of time. A typical result is illustrated in Fig. 21-17*a*. It shows the peak cyclic load as a function of time (cycle number = time/24 sec) for the fatigue test. Note that the peak cyclic load drops from about 6 lb (26.6 N) to 4.5 lb (19.98 N) in the first 100 cycles. However, with further cyclic displacements, the rate of load drop tends to remain constant. Figure 21-17*b* shows the data for another sample subjected to the same displacement boundary condition. It can be seen that the load-drop curves in Fig. 21-17*a* and 21-17*b* are about the same (i.e., the test results are reproducible).

Figure 21-18*a* shows the load–displacement loop measured in the first three cycles. The load–displacement profile is nearly linear in region AB. The slope of the initial elastic region is 2.33 MN/m (13 klb/in.). Beyond point B the inelastic deformation in both the solder and the pin causes a significant nonlinearity in the deformation behavior of the assembly. A similar trend is

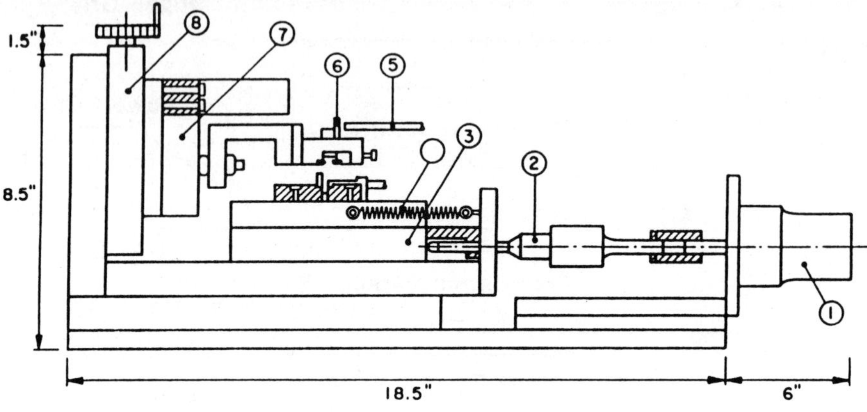

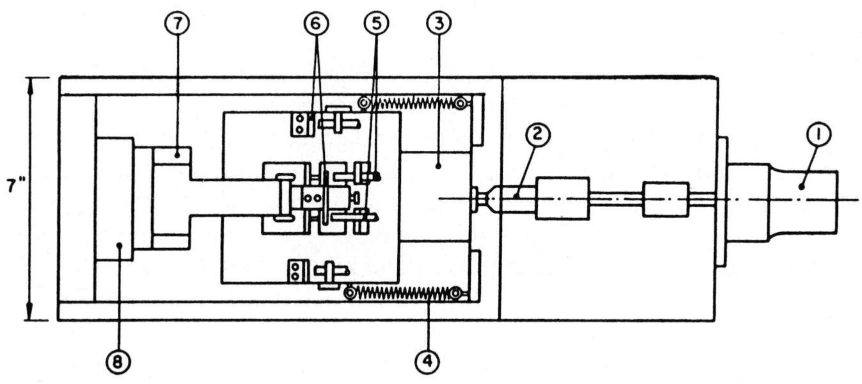

| NO. | PARTICULARS |
|-----|-------------|
| 1 | Motor |
| 2 | Micrometer |
| 3 | Horizontal Stage |
| 4 | Retaining Spring |
| 5 | Contactless Displacement Probe |
| 6 | Probe Target |
| 7 | Load Cell |
| 8 | Vertical Translation Stage |

**Figure 21-16**   Test geometry and loading mode.

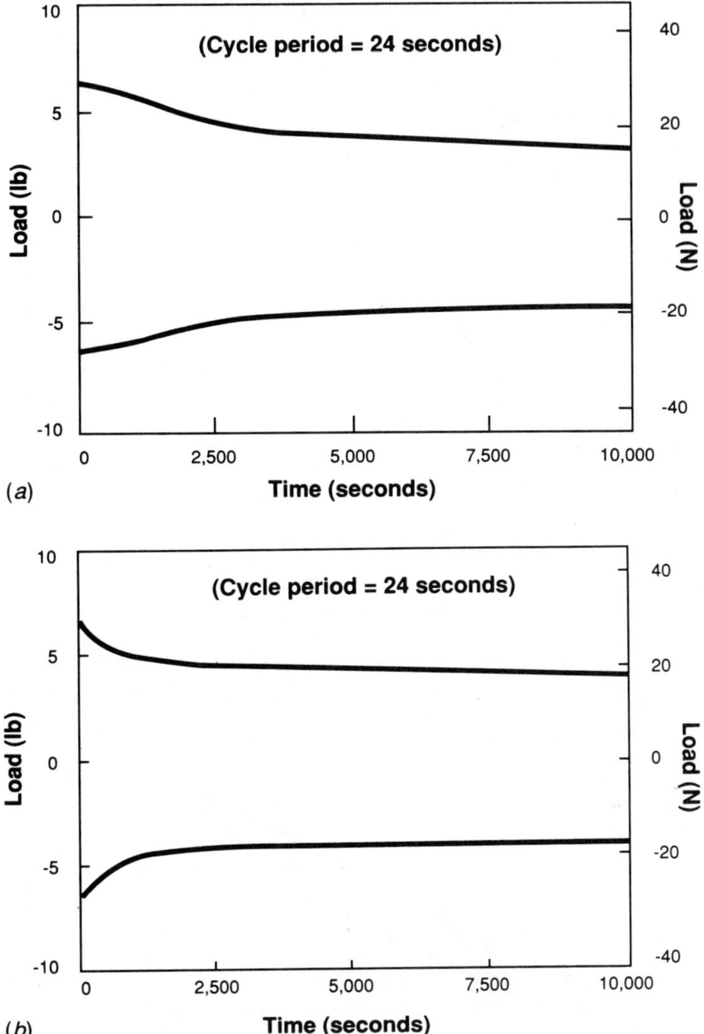


**Figure 21-17**  (a) Load-drop curves (total displacement = 0.0023 in. (b) Load-drop curves (total displacement = 0.0023 in.).

observed upon direction reversal (CDA). The hysteresis loops are found to be reproducible and symmetric.

The slope of the linear region in the load–displacement profile does not significantly change during the early stages of crack growth in the solder. This observation may be reconciled in terms of the relative stiffness of the pin and the solder joint. Since the pin is the most compliant member of the

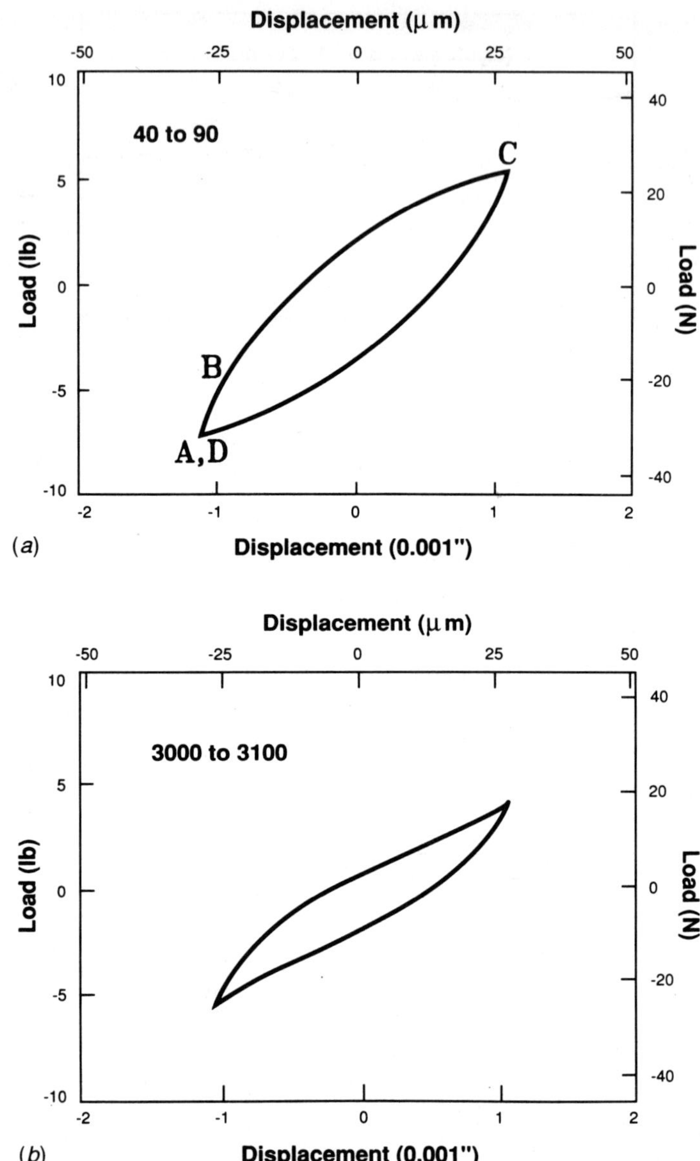


**Figure 21-18**    (a) Load–displacement loop at the first 3 cycles. (b) Load–displacement loop at 125 cycles. (*Continued*)

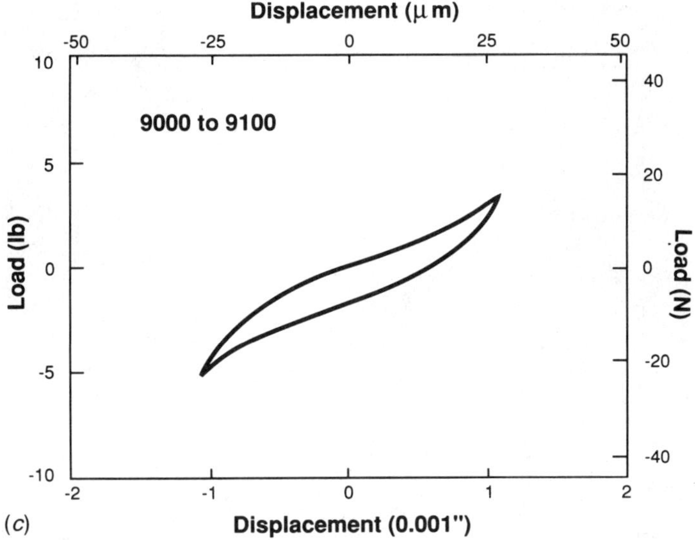



**Figure 21-18** (*continued*)   (*c*) Load–displacement loop at 375 cycles.

soldered assembly, a reduction in solder joint stiffness due to a crack in the solder joint will not be reflected in the overall stiffness.

Figure 21-18*b* shows the load–displacement curve at 125 cycles. Note that a deviation in an otherwise smooth load–displacement curve (Fig. 21-18*a*) is observed. The slope of the load–displacement curve at large displacements increases, while the slope at the smaller displacements decreases. However, it is observed that beyond about 250 cycles, the apparent change in the load–displacement curve is small and both the slope and the shape of the curves appear relatively constant. Figure 21-18*c* shows the load–displacement curve at 375 cycles. The maximum number of cycles applied to these specimens is 1000, and for this number of cycles there is not complete failure of the solder joint.

The increase in slope of the load–displacement curve at large displacements may be explained in terms of a crack closure effect. Upon reaching large displacements, one side of the crack closes. Therefore, the net load is supported by a large area of the solder joint such that the effective stiffness of the specimen increases. This results in an increase in the slope of the load–displacement profile at large displacements.

Cross-sections of all the specimens fatigue-tested in this work indicated the presence of cracks. A typical cross-section is shown in Fig. 21-19. The cracks typically nucleate near the tip of the solder fillet, propagate in the solder joint roughly parallel to the pin, and stop. It appears that crack growth is the primary mechanism responsible for the decrease in peak load, and the

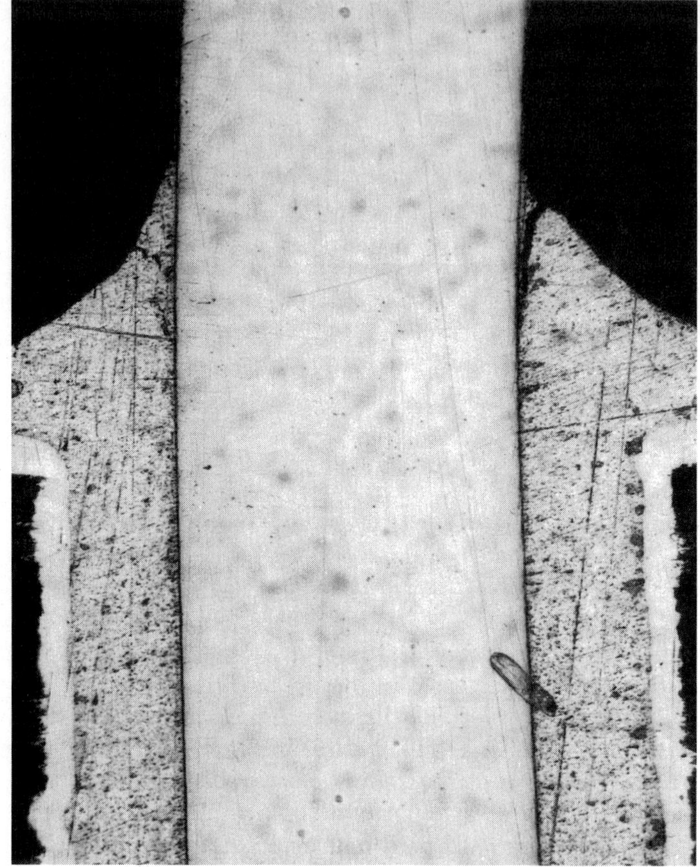

**Figure 21-19**  Cross-section of the fatigue specimen.

crack growth rates decrease with increasing cycle number. The latter is a direct consequence of crack closure. Once the load-bearing area is increased, the crack-tip loading is correspondingly decreased. A decrease in the crack-tip loading will result in smaller plastic strain and crack growth rates, and thus slow down the crack propagation in the solder joint.

## 21.5 FATIGUE LIFE OF PGA SOLDER JOINT AND PTH COPPER

As mentioned earlier, there are two major systems of thermal stresses/strains acting at the solder joint. One is the horizontal normal stress/strain due to the global thermal expansion mismatch between the ceramic PGA and the FR-4 PCB. This system of stress and strain was addressed by fatigue experiments and was found to be no threat to the reliability of solder joints due to the crack closure effect.

The other system of thermal stress and strain is due to the local thermal expansion mismatch between the solder, pin, copper, and FR-4, and was determined by a nonlinear 3D finite element method. It has been shown that the largest plastic strain component in the solder joint is the plastic shear strain component acting in the transverse direction $(\gamma_{rz}^{p})$ and this is shown in Figs. 21-11 and 21-12. In this section, the thermal fatigue life of the PGA solder joint is estimated based on the plastic shear strain (Figs. 21-11 and 21-12) and Solomon's fatigue data of solders.[57–61]

In view of Fig. 21-11, it can be seen that the maximum plastic shear strain is 0.021. Consequently, the thermal fatigue life under the current condition $(\Delta T = 100°C)$, is 2084 cycles.[57–61] Following all the suggestions provided by Solomon, it is safe to predict that the solder joint will last for 520 cycles. This means that after 520 cycles there will be a crack running at mark A of Fig. 21-11.

As mentioned earlier, since most of the thermal fatigue life is spent in propagating the crack (i.e., fatigue crack growth), and the plastic shear strain near the middle section of the solder joint is small (Fig. 21-12), it is reasonable to relax the conservatism and extend the crack to marks J and K. In that case, the plastic shear strain is 0.007 69 and consequently, the solder joint of the PGA interconnection should have an average fatigue life of 4500 cycles.[57]

The maximum values of the total strains in the PTH copper acting in the radial direction $(\varepsilon_{r}^{t})$, vertical direction $(\varepsilon_{z}^{t})$, circumferential direction $(\varepsilon_{\theta}^{t})$, and shear transverse direction $(\gamma_{rz}^{t})$ are tabulated in Table 21-2. The PTH copper strains in the second column are for the case of a typical PGA solder joint; those in the third column are for the case of a PGA solder joint without solder fillets; and those in the fourth column are for the case of a PGA with half-filled solder joints. It can be seen that their values are small compared with the elongation of the PTH copper (0.12). Furthermore, based on the experimental fatigue data of electroless copper[5,6] and the calculated largest maximum total strain (0.009), an average fatigue life of 10 700 cycles was estimated. However, it should be pointed out that: (1) the ductility of electroless copper is higher than that of electroplated copper; (2) a perfect structure has been modeled; and (3) surface roughness, surface activation, contamination, etc., were not considered in the present study. Thus, a safety factor should be adopted in order to give a conservative life estimation. If a safety factor of 3 is chosen, then the copper will last for 3560 cycles.

## 21.6 EFFECTS OF SOLDER GEOMETRY ON THE FATIGUE LIFE OF PGA SOLDER JOINTS

Cross-sections of more than 100 PGA solder joints show that the typical solder joint geometry is the one shown in Figs. 21-4, 21-6, and 21-19. However, it is not uncommon to observe some PGA solder joints without the solder fillet if the PGA is assembled to a very thick PCB with reflow-solder methods. In this study, two other solder joint geometries will be considered.

### 21.6.1 PGA Solder Joints without Solder Fillet

Figure 21-20 shows the finite element model for the analysis of a PGA assembly without solder fillet. Due to symmetry, only the upper half of the assembly is analyzed. The whole-field deflections of the PGA assembly are shown in Fig. 21-21. The dotted lines are for the original mesh and the solid lines are for the displaced mesh. It can be seen that the PGA solder joint is subjected to very large shear deformation, especially at the corner right next to the Alloy-42 pin.

The incremental plastic shear strain contours acting in the PGA solder

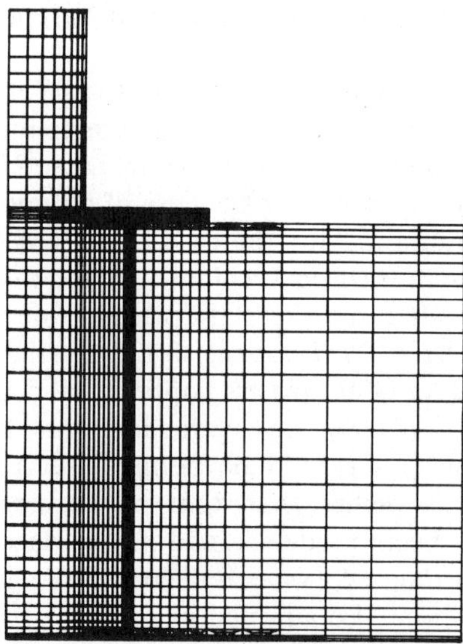

**Figure 21-20**   Finite element model of $\frac{1}{4}$ of a PGA solder joint without fillet.

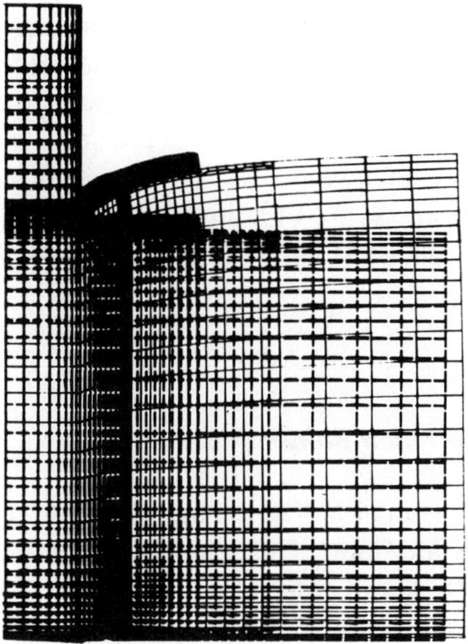

**Figure 21-21**  Displacement of $\frac{1}{4}$ of the PGA solder joint without fillet.

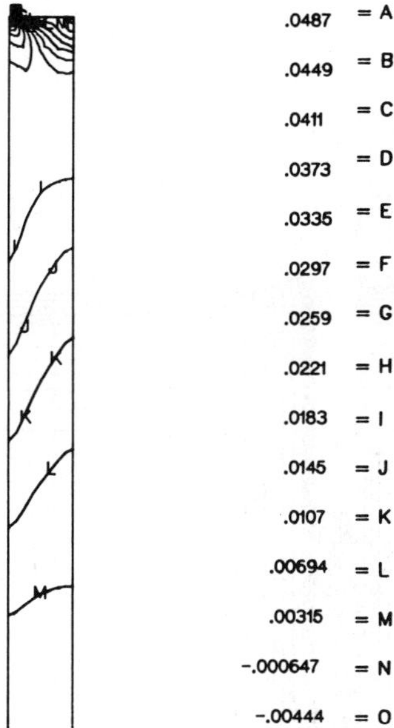


**Figure 21-22**  Incremental plastic shear strain in the solder joint.

joint are shown in Fig. 21-22. It can be seen that very large plastic shear strain (0.05) is concentrated at the corner of the solder joint. This plastic shear strain decreases rapidly as soon as it is away from the corner and is very small near the middle of the solder joint. Consequently, it is reasonable to assume the crack extends to marks K and L with a plastic shear strain equal to 0.008 84 (Fig. 21-22). With Solomon's fatigue data,[57-61] the average thermal fatigue life of the PGA solder joint without solder fillet is estimated to be 3400 cycles.

### 21.6.2 PGA with Half-Filled Solder Joints

Figure 21-23 shows the upper half of the finite element model for the analysis of a half-filled PGA solder joint. The deformation of the PGA assembly and the incremental plastic shear strain contours in the PGA solder joint are shown in Figs. 21-24 and 21-25, respectively. Again, the solder joint is dominated by the shear deformation in the transverse direction (Fig. 21-24) due to the large thermal expansion mismatch between the pin, solder, copper, and PCB. For this case, the average thermal fatigue life is 1200 cycles. This estimate was based on a plastic shear strain value of 0.015 (Fig. 21-25) and Solomon's experimental data.[57-61]

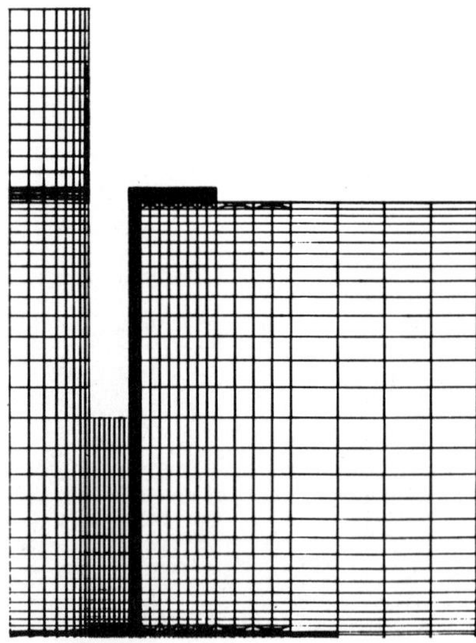

**Figure 21-23**   Finite element model of $\frac{1}{4}$ of the PGA with half-filled solder joint.

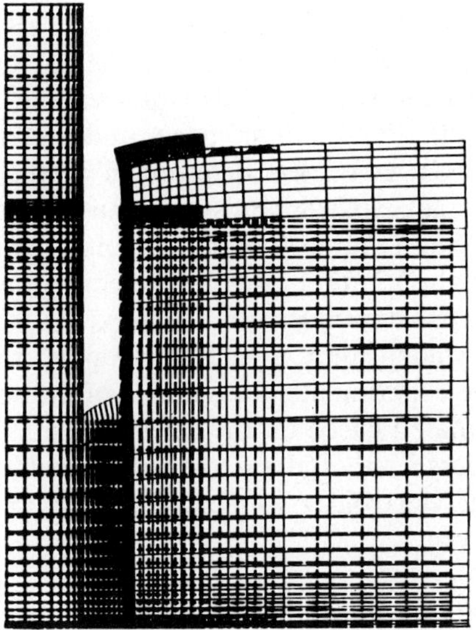

**Figure 21-24**    Displacement of $\frac{1}{4}$ of the half-filled PGA assembly.

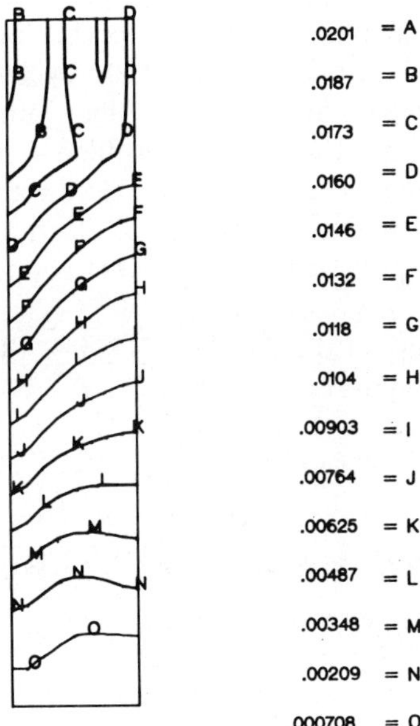



**Figure 21-25**    Incremental plastic shear strain in the solder joint.

## 21.7  SUMMARY

Thermal stresses and strains in the solder joints, PTH copper pads/barrels, and FR-4 PCB of the PGA interconnect under a temperature cycling condition ($\Delta T = 100°C$) have been determined in the present study. The effects of the local thermal expansion mismatch between the copper, Alloy-42, solder, and FR-4 on the reliability of solder joints and PTH copper have been determined using a 3D nonlinear finite element method. The effects of the global thermal expansion mismatch between the ceramic PGA and the FR-4 PCB on the solder joint and PTH copper reliability have been determined by fatigue experiments. Fatigue life of the solder joint and PTH copper was then estimated based on the calculated plastic shear strains and fatigue data on solders,[57–61] and on the calculated total normal strains and fatigue data on electroless coppers.[5,6] Some important results are summarized as follows.

1. The results of fatigue experiments show that they are reproducible and the load–displacement loops are symmetrical. Also, cross-sections of all the specimens show that the cracks typically nucleate near the tip of the solder fillet, propagate in the solder joint roughly parallel to the PGA pin, and stop. Thus, due to the crack closure of the PGA solder joints, the effect of global thermal expansion mismatch on the solder joint reliability may not be of concern.
2. The average thermal fatigue life due to the local thermal expansion mismatch has been estimated to be 4500 cycles for the typical PGA solder joints, 3400 cycles for the PGA solder joints without solder fillet, and 1200 cycles for the PGA with half-filled solder joints.
3. The values of the total strains in the PTH copper pads/barrels are very small compared with the elongation of electrodeposited copper, and the average thermal fatigue life of the PTH copper should last for more than 3560 cycles.
4. The Von Mises stress in the FR-4 PCB is less than the failure strength of the FR-4 epoxy–glass.


# Part II
# Effect of Rework on the Reliability of Pin Grid Array Interconnects

The use of statistical process control and total quality control has minimized the need for rework of less dense components. However, as the number of I/Os of the component increases, the probability of a high-yield is reduced and the probability of failure increases. Consequently, the PGA components must be removed and replaced. In this study, the effects of rework (remove and replace) on the reliability of the solder joint and PTH copper of the PGA assembly will be addressed by fatigue testing. Up to three episodes of rework (i.e., 1-rework, 2-rework, and 3-rework) will be considered and the rework procedures will be briefly stated. Cross-sections of the reworked assemblies before and after the fatigue testing will be provided and their fatigue life will be discussed.

## 21.8 REWORK PROCEDURES

The rework procedures of the 408-pin cavity-down ceramic PGA will be briefly discussed. It consists of three major operations, namely, equipment preparation, board preparation, and rework. The important steps of each operation are stated in the following.

### Equipment Preparation

1. Turn on the M97A Wenesco solder system (Fig. 21-26).
2. Warm up the solder to $254 \pm 6°C$ ($490 \pm 10°F$).
3. Attach the rubber solder cleaning vacuum nozzle to a stainless steel handle.
4. Place the solder nozzle in the solder fountain. The solder nozzle is a specially designed piece of equipment for rework of very dense PCBs. Unlike the traditional solder nozzle, the new design has side slots that are cut into the nozzle (Fig. 21-27). These slots provide exhaust paths for the molten solder so that the nearby solder joints will not be reflowed/bridged. A typical design of the solder nozzle is shown in Fig. 21-28 from the side and bottom views.
5. Adjust the flow rate of the molten solder by adjusting the air pressure pump such that solder can flow smoothly and evenly on four sides of the solder nozzle (Figs. 21-27 and 21-28).

### Board Preparation

1. Desolder all the components within 0.3 in. (7.62 mm) of the PGA from the board.
2. Remove heat sinks from the PGA.
3. Apply liquid flux to the board where the PGA is to be reworked.

**Figure 21-26** M97A Wenesco solder system.

*Existing Design*

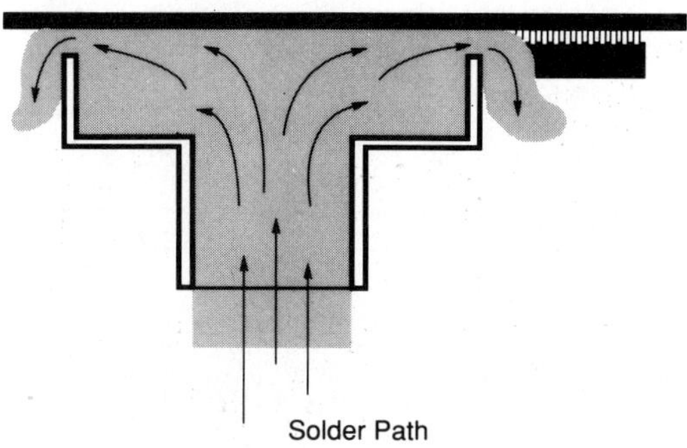


*Proposed Design*

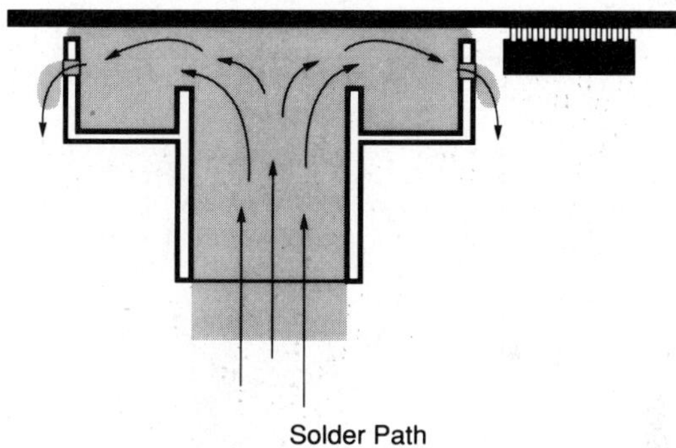


**Figure 21-27**    Existing and proposed designs for the solder nozzle.

## Rework

1. Place the board on the board carrier of the M97A Wenesco rework station.
2. Align the board such that the PGA is right above the solder nozzle.
3. Preheat the board until its top surface reaches 93 ± 6°C (200 ± 10°F).
4. Place and adjust the height of the board at about 0.05 in. (1.27 mm) above the solder nozzle.
5. Pump the solder and remove the PGA from the board within 15 seconds.

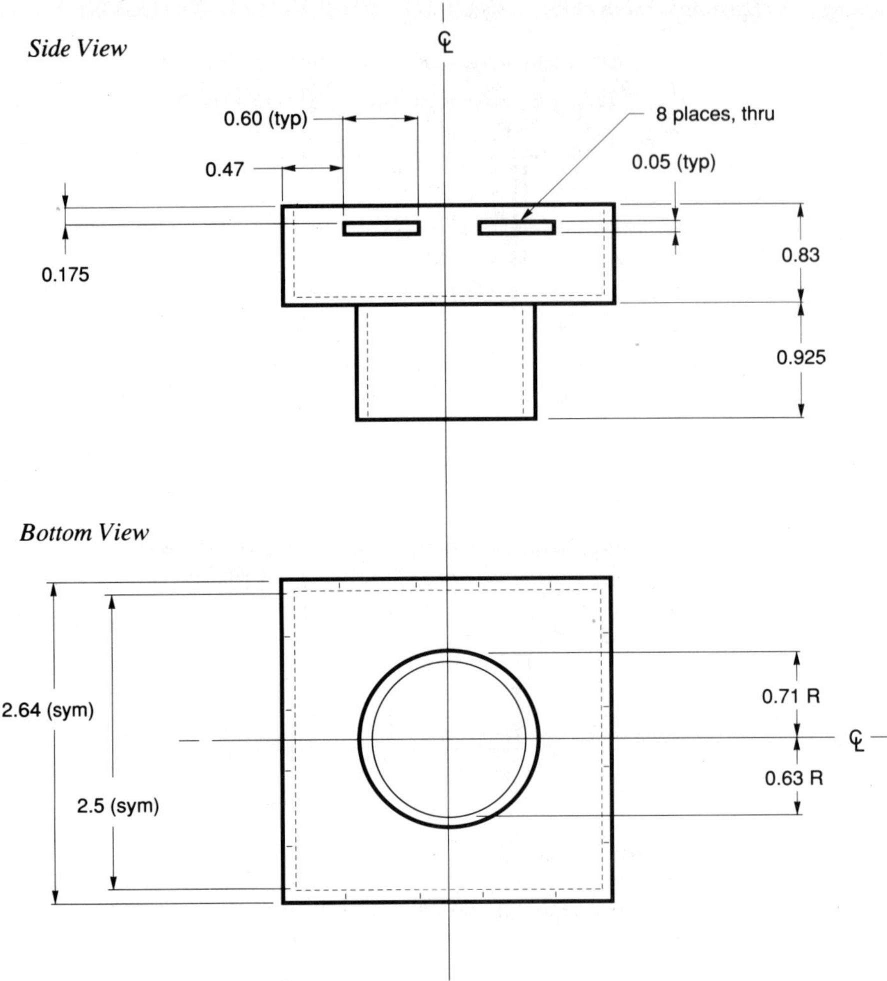


**Figure 21-28**   Dimensions of the new design for the solder nozzle.

6. Remove the solder from the surfaces and holes of the board with the rubber solder cleaning vacuum nozzle.
7. Wash and inspect the board. Manually clear the uncleared holes.
8. Insert a new PGA into the board and repeat steps (1)–(4).
9. Pump the solder to the new PGA for 7 seconds.
10. Wash and inspect the board. Touch up the bad joints with a soldering iron.

## 21.9  CROSS-SECTIONS OF REWORKED PGA INTERCONNECTS (BEFORE TEST)

More than 100 reworked PGA solder joints were inspected. Typical cross-sections of the reworked PGA solder joint and copper are shown in Figs. 21-29 through 21-32 for 0-rework, 1-rework, 2-rework, and 3-rework, respectively (0-rework means the interconnect has never been reworked and 3-rework means the interconnect has been reworked three times). It can be seen that the solder joint and copper for all the cases are almost the same, except the rotation of the copper pad is slightly larger for the specimens with higher numbers of reworks. Furthermore, because of the current rework procedures, copper dissolved into the solder is minimal and the number of reworks does not affect the quality of the solder joint.

## 21.10  FATIGUE TEST OF REWORKED PGA INTERCONNECT

The effect of rework on the reliability of solder joint and copper is further investigated by fatigue experiments. In the tests of this section, three displacement boundary conditions ($\pm 0.0005$ in. or $\pm 0.013$ mm; $\pm 0.001$ in. or $\pm 0.025$ mm; and $\pm 0.0015$ in. or $\pm 0.038$ mm) were imposed on the solder

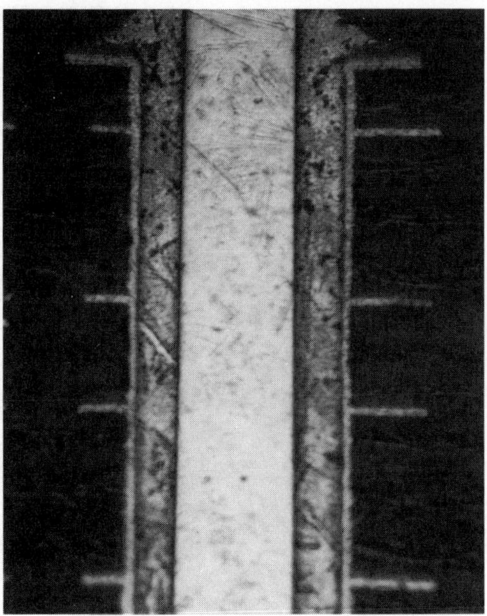

**Figure 21-29**  Cross-section of the 0-rework PGA interconnect (before fatigue test).

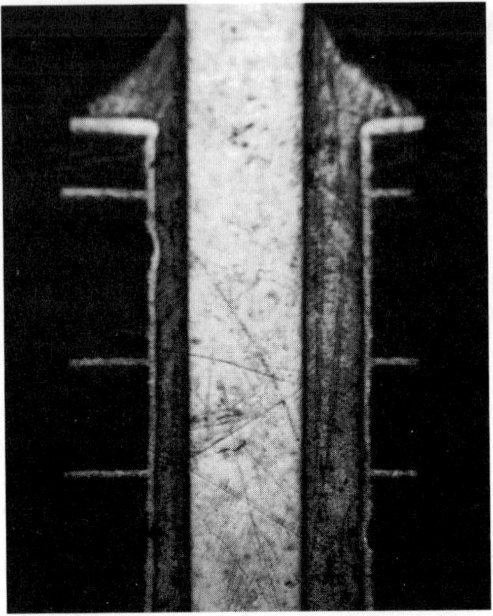

**Figure 21-30**  Cross-section of the 1-rework PGA interconnect (before fatigue test).

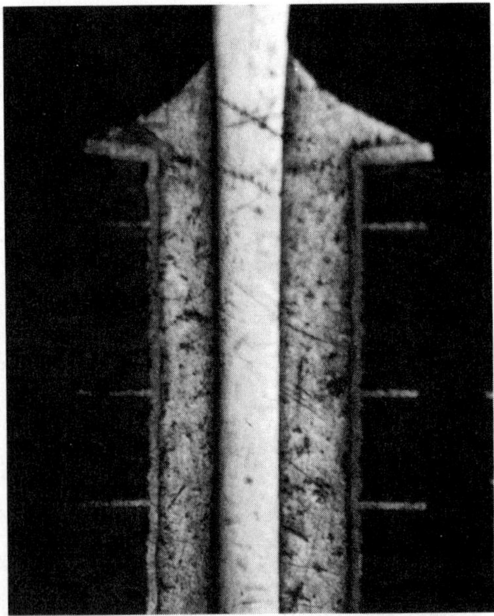

**Figure 21-31**  Cross-section of the 2-rework PGA interconnect (before fatigue test).

**Figure 21-32**  Cross-section of the 3-rework PGA interconnect (before fatigue test).

joint. These displacement boundary conditions could be generated from the global thermal expansion mismatch between the PGA and the PCB under rework conditions.

## 21.10.1 Fatigue Test Setup and Preparation

The specimen preparation and test setup are exactly the same as those mentioned in Section 21.4.1. However, in this section, the pin is loaded (in bending, shear, and axial force) with $\pm 0.0005$ in. (0.0127 mm), $\pm 0.001$ in. (0.0254 mm), and $\pm 0.0015$ in. (0.0381 mm) displacements at 150 cycles/hour, 150 cycles/hour, and 90 cycles/hour, respectively. The test temperature is 25°C and the ambient humidity is 65 $\pm$ 5%.

## 21.10.2 Experimental Results and Discussion

The loads and displacements are continuously monitored as functions of time. It has been shown in Section 21.4.2 that all the experimental results are reproducible and the load–displacement loops are symmetrical. Thus, they will not be repeated here.

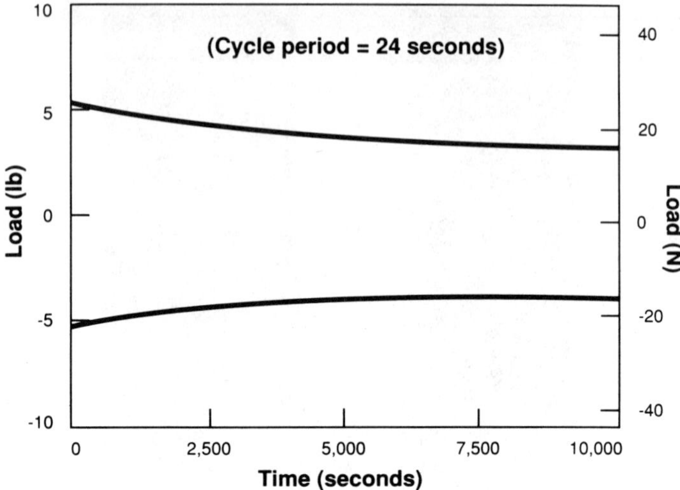


**Figure 21-33**   Load-drop curves for 0-rework specimen (total displacement = 0.051 mm).

Figures 21-33 and 21-34 show the load drops vs. time for the 0-rework and 3-rework specimens, respectively. The displacement boundary condition imposed on these specimens was $\pm 0.001$ in. It can be seen that the load-drop curves for these two cases are almost the same. Thus, there is almost no effect of the rework on the solder joint reliability. This is due to the present rework processes; the solder joint is formed by "fresh" solder at each rework.

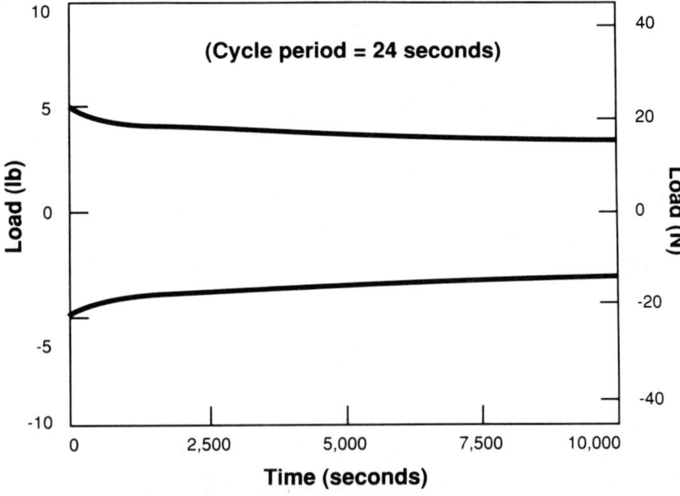


**Figure 21-34**   Load-drop curves for 3-rework specimen (total displacement = 0.051 mm).

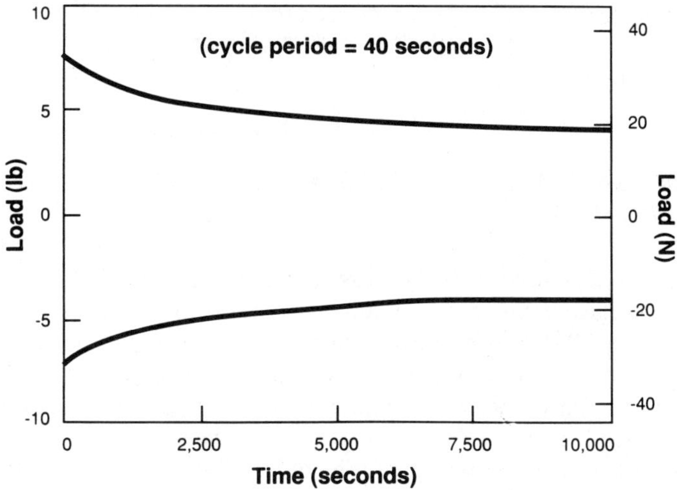


**Figure 21-35**   Load-drop curves (total displacement = 0.076 mm).

Figure 21-35 shows the load drops vs. time for a specimen subjected to a $\pm 0.0015$ in. displacement boundary condition. Note that the peak cyclic load drops from about 7.5 lb to 5 lb in the first 100 cycles. Similarly to all the other cases, however, the rate of load drops tends to remain constant with further cyclic displacements.

Figure 21-36 shows the load-drop curves of a specimen subjected to a

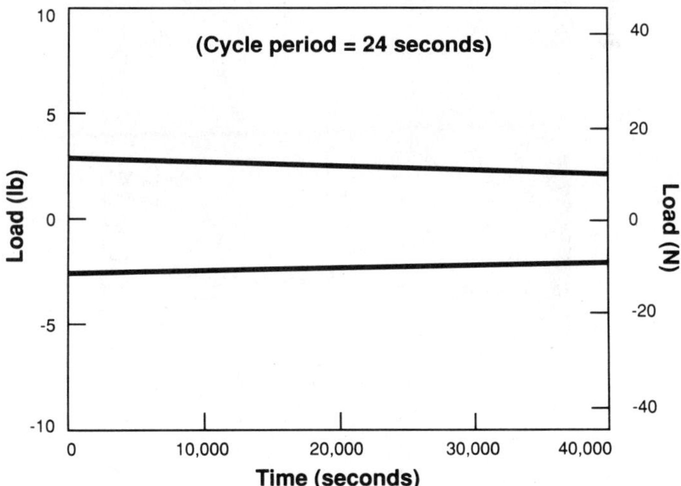


**Figure 21-36**   Load-drop curves (total displacement = 0.025 mm).

displacement of $\pm 0.0005$ in. It can be seen that the initial peak load is 2.87 lb and drops to about 2 lb in 1777 cycles. The load-drop rate is very small, especially at higher cycles.

## 21.11 CROSS-SECTIONS OF REWORKED PGA INTERCONNECTS (AFTER TEST)

Figures 21-37, 21-38, and 21-39 show cross-sections of specimens fatigue-tested at displacements of $\pm 0.0005$ in. (load-drop curves are shown in Fig. 21-36), $\pm 0.001$ in. (load-drop curves are shown in Fig. 21-34), and $\pm 0.0015$ in. (load-drop curves are shown in Fig. 21-35), respectively. It can be seen that the initial peak load and crack opening are in direct relation to the applied displacements. However, it can also be seen that in all the cases the cracks nucleate near the tip of the solder fillet, propagate in the solder joint parallel to the Alloy-42 pin, and stop. It appears that crack growth is the primary mechanism responsible for the decrease in peak load, and the crack growth rates decrease with increasing cycle number. The latter is a direct consequence of crack closure. Once the load-bearing area is increased, the crack tip loading is correspondingly decreased. A decrease in the

**Figure 21-37**  Cross-section of tested specimen (total displacement = 0.025 mm).

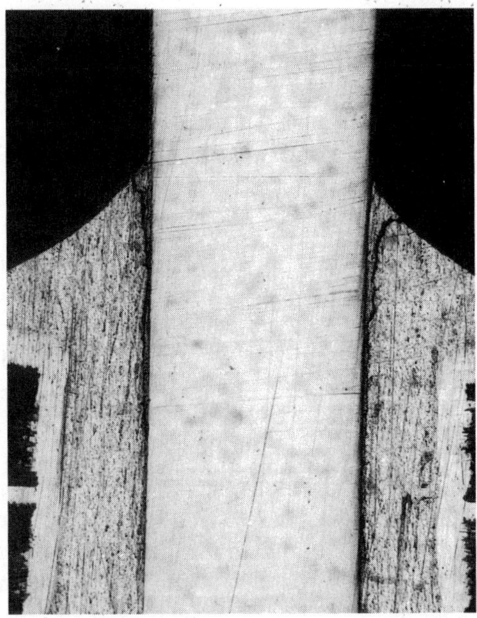

**Figure 21-38**   Cross-section of tested specimen (total displacement = 0.051 mm).

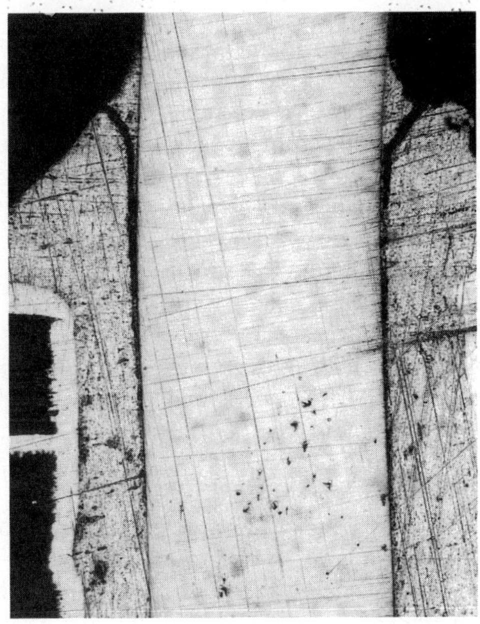

**Figure 21-39**   Cross-section of tested specimen (total displacement = 0.076 mm).

crack-tip loading will result in smaller crack growth rates, and slow down the crack propagation in the solder joint.


## 21.12 RELIABILITY OF REWORKED PGA SOLDER JOINTS AND PTH COPPER

With the present rework process and fatigue experiments, it has been shown that: (1) the amount of copper pads/barrels on the FR-4 board dissolved into the solder during rework is very small; (2) the number of reworks does not affect the quality of the solder joint; and (3) the load-drop curves for the 0-rework and 3-rework specimens are almost the same. Thus, it is reasonable to assume that the stress and strain in the solder joint and copper pad/barrel of the 3-rework assembly are almost the same as those of the 0-rework assembly (i.e., they should have the same level of reliability). Furthermore, due to the crack closure effect, the thermal stresses and strains caused by the global thermal expansion mismatch between the PGA and the PCB under rework condition have been found to be no threat to the reliability of solder joints.


## 21.13 SUMMARY

The effects of rework of the PGA component on the reliability of the solder joint and PTH copper pad/barrel have been determined by fatigue experiments. Cross-sections of the PGA assemblies (before and after fatigue tests) and load-drop curves of the specimens for up to three reworks have also been provided for a better understanding of the failure mechanisms. Furthermore, a workable rework procedure has been briefly stated. Some important results are summarized as follows.

1. Cross-sections of all the specimens show that the cracks typically nucleate near the tip of the solder fillet, propagate in the solder joint roughly parallel to the PGA pin, and stop. Thus, due to the crack closure of the PGA solder joints, the effect of global thermal expansion mismatch between the PCB and PGA under rework condition on the solder joint reliability may not be of concern.
2. The load-drop curves for the 0-rework and 3-rework cases are very similar in shape and are very close in magnitude.
3. There is almost no effect of rework on the reliability of the PGA solder joints. The PGA solder joint is formed by "fresh" solder at each rework.

4. There is almost no effect of up to three reworks on the reliability of the PTH copper provided that the specified rework process is followed. PTH copper dissolved into the solder during rework is minimal.

## ACKNOWLEDGMENTS

The authors thank Tara Bunch, Ann Hogsett, Albert Jeans, Ted Lancaster, Sherman Leung, Jian Miremadi, Larry Moresco, Dong Nguyen, Andre Shaari, Tom Tiernan, and Dan Uno of Hewlett-Packard Company, and Boris Yost of Cornell University for their effective help on this project. Subrahmanyan and Li also would like to thank Semiconductor Research Corporation for their support of this project.

## REFERENCES


1. Tummala, R. R., and E. Rymaszewski, *Microelectronics Packaging Handbook*, Van Nostrand Reinhold, New York, 1989.
2. Seraphim, D. P., R. Lasky, and C. Y. Li, *Principles of Electronic Packaging*, McGraw-Hill, New York, 1989.
3. *Electronic Materials Handbook, Vol. 1, Packaging*, ASM International, Materials Park, OH, 1989.
4. Sinnadurai, F. N., *Handbook of Microelectronics Packaging and Interconnection Technologies*, Electrochemical Publications, Isle of Man, 1985.
5. Vecchio, K. S., and R. W. Hertzberg, "Analysis of Long Term Reliability of Plated-Through Holes in Multilayer Interconnection Boards. Part A: Stress Analyses and Material Characterization," *J. Microelectronics and Reliability*, 1986, pp. 715–732.
6. Vecchio, K. S., and R. W. Hertzberg, "Analysis of Long Term Reliability. Part B: Fatigue Results and Fracture Mechanisms," *J. Microelectronics and Reliability*, 1986, pp. 733–751.
7. Seraphim, D. P., "A New Set of Printed-Circuit Technologies for the IBM 3081 Processor Unit," *IBM J. Research and Development*, **26**(1), January 1982, pp. 37–44.
8. Oien, M. A., "Methods for Evaluating Plated-Through Hole Reliability," *Proc. 14th IEEE Reliability Physics Symposium*, 1976, pp. 129–131.
9. Oien, M. A., "A Simple Model for the Thermo-Mechanical Deformations of Plated-Through-Holes in Multilayer Printed Wiring Boards," *Proc. 14th IEEE Reliability Physics Symposium*, 1976, pp. 121–128.
10. Hines, J. N., "Measurement of Land to Plated-Through-Hole Interface Resistance in Multilayer Boards," *Proc. 14th IEEE Reliability Physics Symposium*, 1976, pp. 132–134.
11. Matsui, A., K. Kawai, and E. Asada, "Experimental Research on Fatigue and Fracture of Solder," *Proc. ISHM International Symposium on Microelectronics*, 1989, pp. 544–551.

12. Rudy, D. A., "The Detection of Barrel Cracks in Plated Through Holes Using Four Point Resistance Measurements," *Proc. 14th IEEE Reliability Physics Symposium*, 1976, pp. 135–140.

13. Boddy, P. J., R. H. Delaney, J. N. Lahti, and E. F. Landry, "Accelerated Life Testing of Flexible Printed Circuits," *Proc. 14th IEEE Reliability Physics Symposium*, 1976, pp. 108–117.

14. Ammann, H. H., and R. W. Jocher, "Measurement of Thermo-Mechanical Strains in Plated-Through-Holes," *Proc. 14th IEEE Reliability Physics Symposium*, 1976, pp. 118–120.

15. Lee, L. C., V. S. Darekar, and C. K. Lim, "Micromechanisms of Multilayer Printed Circuit Boards," *IBM J. Research and Development*, **28**(6), November 1984, pp. 711–718.

16. Kelly, J. H., C. K. Lim, and W. T. Chen, "Optimization of Interconnections Between Packaging Levels," *IBM J. Research and Development*, **28**(6), November 1984, pp. 719–725.

17. Marsh, L. L., R. Lasky, D. P. Seraphim, and G. S. Springer, "Moisture Solubility and Diffusion in Epoxy and Epoxy-Glass Composites," *IBM J. Research and Development*, **28**(6), November 1984, pp. 655–661.

18. Chen, W. T., L. C. Lee, C. K. Lim, and D. P. Seraphim, "Mechanical Modeling for Printed Circuit Boards," *Proc. 3rd Printed Circuit World Conference*, Washington, DC, 1984.

19. Crocombe, A. D., and R. D. Adams, "Peel Analysis Using the Finite Element Methods," *J. Adhesion*, **12**, 1981, pp. 127–139.

20. Nakahara, S., and Y. Okinada, "Microstructure and Ductility of Electroless Copper Deposits," *Acta Metallurgica*, **31**(5), 1983, pp. 713–724.

21. Honma, H., and S. Mizushima, "Applications of Ductile Electroless Copper Deposition of Printed Circuit Boards," *Japan Metal Finishing Technology*, **34**(290), January 1983, pp. 47–52.

22. Fox, A., "Mechanical Properties at Elevated Temperature of Cu Bath Electroplated Copper for Multilayer Boards," *J. Testing and Evaluation*, **4**(1), January 1976, pp. 74–84.

23. Munikoti, R., and P. Dhar, "A New Power Cycling Technique for Accelerated Reliability Evaluation of Plated-Through-Holes and Interconnects in PCBs," *Proc. 40th IEEE Electronic Components and Technology Conference*, May 1990, pp. 426–435.

24. Hagge, J. K., "Strain-Induced Failures in Plated Through Holes," *Proc. Printed Circuit World*, 1980, pp. 32–36.

25. Engel, P. A., C. K. Lim, and M. D. Toda, "Thermal Stress Analysis of Soldered Pin Connectors for Complex Electronics Modules," *Computers in Mechanical Engineering*, May 1984, pp. 56–69.

26. Engel, P. A., "Stress Analysis for Soldered Pin-Through-Hole Connectors for Electronic Packaging Structures," *Proc. PACAM II Conference*, Vina del Mar, Chile, 1991.

27. Mirman, B. A., "Mathematical Model of a Plated-Through Hole Under a Load Induced by Thermal Mismatch," *IEEE Trans. Components, Hybrids, and Manufacturing Technology*, **11**(4), December 1988, pp. 506–511.

28. Engelmaier, W., and T. Kessler, "Investigation of Agitation Effects on Electroplated Copper in Multilayer Board Plated-Through Holes in a Forced Flow Plating Cell," *J. Electromechanical Society*, **125**(1), 1978, pp. 36–43.

29. Arthur, D. J., and E. L. Kozij, "PTH Reliability of High Performance PWB Material," *Proc. International Electronics Packaging Society Conference*, 1988, pp. 74–86.

30. Lea, C., *A Scientific Guide to Surface Mount Technology*, Electrochemical Publications, Isle of Man, 1988.

31. Wassink, R. J. K., *Soldering in Electronics*, Electrochemical Publications, Isle of Man, 1989.

32. Lea, C., and F. H. Howie, "Blowholing in PTH Solder Fillets, Part 1: Assessment of the Problem," *Circuit World*, **12**(4), 1986, pp. 14–19.

33. Howie, F. H., and C. Lea, "Blowholing in PTH Solder Fillets, Part 2: The Nature, Origin and Evolution of the Gas," *Circuit World*, **12**(4), 1986, pp. 20–25.

34. Seah, M. P., F. H. Howie, and C. Lea, "Blowholing in PTH Solder Fillets, Part 3: "Moisture and the PCB," *Circuit World*, **12**(4), 1986, p. 14.

35. Lea, C., M. P. Seah, and F. H. Howie, "Blowholing in PTH Solder Fillets, Part 4: The Plated Copper Barrel," *Circuit World*, **13**(1), 1986, pp. 28–34.

36. Lea, C., and F. H. Howie, "Blowholing in PTH Solder Fillets, Part 5: The Role of the Electroless Copper," *Circuit World*, **13**(1), 1986, pp. 35–42.

37. Lea, C., and F. H. Howie, "Blowholing in PTH Solder Fillets, Part 6: The Laminate, The Drilling and the Hole-Wall Preparation," *Circuit World*, **13**(1), 1986, pp. 43–50.

38. Howie, F. H., D. Tilbrook, and C. Lea, "Blowholing in PTH Solder Fillets, Part 7: Optimising the Soldering," *Circuit World*, **13**(2), 1986, pp. 42–45.

39. Lea, C., F. H. Howie, and M. P. Seah, "Blowholing in PTH Solder Fillets, Part 8: The Scientific Framework Leading to Recommendations for Its Elimination," *Circuit World*, **13**(3), 1986, pp. 11–20.

40. Lea, C., and F. H. Howie, "Controlling the Quality of Soldering of PTH Solder Joints," *Circuit World*, **11**(3), 1985, pp. 5–7.

41. Barker, D., M. Pecht, A. Dasgupta, and S. Naqvi, "Transient Thermal Stress Analysis of a Plated Through Hole Subjected to Wave Soldering," *Trans. ASME, J. Electronic Packaging*, **113**, 1991, pp. 149–155.

42. Lau, J. H., R. Subrahmanyan, D. Rice, S. Erasmus, and C. Li, "Fatigue Analysis of a Ceramic Pin Grid Array Soldered to An Orthotropic Epoxy Substrate," *Trans. ASME, J. Electronic Packaging*, **113**, June 1991, pp. 138–148.

43. Lau, J. H., S. Leung, R. Subrahmanyan, D. Rice, S. Erasmus, and C. Y. Li, "Effects of Rework on the Reliability of Pin Grid Array Interconnects," *Circuit World*, **17**(4), 1991, pp. 5–10.

44. Vaynman, S., "Effect of Strain Rate on Fatigue of Low-Tin Lead-Base Solder," *IEEE Trans. Components, Hybrids, and Manufacturing Technology*, **12**(4), December 1989, pp. 469–472.

45. Lau, J. H., L. Powers, J. Baker, D. Rice, and W. Shaw, "Solder Joint Reliability of Fine Pitch Surface Mount Technology Assemblies," *IEEE Trans. Components, Hybrids, and Manufacturing Technology*, **13**(3), September 1990, pp. 534–544.

46. Lau, J. H., W. D. Rice, and G. Harkins, "Thermal Stress Analysis of TAB Packages and Interconnections," *IEEE Trans. Components, Hybrids, and Manufacturing Technology*, **13**(1), March 1990, pp. 182–187.

47. Lau, J. H., "Thermal Stress Analysis of SMT PQFP Packages and Interconnections," *Trans. ASME, J. Electronic Packaging*, **111**, March 1989, pp. 2–8.

48. Lau, J. H., G. Harkins, D. Rice, J. Kral, and B. Wells, "Experimental and Statistical Analyses of Surface-Mount Technology PLCC Solder-Joint Reliability, *IEEE Trans. Reliability*, **37**(5), December 1988, pp. 524–530.

49. Lau, J. H., and G. Harkins, "Stiffness of 'Gull-Wing' Leads and Solder Joints for a Plastic Quad Flat Pack," *IEEE Trans. Components, Hybrids, and Manufacturing Technology*, **13**(1), March 1990, pp. 124–130.

50. Lau, J. H., and D. W. Rice, "Solder Joint Fatigue in Surface Mount Technology: State of the Art," *Solid State Technology*, **28**, October 1985, pp. 91–104.

51. Hall, P. M., "Creep and Stress Relaxation in Solder Joints in Surface-Mount Chip Carriers," *IEEE Trans. Components, Hybrids, and Manufacturing Technology*, **10**(4), December 1987, pp. 556–565.

52. Ross, R. G., "A Systems Approach to Solder Joint Fatigue in Spacecraft Electronic Packaging," *Trans. ASME, J. Electronic Packaging*, **113**, 1991, pp. 121–128.

53. Lau, J. H., S. Golwalkar, D. Rice, S. Erasmus, R. Surratt, and P. Boysan, "Experimental and Analytical Studies of 28-Pin Thin Small Outline Package (TSOP) Solder-Joint Reliability," *1991 ASME Winter Annual Meeting*, Paper No. 91-WA-EEP-30.

54. Guo, Z., A. Sprecher, D. Jung, and H. Conrad, "Influence of Environment on the Fatigue of Pb–Sn Solder Joints," *IEEE Trans. Components, Hybrids, and Manufacturing Technology*, **14**(4), December 1991, pp. 833–837.

55. Liu, K., and J. Duh, "Microstructural Evolution in Sn/Pb Solder and Pd/Ag Thick Film Conductor Metallization," *IEEE Trans. Components, Hybrids, and Manufacturing Technology*, **14**(4), December 1991, pp. 703–707.

56. Satoh, R., K. Arakawa, M. Harada, and K. Matsui, "Thermal Fatigue Life of Pb–Sn Alloy Interconnections," *IEEE Trans. Components, Hybrids, and Manufacturing Technology*, **14**(1), March 1991, pp. 224–232.

57. Solomon, H. D., "Fatigue of 60/40 Solder," *IEEE Trans. Components, Hybrids, and Manufacturing Technology*, **9**, December 1986, pp. 423–432.

58. Solomon, H. D., "Strain-Life Behavior in 60/40 Solder, *Trans. ASME, J. Electronic Packaging*, **111**, June 1989, pp. 75–82.

59. Solomon, H. D., "Room Temperature Low Cycle Fatigue of a High Pb Solder (Indalloy 151)," *Proc. 2d ASM International Electronic Materials and Processing Congress*, April 1989, pp. 135–146.

60. Solomon, H. D., "Low Cycle Fatigue of Surface Mounted Chip Carrier/Printed Wiring Board Joints," *IEEE Trans. Components, Hybrids, and Manufacturing Technology*, **12**(4), December 1989, pp. 473–479.

61. Solomon, H. D., "High and Low Temperature Strain-Life Behavior of a Pb Rich Solder," *Trans. ASME, J. Electronic Packaging*, **112**, June 1990, pp. 123–128.

62. Subrahmanyan, R., "A Damage Integral Approach for Low-Cycle Isothermal and Thermal Fatigue," Ph.D. Thesis, Cornell University, Ithaca, New York, 1990.

63. Subrahmanyan, R., J. R. Wilcox, and C. Y. Li, "A Damage Integral Approach to Thermal Fatigue of Solder Joints," *IEEE Trans. Components, Hybrids, and Manufacturing Technology*, **12**(4), December 1989, pp. 480–491.

64. Wilcox, J. R., R. Subrahmanyan, and C. Y. Li, "Assembly Stiffness and Failure Criterion Considerations in Solder Joint Fatigue," *Trans. ASME, J. Electronic Packaging*, **112**, June 1990, pp. 115–122.

65. Stone, D., H. Wilson, R. Subrahmanyan, and C. Y. Li, "Mechanisms of Damage Accumulation in Solders During Thermal Fatigue," *Proc. 36th IEEE Electronic Components Conference*, 1986, pp. 630–635.

66. Lau, J. H., *Solder Joint Reliability: Theory and Applications*, Van Nostrand Reinhold, New York, 1991.

67. Lau, J. H., *Handbook of Tape Automated Bonding*, Van Nostrand Reinhold, New York, 1991.

68. Suhir, E., "Analytical Modeling in Electronic Packaging Structures: Its Merits, Shortcomings and Interaction With Experimental and Numerical Techniques," *Trans. ASME, J. Electronic Packaging*, **111**, June 1989, pp. 157–161.

69. Lau, J. H., "A Note on the Calculation of Thermal Stresses in Electronic Packaging by Finite Element Methods," *Trans. ASME, J. Electronic Packaging*, **111**, December 1989, pp. 313–320.

70. Suhir, S., "Interfacial Stresses in Bimetal Thermostats," *Trans. ASME, J. Applied Mechanics*, **55**, 1989, pp. 595–600.

71. Evans, A. G., and J. W. Hutchinson, "On the Mechanics of Delamination and Spalling in Compressed Films," *Int. J. Solids and Structures*, **20**, 1984, pp. 455–466.

72. Eischen, J. W., and J. S. Everett, "Thermal Stress Analysis of a Bimaterial Strip Subject to an Axial Temperaure Gradient," *Trans. ASME, J. Electronic Packaging*, **111**, December 1989, pp. 282–288.

73. Engel, P. A., "Structural Analysis for Circuit Card Systems Subjected to Bending," *Trans. ASME, J. Electronic Packaging*, **112**, March 1990, pp. 2–10.

74. Lau, J. H., "Thermoelastic Solutions for a Finite Substrate With an Electronic Device," *Trans. ASME, J. Electronic Packaging*, **113**, March 1991, pp. 84–88.

75. Lau, J. H., "Thermoelastic Problems for Electronic Packaging," *J. International Society for Hybrid Microelectronics*, No. 25, May 1991, pp. 11–15.

76. Kuo, A., "Thermal Stresses at the Edge of a Bimetallic Thermostat," *Trans. ASME, J. Applied Mechanics*, **56**, 1989, pp. 585–589.

77. Chen, W. T., and C. W. Nelson, "Thermal Stresses in Bonded Joints," *IBM J. Research and Development*, **23**(2), 1979, pp. 179–187.

78. Weaver, W., and P. R. Johnston, *Finite Elements for Structural Analysis*, Prentice-Hall, Englewood Cliffs, NJ, 1984.

79. Oden, J. R., *Finite Elements of Nonlinear Continua*, McGraw-Hill, New York, 1973.

80. Hughes, T., *The Finite Element Method*, Prentice-Hall, Englewood Cliffs, NJ, 1987.

81. Lau, J. H., and A. H. Jeans, "Nonlinear Analysis of Elastomeric Keyboard Domes," *Trans. ASME, J. Applied Mechanics*, **56**, December 1989, pp. 751–755.

82. Lau, J. H., and L. L. Moresco, "Mechanical Behavior of Microstrip Structures Made of Y-Ba-Cu-O Superconducting Ceramics," *IEEE Trans. Components, Hybrids, and Manufacturing Technology*, **11**(4), December 1988, pp. 419–426.

83. Lau, J. H., and L. B. Lian-Mueller, "Finite Element Modeling for Optimizing Hermatic Package Reliability," *Trans. ASME, J. Electronic Packaging*, **111**, December 1989, pp. 255–260.

84. Lau, J. H., and C. A. Keely, "Dynamic Characterization of Surface Mount Component Leads for Solder Joint Inspection," *IEEE Trans. Components, Hybrids, and Manufacturing Technology*, **12**(4), December 1989, pp. 594–602.

85. Lau, J. H., D. W. Rice, and P. A. Avery, "Elasto-Plastic Analysis of Surface-Mount Solder Joints," *IEEE Trans. Components, Hybrids, and Manufacturing Technology*, **10**(3), September 1987, pp. 346–357.

86. Lau, J. H., and G. Harkins, "Thermal Stress Analysis of SOIC Packages and Interconnections," *IEEE Trans. Components, Hybrids, and Manufacturing Technology*, **11**(4), December 1988, pp. 380–389.

87. Lau, J. H., "Elasto-Plastic Large Deflection Analysis of a Copper Film," *1991 ASME Winter Annual Meeting*, Paper No. 91-WA-EEP-29.

88. Lau, J. H., and G. H. Barrett, "Stress and Deflection Analysis of Partially Routed Panels for Depanelization," *IEEE Trans. Components, Hybrids, and Manufacturing Technology*, **10**(3), September 1987, pp. 411–419.

# 22

# Mechanics of Wirebond Interconnects

*Michael Pecht and Pradeep Lall*

## 22.1 INTRODUCTION

This chapter discusses the thermomechanical considerations in the manufacture and design of wirebond interconnects and presents guidelines for design of reliable wirebond interconnects. This section gives an overview of wirebonding technology, including bond types and representative geometries, typical metallurgical systems, considerations in wire selection, common methods of wirebond interconnect evaluation, and a list of observed wirebond failures. Section 22.2 discusses thermomechanical considerations during wirebond manufacture. Thermomechanical considerations concerning bonding process and common bonding problems are discussed, including cratering, pad cleanliness, intermetallic formation, metallization problems, pad lifting, bonding parameters, and geometries. The common post-bond test techniques used to assess the effect of thermomechanical stresses on bond performance are then discussed. Section 22.3 discusses thermomechanical considerations during the wirebond operation, including potential failure mechanisms and physics of failure models. Based on thermomechanical considerations during manufacture and use, outlined in Section 22.2 and 22.3, guidelines for design of reliable wirebond interconnects are presented in Section 22.4.

### 22.1.1 What is Wirebonding?

Wires are often used to connect the input/output pads of a device to other devices or to leads. More specifically, the wire often connects the first-level package (i.e., the chip or the die) to the second level package (i.e., the chip carrier or the lead frame in a single chip carrier). The wires are bonded to the interconnecting pads at each end using thermal or ultrasonic energy or a combination of both. The resulting interconnections are called wirebonds. Most wires used for wirebonding are between 0.5 and 8 mils in diameter. The bonded spans of typical wirebond interconnections are in the range 40 to 150 mils.

### 22.1.2  Wirebond Types and Representative Geometries

The bonded wire at each end of the interconnection has a characteristic shape depending on the bonding tool and the manufacturing process. The manufacturing processes in wirebonding include thermocompression bonding, ultrasonic bonding, and thermosonic bonding. The wirebonds resulting from these manufacturing processes can be grouped into three characteristic shapes. These include ball bonds, wedge bonds, and stitch (crescent) bonds. Figure 22-1 shows the representative geometries of these bonds.

Thermocompression bonds occur when two metal surfaces are brought into intimate contact over a controlled time, temperature, and pressure cycle. During the process of thermocompression bonding (Fig. 22-2), the wire and the underlying metallization undergo a certain amount of plastic deformation and atomic interdiffusion. This process of interdiffusion can result in a uniform interface. The plastic deformation at the mating surfaces ensures intimate contact between the bonding elements, an increase in the bonding area, and breakdown of the interface film layers. Surface roughness, oxide layer formation, and absorbed chemical species or moisture layers can prevent the intimate metal-to-metal contact, thus inhibiting interfacial welding and reducing the inherent strength of the bond. The typical bonding temperatures during thermocompression bonding range from 300 to 400°C. Heat is generated during the manufacturing process either by a heated capillary feeding the wire or by a heated pedestal on which the assembly is placed for the purpose of bonding. The capillary is made of alumina, tungsten carbide, ruby, or other refractory materials.

Ultrasonic bonding is a low-temperature process in which the source of energy for the metal welding is a transducer vibrating the bonding tool in the frequency range from 20 to 60 kHz. In ultrasonic bonding, the wire is threaded through a hole in the wedge and trailed under the bonding tip (Fig. 22-3). The wire is secured with a wire clamp and the bonding tool is positioned over the first bond. The wedge is lowered and the wire is pressed tightly between the wedge and the bond pad. With the interface under firm compressive stress, a burst of ultrasonic energy is applied and a combination

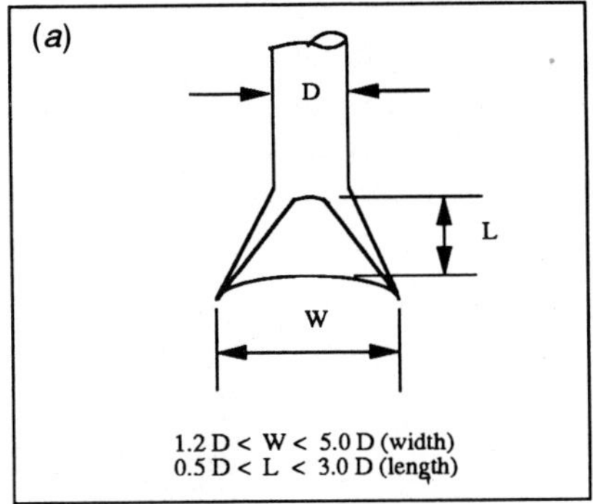


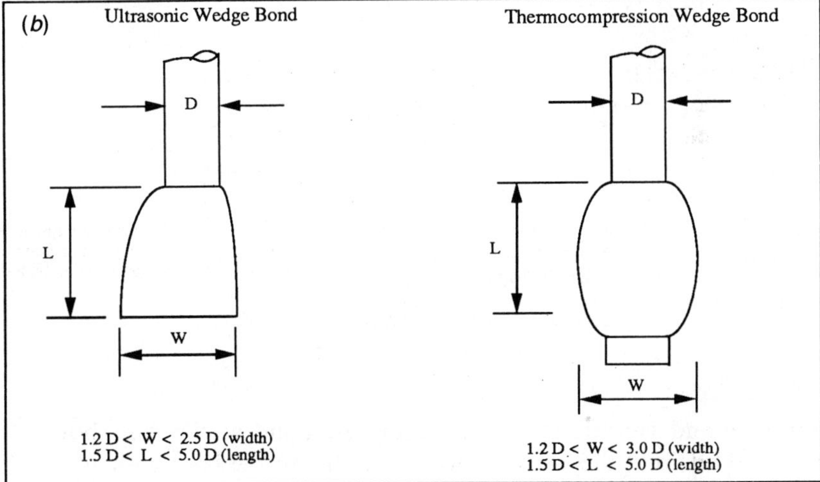


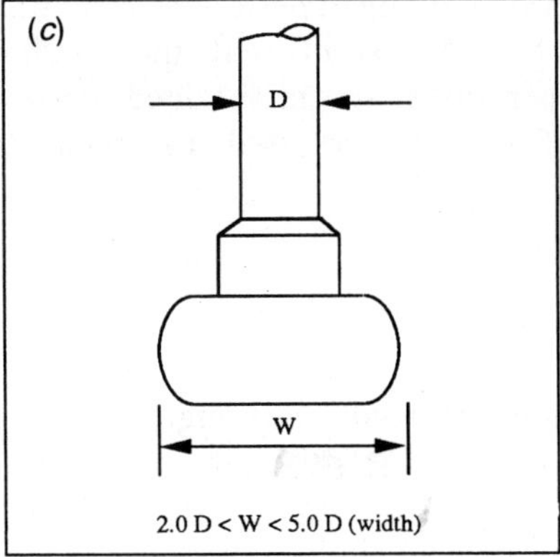


**Figure 22-1** Representative bond geometries (MIL-STD-883). (a) Crescent bond; (b) wedge bond; (c) ball bond.

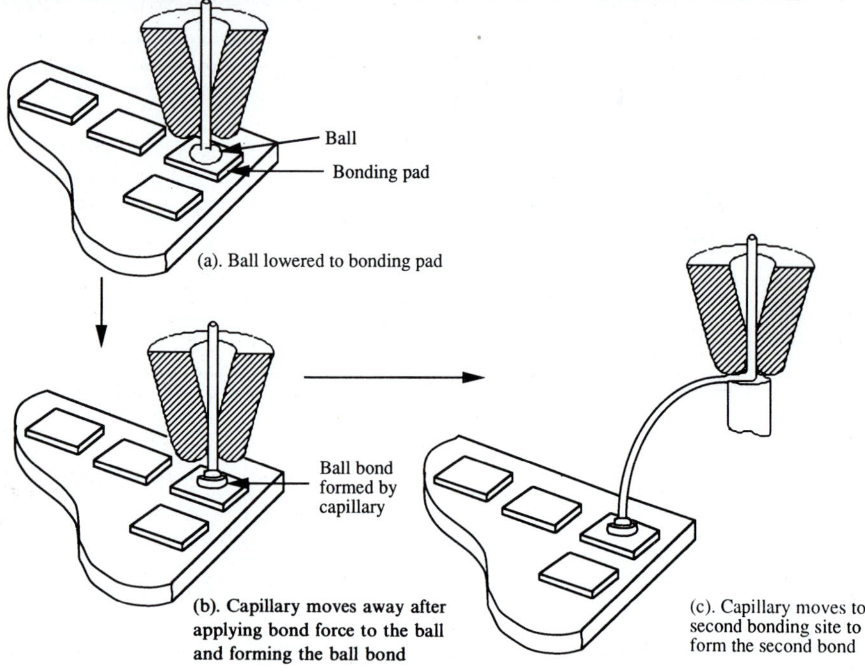



**Figure 22-2**   Thermocompression bonding.

of pressure and vibrational energy forms the bond at the wire–bond pad interface. The tip of the wedge vibrates parallel to the bonding pad.

Thermosonic bonding combines ultrasonic energy with the ball bonding capillary technique of thermocompression bonding. Thermosonic bonding is similar to the thermocompression process except that the capillary is not heated and the substrate temperatures are maintained between 100 and 150°C. Ultrasonic bursts of energy are used to make the bond.

### 22.1.3 Typical Metallurgical Systems

Gold and aluminum are common wire materials, although copper[1–6] and silver[7–11] have also been used. Gold is mostly used in thermocompression bonding while aluminum is mostly used in ultrasonic bonding. Characteristics of wire materials that are of vital importance to the strength of the wire bond include wire dimensions, tensile strength, and elongation.

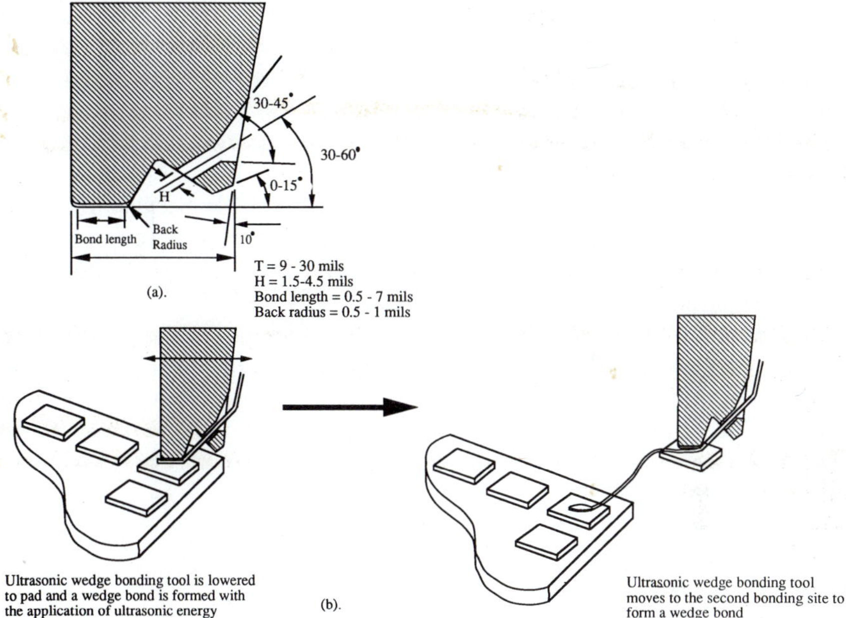



**Figure 22-3**    Ultrasonic wedge bonding. (*a*) Ultrasonic wedge bonding tool with typical dimensions. (*b*) Bonding process.

## Gold Bond Wire Systems

Gold wire is used extensively for thermocompression bonding, and can also be used for thermosonic bonding. In producing gold bonding wires, control of the surface finish, and surface cleanliness, are of the highest priority to ensure the formation of a strong bond and to prevent clogging of bonding capillaries. Pure gold can usually be drawn to produce an adequate breaking strength (ultimate tensile strength of the wire) and proper elongation (ratio of the increase in wire length at rupture to the initial wire length given as a percentage) for use as a bond wire. Ultrapure gold is very soft, but even small amounts of impurities such as 5–10 p.p.m. by weight of Be or 30–100 p.p.m. by weight of Cu make the gold wire workable. Be-doped wire is stronger than Cu-doped wire by about 10–20% under most conditions. The increased strength of the Be-doped wire is advantageous for automated thermosonic bonding, where high-speed capillary movements generate higher stresses than in slow or manual bonders.

Gold wires can be bonded to copper lead frames and can lead to the formation of ductile intermetallic phases ($Cu_3Au$, $AuCu$, $Au_3Cu$) that decrease the bond strength at higher temperatures (in the neighborhood of 200–325°C) as a result of void formation.[12] Temperatures above 300°C accelerate the rate of intermetallic compound formation. However, Pitt et al.[13] studied gold thermosonic bonds to thick-film copper and found little strength degradation at 150°C for up to 3000 hours, and no failures at 250°C for over 3000 hours. Cleanliness of the bonding surface is extremely important to ensure good bondability in Cu–Au systems.[14,15]

Gold wire bonded to a gold bond pad is very reliable because the bond is not subject to interface corrosion, intermetallic formation, or other bond-degrading conditions, and even a poorly welded gold–gold bond will increase in strength with time and temperature.[9]

Gold bonded to silver metallization is very reliable for very long times at high temperatures.[11] This bond system does not form intermetallic compounds and does not exhibit interface corrosion. Contaminants like sulfur reduce the bondability of silver films to gold. High temperature (approximately 250°C) thermosonic bonding may be needed to increase bondability of silver.[16,17]


## Aluminum Bond Wire Systems

Pure aluminum is typically too soft to be drawn into a fine wire. For this reason aluminum is often alloyed with 1% Si or 1% Mg to provide a strengthening mechanism. The equilibrium solid-state solubility of Si in Al at 20°C is of the order of 0.02% by weight. Only at temperatures above about 500°C does Si attain 1% solubility in equilibrium solid solution. Thus 1% silicon exceeds the solubility of silicon in aluminum at room temperature by a factor of 50, so there is always a tendency for Si to precipitate, forming a silicon second phase. This second phase can become a potential fatigue crack nucleation site. The amount of silicon precipitated and the size of silicon nodules, depends on the rate of cooling from higher temperatures. Slower cooling rates result in more precipitation and larger nonuniform silicon nodules, while faster cooling rates do not allow sufficient time for silicon precipitation resulting is uniformly dispersed nodules. Silicon grain size has been found to affect wire ductility.[18,19]

Aluminum alloyed with 1% magnesium can be drawn into a fine wire that exhibits a breaking strength similar to that of Al–1% Si. The Al–1% Mg alloy wire bonds satisfactorily and is superior to Al–1% Si in resistance to fatigue failure and to degradation of ultimate strength after exposure to elevated temperatures. These advantages of Al–1% Mg wire over its silicon substitute occur because the equilibrium solid solubility of Mg in Al is about 2% by weight, and thus at 0.5–1% Mg concentration there is no tendency towards second-phase segregation as is the case with Al–1% Si.

Aluminum wire bonded to a silver-plated lead frame is often used in thick-film hybrids. The Ag–Al phase diagram is very complex, with many intermetallic phases. Kirkendall voids can occur in this metal system, but typically at temperatures higher than the operating range of the micro-circuits. Ag–Al bonds are seldom used because of their tendency to degrade due to interdiffusion and to oxidize in the presence of humidity. An aluminum–chlorine corrosion mechanism is responsible for the bond degra-dation as seen by the formation of aluminum hydroxide, $Al(OH)_3$, in the zeta phase of the Ag–Al intermetallic.[7–11]

Harman[16,17] found that aluminum wires bonded to a nickel coating (used as a substitute for gold) are more reliable than Al–Ag or Al–Au bonds. Large diameter, $>75$ μm (3 mil), aluminum wires bonded to nickel platings have been used in power devices.[16,17] The Al–Ni system is well-suited to high-temperature applications and is less prone to Kirkendall voiding and galvanic corrosion.

Aluminum wire bonded to aluminum metallization is extremely reliable because it is not prone to intermetallic formation and corrosion. Aluminum wire on aluminum metallization welds best ultrasonically, although a thermocompression bond can be produced by high deformation.


## Copper Bond Wire Systems

Copper wires are used primarily because of their economy and their resistance to sweep (tendency of the wire to move in the plane perpendicular to its length) during plastic encapsulation.[1–6] Copper is harder than gold and thus greater attention is needed during the bonding operation to prevent cratering of the silicon die.[2,6] The harder copper wire tends to push the softer metallization aside during the bonding operation. Bonding systems using copper wires therefore require harder metallization.[2,6] Bonding systems using copper wires require inert atmosphere for bonding because of the tendency of copper to oxidize readily. Olsen and James[20] observed that for Al wire bonded to oxygen-free high-conductivity copper, thermal aging at 150°C in ambient did not effect the bond strength, while thermal aging in a vacuum resulted in a weaker bond.

Intermetallic growth in Cu–Al bonds is slower than in Au–Al bonds. The intermetallic growth in Cu–Al bonds does not result in Kirkendall voiding but produces a lower shear strength at 150–200°C due to growth of a brittle $CuAl_2$ phase.[3,5] Cu–Al bonds are stronger than Au–Al bonds in the presence of brominated flame retardants. Cu–Al bonds were strong after 1245 hours at 200°C but Au–Al failed after 700 cycles.[21] Aluminum metallization containing copper–aluminum intermetallics corrode with Cl contamination and water.[22–25]

**Typical Bonding Surfaces**

Bonding surfaces commonly used on the die surface include aluminum, gold, silver, nickel, and copper. Aluminum is the most commonly used bond pad material, though gold bond pad is used to avoid direct contact between gold wire and aluminum bond pad to prevent the formation of gold intermetallics. Because gold does not adhere well to silicon dioxide, other metals are used for form a multilayer metallization to avoid direct contact between gold and silicon.

Silver is used as a bond plating material on lead frames and as a metallization in commercial thick-film hybrids, usually in alloy form with platinum or palladium.[26] At temperatures in the neighborhood of 400°C, silver undergoes selective oxidation, and is therefore not used in high-temperature applications.

Nickel has been widely used in power devices as a substitute for gold in various environments with no significant reliability problems. Nickel is deposited from electroless boride or sulfamate solutions. Low-stress films electroplated from sulfamade baths result in reliable bonds. Bonding to the die and the terminal is affected by many film-related factors, which include surface finish, film hardness and thickness, film preparation, and surface contamination. Nickel surfaces are prone to oxidation and should be bonded soon after being nickel-plated, and should be chemically cleaned before bonding and protected by an inert atmosphere.[16,17] Phosphide electroless nickel solutions used to deposit more than 6–8% of phosphorus can result in reliability and bondability problems.

### 22.1.4  Wire Selection

The properties of various wire and bond pad materials are listed in Tables 22-1($a$) and 22-1($b$). Critical wire material properties include resistivity, shear strength, tensile strength (yield and ultimate), elastic modulus, Poisson's ratio, hardness, thermal coefficient of expansion, percent wire elongation, and endurance limit for the bond pad and chip.

The wire material must be highly electrically conductive metal. The resistivity of wire material must be low because resistance in signal propagation path will cause a voltage drop across the interconnect and deteriorate signal integrity. The interconnect's electrical properties have to be such that the interconnect does not degrade the waveform coming from the on-chip circuitry. Distortion in the signal limits the maximum allowable signal rise time and decreases the maximum allowable clock period. Bandwidth of transmitted signal decreases with increase in distortion. Rise time of the signal which is inversely proportional to bandwidth and clock frequency

**Table 22-1(a)**   Properties of Wire Materials at 27°C

| Property | Aluminum (Al) | Copper (Cu) | Gold (Au) |
| --- | --- | --- | --- |
| Specific heat (watt-sec/(gram-deg K) | 0.9 | 0.385 | 0.1289 |
| Thermal conductivity (watts/(cm-deg K) | 2.37 | 4.03 | 3.19 |
| Specific gravity (gm/cm$^3$) | 2.6989 | 8.96 | 19.32 |
| Melting point (°C) | 660.37 | 1083.4 | 1064.43 |
| Electrical resistivity (ohm-cm) | $2.65 \times 10^{-6}$ | $1.73 \times 10^{-6}$ | $2.249 \times 10^{-6}$ |
| Temperature coefficient of electrical resistivity (ohm-cm/deg C) | $4.3 \times 10^{-9}$ | $6.8 \times 10^{-9}$ | $4 \times 10^{-9}$ |
| Elastic modulus (psi) | $0.5 \times 10^7$ | $19.2 \times 10^7$ | $1.12 \times 10^7$ |
| Yield strength (psi) | 1500 | 10 000 | 25 000 |
| Ultimate tensile strength (psi) | 6500 | 32 000 | 30 000 |
| Coefficient of thermal expansion (1/K) | $46.4 \times 10^{-6}$ | $16.12 \times 10^{-6}$ | $14.2 \times 10^{-6}$ |
| Poisson's ratio | 0.346 | 0.339 | 0.291 |
| Hardness (Brinell) | 17 | 37 | 18.5 |
| Percentage elongation (%) | 50 | 51 | 4 |

**Table 22-1(b)**   Properties of Bond Pad Materials at 27°C

| Property | Aluminum (Al) | Copper (Cu) | Gold (Au) |
| --- | --- | --- | --- |
| Specific heat (watt-sec/(gram-deg K) | 0.9 | 0.385 | 0.1289 |
| Thermal conductivity (watts/(cm-deg K) | 2.37 | 4.03 | 3.19 |
| Specific gravity (gm/cm$^3$) | 2.6989 | 8.96 | 19.32 |
| Melting point (°C) | 660.37 | 1083.4 | 1064.43 |
| Electrical resistivity (ohm-cm) | $2.65 \times 10^{-6}$ | $1.73 \times 10^{-6}$ | $2.249 \times 10^{-6}$ |
| Temperature coefficient of electrical resistivity (ohm-cm/deg C) | $4.3 \times 10^{-9}$ | $6.8 \times 10^{-9}$ | $4 \times 10^{-9}$ |
| Elastic modulus (psi) | $0.5 \times 10^7$ | $19.2 \times 10^7$ | $1.12 \times 10^7$ |
| Yield strength (psi) | 1500 | 10 000 | 25 000 |
| Ultimate tensile strength (psi) | 57 000 | 50 000 | 70 000 |
| Coefficient of thermal expansion (1/K) | $46.4 \times 10^{-6}$ | $16.12 \times 10^{-6}$ | $14.2 \times 10^{-6}$ |
| Poisson's ratio | 0.346 | 0.339 | 0.291 |
| Hardness (Brinell) | 17 | 37 | 18.5 |
| Percentage elongation (%) | | | 2–3 |

thus increases with distortion. The distortion and clock skew from the interconnect can be enough to cause false triggering. The rise time ($t_{rise}$), clock frequency ($F_{clock}$), and the bandwidth are related by

$$\text{Bandwidth} = \frac{0.35}{t_{rise}}$$

$$t_{rise} = \frac{0.07}{F_{clock}} \qquad (22\text{-}1)$$

$$\text{Bandwidth} = 5F_{clock}$$

Thus the bonded materials should have interdiffusion constants such that they permit formation of a strong bond while at the same time preventing the formation of excess intermetallics during the expected operational life. The wirebonding operation is a diffusion phenomenon accomplished either by a combination of high temperature and pressure or by a combination of ultrasonic energy and pressure. The bonded surfaces are subject to subsequent interdiffusion, forming intermetallics at the bonded interface, at a rate depending on the interdiffusion constants of the component metals. The intermetallics increase the interconnect resistance and degrade the propagating signal. The interdiffusion is further accelerated by higher temperatures.

Shear strength and coefficient of thermal expansion of the wire and the bond pad materials are critical material properties. A wire bond subjected to temperature change experiences shear stresses between the wire, the bond pad, and the substrate as a result of differential thermal expansion.[27,28] These stresses are worsened when intermetallics are formed at the interface due the interdiffusion of the bonded materials and their growth is enhanced by increased temperature and temperature cycling.[27] Differences in the hardness and the material properties (including the coefficient of thermal expansion between the intermetallic and the surrounding metal) make the interface a potential site for failure. The shear strength of the wire or the bond pad, if exceeded during these temperature cycles, will cause a failure.

The yield strength, ultimate tensile strength, and endurance limit of the wire in wire bonded interconnections are critical properties because the wires are subjected to flexure during temperature cycling, causing metal fatigue if the endurance limit is exceeded. The endurance limit of the wire material should be greater than the stresses produced in the wire during temperature cycling. The bending or flexural stress in the wire during temperature cycling may equal the breaking load of the wire. The wire will fail in the first cycle if the stresses generated during temperature cycling exceed the yield strength of the wire.

Wire hardness is important because of its effect on cratering. If the wire is harder than the bond pad, it inhibits cratering by absorbing the energy from the bonding process. A wire softer than the bond pad, on the other hand, would readily transmit the energy from the bonding process to the substrate. Ravi[18,19] and Hirota[2] found that the best bonds were formed when the hardness of the wire and pad were reasonably matched.

### 22.1.5 Methods of Wirebond Evaluation

The wirebond evaluation methods are listed in MIL-STD-883. These include:

Nondestructive bond pull test (Method 2023)
Ball bond shear test
Destructive bond-pull test (Method 2011)
Internal visual (Method 2010; Test Condition A and B)
Constant acceleration (Method 2001; Test Condition E)
Random vibration (Method 2026)
Delay measurements (Method 3003)
Mechanical shock (Method 2002)
Moisture resistance (Method 1004)

Each of these methods is discussed in more detail in Section 22.2.8.

### 22.1.6 Observed Wirebond Failures

Common wirebond failure mechanisms include:

Wire flexure fatigue
Wire–bond pad shear failure
Bond pad–substrate shear failure
Axial fatigue of the wire in plastic encapsulated devices
Interdiffusion between the wire and the bond pad leading to intermetallic
    formation and Kirkendall voiding
Corrosion
Dendritic growth
Electrical noise
Vibration fatigue
Resistance change
Bond pad cracking

Each of these failure mechanisms is discussed in more detail in Section 22.3.

### *22.1.7 Role of Thermomechanical Properties*

The bonded interfaces between the wire and the bond pad undergo shear stresses during temperature cycling. For this reason coefficients of thermal expansion of wire and bond pad should be reasonably matched. Wire–bond pad material combinations having widely differing coefficients of thermal expansions usually lead to bond pad lifts.

Differential expansion can also result in chip craters if the fracture toughness or the critical size for the chip is exceeded. Small cracks may be introduced in the chip during the bonding process. Two situations can result from such cracks. If the cracks lie in the subthreshold region, the crack may grow very slowly to the threshold crack size and the propagation of the crack to failure may be well beyond the operational life of the device. A crack is said to be a threshold crack when the relationship between rate of crack growth per cycle versus cyclic stress magnitude is not linear, such that the rate of crack growth is negligible per cycle. On the other hand, if the crack size is larger than the threshold crack size, the crack may propagate during temperature cycling to a critical crack size. The critical crack size is the size of the largest crack after which any loading would cause it to catastrophically progress to failure. Susceptibility to cratering can be assessed by using the Paris's power law or Foreman's law, relating the crack size to fatigue life.

Bond pad lifts during temperature cycling also result from surface oxidation of the bond pads. Nickel surfaces have a tendency to oxidize; for this reason bonds to surfaces such as nickel are made soon after plating. Nickel surfaces are usually chemically cleaned before bonding to remove any oxidizing impurities, and protected by an inert atmosphere.

Wire and bond pad materials have a tendency to interdiffuse, resulting in the formation of intermetallics. The growth of these intermetallics is enhanced at high temperatures in the neighborhood of 160°C for common material combinations including gold–aluminum. Temperature cycling further accelerates the growth of intermetallics, which serve as sites for crack initiation and propagation.

The work hardening characteristics of the wire material when heated strongly influence the life under temperature cycling. During the thermocompression and sometimes during the thermosonic bonding process the capillary feeding the bonding wire is heated. The heating of the wire causes the wire material to work harden. The problem is worsened by impurities from the plating process, including thallium, which wick up to the wire above the ball, resulting in the embrittlement of the wire. Bonded wires with embrittlement above the ball are more susceptible to failure in temperature cycling.

## 22.2 THERMOMECHANICAL CONSIDERATIONS DURING THE BONDING PROCESS

### 22.2.1 Cratering

Cratering is a failure mechanism that occurs during manufacturing predominantly as a result of improper control of bonding parameters. Cratering can be induced in thermocompression bonds by high bonding force, an excessive tool-to-substrate impact velocity, or too small a ball, which allows the hard bonding tool to contact the metallization. In studying the ultrasonic process, Winchell[29,30] found that even though metal flow is equal in all directions, stacking faults in silicon occur perpendicular to the direction of the ultrasonic bonding tool motion, thus verifying that ultrasonic energy is capable of introducing defects into the single-crystal silicon. Cratering is typically noticed in ultrasonic bonding and is uncommon in thermocompression bonding. This was verified by Koyama,[31] who found that force and temperature could not cause silicon nodule cratering without ultrasonic energy. Contamination of the bond pad can cause an improper setup of the bonding parameters, including the bonding energy required. Contaminated bond pads are found to require more ultrasonic energy and higher temperature for making strong bonds. Further, too high or too low a static bonding force can result in cratering in wedge bonds (Fig. 22-4).[32] Kale suggested

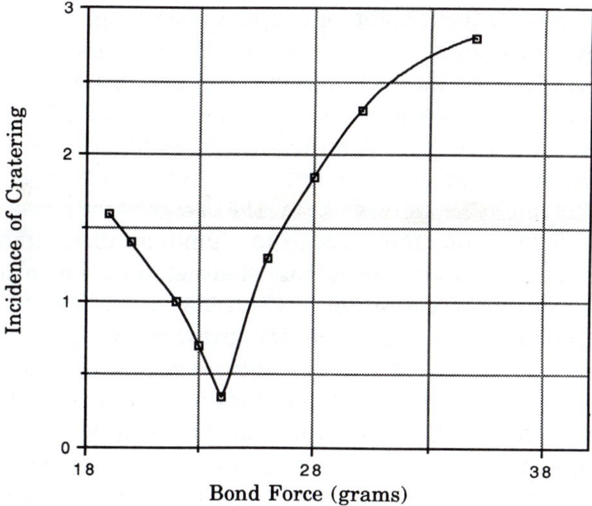


**Figure 22-4**  The incidence of cratering versus bond force for the ultrasonic bonding of 25-μm diameter Al–1% Si of 15–16 gf breaking load. The data were obtained for various silicon devices.[32]

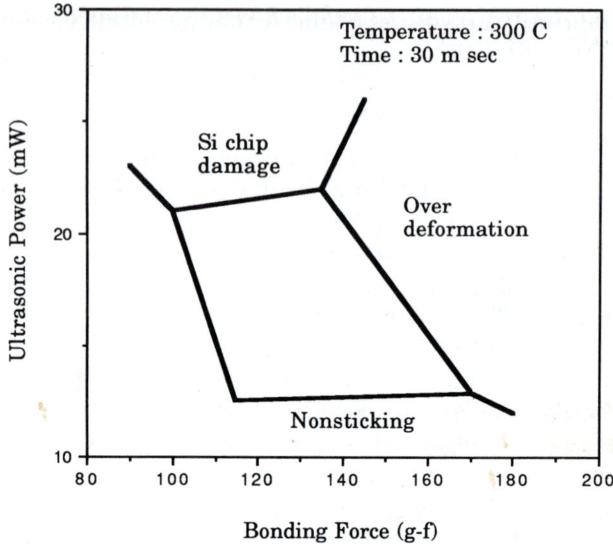


**Figure 22-5**  The preferred ranges of ultrasonic power and bonding force for copper ball bonding to 1.3-μm diameter Al–1% Si over silicon dioxide. The ball diameter was 62 μm.[33]

that the optimum bonding force resulted in a more efficient energy transfer, thus lowering the total energy requirements for the bond. Figure 22-5 shows a two-dimensional parameter plot for copper ball bonding, showing the areas of nonsticking through cratering and overdeformation.[33] Winchell[29,30] noted the effect of bond pad thickness on the phenomenon of cratering. The bond pad was found to serve as a cushion protecting the underlying Si, $SiO_2$, polysilicon, and GaAs from the stresses of the bonding process. The tendency to crater was found to be predominant in thin metallizations.

Silicon added to aluminum has been cited as a possible cause of cratering. One percent silicon is sometimes added to aluminum metallization in order to prevent back diffusion of silicon from shallow junctions into that metal, which would result in deterioration of electrical properties of the device. It has been reported that silicon nodules grow by annealing after metal deposition, and are distributed uniformly before bonding. These micrometer-sized silicon nodules in aluminum bond pads form and act as stress raisers, and crack the underlying glass during thermosonic gold ball bonding.[34] After the bonding process is complete, these nodules decrease in the bonding region and damage is observed on the insulation. Corrective action for silicon nodule cratering includes bonding at higher temperature (250°C), lower ultrasonic power, reducing molding stress, and removing a fracture-prone phosphorus glass layer from under the pad. Ching[35], and Koyama[31] have generally confirmed the silicon nodule cratering effect. Ching solved this problem by changing the structure under the pad (using a hard Ti–W submetallization), by modifying the schedules, and by using rapid ball touchdown. Koyama[31] experienced silicon nodules cratering during surface mount soldering of plastic encapsulated devices. Pramanik,[36] and Umemura[37] considered thermal effects on silicon nodules in metallization. Harder bond pads readily transmit energy from the bonding process to the substrate, which increases the susceptibility to cratering.[2,18,19]

Ching and Schreon[35] developed a theoretical model for contacting surfaces based on the Hertzian contact theory between a sphere and a flat surface. This model was used to study the stresses induced at the surfaces during the bonding process. The assumptions made were that both the surfaces were elastic, that the intermetallic region possessed the same properties as the gold ball, and that the contact was made between a part of the sphere of the intermetallic and the flat semiconductor bond pad. Ultrasonic effects were neglected. The manufacturing variables included time to touchdown after ball formation (i.e., the time between the formation of the ball at the end of the bonding wire by EFO* and the time of wire contact with the bond pad during bonding operation), and the moisture content and the hardness of the gold bond. The conclusion drawn from this model was that to reduce bond pad cracking, the contact area at touchdown should be maximized while the forced exerted on the bond pad is minimized. The Hertz model when applied to cratering starts with a circle being pressed against a bond pad. The radius of curvature establishes the contact area and the initial force creates stresses many times the yield strength or the fracture strength of the bond pad or the wire, implying that the stress is applied to the underlying semiconductor, thus initiating the crater. The metal has a yield strength far lower than that of the semiconductor, deforming the wire and the bond pad metal. As the wire flattens, the stress drops rapidly below the metal yield strength. Another problem with the Hertz model is that it predicts the initiation of cratering at the center of the bond pad area, which is contrary to the observations that the worst damage occurs along the perimeter of the bond pad. The Hertz model is also pessimistic since it assumes elastic properties, whereas bonding results in major plastic deformation of the metals.

Clatterbaugh et al.[38] studied gold–aluminum intermetallics using the ball-shear test and found the variation in probability of cratering versus time at 250°C to vary as shown in Fig. 22-6. The probability of cratering was found to peak (peak value ≈ 20%) at 4 hours. Continued heat treatment at

---

* EFO stands for Electrical Flame-off and is used to melt the ball. EFO is typically classified as positive EFO and negative EFO. Positive EFO involves charging the electrode positive and the wire negative. Negative EFO involves charging the electrode negative and the wire positive.

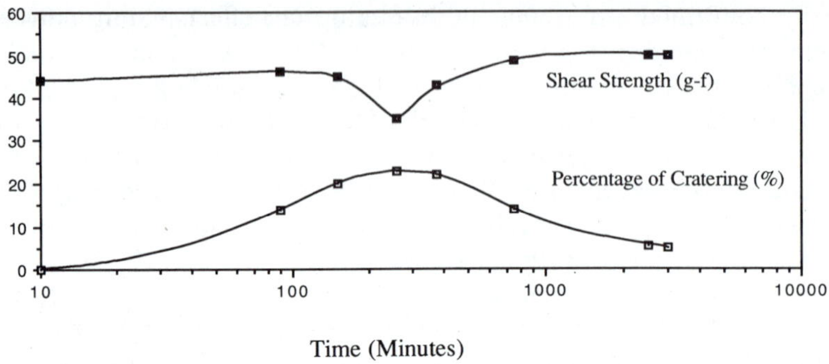


**Figure 22-6** Percentage of cratering and shear strength versus aging time at 250°C for gold bonds on 1-μm thick gold film for various plating densities.[38]

250°C, however, was found to decrease the probability of cratering after 35 hours. This increase in the probability of cratering was attributed to increase in the volume of the original aluminum metallization due to formation intermetallic phases, including $Au_2Al$, causing a high stress under the bond. When subjected to additional time at higher temperature, the reduction in the probability of cratering is due to reduction in stress under the bond due to recrystallization of the intermetallic phases, and formation of other intermetallic compounds with smaller volumes.

### 22.2.2 Pad Cleanliness

Impurities are a major cause of the loss in the bondability of a surface and the premature bond failure during the operational life of the device. Impurities along with interdiffusion in wirebonds can cause Kirkendall voids.[39] For Au–Al bonds, the interdiffusion front moves through the gold to the nickel underplating, forming intermetallic alloy phases. The impurities in gold are swept ahead of the intermetallic diffusion front, and because of their lower solubility in the intermetallic they precipitate ahead of the diffusion front. These impurities act as sinks for the vacancies produced during the diffusion reaction.[39]

Clatterbaugh et al.[40] investigated the effect of contamination on the shear strength of bonds versus bond strength after UV/ozone cleaning, acid cleaning, plasma cleaning, and solvent cleaning. Solvent cleaning was found to be the least effective, and UV/ozone cleaning the most effective in restoring the bond strength to uncontaminated levels. Surface contamination with photoresist was found to have significant effects on the bond strength

($\sim$20%) in both aluminum and gold metallizations. In general bonds made to aluminum metallization were more affected by contaminants than were gold metallizations.

Plating baths are a major source of the impurities in the plating. Impurities include potassium–gold–cyanide, buffers, lactates, citrates, carbonates, and phosphates. Other impurities like thallium, lead, or arsenic are added to the bath to increase plating speed and reduce grain size. Thallium has been identified as a major source of contamination-induced wirebond failures.[41,42] Thallium can be transferred to gold wires from gold-plated lead frames during crescent bond breakoff.[43] Thallium diffuses rapidly during bond formation and concentrates above grain boundaries above the neck of the ball, where it forms a low-melting eutectic. The forces and temperature applied during plastic encapsulation, or during temperature cycling in operational life, lead to wire breaks and device failures.[43] Thallium and lead can cause premature wirebond failures during burn-in by accelerating cracks or Kirkendall-like void formation under the bond.[44,45] The effects of thallium, lead, and arsenic on the bond strength, at concentrations normally used as grain refiners, compared to platings in pure solutions is shown in Fig. 22-7. Increasing plating current density increases the co-deposited impurity level exponentially (Fig. 22-8).[46] Variations in the plating process parameters such as bath temperature, and bath concentration also cause bondability problems.

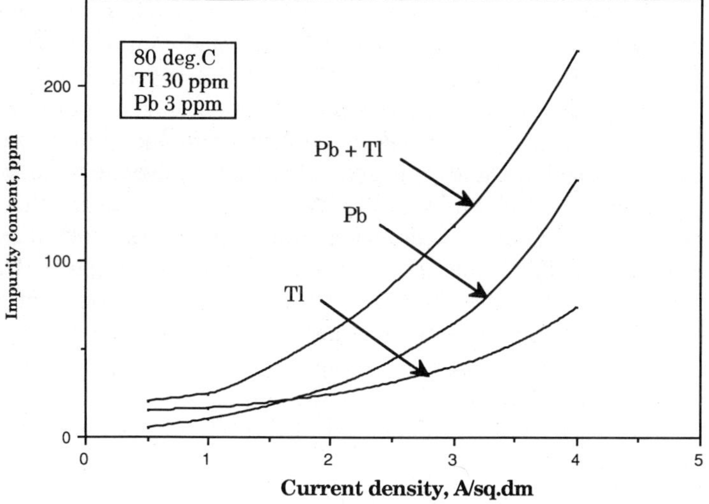


**Figure 22-7**    Thallium and lead content in gold deposits as a function of current density. Initial bath concentration was 10 p.p.m. for lead and 30 p.p.m. for thallium.[44]

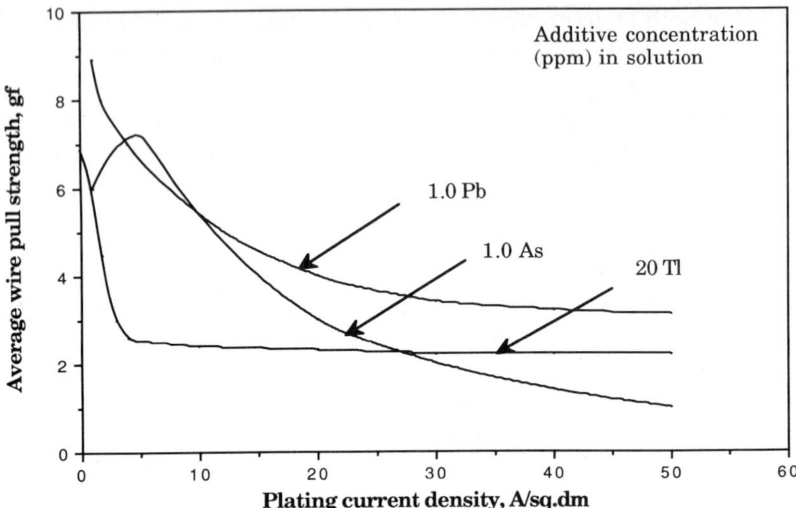


**Figure 22-8**   Wirebond pull strength for 1.3-mil diameter Al–1% Si wire made to 1.25 μm thick gold film.

Thermocompression gold bonding has been found to be highly steady-state-temperature-dependent in the presence of organic films on the bond pad before bonding.[9] Gold does not form a stable oxide film at room temperature, and it has been found that gold–gold bonds are very difficult to obtain when the deformation is low. This problem has been attributed to the inability of gold to destroy contaminant surface films by absorption during solid-phase welding at moderately elevated temperatures (in the neighborhood of 192°C).[9] Jellison investigated the temperature dependence of both conventional thermocompression bonding (using a heated substrate during the bonding process) and pulsed thermocompression bonding process (using a heated capillary and unheated substrates, subjecting the bonding interface to a very small temperature cycle). The bond strength of gold bonds having contaminated surfaces was found to increase for all bonding temperatures after UV-cleaned surfaces were bonded (Fig. 22-9). The time of application of substrate temperature during the bonding process was found to affect the bond strength. Shorter application of substrate temperature resulted in stronger bonds than longer times at substrate temperatures (Fig. 22-10).[9] However, using pulsed thermocompression bonding, the shear force at failure was found to increase with bonding interface temperature both with and without contaminated surfaces (Fig. 22-11). Post-heat treatment of the gold–gold bonds increased the bond strength for pulse-bonded thermocompression bonds while pre-heat treatments were found to decrease the bond strength.[9,47]

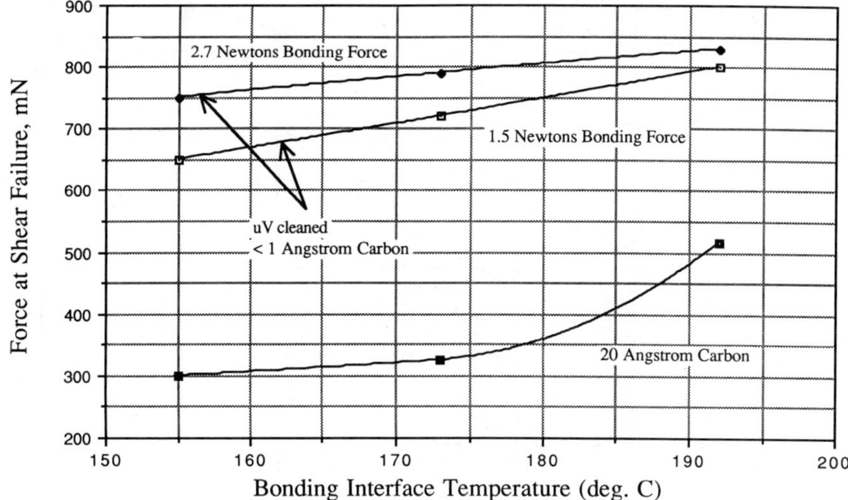


**Figure 22-9**   The bond strength of gold bonds having contaminated surfaces was found to increase for all bonding temperatures after UV-cleaned surfaces were bonded.[9]

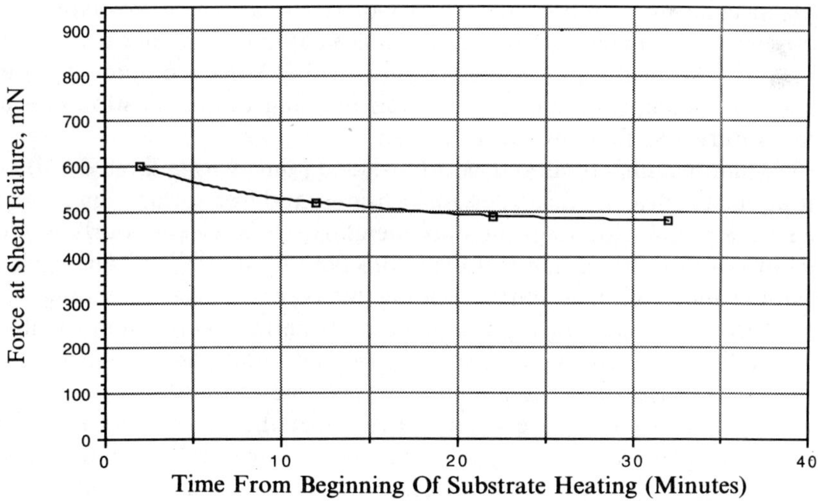


**Figure 22-10**   Shorter application of substrate temperature resulted in stronger bonds than longer times.[9]

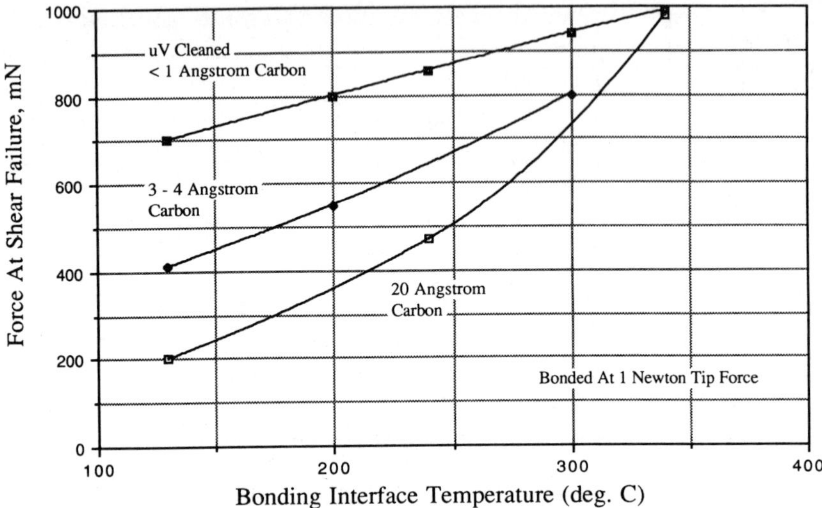


**Figure 22-11**  Shear force at failure was found to increase with bonding interface temperature for surfaces without contamination during pulsed thermocompression bonding.[9]

### 22.2.3 Intermetallic Formation

This failure mechanism shows up in wirebonds in the form of intermetallic compound formation and the associated Kirkendall void formation. This failure mechanism is more an impurity-driven mechanism involving corrosion reactions. In gold–aluminum systems a compound often called purple plague, because of the purple appearance of the $AuAl_2$ compound, forms during thermocompression or thermosonic bonding, or qualification screens, if temperature and time are large enough.

Thick-film metallizations and poorly welded joints with a large density of grain boundary defects, vacancies, dislocations, and free surfaces have much higher rates of diffusion than the bulk metal. For this reason poorly welded bonds or bonds to thick-film metallizations fail rapidly.[10,48–50] The basis of this failure mechanism is initial welding between surfaces consisting of a series of isolated welds during the process of thermocompression or ultrasonic bonding. The deformation of the wire and bond pad during bonding sweeps the surface oxide and contaminants aside into the debris zones, allowing intimate contact between the two metallic parts.[51] As the bond matures, the deformation energy imparted to the wire during bonding causes the microwelds to grow, join together, and spread inward in ultrasonic bonds and outward in thermocompression bonds.[51] Immature bonds consisting of isolated microwelds may result from contaminants on the surface that prevent intimate contact between surfaces. Harman generated a two-dimensional finite element model for gold–aluminum diffusion in structures with specified microweld dimensions[16,17] and found that rate of diffusion was dependent on the shape of microwelds. Diffusion in microweld geometries was found to be much more rapid than in completely welded couples, supporting observations that poorly welded Au–Al bonds fail more rapidly.

### 22.2.4 Metallizations

Metallization material should have a reasonably matched coefficient of thermal expansion (CTE) with the wire material so as to reduce the residual shear stresses introduced into the bonded interface after cooldown from the bonding temperature during the bonding process. The bonded materials should be such that they allow diffusion for bonding but do not allow formation of weak intermetallics in very short periods of operational life.

Nickel degrades bondability and reduces reliability. Nickel may be introduced into the gold plating by accident or by thermally induced grain boundary diffusion upward from the thin underplating, during die attachment or heat treatment.[52,53] Nickel spreads on the surface through surface diffusion, oxidizes, and renders the surface unbondable. Chromium from plating bath contamination, and lead frames, can follow the same route as nickel and oxidizes on the surface, resulting in decreased bondability. Chromium is often used in microelectronic components to promote adhesion between substrates and vacuum-deposited gold films, and can rapidly diffuse through grain boundaries to the surface and oxidize.

Metallizations with hydrogen bubbles in the film are often responsible for degradation of bondability. Hydrogen bubble concentration is a function of plating current density and bath impurity level. Huettner and Sanwald[54] found that during bonding of gold-plated copper wire to gold films, the lowest bondability occurred for plating currents in the range 1.6–2.7 A/dm$^2$, which corresponded to the onset of hydrogen evolution at the cathode and dendritic-like surface morphology. Hydrogen entrapment can be caused by too high a current density, low bath concentration, or low bath agitation. Gold films containing hydrogen are harder than pure films, but this hardness decreases rapidly during high-temperature annealing at 350°C for 8 hours.[55] Gas entrapment decreases the film bondability by increasing the ultrasonic energy absorption. These bubbles coalesce and grow by the process of structural rearrangement.[55] The gas bubbles are ruptured during bonding, preventing the surfaces from coming into intimate contact. However, free hydrogen inside the package has been found to inhibit the formation of

Au–Al intermetallic compounds by filling the vacancies in aluminum and its grain boundaries.[56]

### 22.2.5 Pad Lifting

Pad lifting in wirebonding is generally a result of poor bonds due to lack of surface cleanliness, operational stress, or poorly adjusted bonding parameters.

Impurities are a major cause of the loss in bondability of a surface and premature bond failure during the operational life of the device. Impurities including chromium, nickel, and titanium have a tendency to diffuse to the surface through grain boundaries and oxidize, degrading bondability. Impurities along with interdiffusion in wirebond can cause Kirkendall voids.[39] Intermetallics serve as sites for crack initiation and propagation during temperature cycling.

The bonded interfaces between the wire and the bond pad undergo shear stresses during temperature cycling. Failures due to wire–bond-pad material combinations having widely differing coefficients of thermal expansions usually consist of bond pad lifts.

Improper bonding parameters, including bonding force, bonding temperature, bonding time, bonding power, tool size, tool pressure uniformity, preheat temperature, contact area, and vibration amplitude and frequency in ultrasonic bonds, may result in improper bonds, resulting bond pad lifts in operational life. If the bonding temperature is not large enough to produce contaminant dispersal and the load not large enough to produce good surface interfacial conformity, the bond pad lifts may result fairly early in operational life. On the other hand, if the bonding force is excessive, bond pad lifts due to craters may result from small cracks that may be introduced into the chip during the bonding process.

### 22.2.6 Bonding Parameters

The bond strength is a function of materials and process variables associated with the substrate–metallization–wire composite structure. The key manufacturing wire bonding variables include bonding force, bonding temperature, bonding time, bonding power, tool size, tool pressure uniformity, preheat temperature, contact area, and vibration amplitude and frequency in ultrasonic bonds.

An adequate bond requires a bonding load large enough to increase the area of contact and produce bonding between fresh layers of metal. The bonding process requires a temperature large enough to produce contaminant dispersal and a load large enough to produce good surface interfacial conformity. In the process of thermocompression bonding, English[47] found that less than optimum bonding temperatures and low tool loads resulted in inadequate bonds. Present-day wirebonders address the issue of bond force during the bonding process, using bond process analyzers (BPA).[58,59] A BPA consists of a sensor inside the bonding head that monitors the force developed by the capillary during the bonding operation. As soon as the bonding force drifts out of tolerance, the BPA stops the machine immediately. Figure 22-12 shows the monitoring of the bond force during the bonding operation using the bond process analyzer.

Wire loop profile is of extreme importance in high-frequency circuits requiring strict tolerances on maximum allowable inductance of interconnects. Wirebonders used to manufacture circuits for high-frequency applications are equipped with optically controlled bonding heads that check the $X, Y, Z$ coordinates up to four times per millisecond.[59] The capability allows a 100 mil wire length to be checked 300 times.

The surface roughness effects the maximum achievable bond strength. Seshan and Ray,[143] while studying the effects of surface roughness in evaporated Au–Ni films with an underlying adhesion layer, noticed that the bond strength increased with decrease in surface roughness. Surface roughness was characterized using surface stylus measurement. The output

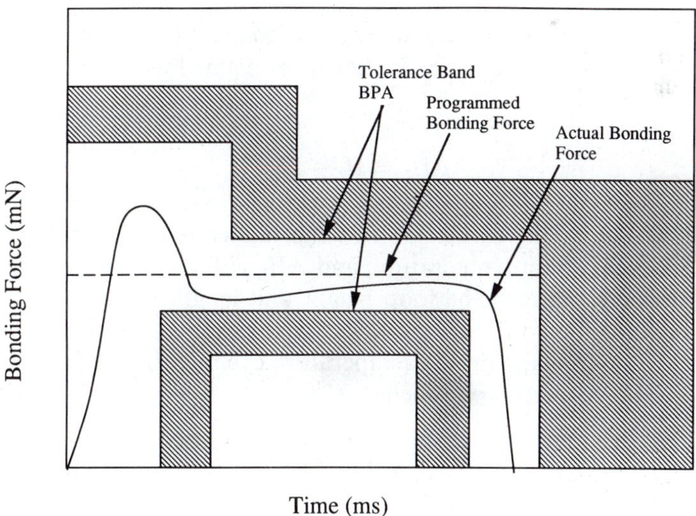


**Figure 22-12**  A typical plot of bonding force versus time on a bond process analyzer.[58,59]

was used to calculate a centerline average (CLA). The CLA average number was higher for rough and lower for smooth surfaces. After heating the wirebonds to 350–400°C and cooling down several times, the bond strength was found to decrease faster on rough films compared to smooth films. The higher bond strength for smooth films was attributed to their superior mechanical properties and lower interdiffusion rates.

Lang and Pinamaneni[14] identified the parameters affecting bond strength as cleaning and copper plating of the lead frames, die-attach cure conditions, atmosphere during bonding, surface of the lead frame, bonding time, bonding force, bonding pressure, and temperature. The presence of inert atmosphere was found to be essential to prevent oxidation of the lead frame.

Clatterbaugh et al.[40] defined the machine parameters that could affect the shear strength of the bonds. They found that an increase in ultrasonic power setting of the first bond improved the shear strength of the bonds. The increase in substrate temperature was also found to increase the shear strength of the bond by 0.13 g/°C for aluminum metallizations. The increase in substrate temperature for gold metallizations was found to have negligible effect on the bond strength. The second bond power setting, residence time of substrate on the heated bonder pedestal, and metallization thickness was found to have no systematic effect on the bond shear strength for both gold and aluminum metallizations.

### 22.2.7  Geometries

In thermal cycling environments wires undergo flexure, resulting in a change in the loop height. Mathematically, flexure is defined as the change in loop height[60]

$$\Delta H_L = H_L - H_{L0} \tag{22-2}$$

where $H_L$ is the height of the loop at the higher temperature, $H_{L0}$ is the height of loop at room temperature, and $\Delta H_L$ is the change in loop height. The relationship between the loop height and length on the wire depends on the shape of the wire. For a circular wire loop, assuming that the wire retains its circular shape during temperature cycling, the height of the loop in terms of the geometric parameters is[60]

$$H_L = \frac{\sqrt{3(3S + 5L)(S - L)}}{8} \tag{22-3}$$

where $S$ is the total wire length, and $L$ is the bond-to-bond spacing.

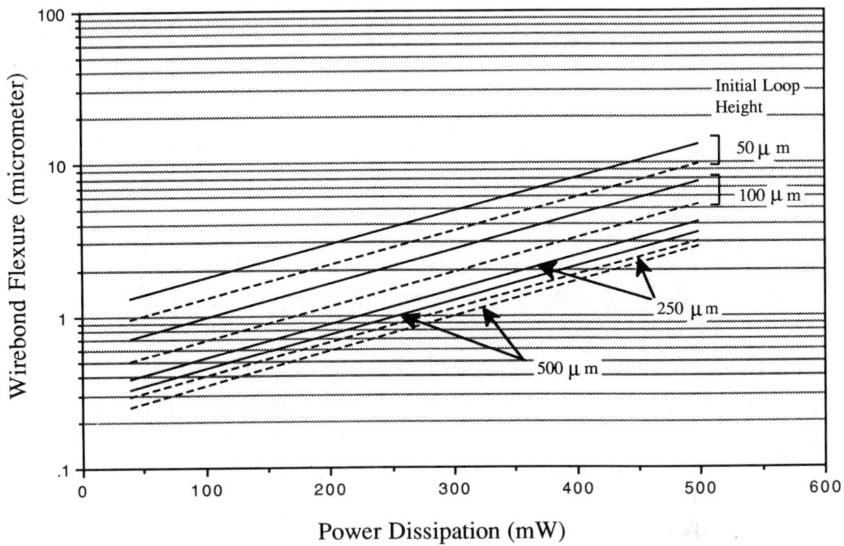


**Figure 22-13**   Wirebound flexure versus power dissipation for various values of initial loop height. (Solid lines for triangular loops; dashed lines for circular loops.)[60]

For a triangular loop, the loop height in terms of its geometric parameters is expressed as[60]

$$H_L = 2 \frac{\sqrt{s(s-a)(s-b)(s-L)}}{L} \tag{22-4}$$

where wire length $S = a + b$, and $s = (a + b + L)/2$. The curves for flexure versus power are plotted for circular and triangular loops in Fig. 22-13. From the curves it is evident that flexure is approximately proportional to power dissipated and inversely proportional to initial loop height. Thus large loop heights increase the wirebond reliability during temperature cycling. This result is valid for geometries with (loop height/bond length) ratio greater than 25%. However, the result does not hold true for larger wire sizes, where the stiffness of the wire becomes a significant factor in the flexibility of the loop.

### 22.2.8 Post-Bond Testing

A summary of screening and test procedures of wire–wire bond assemblies can be found in MIL-STD-883C. This section outlines the applicability of existing test techniques for assessing wire and wirebond integrity. Typically, the same test method may be used for quality assurance and qualification at different stress levels.

## Nondestructive Bond Pull Test (Method 2023)

This screen is supposed to precipitate the weak bonds resulting from excessive neckdown at the heel, or excessive embrittlement of the wire, during the bonding operation or poor bondability due to surface roughness, presence of contaminants, and improper setup of the bonding machine parameters. The purpose of this test is to detect unacceptable wirebonds while avoiding damage to acceptable wirebonds, and can be used for thermocompression or ultrasonic bonds. Bonds with such defects, if allowed to enter operational life, will exhibit failure mechanisms including flexure failures, interfacial de-adhesion, cratering, and shear fatigue at bonded interfaces during temperature cycling. The weak bonds due to excessive neckdown or embrittlement will break at the heel of the wire, while bonds with improper bonding parameters will exhibit interfacial de-adhesion (indicating poor bondability) or cratering (indicating improper bonding parameters such as too much impact force, excess approach velocity of tool, excess hardness of wire).

The wirebond pull test is the most universally accepted test method to evaluate bond strengths of wirebond interconnects. This test method involves pulling the bond wire with a hook. The force exerted by hook on the wire is represented by[16,17]

$$F_{wt} = F\left[ \frac{(h^2 + \varepsilon^2 d^2)^{1/2}\left((1 - \varepsilon)\cos\phi + \dfrac{(H + h)}{d}\sin\phi\right)}{h + \varepsilon H} \right] \tag{22-5}$$

$$F_{wd} = F\left[ \frac{\left(1 + \dfrac{(1 - \varepsilon)^2 d^2}{(H + h)^2}\right)^{1/2}(H + h)\left(\varepsilon\cos\phi - \dfrac{h}{d}\sin\phi\right)}{h + \varepsilon H} \right] \tag{22-6}$$

where $F_{wt}$ is the component of force acting along the wire at the terminal or the lead frame, and $F_{wd}$ is the component of force along the wire at the die (Fig. 22-14). If both the bonds are at the same level and the hook is applied at the center, the forces can be represented as

$$F_{wt} = F_{wd} = \frac{F}{2}\sqrt{1 + \left(\frac{d}{2h}\right)^2} = \frac{F}{2\sin\theta} \tag{22-7}$$

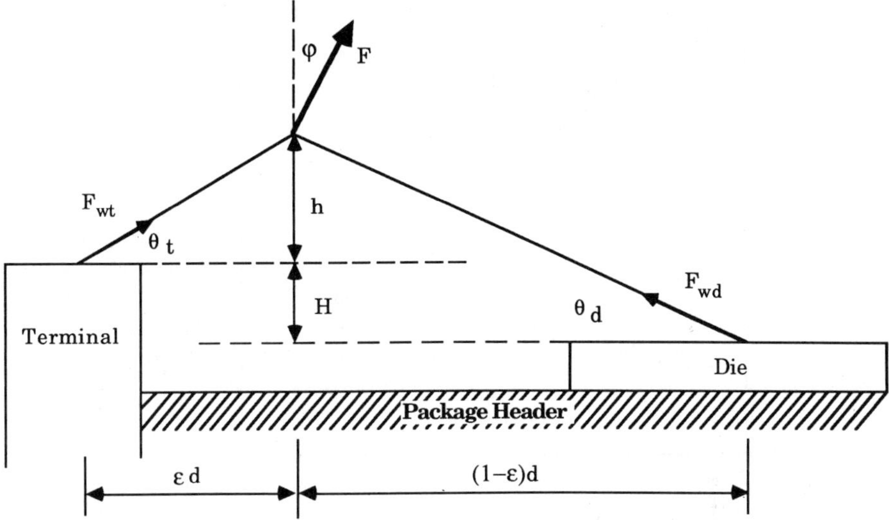


**Figure 22-14**    Geometric variables for wirebound pull test in the plane of bond loop.[60]

where $\theta_t = \theta_d = \theta$. For bonds of a given strength, larger $h/d$ ratios will result in larger pull values. The placement of the wire hook determines the values for $\varepsilon$ and $\phi$. The hook is usually positioned in such a way that the forces on both the bonds are equal.[16,17]

The bond pull test can potentially result in misleading conclusions. This can result from the hook trying to slip to the highest portion of the loop. Hook slippage can result in peel mode failures giving a lower bond pull value. In wirebond interconnects consisting of a ball bond and a wedge bond, the wire rises vertically from the center of the ball bond. If the pull force is applied to the peak of the loop, most of the force is directly applied to the ball bond, which because of its larger area is stronger than the wire. Figure 22-15 shows the effect of hook position on the bond pull force. The second bond, which is usually weaker than the ball bond, is subjected to relatively little force, thus testing the stronger bond only. On the other hand, if the hook rises to the peak of the loop, more force is exerted on the weaker bond and the stronger bond remains untested. Elongation of the wire material also affects the bond pull test results. Figure 22-16 shows the effect of elongation on the bond pull strength. During pulling the $h/d$ ratio may increase significantly, yielding a pull force higher than what might be expected if only the breaking load of the wire and the initial bond geometry wire is considered. The effect of the change in $h/d$ ratio is increased if the value of $h$ is low.[16,17]

The method requires that all bond wires on the device be pulled and counted. Such requirements are feasible for present-day devices, but with the

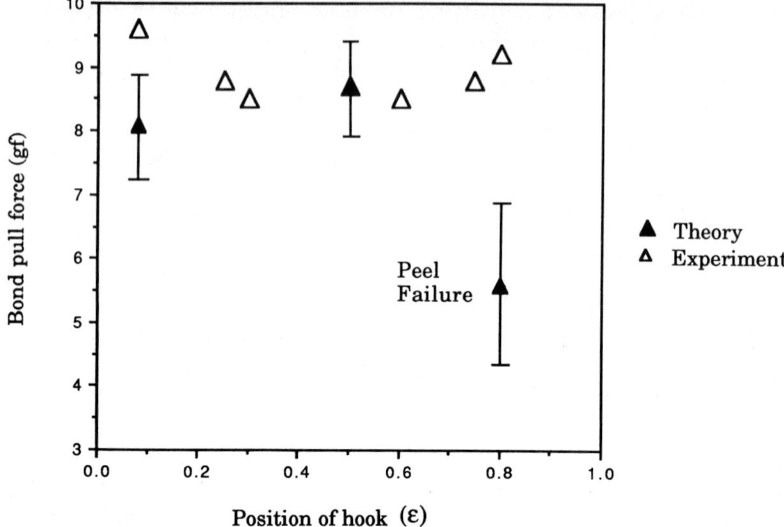



**Figure 22-15**    Measured and calculated values of the pull force as a function of hook position for single-level, 25-μm diameter ultrasonic aluminum bond pairs having first and second bonds of equal breaking strengths. Each point is the mean of 25–30 bonds pulled at the indicated hook positions.[93]

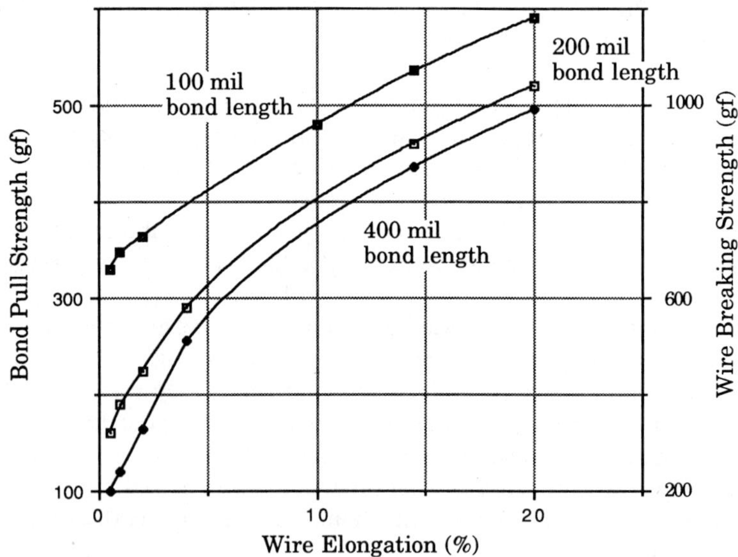



**Figure 22-16**    The variation in bond pull strength and wire breaking strength versus wire elongation.[16,17]

advent of ULSI devices consisting of hundreds of bond wires, such a procedure may not be feasible. Besides, a decrease in the pitch of the interconnections increases the probability of damaging adjacent wires during qualification.

## Internal Visual (Method 2010: Test Condition A and B)

The purpose of this test is to check the internal materials, construction, and workmanship of microcircuits for compliance with requirements of the applicable procurement document. This screen precipitates defects such as improper bond geometries, including deformed ball bonds, small ball bonds, large ball bonds, golfball bonds, necked ball bonds, ball off the bond pad; deformed wedge bonds with excessive deformation at the heel, or where the bond has damaged the surrounding passivation or semiconductor device circuitry; bonds with excessively large tails touching adjacent bond pads or surrounding metallization; excessive intermetallics; and visible corrosion. Such defects may show up in operational life in the form of interfacial de-adhesion failures, craters, flexure failures in temperature cycling, vibration fatigue failures, electrical shorts, electrical noise, electrical leakage, dendritic growth, or corrosion in presence of contaminants.

The test is done prior to capping or encapsulation to detect and eliminate devices with an internal defect, which could lead to device failure in normal application. According to MIL-STD-883 the following defects are not acceptable:

Voids in the bonding pad or fillet area that reduce the metallization path width connecting the bond to the interconnecting metallization to less than 75% of the narrowest entering metallization stripe width.

Gold ball bonds on the die or package post wherein the ball bond diameter is less than 2 times or greater than 5 times the wire diameter.

Gold ball bonds where the wire exit is not completely within the periphery of the ball.

Gold ball bonds where the wire center exit is not within the boundaries of bonding pad.

Intermetallic formation extending radially more than 0.1 mil completely around the periphery of any gold ball bond for that portion of the gold ball bond located on the metal.

Ultrasonic wedge bonds on the die or package post that are less than 1.2 times or more than 3 times the wire diameter in width, or are less than 1.5 times or more than 5 times the wire diameter in length.

Thermocompression wedge bonds on the die or package post that are less than 1.5 times or more 3 times the wire diameter in width, or are less than 1.5 times or are more than 5 times the wire diameter in length.

Crescent bonds on the die or the package post that are less than 1.2 times or more than 5 times the wire diameter in width, or less than 0.5 times or more than 3 times the wire diameter in length.

Crescent bonds where the bond impression does not cover the entire width of the wire.

Bonds where less than 75% of the bond is within the unglassivated bond pad area.

Wirebond tails that extend or make contact with any unglassivated metallization.

Wirebond tails that extend more than two wire diameters in length.

Any wire that comes closer than two wire diameters to unglassivated operating metallization, another wire, package, post, unpassivated die area, or the package lid.

Nicks, bends, cuts crimps, scoring, or neckdown that reduces the wire diameter by more than 25%.

### Temperature Cycling (Method 0101; Test Condition C)

The purpose of this test is to determine the resistance of wire bond interconnect to exposures at high and low temperatures and the effects of alternate exposure to these temperatures. The failure mechanisms addressed by the test include flexure failure of the wire at the heel, bond pad–substrate shear failure, wire–substrate shear failure. Susceptibility to these failure mechanisms may be actuated by excessive neckdown at the heel or excessive embrittlement of the wire during bonding, or poor bond shear strength resulting from surface roughness and the presence of contaminants.

### Constant Acceleration (Method 2001; Test Condition E)

This screen is supposed to precipitate defects such as improper interconnect material for applications subjected to high accelerations of the order of $10\,000g$ loading. Improper hardness of the wire material, or improper stiffness of the wirebond structure in a high-acceleration application such as cannon-launched devices can result in plastic collapse under impact during operation. This test simulates the plastic collapse of the wirebond at high acceleration levels encountered in cannon-launched devices.

The failures during the screen will be in the form of plastically deformed wires, resulting in shorts to adjacent wirebonds. Constant acceleration test should be used with caution because high stress levels used in the screen may introduce residual stress, potentially leading to failure in operational life.

Screening transforms based on the physics of failure of the package elements should be used in determining screen stress levels. Sometimes the results of the constant-acceleration test can be more conveniently achieved by subjecting the components to mechanical shock, which is an easier process.

### Random Vibration (Method 2026)

This test detects the suitability of wirebond interconnects to vibrational loads in operational life. The device is rigidly fastened on a vibration platform and the leads are adequately secured. The device is subjected to random frequencies and intensities of vibration with the total power applied distributed among different frequencies as represented by the spectral density curves. The spectral density curves plot the output of the vibration machine as power spectral density versus frequency. The power spectral density is the mean square value of an oscillation passed by a narrow-band filter per unit filter bandwidth.

When used as a screen, random vibration aims at detecting cracks in the die beneath the bond, and detecting the presence of brittle intermetallics at the bonded interfaces of wirebond interconnects. Such interconnects, if allowed to enter operational life, will fail as a result of thermomechanical fatigue in temperature cycle or vibration. The failures during this screen will be interfacial de-adhesion and delamination failures at the bonded interfaces in wirebonded interconnections.

### Delay Measurement (Method 3003)

This method establishes the means for measuring the propagation delay of microelectronic devices. The interconnect's electrical properties have to be such that the interconnect does not degrade the pristine waveform coming from the on-die circuitry. The distortion and clock skew from the interconnect can be so great that they could cause false triggering. The rise time, clock frequency, and the bandwidth are related by

$$\text{Bandwidth} = \frac{0.35}{t_{\text{rise}}}$$

$$t_{\text{rise}} = \frac{0.07}{F_{\text{clock}}} \qquad (22\text{-}8)$$

$$\text{Bandwidth} = 5F_{\text{clock}}$$

The defects in the form of increased bond resistance due to improper bonding, or precipitation of impurities such as Si (as in Al–1% Si wires) due to improper bonding temperature, can be detected with this screen. These defects, if allowed to enter operational life, will exhibit failures in the form of electrical noise in the output, signal distortion, and false triggering.


## Ball Bond Shear Test

This test has found limited application as a qualification test, to certify shear strength of the bond at the wirebond assembly's bonded interface. The shear strength value is used as a life estimate of the bond subjected to temperature cycling or vibration and is a measure of the resistance to interfacial de-adhesion or cratering at the wirebonded interface. This test serves as a complement to the bond pull test, to determine parameters that are neglected due to the limitations of the bond pull test. Most ball bonds have interfacial welds of the order of 6 to 10 times the cross-sectional area of the wire. Therefore, it is apparent that the wire will fail in pull testing before even a poorly attached bond is lifted. In addition the wire near the neck of the ball bond is totally annealed, recrystallized, and becomes the weakest part of the ball bond wire wedge system. In a pull test, the wire often breaks at this point, depending on where the hook is placed and the ball geometry, even if the ball is welded even a little over 10% of its area.[16,17] A shear test gives the shear strength of the bond, and serves as a check for bondability and appropriateness of the machine setup parameters. Equipment for bond shear testing ranges from tweezers and other hand-held probes to dedicated shear test machines with strain-gauge force sensors.

There is no documentation on the physics of failure-based acceleration transform, correlating the test life and operational life. However, the test results are often related to operational life based on experimentation. The logic in the approach is that if a bond will not fail under a particular stress level, it will last reasonably long. Based on the same logic, strong ball bonds will have a shear strength of greater than 50 gf.[16,17]

Unfortunately, the shear test holds the potential to provide misleading data. Mispositioning of the shear tool can give wrong results. In an ideal situation the tool should not drag the surface, and should approach the ball bonds about 2.5–5 μm above the surface for normally deformed balls, and about 13 μm above the surface for large balls. If the tool is positioned higher it will tend to ride or smear over the top of the ball.[16,17] Since the tool is positioned less than 13 μm from the surface, metallic smears, bits of aluminum, gold, glassivation, and silicon from the chip or the substrate sticking to the tip of the tool may prevent the correct positioning of the tool. Substrate dragging increases the shear force value by 10–20 gf. Gold ball bonds are susceptible to friction rewelding to gold surfaces, giving wrong values for the shear force. Friction rewelding seldom occurs in aluminum-metallized IC pads.[16,17] Ball bonds on thick-film metallization also hold the potential for problems, since they are themselves higher than the vertical position of the tool above the substrate, causing it to drag across the thick-film surface. Even if the positioning of the tool is correct, shear tests made on bonds welded to thick films will yield lower values than expected if the metallization adheres poorly to the substrate. Intermetallic compounds formed in Au–Al bonds can trigger exceptionally high shear test values. If the intermetallic interface is void-free, compounds could increase the strength since intermetallics form irregularly in depth across the interface and effectively increase the surface area of both the gold and aluminum, thus resulting in an increase in the resistance to shear.

**Destructive Bond Pull Test (Method 2011)**

The purpose of this test is to certify the proper setup of the bonding machine parameters. This test is supposed to certify the bond geometry against failure mechanisms including flexure failures, interfacial de-adhesion, cratering, and shear fatigue at bonded interfaces during temperature cycling, and to determine bond strength, bond strength distribution, and compliance with specified bond strength, The failures may include:

Wire break at neckdown point. (The neckdown point is the area where
   the cross-section has been reduced due to the bonding process.)
Failure in bond at the die.
Lifted metallization from the die.
Lifted metallization from substrate or package post.

There is no documentation on the physics of failure-based acceleration transform, correlating the test life and operational life. However, the test results are often related to operational life based on experimentation The logic in the approach is that if a bond will not fail under a particular stress level, it will last reasonably long. Based on the same logic, strong ball bonds will have a bond pull strength of greater than 10–13 gf.[16,17]

**Mechanical Shock (Method 2002)**

This test is intended to certify the suitability of the wirebond geometries and materials for use in electronic equipment that may be subjected to moderately severe shocks as a result of suddenly applied forces or abrupt changes in motion produced by rough handling, transportation, or field operation. Shocks of this type may disturb operating characteristics or cause damage resulting from excessive vibration, particularly if the shock pulses are repetitive.

## Stabilization Bake (Method 1008)

The purpose of this test is to determine the effect on microelectronic devices of storage at high temperatures without electrical stress. This can be used as a preconditioning sequence to other tests, and is not a life test. These defects if allowed to enter operational life may show up in the form of increased bond resistance, electrical noise, and shear fatigue of the bond in temperature cycling due to increased brittleness of the bonded interface as a result of excessive intermetallic formation.

There is no documentation on the physics of failure-based acceleration transform, correlating the test life and operational life. However, the test results are related to operational life based on experimentation. The logic in the approach is that if a bond will not fail under a particular stress level, it will last reasonably long. Some of the correlation results from various investigators are listed in Table 22-2.

## Moisture Resistance (Method 1004)

The moisture resistance test is performed to evaluate in an accelerated manner the resistance of the packages to deteriorative effects of high

**Table 22-2**  Stabilization Bake Results

| Time (hours) | Temperature (°C) | Life Estimate Criteria | Reference |
|---|---|---|---|
| 1 | 390 | If the bond is strong after the specified thermal stress, it is acceptable. | Horsting[39] |
| 3000 | 150 | If the bond passes MIL-STD-883, Method 2011, after the specified treatment, it is acceptable. | Ebel[61] |
| 1 | 300 | The bond is acceptable if after the thermal treatment, it has a pull strength of<br>1 gf for 1-mil wire Au or Al,<br>0.5 gf for <1 mil Au or Al<br>wires. | MIL-STD-883 (Method 5008) |

humidity and heat. The failure mechanism addressed by this test is corrosion of the wirebonded assembly. The test certifies the corrosion resistance of the materials in a high-humidity environment.


## 22.3  THERMOMECHANICAL  CONSIDERATIONS  DURING  USE

### 22.3.1 Intermetallic Formation and Kirkendall Voiding

Intermetallic compound formation can be accelerated during operational life if the temperature and time are large enough. A few monolayers of these compounds can also form at room temperature. Intermetallic formation has been most widely cited in gold–aluminum systems. The compound, often called purple plague because of the purple appearance of the $AuAl_2$ compound, occurs between gold wirebonded to aluminum metallization, and vice versa. There are five intermetallic compounds which are all colored, with $AuAl_2$ being purple and the remaining being white or tan (Fig. 22-17).[62] The Al-rich gold–aluminum compound has a high melting point and is relatively stable. Prolonged thermal exposure results in continued diffusion until gold or aluminum is consumed. When bonded together, Au–Al present a thermal cycle problem because of the formation of Au–Al intermetallics, which are more susceptible to flexure damage than pure Au and Al wires. Intermetallics are stronger and more brittle than pure aluminum or gold, causing the bond to be much more susceptible to brittle fracture in temperature cycling.[63,64] Further intermetallic brittleness and growth is enhanced by temperature cycling. The growth rate of these compounds follows a parabolic relationship:[65]

$$x = k_D t^{1/2}$$
$$k_D = D_0 e^{-E/kT}$$

$$(22\text{-}9)$$

where $x$ is the intermetallic layer thickness, $t$ is the time, $D_0$ is a constant, $E$ is the activation energy, $k$ is the Boltzmann constant, and $T$ is the absolute temperature (K). The relative growth rates for each of these intermetallic compounds are shown in Fig. 22-18,[64] showing that $Au_5Al_2$ grows much faster than the other phases. Figure 22-19 shows that the formation of significant amounts of purple plague can take years at 150°C and below.

The diffusion rate for intermetallic formation is a function of the lattice defects. Thick-film metallizations and poorly welded joints with a large density of grain boundary defects, vacancies, dislocations, and free surfaces have much higher rates of diffusion than the bulk metal. For this reason poorly welded bonds or bonds to thick-film metallizations fail rapidly.[10,48–50] The basis of this failure mechanism is initial welding between

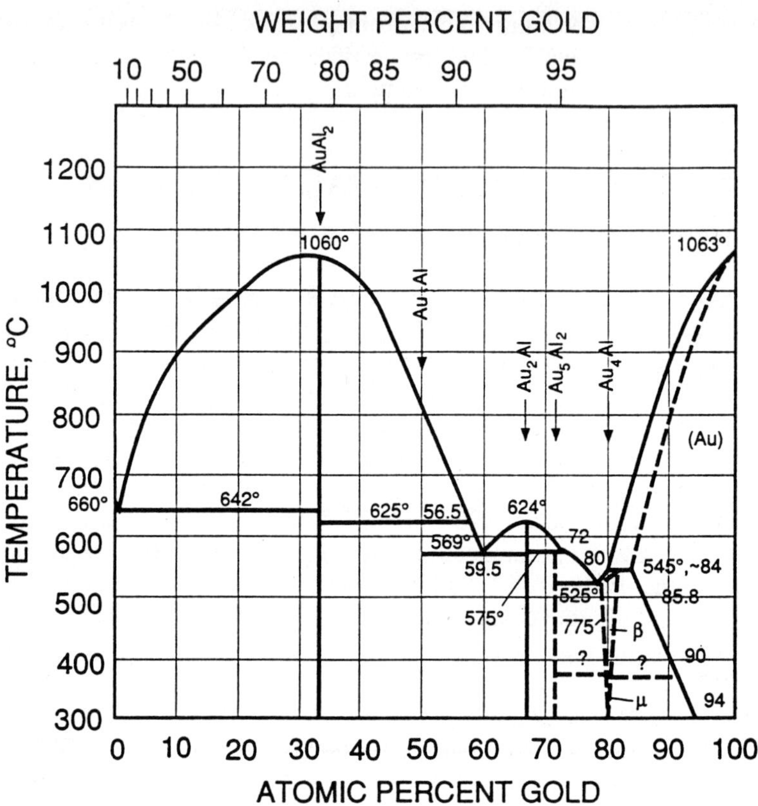


**Figure 22-17**   Five intermetallic compounds can form when gold and aluminum are bonded.[62]

surfaces consisting of a series of isolated welds during the process of thermocompression or ultrasonic bonding. The deformation of the wire and bondpad during bonding sweeps the surface oxide and contaminants aside into the debris zones, allowing intimate contact between the two metallic parts.[51]

Kirkendall voiding around the periphery results in a bond mechanically strong but having a high electrical resistance. Kirkendall voiding beneath the bond causes the bond to fail as a result of mechanical weakness. Formed by the different interdiffusion rates of gold and aluminum, under various circumstances the voids may appear either in the gold or in the aluminum. Vacancies pile up and condense to form voids, normally on the gold-rich side along the Au–Al interface. The rates of diffusion vary with temperature, and with different phases, and are dependent upon the adjacent phases as

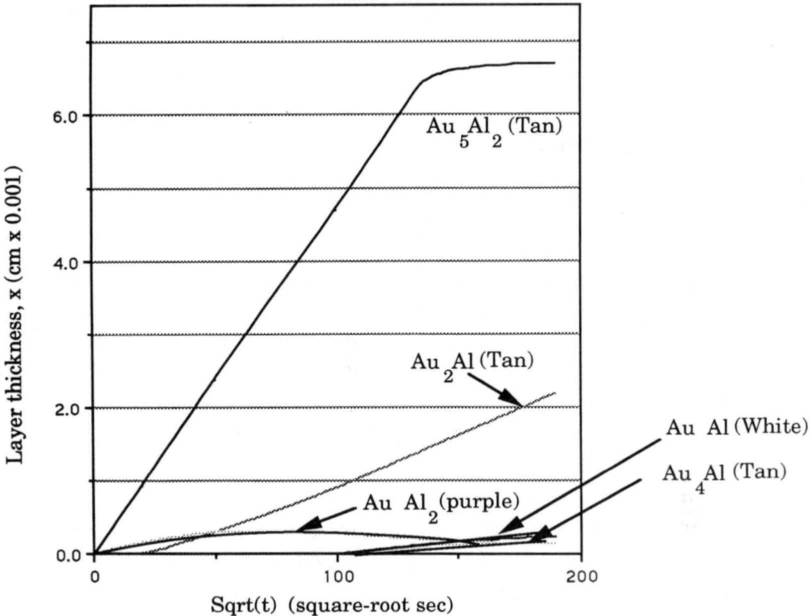


**Figure 22-18**   Layer thickness of five intermetallic phases versus the square-root of time at 400°C.[64,67]

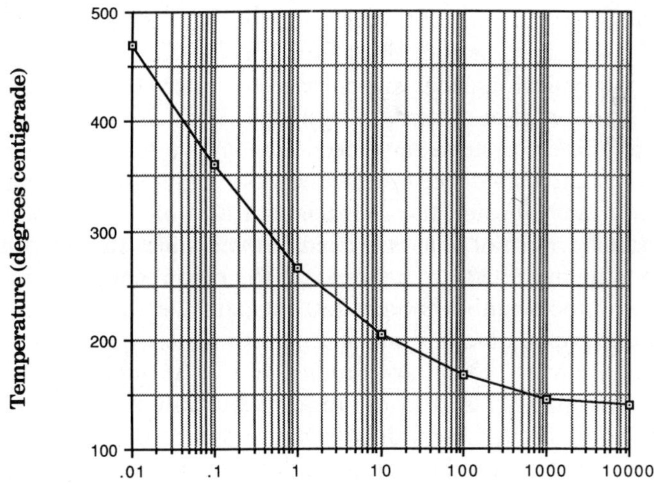


**Figure 22-19**   The formation of significant amounts of "purple plague" can take years in equipment operating range −5 to 125°C. DELCO data (as of 11-2-88) EDTL LABCOM— from 636 nos. ICs analyzed, 513 nos. Bipolar and 123 nos. MOS.

well as on the number of vacancies in the original metals. Kashiwabara and Hattori,[66] and later Philosky,[64,67] noted that thinner metallization restricted voiding by restricting the availability of one of the intermetallic elements, and intermetallic failures could be prevented to a large extent if the ratio of the width of the aluminum wedge bond to the thickness of the gold film was greater than 4 and the storage temperature less than 350°C. Shih and Ficalora[56] noted that the presence of hydrogen inside the package reduced the rate of intermetallic formation. Hydrogen inside the package, however, presents a potential problem, because it reacts with iron, nickel, and palladium inside the package, causing embrittlement. Hydrogen also combines with oxygen, producing water that can result in corrosion.[16,17] Palladium is added to gold thick film ultrasonically bonded to aluminum to reduce interdiffusion. This configuration results in a stable ternary compound of Au–Al–Pd, which slows down the rate of Au–Al diffusion and results in longer bond life.[48,68]

Interdiffusion can be enhanced by impurities in the gold-plated film.[39] These impurities include nickel, iron, cobalt, boron, chlorine, and bromine. During high-temperature bake for a gold–aluminum bond, Horsting noticed that the intermetallic diffusion front moved through gold plating down to the nickel underplating. The impurities became concentrated ahead of the intermetallic diffusion front, and precipitated there, acting as sinks for vacancies produced by the diffusion reaction and resulting in Kirkendall-like voids. Halogens also corrode aluminum metallizations in integrated circuits. Epoxies containing brominated flame retardants have been found to cause wirebond failures in a period of 24 hours at 200°C.[21] These failures are marked by the typical lamellar structure of the intermetallic produced as a result of reaction with outgassed halogenated products.

Copper–aluminum systems have been also known to form intermetallics. In spite of substantial evidence for the formation of intermetallic phases, and the significant interdiffusion between temperatures of 100 and 500°C, there is no consistent pattern of phases formed in the couples studied. Pitt et al.[69] found that aluminum alloys bonded to thick-film copper and aged at temperatures in the neighborhood of 150 and 250°C, exhibited little evidence of intermetallic growth. No intermetallic corrosion occurred during exposure to high humidity over 2000 hours. Studies on Cu–Al bonds in the temperature range 400–535°C,[70] and in the range 300–500°C[71] show a significant decrease in bond strength and ductility after aging. The decrease in bond strength was directly related to the total thickness of the intermetallics. The critical thickness of the intermetallic range for most studies has been reported to be in the neighborhood of 16 µm. The intermetallic compounds formed include $CuAl$, $CuAl_2$, and $Cu_9Al_4$. Lower temperature studies by Gershinskii et al.[72] and Campisano et al.[73] reveal that the above-mentioned intermetallic compounds occur in mixtures of pairs. The identified mixtures include ($CuAl$,

$CuAl_2$) and ($CuAl_2$, $Cu_9Al_4$). The ambient atmosphere composition has been shown to play a major role in the rate of intermetallic formation.[20] Olsen and James aged Al–Cu bonds in air, vacuum, and nitrogen and found that intermetallic formation occurred under all aging conditions. The formation of oxides in air-aged samples retarded the intermetallic formation by a factor of 2.8.[20]

The strengths of gold–copper bonds have been found to decrease when aged at temperatures in the neighborhood of 200–300°C. The rate of decrease in bond strength versus aging time was found to increase with the increase in aging temperature.[12] Figure 22-20 shows the temperature dependence of time to decrease the copper to gold bond strength 40% below the "as-bonded" strength for temperatures above 300°C.[12] The decrease in strength was attributed to the formation of intermetallics. The observed intermetallic phases include $Cu_3Au$, $CuAu$, and $Au_3Cu$. Oxidation of copper in air with formation of copper oxide was found to have no effect on the time to decrease in bond strength at elevated temperatures.[12] Johnson[74] while working on gold–copper bonds at lower temperatures found no intermetallic formation. The use of a nickel barrier of 5000 ± 2500 Å between copper and gold prevents the degradation of wirebonds due to interdiffusion between gold and copper. The extrapolated life at 50°C in the presence of nickel barrier was found to be in the neighborhood of 40 years.[12] The critical intermetallic thickness has been found by Onuki et al.[5] to be in the neighborhood of 2 μm.

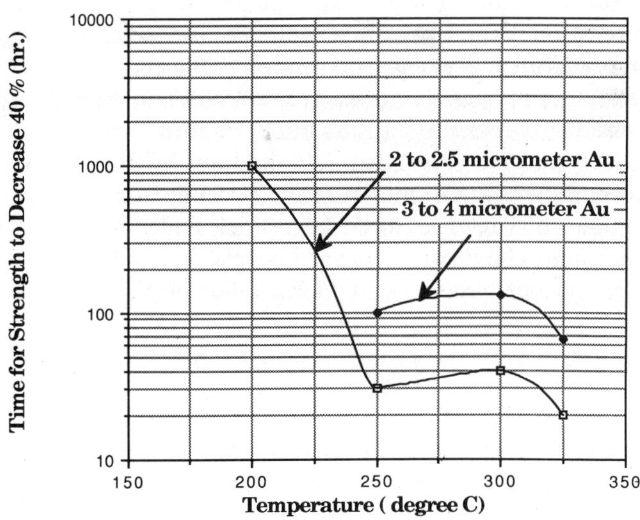


**Figure 22-20**   Temperature dependence of time to decrease copper–gold bond strength 40% below the as-bonded strength.

The estimated degradation time at 150°C was found to be in the neighborhood of 4.5 years.


### 22.3.2  Wire Flexure Fatigue

Wirebond interconnects fail as a result of metallurgical cracks in the heel.[60] Cracks can arise as a result of excessive flexing of the wire during loop formation especially when the second bond is significantly lower than the first bond. The sharp metallurgical microcracks can propagate through the wire and cause failure during device operating life, when temperature changes cause the wire to expand and contract. Flexure of the wire produces stress reversal at the heel of the bond in wedge bonds and stitch bonds, and causes eventual fatigue failure of the wire.

Gaffney,[75] Villela,[76] Ravi,[77] and Phillips,[78] while examining 0.001-in. diameter aluminum wires with 1% silicon impurities for failures under temperature cycling, found that the failures occurred due to repeated wire flexing resulting from different coefficients of thermal expansion between the aluminum wire and the header as the device heated up and cooled down. The maximum flexure and therefore the failure was found to occur at the thinned bond heels. The heel of the die bond was found to experience a greater temperature excursion and therefore was more prone to fail than the heel of the lead frame bond pad. Villela[76] conducted statistical tests with temperature-cycled devices and determined that aluminum ultrasonic bonds were more reliable under temperature cycling conditions than aluminum thermocompression bonds. Ravi[77] experimentally investigated the metallurgical flexure fatigue of a number of aluminum alloy wires and showed that aluminum wires with 0.1% magnesium was superior to that of the commonly used aluminum–1% silicon alloy. Phillips[78] calculated wire bond geometry effects and recommended that the loop height be at least 25% of the bond-to-bond spacing to minimize the bond flexure.

A wedge bonded wire is modeled as a beam under pure bending due to thermal expansion. The theory of curved beams[79] gives the maximum strain expression of a critical section at the heel of the wire[80]

$$\varepsilon = \left( \frac{R}{\rho \pm r} - 1 \right)\left( \frac{dR}{R} \right) \tag{22-10}$$

where $\rho$ is initial radius of curvature measured to the centroid of the cross-section, $r$ is the cross-sectional radius of the wire, and $R$ is the distance between the center of the curvature and the neutral axis of the wire. The positive or negative sign corresponds to the inner or outer portion of the wire toward the center of curvature. Generally, the neutral axis for a curved beam is not at the centroid of the cross-section. For a beam with a circular cross-section under pure bending, the relationship between $R$ and $\rho$ is expressed as[80]

$$R = \frac{\frac{1}{2}(r/\rho)^2}{1 - (1 - (r/\rho)^2)^{1/2}}\,\rho \tag{22-11}$$

In the case of wire bonds, $r < \frac{1}{5}\rho$ which corresponds to $R \approx 0.99\rho$, and, thus, the strain of the critical point at the heel is simplified from the above equation[80]

$$\varepsilon \approx \frac{r}{\rho}\left(\frac{d\rho}{\rho}\right) \tag{22-12}$$

The corresponding tensile stress is[80]

$$\sigma = E_w \frac{r}{\rho}\left(\frac{d\rho}{\rho}\right) \tag{22-13}$$

where $E_w$ is the modulus of elasticity of the material. Determination of the shape and radius of the curvature at the heels, $\rho$, has been solved numerically by a variational scheme for the wire with length of $2L$ bonded over a span of $2D$. The process of analysis and numerical results showed that the curvature at the heels and the span length can be described by an empirical power law relationship[80]

$$\frac{\rho}{D} = \frac{1}{6}\left(\frac{L}{D} - 1\right)^{-1/2} \tag{22-14}$$

The first-order derivative of the equation gives[80]

$$\frac{d\rho}{\rho} = \left(\frac{1}{D} + \frac{L}{2D(L - D)}\right)dD + \frac{1}{2(L - D)}\,dL \tag{22-15}$$

Substituting $dD = 2D\alpha_s\,\Delta T$, $dL = 2L\alpha_w\,\Delta T$, and for $\rho$ and $d\rho/\rho$ in the equation for stress gives the variation of bending stress at the heels of the wire due to thermal expansion during temperature cycles, $\Delta T$, as[80]

$$\sigma = 6E_w\frac{r}{D}\left(\frac{L}{D} - 1\right)^{1/2}\left(2\alpha_s + \frac{\alpha_s - \alpha_w}{1 - (D/L)}\right)\Delta T \tag{22-16}$$

where $E_w$ is the elastic modulus of the wire, $\alpha_w$ and $\alpha_s$ are the thermal expansion coefficients of the wire and the substrate (or the chip) materials, respectively, and $\Delta T$ is the temperature change encountered during operation. The number of cycles to failure, $N$, in flexure is related to the stress change in fatigue (Basquin's relation) and is[80]

$$N = C_w \sigma^{-m_w} \tag{22-17}$$

where $\sigma$ is the bending stress (already calculated), and $C_w$ and $m_w$ are fatigue properties determined by tensile fatigue tests of the wire material.

### 22.3.3 Wirebond Fatigue

A wirebond subjected to temperature change experiences shear stresses between the wire, the bond pad, and the substrate (or the chip) as a result of differential thermal expansion.[27,28,80] Occasionally when gold and aluminum are bonded together, intermetallics are formed at the interface due to the interdiffusion of gold and aluminum. Gold–aluminum intermetallics are stronger than pure metals and their growth is enhanced by increased temperature and temperature cycling.[27] The difference in the hardness and the material properties, including the coefficient of thermal expansion, between the intermetallic and the surrounding metal, make the interface a potential site for failure.

***Bond Shear***   As the temperature changes, bimetal bonds experience shear stresses as a result of differential thermal expansion between the wire and the bond pad or between the bond pad and the substrate. Since the bond pad is typically very thin compared to the substrate, it is assumed to act as an interlayer between the wire and the substrate, and the "shear lag" model[81] is used in this study as a first-order approximation. Ignoring the bending deformation in the thickness direction of the bond pad, the in-plane thermal deformation $d\delta$ in the $x$-direction for the wire is[80]

$$d\delta_w = \frac{P_w(x)\,dX}{E_w A_w} + \alpha_w \,\Delta T\, dX \tag{22-18}$$

and for the substrate strip, it is written as[80]

$$d\delta_s = \frac{P_s(x)(1 - v_s)\,dX}{E_w A_w} + \alpha_s \,\Delta T\, dX \tag{22-19}$$

where $P_w(x)$ and $P_s(x)$ are the mechanical forces in the $x$-direction as a result of the thermal expansion mismatch; $E_w$ and $E_s$ are modulus of elasticity of the wire and the substrate materials, respectively, and $v_s$ is Poisson's ratio for the substrate materials. $A_w$ is the cross-sectional area of the wire and $A_s$ is the effective cross-sectional area of the substrate, which is assumed to be equal to $b_s(W_p + W_s)/2$, where $W_p$ is the width of the bond pad. It is assumed that the in-plane normal stress and thermal deformations are uniform through the $z$-direction.

The thermal expansion mismatch between the wire and the chip is assumed to cause only shear deformation and shear stress in the bond pad in accordance with the simplification of "shear lag" theory. Assuming that half the diameter of the wire is contacted on the bond pad, the equilibrium conditions give[80]

$$dP_w - \tau r\, dx = 0, \qquad -l_w \leqslant x \leqslant l_w \tag{22-20}$$

$$dP_s - \tau W_p\, dx = 0, \qquad -l_w \leqslant x \leqslant l_w \tag{22-21}$$

where $\tau$ is the shear stress on two surfaces of the bond pad due to temperature change, and $r$ is the radius of the wire. The compatible shear strain and resulting stress in the bond pad is[80]

$$\gamma = \frac{(\delta_w - \delta_s)}{b_p} = \frac{\tau}{G_p}, \qquad l_w \leqslant x < l_w \tag{22-22}$$

where $G_p$ is the shear modulus of the bond pad material and $b_p$ is the thickness of the bond pad. The strain field is uniform in the thickness and width directions and thus reduces the problem to one dimension.

Differentiating the above equation and substituting for $d\delta_w$ and $d\delta_s$ gives[80]

$$\frac{d\tau}{dx} = \frac{G_p}{b_p}\left(\frac{P_w}{E_w A_w} - \frac{(1 - v_s)P_s}{E_s A_s} + (\alpha_w - \alpha_s)\,\Delta T\right) \tag{22-23}$$

Differentiating and using the equilibrium conditions, gives a second-order, homogeneous, ordinary differential equation[80]

$$\frac{d^2\tau}{dx^2} - Z^2\tau = 0, \qquad -l_w \leqslant x < l_w \tag{22-4}$$

where

$$Z^2 = \frac{G_p}{b_p}\left(\frac{r}{E_w A_w} + \frac{(1 - v_s)W_p}{E_s A_s}\right) \tag{22-25}$$

The solution of the one-dimensional differential equation, subject to the symmetric boundary condition that $\tau = 0$ at $x = 0$, is[80]

$$C \sinh(Zx) \qquad -l_w \leqslant x < l_w \tag{22-26}$$

where the constant $C$ can be determined from the boundary conditions at the ends of the wire and the bond pad. For a two-layer connection where $l_w \leqslant x < l_p$, compatibility conditions require that $d\delta_p = d\delta_s$, i.e.,[80]

$$\frac{P_p(x)\,dx}{E_p A_p} - \frac{P_s(x)(1 - v_s)\,dx}{E_s A_s} + (\alpha_p - \alpha_s)\,\Delta T\,dx = 0, \qquad l_w \leqslant x < l_p \tag{22-27}$$

where $E_p$ and $A_p$ are the modulus of elasticity and cross-sectional area of the bond, and $P_p$ is the mechanical force in the $x$-direction as a result of the thermal expansion mismatch. The equilibrium equation gives[80]

$$dP_p(x) + dP_s(x) = 0; \qquad P_p(x) + P_s(x) = c, \qquad l_w \leqslant x < l_p \tag{22-28}$$

where $c$ is a constant. At $x = l_p$, $P_p = P_s = 0$, which leads to $c = 0$,[80]

$$P_p(x) = -P_s(x), \qquad l_w \leqslant x \leqslant l_p \tag{22-29}$$

Substituting for $P_p$ into the compatibility conditions gives a constant value of uniaxial force, which provides a boundary condition[80]

$$P_s = \left( \frac{(\alpha_w - \alpha_s)\,\Delta T}{\dfrac{(1 - v_s)}{E_s A_s} + \dfrac{1}{E_p A_p}} \right), \qquad l_w \leqslant x \leqslant l_p \tag{22-30}$$

Because the wire is subjected to a pure bending moment at the heel, the value of $P_w(x)$ at $x = l_w$ is zero, i.e., $P_w(l_w) = 0$. Substituting $P_s$ and $P_w(l_w) = 0$ into the right side of the equation for $d\tau/dx$ and substituting $\tau = C \sinh(Zx)$, with the value of $x = l_w$ into the left side of the equation for $d\tau/dx$, gives[80]

$$CZ \cosh(zw) = \left( \frac{G_p}{b_p} \right) \left( (\alpha_w - \alpha_s) - \frac{(\alpha_s - \alpha_p)}{\left( 1 + \dfrac{(E_s A_s / E_p A_p)}{(1 - v_s)} \right)} \right) \Delta T \tag{22-31}$$

This equation is used to determine the constant $C$,[80]

$$\tau = \left(\frac{G_p \Delta T}{b_p Z}\right)\left((\alpha_w - \alpha_s) - \frac{(\alpha_s - \alpha_p)}{\left(1 + \dfrac{(E_s A_s / E_p A_p)}{(1 - v_s)}\right)}\right)\left(\frac{\sinh(zx)}{\cosh(zl_w)}\right) \qquad (22\text{-}32)$$

This equation shows that shear stresses in the pad is a function of position along the interface layer, and they are maximum at $x = l_w$ and decrease towards the center of the bond pad–wire, or bond pad–substrate interface. Notice that $Zl_w \gg 1$, and therefore $\tanh(Zl_w) \approx 1$; the amplitude of maximum shear stress due to the temperature cycles, $\Delta T$, at the critical point $x = l_w$ is determined as[80]

$$\tau_{\max} = Q\,\Delta T \qquad (22\text{-}33)$$

where

$$Q = \left(\frac{G_p}{b_p Z}\right)\left((\alpha_w - \alpha_s) - \frac{(\alpha_s - \alpha_p)}{\left(1 + \dfrac{(E_s A_s / E_p A_p)}{(1 - v_s)}\right)}\right) \qquad (22\text{-}34)$$

Once the maximum amplitude of the shear stress is determined, the numbers of cycles to the shear fatigue failure of the bond pad material due to differential thermal expansion between wire and chip can be predicted by Basquin's relation[80]

$$N = C_{p'}\,\tau_{\max}^{-m_{p'}} \qquad (22\text{-}35)$$

where $C_{p'}$ and $m_{p'}$ are the shear fatigue properties for the bond pad materials.

**Shear of Wire and Substrate**   In accordance with shear lag theory, it is assumed that the wire and substrate near the bonding layer are subjected to longitudinal normal stress. The normal traction in wire and substrate near the bondline can be determined by integrating equilibrium equations previously derived, and substituting for $\tau$ gives[80]

$$P_w = \frac{rQ\,\Delta T}{Z}\left(\frac{\cosh(Zx)}{\cosh(Zl_w)} - 1\right) \qquad (22\text{-}36)$$

and

$$P_s = \frac{W_p Q\,\Delta T}{Z}\left(1 - \frac{\cosh(Zx)}{\cosh(Zl_w)}\right) + \frac{(\alpha_s - \alpha_p)\,\Delta T}{\dfrac{(1 - v_s)}{(E_s A_s)} + \dfrac{1}{E_p A_p}} \qquad (22\text{-}37)$$

where $Q$ has been derived previously. The boundary conditions are obtained from the equation for $P_s$ and from the condition that $P_w(l_w) = 0$. From the above equations the average longitudinal normal stress in the wire and substrate is given by[80]

$$\sigma_w = \frac{P_w}{A_w} = \frac{rQ\,\Delta T}{ZA_w}\left(\frac{\cosh(Zx)}{\cosh(Zl_w)} - 1\right) \tag{22-38}$$

and

$$\sigma_s = \frac{P_s}{A_s} = \frac{W_pQ\,\Delta T}{ZA_s}\left(1 - \frac{\cosh(Zx)}{\cosh(Zl_w)}\right) + \frac{(\alpha_s - \alpha_p)\,\Delta T}{\dfrac{(1 - v_s)}{(E_sA_s)} + \dfrac{1}{E_pA_p}} \tag{22-39}$$

where $A_s = b_s(W_s + W_p)/2$ is the effective cross-sectional area as defined previously. This shows that the longitudinal normal stress has a maximum value at $x = 0$ in the bond pad–wire or bond pad–substrate interface and decreases to zero towards the free ends. Assuming that the normal stress components perpendicular to the interface are negligible, the maximum shear stress in the wire and substrate material are given according to the theory of elasticity:[80]

$$\tau_w(x) = (\tfrac{1}{4}\sigma_w(x)^2 + \tau(x)^2)^{1/2} \tag{22-40}$$

$$\tau_s(x) = (\tfrac{1}{4}\sigma_s(x)^2 + \tau(x)^2)^{1/2} \tag{22-41}$$

where $\tau(x)$, $\sigma_w(x)$, and $\sigma_s(x)$ have been calculated previously. As shown, the shear stress increases and longitudinal normal stress decreases towards the free ends. Hence, the maximum shear stress at each point along the interface layers is a function of $x$. The location of the maximum shear stresses can be determined by solving the following differential equations:[80]

$$\frac{d\tau_w(x)}{dx} = 0 \tag{22-42}$$

and

$$\frac{d\tau_s(x)}{dx} = 0 \tag{22-43}$$

Solving these two equations for $x$ gives the locations of largest maximum shear stress, $x_w$, in the wire, and $x_s$, in the substrate:[80]

$$x_w = (\pm)\,\text{arctanh}\left(\frac{A_w}{r}\right) \tag{22-44}$$

and

$$x_s = (\mp)\,\mathrm{arctanh}\left(\frac{A_s}{W_p}\right) \tag{22-45}$$

The magnitude of the largest maximum shear stress at these points is therefore determined by substituting $x_w$ and $x_s$ into the equations for $\tau_w(x)$, and $\tau_s(x)$, giving[80]

$$\tau_{w,M} = \left(\frac{r^2}{4Z^2 A_w^2}\left(\frac{\cosh(zx)}{\cosh(zl_w)}-1\right)^2 + \frac{\sinh^2(Zx_w)}{\cosh^2(Zl_w)}\right)^{1/2} Q\,\Delta T \tag{22-46}$$

$$\tau_{s,M} = \left[\left(\frac{W_p Q}{2Z^2 A_s^2}\left(1 - \frac{\cosh(Zx_w)}{\cosh(Zl_w)}\right) + \frac{(\alpha_s - \alpha_p)}{\dfrac{(1-v_s)}{(E_s A_s)} + \dfrac{1}{(E_p A_p)}}\right)^2 + Q^2\,\frac{\sinh^2(Zx_w)}{\cosh^2(Zl_w)}\right]^{1/2} \Delta T \tag{22-47}$$

The numbers of cycles to shear fatigue failure of wire and substrate materials can be predicted as (Basquin's relation)[80]

$$N = C_{w'}\,\tau_{w,M}^{-m_{w'}} \tag{22-48}$$

$$N = C_{s'}\,\tau_{s,M}^{-m_{s'}} \tag{22-49}$$

where $C_{w'}$, $m_{w'}$, $C_{s'}$, and $m_{s'}$ are the shear fatigue properties of wire material and substrate material.

### 22.3.4 Axial Fatigue of the Wire in Plastic-Encapsulated Devices

Another wirebond metallurgical failure mode was identified by Adams[82] for gold wire in plastic-encapsulated devices where axial stresses in the wire result from the differential expansion of the epoxy encapsulation and the wire. A typical case of metal fatigue was encountered when the device was made to undergo thermal cycling. Adams calculated that for a $\Delta T$ of 100°C the axial stress in the wire due to different expansion coefficients of the wire and plastic would almost equal the breaking load of the wire, assuming that the plastic was perfectly bonded to the wire. Repeated temperature fluctuations cause axial fatigue of the wire. Assuming that there is no slippage between the wire and encapsulant, the value of the forces presented in the wire, $P_w$, are just equal to those of the encapsulant, $P_e$, i.e.,

$$P_w = P_e \tag{22-50}$$

The total axial deformation in the wire and encapsulant is the deformation due to the rise of temperature, $\Delta T$, plus the deformation due to the mechanical forces experienced by the wire and the encapsulant. The compatibility condition requires that total deformations in the wire and encapsulant should be equal, which gives

$$\alpha_w \, \Delta T \, l_0 + \frac{P_w l_0}{E_w A_w} = \alpha_e \, \Delta T \, l_0 - \frac{P_e l_0}{E_e A_e} \tag{22-51}$$

where $l_0$ is the length of wire under consideration. Solving the equations for the force $(P_w)$ in the wire gives

$$P_w = \frac{(\alpha_e - \alpha_w) \, \Delta T}{\dfrac{1}{(E_w A_w)} + \dfrac{1}{(E_e A_e)}} \tag{22-52}$$

where $\alpha_e$ is the coefficient of thermal expansion, $E_e$ is the modulus of elasticity of encapsulant material, and $A_e$ is the cross-sectional area of encapsulant. Generally, for an encapsulant package, $A_e > 5 \, \text{mm}^2$, $A_w < 10^{-3} \, \text{mm}^2$, $E_w \approx 50 \, \text{GN/m}^2$, and $E_e \approx 10 \, \text{GN/m}^2$. Since $E_w A_w > 1000 E_e A_e$, the second term in the denominator can be neglected; the stress variation in the wire is therefore evaluated as

$$\sigma_w = E_w(\alpha_e - \alpha_w) \, \Delta T \tag{22-53}$$

Substituting into the equation for the number of cycles to failure gives

$$N = C_w \sigma_w^{-m_w} \tag{22-54}$$

where $C_w$ and $m_w$ are fatigue properties for the wire material.

### 22.3.5 Corrosion

Forrest[83] noted that another failure mechanism was that of corroded wirebonds. Corrosion opened one end of the wire completely and occasionally both ends of the wire, permitting the wire to move freely within the package volume and causing intermittent electrical short circuits. It was found that chlorine ions concentrated around wire bonds during the high-purity water rinse. Capillary action of the wirebond to water interface concentrated any dissolved chlorine at that point, causing the formation of $AlCl_3$ during elevated temperatures encountered during burn-in. Exposure of the conductor material to a chlorine environment caused a replacement chemical reaction, converting copper oxide to copper chloride at the substrate interface, the presence of which caused the Al wirebonds to corrode or develop high resistance intermetallics. The corrosion mechanism consists of the adsorption of $Cl^-$ on the oxide–solution interface under the influence of an electric field, caused by the electric double layer at the oxide–solution interface or the galvanic couple of the bond. The $Cl^-$ ion competes with the $OH^-$ and $H_2O$ molecules for surface sites on the hydrated surface. Aluminum hydroxide reacts with the chloride ions to form a basic hydroxy-chloride aluminum salt. The reaction is represented as[84,85]

$$Al(OH)_3 + Cl^- \rightarrow Al(OH)_2Cl + OH^- \tag{22-55}$$

Once the surface oxide is dissolved, the underlying Al reacts with the $Cl^-$ as[84,85]

$$Al + 4Cl^- \rightarrow Al(Cl)_4^- + 3e^- \tag{22-56}$$

The $Al(Cl)_4^-$ then reacts with the available water as[84,85]

$$2Al(Cl)_4^- + 6H_2O \rightarrow 2Al(OH)_3 + 6H^+ 8Cl^- \tag{22-57}$$

This reaction liberates a chloride ion, which is then available to continue the corrosion process by the reactions listed above. Gold on aluminum produces a galvanic couple that can accelerate the corrosion process by providing the driving force for the aluminum oxidation reaction. The bond corrosion may not directly result in a failure but increases the electrical bond resistance to a value that renders the device nonfunctional.

Bond corrosion in Au–Al bonds can also result from the liberation of bromine from brominated flame retardants at high temperatures. The mechanism proposed by Ritz included the liberation of bromine from $CH_3Br$ or HBr from the high-temperature breakdown products of the resin:

$$HBrCH_3Br \rightarrow CH_3 + Br^- + HBr$$
$$4HBr + 2O \rightarrow 4Br^- + 2H_2O \tag{22-58}$$

$Br^-$ reacts with the Al in $Au_4Al$ intermetallic phase forming $AlBr_3$ and Au. The aluminum bromide oxidizes and this oxidation reaction provides the driving force till the $Au_4Al$ intermetallic is consumed. The reactions are

$$Au_4Al + 3Br^- \rightarrow 4Au + AlBr_3$$
$$2AlBr_3 + 3O \rightarrow Al_2O_3 + 6Br^-$$

(22-59)

The bromide ions liberated in the above reactions sustain the reactions and completely corrode the bond pad.


### 22.3.6 Dendritic Growth

Another failure mechanism was that of silver dendrite growth from the wire bond pads of an integrated circuit. This is essentially an electrolytic process. It is observed in biased tests where the metal from the anode region migrates to the cathodic areas and forms dendrites there. The metal migration phenomenon leads to an increase in leakage current between the bridged regions or causes a short if complete bridging occurs (migrative resistance shorts, MGRS). Although Ag migration has been most widely reported, depending on environmental conditions many other metals like Pb, Sn, Ni, Au, and Cu have been found to undergo migration.[86] Being a time-dependent phenomenon, this is a wearout mechanism.

Metal migration is governed by availability of metal, presence of condensed water and ionic species, and the existence of a voltage differential. As the migration phenomenon is an electrolytic process, it is essential to have a conducting medium, i.e., water with dissolved ionic species. The ionic species could be impurities like chlorides or products generated during corrosion. Finally, the driving force necessary to cause metal migration is a potential difference. This condition exists in a biased condition of the package. To mitigate the phenomenon, metals known to show metal migration (Pb, Sn, Ni, Au, Cu, and Ag) should be exposed to the atmosphere.


### 22.3.7 Electrical Noise

Electrical leakage failure is the result of poor insulation from the Si substrate due to the lack of an MLO (multilayer oxide) layer underneath the bond pad. Bonds with no visible evidence of damage or mechanical weakness may have intermittent electrical leakage, especially in devices with an MLO-free bond pad. Failure analysis of these leakage failures showed that there were no cracks on pads using the same bonding conditions.

### 22.3.8 Vibration Fatigue

Harman[60] found that vibration forces that occur in the field are seldom severe enough to cause metallurgical fatigue or other bond damage. In general, large components of assembled systems were found to fail before such forces were sufficient to damage the bonds. Schafft[87] calculated the resonant frequency as well as centrifuge-induced forces for gold and aluminum wire bonds having various geometries. The minimum excitation frequency that might induce resonance and thus damage gold wire bonds having typical geometries was found to be in the range 3–5 kHz. For most aluminum wirebond geometries, the resonant frequencies required to damage the bonds were found to be greater than 10 kHz. Winchell[29,30] found that the natural frequency of vibration increased with the increase in the hardness of the aluminum wire.

Gold wirebonds have been found to fail due to vibration fatigue during ultrasonic cleaning.[60,88] The lowest frequency mode is the lateral mode (side to side). Maximum wire movement is experienced at this mode at the lowest frequency. The potential danger of wirebond failure occurs because of many wirebond configurations encountered in a package. A given package can contain bonds made with gold and/or aluminum wire with diameters ranging from 0.7 to 1.5 mils. The bond spacing may vary from 25 to 150 mils, and the loop height may also vary. When a large number of devices are being cleaned simultaneously, there is a possibility that reflections can cause energy concentrations in some areas, thus damaging only specific bond dimensions. Previously, ultrasonic cleaners have been designed with frequencies in the neighborhood of 20 kHz, which falls in the range of resonant frequency for wirebonds. More recent ultrasonic cleaners have resonant frequencies in a range of 20–100 kHz, making them unlikely to damage most wirebond geometries.

### 22.3.9 Resistance Change

This failure mechanism is typical of the Ag–Al bond system and is found to occur in dry air due to selective oxidation of the Ag–Al alloy. If the package absorbs water before soldering, soldering heat stress causes a peeling-off phenomenon of the wire ball from the substrate or chip. Shukla and Deo[8] found that silver metallization in CERDIPs failed catastrophically by high resistance change. The Ag–Al substrate bond system showed a high resistance, as a result of a thermally activated process that causes the resistance of the bond to change from negligible (0.1 ohm) to 20 ohms or higher. The mechanism was attributed to the selective oxidation of the Ag–Al intermetallic layer that resulted in an insulating oxide barrier at the interface.

The mechanism was activated at temperatures above 400°C. The resistance change is typically used to assess the long-term reliability of IC devices. Forrest[83] found that there is no discernible resistance change until a critical time is reached, at which time it rises in a dramatic manner to values ranging as high as 20 ohms or more. Mantese and Alcini[89] found that accelerated Al oxidation occurs as the temperature of the bond material is elevated, causing degradation of contact. Al melts at 660°C and oxidizes readily, making it unsuitable for devices that experience high temperature. Other parameters found to affect bond quality are the bonding time, ultrasonic power, tool length, tool wear, and type of substrate material.

## 22.4 WIREBOND INTERCONNECT DESIGN GUIDELINES

The aim of the wirebond design is to prevent the potential failure mechanisms of wire flexure fatigue, wire–bond-pad shear fatigue, bond-pad–substrate shear fatigue, interdiffusion and Kirkendall voiding, corrosion, dendritic growth, electrical noise, vibration fatigue, resistance change, and bond pad cracking. The inputs for the design process include chip technology (bipolar or CMOS); chip material and thickness; bond pad material, pitch, length, width, and thickness; clock frequency; output high and low voltages; maximum allowable interconnect resistance per unit length; designed-for output capacitive load; transistor conduction channel resistance; maximum allowable interconnect inductance; and mission profile loads.

The process of geometry design of wirebonded interconnects is divided into the design of wire diameter and wire aspect ratio. The minimum allowable wire diameter is based on maximum allowable resistance per unit length. The bond pad dimensions including the length, width, and pitch are used to calculate the maximum for interconnections. A plot of the resistance per unit length of the candidate wire materials versus wire diameters gives the allowable wire diameters for various materials.

The wire profiles are determined to satisfy noise requirements. The constraint on noise voltage is a function of the device technology. For example, bipolar devices can accommodate a noise of 0.4 V on an input of 5 V, while CMOS chips can take a noise of 1.06 V. The electrical parameters including the clock frequency, output high and low voltages, designed-for output capacitive load, transistor conduction channel resistance, and maximum allowable interconnector inductance are used to calculate noise voltages for various aspect ratios. The noise voltage due to simultaneous switching noise and signal integrity noise for various geometries are used to determine the allowable wire aspect ratios that satisfy the noise requirements for the packaged elements.

The geometry design of the wire is followed by the material selection for the wire. In this design step the undesirable material combinations based on design guidelines are eliminated from the list of candidate wire materials. Using the design guidelines, the designer at this step chooses a wire diameter, aspect ratio, and wire materials from the allowable ranges isolated. The design is then subjected to a series of checks using physics of failure models documented in the design guidelines, to ensure adherence to mission profile requirements.

### 22.4.1 Input Information and Design Goals

Table 22-3 summarizes the inputs to the guidelines for design of reliable wirebond interconnects. The outputs listed in the table are the design variables determined by the design guidelines.

**Table 22-3**   Wirebond Interconnect Design

| Inputs | Outputs |
| --- | --- |
| *Design invariants* | *Material parameters* |
| Chip material | Wire material |
| Chip thickness | |
| Chip technology (bipolar or CMOS) | *Geometry parameters* |
| Bond pad material | Allowable wire diameter range |
| Bond pad pitch | consistent with performance |
| Bond pad length and width | requirements |
| Clock frequency | Allowable wire aspect ratio |
| Output high and low voltages | consistent with performance |
| Designed-for output capacitive load | requirements |
| Transistor conduction channel resistance | |
| | |
| *Previously determined design variables* | |
| Bond span | |
| Maximum interconnect resistance per unit length | |
| Allowable junction temperature ($T_j$) | |
| Average junction temperature ($T_{j,\mathrm{av}}$) | |
| Average temperature cycle ($\Delta T_{\mathrm{average}}$) | |
| Required number of operational temperature cycles ($N_{f,r}$) | |
| Manufacturing capabilities | |
| Costs of materials and manufacturing processes | |

### 22.4.2 Wirebond Interconnect Design Process

This section presents the design guidelines for design of reliable wire bond interconnects. The subsections from A to F represent the design steps in the design process. The details of design decisions involved in each design step are discussed in each of the respective subsections. The guidelines in each design step are presented in *italics*. The physics of failure basis is discussed under each of the italicized design guidelines.

## A. Determination of Allowable Wire Diameters

1. *Any material that has a resistance per unit length of wire less than the required electrical value is a candidate wire material*

$$R_l < R_{l_{\text{required max.}}}$$

where the resistance per unit length, $R_l$, is

$$R_l = \frac{\rho}{A_{\text{wire}}} \tag{22-60}$$

where $\rho$ is the resistivity, and $A_{\text{wire}}$ is the cross-sectional area of the wire. This requirement establishes the minimum acceptable wire diameter for all candidate wire materials (Fig. 22-21). An example plot of resistance per unit length vs. wire diameter for common wire materials is shown in Fig. 22-21. Most wire materials have resistivity less than $3 \times 10^{-6}$ $\Omega$-cm.

2. *The wire diameter should not exceed 1/4 of the pad size in case of ball bonds and 1/3 of the pad size in case of wedge bonds.*

3. *The bond size should not exceed 3/4 of the pad size.*

## B. Determination of Allowable Wire Aspect Ratios

1. The interconnects have to be designed such that they do not degrade the pristine wave form coming from the on-die circuitry because distortion in the signal limits the rise time of the signals that can propagate in the interconnect environment, and increases the minimum possible clock period.

There are two sources to noise in the interconnects: simultaneous switching noise, and signal integrity noise.

***Simultaneous Switching Noise***  All CMOS logic devices operate by charging and discharging capacitive loads. The load capacitor makes the

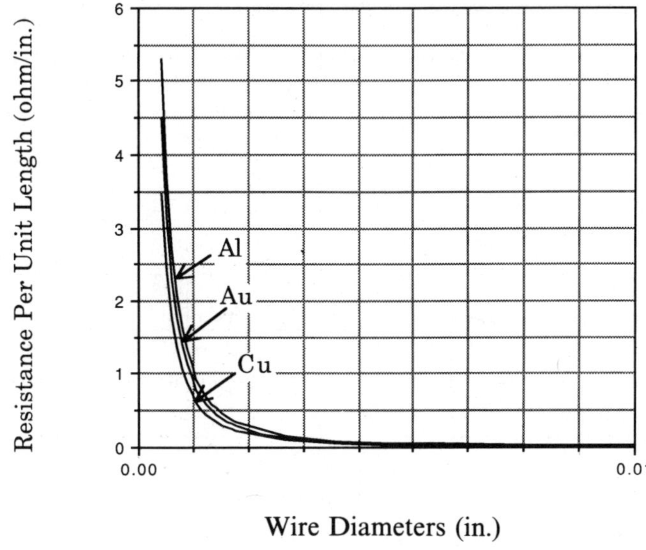



**Figure 22-21**    Allowable resistance per unit length vs. wire diameter.

instantaneous switching of logic levels impossible. The output structure on a die for a single logic high to low transition can be modeled by an *RC* network. The value of the resistor in the *RC* network is determined by the size of the output transistor. The capacitance is the load capacitance for which the die is optimized. The current flow and the interaction of this current with package parasitic impedances are determining factors in the performance degradation observed when several outputs are simultaneously switching on a multiple-output device. The package parasitics that have most effect are the self and mutual inductances in the bond wires. For a pair of parallel current-carrying bond wires,[142]

$$L_i = \frac{\mu l}{8\pi}$$

$$M = 5l\left[\ln\left(\frac{1}{d}\sqrt{1 + \left(\frac{l}{d}\right)^2}\right) - \sqrt{1 + \left(\frac{d}{l}\right)^2} + \frac{d}{l}\right]$$

(22-61)

where $\mu$ is the permittivity, $L$ is the self inductance, $M$ is the mutual inductance, $l$ is the length of the bond wires, and $d$ is the distance between the bond wires. The inductor has the property that a voltage is generated across it as the current through it changes, and for short interconnect lengths it is highly dependent on the location of the neighboring interconnects. The inductance of a wirebond can be 2 nH if the return path is an adjacent

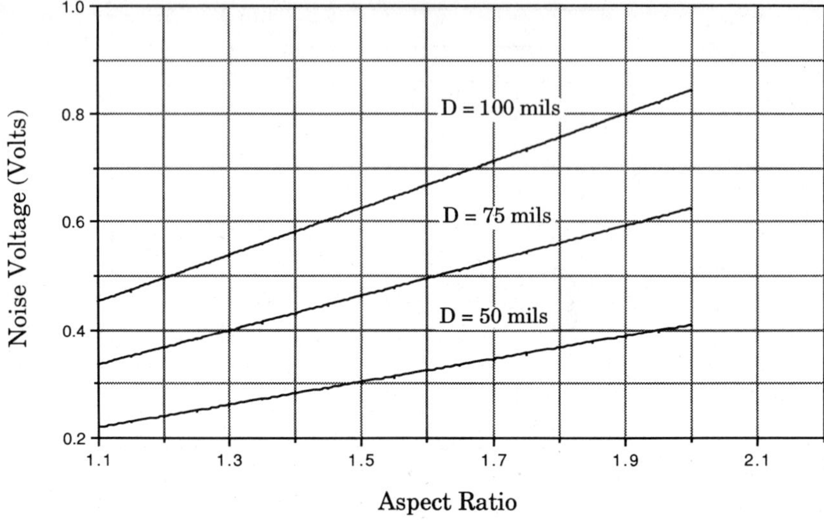


**Figure 22-22**   Simultaneous switching noise vs. aspect ratio (aspect ratio = wire length/bond span).

interconnect, and may be 20 nH if the return path is farther away. The inductance per length of the wire (when the adjacent wire is carrying current in the opposite direction) is defined by

$$L = 5 \ln\left(\frac{D}{r}\right) nH/\text{inch} \tag{22-62}$$

where $L$ is the inductance per unit length, $D$ is the distance between the two conductors, and $r$ is the radius of the wire.

After calculating the inductance, the designer should plot the simultaneous switching noise voltage ($V_L$) versus aspect ratio (wire length/bond span) as shown in Fig. 22-22. The allowable noise voltage for the element technology should then be used to calculate the allowable range of aspect ratios for the wire.

***Signal Integrity Noise***   Signal integrity noise is generated by short segments of nonuniform geometry that behave as *LRC* circuits. The inductance arises from bond wires and capacitance elements can be gate input capacitances or pads. If there is no series resistance the circuit will ring with a frequency of

$$f_{\text{ringing}} = \frac{1}{2\pi\sqrt{LC}} \tag{22-63}$$

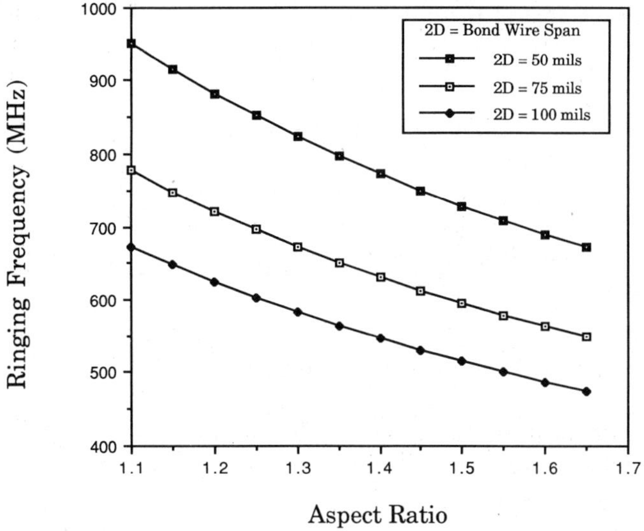


**Figure 22-23**    Signal integrity noise ringing frequency vs. aspect ratio (aspect ratio = wire length/bond span).

where $L$ is the inductance, $C$ is the die capacitance, and $f_{\mathrm{ringing}}$ is the ringing frequency. As the frequency components in the signal get closer to the ringing frequency, the interconnects more readily excite ringing signal integrity noise.

After calculating the inductance, the designer should plot the signal integrity noise ringing frequency versus aspect ratio (wire length/bond span) as shown in Fig. 22-23. The allowable range of aspect ratios for the wire should be such that the ringing frequency of the interconnect is far from the clock frequency of the circuit. If the ringing frequency is close to the clock frequency of the signal, the interconnects will excite ringing signal integrity noise.

2. *In plastic encapsulated packages, use low-dielectric-constant materials in the interconnect medium.* The signal that propagates from one gate to another in a circuit is the voltage difference between the signal line and the ground return line. There exists an electric field in the medium surrounding the conductors, which propagates down the line. The speed of propagation of the electric field is inversely proportional to the square root of the dielectric constant of the surrounding material of the interconnect:

$$V_{\mathrm{medium}} = \frac{V_{\mathrm{air}}}{\sqrt{\varepsilon_{\mathrm{medium}}}} \qquad (22\text{-}64)$$

where $V_{\mathrm{air}}$ is the velocity of the signal in air, $V_{\mathrm{medium}}$ is the propagation velocity of the signal in the medium, and $\varepsilon_{medium}$ is the dielectric constant of the medium.

## C. User-Specified Wire Diameter and Aspect Ratio

After the allowable ranges of wire diameters and wire aspect ratios have been identified, the user specifies a wire diameter and a wire aspect ratio in the allowable range. Further design steps for the wirebond interconnect are carried on with the user-specified dimensions.

## D. Identification of Allowable Wire Materials (see Table 22-4)

In this design step the designer identifies the allowable wire materials. The design guidelines are used to eliminate the unacceptable materials, and identify allowable materials.

1. *Wire and bond pad hardness should be reasonably matched*. A softer bond pad inhibits cratering by absorbing the energy from the bonding process. A harder bond pad, on the other hand, would readily transmit the energy from the bonding process to the substrate.[18,19]

2. *The bond pad material should be free of impurities to ensure good bondability*. Impurities are a major cause for the loss in the bondability of a surface and the premature bond failure during the operational life of the device. Impurities along with interdiffusion in wirebond can cause Kirkendall voids.[89] Impurities such as thallium, lead, or arsenic are added to the bath to increase plating speed and reduce grain size. Thallium and lead can cause premature wirebond failures during burn-in by accelerating cracks or Kirkendall-like void formation under the bond.[44,45] The contaminants that should be avoided to ensure good bondability, and their sources are outlined in Table 22-4.

3. Contamination of the bond pad can cause an improper setup of the bonding parameters, including the bonding energy required. Contaminated bond pads are found to require more ultrasonic energy and higher temperature for making strong bonds. Further, too high or too low a static bonding force can result in cratering in wedge bonds.

4. *Nickel, copper, and chromium should be used with caution on bonding surfaces*. Nickel degrades bondability and reduces reliability. Nickel may be introduced into the gold plating by accident or by thermally induced grain boundary diffusion upward from the thin underplating, during die attachment or heat treatment.[52,53] Nickel spreads on the surface through surface diffusion, oxidizes, and renders the surface unbondable. Copper from plating

**Table 22-4**   Impurities to be Avoided to Ensure Good Bondability and Their Possible Sources of Origin

| Impurity | Possible Source of Impurity |
| --- | --- |
| Halogens | Plasma etching, epoxy outgassing, silox etch, photoresist stripper, solvents (TCA, tetrachloroethylene, chlorofluoro compounds) |
| Thallium, lead, chromium, nickel, iron, copper, hydrogen | Plating |
| Sulfur | Packing containers, ambient air, cardboard and paper, rubber bands |
| Sodium, phosphorus, moisture, carbon, copper, bismuth, cadmium, tin, glass | Sources can be randomly distributed |

bath contamination, and lead frames, can follow the same route as nickel and oxidize on the surface, resulting in decreased bondability. Chromium is often used in microelectronic components to promote adhesion between substrates and vacuum-deposited gold films, and can rapidly diffuse through grain boundaries to the surface and oxidize, reducing bondability.

5. *Plating bath parameters for depositing the bonding surfaces should be controlled to prevent gas entrapments in the films.* Hydrogen bubbles in the film are often responsible for degradation of bondability. Hydrogen bubble concentration is a function of plating current density and bath impurity level. Excessive hydrogen entrapment can be caused by too high a current density, low bath concentration, or low bath agitation. Gas entrapments decrease the film bondability by increasing the ultrasonic energy absorption. These bubbles coalesce and grow by the process of structural rearrangement.[55] The gas bubbles are ruptured during bonding, preventing the surfaces from coming into intimate contact. Huettner and Sanwald[54] found that during bonding of gold-plated copper wire to gold films the lowest bondability occurred for plating currents in the range 1.6–2.7 A/dm$^2$, which corresponded to the onset of hydrogen evolution at the cathode and dendritic-like surface morphology. Hydrogen entrapment can be caused by too high a current density, low bath concentration, or low bath agitation. Gold films containing hydrogen are harder than pure films, but this hardness decreases rapidly during high-temperature annealing at 350°C for 8 hours.[55] However, free hydrogen inside the package has been found to inhibit the formation of

Au–Al intermetallic compounds by filling the vacancies in aluminum and its grain boundaries.[56]

6. *The bond pad material should not be able to form intermetallics with the wire material at operating temperatures* (information obtained from the mission profile). Interdiffusion shows up in wirebonds in the form of intermetallic compound formation and associated Kirkendall void formation. In gold–aluminum bonds, prolonged thermal exposure to temperatures higher than 175°C results in continued diffusion until gold or aluminum is consumed. Gold–aluminum intermetallics are stronger than pure aluminum or gold causing the bond to be most susceptible to flexure failure or brittle fracture in temperature cycling.[63,64] Intermetallic brittleness and growth is enhanced by temperature cycling. The growth rate of these compounds follows a parabolic relationship[65]

$$x = k_D t^{1/2}, \qquad k_D = D_0 \, e^{-E/kT} \tag{22-65}$$

where $x$ is the intermetallic layer thickness, $t$ is the time, $D_0$ is a constant, $E$ is the activation energy, $k$ is the Boltzmann constant, and $T$ is the absolute temperature (K). The diffusion rate is a function of the lattice defects. Thick-film metallizations and poorly welded joints with a large density of grain boundary defects, vacancies, dislocations, and free surfaces have much higher rates of diffusion than the bulk metal. For this reason poorly welded bonds or bonds to thick-film metallizations fail rapidly.[10,48–50]

7. *Gold–copper bonds should not be used for high-temperature applications.* Operating temperatures above 300°C accelerate the rate of Cu–Au intermetallic compound formation. Gold wires bonded to bare copper lead frames and to copper thick films in hybrids react, forming three intermetallic phases $Cu_3Au$, $AuCu$, and $Au_3Cu$. The "time to decrease bond strength" is found to decrease with an increase in temperature.[12] Temperature–time studies on thermocompression lead frame bonds indicate a decrease in strength as a result of void formation.[90–92] Cleanliness of the surfaces is extremely important to ensure good bondability in Cu–Au systems.

8. *Gold wirebonded to gold is very reliable.* Gold–gold bond is not subject to interface corrosion, intermetallic formation, or other bond-degrading conditions. Poorly welded gold–gold bonds improve with time and temperature.[9]

9. *Gold bonded to silver is very reliable for very long times at high temperatures.*[11] The gold–silver bond system does not form intermetallic compounds and does not exhibit interfacial corrosion. Presence of sulfur reduces the bondability of silver to gold films.[16,17]

10. *Silver–aluminum bond systems should be used with caution.* Ag–Al bonds have a tendency to degrade by interdiffusion, and oxidize in the presence of humidity. The Ag–Al substrate bond system showed a high resistance as a result of a thermally activated process that causes the resistance of bond to change from negligible (0.1 ohm) to 20 ohms or higher.[8] The mechanism was attributed to the selective oxidation of the Ag–Al intermetallic layer which resulted in an insulating oxide barrier at the interface. Typically in Ag–Al no discernible resistance change occurs until a critical time is reached, at which time it rises in a dramatic manner to values ranging as high as 20 ohms or more.[83] Accelerated Al oxidation occurs as the temperature of the bond material is elevated, causing degradation of contact.

Intermetallics are also responsible for electrical noise in the output of the circuit.[83] This type of bond exhibits high mechanical strength in conjunction with low conductivity due to formation of resistive compounds at the interface.

11. *In general, silver in bond systems should be used with extreme caution, after giving consideration to operating environment and device architecture.* Silver is the most widely reported material forming migrative resistive shorts (MGRS). This is essentially an electrolytic process. It is observed in biased tests where the metal from the anode region migrates to the cathodic areas and forms dendrites there. The metal migration phenomenon leads to an increase in leakage current between the bridged regions or causes a short if complete bridging occurs.

12. *Aluminum wires bonded to nickel coatings are reliable under various environments.* Harman[16,17] found that aluminum wires bonded to nickel coating (used as a substitute for gold) are reliable under various environments and are more reliable than Al–Ag or Al–Au bonds. Large diameter, > 75 µm (3 mil), aluminum wires bonded to nickel platings have been used in power devices.[16,17] Al–Ni bonds are more reliable than Al–Ag or Al–Au bonds. The Al–Ni phase diagram is more complex with various intermetallic phases. The system is well suited to high-temperature applications, and is less prone to Kirkendall voiding and galvanic corrosion.

13. *Aluminum wire bonded to aluminum metallization is extremely reliable.* Aluminum wire bonded to aluminum metallization is extremely reliable because it is not prone to intermetallic formation and corrosion. Aluminum wire on aluminum metallization welds best ultrasonically, even though a thermocompression bond can be produced by high deformation.

14. *Copper wires are preferable during plastic encapsulation because of their economy and their resistance to sweep* (tendency of the wire to move in the plane perpendicular to its length).[1–6] The resistance to sweeps results from

**Table 22-5**  Relative Reliability Rating of Wirebond Interface for Different Material Combinations

| Metallurgical System | Relative Reliability Rating |
| --- | --- |
| Gold–gold | 1 |
| Aluminum–aluminum | 2 |
| Gold–silver | 3 |
| Aluminum–nickel | 4 |
| Gold–aluminum | 5 |
| Gold–copper | 6 |
| Silver–copper | 7 |
| Aluminum–copper | 8 |
| Aluminum–silver | 9 |

the fact that copper is harder than gold, but greater attention is needed during the copper bonding operation to prevent cratering of the silicon die. The harder copper wire tends to push the softer metallization aside during the bonding operation.

15. *Wires and substrates stored in air for sometime before bonding should be bonded at higher than required temperatures and bake times.* Leads and metallized substrates stored in air for a few days prior to interconnection carry an adsorbed layer of water vapor and contaminants.

16. *An optimum selection of the bonding parameters is essential to making a good bond.* High bonding force, an excessive tool-to-substrate impact velocity, or too small a ball that allows the hard bonding tool to contact the metallization can induce cratering in thermocompression bonds.


**E. Fatigue Life Check**

1. *The life prediction estimate for bending stress at the heels of the wire due to thermal expansion during temperature cycles, $\Delta T$, should be greater than the required life from the mission profile, that is,*[80]

$$N_S > N_R \tag{22-66}$$

where $N_R$ is the number of cycles the component is expected to withstand (data from the mission profile). The fatigue life of the component can be empirically related to the flexure stress amplitude by the relation (Basquin's relation)

$$N = C_w \sigma^{-m_w} \tag{22-67}$$

Using the theory of curved beams, the stress amplitude can be estimated as

$$\sigma = 6E_w \frac{r}{D}\left(\frac{L}{D} - 1\right)^{1/2}\left(2\alpha_s + \frac{\alpha_s - \alpha_w}{1 - (D/L)}\right)\Delta T \qquad (22\text{-}68)$$

where $\alpha_w$ and $\alpha_s$ are the thermal expansion coefficients of the wire and the substrate materials, respectively, $\Delta T$ is the temperature change encountered during operation, and $C_w$ and $m_w$ are fatigue properties determined by tensile fatigue tests of the wire material.

2. *The life prediction estimate for the axial fatigue due to differential thermal expansion between the wire and the encapsulant should be greater than the required life from the mission profile,*

$$N_S > N_R \qquad (22\text{-}69)$$

where $N_R$ is the number of cycles the component is expected to withstand (data from the mission profile). The fatigue life of the component can be empirically related to the flexure stress amplitude by the following relation (Coffin–Manson relation):

$$N = C_w \sigma_w^{-m_w} \qquad (22\text{-}70)$$

Equating the stresses due to thermal expansion mismatch between the encapsulant and the wire, the stress amplitude can be estimated to be

$$\sigma_w = E_w(\alpha_e - \alpha_w)\,\Delta T \qquad (22\text{-}71)$$

where $C_w$ and $m_w$ are fatigue properties determined by tensile fatigue tests of the wire material.

3. *The life prediction estimate for the number of cycles to failure as a result of shear between the bond pad and the substrate under thermal cycling should be greater than the required life estimate from the mission profile, that is,*[80]

$$N_S > N_R \qquad (22\text{-}72)$$

where $N_R$ is the number of cycles the component is expected to withstand (data from the mission profile). The fatigue life of the component can be empirically related to the shear stress amplitude by the following relation (Coffin–Manson relation):

$$N_S = C_{p'}\tau_{\max}^{-m_{p'}} \qquad (22\text{-}73)$$

Assuming the bond pad to be a thin interlayer between the wire and the substrate, ignoring the bending deformation in the thickness direction of the bond pad, and equating the in-plane thermal deformation of the wire and the substrate, using the shear lag theory, the maximum value of shear stress between the bond pad and substrate can be estimated by the relations

$$\tau_{max} = Q\,\Delta T \tag{22-74}$$

and

$$Q = \left(\frac{G_p}{b_p Z}\right)\left((\alpha_w - \alpha_s) - \frac{(\alpha_s - \alpha_p)}{\left(1 + \dfrac{(E_s A_s/E_p A_p)}{(1 - \nu_s)}\right)}\right) \tag{22-75}$$

where $G_p$ is the shear modulus of the bond pad material; $\Delta T$ is the temperature cycle magnitude; $b_p$ is the bond pad thickness; $Z$ is as defined by Eq. (22-81); $\alpha_s$, $\alpha_p$, $\alpha_w$, are the thermal coefficients of expansion for the substrate, pad, and wire, respectively; $E_p$ and $E_s$ are modulus of elasticity of the pad and the substrate materials, respectively; $\nu_s$ is the Poisson ratio for the substrate materials; $A_p$ is the cross-sectional area of the pad; $A_s$ is the effective cross-sectional area of the substrate, which equals $b_s(W_p + W_s)/2$, where $W_p$ is the width of the bond pad; and $C_{p'}$ and $m_{p'}$ are the shear fatigue properties for the bond pad materials.

4. *The life prediction estimate for the number of cycles to failure as a result of shear between the wire and the bond pad under thermal cycling should be greater than the required life estimate from the mission profile, that is,*[80]

$$N_S > N_R \tag{22-76}$$

where $N_R$ is the number of cycles the component is expected to withstand (data from the mission profile). The fatigue life of the component can be empirically related to the shear stress amplitude by the following (Basquin's relation)

$$N_S = C_{w'}\,\tau_{w,M}^{-m_{w'}} \tag{22-77}$$

Assuming that the wire and substrate near the bonding layer are subjected to longitudinal normal stress, the normal traction in the wire and substrate near the bondline can be determined by integrating the equations of equilibrium at the wire–substrate interface. The shear stress under thermal expansion mismatch is then given by

$$N_S = C_{s'}\,\tau_{s,M}^{-m_{s'}} \tag{22-78}$$

where

$$\tau_{w,M} = \left( \frac{r^2}{4Z^2 A_w^2} \left( \frac{\cosh(zx)}{\cosh(zl_w)} - 1 \right)^2 + \frac{\sinh^2(Zx_w)}{\cosh^2(Zl_w)} \right)^{1/2} Q \, \Delta T \quad (22\text{-}79)$$

and

$$\tau_{s,M} = \left\{ \left[ \frac{W_p Q}{2Z^2 A_s^2} \left( 1 - \frac{\cosh(Zx_w)}{\cosh(Zl_w)} \right) + \frac{(\alpha_s + \alpha_p)}{\dfrac{(1 - \nu_s)}{(E_s A_s)} + \dfrac{1}{(E_p A_p)}} \right]^2 \right. $$
$$\left. + Q^2 \frac{\sinh^2(Zx_w)}{\cosh^2(Zl_w)} \right\}^{1/2} \Delta T \qquad\qquad (22\text{-}80)$$

where

$$Z^2 = \frac{G_p}{b_p} \left( \frac{r}{E_w A_w} + \frac{(1 - \nu_s)W_p}{E_s A_s} \right) \qquad\qquad (22\text{-}81)$$

where $E_p$ and $E_s$ are the modulus of elasticity of the pad and the substrate materials, respectively; $\nu_s$ is the Poisson ratio for the substrate materials; $A_p$ is the cross-sectional area of the pad; and $A_s$ is the effective cross-sectional area of the substrate, which equals $b_s(W_p + W_s)/2$, where $W_p$ is the width of the bond pad; $x_w$ gives the location of the maximum stress in the wire; $r$ is the radius of the wire; $\Delta T$ is the temperature cycle; $Q$ is as defined by Eq. (22-75); and $C_{w'}$, $m_{w'}$, $C_{s'}$, and $m_{s'}$ are the shear fatigue properties of wire material and substrate material.

### Satisfying Mission Life

If the fatigue life estimate obtained in the preceding design step is more than the mission life requirement from the design constraints, then this step marks the end of the design process. On the other hand, if the fatigue life estimate is less than the mission life requirement, the designer iterates the design process with longer wire lengths and different wire materials, until the mission life is satisfied.

## 22.5 SUMMARY

Thermal considerations in the manufacture, design, and operation of wire-bond interconnects have been addressed. The aspects of the bonding process discussed include cratering, pad cleanliness, intermetallic formation, metallizations, pad lifting, bonding parameters, geometries, and post-bond testing. Further, thermomechanical considerations during use have been outlined for intermetallic formation and Kirkendall voiding, wire flexure fatigue, wirebond fatigue, axial fatigue of the wire, corrosion, dendritic growth, electrical noise, vibration fatigue, and resistance change. The physics of failure models quantifying the concerns regarding the influence of temperature on wirebond interconnect reliability have been identified for both manufacture and operation.

Wirebond design guidelines have been presented for design of reliable wirebond interconnects for IC packages, hybrids, and multichip modules. The guidelines take into account environmental conditions, operational requirements in terms of clock frequency, propagation delay, signal rise times, noise, and maximum allowable inductance, and contractual requirements on mission life in terms of number of cycles to failure. The output from the design guidelines is in the form of allowable ranges of geometries and material properties.

## REFERENCES


1. Kurtz, J., D. Cousens, and M. Dufour, "Copper Ball Bonding," *34th IEEE Electronic Components Conference*, New Orleans, Louisiana, May 1984, pp. 1–5.
2. Hirota, J., K. Machida, T. Okuda, M. Shimotomai, and R. Kawanaka, "The Development of Copper Wire Bonding for Plastic Molded Semiconductor Packages," *35th Electronic Component Conference Proceedings*, Washington, DC, 1985, pp. 116–121.
3. Atsumi, K., T. Ando, M. Kobayashi, O. Usuda, "Ball Bonding Technique, For Copper Wire," *36th Proc. IEEE Electronic Components Conference*, Seattle, Washington, 1986, pp. 312–317.
4. Levine, L., and M. Schaeffer, "Copper Ball Bonding," *Semiconductor International*, August 1986, pp. 126–129.
5. Onuki, J., M. Koizumi, and I. Araki, "Investigation on Reliability of Copper Ball Bonds to Aluminum Electrodes," *IEEE Trans. Components, Hybrids, and Manufacturing Technology*, **CHMT-10**, 1987, pp. 550–555.
6. Riches, S. T., and N. R. Stockham, "Ultrasonic Ball/Wedge Bonding of Fine Cu Wire," *Proc. 6th European Microelectronic Conference (ISHM)*, Bournemouth, England, June 1987, pp. 27–33.
7. Hermansky, V., "Degradation of Thin Film Silver-Aluminum Contacts," *Fifth Czech. Conference on Electronics and Physics*, Czechoslovakia, October 1972, p. IIC-11.
8. Shukla, R., and J. Singh-Deo, "Reliability Hazards of Silver Aluminum Substrate Bonds in MOS Devices," *20th Annu. Proc. Reliability Physics Symposium*, San Diego, CA, 1982, pp. 122–127.
9. Jellison, J. L., "Susceptibility of Microwelds in Hybrid Microcircuits to Corrosion Degradation," *13th Ann. Proc. Reliability Physics Symposium*, Las Vegas, NE, 1975, pp. 70–79.

10. Kamigo, A., and H. Igarashi, "Silver Wire Ball Bonding and Its Ball/Pad Interface Characteristics," *35th Proc. IEEE Electronic Components Conference*, Washington, DC, 1985, pp. 91–97.

11. James, K., "Reliability Study of Wirebonds to Silver Plated Surfaces," *IEEE Trans. Parts, Hybrids, and Packaging*, **PHP-13**, 1977, pp. 419–425.

12. Hall, P. M., N. T. Panousis, and P. R. Manzel, "Strength of Gold Plated Copper Leads on Thin Film Circuits Under Accelerated Aging," *IEEE Trans. Parts, Hybrids, and Packaging*, **PHP-11**(3), 1975, pp. 202–205.

13. Pitt, V. A., and C. R. S. Needes, "Thermosonic Gold Wire Bonding to Copper Conductors," *IEEE Trans. Components, Hybrids, and Manufacturing Technology*, **CHMT-5**, 1982, pp. 435–440.

14. Lang, B., and S. Pinamaneni, "Thermosonic Gold Wire Bonding to Precious-Metal-Free Copper Leadframe," *38th Proc. IEEE Electronic Components Conference*, Los Angeles, CA, 1988, pp. 546–551.

15. Fister, J., J. Breedis, and J. Winter, "Gold Leadwire Bonding of Unplated C194," *20th Proc. IEEE Electronic Components Conference*, San Diego, CA, 1982, pp. 249–253.

16. Harman, G. G., *Reliability and Yield Problems of Wire Bonding in Microelectronics*, Technical Monograph of the ISHM, 1989.

17. Harman, G. G., and C. L. Wilson, "Materials Problems Affecting Reliability and Yield of Wire Bonding in VLSI Devices," *Proc. 1989 Materials Research Society, Electronic Packaging Materials Science IV*, Vol. 154, San Diego, CA, 1989; quoted from ref. 16.

18. Ravi, K. V., and R. White, "Reliability Improvement in 1-mil Aluminum Wire Bonds for Semiconductors," Final Report, Motorola SPD, NASA Contract NAS8-26636, December 6, 1971.

19. Ravi, K. V., and E. Philosky, "The Structure and Mechanical Properties of Fine Diameter Alumina–1% Si Wire," *Metallurgical Transactions*, **2**, March 1971, pp. 712–717.

20. Olsen, D. R., and K. L. James, "Evaluation of the Potential Reliability Effects of Ambient Atmosphere on Aluminum–Copper Bonding in Semiconductor Products," *IEEE Trans. Components, Hybrids, and Manufacturing Technology*, **CHMT-7**, 1984, pp. 357–362.

21. Thomas, R. E., V. Winchell, K. James, and T. Scharr, "Plastic Outgassing Induced Wirebond Failure," *27th Annu. Proc. IEEE Electronics Components Conference*, Arlington, VA, May 1977, pp. 182–187.

22. Nesheim, J. K., "The Effects of Ionic and Organic Contamination on Wirebond Reliability," *Proc. 1984 Int. Symposium on Microelectronics (ISHM)*, Dallas, TX, 1984, pp. 70–78.

23. Thomas, S., and H. M. Berg, "Micro-Corrosion of Al–Cu Bonding Pads," *23d Annu. Proc. Reliability Physics Symposium*, Orlando, FL, March 1985, pp. 153–158.

24. Totta, P., "Thin Films: Interdiffusion And Reactions," *J. Vacuum Science and Technology*, **14**, 1977, pp. 26.

25. Zahavi, J., M. Rotel, H. C. Huang, and P. A. Totta, "Corrosion Behavior of Al–Cu Alloy Thin Films in Microelectronics," *Proc. Int. Congress on Metallic Corrosion*, Toronto, Canada, June 1984, pp. 311–316.

26. Baker, J. D., B. J. Nation, A. Achari, and G. C. Waite, "On the Adhesion of Palladium Silver Conductors Under Heavy Aluminum Wire Bonds," *Int. J. Hybrid Microelectronics*, 1981, pp. 155–160.

27. Philosky, E. V., and K. V. Ravi, "On Measuring the Mechanical Properties of Aluminum Metallization and Their Relationship to Reliability Problems," *IEEE 11th Annu. Proc. Reliability Physics*, 1973, pp. 33–40.

28. Pecht, M., A. Dasgupta, and P. Lall, "A Failure Prediction Model For Wire Bonds," *Proc. 1989 Int. Microelectronic Symposium, ISHM*, 1989.

29. Winchell, V. H., "An Evaluation of Silicon Damage Resulting from Ultrasonic Wire Bonding," *14th Annu. Proc. Reliability Physics Symposium*, Las Vegas, NE, 1976, pp. 98–107.

30. Winchell, V. H., and H. M. Berg, "Enhancing Ultrasonic Bond Development," *IEEE Trans. Components, Hybrids, and Manufacturing Technology*, **CHMT-1**, 1978, pp. 211–219.

31. Koyama, H., H. Shiozaki, I. Okumura, S. Mizugashira, H. Higuchi, and T. Ajiki, "A Bond Failure Wire Crater in Surface Mount Device," *26th Annu. Proc. Reliability Physics*, Monterey, California, 1988, pp. 59–63.

32. Kale, V. S., "Control of Semiconductor Failures Caused by Cratering of Bond Pads," *Proc. 1979 Int. Microelectronics Symposium*, Los Angeles, CA, 1979, pp. 311–318.

33. Mori, S., H. Yoshida, and N. Uchiyama, "The Development of New Copper Ball Bonding-Wire," *38th Electronics Components Conference*, Los Angeles, CA, 1988, pp. 539–545.

34. Koch, T., W. Richling, J. Whitlock, and D. Hall, "A Bond Failure Mechanism," *Int. Reliability Physics Symposium 1986*, Anaheim, CA, 1986, pp. 55–60.

35. Ching, T. B., and W. H. Schreon, "Bond Pad Structure Reliability," *International Reliability Physics Symposium*, Monterey, CA, 1988, pp. 64–70.

36. Pramanik, D., and A. N. Saxena, *VLSI Metallization and its Alloys, Part I, Solid State Technology*, 1983, pp. 127–133.

37. Umemura, E., H. Onoda, and S. Madokoro, "High Reliable Al–Si Alloy/Si Contacts by Rapid Thermal Sintering," *26th Annu. Proc. Reliability Physics Symposium*, Monterey, CA, 1988, pp. 230–233.

38. Clatterbaugh, G.V., J. A. Weiner, and H. K. Charles, "Gold–Aluminum Intermetallics: Ball Bond Shear Testing and Thin Film Reaction Couples," *IEEE Trans. Components, Hybrids, and Manufacturing Technology*, **CHMT-7**, 1984, pp. 349–356.

39. Horsting, C. W., "Purple Plague and Gold Purity," *10th Annu. Proc. Int. Reliability Physics Symposium*, San Diego, CA, 1972, pp. 155–158.

40. Clatterbaugh, G. V., J. A. Weiner, H. K. Charles, and B. M. Romenesko, "Gold–Aluminum Intermetallics: Ball Bond Shear Testing and Thin Film Reaction Couples," *IEEE Trans. Components, Hybrids, and Manufacturing Technology*, **CHMT-7**, 1984, pp. 349–356.

41. McDonald, N. C., and P. W. Palmberg, "Application of Auger Electron Spectroscopy for Semiconductor Technology," *Int. Electron Device Meeting*, Washington, DC, October 1971, pp. 42.

42. McDonald, N. C., and G. E. Riach, "Thin Film Analysis for Processes Evaluation," *Electronic Packaging and Production*, 1973, pp. 50–56.

43. James, H. K., "Resolution of the Gold Wire Grain Growth Failure Mechanism in Plastic Encapsulated Microelectronic Devices," *IEEE Trans. Components, Hybrids, and Manufacturing Technology*, **CHMT-3**, September 1980, pp. 370–374.

44. Wakabayashi, S., A. Murata, and N. Wakobauashi, "Effects of Grain Refiners in Gold Deposits on Aluminum Wire-Bond Reliability," *Plating and Surface Finishing*, **V**, 1982, pp. 63–68.

45. Evans, K. L., T. T. Guthrie, and R. G. Hays, "Investigation of the Effect of Thallium on Gold/Aluminum Wirebond Reliability," *Proc. ISTFA*, Los Angeles, CA, 1984, pp. 1–10.

46. Endicott, D. W., H. K. James, and F. Nobel, "Effects of Gold Plating Additives on Semiconductor Wire Bonding," *Plating and Surface Finishing*, **V**, 1981, pp. 58–61.

47. English, A. T., and J. L. Hokanson, "Studies of Bonding Mechanisms and Failure Modes in Thermocompression Bonds of Gold Plated Leads to Ti–Au Metallized Substrates," *9th Annu. Int. Reliability Physics Symposium*, Las Vegas, NE, 1971, pp. 178–186.

48. Hund, T. D., and P. V. Plunkett, "Improving Thermosonic Gold Ball Reliability," *IEEE Trans. Components, Hybrids, and Manufacturing Technology*, **CHMT-8**, 1985, pp. 446–456.

49. Khan, M. M., T. S. Tarter, and H. Fatemi, "Aluminum Bond Pad Contamination by Thermal Outgassing of Organic Material from Silver Filled Epoxy Adhesives," *IEEE Trans. Components, Hybrids, and Manufacturing Technology*, **CHMT-10**, 1987, pp. 586–592.

50. Ahmad, S. S., "Impact of Residue on Al/Si Pads on Gold Bonding," *38th Proc. IEEE Electronic Components Conference*, Los Angeles, CA, 1987, pp. 534–538.

51. Harman, G. G., and K. O. Leedy, "An Experimental Model of the Microelectronic Ultrasonic Wire Bonding Mechanism," *10th Annu. Proc. Reliability Physics Symposium*, Las Vegas, NE, 1972, pp. 49–56.

52. Hall, P. M., and J. M. Morabito, "Diffusion Problems in Microelectronics Packaging," *Thin Solid Films*, **53**, 1978, pp. 175–182.

53. Loo, M. C., and K. Su, "Attach of Large Dice With Big Glass in Multilayer Packages," *Hybrid Circuits (UK)*, No. 11, September 1986, pp. 8–11.

54. Huettner, D. J., and R. C. Sanwald, "The Effect of Cyanide Electrolysis Products on the Morphology and Ultrasonic Bondability of Gold, Plating and Surface Finishing," August 1972, pp. 750–755.

55. Joshi, K. C., and R. C. Sanwald, "Annealing Behavior of Electrodeposited Gold Containing Entrapments," *J. Electronic Materials*, **2**(4), 1973, pp. 533–551.

56. Shih, D. Y., and P. J. Ficalora, "The Reduction of AuAl Intermetallic Formation and Electromigration in Hydrogen Environments," *16th Annu. Proc. Int. Reliability Physics Symposium*, San Diego, CA, 1978, pp. 268–272.

57. Newsome, J. L., R. G. Oswald, and W. R. Rodrigues de Miranda, "Metallurgical Aspects of Aluminum Wire Bonds to Gold Metallization," *14th Annu. Proc. IEEE Electronics Components Conference*, Las Vegas, NE, 1976, pp. 63–74.

58. Nehl, W., "Gold Wire Bonders For The Nineties," *Solid State Technology*, June 1991, pp. 59–62.

59. DiOrio, M., and T. Riesterer, "Gold Thermosonic Wire Bonding Technology in the 90s," *Electronics and Semiconductor Production and Packaging Conference; First Technical Conference on INTERNEPCON/Semiconductor*, Thailand, December 1991.

60. Harman, George G., "Metallurgical Failure Modes of Wirebonds," *12th Int. Reliability Physics Symposium*, 1974, pp. 131–141.

61. Ebel, G. H., "Failure Analysis Techniques Applied in Resolving Hybrid Microcircuit Reliability Problems," *15th Annu. Proc. Reliability Physics*, Las Vega, NE, 1977, pp. 70–81.

62. Hansen, M., "The Constitution of Binary Phase Diagrams," 2d edn., McGraw-Hill, New York, 1958.

63. Philosky, E., "Design Limits When Using Gold Aluminum Bonds," *9th Annu. Proc. Int. Reliability Physics Symposium*, Las Vegas, NE, 1971, pp. 11–16.

64. Philosky, E., "Intermetallic Formation in Gold–Aluminum Systems," *Solid State Electronics*, **13**, 1970, pp. 1391–1399.

65. Kidson, G. V., "Some Aspects of the Growth of Diffusion Layers in Binary Systems," *J. Nuclear Materials*, **3**(1), 1961, pp. 21–29.

66. Kashiwabara, M., and S. Hattori, "Formation of Al–Au Intermetallic Compounds and Resistance Increase for Ultrasonic Al Wire Bonding," *Review of the Electrical Communication Laboratory*, **17**, 1969, pp. 1001–1013.

67. Philosky, E., "Purple Plague Revisited," *8th Annu. Proc. Int. Reliability Physics Symposium*, Las Vegas, NE, 1970, pp. 177–185.

68. Horowitz, S. J., J. J. Felton, D. J. Gerry, J. R. Larry, and R. M. Rosenberg, "Recent Developments in Gold Conductor Bonding Performance and Failure Mechanisms," *Solid State Technology*, **22**, March 1979, pp. 37–44.

69. Pitt, V. A., C. R. S. Needes, and R. W. Johnson, "Ultrasonic Aluminum Wire Bonding to Copper Conductors," *Electronics Components Conference*, 1981, pp. 18–23.

70. Funamizu, Y., and K. Watanabe, "Interdiffusion In The Aluminum–Copper System," *Trans. Japan Institute of Metals*, **12**, May 1971, pp. 147–152.

71. Wallach, E. R., and G. J. Davies, "Mechanical Properties of Al–Cu Solid-Phase Welds," *Metals Technology*, **4**, April 1971, pp. 183–190.

72. Gershinskii, A. E., B. I. Formin, E. J. Cherepov, and F. L. Edelman, *Thin Solid Films*, **42**, 1977, pp. 269–275.

73. Campisano, S. U., E. Costanzo, F. Scaccianoce, and R. Cristofollini, *Thin Solid Films*, **52**, June 1978, pp. 97–101.

74. Johnson, D. R., *Plating in the Electronic Industry Symposium*, American Electroplating Society Inc., 1973, pp. 272.

75. Gaffney, J., "Internal Lead Fatigue Through Thermal Expansion in Semiconductor Devices," *IEEE Trans. Electronic Devices*, **ED-15**, 1968, pp. 617.

76. Villela, F., and M. F. Nowakomski, "Thermal Excursion Can Cause Bond Problems," *9th Annu. Proc. IEEE Reliability Physics Symposium*, 1971, pp. 172–177.

77. Ravi, K. V., and E. M. Philofsky, "Reliability Improvement of Wire Bonds Subjected to Fatigue Stresses," *10th Annu. Proc. IEEE Reliability Physics Symposium*, 1972, pp. 143–149.

78. Phillips, W. E., "Microelectronic Ultrasonic Bonding," National Bureau of Standards Special Publication 400-2, 1974, pp. 80–86.

79. Volterra, E., and J. H. Gaines, *Advanced Strength of Material*, Prentice-Hall, Englewood Cliffs, NJ, 1971.
80. Hu, J. M., M. Pecht, and A. Dasgupta, "A Probabilistic Approach for Predicting Thermal Fatigue Life of Wirebonding in Microelectronics," *ASME J. Electronics Packaging*, **113**, 1991, pp. 275–285.
81. Jones, R. M., *Mechanics of Composite Material*, McGraw-Hill, New York, 1975.
82. Adams, C. N., "A Bonding Wire Failure Mode in Plastic Encapsulated Integrated Circuits," *IEEE 11th Annu. Proc. Reliability Physics*, 1973, pp. 41–44.
83. Forrest, N. H., "Reliability Aspects of Minute Amounts of Chlorine on Wire Bonds Exposed to Pre-seal Burn-in," *Int. J. Hybrid Microelectronics*, **5**, 1982, pp. 549–551.
84. Paulson, W. M., and R. P. Lorigan, "The Effect of Impurities on the Corrosion of Aluminum Metallization," *14th Annu. Proc. Reliability Physics Symposium*, Las Vegas, NE, 1976, pp. 42–47.
85. Iannuzi, M., "Bias Humidity Performance and Failure Mechanisms of Non-Hermetic Aluminum SICs in an Environment Contaminated with $Cl_2$," *20th Annu. Proc. Reliability Physics Symposium*, San Diego, California, 1982, pp. 16–26.
86. Dumoulin, P., J. P. Seurin, and P. Marce, "Metal Migrations Outside the Package During Accelerated Life Tests," *IEE Trans. Components, Hybrids, and Manufacturing Technology*, **CHMT-5**(4), 1982, p. 479.
87. Schafft, H. A., "Testing and Fabrication of Wirebond Electrical Connections—A Comprehensive Survey," National Bureau of Standards, Technical Note 726, 1972.
88. Ramsey, T. H., "Metallurgical Behavior of Gold Wire in Thermal Compression Bonding," *Solid State Technology*, **16**, 1973, pp. 43–47.
89. Mantese, Joseph H., and William V. Alcini, "Platinum Wire Wedge Bonding: A New IC and Microprocessor Interconnect," *J. Electronic Materials*, **17**(4), 1988, pp. 285–289.
90. Pinnel, M. R., and J. E. Bennett, "Mass Diffusion in Polycrystalline Copper/ Electrodeposited Gold Planar Couples," *Metallurgical Transactions*, **3**, July 1972, pp. 1989–1997.
91. Feinstein, L. G., and J. B. Bindell, "The Failure of Aged Cu–Au Thin Films by Kirkendall Porosity," *Thin Solid Films*, **62**, 1979, pp. 37–47.
92. Feinstein, L. G., and R. J. Pagano, "Degradation of Thermocompression Bonds to Ti–Cu–Au and Ti–Cu by Thermal Aging," *29th Proc. Electronic Components Conference*, Cherry Hill, NJ, 1979, pp. 346–354.
93. Harman, G. G., and C. A. Cannon, "The Microelectronic Wire Bond Pull Test, How to Use It, How to Abuse It," *IEEE Trans. Components, Hybrids, and Manufacturing Technology*, **CHMT-1**, September 1978, pp. 203–210.
94. Albers, J. H., "The Destructive Bond Pull Test," NBS Special Publication, 1976, pp. 400–418.
95. Ang, A. H., and W. H. Tang, *Probability Concepts in Engineering Planning and Design*, Wiley, New York, 1984.
96. ASCE Committee on Fatigue and Fracture Reliability, "Fatigue Reliability," *J. Structural Division, ASCE*, **108**, No. ST1, 1982, pp. 1–104.
97. ASME Handbook, *Metal Properties*, McGraw-Hill, New York, 1954.

98. Casey, G. J., and D. W. Endicott, "Control of Surface Quality of Gold Electrodeposits Utilizing Auger Electron Spectroscopy, *Plating and Surface Finishing*, **67**(7), 1980, pp. 39–42.
99. Chevez, C., Westinghouse Electric Corporation, private communication, 1991.
100. Coucoulas, A., "Ultrasonic Welding of Aluminum Leads to Tantalum Thin Films," *Trans. Met. Soc. AIME*, 1966, pp. 587–589.
101. Cunnigham, J. A., "Expanded Contacts and Interconnections to Silicon Monolithic Integrated Circuits," *Solid State Electronics*, April 1965, pp. 735–745.
102. Devaney, J. R., "Failure Mechanisms in Active Device," *Electronic Materials Handbook*, Vol. 1, ASM International, 1989.
103. Ebel, G. H., J. A. Jeffery, and J. P. Farrell, "Wirebonding Reliability Techniques and Analysis," *IEEE Trans. Components, Hybrids, and Manufacturing Technology*, **CHMT-5**(4), 1982, pp. 441–445.
104. Fuchs, H. O., and R. I. Stephens, *Metal Fatigue in Engineering*, Wiley, New York, 1980.
105. Gale, R. J., "Epoxy Degradation Induced Au–Al Intermetallic Void Formation in Plastic Encapsulated MOS Memories," *22d Annu. Proc. Int. Reliability Physics Symposium*, Las Vegas, NE, 1984, pp. 37–47.
106. Gerling, W., "Electrical and Physical Characterization of Gold-Ball Bonds on Aluminum Layers," *34th Proc. IEEE Electronic Components Conference*, New Orleans, LO, May 1984, pp. 13–20.
107. Harper, C. A. (ed.), *Handbook of Components for Electronics*, McGraw-Hill, New York, 1977.
108. Harper, C. A. (ed.), *Handbook of Materials and Processes for Electronics*, McGraw-Hill, New York, 1970.
109. Hill, P., "Uniform Metal Evaporation," *Proc. Conf. Reliability of Semiconductor Devices and Integrated Circuits*, Vol. 2, Sect. 27, 1964, pp. 1–29.
110. Hinton, E., and D. R. J. Owen, *Finite Element Programming*, Academic Press, Orlando, FL, 1977.
111. Howell, J. R., and J. W. Slemmons, "Evaluation of Thermocompression Bonding Processes," presented to the 9th Welded Electric Packaging Association Symposium, Santa Monica, CA, 1964; Autonetics Report No. T4-240/3110, (13), March 1964, pp. 16, 19, 27, 28, 30.
112. Kawanobe, T., K. Miyamoto, M. Sieno, and S. Shoji, "Bondability of Silver Plating on IC Leadframe," *35th Proc. IEEE Electronic Components Conference*, Washington, DC, 1985.
113. Khan, M. M., and H. Fatemi, "Gold Aluminum Bond Failure Induced by Halogenated Additives in Epoxy Molding Compounds," *Proc. 1986 Int. Symposium on Microelectronics (ISHM)*, Atlanta, GA, October 1986, pp. 420–427.
114. Lall, P., D. Barker, A. Dasgupta, M. Pecht, and S. Whelan, "Practical Approaches to Microelectronic Package Reliability Prediction Modelling," *ISHM Proc.*, 1989, pp. 126–130.
115. Lall, P., M. Pecht, and A. Dasgupta, "A Failure Prediction Model for Wire Bonds," *IHM Proc.*, 1989, pp. 607–612.
116. Langhaar, H. L., *Energy Methods in Applied Mechanics*, Wiley, New York, 1962.

117. Lee, J. D., "Three Dimensional Finite Element Analysis of Damage Accumulation in Composite Laminate," *Computers and Structures*, **15**(3), 1982, pp. 335–350.

118. Lin, R., E. Blackshear, and R. Serisky, "Moisture Induced Package Cracking in Plastic Encapsulated Surface Mount Components During Solder Reflow Process," *26th IEEE Ann. Proc. Reliability Physics*, 1988, pp. 83–89.

119. Lipson, C., and J. Sheth, *Statistical Design and Analysis of Engineering Experiments*, McGraw-Hill, New York, 1973.

120. Lum, R. M., and L. G. Feinstein, "Investigation of the Molecular Processes Controlling Corrosion Failure Mechanisms in Plastic Encapsulated Devices," *30th Annu. Proc. IEEE Electronics Components Conference*, San Francisco, CA, 1980, pp. 113–120.

121. Majni, G., and G. Ottaviani, "AuAl Compound Formation by Thin Film Interactions," *J. Crystal Growth*, **47**, 1979, pp. 583–588.

122. Moore, Kevin D., "Interconnection Failures in Circuit Assemblies," *38th Electronic Component Conference*, Los Angeles, CA, pp. 521–526.

123. Nelson, G. C., and P. H. Holloway, "Determination of the Low Temperature Diffusion of Chromium Through Gold Films by Ion Scattering Spectroscopy and Auger Electron Spectroscopy," *ASTM Special Technical Publication 596, Surface Analysis Techniques*, 1976, pp. 68–77.

124. Zienkiewicz, O. C., *The Finite Element Method*, McGraw-Hill, New York, 1977.

125. Onishi, M., and K. Fukumoto, "Diffusion Formation of Intermetallic Compounds on Au–Al Couples by Use of Evaporated Al Films," *Jap. J. Met. Soc.*, 1974, pp. 38.

126. Panousis, N. T., and H. B. Bonham, "Bonding Degradation in Tantalum Nitride Chromium–Gold Metallization System," *11th Annu. Proc. IEEE Reliability Physics Symposium*, 1973, pp. 21–25.

127. Freudenthal, A. M., "Physical and Statistical Aspects of Fatigue," *Advances in Applied Mechanics*, Vol. 4, Academic Press, Orlando, FL, 1956, pp. 116–159.

128. Plunkett, P. V., and J. F. Dalporto, "Low Temperature Void Formation in Gold Aluminum Contacts," *32d Annu. Proc. IEEE Electronics Components Conference*, San Diego, CA, 1982, pp. 421–427.

129. Polcari, S. M., and J. J. Bowe, "Evaluation of Non-destructive Tensile Testing," Report No. DOT-TSC-NASA-71-10, June 1971, pp. 1–46.

130. Poonawala, M., "Evaluation of Gold Wirebond's in IC's Used in Cannon Launched Environment," *33rd Electronic Components Conference*, Orlando, FL, 1983, pp. 189–192.

131. Riches, J. W., O. D. Sherby, and J. E. Dorn, "The Fatigue Properties of Some Binary Alpha Solid Solutions of Aluminum," *Trans. ASM*, **44**, 1952, pp. 882–895.

132. Ritz, K. N., W. T. Stacy, and E. K. Broadbent, "The Microstructure of Ball Bond Corrosion Failures," *25th Annu. Proc. Int. Reliability Physics Symposium*, San Diego, CA, 1987, pp. 28–33.

133. Singh, M. P., and P. A. Heller, "Random Thermal Stress in Unclear Containments," *Proc. Conf. Probabilistic Mechanics and Statistical Reliability*, A. H. Ang, ed., ASCE, 1979, pp. 16–19.

134. Smith, J. R. H., "Fatigue Strength of Elevated Temperature of L65 Aluminum Alloy Notched and Lag Specimens," *Current Aeronautical Fatigue Problems*, Pergamon Press, Oxford, 1967, pp. 103–130.

135. Smith, J. M. et al., "Hybrid Microcircuit Tape Chip Carrier Materials/Processing Trade-offs," *IEEE Trans. Parts Hybrids Packaging*, **PHP-13**(3), September 1977, pp. 257–268.

136. Solomon, Harvey D., "Influence of Hold Time and Fatigue Cycle Wave Shape on the Low Cycle Fatigue of 60/40 Solder," *38th Electronic Components Conference*, Los Angeles, CA, 1988, pp. 7–12.

137. Thompson, R. J., D. R. Cropper, and B. Whitaker, "Bondability Problems Associated with the Ti–Pt–Au Metallization of Hybrid Microwave Thin Film Circuits," *IEEE Trans. Components, Hybrids, and Manufacturing Technology*, **CHMT-4**, 1981, pp. 439–445.

138. Villela, F., and M. F. Nowakowaski, "Investigation of Fatigue Problems in 1 Mil Diameter Thermocompression and Ultrasonic Bonding of Al-Wire," NASA TM-X-64566, 1970.

139. Wallance, W. E., and C. Sontz, "Introduction on MIL-STD-781D and MIL-HDBK-781," *IEEE Proc. Annual Reliability and Maintainability Symposium*, pp. 418–425, 1985.

140. Weaver, C., and L. C. Brown, "Diffusion in Evaporated Films of Gold–Aluminum, *Philosophical Magazine*, **7**, 1961, pp. 1–16.

141. Fung, Y. C., *Foundations of Solid Mechanics*, Prentice-Hall, Englewood Cliffs, NJ, 1965.

142. Grover, F. W., *Inductance Calculations: Working Formulas and Tables*, Van Nostrand, New York, 1946.

143. Sesham, K., and S. K. Ray, "Effects of Surface Roughness on Wire Bondability," *Proc. Int. Electronic Manufacturing Technology Symposium*, San Francisco, CA, 1986, pp. 109–113.

# 23

# Corrosion in Microelectronics Packages

*X. Shan and M. Pecht*

## 23.1 INTRODUCTION

The potential for failure in a microelectronic device due to corrosion has increased as the device metallization tracks have become narrower and thinner, the separation between metallization tracks has become closer, and bond pads have become smaller. Although corrosion problems are well documented in the literature and preventive measures are known, corrosion failures continue to occur. About 20% of electronic failures are caused by corrosion.[1,2] Most of the problems result from a violation of one or more of the fundamental design or corrosion-prevention guidelines, caused by improper material selection, improper design of geometry, or inadequate control of the manufacturing and assembly processes.

Corrosion is the deterioration of materials from their interaction with the environment through chemical or electrochemical mechanisms, resulting in the conversion of materials into oxides, salts, or other compounds. All materials corrode under certain conditions.

Corrosion occurs in most metals because they are not in their natural states. Most metals found in nature are in the form of an oxide, sulfide, or some other metal compound. Only a few metals, such as gold, exist in nature in the metallic form. All metals derived from ores have a tendency to return to the original form. Corrosive attack frequently occurs in combination with other mechanisms of failure, such as corrosion fatigue and stress corrosion cracking. Metals and their alloys that have undergone corrosion typically lose their strength, ductility, and other desirable mechanical and physical properties. Corrosion in microelectronic devices may result in an open circuit, a short circuit, or a change in an electrical parameter due to changes in resistance and capacitance.

The metal structures in microelectronics packages, such as electrical contacts to the semiconductor, interconnect lines, and bonding pads for wire or solder connections, typically consist of aluminum, gold, copper, or their alloys. The metal structures are usually coated with inorganic or organic protective films to provide both mechanical and chemical protection for the device. The passivating films, if defect-free, are effective in preventing corrosion of underlying thin-film resistors or conductors. However, localized structural defects in the protective coatings, or the presence of unprotected metal areas, such as the bonding wires and pads, make the metals vulnerable to corrosive attack. Bonding wires and pads cannot easily be passivated because of the assembly processes and are thus subjected to corrosion attack. Figures 23-1 through 23-6 illustrate corrosion phenomena in microelectronics packages.

Corrosion is generally a failure mechanism of concern in plastic-encapsulated packages where porosity, permeability, and inherent contamination are corrosion-accelerating factors. This, however, is not meant to

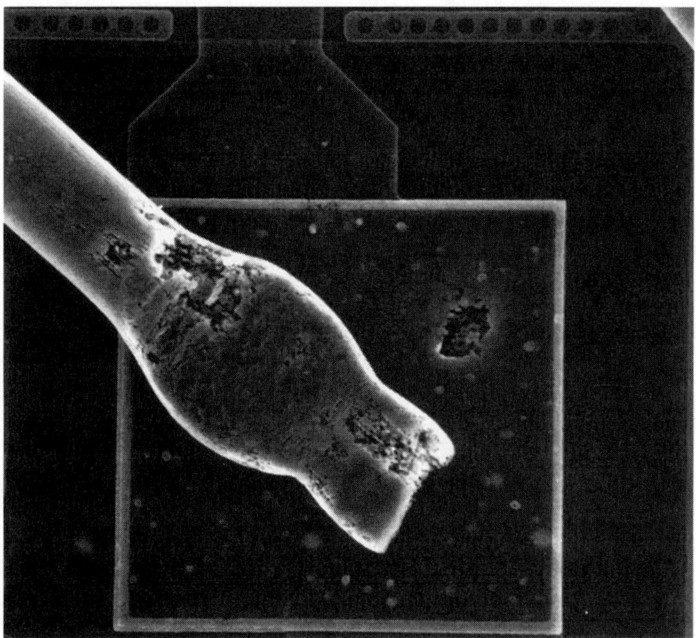

**Figure 23-1** SEM micrograph of bond pad and wire bond corrosion (courtesy of UNISYS-NASA).

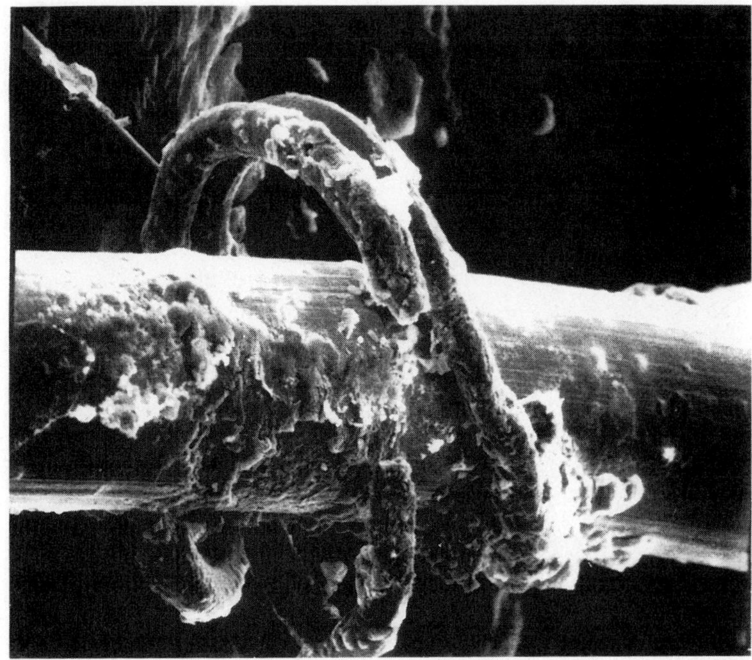

**Figure 23-2**   SEM micrograph of corrosion of relay coil interconnection wire joint (courtesy of UNISYS-NASA).

**Figure 23-3**   SEM micrograph of dendritic growth in SN 14, ×9620 (courtesy of UNISYS-NASA).

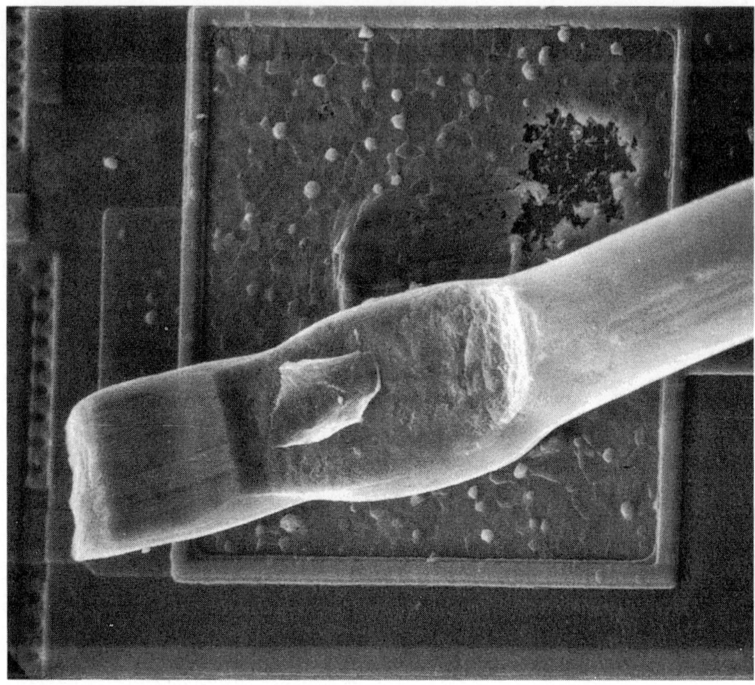

**Figure 23-4**  SEM micrograph of die bond pad corrosion (courtesy of UNISYS-NASA).

exclude metal and ceramic packages, which may, in fact, be more susceptible to corrosion than plastic packages because of sealed-in moisture and contaminants, increased handling, and moisture ingress through the lead–lead seal interface. In fact, a metal case itself corrodes if the plating has pinholes or scratches.

Corrosion may occur during manufacturing, storage, shipping, and service. Halogen ions ($Cl^-$, $Br^-$, etc.), potassium ion ($K^+$), hydrogen sulfide ($H_2S$), sulfur dioxide ($SO_2$), nitrogen compounds, and other airborne contaminants are the common contaminants. The sources of these contaminants and the methods of controlling them have been investigated and identified.[3] Studies show that part-per-billion levels of selected pollutants are sufficient under proper conditions of temperature and relative humidity to initiate and accelerate corrosion processes in microelectronics packages.[4]

Corrosion occurs under large variety of conditions. Numerous books have been devoted to the study of corrosion.[6–9] The interaction among electrical, metallurgical, and environmental conditions, and small dimensions lead to a unique set of corrosion problems for microelectronic packages. In this chapter, we overview the forms, reaction mechanisms, thermodynamics, and

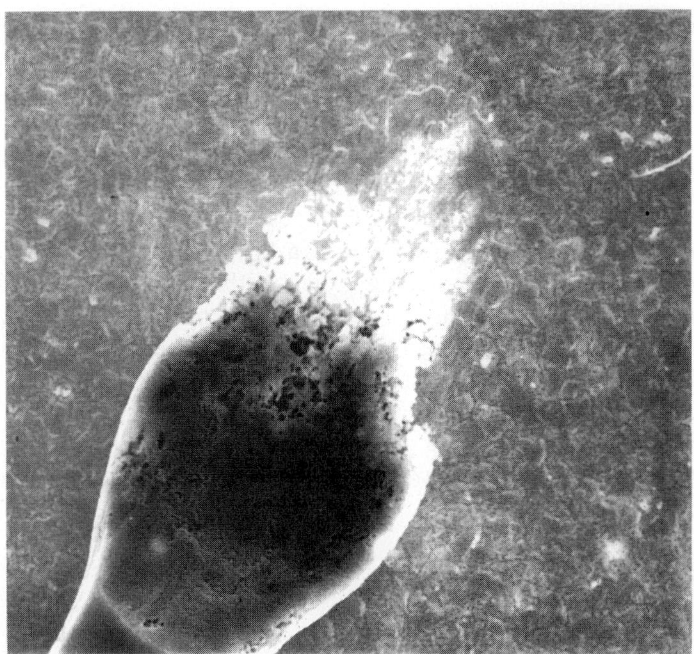

**Figure 23-5**    SEM micrograph of a corroded wire bond at the lead frame (courtesy of UNISYS-NASA).

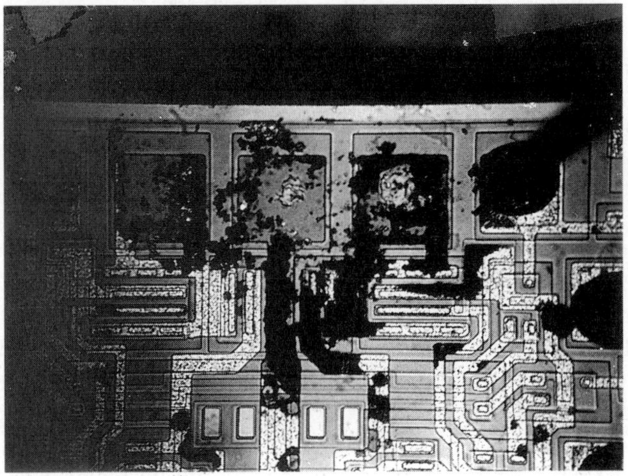

**Figure 23-6**    SEM micrograph of die metallization corrosion near $V_{dd}$ pad (courtesy of Edward Hakim).

kinetics of corrosion, discuss the governing equations for generic corrosion processes, and present general approaches to modeling and simulating various forms of corrosion in microelectronics packages.


## 23.2 FORMS OF CORROSION

The classification of corrosion forms is arbitrary because many forms of corrosion are interrelated, making exact distinction impossible. For the purpose of simplifying discussion, corrosion in microelectronic devices is classified into forms according to the reaction mechanisms and the visual characteristics of the morphology of the corrosion attack:

| | |
|---|---|
| Uniform corrosion | Intergranular corrosion |
| Galvanic corrosion | Selective leaching |
| Electrolytic corrosion | Stress corrosion fatigue |
| Crevice corrosion | Stress corrosion cracking |
| Pitting | Fretting |

Reactions in which the reactant gains electrons is known as cathodic reactions. Examples of cathodic reactions are

$$Al^{3+} + 3e^- \rightarrow Al \tag{23-1}$$

$$H^+ + e^- \rightarrow \tfrac{1}{2}H_2 \tag{23-2}$$

$$O_2 + 2H_2O + 4e^- \rightarrow 4OH^- \tag{23-3}$$

Reactions in which the reactant loses electrons are known as anodic reactions. Examples of anodic reactions are

$$Al \rightarrow Al^{3+} + 3e^- \tag{23-4}$$

$$Fe \rightarrow Fe^{3+} + 3e^- \tag{23-5}$$

$$2Cl^- \rightarrow Cl_2 + 2e^- \tag{23-6}$$

The electrode at which an electrochemical reduction occurs is defined as the cathode. The electrode at which an electrochemical oxidation occurs is defined as anode. The medium supporting ionic mass transport is known as the electrolyte. Ions such as $Na^+$, $Cu^+$, and $Al^{3+}$, which migrate toward the cathode when electricity flows through the cell, are called cations. Ions such as $Cl^-$ and $OH^-$, which migrate toward the anode when electricity flows through the cell, are referred to as anions.

Uniform corrosion, also called direct corrosive attack, is characterized by

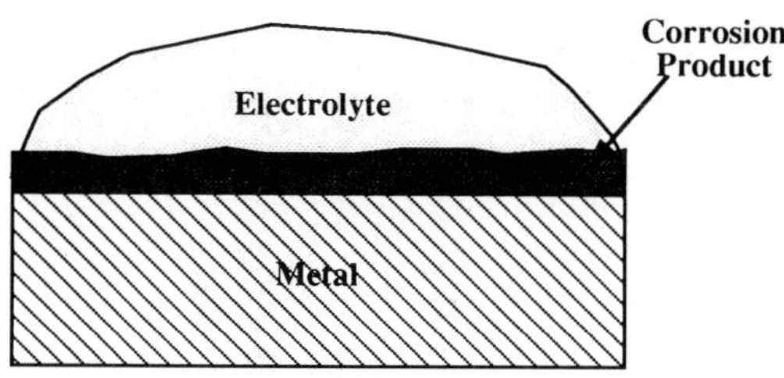



**Figure 23-7**   Uniform corrosion.

chemical or electrochemical reactions that proceed uniformly over the exposed surface (Fig. 23-7). For electrochemical reactions, the anodic and cathodic sites shift constantly so that corrosion spreads over the entire metal surfaces. The corroding metal becomes thinner and eventually fails. Weathering steels and copper alloys are examples of materials that typically exhibit general attack, while passive materials such as stainless steels or nickel-chromium alloys are generally subject to localized attack.

Galvanic corrosion is a form of electrochemical corrosion that occurs when two dissimilar metals are in contact in the presence of an electrolyte (Fig. 23-8). When two dissimilar metals in contact are electrically connected in an electrolyte, an electrochemical cell is formed. The more active metal, the one with the more negative potential, becomes the anode and is corroded by electrochemical reactions. The more noble electrode, the one with more positive potential, becomes a cathode. Galvanic corrosion may accelerate any other form of corrosion. Typically, the failure mode of galvanic corrosion is an open circuit. The reaction rate of the galvanic corrosion is dependent on temperature, contaminant concentrations, and electrochemical potential difference between the two metals in contact.

Electrolytic corrosion is a form of electrochemical corrosion that requires

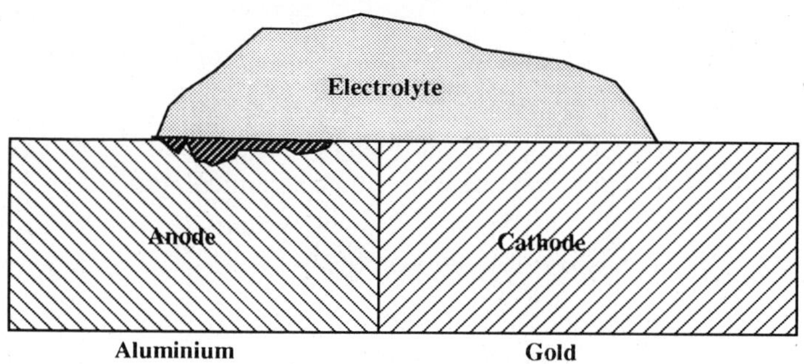



**Figure 23-8**   Galvanic corrosion.

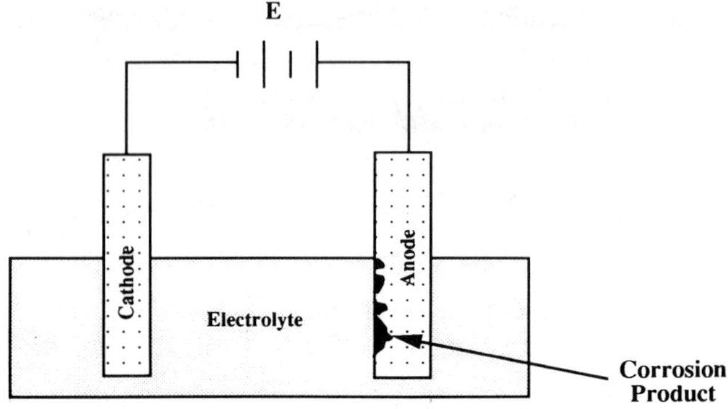



**Figure 23-9**   Electrolytic corrosion.

a potential gradient across an electrolyte between adjacent bond pads/ metallization tracks (Fig. 23-9). Corrosion of this type typically results in a short circuit (dendritic growth) or an open circuit (corrosion of metal). The process is accelerated by increasing temperature, contaminant concentration, and applied potential. Of all forms of corrosion, electrolytic corrosion is most prominent in microelectronic devices. In electrolytic corrosion, the cathode is the negative pole; the anode is the positive pole.

Crevice corrosion is the local attack in a crevice between metal-to-metal interface or between metal-to-nonmetal interface (Fig. 23-10). One side of the crevice is exposed to the corroding environment and the corroding ions must be in the crevice. Crevice corrosion is destructive because the damage is localized. The mechanism of crevice corrosion is illustrated with a metal in an aqueous solution. The overall reaction is the dissolution of metal M and the reduction of oxygen to hydroxide ions:

$$\text{Anode:} \qquad M \rightarrow M^{n+} + ne$$
$$\text{Cathode:} \qquad O_2 + 2H_2O + 4e \rightarrow 4OH^- \tag{23-7}$$

Initially, the anodic and cathodic reactions occur uniformly over the entire

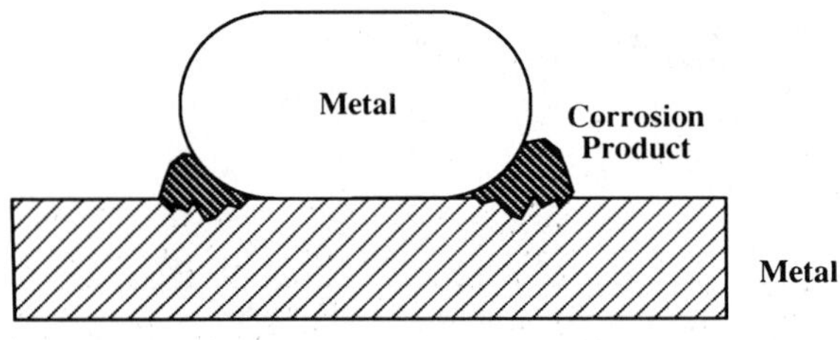



**Figure 23-10**   Crevice corrosion.

surface. Charge conservation is maintained in both the metal and the solution. When oxygen within the crevice is depleted because of the restricted convection, oxygen reduction stops in this area. No further oxygen reduction occurs, although the dissolution of metal M continues, producing an excess of positive charge in the solution. To balance the charge, negatively charged ions, for example, chloride ions, migrate into the crevice. As a result of hydrolysis, some form of acid is produced. The acid produced destroys the protective films or corrosion products, exposing more fresh metal for further corrosion attack. Crevice corrosion causes failures rapidly because of the autocatalytic nature of the process.

Pitting is a form of localized corrosion characterized by surface cavities, as shown in Fig. 23-11. Pitting is one of the most destructive forms of corrosion since only a small percentage weight loss may cause failures such as loss of hermeticity. The mechanism of pitting is similar to that of crevice corrosion in that it is autocatalytic, that is, the corrosive ions such as $Cl^-$ are both reactants and products. There are numerous theories of pitting. No complete agreement exists on the essential causes and mechanisms of pit initiation and growth. The theories were organized into two groups: adsorption theory and anion penetration and migration theory. Detailed descriptions of the theories of pitting corrosion can be found in ref. 10.

One of the characteristic features of pitting is the existence of a threshold value of the anodic potential below which pitting does not occur in the given corrosive environment. The most common and important type of pitting occurs on passivated metals in contact with chloride or other halide solutions, although other chemical species can also cause pitting. Pitting also occurs in nonpassivating metals and alloys with physical or chemical inhomogeneities or with defective protective coatings, or in certain heterogeneous low-conductivity corrosive media.

Intergranular corrosion is the attack that occurs at the grain boundaries of

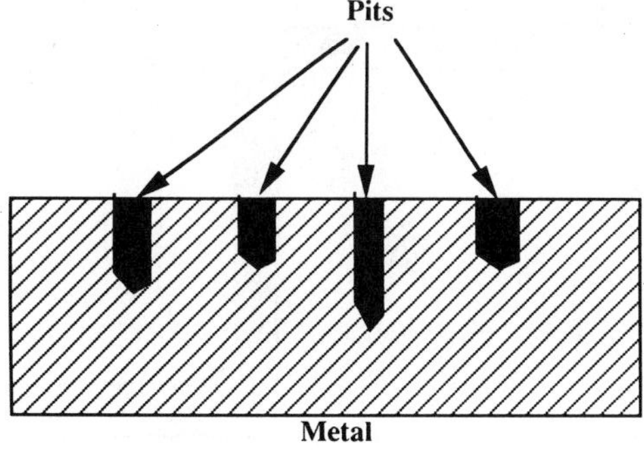


**Figure 23-11**   Pitting corrosion.

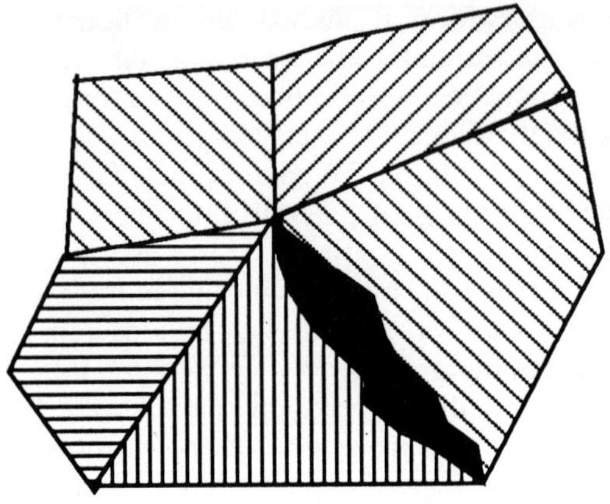

**Figure 23-12**    Intergranular corrosion.

metals and alloys. Improper heat treatment or welding causes metals to become highly susceptible to intergranular corrosion. Frequently, the grain boundaries are anodic to the main body of the grain so that selective corrosion of the grain boundaries occurs. Figure 23-12 illustrates intergranular corrosion.

Selective corrosion of alloy is the preferential dissolution of one component in an alloy. The overall results of selective corrosion is reduced mechanical strength and increased permeability because the degraded metal attains a porous structure (Fig. 23-13).

Corrosion fatigue is the reduction of fatigue resistance due to the presence of a corrosion medium and cyclic or fluctuating stresses. The fatigue resistance is reduced in the presence of corrosion because corrosion pits act as stress raisers and initiate cracks (Fig. 23-14).

Stress corrosion cracking (SCC) is the corrosion-induced cracking of a

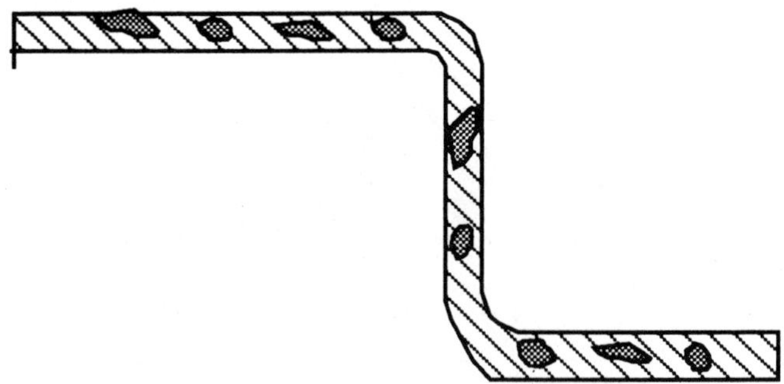

**Figure 23-13**    Selective corrosion (leaching).

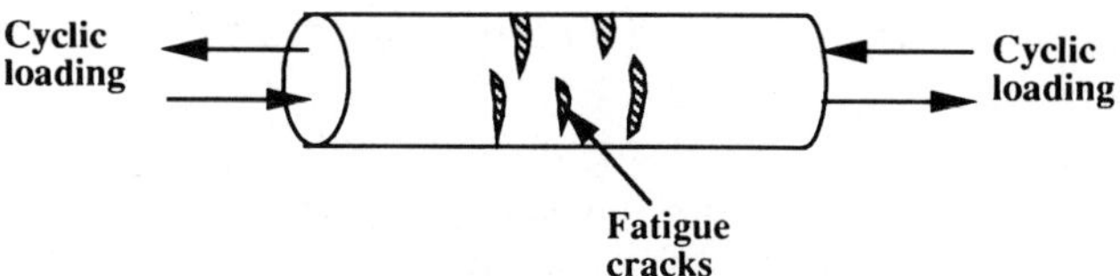



**Figure 23-14**   Corrosion fatigue.

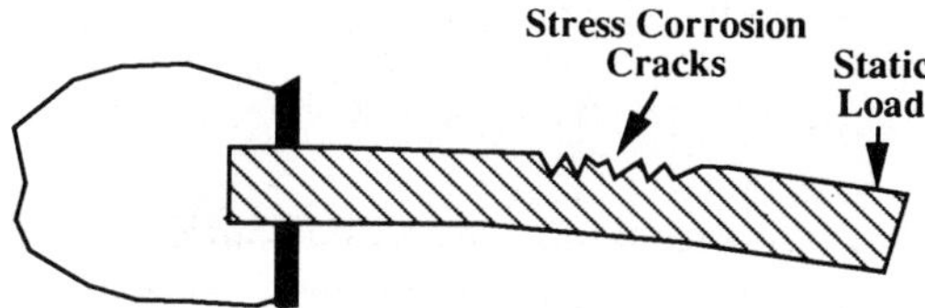



**Figure 23-15**   Stress corrosion cracking.

material under static stress, either applied or residual (Fig. 23-15). The resulting cracks are branched and can propagate transgranularly and intergranularly. Stress corrosion cracking produces greater deterioration in mechanical properties through the simultaneous action of static tensile stress and corrosion. Corrosion or mechanical stress acting separately will not produce the same degree of damage as the simultaneous actions of mechanical stress and corrosion. The stresses required to cause SCC are small, usually below the macroscopic yield stress, and are tensile stresses. Static loading is usually considered responsible for SCC, while environmentally induced crack propagation due to cyclic loading is defined as corrosion fatigue. Cold-working or straining, quenching, grinding, or welding may produce internal stresses.

Fretting corrosion is a combined wear and corrosion process in which material is removed from contacting surfaces even when motion between the surfaces is restricted to small-amplitude oscillation. Fretting corrosion occurs in connectors and socketed microelectronic packages. Factors affecting fretting include contact load, amplitude, number of cycles, relative humidity, and temperature. Fretting decreases the fatigue life of parts operating under fatigue loading. The overall effects from fretting corrosion are fatigue and structural failures.

## 23.3 THERMODYNAMICS OF CORROSION REACTIONS

This section introduces the fundamentals of thermodynamics and derives the thermodynamic relationships used to describe a corroding system.

Thermodynamics is useful in predicting whether a metal will corrode in a given environment.

The tendency for a chemical or electrochemical reaction to occur is measured by the change of Gibbs free energy of the reaction, $\Delta G$. The more negative the value of the free energy change, the greater the tendency for the reaction to occur.

From the first law of thermodynamics, the total change in internal energy, $\Delta U$, of a closed system is equal to the heat added to the system, $q$, minus the work done by the system, $w$:

$$\Delta U = q - w \tag{23-8}$$

The work is divided into mechanical, electrical, surface, and magnetic work. In electrochemical systems (corrosion systems), electrical and mechanical work are usually considered.

Consider a system that undergoes a reversible process at constant temperature and pressure in which both mechanical work and electrical work are done. The mechanical work at constant temperature and pressure is

$$w_m = \int_0^L \mathbf{F} \cdot d\mathbf{l} = \int_0^L (P\mathbf{S}) \cdot d\mathbf{l} = \int_0^V P \, dV = P \, \Delta V \tag{23-9}$$

The total work done by the system is the sum of mechanical and electrical work, $w_e$:

$$w = w_e + w_m \tag{23-10}$$

The reversible heat transferred to the system is given by

$$q = T \, \Delta S \tag{23-11}$$

where $\Delta S$ is the entropy change. The change in Gibbs free energy, $\Delta G$, is

$$\Delta G = \Delta H - T \, \Delta S \tag{23-12}$$

where $\Delta H$ is the enthalpy change and is given by

$$\Delta H = \Delta U + P \, \Delta V \tag{23-13}$$

Combining Eqs. (23-8) through (23-13) gives

$$\Delta G = -w_e \tag{23-14}$$

The maximum electrical work that can be done by the system is

$$w_e = qE = nFE \tag{23-15}$$

where $n$ is the number of electrons involved in the electrochemical reaction, and $F$ is Faraday's constant. Substituting Eq. (23-15) into Eq. (23-14) gives

$$\Delta G = -nFE \tag{23-16}$$

Equation (23-16) is a fundamental equation that relates electrochemical potential to the change of a thermodynamic state variable, $\Delta G$, the change of Gibbs free energy. According to the second law of thermodynamics, a spontaneous process at constant temperature and pressure results in a decrease in Gibbs free energy. Thus a positive potential is expected when the cell reaction is spontaneous.

The Gibbs free energy change at standard conditions is calculated from the free energy of formation, $\Delta G_{f,i}$ by

$$\Delta G^0 = \sum_i^n v_i \, \Delta G_{f,i} \tag{23-17}$$

where $v_i$ is the stoichiometric coefficient, which is positive for products and negative for reactants. A list of standard electrode potentials is given in Table 23-1.

The change of the Gibbs free energy at nonstandard conditions is given by

$$\Delta G = \Delta G^0 + RT \ln(\textstyle\prod a_i^{v_i}) \tag{23-18}$$

where $a_i$ is the activity of $i$th species. Combining Eqs. (23-16) and (23-18) gives

$$E = E^0 - \frac{RT}{nF} \ln(\textstyle\prod a_i^{v_i}) \tag{23-19}$$

Equation (23-19) is known as the Nernst equation. The Nernst equation relates the potential of an electrochemical cell to the standard potential and the concentrations (activities) of the chemical species involved in the electrochemical reaction. One common method of representing the Nernst equation is to plot reversible potential versus pH (Pourbaix diagram). Figure 23-16 illustrates a potential–pH diagram for the aluminum–water system.[39] Lines separating regions represent the potential and pH at which the various thermodynamically stable species are in equilibrium at unit activity.

**Table 23-1**   Standard Electrochemical Potentials

| Reaction | $E^0$ (V) | Reaction | $E^0$ (V) |
|---|---|---|---|
| $Fe^{3+} + 3e \rightleftharpoons Fe$ | $-0.037$ | $2H^+ + 2e \rightleftharpoons H_2$ | $0.0$ |
| $2Cu(OH)_2 + 2e \rightleftharpoons Cu_2O + 2OH^- + H_2O$ | $-0.080$ | $N_2 + 2H_2O + 6H^+ + 6e \rightleftharpoons 2NH_4OH$ | $0.092$ |
| $Pb^{2+} + 2e \rightleftharpoons Pb$ | $-0.126$ | $HgO + H_2O + 2e \rightleftharpoons Hg + 2OH^-$ | $0.0977$ |
| $CrO_4^{2-} + 4H_2O + 3e \rightleftharpoons Cr(OH)_3 + 5OH^-$ | $-0.130$ | $Pt(OH)_2 + 2e \rightleftharpoons Pt + 2OH^-$ | $0.14$ |
| $Sn^+ + 2e \rightleftharpoons Sn$ | $-0.138$ | $S + 2H^+ + 2e \rightleftharpoons H_2S(aq)$ | $0.142$ |
| $In^+ + e \rightleftharpoons In$ | $-0.140$ | $IO_3^- + 2H_2P + 4e \rightleftharpoons IO^- + 4OH^-$ | $0.15$ |
| $Cu(OH)_2 + 2e \rightleftharpoons Cu + 2OH^-$ | $-0.222$ | $Sn^{4+} + 2e \rightleftharpoons Sn^{2+}$ | $0.151$ |
| $Ni^{2+} + 2e \rightleftharpoons Ni$ | $-0.257$ | $Sb_2O_3 + 6H^+ + 6e \rightleftharpoons 2Sb + 3H_2O$ | $0.152$ |
| $PbCl_2 + 2e \rightleftharpoons Pb + 2Cl^-$ | $-0.268$ | $Cu^{2+} + e \rightleftharpoons Cu^+$ | $0.153$ |
| $Co^+ + 2e \rightleftharpoons Co$ | $-0.280$ | $AgCl + e \rightleftharpoons Ag + Cl^-$ | $0.222$ |
| $PbBr_2 + 2e \rightleftharpoons Pb + 2Br^-$ | $-0.284$ | Calomel electrode, saturated NaCl | $0.236$ |
| $Cu_2O + H_2O + 2e \rightleftharpoons 2Cu + 2OH^-$ | $-0.360$ | Calomel electrode, saturated KCl | $0.241$ |
| $Ti^{3+} + e \rightleftharpoons Ti^{2+}$ | $-0.368$ | $Cu^{2+} + 2e \rightleftharpoons Cu$ | $0.342$ |
| $Fe^{2+} + 2e \rightleftharpoons Fe$ | $-0.447$ | $Ag_2O + H_2 + 2e \rightleftharpoons 2Ag + 2OH^-$ | $0.342$ |
| $H_3PO_3 + 3H^+ + 3e \rightleftharpoons P + 3H_2O$ | $-0.454$ | $Cu^{2+} + 2e \rightleftharpoons Cu(Hg)$ | $0.345$ |
| $Fe(OH)_3 + e \rightleftharpoons Fe(OH)_2 + OH^-$ | $-0.56$ | $O_2 + 2H_2O + 4e \rightleftharpoons 4OH^-$ | $0.401$ |
| $PbO + H_2O + 2e \rightleftharpoons Pb + 2OH^-$ | $-0.580$ | $H_2SO_3 + 4H^+ + 4e \rightleftharpoons S + 3H_2O$ | $0.449$ |
| $Ni(OH)_2 + 2e \rightleftharpoons Ni + 2OH^-$ | $-0.720$ | $IO^- + H_2O + 2e \rightleftharpoons I^- + 2OH^-$ | $0.485$ |
| $Cr^{3+} + 3e \rightleftharpoons Cr$ | $-0.744$ | $NiO_2 + 2H_2O + 2e \rightleftharpoons Ni(OH)_2 + 2OH^-$ | $0.49$ |
| $Zn^{2+} + 2e \rightleftharpoons Zn$ | $-0.762$ | $Cu^+ + e \rightleftharpoons Cu$ | $0.521$ |
| $2H_2O + 2e \rightleftharpoons H_2 + 2OH^-$ | $-0.828$ | $Sb_2O_5 + 6H^+ + 4e \rightleftharpoons 2SbO^+ + 2H_2O$ | $0.581$ |
| $Cr^{2+} + 2e \rightleftharpoons Cr$ | $-0.913$ | $O_2 + 2H^+ + 2e \rightleftharpoons H_2O_2$ | $0.695$ |
| $Cr(OH)_3 + 3e \rightleftharpoons Cr + 3OH^-$ | $-1.480$ | $ClO_2^- + 2H_2O + 4e \rightleftharpoons Cl^- + 4OH^-$ | $0.760$ |
| $Al^{3+} + 3e \rightleftharpoons Al$ | $-1.662$ | $2No + H_2O + 2e \rightleftharpoons N_2 + 2OH^-$ | $0.760$ |
| $Sc^{3+} + 3e \rightleftharpoons Sc$ | $-2.077$ | $Fe^{3+} + e \rightleftharpoons Fe^{2+}$ | $0.771$ |
| $H_2 + 2e \rightleftharpoons 2H^-$ | $-0.230$ | $Ag^+ + e \rightleftharpoons Ag$ | $0.8$ |
| $Na^+ + e \rightleftharpoons Na$ | $-2.710$ | $2NO_3^- + 4H^+ + 2e \rightleftharpoons N_2O_4 + 2H_2O$ | $0.803$ |
| $Ca^{2+} + 2e \rightleftharpoons Ca$ | $-2.868$ | $Pd^{2+} 2e \rightleftharpoons Pd$ | $0.951$ |
| $K^+ + e \rightleftharpoons K$ | $-2.931$ | $Pt^{2+} + 2e \rightleftharpoons Pt$ | $1.118$ |
| $Ca^+ + e \rightleftharpoons Ca$ | $-3.80$ | $O_2 + 4H^+ + 4e \rightleftharpoons 2H_2O$ | $1.229$ |
| $2H^+ + 2e \rightleftharpoons H_2$ | $0.00$ | $Au^{3+} + 3e \rightleftharpoons Au$ | $1.498$ |

Potential–pH diagrams usually include the region of water stability (lines a and b). The potential–pH diagram is divided into three regions:

1. *Immunity.* The metal is thermodynamically stable and dissolution or corrosion cannot occur.
2. *Corrosion.* Ions of the metal are thermodynamically stable and corrosion will occur.
3. *Passivity.* A corrosion product is thermodynamically stable. The corrosion product forms a protective film on surfaces and prevent metals from further corrosion.

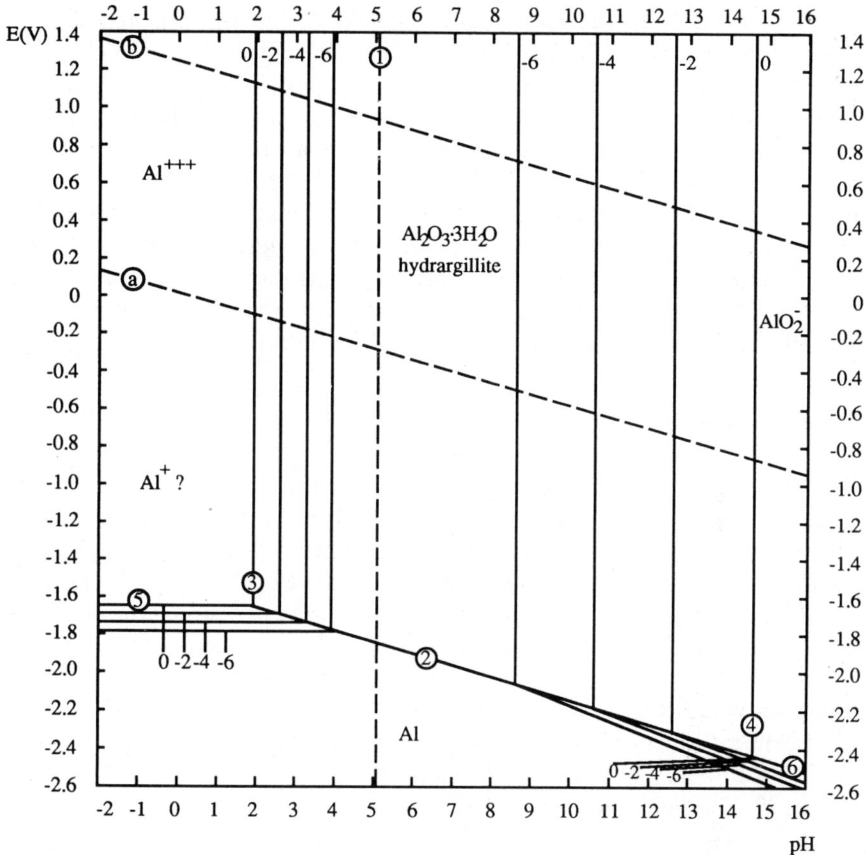



**Figure 23-16**  Potential–pH diagram for aluminum–water system.[39]

Potential–pH diagrams are practical tools that aid in corrosion studies. These diagrams show under what conditions corrosion may occur or may be prevented and thus provide guidance in the design of electronics packages. Because thermodynamics does not provide information on corrosion rate, experiments are required to determine whether corrosion reactions predicted to occur thermodynamically actually occur at a rapid enough rate that metal degradation becomes a problem.

The orderly arrangement of the standard reduction potentials versus a standard hydrogen–platinum electrode for metals is referred to as the electromotive force (EMF) series. The position of a metal in the EMF series is determined by the equilibrium potential of a metal in contact with its ions at unit activity. Of two metals composing a galvanic cell, the anode is the more active metal in the EMF series, provided that the ion activities in equilibrium are unity. The EMF series is limited in predicting which metal will corrode since the activities of ions vary greatly with the environment. Because of the limitations of the EMF series for predicting galvanic relations of metals, the galvanic series is established. The galvanic series is an orderly arrangement of metals and alloys according to their EMFs in a given environment. The ranking of a metal may be different from one electrolyte to another. Table 23-2 lists a galvanic series in seawater. In electronic design, the galvanic series in a given environment must be used instead of the standard EMF series. ASTM Committee G1 has prepared a Standard Guide G82 for the development and use of a galvanic series.[11]

## 23.4 KINETICS OF CORROSION REACTIONS

Corrosion rate depends not only on the potential difference (the driving force) between the anode and the cathode but also on the kinetics, the relative areas of two electrodes, and the rate of the mass transport. Thermodynamics provides information on the possibility or tendency of a corrosion process occurring but gives no information regarding the corrosion rate. Kinetics deals with how fast a reaction can proceed. This section introduces the fundamentals of kinetics of a corroding system and derives equations governing the corrosion reactions.

In chemical kinetics, the progress of a reaction is frequently described by plotting the energy of the species against the reaction coordinate as shown in Fig. 23-17. Electrochemical reactions are treated analogously, but the

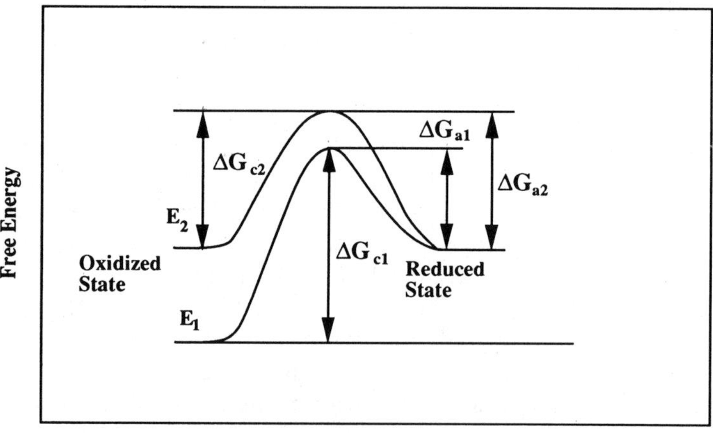


**Figure 23-17**    Reaction energy versus reaction coordinate.

**Table 23-2**   Galvanic Series in Sea Water

| | |
|---|---|
| **Anodic** ↑ | Magnesium alloys |
| | Beryllium |
| | Zinc |
| | Galvanized steel |
| | Aluminum alloys |
| | Chromium |
| | Galium |
| | Cadmium |
| | Mild steel |
| | Wrought iron |
| | Indium |
| | Low-alloy steels |
| | Cast iron |
| | Low-alloy cast iron |
| | 4 6^ Cr steel |
| | Ni cast iron |
| | 12–14 Chromium steel and 25–30% |
| | Lead–tin solders |
| | 16–18% Chromium steel |
| | Austenitic Cr–Ni stainless steel |
| | Austenitic Cr–Ni–Mo stainless steel |
| | Lead |
| | Tin |
| | Manganese bronze |
| | Naval brass |
| | Cobalt |
| | Nickel |
| | Inconel (13% Cr, 6.5% Fe, Bal. Ni) |
| | Yellow brass |
| | Admiralty brass |
| | Aluminum bronze |
| | Red brass |
| | Antimony |
| | Copper |
| | Silicon bronze |
| | Nickel silver |
| | 70–30% Copper nickel |
| | Titanium |
| | Monel |
| | Composition G bronze |
| | Composition M bronze |
| | Silver solder |
| | Nickel (passive) |
| | 70–30 Nickel–copper |
| | Stainless steels (passive) |
| | Silver |
| | Palladium |
| | Gold |
| | Rhodium |
| | Platinum |
| **Cathodic** ↓ | Carbon |

effects of potential on the reaction rate must be considered since the charge transfer process is involved. The following reaction, the reduction of species $O$ to species $R$ is used to derive the equation governing the electrochemical reaction at the electrode interface.

$$O^{n+} + ne^- \rightleftharpoons R \qquad (23\text{-}20)$$

To develop a quantitative description of the kinetics of a reaction, consider the case where we begin an experiment at the potential $E_1$ and change it to $E_2$ as shown in Fig. 23-17. The activation energy for the first cathodic and anodic processes are denoted by $\Delta G_{c1}$ and $\Delta G_{a1}$, respectively; $\Delta G_{c2}$ and $\Delta G_{a2}$ denote the second cathodic and anodic processes, respectively. Since the reaction is reversible, even though the energy has been raised by the amount $[-nF(E_2 - E_1)]$, only a fraction of this amount has been used to promote the cathodic reaction. If we denote this fraction as $\beta$, then

$$\Delta G_{c2} = \Delta G_{c1} + \beta nF(E_2 - E_1) \qquad (23\text{-}21)$$

where the subscript $c$ denotes a cathodic process. The activation energy for the anodic process is raised by the amount of $(1 - \beta)nF(E_2 - E_1)$:

$$\Delta G_{a2} = \Delta G_{a1} - (1 - \beta)nF(E_2 - E_1) \qquad (23\text{-}22)$$

where the subscript $a$ denotes the anodic process. Note that the potential $E_2$ is more negative than the potential $E_1$. The fraction $\beta$ is commonly known as the asymmetry factor and varies between 0 and 1. It is the fraction of the energy promoting the cathodic reaction. The experimental values are generally close to 0.5.

The form of the reaction rate constant used is the same as that for chemical reactions:

$$k = k^0 \exp\left(-\frac{\Delta G}{RT}\right) \qquad (23\text{-}23)$$

where $\Delta G$ is the free energy of activation and $k^0$ is a rate constant. The rate of an electrochemical reaction is directly proportional to the current density as shown below:

$$r = \frac{i}{nF} = k^0 c^v \exp\left(-\frac{\Delta G}{RT}\right) \qquad (23\text{-}24)$$

where $r$ is the reaction rate, $c$ is the reactant concentration, and $v$ is the coefficient of the reactant. Substituting Eq. (23-21) into Eq. (23-24) gives

$$r_c = k_c c_o \exp\left(\frac{-\beta n F E}{RT}\right) \qquad (23\text{-}25)$$

where $E = E_2 - E_1$ and

$$k_c = k_c^0 \exp\left(\frac{-\Delta G_{c1}}{RT}\right) \qquad (23\text{-}26)$$

Similarly, the rate equation for the anodic process is

$$r_a = k_a c_r \exp\left[\frac{(1 - \beta) n F E}{RT}\right] \qquad (23\text{-}27)$$

The anodic current is conventionally defined as positive in electrochemistry. The net current density is the difference between the current density for the anodic process and the current density for the cathodic process:

$$i = i_a - i_c$$

$$= nF\left\{k_a c_r \exp\left[\frac{(1 - \beta) n F E}{RT}\right] - k_c c_o \exp\left(\frac{-\beta n F E}{RT}\right)\right\} \qquad (23\text{-}28)$$

At equilibrium the net current density is zero while the current densities for the anodic and cathodic processes are not. The magnitude of this equal and opposite current density is called the exchange current density, $i_0$:

$$i_0 = nF k_c c_o \exp\left(\frac{-\beta n F E_0}{RT}\right)$$

$$= nF k_a c_r \exp\left[\frac{(1 - \beta) n F E_0}{RT}\right] \qquad (23\text{-}29)$$

Because the net current at equilibrium is zero, Eq. (23-18) becomes

$$nF k_a c_r \exp\left[\frac{(1 - \beta) n F E_0}{RT}\right] - nF k_c c_o \exp\left(\frac{-\beta n F E_0}{RT}\right) = 0 \qquad (23\text{-}30)$$

Rearranging Eq. (23-30) gives

$$E_0 = \frac{RT}{nF} \ln\left(\frac{k_c}{k_a}\right) - \frac{RT}{nF} \ln\left(\frac{c_r}{c_o}\right)$$

$$= E^0 - \frac{RT}{nF} \ln\left(\frac{c_r}{c_a}\right) \tag{23-31}$$

where $E_0$ is the potential at equilibrium and $E^0$ is the potential at standard state (unit activity, 25°C, and 1 atm). As we expected, the kinetic equation collapses into the thermodynamic equation, the Nernst equation (23-19), when the reaction is at equilibrium.

The overpotential, the departure of the electrode potential from the equilibrium potential, is customarily used in kinetic studies. Substituting $\eta = E - E_0$ and Eq. (23-31) into Eq. (23-18) and rearranging gives

$$i = i_0 \left[ \exp \frac{\alpha_a F}{RT} \eta - \exp \frac{-\alpha_c F}{RT} \eta \right] \tag{23-32}$$

where

$$i_0 = nF k_c^{(1-\beta)} k_a^{\beta} c_o^{(1-\beta)} c_r^{\beta} \tag{23-33}$$

$$\alpha_a = (1 - \beta)n \tag{23-34}$$

$$\alpha_c = \beta n \tag{23-35}$$

Equation (23-32) is the Butler–Volmer equation used to describe the current–potential relationship for an electrode at a given temperature, pressure, and concentration of species involved in the reaction. The $\alpha_a$ and $\alpha_c$ are the transfer coefficients. The Butler–Volmer equation is often used in a strictly empirical way to analyze corrosion data. When the overpotential, $\eta$, is sufficiently large, one of the exponential terms in Eq. (23-32) will be negligible compared with the other term. For example, when the overpotential is large and positive, the cathodic component of the current is negligible. The total current is

$$i = i_0 \exp \frac{\alpha_a F}{RT} \eta \tag{23-36}$$

Equation (23-36) is known as the Tafel equation, discovered empirically by Tafel.

## 23.5  GENERIC CORROSION PROCESSES AND REACTION MECHANISMS OF CORROSION

Understanding the fundamental corrosion processes and corrosion mechanisms is essential in corrosion modeling. Figure 23-18 shows the generic corrosion processes in a plastic package. Moisture and contaminants get into a metal or ceramic package through different mechanisms, although the other corrosion steps are similar. Plastic packages are used to illustrate the approach to corrosion modeling in this paper.

The generic corrosion processes include the adsorption of the reactants to the plastic, diffusion of reactants (contaminants and moisture) from the environment to the electrolyte–plastic interface, desorption, diffusion and/or migration from the bulk electrolyte to the surface film, diffusion and/or migration across the surface film to the electrode interface, adsorption and reaction, diffusion and/or migration of products from the electrode interface back to the bulk electrolyte, and diffusion of products across the plastic to the environment.

The overall corrosion rate of a corrosion system is determined by the slowest reaction step. The first step of corrosion modeling is the identification of the reaction mechanisms and the rate-determining step. To achieve charge balance, the total current at the anode must equal the total current at the cathode. Hence, reactions at both anode and cathode must be identified to find the rate-determining step.

Aluminum is widely used as die metallization in microelectronic packages.

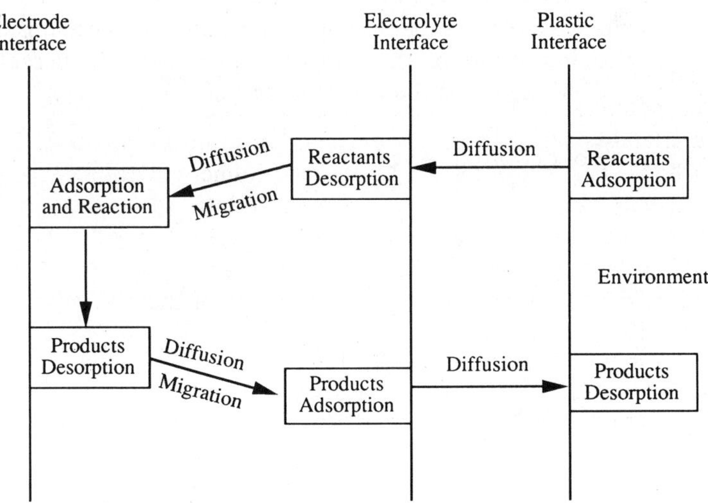



**Figure 23-18**  Generic corrosion process.

Nguyen and Foley[12] postulated that the mechanism of aluminum corrosion in the presence of chloride or other halide ions is as follows.

1. The adsorption of the aggressive anion ($Cl^-$) on the oxide film and the chemical reaction of the adsorbed anion with the $Al^{3+}$ in the oxide lattice:

$$Al^{3+} + Cl^- + 2OH^- \rightarrow Al(OH)_2Cl \qquad (23\text{-}37)$$

or

$$Al^{+3} + 2Cl^- + 2OH^- \rightarrow Al(OH)_2Cl_2^- \qquad (23\text{-}38)$$

2. The thinning of the oxide film by dissolution and the direct attack of the exposed metal by the aggressive anion with the formation of transient complexes:

$$Al \rightarrow Al^{+3} + 3e \qquad (23\text{-}39)$$

$$Al^{+3} + 4Cl^- \rightarrow AlCl_4^- \qquad (23\text{-}40)$$

3. The transient complexes hydrolyze, producing the aggressive anion, $Cl^-$:

$$AlCl_4^- + 2H_2O \rightarrow Al(OH)_2Cl + 2H^+ + 3Cl^- \qquad (23\text{-}41)$$

The chlorine ions are recycled to continue the corrosion process so that a small amount of chlorine ions may cause considerable corrosion damage.

By the law of charge conservation, the total current at the anode equals the total current at the cathode. Hence, the reaction mechanism at the cathode is also important in the corrosion processes of microelectronic devices. In aqueous solutions the common cathodic reactions are

1. Reduction of dissolved oxygen

$$O_2 + 4e^- + 2H_2O \rightarrow 4OH^- \qquad (23\text{-}42)$$

2. Hydrogen evolution as a result of the reduction of hydrogen ions:

$$2H^+ + 2e^- \rightarrow H_2 \qquad (23\text{-}43)$$

3. Deposition of metal ions

$$M^{n+} + ne^- \rightarrow M \qquad (23\text{-}44)$$

Several cathodic reactions may occur simultaneously depending on the electrode potential and the concentrations of the reactants. In neutral solutions, the reduction of oxygen is of considerable importance for many metals and in such cases the amount of oxygen dissolved in the water may determine the overall corrosion rate.

## 23.6 CORROSION MODELING

Most corrosion models in the literature concentrate on phenomenological correlations of time-to-failure with various stresses (temperature, relative humidity, bias, contaminant concentration, and duty cycles) and do not emphasize the reaction mechanisms and the rate-determining step. These approaches do not provide any insights into the physical processes of corrosion and that the models apply only to specific systems under specific conditions (device type, molding compound, testing time, and aging conditions) for which the equations were derived. The models in the literature yield a wide range of acceleration factors. Mathematical models based on the understanding of physical processes of corrosion systems are useful for interpreting data and justifying the application of a particular model to obtain a time to corrosion failure of microelectronic devices.

Identification of the corrosion sites, corrosion products, and corrosion mechanisms is the first step in corrosion modeling. The physical and chemical analyses of corrosion sites and products have been studied using techniques such as optical and electron microscopy, energy-dispersive X-ray spectroscopy (EDS), Auger electron spectroscopy (AES), secondary-ion mass spectrometry (SIMS), X-ray photoelectron spectroscopy (XPS or ESCA), scanning acoustic microscopy (SAM), electrochemical impedance spectroscopy (EIS), and potentiostatic and galvanostatic polarizations.

This section discusses the governing equations for generic corrosion processes and general methods of modeling and simulation. Models for electrolytic corrosion, stress corrosion fatigue, stress corrosion cracking, and mass transport-controlled corrosion processes are discussed.

### 23.6.1 Governing Equations of Corrosion Processes

**Mass Transport Equation**

The flux of mass transport in dilute solutions of an electrochemical system is described by[13]

$$N_i = -D_i \nabla c_i - n_i u_i F c_i \nabla \phi + v c_i \qquad (23\text{-}45)$$

where $N_i$ is the flux of the $i$th species, $D_i$ is the diffusivity, $\nabla c_i$ is the concentration gradient, $n_i$ is the charge carried by the $i$th species, $\nabla\phi$ is the potential gradient, $v$ is the velocity of the bulk electrolyte, and $u_i$ is the mobility that is related to the diffusion coefficient, $D_i$, and temperature by

$$u_i = \frac{D_i}{RT} \tag{23-46}$$

The first term of the right-hand side of Eq. (23-45) is the contribution of diffusion due to the concentration gradient; the second term is the contribution of migration due to potential gradients; and the third term is the contribution of convection due to the movement of the bulk electrolyte. In microelectronics packages, the third term due to convection is neglected since there is negligible movement of the bulk electrolyte.

**Continuity Equation**

By the law of mass conservation, a continuity equation can be derived:

$$\frac{\partial c_i}{\partial t} = -\operatorname{div} N_i + R_i$$

$$= D_i \nabla^2 c_i + n_i u_i F \, \nabla \cdot (c_i \, \nabla\phi) + R_i \tag{23-47}$$

where $R_i$ is the species $i$ generated in a control volume, and div is the divergence operator:

$$\operatorname{div} = \frac{\partial}{\partial x}\mathbf{i} + \frac{\partial}{\partial x}\mathbf{j} + \frac{\partial}{\partial z}\mathbf{k} \tag{23-48}$$

**Equation for Electrochemical Reactions**

The reaction rate for an electrochemical reaction is described by the Butler–Volmer equation:[13]

$$i = i_0 \left[ \exp\left(\frac{\alpha_a F}{RT}\eta\right) - \exp\left(-\frac{\alpha_c F}{RT}\eta\right) \right] \tag{23-49}$$

where $i$ is the current density; $i_0$ is the exchange current density depending on the concentrations and rate constant; $\eta$ is the overpotential, that is, the difference between the electrode potential and the equilibrium electrode potential; $R$ is the gas constant; and $T$ is the absolute temperature.

## Poisson's Equation

The potential distribution is governed by the Poisson equation:

$$\nabla^2\phi = -\frac{q}{\varepsilon\varepsilon_0} \tag{23-50}$$

where $q$ is the charge density, $\varepsilon$ is the dielectric constant of the electrolyte, and $\varepsilon_0$ is the permittivity of free space. Poisson's equation is usually approximated by the equation of local charge neutrality:

$$\sum_i^n z_i c_i = 0 \tag{23-51}$$

The above equations are coupled and highly nonlinear. The most common approximations used to solve the problems are

Reducing the number of dimensions (usually to one).
Considering only one of the mass transport mechanisms (diffusion, migration, or convection).
Neglecting the moving boundary that may characterize the system.
Simplifying the chemical and electrochemical reactions.
Assuming the electrolyte is neutral to transform the Poisson equation to the Laplace equation.
Assuming a steady-state condition.

### 23.6.2 Calculation of Time-to-Failure

The purpose of corrosion modeling is to predict a time-to-failure to aid in the design and materials selection of an electronics package. From Pecht and Ko,[14] the time-to-failure is the sum of an induction time, $t_i$, and a time for corrosion failure, $t_c$, and is expressed as

$$t_f = t_i + t_c \tag{23-52}$$

The induction time is the time required for the internal relative humidity to reach a critical value to condense and to support ionic transport. It was estimated by solving the diffusion equation with appropriate boundary and initial conditions.[14]

To determine a time to corrosion failure, we first define the failure criterion,

$F_c$, as the fraction of the total mass corroded:

$$F_c = \frac{w_c}{w_t} \tag{23-53}$$

where $w_c$ is the mass corroded, and $w_t$ is the total mass of metal. Faraday's law states that the total charge passed (in moles) equals the reactant (metal) consumed in chemical equivalents. That is,

$$Q_t = \int_0^{t_c} \int_0^{A} i \, ds \, dt = \frac{w_t F_c}{M} nF \tag{23-54}$$

where $Q_t$ is the total charge, $M$ is the molecular weight of the metal, $w_t$ is the total mass of the metal, and $n$ is the valence of the metal. Solving Eq. (23-54), the time to corrosion failure is

$$t_c = \left(\frac{w_t F_c}{M}\right) nF \left(\frac{1}{i_{ave} A}\right) = C \frac{1}{i_{ave}} \tag{23-55}$$

where $i_{ave}$ is the average corrosion current density, and $A$ is the area of the active surface. The current density in an electrolyte solution is due to the movement of charged species:

$$i = F \sum n_i N_i$$
$$= F\left(-\sum n_i D_i \nabla c_i - F \nabla\phi \sum n_i^2 u_i c_i\right) \tag{23-56}$$

If the electrochemical reaction is the rate-controlling step, the current density described by the Butler–Volmer equation (23-49) is used directly to calculate the current density. If the reaction (23-41) is the rate-determining step of the electrochemical reaction, from reaction (23-40), we get

$$K_2 = \frac{c_{AlCl_4^-}}{c_{Cl^-}^4} \tag{23-57}$$

where $K_2$ is the equilibrium constant, $c_{AlCl_4^-}$ is the concentration of $AlCl_4^-$, and $c_{Cl^-}$ is the concenttation of $Cl^-$. The current density is

$$i = nFr = 3Fk_0[\exp(-\Delta G/RT)\exp(\alpha_a FE/RT)]c_{H_2O}^2(K_2 c_{Cl^-}^4) \tag{23-58}$$

Assuming Raoult's law $p = p_0 x$ holds where $p$ is the partial pressure of the water, $p_0$ is the saturation pressure of the water, and $x$ is the mole fraction of the water; from the definition of the relative humidity, $RH = p/p_0$, we get $RH = x$. Therefore, the concentration of the water is related to the relative humidity by

$$c_{H_2O} \cong \frac{x}{M_{H_2O}} 1000 = \frac{1000RH}{M_{H_2O}} \tag{23-59}$$

Substituting Eq. (23-59) into Eq. (23-58) gives

$$i = C_1[\exp(-\Delta G/RT) \exp(\alpha_a FE/RT)]RH^2 c_{Cl^-}^4 \tag{23-60}$$

where

$$C_1 = 3Fk_0 \left(\frac{1000}{M_{H_2O}}\right)^2 K_2 \tag{23-61}$$

Substituting Eq. (23-60) into the time to corrosion failure Eq. (23-55) gives

$$t_c = C_3[\exp(\Delta G/RT) \exp(-\alpha FE/RT)]RH^{-2} c_{Cl^-}^{-4} \tag{23-62}$$

The exponent of RH determined by Peck[15] ranges from $-2.50$ to $-3.0$.

If the rate-determining step is a diffusion process, either diffusion from ambient to the bulk electrolyte or diffusion from the bulk electrolyte to the electrode interface, then the diffusion equation and the equation of continuity must be solved with appropriate initial and boundary conditions. If the diffusion is one-dimensional and there is no potential gradient, the continuity equation becomes

$$\frac{\partial c_i}{\partial t} = D_i \frac{\partial^2 c_i}{\partial^2 x} \tag{23-63}$$

By the method of separation of variables, the general solution of Eq. (23-63) for initial and boundary conditions

$$\begin{aligned}
c_i &= c_1, & x &= 0, & t &\geqslant 0 \\
c_i &= c_2, & x &= l, & t &\geqslant 0 \\
c_i &= f(x), & 0 &< x < l, & t &= 0
\end{aligned} \tag{23-64}$$

is[16]

$$c_i = c_1 + (c_2 - c_1)\frac{x}{l} + \frac{2}{\pi}\sum_{n=1}^{\infty}\frac{c_2 \cos n\pi - c_1}{n}\sin\frac{n\pi x}{l}e^{-Dn^2\pi^2 t/l^2}$$

$$+ \frac{2}{l}\sum_{n=1}^{\infty}\sin\frac{n\pi x}{l}e^{-Dn^2\pi^2 t/l^2}\int_0^l f(t)\sin\frac{n\pi t}{l}\,dt \qquad (23\text{-}65)$$

where $l$ is the thickness of the diffusion barrier. The current density is

$$i = -nFD_i\left.\frac{\partial c_i}{\partial x}\right|_{x=l} \qquad (23\text{-}66)$$

If concentration variation is negligible, the current density equation is simplified to Ohm's law:

$$i = -\nabla\phi F^2\sum n_i^2 u_i c_i = -k\,\nabla\phi = -\frac{1}{\rho}\nabla\phi \qquad (23\text{-}67)$$

where $k$ is the conductivity and is equal to $F^2\sum n_i^2 u_i c_i$, $\rho$ is the resistivity, and $\nabla\phi$ is the potential gradient. Note that Ohm's law holds when the concentration variation is negligible. The time-to-failure under this condition is

$$t_c = \frac{w_t F_c}{M}nF\frac{1}{-k\,\nabla\phi\,A} = \frac{w_t F_c}{MV}nF\frac{S}{L}\frac{\rho}{z} \qquad (23\text{-}68)$$

where $V$ is the potential difference, $L$ is the length of the metal track exposed to the corroding electrolyte, and $S$ is the separation of the metal tracks. The result under this condition is the same as the model proposed by Howard[17] and modified by Pecht and Ko.[14]

### 23.6.3 Pitting and Crevice Corrosion

When a metal is exposed to a corrosive environment, corrosion may occur at a number of isolated sites. If the corrosion sites (anodes and cathodes) do not shift in position and the total corrosion area is much smaller than the total exposed area, the corrosion is referred to as localized corrosion. The rate of localized corrosion is much greater than that of uniform corrosion and it is more destructive because a perforation resulting from a single pit may result in a complete failure. The mechanisms of propagation of pitting and crevice corrosion are similar, so that the modeling of pitting and crevice corrosion is discussed together. Other types of localized corrosion include stress corrosion cracking, corrosion fatigue, and intergranular attack, and will be considered in the following sections.

Pitting processes consist of pit initiation and pit propagation. The factors affecting pit propagation are different from those that cause pit initiation. The initiation of crevice corrosion is different from pitting processes, but the propagation process of crevice corrosion is similar to that of pitting corrosion. The aim of this section is to describe a general approach to modeling the crevice, pitting, and galvanic corrosion based on the mathematical representation of the physical mechanisms governing the process. Detailed solutions of the mathematical model can be found in the references given.

The corrosion current at any point on a metal surface in an electrolyte can be predicted if the electrochemical potential in the electrolyte around that point is known. The equation governing the potential distribution in an electrochemical system can be derived from the mass transport, continuity, and Poisson equations. In addition to the processes shown in Fig. 23-18, a full description of crevice or pitting corrosion must take into account the solution chemistry within crevice/pit, the chemical/electrochemical reaction rate, the mass transport of ions into the crevice/pit, the effect of the moving boundary of the crevice/pit as propagation proceeds and the effect of the blocking of the crevice with solid corrosion product. These processes are shown in Fig. 23-19.

In a dilute electrolyte, the flux of a species, $N_i$, is given by Eq. (23-45).

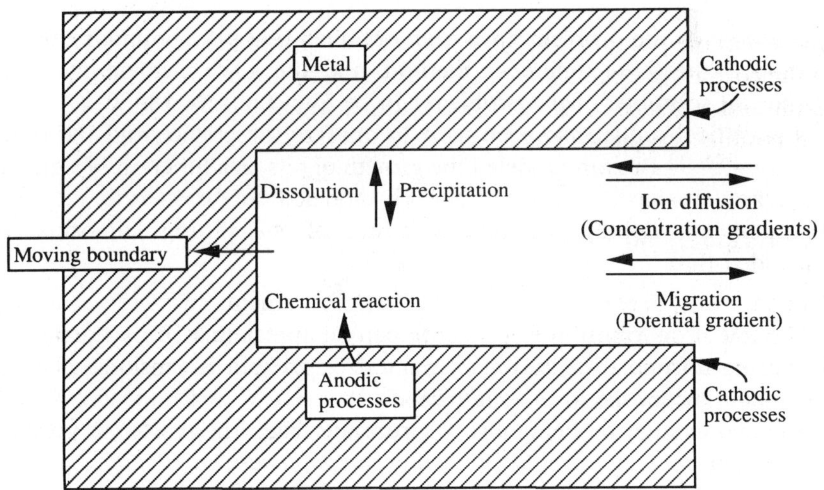


**Figure 23-19**    Schematic illustration of the crevice/pit model.

The continuity equation is

$$\frac{\partial c_i}{\partial t} = -\operatorname{div} N_i + R_i$$

$$= D_i \nabla^2 c_i + n_i u_i F \, \nabla \cdot (c_i \, \nabla \phi) + R_i \qquad (23\text{-}69)$$

The potential in the electrolyte is governed by the Poisson equation

$$\nabla^2 \phi = -\frac{q}{\varepsilon \varepsilon_0} \qquad (23\text{-}70)$$

where $q$ is the charge density, $\varepsilon$ is the dielectric constant of the electrolyte, and $\varepsilon_0$ is the permittivity of free space. The corrosion rate is related to fluxes of charged species by

$$i = F \sum n_i N_i \qquad (23\text{-}71)$$

The boundary conditions are that:

The concentrations are constant at the cavity mouth and are equal to the values in the bulk electrolyte outside the pit or crevice.
The flux of the $i$th species equals the rate of electrochemical reactions of the $i$th species produced or consumed on the active metal surface.
The fluxes of the other species on the active surface are zero.

Solutions of the problem have been obtained using the boundary element method,[19] the finite difference method, and the finite element method.[18] Numerical solutions of the problem can be found in refs. 18–20.

A statistical model has been used to analyze the crevice and the pitting corrosion.[21–23] Provan modeled the growth of pits as a function of time with a Markov stochastic process.[21] Sheikh modeled corrosion rate, pit depth, maximum pit depth, and time-to-failure of the pitting corrosion and concluded that:

The initial distribution of pit depth can be approximated by normal law.
The log-normal distribution characterizes the pit dimensions when the metal is exposed for a long period.
The exponential model is also appropriate when the scatter in pit depth is high.
The maximum pit depths are characterized by extreme value distribution.[22]

### 23.6.4 Stress Corrosion Cracking

Study of stress corrosion cracking has been an area of considerable interest. Fracture mechanics and corrosion engineering are used to study the phenomena. One of the models used to calculate corrosion crack rate assumed that the process is dissolution-controlled. From Faraday's law, we have

$$\frac{da}{dt} = i_a \frac{M}{nFd} \tag{23-72}$$

where $da/dt$ is crack velocity, $i_a$ is anodic current density, $M$ is atomic weight of the metal, $n$ is the number of electrons involved in the electrochemical reaction, and $F$ is Faraday's constant.

Equation (23-72) does not take into account film growth and crack-tip strain rate effect. To include crack-tip strain rate effect, Eq. (23-72) is modified as[24]

$$\frac{da}{dt} = \frac{Q_f}{\epsilon_f} \epsilon_{CT} \frac{M}{nFd} \tag{23-73}$$

where $Q_f$ is the charge passed, $\epsilon_f$ is the strain at which fracture occurs, and $\epsilon_{CT}$ is the crack-tip strain rate. The relationship between strain rate and environment-sensitive crack growth is known for a variety of systems. Various expressions for strain rate are summarized in Table 23-3.

Various researchers have used Eq. (23-73), or some modified version thereof, to compare observed with calculated crack velocities as a function of strain rate.[24,25] Some element of fitting was frequently involved in obtaining an acceptable coincidence of the calculated and observed results. Detailed discussions of stress corrosion cracking can be found in refs. 24–26.

### 23.6.5 Stress Corrosion Fatigue

The study of corrosion fatigue of metals involves several scientific disciplines including metallurgy, corrosion engineering, electrochemistry, and mechanical engineering. The development of corrosion fatigue failure can be divided into crack initiation, crack propagation, and fracture stages. The corrosive environment affects the initiation and propagation of fatigue cracks, although it might not affect fracture. The eletrochemical conditions within the cracks have important effects on the rate of crack propagation in stress corrosion cracking and corrosion fatigue. Turnbull[30] calculated the oxygen concentration in a corrosion fatigue crack by assuming that oxygen was consumed

**Table 23-3**  Mathematical Models of Strain Rate

| Crack Tip Strain Rate* | Reference |
| --- | --- |
| $\dfrac{1}{2T}\dfrac{(\Delta K)^2}{G\sigma_y}$ | Parkins[24] |
| $-\dfrac{1}{T}\ln[1 - 0.5(1 - R)^2]$ | Scott[27] |
| $\dfrac{2\gamma_0}{T}\left[\dfrac{\beta}{\beta - 1}\ln(\beta) - 1\right]$ | Lidbury[28] |
| $\dfrac{1}{T}\dfrac{\alpha(\Delta K)^2}{2WE\sigma_y}$ | Tomkins[29] |
| $\dfrac{2\gamma_0}{T}\beta$ | Lidbury[28] |
| $\dfrac{2\gamma_0}{T}[1 + \ln(\beta)(1 + 0.5\ln(\beta))]$ | Lidbury[28] |

* $T$ = loading time; $\Delta K$ = stress intensity factor range; $G$ = shear modulus; $\sigma_y$ = yield stress; $R = K_{min}/K_{max}$; $K$ = stress intensity factor; $\gamma$ = shear strain; $\beta = (\Delta K/\Delta K_{th})^2$; $\Delta K_{th}$ = stress intensity factor range threshold; $\alpha$ = constant; $W$ = width of active shear band; $E$ = Young's modulus.

by cathodic reduction on the walls of the crack and mass transport occurred by diffusion and forced convection. He concluded that the cyclic variation of the crack walls significantly enhanced the mass transport of oxygen in the crack.

Many variables influence the corrosion fatigue behavior of materials.[31] These variables include

Maximum stress or stress intensity factor
Cyclic stress or stress intensity range
Stress ratio
Cyclic loading frequency
Alloy composition
Distribution of alloying elements and impurities
Microstructure and crystal structure
Heat treatment
Mechanical working
Orientation of grains and grain boundaries

Mechanical properties
Temperature
Types of environment
Concentration of damaging species
Electrochemical potential
pH
Viscosity of the environment
Coatings
Inhibitors

Modeling of environmentally assisted fatigue crack growth in pure gases and binary gas mixtures has been done.[31,32] The model assumed that the rate of crack growth, $(da/dN)_e$ in a corrosive environment is the sum of three components:

$$\left(\frac{da}{dN}\right)_e = \left(\frac{da}{dN}\right)_r + \left(\frac{da}{dN}\right)_{cf} + \left(\frac{da}{dN}\right)_{scc} \qquad (23\text{-}74)$$

where $(da/dN)_r$ is the rate of fatigue crack growth in an inert or reference environment, representing the contribution of mechanical fatigue, $(da/dN)_{cf}$ is the cycle-dependent contribution, and $(da/dN)_{scc}$ is the contribution due to static stresses. Weir et al. assumed molecular flow and first-order reaction kinetics and derived equations for transport-controlled and surface reaction-controlled fatigue crack growth.[32]

*Transport control:*

$$\left(\frac{da}{dN}\right)_{cf} = \left(\frac{da}{dN}\right)_{cf,s} \left(\frac{p_0}{2}f\right) \Big/ \left(\frac{p_0}{2}f\right)_s \qquad (23\text{-}75)$$

*Surface reaction control:*

$$\left(\frac{da}{dN}\right)_{cf} = \left(\frac{da}{dN}\right)_{cf,s} [1 - \exp(-k_c p_0/2f)] \qquad (23\text{-}76)$$

where $p_0$ and $k_c$ are the gas pressure in the external environment and the reaction rate constant, respectively, and $(da/dN)_{cf,s}$ is the maximum enhancement in the rate of cycle-dependent fatigue crack growth.


## 23.7 METHODS OF PREVENTING OR MINIMIZING CORROSION

Corrosion is the result of the interaction of a given material and a particular environment. The following conditions must be fulfilled for the electrochemical corrosion process to occur:

An anode where the oxidation reactions occur
A cathode where the reduction reactions occur
A potential gradient between the anode and the cathode
An electrolyte to support ionic conduction
A connection between the anode and the cathode to complete the electrical
   path

Controlling or eliminating any one of the conditions can retard or avoid corrosion. The following techniques can be used to prevent or retard corrosion:

Material selection
Control of the environment
Use of corrosion inhibitors
Use of cathodic and anodic protection
Special design of geometry

The primary method of corrosion prevention is through material selection. In the selection and use of materials, it is important to consider not only the electrical and mechanical properties, but also factors such as basic corrosion resistance, compatibility between metals, stability of organic materials, and maintainability. The corrosion tendency is dependent on the electrochemical potential of a material. The more positive the electrochemical potential of a material, the lower the tendency to corrode. A wide variety of metals are used in electronic design because of particular physical, electrical, and mechanical properties, and cost. When two metals are in contact, a galvanic cell is formed. In the presence of moisture and contaminants, destructive galvanic corrosion can take place. To minimize galvanic corrosion, the contact of dissimilar metals as defined in MIL-STD-889,[33] and MIL-STD-1250[34] must be avoided whenever possible. When such contacts are unavoidable because of design constraints, the following methods can be used to minimize the corrosion rate:

1. Design the couple so that the area of the more noble metal is appreciably smaller than the area of the more active metal. Decrease the cathode area by painting or coating.
2. Plate the cathode and/or anode with a compatible material to reduce the potential difference.
3. Clean and seal the junctions to minimize the presence of an electrolyte at the bimetallic interface.

4. Place the electronics in a hermetically sealed enclosure pressurized with an inert gas. If impossible to seal hermetically, use elastomeric seals and maintain relative humidity below 40% at room temperature.
5. Apply corrosion inhibitors.

Controlling moisture and contamination level is an effective means of corrosion control. The source of the moisture, halide ions, and the other contaminants have been well documented[3] and are summarized in three groups: contaminants from manufacturing and assembly process, contaminants from the environment, and contaminants contained in the materials used. Major sources of occurrence include:

Molding and coating materials
Substrate and substrate fabrication
Environment
Die- and substrate-attach materials
Solder flux residue
Manufacturing and assembly process (receiving, storage, soldering, cleaning, coating, encapsulation, testing)
Repair and rework

The following methods are used to minimize the corrosion effects by controlling the moisture and contaminant levels:

1. Design a configuration that does not provide moisture traps.
2. Encapsulate components with moisture-resistant materials so that moisture and contaminants cannot get in.
3. Mount printed circuit boards vertically and well above the bottom of the housing to prevent moisture and debris from collecting on the boards.
4. Locate edge connectors on the vertical sides, not the bottom, of the printed circuit board so that moisture and debris will not collect in the mated connectors and degrade the contacts.
5. Avoid materials that emit corrosive vapors.
6. Use a volatile corrosion inhibitor that is carefully selected to protect those metals of concern in a particular application.

Design for corrosion control must take into account the geometry. The geometry of a structure is basically determined by its functional, material, and fabrication requirements. However, a designer does have choices in selecting geometries to reduce corrosion attack. The following general principles should be observed:

1. Avoid dissimilar metal contacts.
2. Provide drainage paths for extraction of contaminants and condensed moisture away from the critical areas.
3. Keep spacing between conductors of different potential as wide as possible to reduce electrolytic corrosion.

Cathodic and anodic protection techniques are used to protect the package case from corrosion. Cathodic protection is based on electrochemical theory, using applied current or voltage to retard corrosion processes. Current generated by sacrificial anodes or a DC power source is applied in such a way that the electrochemical potential of the metal structure to be protected is shifted to a position at which corrosion does not occur. Anodic protection is also an electrochemical method of controlling corrosion. Anodic protection is based on the phenomenon of passivity. The electrochemical potential and/or environment of a metal must be controlled to achieve passivity. Anodic protection can only be applied to those metals and alloys that exhibit passivity. Since both cathodic and anodic protection involve controlling the potential, these two techniques can only be applied where there are no requirements of specific potential to perform electrical functions. Cathodic and anodic protection are effective but also expensive.

## 23.8 LABORATORY EXPERIMENTAL TECHNIQUES AND STANDARD TESTS

Laboratory corrosion tests are used to predict corrosion behavior when service data and time are short or budget constraints prohibit field testing. These tests are useful for quality control, material selection, predicting service life, determining the most economical means of reducing corrosion, and studying corrosion mechanisms. Most of the laboratory tests are accelerated tests by design. Standard test methods are described in the Annual Book of ASTM Standards and in military standard MIL-STD-883.

The probes for corrosion research can be voltage, current, charge, electron beam, light, or ultrasound. The techniques for thermodynamic and kinetic studies are classified into steady-state techniques, including potentiostatic and galvanostat polarizations, and transient techniques, including cyclic voltammetry, electrochemical impedance spectroscopy, and Tafel measurements. Table 23-4 lists the common transient techniques used for corrosion studies.

The physical and chemical analyses of corrosion sites and products can be studied using techniques such as optical and electron microscopy, energy-dispersive X-ray spectroscopy (EDS), Auger electron spectroscopy (AES), secondary-ion mass spectrometry (SIMS), X-ray photoelectron

**Table 23-4**   Common Transient Techniques for Corrosion Studies

| Method | Perturbation | Typical Response |
| --- | --- | --- |
| 1. Single potential step | 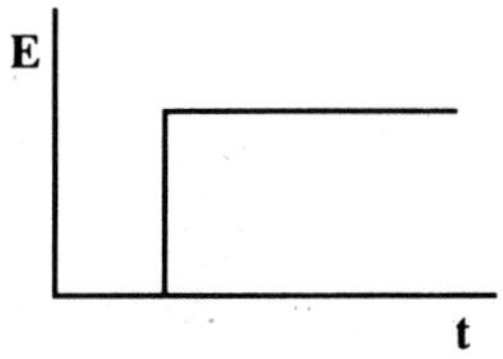 | 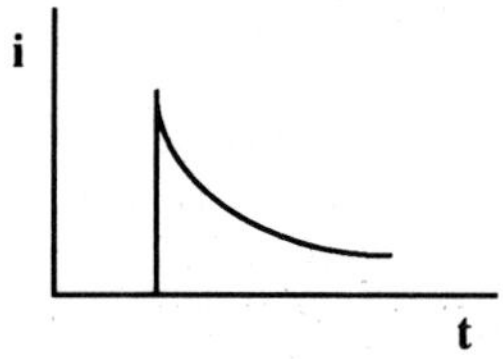 |
| 2. Double potential step | 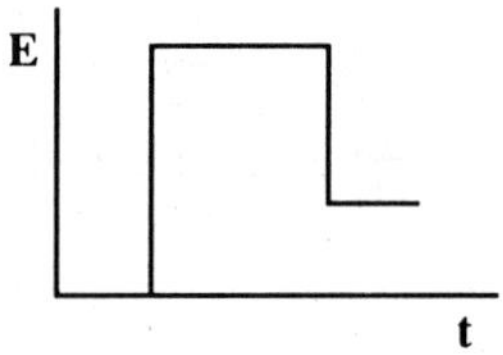 | 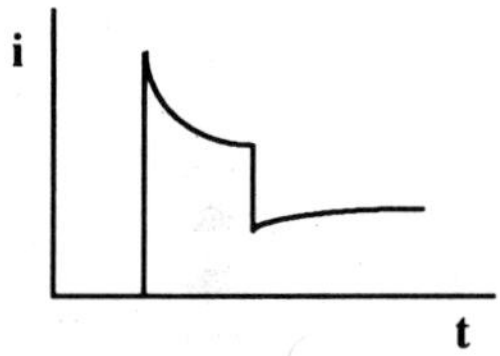 |
| 3. Linear sweep voltammetry | 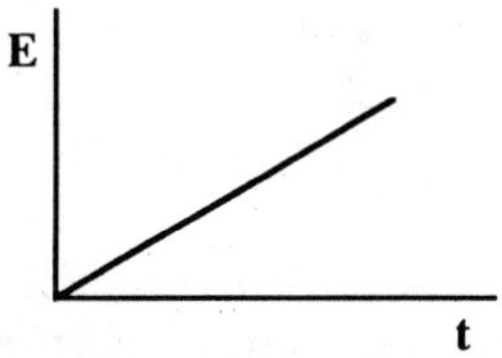 | 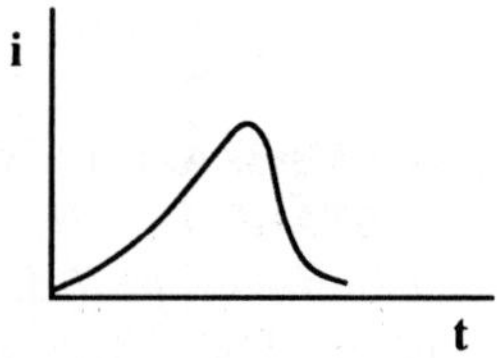 |
| 4. Cyclic voltammetry | 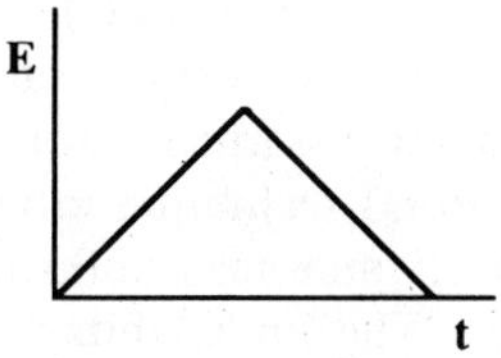 | 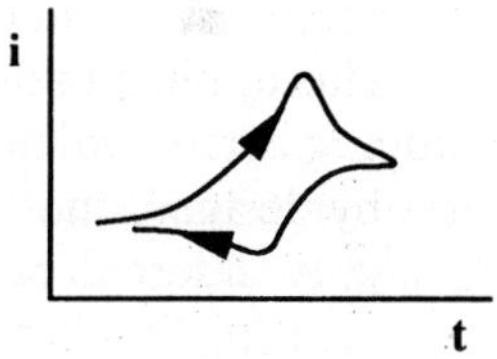 |
| 5. Impedance spectroscopy | 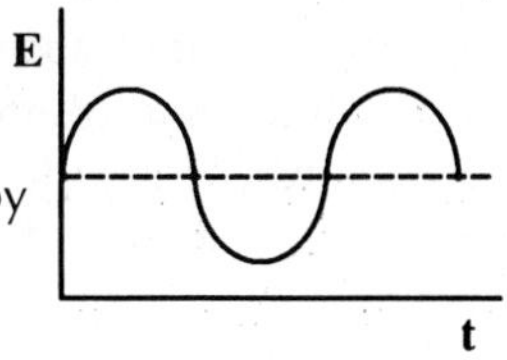 | 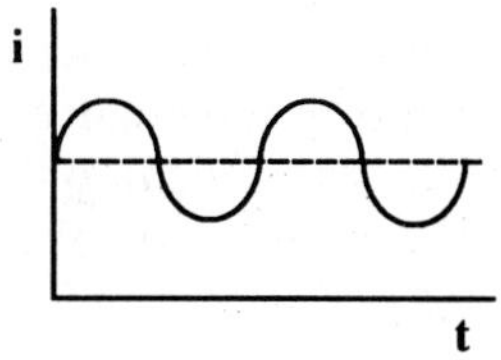 |

(*continued*)

**Table 23-4** (*continued*)

| Method | Perturbation | Typical Response |
|---|---|---|

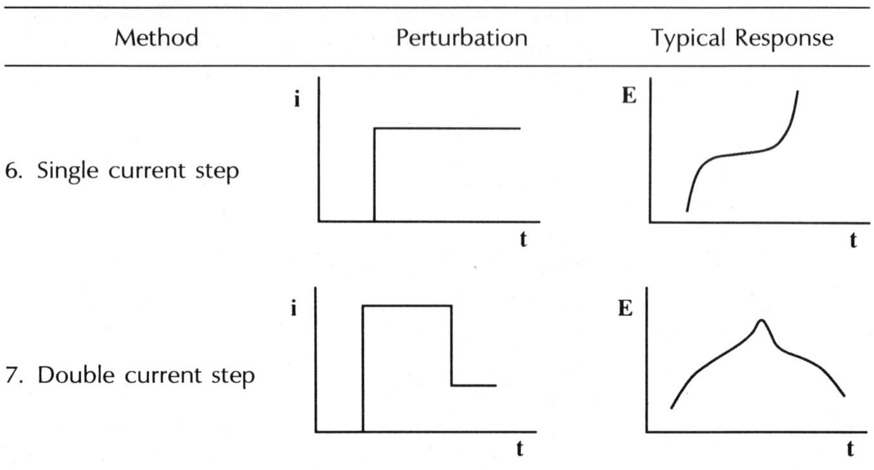


spectroscopy (XPS or ESCA), and scanning acoustic microscopy (SAM). The common techniques for corrosion studies—electrochemical impedance spectroscopy (EIS), cyclic voltammetry, polarization curve, and polarization resistance measurement—are briefly introduced. Detailed discussions on most corrosion study techniques can be found in refs. 7 and 35.

***Electrochemical Impedance Spectroscopy*** EIS uses an AC or a step-function DC signal with small amplitude to perturb a system at steady state and the response is measured to calculate the impedance.[36] In transient measurements, a step function of voltage [$V(t) = V_0$ for $t > 0$, $V(t) = 0$ for $t < 0$] is applied and the resulting time-varying current $i(t)$ is measured. The ratio $V_0/i(t)$ is calculated and transformed into the frequency domain by Fourier transformation. A second technique in impedance spectroscopy is to apply a signal composed of random (white) noise to the interface and measure the resulting current. The impedance in the time domain is then transformed to frequency domain with the fast Fourier transform (FFT). This technique offers the advantage of fast data collection because only one signal is applied to the interface for a short time. The third technique is to measure impedance directly in the frequency domain by applying a single-frequency signal to the interface and measuring the phase shift and amplitude. The advantages of this technique are the availability of these instruments, the ease of use, and high accuracy.

By combination of single-frequency and white-noise techniques, the electrochemical impedance in a wide spectrum of frequencies ($10^{-4}$ to $10^6$ Hz) can be obtained with current instruments such as PAR 386 impedance measurement systems and Solartron frequency analyzers. Because of this wide frequency band, corrosion processes with different relaxation time constants can be differentiated. Electrochemical impedance spectroscopy has been widely used to study corrosion mechanisms, estimate corrosion rate, evaluate the effectiveness of coating, and study defects in semiconductor. The corrosion rate can be estimated with an equivalent electrical circuit model or a physical model.

The advantages of EIS are that it can be applied in media of high resistivity, it is nondestructive, and the capacitive hysteresis is eliminated. The information obtained from impedance measurements includes corrosion rate, solution resistance, double-layer capacitance, and reaction mechanisms. The interpretation of data is usually based on electrical equivalent circuit models. Figure 23-20 shows a simple equivalent circuit model of metal–electrolyte interface and the complex-plane plot of the electrochemical impedance.

***Cyclic Voltammetry***    This uses a potential sweep to probe the corrosion system (Fig. 23-21). The ASTM standard G61[37] describes a procedure for conducting cyclic potentiodynamic polarization measurements to determine relative susceptibility to localized corrosion. The potential sweep rates ranging from a few millivolts per hour to tens of volts per second have been employed. The slow sweep rates are frequently used to measure "steady-state" current–voltage curves on the assumption that the surface relaxes sufficiently rapidly that the system is negligibly different from a true steady state. The validity of this assumption can be checked by measuring the current–voltage curves over a range of sweep rates. From cyclic voltammetry, corrosion reaction mechanism can be deduced and corrosion rate can be estimated. Particular attention is focused on two features of the cyclic anodic polarization behavior diagram: the potential at which the anodic current increases considerably with applied potential or the breakdown potential and the potential at which the hysteresis loop is completed upon reverse polarization scan (protection potential). In general, the more positive the potential, the less susceptible the alloy to the initiation of localized attack.

***Polarization Technique (Tafel Plot)***    This applies either a slow potential sweep or a fixed potential at a given time to probe the system, record the resulting current, and plot the applied potential versus the logarithm of current density. Figure 23-22 shows schematic anodic and cathodic polarization curves. Several methods can be used to estimate corrosion currents from the polarization curve. The specimen can be polarized to a potential at least 50 mV more negative than the open circuit potential to obtain a reliable Tafel line. Then the current density can be extrapolated back to corrosion potential to obtain the corrosion current density value. The

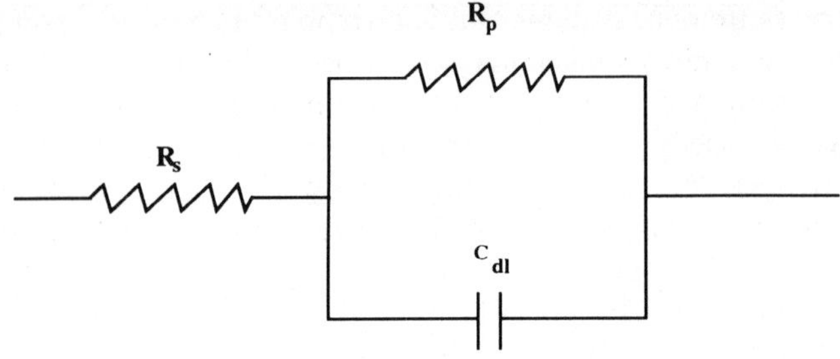



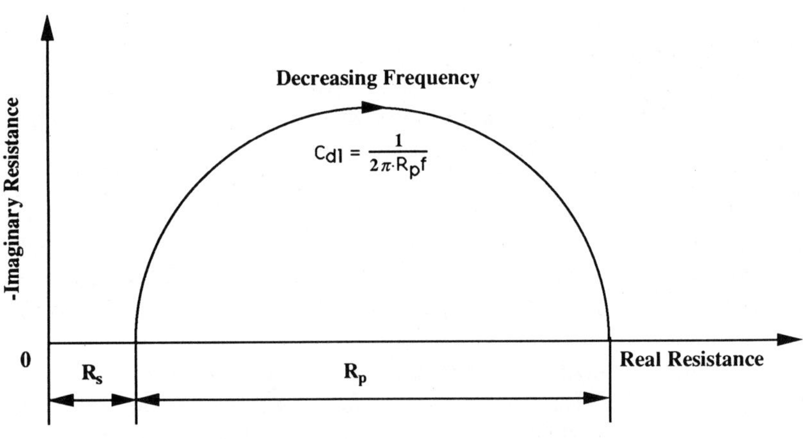



$$C_{dl} = \frac{1}{2\pi \cdot R_p f}$$

**Figure 23-20**   Equivalent circuit model of metal–electrolyte interface and complex-plane plot of electrochemical impedance.

second method is analogous to the first. The anodic Tafel line is used instead of cathodic Tafel line. The third method uses both cathodic and anodic Tafel lines. The corrosion current density is obtained at the intersection of the anodic and cathodic extrapolated lines. The three methods to obtain corrosion current density are shown in Fig. 23-23. In addition to calculating corrosion current density, polarization curves can be used to obtain breakdown potential, critical pitting potential, critical crevice potential, and corrosion potential.

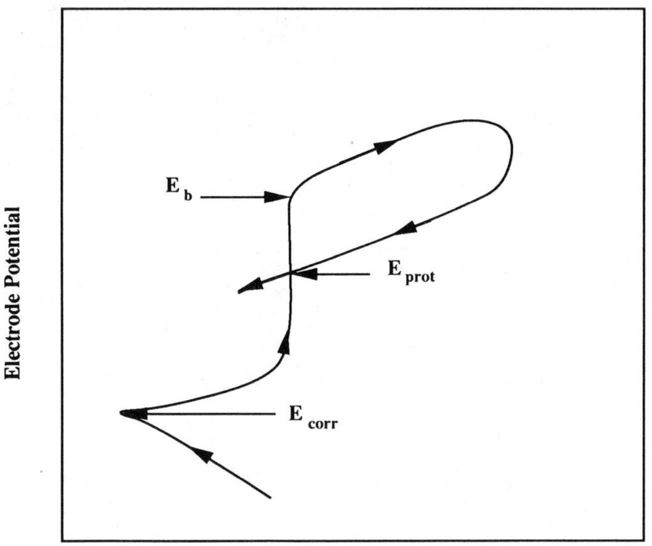



**Figure 23-21**    Cyclic polarization to examine breakdown of passivity.

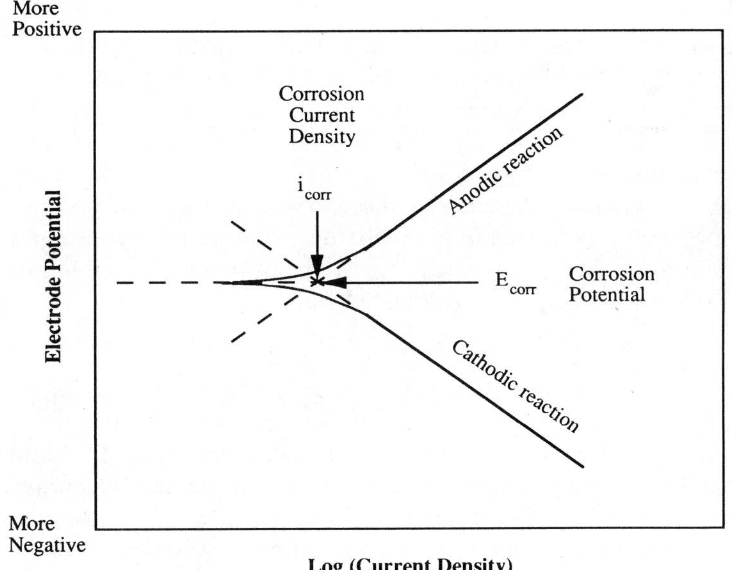



**Figure 23-22**    Schematic anodic and cathodic polarization curves.

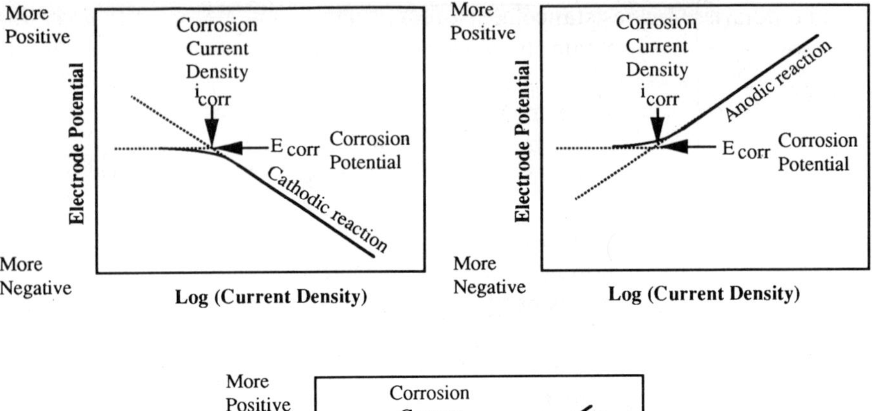


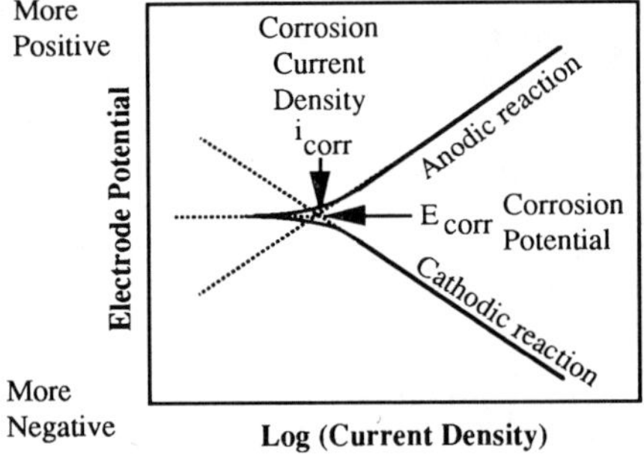


**Figure 23-23**   Tafel extrapolation methods for estimating corrosion rate.

***Polarization Resistance Measurement***   This is used to estimate corrosion rate based on the Stearn–Geary equation.[5] This technique is a well-developed method for determining corrosion rates, and its theoretical background is well understood. The theory assumes that the corrosion rate of a specimen is inversely proportional to its polarization resistance, that is, the slope of the potential–current response curve near the corrosion potential. A test specimen is polarized to about 10 mV around the corrosion potential. The resulting current is recorded as a measure of corrosion rate. The polarization resistance, $R_p$, is calculated from

$$R_p = \left(\frac{1}{2.3 i_{\text{corr}}}\right)\left(\frac{\beta_a \beta_c}{\beta_a + \beta_c}\right) \tag{23-77}$$

where $i_{\text{corr}}$ is the corrosion current density, and $\beta_a$ and $\beta_c$ are the anodic and cathodic Tafel slopes, respectively. In order to use this technique, it is necessary to obtain Tafel slopes independently. The ASTM standard G59 describes procedures for conducting polarization resistance measurements.[41] The electrochemical impedance technique described above can also be used to obtain polarization resistance.

The polarization resistance measurement is a quick method of estimating the uniform corrosion rate. It measures the corrosion current density with small polarizations about $E_{corr}$ and thereby disturbs the electrode–electrolyte interface very little. If the measurement has been made correctly and the corrosive environment has been simulated well, the polarization resistance measurement is a valuable method for determining the uniform corrosion rate.

There are several limitations to the polarization resistance method. The corroding environment must be an electrolyte with reasonably low resistivity. High-resistivity electrolytes produce erroneously low corrosion rates. Polarization resistance measurements do not provide information on localized corrosion such as pitting and stress corrosion cracking.

The following standard tests listed in MIL-STD-883 can be used to test corrosion resistance of the electronic devices.

1. *Salt atmosphere* (Method 1009). This test is an accelerated laboratory corrosion test simulating the effects of a seacoast atmosphere on devices and package elements. Salt atmosphere test is considered destructive. It is intended for device qualification and process monitoring.
2. *Thin film corrosion test* (Method 1031). The purpose of this test is to demonstrate the quality or reliability of devices subjected to specified conditions over a specified time period. This test is used either as a short-term specialized quality assurance test or as a long-term acceleration test to assure device reliability. It is for devices containing thin film conductors, resistors, or fuses that are susceptible to corrosion. The intention of this test is to reveal time-, temperature- and stress-dependent moisture-related failure modes resulting from metallization corrosion.
3. *Moisture resistance* (Method 1004). The purpose of this test is to evaluate, in an accelerated manner, the resistance of component parts and constituent materials to the deteriorative effects of the high humidity and heat conditions typical of tropical environments.

## 23.9 SUMMARY

Corrosion is one of the major failure mechanisms in microelectronic devices. As packaging density become higher and signal speed becomes faster, microelectronic devices are more prone to corrosion attack because of severe dimensional constraints. To prevent or reduce corrosion, the general principle of corrosion prevention must be incorporated into the design, manufacturing, assembly, storage, and shipping processes.

This chapter introduces general principles of corrosion—thermodynamics and kinetics of electrochemical reactions and mass transport in dilute electrolytes—to understand why corrosion occurs, factors that affect corrosion, and methods to prevent or reduce corrosion. Common forms of corrosion and their remedial measures in microelectronic devices are discussed. Mathematical models for various forms of corrosion are reviewed.


## NOMENCLATURE

| | |
|---|---|
| $\alpha_a$ | Transfer coefficient of anode |
| $\alpha_c$ | Transfer coefficient of cathode |
| $\beta$ | Transfer coefficient |
| $\Delta G$ | Gibbs free energy |
| $\Delta H$ | Entropy |
| $\Delta S$ | Enthalpy |
| $\Delta U$ | Internal energy |
| $\epsilon_{CT}$ | Crack tip strain rate |
| $\epsilon_f$ | Strain at which fracture occurs |
| $\varepsilon$ | Permittivity of electrolyte |
| $C$ | Proportional constant |
| $c_i$ | Concentration of $i$th species |
| $D_i$ | Diffusion coefficient of $i$th species |
| div | Divergence operator |
| $E$ | Electrochemical potential |
| $E_a$ | Activation energy |
| $E^0$ | Standard electrochemical potential |
| $E_0$ | Equilibrium electrochemical potential |
| $F$ | Faraday's constant (96 500 C/eq) |
| $F_c$ | Failure criterion, ratio of corroded weight to the total weight |
| $i$ | Current density |
| $i_{\text{ave}}$ | Average corrosion current density |
| $i_0$ | Exchange current density |
| $k$ | Reaction rate constant |
| $k$ | Boltzmann constant |
| $k^0$ | Reaction rate constant at reference state |
| $n$ | Number electrons involved in an electrochemical reaction |
| $N_i$ | Flux of $i$th species |
| $q$ | Charge density |
| $Q_f$ | Charge |
| $r$ | Reaction rate |
| $R$ | Gas constant |
| $r_a$ | Anodic reaction rate |
| $r_c$ | Cathodic reaction rate |
| RH | Relative humidity |
| $R_i$ | Rate of generation of $i$th species |
| $T$ | Absolute temperature |

$t_c$    Time-to-failure due to corrosion
$t_f$    Time-to-failure
$t_i$    Induction time to corrosion
$u_i$    Mobility of $i$th species
$V$    Potential difference
$w_c$    Amount of metal corroded
$w_e$    Electrical work
$w_m$    Mechanical work
$w_t$    Total mass of a metal
$z_i$    Charge carried by $i$th species
$\nabla\phi$    Potential gradient
$\nabla c_i$    Concentration gradient of $i$th species

# REFERENCES


1. Dobbs, B., and G. Sleask, "Investigation of Corrosion Related Failures in Electronic Systems," *Materials Performance*, **23**(3), 1984, pp. 35–38.
2. Munn, R. S., "Application of Numerical Analysis and Design of Microelectronic Components Under Conditions of Atmospheric Corrosion," *Corrosion/88*, Paper No. 113, NACE, Houston, TX, 1988.
3. Tautscher, C. J., *Contamination Effects on Electronic Products*, Marcel Dekker, New York, 1991.
4. Abbott, W. H., "Field Versus Laboratory Experience in the Evaluation of Electronic Components and Materials," Paper Presented at Corrosion/83, Anaheim, CA, NACE, April 1983.
5. Uhlig, H. H., and R. W. Revie, *Corrosion and Corrosion Control*, Wiley, New York, 1985.
6. Mansfield, F., *Corrosion Mechanisms*, Marcel Dekker, New York, 1987.
7. Baboian, R., *Electrochemical Techniques for Corrosion Engineering*, NACE, Houston, TX, 1986.
8. Pludek, V. R., *Design and Corrosion Control*, Wiley, New York, 1977.
9. Crooker, T. W., and B. N. Leis, *Corrosion Fatigue: Mechanics, Metallurgy, Electrochemistry and Engineering*, ASTM, Philadelphia, PA, 1983.
10. Szklarska-Smialowska, Z., *Pitting Corrosion of Metals*, NACE, Houston, TX, 1986.
11. ASTM G82, "Standard Guide for Development and Use of a Galvanic Series for Predicting Galvanic Corrosion Performance," *Annual Book of ASTM Standards*, ASTM, Philadelphia, PA, 1990.
12. Nguyen, T. H., and R. T. Foley, "The Chemical Nature of Aluminum Corrosion," *J. Electrochemical Society*, **127**(12), 1980, pp. 2563–2566.
13. Newman, J., *Electrochemical Systems*, Prentice-Hall, Englewood Cliffs, NJ, 1991.
14. Pecht, M., and W. Ko., "A Corrosion Rate Equation for Microelectronic Die Metallization," *Int. J. Hybrid Microelectronics*, **13**(2), 1990, p. 41.
15. Peck, D. S., "Comprehensive Model for Humidity Testing Correlation," *Proc. 24th Int. Reliability Physics Symposium*, 1986, pp. 44–50.
16. Crank, J., *The Mathematics of Diffusion*, Oxford University Press, Oxford, UK, 1967.

17. Howard, R. T., "Electrochemical Model for Corrosion of Conductors on Ceramic Substrates," *IEEE Trans. Components, Hybrids, and Manufacturing Technology*, **CHMT-4**(4), 1981, pp. 520–525.

18. Fu, J. W., and S. K. Chan, "A Finite Element Method for Modeling Localized Corrosion Cells," *Corrosion*, **40**(10), October 1984, pp. 540–544.

19. Aoki, S., "Prediction of Galvanic Corrosion Rates by the Boundary Element Method," *Mathematical and Computer Modeling*, **15**(3), 1991, p. 11.

20. Sharland, S. M. et al., "A Finite-Element Model of the Propagation of Corrosion Crevices and Pits," *Corrosion Science*, **29**, 1989, pp. 1149–1166.

21. Provan, J. W., and E. S. Rodriguez III, "Part I: Development of a Markov Description of Pitting Corrosion," *Corrosion*, **45**, 1989, pp. 178–192.

22. Sheikh, A. K. et al., "Statistical Modeling of Pitting Corrosion and Pipeline Reliability," *Corrosion*, **46**(3), 1990, pp. 190–197.

23. Watson, M., and J. Postlethwaite, "Numerical Simulation of Crevice Corrosion of Stainless Steels and Nickel Alloys in Chloride Solutions," *Corrosion*, **46**(7), 1990, pp. 522–530.

24. Parkins, R. N., "Factors Influencing Stress Corrosion Crack Growth Kinetics," *Corrosion*, **43**(3), March 1987, pp. 130–139.

25. Parkins, R. N., "The Application of Stress Corrosion Crack Growth Kinetics to Predicting Lifetimes of Structures," *Corrosion Science*, **29**(8), 1989, pp. 1019–1038.

26. Galvele, J., "A Stress Corrosion Cracking Mechanism Based on Surface Mobility," *Corrosion Science*, **27**(1), 1987, pp. 1–33.

27. Scott, P. M., and A. E. Truswell "Corrosion Fatigue Crack Growth in Reactor Pressure Vessel Steels—Measurement and Application," in *Structural Integrity of Light Water Reactor Components*, L. E. Steels, ed., Applied Science Publications, London, 1982, pp. 287–309.

28. Lidbury, D. P., "The Estimation of Crack Tip Strain Rate Parameters Characterizing Environment Assisted Crack Growth Data," in *Embrittlement by the Localized Crack Environment*, R. P. Gangloff, ed., American Institute of Mining, Metallurgical, and Petroleum Engineers, New York, 1983, pp. 149–172.

29. Tomkins, B., and J. Waring, *Metal Science*, **11**, 1977, p. 414.

30. Turnbull, A., and D. H. Ferriss, "Mathematical Modelling of the Electrochemistry in Corrosion Fatigue Cracks in Structural Steel Cathodically Protected in Sea Water," *Corrosion Science*, **26**(8), 1986, pp. 601–628.

31. Wei, R. P., *Fatigue Mechanisms*, ASTM STP 675, ASTM, Philadelphia, PA, 1979, pp. 816–831.

32. Weir, T. W., G. W. Simmons, R. G. Hart, and R. P. Wei, *Scripta Metallurgica*, **14**(3), March 1980, pp. 357–364.

33. MIL-STD-889, *Dissimilar Metals*, Naval Publications and Forms Center, Philadelphia, PA.

34. MIL-STD-1250, *Corrosion Prevention and Deterioration Control in Electronic Component and Assemblies*, Naval Publications and Forms Center, Philadelphia, PA, June 1992.

35. Haynes, G., and R. Baboian, *Laboratory Corrosion Tests and Standards*, ASTM, Philadelphia, PA, 1985.

36. Macdonald, J. R., *Impedance Spectroscopy*, Wiley, New York, 1987.

37. ASTM G61, "Standard Practice for Conducting Cyclic Potentiodynamic

Measurements for Localized Corrosion," *Annual Book of ASTM Standards*, ASTM, Philadelphia, PA, 1990.

38. Pecht, M., "A Model for Moisture Induced Corrosion Failures in Microelectronic Packages," *IEEE Trans. Components, Hybrids, and Manufacturing Technology*, **13**, 1990, pp. 383–389.

39. Pourbaix, M., *Atlas of Electrochemical Equilibria in Aqueous Solutions*, 2d ed., NACE, Houston, TX, 1974.

40. Wei, R. P., and G. W. Simmons, "Surface Reactions and Fatigue Crack Growth," *Proc. 27th Army Materials Research Conference*, Bolton Landing, New York, July 1980.

41. ASTM G59, "Standard Practice for Conducting Potentiodynamic Polarization Resistance Measurements," *Annual Book of ASTM Standards*, ASTM, Philadelphia, PA, 1984.

# Author Biographies

**John H. Lau** is a senior engineer of Hewlett-Packard Company located in Palo Alto, California. His research activities cover a broad range of microelectronics packaging and manufacturing technology.

He received the BE degree in civil engineering from the National Taiwan University, the MASc degree in structural engineering from the University of British Columbia, the MS degree in engineering mechanics from the University of Wisconsin, the MS degree in management science from Fairleigh Dickinson University, and the PhD degree in theoretical and applied mechanics from the University of Illinois.

Prior to joining Hewlett-Packard Laboratories in 1984, he worked for Exxon Production and Research Company and Sandia National Laboratory. He has more than 20 years of research and development experience in applying the principles of engineering and physics to the petroleum, nuclear, defense, and electronics industries. He has published extensively in these areas and is the author and editor of the books, *Solder Joint Reliability: Theory and Applications* and *Handbook of Tape Automated Bonding* published by Van Nostrand Reinhold.

He is an associate editor of the *ASME Transactions, Journal of Electronic Packaging*, the *IEEE Transactions on Components, Hybrids, and Manufacturing Technology*, and the *International Journal of Microelectronics Packaging*. He has also served as session chairman, program chairman, general chairman, and invited speaker of several ASM/ASME/IEEE/ISHM/MRS/NEPCON/ TMS conferences and symposiums.

**Peter M. Hall** is a Distinguished Professor of Physics at Johnson C. Smith University, Charlotte, North Carolina. Dr. Hall teaches physics, is in charge of a dual-degree engineering program, and serves on the faculty of the Honors College. He has 33 years experience at AT&T Bell Labs in the fields of thin films, superconductivity, electronics packaging, and solder mechanics. Dr. Hall received his BA in physics from Hobart College, and his PhD in physics from Iowa State University. He is a Fellow of IEEE, a Bell Labs Fellow, and a registered Professional Engineer. He serves on the program committee of the Electronic Components and Technology Conference and is program chairman of the 1990 VLSI and GaAs Packaging Workshop. He also serves as Associate Editor of the ASME Transactions of the *Journal of Electronic Packaging*. Dr. Hall has published over 45 papers, and is coauthor of *Thin Film Technology*.

**Wan-Lee Yin** is a Professor with the Engineering Science and Mechanics Program, School of Civil Engineering, Georgia Institute of Technology, Atlanta, Georgia. Dr. Yin received his MS in Mechanics and Materials from the University of Minnesota and his PhD in Applied Mathematics from Brown University. He held academic positions at the Johns Hopkins University, Carnegie-Mellon University, the University of Singapore, and the National Tsinghua University in Taiwan. Before joining the Georgia Institute of Technology in 1979, he was a Structural Engineer in Stone and Webster Engineering Corporation in Cherry Hill, New Jersey. He has published papers in various research areas including continuum mechanics, flow of nonNewtonian fluids, nonlinear elasticity plasticity and mechanics of composite materials. His current research interests include the stability of plates, micromechanics, delamination, and interlaminar stresses in composite laminates, and computational mechanics. Dr. Yin is a member of AIAA and the Society for Natural Philosophy. He has served as a consultant for the aerospace industry over a number of years.

**An-Yu Kuo** is an Associate of Structural Integrity Associates, Inc., San José, California. Dr. Kuo received his BS degree in Power Mechanical Engineering from the National Tsing-Hua University, Taiwan, his MS degree in Mechanical Engineering from the National Taiwan University, Taiwan, and his PhD degree in Theoretical and Applied Mechanics from the University of Illinois at Urbana-Champaign. Dr. Kuo's major research areas have been in fatigue and fracture mechanics evaluation of structural components, mechanics of composite materials, thermal fatigue/fracture of power plant and aerospace components, and damage and/or crack detection and characterization of electronic materials. He is an active member of ASME and ASTM.

**Kuan-Luen Chen** is a Senior Engineer with Structural Integrity Associates, Inc., San José, California. Dr. Chen received his BS degree in Power Mechanical Engineering from the National Tsing-Hua University, Taiwan, and his MS and PhD degrees in Mechanical Engineering and Civil Engineering, respectively, from the Georgia Institute of Technology, Atlanta, Georgia. Dr. Chen's major research areas have been in fracture mechanics evaluation of structural components, manufacturing process simulation, residual life estimation of power plant components, and thermal fatigue of electronic devices. He is an active member of ASME and ASTM.

**Rajen Chanchani** received his BTech degree from the Indian Institute of Technology in 1973, and his MS and PhD degrees in Materials Science and Engineering from the University of Florida in 1984. He is currently a senior member of technical staff at Sandia National Laboratories. Prior to his present assignment, he was with AT&T Bell Labs, 1985–1990, and AVX Corporation, 1984–1985. His current interests include electronic packages and substrates, multichip modules, surface mount technology, ceramic capacitors and thermal modeling and analysis. He has published several papers in these areas.

**Goran S. Matijasevic** is a Post-Doctoral Researcher at the University of California, Irvine. Dr. Matijasevic has been involved in work on microelectronics packaging, with special interest in die-attach technology. His research interests also involve nondestructive testing techniques, especially acoustic microscopy, as well as failure analysis and reliability issues. He has presented his work at the International Reliability Physics Symposium in 1989 and the IEEE Electronic Components and Technology Conference in 1990. He received his BEng degree in Applied Physics from the Faculty of Electrical Engineering of the University of Belgrade, Yugoslavia in 1984 and his MS and PhD degrees in Electrical and Computer Engineering from the University of California, Irvine in 1985 and 1991, respectively. Dr. Matijasevic is a member of the IEEE, MRS, and Eta Kappa Nu.

**Chen Yu Wang** is a Research Assistant at the University of California, Irvine. He received his BS degree in Physics from the National Taiwan University, Taiwan in 1986, his MS degree in Electrical Engineering from the University of California, Irvine in 1989, and his PhD degree from the University of California, Irvine in 1992. He has worked as a Teaching Assistant and Research Assistant at University of California, Irvine since 1988. His current research interests include bonding technologies of electronic devices, scanning acoustic microscopy, and the thermal and electrostatic analysis of semiconductor devices. He is a student member of the IEEE.

**Chin C. Lee** is an Associate Professor at the University of California, Irvine. Dr. Lee received his BE and MS degrees in Electronics from the National Chiao-Tung University, Hsinchu, Taiwan in 1970 and 1973, respectively, and his PhD degree in Electrical Engineering from Carnegie-Mellon University, Pittsburgh, Pennsylvania in 1979. From 1979 to 1980 he was a research associate with the Electrical Engineering Department of Carnegie-Mellon University. From 1980 to 1983 he was with the Electrical Engineering Department of the University of California, Irvine as a research specialist. In 1984, he joined the same department as an Assistant Professor and later became Associate Professor of Electrical and Computer Engineering in 1990.

His research interests include thermal design and measurement of integrated circuit devices, electromagnetic theory, packaging technologies of semiconductor devices, scanning acoustic microscopy and its application to fiber optics. He is the coauthor of one book chapter and 70 papers in the subject areas mentioned above. Dr. Lee is a senior member of the IEEE and a member of the International Society for Boundary Elements and Tau Beta Pi.


**James N. Sweet** is a Distinguished Member of the Technical Staff, Advanced Packaging Department, Sandia National Laboratories, Albuquerque, New Mexico. Dr. Sweet has recently been working on development of test techniques and methodology for evaluation of packaged part reliability. One area of concentration is the design of Assembly Test Chips or ATCs for reliability and failure mechanism studies. He was previously supervisor of the Thermophysical Properties Division at Sandia. He received his BS in Physics from Stanford University and his PhD in Physics from the University of California at Berkeley. He is a senior member of the IEEE and a member of the American Physical Society.


**Arkady Voloshin** is currently with the Department of Mechanical Engineering and Mechanics at Lehigh University, Bethlehem, Pennsylvania, and received his PhD degree in 1978 from Tel Aviv University, Israel. He has been actively involved in the field of optical methods in experimental mechanics. His main area of research is the utilization of the digital image enhanced processing for problems in experimental mechanics of electronic packages. His pioneering work on the analysis of fractional fringe values was awarded the M. Hetenyi Award from the Society for Experimental Stress Analysis "for the most significant research paper published in Experimental Mechanics during the period January to December 1983."

Professor Voloshin is the author of over 45 papers. He is Associate Editor of the SEM journal *Experimental Mechanics*.

**Chao-Pin Yeh** received his PhD degree from the Georgia Institute of Technology in 1992. He is currently Advisory Engineer at IBM Endicott where he is responsible for (1) developing physics-of-failure based reliability models, (2) performing design for reliability (DFM) tool automation/integration, and (3) performing finite element analysis for various electronic packaging applications (multi-chip modules, printed wiring boards, surface mounted interconnects, etc.). Dr. Yeh's research interests include concurrent engineering, engineering databases, electronic design automation (EDA), reliability modeling, electronic packaging technologies and finite element method. He is also an active member of the ASME Electronic Packaging Division. He has issued more than 20 technical papers.

**Charles Ume** received his BS degree in Mechanical Engineering from the University of North Carolina, Charlotte in 1978 and his MS and PhD degrees from the University of South Carolina in 1981 and 1985, respectively. In 1985, he joined the faculty of the School of Mechanical Engineering at the Georgia Institute of Technology, Atlanta. In 1989, Professor Ume spent 6 months with the System Technology Division Laboratory at IBM in Austin, Texas as a faculty research fellow. He also spent summer 1990 as a research fellow with Corporate Manufacturing Research Center at Motorola, Inc., Schaumburg, Illinois. He is currently an Associate Professor in the School of Mechanical Engineering. He is an active member of ASEE, SME, SEM, ASNT, and ASME. He was Chairman of the Transducer/Sensor Technical Division of SEM. His current research interests are experimental and finite element analyses of thermomechanical effects in printed wiring board designs, laser phased array generation of ultrasound for on-line control of manufacturing processes – weld quality control, sensors with applications to robotic controls, and mechatronics.

**Robert E. Fulton** is Professor of Mechanical Engineering and Co-Director of the CAE/CAD Laboratory at the Georgia Institute of Technology. Dr. Fulton received his BS degree in Civil Engineering from Auburn University in 1953, and MS and PhD degrees in Civil Engineering from the University of Illinois in 1958 and 1960, respectively. He joined the NASA Langley Research Center in 1962 where, until 1984, he conducted or directed research in a broad range of structural mechanics and computer-aided design activities. From 1972 to 1984, he was Manager of the NASA/Navy sponsored $30 million CAD/CAM project denoted IPAD (Integrated Programs for Aerospace Vehicle Design) to develop technology and associated computer software for integrated company management of engineering information. He is the author of over 200 technical publications in such areas as finite element methods, numerical methods, static and dynamics analysis of shell structures, dynamic stability, and the use of computers in engineering analysis, design and manufacturing. He has also served on the faculties of the George Washington University, Old Dominion University, North Carolina State University, University of Illinois, and VPI&SU. His professional society affiliations include membership and active leadership roles in the National Computer Graphics Association; the American Society of Civil Engineers; the American Society of Mechanical Engineers; the American Society for Engineering Education; the American Academy of Mechanics; and the American Institute of Aeronautics and Astronautics. He has served on several National Academy of Sciences and NSF Committees to help identify critical technology needs associated with computer application to engineering. He is a past president of the National Computer Graphics Association, current chairman of the ASME Engineering Database Program, and active in the ASME Electronic Packaging Division.

**Karl W. Wyatt**, a 1976 BSEE graduate of the Department of Electrical Engineering at the Tennessee State University, is the Manager of Technology Applications and Concurrent Engineering, Simulation and Test Laboratory at the Motorola Corporate Manufacturing Research Center (CMRC) in Schaumburg, Illinois. Dr. Wyatt oversees Motorola sponsored manufacturing research at the Hampton Institute, the University of Illinois, the Georgia Institute of Technology, and MIT. His areas of technical activity include: thermal, thermomechanical, and electromagnetic analysis of multi-chip modules (silicon, ceramic, and organic substrates); development and use of chip level behavioral models for digital and analog ICs in board level simulations; development and use of CAD tools for design for testability and manufacturability; and the application of advanced packaging technology and system design tools and methodology to new product development.

Dr. Wyatt's background includes the physical modeling and analysis of continuum electromechanics systems and gas discharges. He has designed an electromagnetic shield for $Nb_3Sn$ superconducting rotor windings of electrical generators, systems for the magnetohydrodynamic precipitation of magnetite ($Fe_3O_4$) from turbulent steam flows in superheated steam power plants, and he has developed the numerical solution for nonlinear, coupled drift-diffusion-Poisson sulfur hexafluoride electron avalanche discharges (from single electron incipience to plasma channel development; cylindrical coordinates and time). In addition, Dr. Wyatt has performed quantum mechanical gallium arsenide heterostructure device design ·and he has managed digital gallium arsenide IC design as a Member of Technical Staff and Supervisor at the AT&T Bell Laboratories.

He received his PhD in Electrical Engineering from MIT in 1985. Prior to joining Motorola in 1990, he supervised the Digital GaAs IC Design group of the AT&T–DARPA III GaAs Pilot Line at Reading in his role as technology and product development manager for GaAs digital IC applications to communication, digital signal processing, and computer (PC to supercomputer) systems.

**John W. Stafford** received his BS in Aeronautical Engineering from MIT, his MS in Applied Mechanics and Electrical Engineering from the Polytechnic Institute of Brooklyn (now PINY) and his MBA from Fairleigh Dickinson University. He has worked on the mechanical design and development of such projects as communication satellites (Telstar), emergency power sources for telephone sites, development of IC maskmaking equipment such as the Electron Beam Exposure System (EBES), etc., and the packaging of integrated circuits developing Direct Chip Attach (DCA) technology to organic printed circuit boards, and is the author of numerous papers. He is a senior member of the IEEE and a member of IEPS. He has worked at the Grumman Aircraft Company, GE Knolls Atomic Power Laboratory, Bell Laboratories, AT&T–IS, and LYTEL, Inc. He was Manager of the Electronic Packaging Processes Laboratory of the Motorola, Inc. Manufacturing Research Center and is now Manager, Packaging Prototype Laboratory of the Motorola Semi-conductor Products Sector Advanced Packaging Development Center.

**Quanxin Guo** is a Research Engineer with Terra Tek, Inc., Salt Lake City, Utah. He received his BS in Mathermatics and Mechanics from Lanzhou University, Lanzhou, China, his MS in Engineering Mechanics from Hua-zhong University of Science and Technology, Wuhan, China, and his PhD in Mechanical Engineering from Northwestern University, Evanston, Illinois. Before he joined Terra Tek, he was a Research Associate at the Department of Materials Science and Engineering, Northwestern University. He has published 15 technical articles.

**Leon M. Keer** is Professor of Civil and Mechanical Engineering and Associate Dean for Research and Graduate Studies, Robert R. McCormick School of Engineering and Applied Science, Northwestern University, Evanston, Illinois. Professor Keer was educated at the California Institute of Technology and the Institute of Technology of the University of Minne-sota. He has been a member of the Technical Staff, Hughes Aircraft Company, and has consulted for the Borg-Warner Corporation, Sandia Laboratories, and the Timkin Company. He has held positions at the University of Minnesota (Research Fellow and Instructor), Columbia Uni-versity (Preceptor), and since 1964 at Northwestern University (currently Professor of Civil and Mechanical Engineering and Associate Dean for Research and Graduate Study in the Technological Institute). He has taught courses in Structural Mechanics, Structural Engineering and Applied Mathe-matics. He is a former North Atlantic Treaty Organization Postdoctoral Fellow (1962–63), a John Simon Guggenheim Memorial Foundation Fellow

(1972–73) and a Japan Society for the Promotion of Science Fellow (1986). He conducts research in the area of solid mechanics and investigates specialized topics of fracture and fatigue, surface mechanics and tribology, wave propagation, theory of layered media and structural dynamics. He has published over 200 technical articles.

**Yip-Wah Chung** is Professor of Materials Science and Engineering, Robert R. McCormick School of Engineering and Applied Science, Northwestern University, Evanston, Illinois. He was involved in establishing the National Science Foundation Industry/University Cooperative Research Center for Engineering Tribology at Northwestern and is currently serving as the Center Director. He received his BS in Physics and Mathematics in 1971, his MPhil degree in Physics in 1973, both from the University of Hong Kong, and his PhD in Physics from the University of California at Berkeley in 1977. He joined Northwestern University in 1977 and has been there ever since. He is on the Executive Committee of the Center for Catalysis and Surface Science and a member of the University's Faculty Planning Committee. Dr. Chung is a life member of the American Physical Society, member of the Society of Tribologists and Lubrication Engineers, Metallurgical Society, and American Vacuum Society. He also serves as the Vice-Chairman of the Illinois Chapter of the American Vacuum Society. He won the 1990 Ralph A. Teetor Engineering Educator Award from the Society of Automotive Engineers and the 1990 Best Paper Award from the Tribology Division of the American Society of Mechanical Engineers. Dr. Chung is author and coauthor of over 80 publications. His research interests include the role of surface chemical reactivity in fatigue crack initiation, environmental effects on friction and wear (including thin film rigid disk systems), deposition of thin films and superhard coatings, thermal fatigue, scanning tunneling microscopy, nanometer lithography, atomic force microscopy and its application as an atomic scale indentor.

**Matt A. Korhonen** researches deformation phenomena and their characterization by mechanical and X-ray methods in the Department of Materials Science and Engineering at Cornell University. A graduate of Helsinki University of Technology, he received his LicTech in 1976 and DTech in 1980 in Materials Science. He has held teaching and research positions at Helsinki University of Technology and the Academy of Finland. His recent research work involves deformation phenomena and their modeling in microelectronic materials, particularly in interconnects.

**Peter Børgesen** is Senior Research Associate in the Department of Materials Science and Engineering at Cornell University, has done extensive research with particle–solid interactions, including the irradiation of solid hydrogen isotopes at cryogenic temperatures, and more recently on fine lines and electronic packaging. He received his PhD in Physics from the University of Århus in Denmark, then spent four years in postdoctoral work at the Max-Planck-Institute for Plasmaphysics in Garching, Germany.

**Che-Yu Li** is Professor of Materials Science and Engineering at Cornell University and has been on the Cornell faculty since 1962. Engaged in electronic packaging research for 10 years, he specializes in solder interconnects, metallization, and Tab-automated bonding. He also studies mechanical reliability and micromechanical testing of components down to the submicron level. He is director of the Industry–Cornell University Alliance for Electronic Packaging and is currently coordinating electronic packaging research sponsored by the Alliance and Semiconductor Research Corporation.

**Ephraim Suhir** is a Member of Technical Staff, Basic Research, AT&T Bell Laboratories, Murray Hill, New Jersey, Division of Materials Science, Engineering and Academic Affairs. Since joining AT&T Bell Laboratories in 1984, Dr. Suhir has performed studies in materials science, and mechanical and reliability engineering in the areas of microelectronics and fiber optics, and supports ongoing projects in physical design, manufacturing and reliability of electronic and fiber optic packages and systems. Dr. Suhir has written over 100 technical papers, research reports, book chapters, and books, and has won several best paper awards in mechanical, structural, reliability, and ocean engineering. Dr. Suhir is cofounder and Senior Associate Editor of the *ASME Journal of Electronic Packaging*, program chairman and organizing committee member of various IEEE, ASME, MRS, and intersociety conferences and symposia, as well as advising committee member of several ongoing book series on mechanical performance of "high-tech" systems. He is a senior member of the IEEE and the SPE.

**Michel Mermet-Guyennet** is Senior Engineer, Packaging, at Advanced Computer Research Institute (Lyon, France) where he is involved in various projects related to packaging for definition and implementation of ACRI's supercomputer. From 1989 to 1990, he was project leader at SGS-Thomson (Grenoble, France) for high pin count packages using TAB technology and in charge of CAD and modeling tools for Corporate Package Development Grenoble. From 1984 to 1988, he was R&D engineer in Packaging and Assembly at Thomson Composants Militaires et Spatiaux where he carried out various projects such as: design of ASIC packages in multilayer cofired ceramic, assembly of MultiChip Module for high reliability applications, set up of assembly process for large chips and high pin count. He received his Thèse in Physics in 1984 from Université de Marseille-Luminy. He is a graduate of Ecole Centrale de Paris (1981). He is a member of the American Society of Mechanical Engineers and the International Electronics Packaging Society.

**Suresh V. Golwalkar** is Manager, Package Development, Assembly Engineering and Reliability, Intel Corporation, Folsom, California. His research activities cover a broad range of electronic materials, processing, applications and reliability of IC packages and systems. His current responsibilities involve development of new packages and technologies for the portable markets.

Dr. Golwalkar joined Intel in 1984, where he worked on various projects in the area of quality, reliability and failure analysis of advanced circuit packaging. He developed several new capabilities and techniques for materials characterization, taught intensive courses related to assembly technologies and facilitated development of Intel's offshore sites. He has conducted research in enhancing the manufacturability of existing packaging technologies, in the development of new thin memory packages and in fine pitch surface mount applications. Prior to joining Intel, he was a faculty member in the Materials Engineering Department of Rensselaer Polytechnic Institute, Troy, New York.

He received his BTech and MTech degrees in Metallurgical Engineering from the Indian Institute of Technology, Bombay and his PhD in Materials Engineering from Rensselaer Polytechnic Institute, Troy, New York. He has served as a member of technical committees, session chairman and workshop speaker of several IEEE, ASME, ASM conferences and symposiums. He has authored numerous technical papers on analytical techniques, high temperature materials and electronic packaging.

**Peter L. Hacke** is a Graduate Student at North Carolina State University. He obtained his BE in Materials and Metallurgical Engineering from the Stevens Institute of Technology. Mr. Hacke is currently working towards his PhD degree in Materials Science and Engineering under the direction of Professor H. Conrad. His thesis work involves thermomechanical fatigue of Pb–Sn solders. Previously, he was an engineer at Emulsitone Co. manufacturing semiconductor dopant sources.

**Arnold F. Sprecher** received his BS degree in Physics in 1981 and his PhD degree in Materials Engineering, under the direction of H. Conrad, from North Carolina State University in 1984. His dissertation topic was "The Electroplastic Effect in Metals."

Since receiving his doctorate, he has worked as a research associate and then as a senior research associate/instructor with the Department of Materials Science and Engineering at NCSU. Areas of research include electrorheology, electroplasticity and effects of electric fields on phase transformations, superplasticity, recrystallization and grain growth and in addition, fatigue of soft solders used in the microelectronics packaging industry.

**Hans Conrad** is a Professor in the Materials Science and Engineering Department, North Carolina State University, Raleigh, North Carolina. Dr. Conrad received his BS in Metallurgical Engineering from the Carnegie Institute of Technology and his Master of Engineering and Doctor of Engineering degrees in Metallurgy from Yale University. Prior to joining the faculty of North Carolina State University as Professor and Department Head in January 1981, Dr. Conrad held the position of Professor and Chairman of the Department of Metallurgical Engineering and Associate Director of the Institute for Mining and Minerals Research at the University of Kentucky. Immediately previous to joining the University of Kentucky, Dr. Conrad was Technical Director of the Materials Science and Engineering Department of The Franklin Institute Research Laboratories. Prior to that appointment he was associated with a number of industrial and research organizations including Aerospace Corporation, Atomics International, Westinghouse Research Laboratories, Chase Brass and Copper Company, R. Wallace and Sons Mfg. Company, the Napier Company, and The Aluminum Company of America.

Dr. Conrad's experience has been in both basic and applied research. In basic research, he has worked principally in the areas of crystal structures, their defects and the effect these have on physical and mechanical properties. He has engaged in research in the general area of the mechanical behavior of materials and in the areas of electroplasticity, electromigration, electrorheology, crystal growth, semiconductivity, superconductivity, laser action, cohesion and adhesion, metastable structures, ultra high vacuum and high pressures. In applied research, he has worked in the areas of metal forming, metal deformation processing, erosion, electrodeposition and metal finishing, shell mold casting, tool and die steels, machinability and alloy development. He has authored or coauthored over 280 scientific papers on his various research activities.

**Ryohei Satoh** received his BS degree in Metallurgical Engineering from Muroran Industrial University, Japan in 1969 and his MS and PhD degrees in Metallurgical Engineering from Hokkaido University, Japan in 1973 and 1988, respectively.

He joined the Production Engineering Research Laboratory, Hitachi Ltd., Japan in 1973 where he has been engaged in research and development of packaging technologies, including soldering technology, for hybrid microelectronics.

Dr. Satoh is a member of the Japan Institute of Metals and the Japan Welding Society.

**Jeng-Gong Duh** is a Professor in the Department of Materials Science and Engineering, National Tsing Hua University, Hsinchu, Taiwan. He received his BS degree in Nuclear Engineering from the National Tsing Hua University in 1974 and his PhD degree in Materials Engineering from Purdue University in 1983. He joined the Faculty of Materials Department at the National Tsing Hua University and was promoted to Professor in 1989. During 1987–88, he was a visiting scientist at the Department of Materials Science and Engineering, Cornell University. His current interests include ceramic thin film, interfacial phenomena, ceramic–metal bonding, surface modification of materials and electron microscopy analysis.

**Kuo-Chuan Liu** is a Research Associate in the Department of Materials Science and Engineering, National Tsing Hua University, Taiwan. He received his BE degree in Chemical Engineering from the National Cheng Kung University in 1987 and his MS degree in Materials Science and Engineering from the National Tsing Hua University in 1989. He is currently working on the development of microwave absorbing ferrite and the interfacial reaction between thick film and substrate material.

**Bi-Shiou Chiou** is a Professor in the Institute of Electronics, National Chiao Tung University, Hisnchu, Taiwan. She received her BS degree in Nuclear Engineering in 1975 and her MS degree in Health Physics in 1977, both from the National Tsing Hua University. She then received her PhD degree in Materials Engineering from Purdue University in 1981. She was a visiting professor at Purdue University and a visiting scientist at Cornell University during 1987–88. Her major areas of research are in the fields of electronic ceramics, thick film technology and ceramic thin film.

**Jim Lynch** received his BA degree in Chemistry from Oxford University, England, in 1966 and his MA and DPhil degrees, also from Oxford University, in 1971. His DPhil thesis examined the oxidation of iron using field ion microscopy. In 1970 he joined GEC–Marconi Materials Technology (formerly Plessey Research Caswell) in Towcester, Northamptonshire working on contact phenomena. In 1975 he became a member of the Quality Assurance Division at GMMT and is currently Quality Assurance Manager.

**Alberto Boetti** received his degree in Electronics Physics from the University of Turin in 1970. From 1965 to 1973 he was with CSELT, a research center for telecommunications in Turin. Since 1974 he has been with the European Space Agency in Noordwijk, The Netherlands, where he is involved with the evaluation and qualification of electronic components and hybrids for space application.

**Ronald G. Ross, Jr.** is Supervisor, Thermal and Cryogenics Technology Group, Jet Propulsion Laboratory, Pasadena, California. Dr. Ross, in addition to his broader management responsibilities, heads JPL's Reliability Physics Research directed at developing NASA design guidelines and test standards for solder attachment of surface-mount electronic components. Over the past 20 years he has conducted extensive research in the fields of reliability, life prediction and qualification of semiconductor products. He has authored or coauthored over 100 publications covering his research, and is JPL's key troubleshooter for solder related failures. He received his PhD from the University of California at Berkeley in 1968, specializing in Mechanical Design and Finite Element Structural Modeling. He is a member of Tau Beta Pi, Pi Tau Sigma and Sigma Xi Honor Societies.

**Liang-Chi Wen** is a Member of Technical Staff, Thermal and Cryogenics Technology Group, Jet Propulsion Laboratory, Pasadena, California. Dr. Wen is a senior researcher with emphasis in modeling the thermal–mechanical response and degradation dynamics of high reliability spacecraft systems. He has been working in the field of reliability physics and thermal–mechanical modeling since 1969. He received his PhD from Purdue University in 1968.

**Donald B. Barker** is Associate Professor, Department of Mechanical Engineering, University of Maryland. He is also Associate Director of the CALCE Electronic Packaging Research Center. The CALCE/EPRC is an industry/university cooperative research center which investigates reliability issues in the design of electronic equipment. Within the Center, Dr. Barker's research involves the experimental and numerical modeling of mechanical failures in electronic components and assemblies and the integration into a concurrent engineering design environment. This research is leading to practical approaches in microelectronic package design and the implementation of reliability prediction methodologies, based on physics of failure concepts. Dr. Barker obtained his PhD degree in Engineering Mechanics from UCLA and joined the faculty at the University of Maryland in 1976. He has published extensively in the general area of experimental mechanics, fracture mechanics, fatigue, dynamic material response, and electronics packaging.

**Abhijit Dasgupta** is Assistant Professor, Department of Mechanical Engineering, University of Maryland. Dr. Dasgupta obtained his PhD in Theoretical and Applied Mechanics from the University of Illinois and is a faculty member with research experience and interests in the areas of mechanics of material behavior, including thermomechanical material constitutive modeling, damage analysis, fracture mechanics and the mechanics of fatigue damage in heterogeneous materials and assemblies. He is currently involved in several research projects dealing with the physics of failure and reliability of electronic components assemblies and interconnects. He has authored over 25 journal papers and conference proceedings and four chapters in textbooks on physics of failure approaches to reliability of electronic packages.

**Ravi Subrahmanyan** received his PhD degree in Materials Science from Cornell University, Ithaca, New York in 1990, and is currently with the Advanced Packaging Development Center, Motorola, Inc., Phoenix, Arizona. Since 1985 he has been actively involved in research and development in flip chip technology, particularly in the area of accelerated testing and reliability assessment. He is currently involved in the development of flip chip interconnect systems in GaAs based packages for RF/microwave applications.

**Stephen J. Erasmus** is a Project Manager at Hewlett-Packard's Product Processes Organization in Palo Alto, California, where he is working on advanced electronics packaging. Prior to this, he developed inspection and measurement techniques for semiconductor lithography, and registration methods for electron lithography. He received his PhD in Engineering from the University of Cambridge in 1982, and then spent a year as a Research Fellow of Trinity College, Cambridge, where he researched digital image processing for electron microscopy.

**Sherman Leung** is Manufacturing Development Engineer with Hewlett-Packard, Networked Computer Manufacturing Operation. Mr. Leung is currently working on assembly processes for new products at HP. He joined the company in 1988 and had previously worked as a process equipment engineer at IBM, San José. Mr. Leung holds a BS degree in Mechanical Engineering from the University of California, Berkeley.

**Michael Pecht** is a tenured faculty member with a joint appointment in Systems Research and Mechanical Engineering, and the Director of the CALCE Electronic Packaging Research Center at the University of Maryland. He has a BS in Acoustics, a MS in Electrical Engineering and a MS and PhD in Engineering Mechanics from the University of Wisconsin. He is a Professional Engineer and an IEEE Fellow. He serves on the board of advisors for various companies. He is the chief editor of the *IEEE Transactions on Reliability*, and a section editor for the Society of Automotive Engineering. He is on the Editorial Board of the *Journal of Electronics Manufacturing* and on the Advisory Board of the Society of Manufacturing

Engineers Electronics Division. He has edited three books on electronics design, reliability assessment and qualification. He has received many "best paper" awards and was a Westinghouse Professor.

**Pradeep Lall** is Chief of the Component Reliability Branch in the CALCE Electronic Packaging Research Center at the University of Maryland. He has published extensively in the area of temperature dependence of microelectronics, reliability assessment and development of failure modeling strategies for microelectronic packages including formulation of physics of failure models and guidelines for design, validation testing, screening and derating for reliable electronic packages. Dr. Lall's research involves: development of physics of failure based software tools for design, reliability assessment, testing, and screening of microelectronics; temperature dependence of microelectronic device failure mechanisms and performance parameters; guidelines for thermal derating; and development of alternative test techniques for first-level interconnects. He is a member of ASME, IEEE, SAMPE, and ISHM. He received his BS in Mechanical Engineering from the University of Delhi, and his MS and PhD in Mechanical Engineering from the University of Maryland at College Park.

**Xuning Shan** is a Research Scientist in the Electronic Packaging Research Center, University of Maryland. Dr. Shan received his PhD in Corrosion and Electrochemical Engineering from The Johns Hopkins University in 1991. His research focuses on corrosion modeling, failure analysis, and reliability in electronic packaging. He is a member of the National Association of Corrosion Engineers (NACE) and the International Society for Hybrid Microelectronics (ISHM).

# Index